third edition

WORLD
Regional Geography

ISSUES FOR TODAY

third edition

WORLD
Regional Geography
ISSUES FOR TODAY

Richard H. Jackson
Lloyd E. Hudman

JOHN WILEY & SONS
New York Chichester Brisbane Toronto Singapore

Cover photo: Zygmunt Malinowski/Colca, Peru

City cover photo: Sheila Granda

Production manager: Lucille Buonocore

Production supervisor: Sandra Russell

Photo research: Stella Kupferberg

Interior and cover design: Dawn L. Stanley

Copyright © 1982, 1986, 1990, by John Wiley & Sons, Inc.

All rights reserved. Published simultaneously in Canada.

Reproduction or translation of any part of
this work beyond that permitted by Sections
107 and 108 of the 1976 United States Copyright
Act without the permission of the copyright
owner is unlawful. Requests for permission
or further information should be addressed to
the Permissions Department, John Wiley & Sons.

Library of Congress Cataloging in Publication Data:

Jackson, Richard H., 1941–
    World regional geography: issues for today/Richard H. Jackson,
  Lloyd E. Hudman.—3rd ed.
    Includes bibliographical references.
    ISBN 0-471-50633-8
    1. Geography. I. Hudman. Lloyd E. II. Title.
  G128.J32 1990
  910—dc20

Printed in the United States of America

10 9 8 7 6 5 4 3 2

Preface

The third edition of *World Regional Geography* reflects the comments and suggestions of numerous colleagues and students who have used the first edition. The underlying theme of the earlier edition which introduced students to the how and why of the geographic factors creating broad global contrasts is still emphasized. The overall structure is basically unchanged; however, coverage of some areas has been increased, along with the number of maps and vignettes.

Specific changes include (1) expansion of regional coverage in Africa, Latin America, and Asia; (2) analysis of China's adoption of elements of the free enterprise economic system, and failed democracy movement; (3) expansion of the number of case studies and boxed material, emphasizing today's issues of conflict, hunger, and environmental degradation; (4) new cartography illustrating world and regional geographic relationships; (5) complete listing in chapter overviews of all new concepts introduced in each chapter; and (6) definitions of key concepts and terms at the bottom of the page where they are first cited; (7) major geographic characteristics and (8) a glossary of terms. Although the difficulty of obtaining both accurate and current statistical data for all areas of the world is recognized, the text is updated to reflect worldwide conditions at the end of the 1980s.

The text is divided into four parts organized to provide an understanding of the major issues facing the world today, and the geographic relationships and patterns related to them.

Part I deals with basic concepts: the contrast between rich and poor, the character and challenge of physical environments, human populations and their growth, and regional variations in economic systems. This part provides the necessary geographic framework for examining the fundamental contrasts of wealth and poverty, levels of economic development, and challenges to development facing the world. The principles and concepts introduced in Part 1 are referred to throughout the balance of the text.

Part 2 examines the regions with a high standard of living and quality of life. These are the highly industrialized nations of Western Europe, Anglo-America, Japan, Australia, New Zealand, and the centrally planned economies of the Soviet Union and the Eastern Bloc European nations. By analyzing the geographic patterns and problems of each of the major areas we attempt to determine the factors that fostered their economic development and gave rise to their world dominance.

Part 3 deals with those regions that have been variously referred to, collectively, as the Third World, the Developing World, the less industrialized world, or even sometimes the region of underdevelopment. Such terms as underdevelopment, less developed, or less industrialized may be used pejoratively and may imply a negative value judgment; nonetheless, it is essential for students to understand the great dichotomy between life in the industrialized, wealthy world described in Part 2 and the life of people in the rest of the world. The plight of this broad region, which has the majority of the world's hunger, raises important issues for the whole world, at present and in the future. In examining this region we contrast the extremely poor countries with what we call the newly industrialized countries and the small group of extremely wealthy countries referred to as surplus-capital oil-exporting nations. The level of economic development in these subregions differs, but the nature of their problems remains the same; only the degree of immediacy differs. The relationships between the physical geography, the cultures, the locations, the technology, and the ties to the industrialized regions illustrate the factors that have had an impact on the present level of development within these countries.

Part 4 consists of Chapter 18, which summarizes the major challenges facing the world in the last two decades of the twentieth century. This chapter discusses and analyzes the geographic and social significance of problems such as continuing population growth, world desertification, loss of tropical forests, and destruction of agricultural land that must be resolved in order to increase the standard of living of the world's poor while maintaining the standard of living of the world's more fortunate minority.

Vignettes of life in the different regions are included to indicate how the geographic factors that are discussed unite to affect the life-styles of the individuals within each region. Maps, graphs, and charts are utilized frequently to show geographic relationships, trends, and changes and should be viewed as an integral part of the text. Each regional chapter has maps illustrating population distribution, climate, landforms, resources, industries, and political patterns.

The intellectual debt associated with the development of a textbook that encompasses the entire world is complex and impossible to fully acknowledge. The co-workers, students, teachers, and fellow geographers who have contributed ideas and insight into our own view of the world are simply too numerous to catalog. Contributions range from comments made in informal conversations that have prompted inclusion of sections of the text, to formal introduction to the geographic viewpoints of our own instructors, to the technical expertise of the numerous people who have been involved in the production of the book. In spite of the difficulty of fully acknowledging the numerous and disparate sources from which ideas used in the book have been culled, certain groups and individuals have played such an important role that they must be singled out for mention.

Of central importance are the hundreds of students who have interacted with us individually as teachers as we have taught courses from which the present text evolved. We are grateful for the patience with which they met our attempts to inculcate in them a sense of curiosity about the world and its peoples. Education is of necessity a two-way street, and their comments, suggestions, and criticisms in class and informal situations have provided us with a wealth of ideas.

Numerous geographers and others who used the first edition offered suggestions that we have incorporated into the second edition of the text. Suggestions have been made to us personally as our paths have crossed, through letters, phone calls, and communication with our editor. All of the suggestions were helpful, and many colleagues will recognize specific places where their ideas have been incorporated in the text. Space prevents a listing of all who took time to offer suggestions, but the following individuals need to be recognized for their critical review of all or part of the text: Joseph Brownell (SUNY-Cortlend); Donald Floyd (Cal. Polytechnic State University); Mike Folsom (Eastern Washington University); Richard MacKinnon (Allan Hancock College); Rickie Sanders (Indiana University of Pennsylvania); and Richard Walesek (University of Wisconsin-Parkside). The critical and timely analyses and recommendations of these individuals contributed greatly to the improvements within the third edition.

Our thanks also to the staff of John Wiley & Sons, Inc. who have provided guidance in the revision of the third edition. Although it is impossible to mention all who were directly or indirectly involved in the production of this edition, a few played such a cardinal role that they must be mentioned. Editor Stephanie Happer, who has maintained her interest and support for *World Regional Geography*, was instrumental in the production of the third edition. Stephanie coordinated our work with that of the Wiley staff and encouraged us to meet the deadlines created by production schedules. Richard Pusey of Charthouse supervised the creation and revision of cartography, and insured that the completed maps reflected the high standards associated with the Wiley organization. Marjorie Shustak supervised the difficult task of copy editing with all that it entails. The excellent quality of the finished book reflects her outstanding work in coordinating the work of Pamela Landau (who edited the entire manuscript) with the comments, deletions, and additions of the authors. Dawn Stanley provided outstanding assistance in both art and design, and her expertise was invaluable to the authors. Sandra Russell supervised the production of the book, insuring that it maintained the Wiley tradition of excellence.

To those and all other individuals who have been involved in the production of this volume we express our gratitude.

Provo, Utah

Richard H. Jackson
Lloyd E. Hudman

Contents

Map List

Part One

PEOPLE, PLACES AND POVERTY: ISSUES IN WORLD GEOGRAPHY 1

1 THE HUMAN FACTOR: FOCUS OF REGIONAL GEOGRAPHY 2
The Geography of Poverty: The Fundamental Issue 5
The Human Dimension: Paco Diaz of Mexico 22
The Human Dimension: Life in the Poorest Countries 27
Geographic Factors Affecting Development 23
Location: Key to Prosperity 24
Applied Geography: What Geographers Do 29

2 THE ENVIRONMENTAL FACTOR: RESOURCES AND DEVELOPMENT 32
The Limits of the Earth 34
The Earth as a Resource 39
The Life-or-Death Issue: Food Resources of the World 50
Resources for Industrialization: Contributors to Wealth 53

3 THE HUMAN EQUATION: THE CENTRAL ISSUE 64
The History of the Population Problem 67
World Distribution of Population 77
Conclusion: Population in Retrospect 83

4 POVERTY AND WEALTH: VARIATION IN ECONOMIC DECISIONS 87
Institutional Framework: Capitalism, Socialism, and Communism 89
Three Approaches: Market, Peasant, and Subsistence Systems 91

The Interconnected World of the Twentieth Century 96
Causes of Wealth and Poverty 100

Part Two

THE WORLD OF WEALTH: REGIONS AND PROBLEMS 111

5 EUROPE: AN INDUSTRIALIZED MARKET ECONOMY 112
The Environmental Setting 118
Cultural Diversity in the Peninsula 126
Europe: Home of Revolution 129
Problems and Conflicts in the Industrial World 139
The Human Dimension: The Evolution of the Common Market 145
The Human Dimension: Preserving Basque Identity 147

6 REGIONAL VARIATIONS IN DEVELOPMENT IN EUROPE IN THE INDUSTRIALIZED CORE AREA OF THE WORLD 150
The Nations of Western Europe 154
The Countries of Northern Europe 164
The Human Dimension: A "Completely Normal Austrian" 164
The Mediterranean Lands of Europe 167
The Nations of Eastern Europe: Alternative to Development 168
Problems in the Industrialized European Core 172
The Human Dimension: The Less Developed World in Europe 172

7 THE ANGLO-AMERICAN REALM 176
The Anglo-American Environment 180
Environmental Issues in Developed Countries 192
American Peoples: Unity or Diversity 193
Cities: Home of Americans 202
The Economy of North America 206
The American Experience in Retrospect: Example or Lesson 210

vii

8 AUSTRALIA AND NEW ZEALAND: OUTPOSTS OF THE DEVELOPED WORLD 214
The Australian Experience 216
New Zealand 225
The Australia-New Zealand Periphery in Retrospect 227

9 THE SOVIET UNION: THE COMMUNIST ALTERNATIVE 228
The Communist Experience 231
The Land 236
The Harsh Lands 241
The Human Dimension: Life in the Subarctic 243
Mineral Resources in the Soviet Union 244
Communism and Economic Development 246
Industry in the Soviet Union 251
Soviet People: Diversity in the Communist Land 257
The Human Dimension: Portrait of a Family 262
The Soviet Union: An Assessment 263
The Human Dimension: Soviet People; Western Veshch (Thing) 265

10 JAPAN: A UNIQUE PATTERN OF DEVELOPMENT 268
The Japanese Experience 270
The Japanese Environment 273
Modern Japan: Miracle in Asia 278
The Japanese Experience in Retrospect 284
The Human Dimension: The Japanese Middle Class 286

Part Three
THE WORLD OF POVERTY: CHARACTERISTICS AND CHALLENGES 289

11 ASIA AND THE ORIENTAL WORLD: POPULATION, POVERTY, AND CHANGE 290
The Population Problem in the Orient 294
The Human Dimension: Notes of an Agency for International Development Adviser in the Philippines 300
The Question of Overpopulation 301
The Geographical Base of Life in the Orient 303
The Diverse Cultures of Asia 307
The Human Dimension: The Tragedy of Cambodia 316
The Peasant Masses of the Orient 317
Issues for the Future 321

12 THE INDIAN SUBCONTINENT: THE DEMOCRATIC REVOLUTION AND ECONOMIC DEVELOPMENT 322
The Physical Basis of the Indian Subcontinent 326
Historic Background of India 330
The Cultural Geography of the Indian Subcontinent 334
Indian Development: Rural and Urban 338
The Human Dimension: An Indian Woman 340
Population and India's Future 345
The Human Dimension: Water for Life 347
India in Retrospect 349
The Indian Perimeter 350

13 CHINA: MODERNIZATION AND COMMUNISM 354
China Before the Revolution 357
The Communist Revolution 361
The Human Dimension: Life in Feudal China 362
The Physical Geography of China 368
Chinese Peoples: The Ultimate Resource 372
China Since Mao: Revolution and Change 374
The Human Dimension: A Contract with Chinese Peasants 379
The Chinese Experience in Retrospect 386
Taiwan: The Other China 387
Hong Kong: China's Window to the World 388
The Two Koreas: A Nation Divided 389

14 SOUTHEAST ASIA AND THE PACIFIC: THE COLONIAL HERITAGE 396
Regional Definitions 399
Geographical Characteristics 402
The Cultural Geography of Southeast Asia and the Pacific 403
Southeast Asia in the Postcolonial Era 410
The Human Dimension: Cheaper than Machines 413
Tomorrow's Issues: The Future of Southeast Asia 415
The Human Dimension: An Indonesian Farmer 416
The Human Dimension: Slowing Population Growth in Indonesia 418
The Pacific World: Variation in Development 418
The Human Dimension: Aiding the Poor 420

15 THE MIDDLE EAST AND NORTH AFRICA: CRADLE OF CIVILIZATION, CENTER OF CONFLICT 424
The World Importance of the Middle East 427
The Cultural Diversity of the Middle East and North Africa 434
The Human Dimension: The Hajj: An introduction 435
The Development of North Africa and the Middle East 442

Middle Eastern Lands: Challenge or Opportunity 445
The Modern Middle East and North Africa: 449
Development and Conflict 449
The Human Dimension: Korean Workers in the Middle East 458
Change and Conflict in the Middle East and North Africa 459
The Human Dimension: From Suqs to Supermarkets 461

16 AFRICA: THE INDUSTRIALIZED SUBSAHARAN DEVELOPED REGION 466

European Influence 468
The African Environment 474
Economies and Potential for Development 481
The Human Dimension: Drought in Africa 483
The Human Dimension: Women of the Less Industrialized World 488
Population Diversity of Africa 488
The Human Dimension: Women of Botswana 491
Developmental and Regional Variation in Africa 494
The Human Dimension: African Farmers 497
The Human Dimension: The Struggle for South Africa's Cities 502
The Human Dimension: Removing the "Black Spots" 504
Africa's Challenges Tomorrow 505

17 LATIN AMERICA: FOCUS OF CHANGE IN THE WESTERN HEMISPHERE 508

The Beginnings of Modernization in Latin America 513
Discovery and Conquest: European Expansion into Latin America 515
Modern Latin America: Variation in Development 520
Cultural and Institutional Settings 525
Latin America Economies 526

The Winds of Change in Latin America 528
Variations in Development: Regional Contrasts 531
The Human Dimension: Latin America: The Children's Voices 532
The Human Dimension: Haiti—"The Problems of Life" 534
The Human Dimension: A Tale of Poverty, Water, and Roses 537
Latin America in Retrospect 540

Part Four

THE WORLD OF TOMORROW: ISSUES FOR TODAY 543

18 TOMORROW'S ISSUES: WORLD SOCIETY OR WORLD CONFLICT 544

Population: The Central Issue 546
Food: The Moral Dilemma 549
Providing Other Basic Human Needs 553
Resource Allocation and Economic Development 554
The Human Dimension: A Windmill and a Farmer 557
Capital Transfer: Aid or Hindrance 558
Technology Transfers and Tomorrow's Poor 560
Political and Social Issues in Tomorrow's World 560
The Environmental Dimension: The Greenhouse Effect and the Ozone Layer 562
Environmental Problems 563
World Conflict or World Prosperity: The Alternative Futures 563

Glossary 567

Photo Credits 579

Index 581

Map and Graph List

Figure 1.1– The traditional division of world economies
Figure 1.2– World distribution of physicians
Figure 1.3– World infant mortality rates
Figure 1.4– World literacy rates
Figure 1.5– World regions based on level of industrial development
Figure 1.6– Food deficit countries of the world
Figure 1.7– World patterns of agricultural employment
Figure 1.8– Growth in income from petroleum in selected surplus-capital oil-producing nations
Figure 1.9– Changes in world birthrates
Figure 1.10– Situation relationships of Rotterdam and the Lower Rhine River
Figure 2.1– World population densities
Figure 2.2– Club of Rome projections of future world quality of life
Figure 2.3– Changes in per capita Gross National Product projected to 1990
Figure 2.4– Proposed sequence of changes in the distribution of the earth's land mass
Figure 2.5– Relationship of the earth's inclination to the sun's rays
Figure 2.6– Earth-Sun relationships
Figure 2.7– Mechanisms for lifting air
Figure 2.8– World climatic distribution based on the Köppen system
Figure 2.9– Cereal production in the industrialized and less industrialized regions (1988)
Figure 2.10– Major agricultural regions
Figure 2.11– World coal and petroleum resources
Figure 2.12– Model of the perception of the environment
Figure 3.1– World population growth
Figure 3.2– Origins and diffusion of domesticated plants and animals
Figure 3.3– Major population migrations related to industrialization and development of Europe
Figure 3.4– Demographic transition
Figure 3.5– Population pyramids
Figure 3.6– Change in birthrates of selected countries
Figure 3.7– Rate of population growth
Figure 4.1– Family relationships in peasant societies
Figure 4.2– The succession of activities in a slash-and-burn agricultural system
Figure 4.3– Level of interconnections as indicated by paved roads
Figure 4.4– Level of interconnections as indicated by railroads
Figure 4.5– Growth of manufactured items in export trade
Figure 4.6– Changes in amounts and types of aid to developing countries projected to 1990
Figure 4.7– The Sahel area on the southern margins of the Sahara
Figure 4.8– Harvest losses in less industrialized nations
Figure 4.9– Increase in fertilizer consumption 1970–1984
Figure 5.1– Location, political divisions, and major cities of Europe
Figure 5.2– Population distribution of Europe
Figure 5.3– Europe's proportion of industrial production
Figure 5.4– Composition of European exports
Figure 5.5– Composition of European imports
Figure 5.6– Migration of Europeans to other parts of the world
Figure 5.7– Relative location and size of Europe compared to North America
Figure 5.8– Major landforms of Europe
Figure 5.9– Climatic types of Europe
Figure 5.10– Volume of freight moved on rivers of Western Europe
Figure 5.11– Mineral resources of Europe
Figure 5.12– Major language groups of Europe
Figure 5.13– Increasing yields of grain in Europe, 1948–1983
Figure 5.14– The colonial possessions of European nations at the start of World War II
Figure 5.15– Representative population pyramids for European countries, 1980
Figure 5.16– Evolution of EEC
Figure 5.17– Territorial claims of the countries that are full or associate members of the EEC
Figure 5.18– EEC compared to other industrial powers
Figure 6.1– Major regional groupings of Europe based on relative level of economic development
Figure 6.2– Major cities and industrial regions of the United Kingdom
Figure 6.3– Energy resources of the British Isles
Figure 6.4– The "Irish Curve" of population growth
Figure 6.5– Major industrial concentrations of Europe
Figure 6.6– Reclamation of land from the former saltwater Zuider Zee
Figure 6.7– Major agricultural regions of Europe
Figure 6.8– The major mineral resources and industrial centers of Eastern Europe

Figure 7.1–	Political subdivisions and major cities of the Anglo-American Realm		Figure 11.7–	The challenge to the developing countries of providing employment for a rapidly growing population
Figure 7.2–	Population distribution of Anglo-America		Figure 11.8–	Changing birth and death rates in Sri Lanka
Figure 7.3–	Climatic patterns of Anglo-America		Figure 11.9–	Per capita consumption of energy
Figure 7.4–	Landforms of Anglo-America		Figure 11.10–	Landforms of Asia
Figure 7.5–	Fossil fuel energy resources of Anglo-America		Figure 11.11–	Climates of Asia
Figure 7.6–	The relative importance of energy resources		Figure 11.12–	Major resources found in Asia
Figure 7.7–	Distribution of major metallic minerals in Anglo-America		Figure 11.13–	Cultural variation in Asia
Figure 7.8–	Dependency of the United States on imported minerals		Figure 11.14–	Concentrations of Chinese in Southeast Asia
Figure 7.9–	The population of American Indians in the United States and Canada		Figure 11.15–	Changing boundaries in Southeast Asia
Figure 7.10–	Changing patterns of immigration into the United States		Figure 11.16–	Crop production vs. energy expenditures
Figure 7.11–	Present and projected areas of metropolitan concentration of population in Anglo-America		Figure 12.1–	Political subdivisions and major cities of the Indian subcontinent
Figure 7.12–	Agricultural regions of Anglo-America		Figure 12.2–	Landforms of the Indian subcontinent
Figure 7.13–	Principal manufacturing regions of Anglo-America		Figure 12.3–	Climates of the Indian subcontinent
Figure 8.1–	Political subdivisions and major cities of Australia and New Zealand		Figure 12.4–	Expansion of British influence in the Indian subcontinent
Figure 8.2–	Climatic Regions of Australia		Figure 12.5–	Division of British India at time of Independence in 1947
Figure 8.3–	Population distribution in Australia		Figure 12.6–	Languages of the Indian subcontinent
Figure 8.4–	Major crop and livestock patterns in Australia		Figure 12.7–	Present political subdivisions of India
Figure 8.5–	Distribution of mineral resources in Australia		Figure 12.8–	Resources of India
Figure 8.6–	Cities, minerals, and agricultural regions in New Zealand		Figure 12.9–	Major industrial regions of the Indian subcontinent
Figure 9.1–	Political subdivisions and major cities of the Union of Soviet Socialist Republics		Figure 12.10–	Comparison of wheat yields between India and the United States
Figure 9.2–	Boundary changes of the Soviet Union since the 1917 communist revolution		Figure 12.11–	Generalized agricultural regions of India
Figure 9.3–	Comparison of the United States and the Soviet Union		Figure 12.12–	Changes in birth and death rates in India in the twentieth century
Figure 9.4–	Landforms of the Soviet Union		Figure 13.1–	Political subdivisions and major citis of The People's Republic of China
Figure 9.5–	Climatic regions of the Soviet Union		Figure 13.2–	Chinese dynasties
Figure 9.6–	Principal resources of the Soviet Union		Figure 13.3–	China's boundaries and foreign influence in the nineteenth and twentieth centuries
Figure 9.7–	Reserves of natural gas and petroleum in the Soviet Union		Figure 13.4–	Land and water buffalo ownership in pre-revolutionary China
Figure 9.8–	Major agricultural regions in the Soviet Union		Figure 13.5–	China's industrial centers
Figure 9.9–	Major industrial districts and cities of the Soviet Union		Figure 13.6–	Landforms of China
Figure 10.1–	Political subdivisions and major cities of Japan		Figure 13.7–	Climate regions of China
			Figure 13.8–	Resources of China
Figure 10.2–	Expansion of the Japanese Empire during the World War II era		Figure 13.9–	Distribution of China's minorities
			Figure 13.10–	Political divisions of China
Figure 10.3–	Landforms of Japan		Figure 13.11–	Agricultural regions of China
Figure 10.4–	Climates of Japan		Figure 13.12–	Major industrial regions of China
Figure 10.5–	Industrial regions of Japan		Figure 13.13–	The island of Taiwan
Figure 11.1–	Political subdivisions and major cities of Asia		Figure 13.14–	Korea
			Figure 14.1–	Political subdivisions and major cities of Southeast Asia
Figure 11.2–	Population density in Asia		Figure 14.2–	Climate regions of Southeast Asia
Figure 11.3–	Urbanization in Asia		Figure 14.3–	Language groups in Southeast Asia
Figure 11.4–	Asia's annual growth rate		Figure 14.4–	Resources in Southeast Asia
Figure 11.5–	Growth of world population and food production		Figure 14.5–	Colonial holdings of European powers in Southeast Asia prior to 1945
			Figure 14.6–	Non-Palestinian refugees
Figure 11.6–	Population growth in India		Figure 14.7–	Political divisions of the Pacific Region
			Figure 14.8–	Territorial claims in Southeast Asia and the Pacific Ocean
			Figure 15.1–	Political subdivisions and major cities of North Africa and the Middle East

Figure 15.2– Distribution of population in the Middle East and North Africa
Figure 15.3– The oil reserves of the Middle East
Figure 15.4– Maximum extent of areas dominated by Islamic religion
Figure 15.5– Distribution of world legal systems
Figure 15.6– Pilgrimage space in Mecca
Figure 15.7– Language groups of North Africa and the Middle East
Figure 15.8– Distribution of major religious systems of the world
Figure 15.9– Territorial extent of the Ottoman Empire
Figure 15.10– Climate Regions in North Africa and the Middle East
Figure 15.11– Highland areas of the Middle East and North Africa
Figure 15.12– Petroleum and natural gas in North Africa and the Middle East
Figure 15.13– Average per capita gross incomes in the Middle East
Figure 15.14– Expansion of area under Israeli control
Figure 15.15– Israeli settlements and settlement areas
Figure 16.1– Political subdivisions and major cities of Africa
Figure 16.2– Representative examples of European Colonial boundaries in Africa
Figure 16.3– Tribal boundaries in Africa
Figure 16.4– Exports of African countries
Figure 16.5– Landforms of Africa
Figure 16.6– Climates of Africa
Figure 16.7– Area of Africa affected by Tsetse flies
Figure 16.8– Resources in Africa
Figure 16.9– Distribution of agricultural economies in Africa
Figure 16.10– Population distribution in Africa
Figure 16.11– World Fertility Rate
Figure 16.12– Major language groups of Africa
Figure 16.13– Location and areal extent of Black homelands established or planned by the South African government
Figure 17.1– Political divisions and major cities of Latin America
Figure 17.2– Standard of living of Latin America compared to that of the United States
Figure 17.3– Changes in population indices in Latin America
Figure 17.4– Population groups in Latin America
Figure 17.5– Landforms in Latin America
Figure 17.6– Climates of Latin America
Figure 17.7– Resources of Latin America
Figure 17.8– Petroleum deposits in Latin America
Figure 17.9– Major transport routes in Latin America
Figure 18.1– Growth in population of females of marriageable age
Figure 18.2– Changes in world food production
Figure 18.3– Percentage of GNP granted as foreign aid
Figure 18.4– Growth in the foreign debt of the less industrialized nations
Figure 18.5– Inflation rates in Ghana
Figure 18.6– Rural-urban migration in Morocco
Figure 18.7– Tropical Rainforests

Part One

PEOPLE, PLACES, AND POVERTY
issues in world geography

I Am Part Of All That I Have Met; Yet All Experience Is An Arch Wherethro' Gleams That Untraveled World Whose Margin Fades For Ever And For Ever When I Move.

ALFRED LORD TENNYSON

Geography Is Selective; It Is Explanatory; It Is Reasonable; It Is Discipline; It Is A Way Of Thinking.... Geography Is A Sequence Of Ideas Or Concepts Which Develop Into A Coherent Subject Looking At The World From A Point Of View.

NEVILLE V. SCARFE

Wad Kowli, Sudan.

THE WORLD PROFILE, 1990

Population (in millions)	5303	Energy Consumption-pounds per Capita (kg)	1707
Growth Rate (%)	1.7	Calories (daily)	2985
Years to Double Population	40	Literacy Rate (%)	74
Life Expectancy (years)	63	Eligible Children in Primary School (%)	98
Infant Mortality (per thousand)	77		
Per Capita GNP ($)	3010	Percent Urban	45
Percent of Labor Force Employed in:		Percent Annual Urban Growth	
Agriculture	48	*Developing*	3.8
Industry	23	*Industrial*	1.65
Service	29		

1

The Human Factor Focus of Regional Geography

MAJOR CONCEPTS

Industrialization / Centrally planned economies / Economic growth / Economic development / Developing/Less developed GNP / Quality of life index / Calorie deficit Holistic / Location / Relative location Spatial interaction / Site / Situation Break-in-bulk / Hinterland Complementarity / Generative

IMPORTANT ISSUES

Distribution of wealth and poverty
Contrasts in life-styles in three worlds
Quality of life in different regions
Allocation of world resources
Ethnocentricity and world poverty
Economic development

Global Geographic Characteristics

1. The central issue in geography revolves around the character of places, their similarities and differences.
2. Global geographic patterns are complex and constantly evolving, reflecting variations in climate, landforms, culture, economies, and other factors.
3. The world can be divided into broad regions sharing common geographic characteristics that represent a unifying element to local geographic variation.
4. The quality of life of the people occupying the major world regions ranges from the high standards found in Europe and North America to the low quality of life in much of Africa.
5. The majority of the world's population lives in regions with lower standards of life.

Geography analyzes the earth as the home of the human population, and in the process reveals numerous patterns and contrasts. Of primary interest to most of us are the differences that combine to make each individual place on the face of the globe somehow unique. Study of the world's geography is an attempt to gain insight into the causes of this uniqueness, which identifies each place and is a combination of the natural, or physical setting of climate, landforms, and resources, and the features created by the residents of that place such as buildings, economy, dress styles, or other people-created, or cultural features. As a result, the study of geography of place is extremely broad-ranging.

Understanding the geography of the world becomes more complicated—and more interesting—because every place on the earth's surface changes over time. Changes in economy, political organization, population size, and the physical environment constantly alter the texture and fabric of the complex mosaic that makes up the geography of the world. We could choose any aspect that contributes to the geographic complexity of the earth's surface—economies, environments, cultures—as the focus of this book. Central to understanding this world, however, is recognition of humans—their numbers and their actions—as the critical factor in the use of the earth and its resources. The actions of people have resulted in a complex pattern of life-styles, each unique to a specific region. This complex pattern can be divided into broad regional groupings in many ways; one basis is the relative well-being of the residents of the various regions. The distribution of wealth among the world's people is critical to understanding the character of the earth's surface.

One of the primary reasons for studying the world is to gain an understanding of the way people live. Unless future generations understand the realities of life in other areas of the world, it will be difficult ever to arrive at a truly peaceful and cooperative world society. Central to an understanding of contrasts in wealth is an understanding of why certain areas have been the scene of wealthy, consumer-oriented societies, while in other areas, basic necessities (food, education, medical care, shelter, and a meaningful occupation) are limited. We could examine these global contrasts in a variety of ways, but because the level of living in a specific place is central to the well-being and outlook of the residents, we have chosen to use the level of economic development as a unifying theme for examining the geography of the world. *Economic development* refers to the nature of a country's economy in terms of level of *industrialization* and modernization. Countries in which the majority of the people grow food only for their own needs are at a low level of economic development. Economic development is not the same as *economic growth*, which refers only to the rate at which the level of economic development increases.

The degree of economic development is central to the distinctive characteristics of a place because it affects the life of the people in a multitude of subtle and complex ways. The level of economic development largely determines a resident's degree of deprivation, job types and demands, and attitudes toward the future. People in areas of the world that are not highly developed economically often lack adequate, basic human needs of food, clothing, medical care, and shelter, whereas those in highly developed industrial economies tend to have a surplus, suffering instead from other problems resulting from the very affluence they enjoy. A person in a *less developed* area has a different impact on the environment from one in an *economically developed* area, and the life-styles of the two are almost opposites.

Wealthy regions with their highly developed industrially based economies are the home of only a minority of the population of the world. The vast majority lives in areas that are less industrialized and less developed economically, with varying degrees of poverty, disease, malnutrition, and other problems. The majority of the readers of this book will be from the wealthy regions of the world, and it is important that we gain an understanding of how the majority populations in the less industrialized regions live and why. Insight into the nature and causes of the poverty of most of the world may help to create a spirit of worldwide concern for the plight of the world's impoverished and develop a commitment to overcoming the world imbalance in allocation of resources and wealth. Such an insight will not only lead to a higher standard of living for the world's poor, but will benefit the world's rich by minimizing future conflicts between rich and poor.

THE GEOGRAPHY OF POVERTY: THE FUNDAMENTAL ISSUE

One of our most significant problems today is the uneven distribution of wealth and resulting impoverishment of certain areas of the world. The effects of these differences are manifest in a range of socio-economic problems, including poor housing, lack of adequate medical care, high infant mortality rates, illiteracy, malnutrition, and unemployment, which consequently are useful indices to the level of economic development in a specific country. Using such

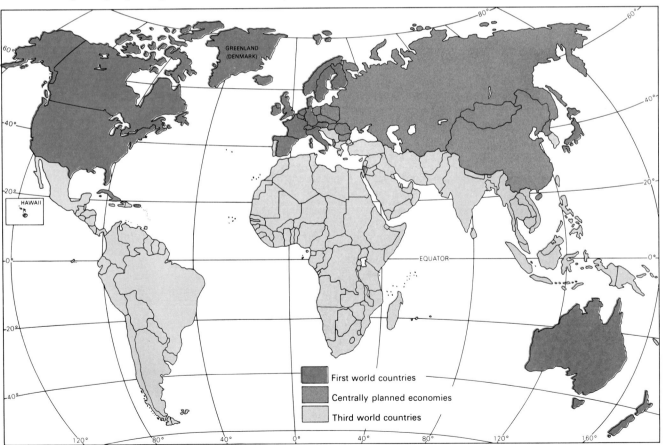

Figure 1.1 The traditional division of world economies utilized a combination of economic and political factors. The categories grouped together countries with major differences in level of economic development.

THE HUMAN FACTOR

Regions: tools for geographical analysis

The specific characteristics that make each place in the world unique are the basis for broader classifications of places that geographers call *regions*. Regions are a tool used by geographers to provide a classification system to facilitate analysis of the world. Regional analysis in geography is based on the identification of a specific area (region) that has some unifying factor or group of factors that make it different from surrounding regions. The regional concept is utilized on an intuitive basis by everyone. For example, we refer to "the West" or "the East" in the United States. Although we may not consciously list the factors that make the eastern region different from the western, we understand that they are distinct and recognizable regions. The daily usage of the regional concept illustrates the basic questions raised by any science, including Geography: What, where, how, and why. When we refer to "the West" as a region in common usage, we mentally answer these questions in general terms.

Geographers attempt to answer what, where, how, and why questions for a variety of types of regions. Regions can be identified on the basis of physical geography, language, politics, economics, or any other criteria that is identifiable and measurable. The challenge to geography involves answering seemingly simple questions about a region. The answer to the question "What is the West?" appears apparent. For most residents of the United States, the West includes the Pacific coast states (Oregon, Washington, California), and the Rocky Mountain states. But what of the plains states of Kansas and Nebraska? Are they part of the West? If so, what of the Dakotas, Texas, and Oklahoma? Are they part of the West? And what is it that makes the West unique? Is it language? Climate? Location? Or is it a combination of elements? In science, the "what" questions are commonly the easiest to answer; the "why" questions the most difficult. If we agree on the boundaries of the West, can we then answer how and why it developed as a unique region? A central focus of geography is its attempt to provide a logical system for dividing the world into regions and supplying answers to the fundamental questions of science relating to these regions, but it should be apparent that doing so is no simple task.

In this text we regionalize the world using the criteria of level of economic development, because economic development affects so many other factors that determine the geographic characteristics of a place. In broad terms there is general recognition in the world that there are two broad regions: one rich, one poor. Terms commonly applied to the rich region include *Developed, Industrial, Scientific-Technological,* or *First World.* Terms applied to the poorer region of the world include *Undeveloped, Underdeveloped, Less developed, Nonindustrial, Less industrial, Peasant,* or *Third World.* These terms are used as a descriptive shorthand, and designation of a country or area as less developed or underdeveloped does not imply inferiority of the people or culture, but indicates only the relative level of industrialization, wealth, and type of economy typically associated with economies that are less industrialized. As with the American West or East, there is general awareness as to what and where these terms apply. This text will endeavor to clarify the what and where of rich and poor regions, and provide insights into the how and why of their geography.

indices as a basis, it has been popular to divide the countries of the world into three groups based on economic development, as shown in Figure 1.1. The First World was historically defined as that of the wealthy, technologically advanced economies of the Western industrialized world. The Second World was historically defined as the Soviet Union, China, Cuba, and other countries with centrally planned economies. The Second World includes countries that are industrialized (Soviet Union), as well as those countries that are less industrialized (China). The remainder of the world was classified since the late 1950s as the Third World.

Defined in this manner the Third World includes the majority of the world's population; these people live in countries where the prevailing level of living is low, there is a relative lack of industry, and the majority of the people live in a perennial condition of near misery and mass poverty. In these Third World countries the majority of the employed population is involved in farming or hunting activities and relies on human or animal power for energy; the countries are overpopulated in terms of the numbers of people who must be supported by the available cultivated land. The problems of these Third World countries are compounded by a high birthrate and a declining death rate, which have resulted in a literal population explosion since the end of World War II. Life expectancy in Third World countries is low in comparison to that of the industrialized regions, diet is often inadequate in nearly all respects, and illiteracy is the rule rather than the exception. The continued expansion of the earth's population, coupled with a declining rate of economic growth, compound these problems in Third World countries.

The degree of severity of socioeconomic problems facing the Third World countries is not uniform; some suffer from mass poverty, illiteracy, malnutrition, and disease, while in others the majority of the population is able to obtain the basic needs of life. Because of

Evidence of diffusion of habits and tastes of the industrialized nations: advertisement for Coca Cola in Guangzhou (Canton) China.

the great difference between the poorest countries of the world and those at the upper end of the continuum of poverty, the traditional three-part division into developed, centrally planned, and Third World countries is inadequate to study the world accurately.

DIVIDING THE WORLD: DEVELOPED TO IMPOVERISHED

A more meaningful division of the world would first separate the wealthy, industrialized countries (commonly referred to as developed) of the world with their high per capita income, near total literacy, low infant mortality rates, and high levels of consumption, from the balance of the world, (commonly called less developed or less industrialized), which is characterized by poverty of greater or lesser degree. Then the second group can be divided into three broad areas: the newly industrialized countries (NICs), the capital-surplus, oil-exporting countries, and the least industrialized countries. Dividing the world in such a fashion is somewhat arbitrary and reflects the difficulty of selecting adequate indices of development. Traditionally, **gross national product** (GNP) per capita has been

Gross National Product (GNP): Refers to the total value of all goods and services produced annually in a country.

Figure 1.2 World distribution of physicians. The relative availability of medical care is one indication of the standard of living in a country. The concentration of medical care in the industrialized countries is apparent.

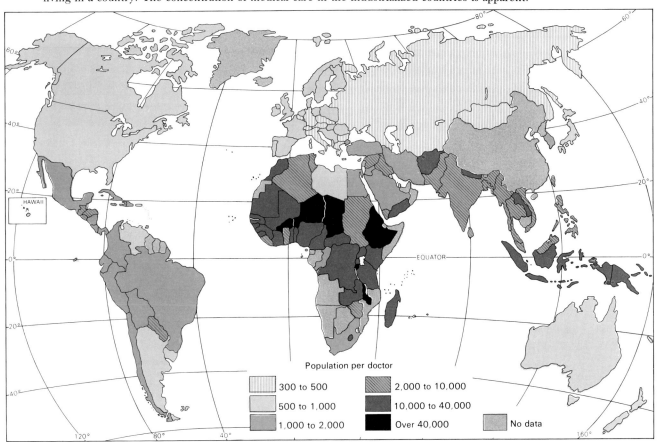

THE HUMAN FACTOR

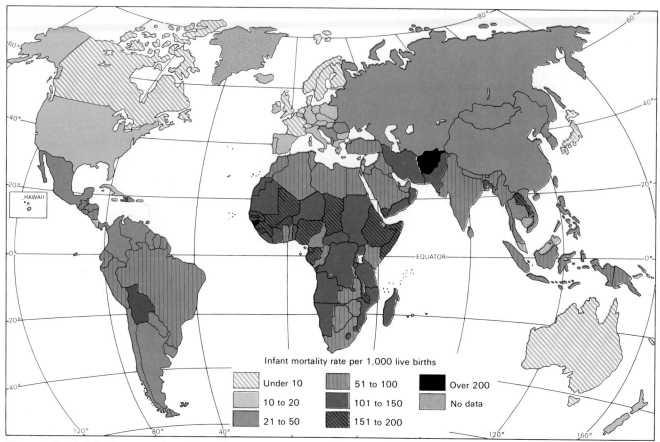

Figure 1.3 World infant mortality rates. The concentration of high infant mortality rates is nearly identical with areas having the least access to medical care.

the primary index of development. Gross national product is useful because it provides a quantitative figure that can be readily compared between countries. It has major weaknesses as the sole criterion for classification because of the different types of economies involved. For example, if the dividing figure between industrialized nations and newly industrialized nations is $5000 GNP per capita per year, the oil-rich country of Brunei shows up as an industrialized country with one of the highest per capita incomes in the world. The approximately $16,000 per capita annual gross national product in Brunei is misleading because the wealth is not evenly distributed and because the high GNP is based on a relatively temporary phenomenon, the export of petroleum.

A further problem in comparing annual per capita incomes among countries is the wide variation in purchasing power. A $300 per capita income in India will purchase more basic foodstuffs than $300 would purchase in an industrialized country. In spite of these difficulties in using GNP when economies and purchasing power differ, the GNP is a useful general indicator of the degree of development in a country.

The difference between countries with per capita GNP of under $400 and those of over $5000 is sufficient to outweigh the weakness of using GNP as a divider.

We can achieve a finer division of economies when we examine the quality of life. The *quality of life index* is a measure based on the standard of living indicated by key indices such as literacy, infant mortality, and access to medical and other social services. The weaknesses associated with measuring quality of life come from the difficulty of obtaining accurate figures. For most areas of the world obtaining more than a general approximation of these statistics is difficult. Nevertheless, world organizations have developed estimates for the physical quality of life for the world's countries. The number of residents per available doctor (Figure 1.2) in a country gives a general indication of the level of medical care available even though doctors in Third World countries primarily treat those with wealth rather than the masses of the poor. The level of infant mortality (Figure 1.3) and literacy levels (Figure 1.4) also provide a general insight concerning the degree of development.

Comparison of literacy rates, infant mortality rates,

PEOPLE, PLACES, AND POVERTY

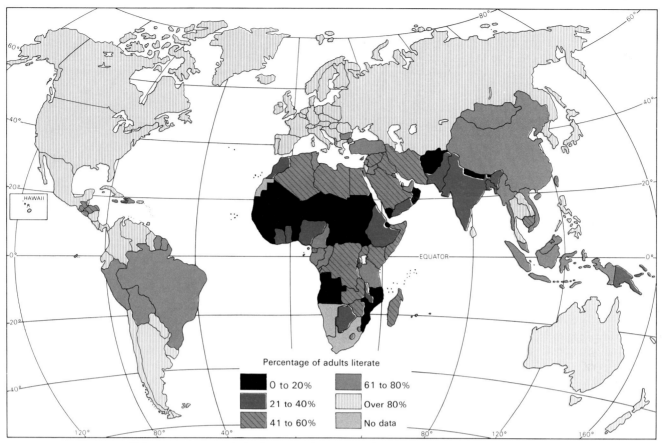

Figure 1.4 World literacy rates parallel infant mortality rates and level of medical care and reinforce the regional contrasts between the industrial countries and the balance of the world.

and gross national product provides a relatively useful basis for division of the world on the basis of quality of life for all but the centrally planned economies. In China, Cuba, and other countries with a centrally planned economy the GNP is more uniformly distributed throughout the entire population, and government efforts have led to higher literacy rates and lower infant mortality rates.

CHARACTERISTICS OF THE WORLD'S REGIONS

The first group, the wealthy industrialized countries (see Figure 1.5), is characterized by a high per capita income, high levels of personal consumption, and a large, well-developed middle class. The majority of the employed population is involved in manufacturing or provision of services. The countries that belong to the industrialized world include Canada, the United States, all countries of Europe (except Albania), the Soviet Union, Japan, Australia, and New Zealand. A casual examination of the countries that comprise the industrialized world shows that the size of a country, either in area or population, is not a necessary determinant of economic strength. Switzerland, Belgium, the Netherlands, Luxembourg, and even Sweden and Norway have per capita gross national products equal to or greater than that of the United States (Table 1.1). There is a wide range in per capita income within the developed regions, but the quality of life as measured by literacy and infant mortality rates reveals a rather uniform level (Table 1.1). The standard of living and quality of life within these industrialized countries may differ in the amount of material possessions available to an individual, but in all, the standard of living is markedly better than it is for the majority of the population in the less industrialized world.

The capital-surplus, oil-exporting counries are a unique situation. Countries include Saudi Arabia, Kuwait, the United Arab Emirates, Brunei, and other oil-rich countries with small populations. In these countries per capita GNP is among the highest in the world. Nevertheless, these countries are in many ways more similar to the rest of the less industrialized world

THE HUMAN FACTOR

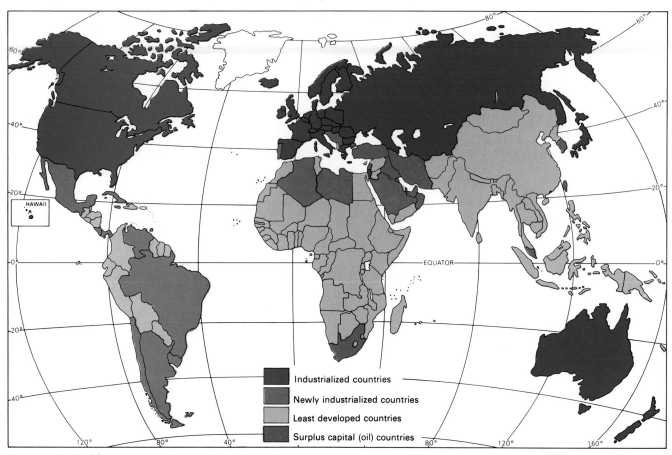

Figure 1.5 World regions based on level of industrial development. Note the relationships between this map and Figures 1.2, 1.3, and 1.4.

than to the industrialized countries. The standard of living for the majority of the population is only beginning to improve rapidly as a result of surplus wealth. Literacy rates remain low, typically below 50 percent for women and below 70 percent for men. The majority of the population is still involved in economic pursuits characteristic of the less industrialized countries rather than the industrial countries. Population characteristics are similar to the less industrialized world with high birthrates and low death rates for the total population, but with high infant mortality rates. The governments of these oil-surplus countries are making a concerted effort to transform their countries through subsidized services of medicine, education, housing, and job opportunities. It is too soon to assess the success of these attempts, because the bulk of their great wealth has been accumulated only since the mid-1970s. Meanwhile, their tremendous wealth separates them from other less industrialized countries, but the persistence of socioeconomic characteristics common to the less industrialized world identifies them as different from the industrialized countries (Table 1.2).

The newly industrializing countries (NICs) are those in which the cultural and economic climate has led to a rapid rate of industrialization and growth since the 1960s. The NICs are in a transition between the least industrialized countries of the world and the mature, industrialized economies of the developed world. Industry is becoming increasingly important, but the majority of the population is still involved in the types of occupations that have persisted with little change for centuries. Those residents who are employed in the emerging industrial sector have more disposable income, higher literacy rates, and better medical care than those who are still engaged in the traditional economy (see Table 1.3). Examples include Singapore, Hong Kong, Brazil, Mexico, Taiwan, South Korea, Argentina, and Chile.

The NICs are consciously attempting to adopt the industrial world's economic practices and are focusing their development efforts in the industrial sector. Consequently, during the decade of the 1980s their annual per capita income grew at a rate greater than that of the United States or the United Kingdom, the traditional representatives of the industrial world. Production of critical industrial items like steel rose dramatically in these countries during the last decade. The NICs are important because they are evidence

TABLE 1.1 Income, Literacy, Birthrate, Mortality, and Quality of Life for Selected Industrialized Countries

Country	Per Capita GNP	Adult Literacy (%)	Birth Rate (per thousand)	Infant Mortality (per thousand)	Quality of Life
Australia	$11,910	99	15	9.8	98
Canada	14,100	99	15	7.9	98
France	10,740	99	14	8.0	98
Germany, Fed. Rep. (West)	12,080	99	10	8.6	98
Greece	3,680	75	11	12.3	89
Italy	8,570	98	10	10.1	96
Japan	12,850	99	13	7.7	97
Luxembourg	15,920	99	10	8.6	98
Netherlands	10,050	99	13	7.7	98
Norway	15,450	99	13	8.5	98
Poland	2,070	98	17	17.5	92
Sweden	13,170	99	11	8.5	98
Switzerland	17,840	99	12	6.8	99
U.S.S.R.	7,400	99	20	2.5	92
United Kingdom	8,920	99	12	9.5	97
United States	17,500	99	16	10.0	97
Average of all industrial countries	10,810	99	13	10	92
Average of countries under $400 per capita	270	52	29	72	46
World average	3,010	74	28	77	68

ªThe Physical Quality of Life Index is a composite index of life expectancy, infant mortality, and literacy, which appears to sum the combined effects of social relations, nutritional status, public health, and family environment.

SOURCE: Morris David Morris, *Measuring the Condition of the World's Poor: The Physical Quality of Life Index* (New York: Pergamon Press, 1979), pp. 128–135 and *1988 World Population Data Sheet* (Washington: Population Reference Bureau, 1988).

TABLE 1.2 Income, Literacy, Birthrate, Mortality, and Quality of Life for Selected Capital-Surplus, Oil-Exporting Countries

Country	GNP Per Capita	Adult Literacy (%)	Birth Rate (per thousand)	Infant Mortality (per thousand)	Quality of Life
Kuwait	$13,890	60	32	18.4	80
Libya	8,640	55	38	93	59
Saudi Arabia	6,930	25	42	85	45
United Arab Emirates	14,410	56	30	38	60
World	3,010	74	28	77	68

SOURCE: World Bank, *World Development Report, 1988*; and Morris David Morris, *Measuring the Conditions of the World's Poor: The Physical Quality of Life Index* (New York: Pergamon Press, 1979), pp. 128–135.

that countries *can* move from the less developed world into the developed, industrial world, given sufficient time and capital. These countries also illustrate that the size of a country in either area or population is no more important for development today than it was historically in the experience of Western Europe.

The majority of the world's countries fall into the category of least developed. In these countries in-

TABLE 1.3 Income, Literacy, Birthrate, Mortality, and Quality of Life for Selected Newly Industrializing Countries

Country	GNP Per Capita	Adult Literacy (%)	Birth Rate (per thousand)	Infant Mortality (per thousand)	Quality of Life
Brazil	$ 1,810	76	28	63	73
Hong Kong	6,720	90	13	7.7	93
Mexico	1,850	76	30	50	76
Portugal	2,230	70	12	15.8	89
Singapore	7,410	83	15	9.4	91
Uruguay	1,860	94	18	29.5	88
Venezuela	2,930	82	30	35.5	83
Yugoslavia	2,300	85	15	27.1	87
Average of countries under $400 per capita	270	52	29	72	46
Industrial countries	10,810	99	13	10	92
World average	3,010	74	28	77	68

SOURCE: World Bank, *World Development Report, 1988;* and Morris David Morris, *Measuring the Conditions of the World's Poor: The Physical Quality of Life Index* (New York: Pergamon Press, 1979), pp. 128–135, and *1988 World Population Data Sheet* (Washington, D.C.: Population Reference Bureau, 1988).

TABLE 1:4 The Least Industrialized Countries of the World

Country	Per Capita GNP	Adult Literacy (%)	Birth Rate (per thousand)	Infant Mortality (per thousand)	Quality of Life
Bangladesh	$ 160	26	43	135	43
Ethiopia	120	15	46	118	27
Nepal	160	19	42	112	34
Zaire	160	55	45	103	45
Burma	200	66	34	103	60
India	270	36	33	104	48
Haiti	330	23	41	117	39
Kenya	300	47	54	76	48
Togo	250	18	47	117	30
Average of countries under $400 per capita	270	52	29	72	46
World average	3,010	74	28	77	68

SOURCE: World Bank, *World Development Report, 1988,* and Morris David Moris, *Measuring the Conditions of the World's Poor: The Physical Quality of Life Index* (New York: Pergamon Press, 1979), pp. 128–135, and *1988 World Population Data Sheet,* (Washington, D.C.: Population Reference Bureau, 1988).

dustry employs a minority, agriculture is the majority occupation, and poverty is the rule. Within this group we can recognize those with a per capita income of $500–1500 as a low-income group, and those with a per capita income of under $400 per year as comprising the least developed and poorest countries of the world (Table 1.4). The countries of the least developed realm vary widely in political ideologies, religious beliefs, and geographic settings. The constant that unites them is their lack of industrialization and the high incidence of illiteracy, infant mortality, malnutrition, and general misery of their population.

TABLE 1.5 GNP Per Capita by Regions, 1950–1988

Region	Population 1988 (in millions)	GNP Per Capita 1950	1975	1988
Africa	623	170	308	1,000
China, People's Rep. of	1,087	113	220	300
East Asia	215.1	130	341	1,425
Latin America	429	495	944	1,720
Middle East	124	460	1,660	2,750
South Asia	1,137	85	132	780
Less industrial countries	3,931	160	375	780
Industrial countries	1,198	2,378	5,238	10,700

SOURCES: World Bank. Col. 3: Data tapes, World Bank Atlas, March, 1977. Col. 2: Estimated by applying growth rate of GDP per capita 1950–1960 (World Bank, World Tables, 1966) to figures for 1960 GNP per capita (Atlas tapes, March 1977). Col. 1, 4: *1988 World Population Data Sheet*, (Washington, D.C.: Population Reference Bureau, 1988).

People in these poorest countries of the world are aware of their plight, and many are hopeful that world developments will lead to an increase in their standard of living. But their rising expectations are only partially fulfilled. In the 1980s a large number of countries in the least industrialized world experienced rising gross national products, coupled with an increasing share of the gross national product going to the poorest segment of the population (Table 1.5). The resulting increase in quality of life is measurable, but in most areas it has been only marginal. For example, there has been relatively little increase in the availability of food in all developing countries since 1950. Many of the countries of the least industrialized world produce less food than their people need. Grain and other food imports are costly and often unavailable to the poorest segments of the population (Figure 1.6). The tragedy is that the existing **calorie deficit** of the poorest countries in the world represents no more than 2 percent of current world production of food grains, and could easily be met by better distribution of food surpluses from wealthy countries. The lack of improvement in food supply among the world's poor is in part the result of inadequate purchasing power.

Per capita gross national product has increased in least industrialized countries since World War II, but the purchasing *value* has remained stable or declined as a result of worldwide inflation. Among the largest of the least industrialized countries in the world, only in the People's Republic of China and Mexico is the poorest segment of the population clearly better off. In India, Indonesia, Pakistan, Bangladesh, and portions of Africa, the evidence is unclear as to whether they have remained stable or actually become worse off. What is clear is that for the majority of the world's population the improvements in standard of living in the industrialized countries, the newly industrialized countries, and the oil-rich, capital-surplus countries contribute to rising expectations without significantly affecting the quality of life of the world's poorest residents in the least industrialized countries.

Only in the area of health standards has there been any significant improvement in the least industrialized countries since World War II. If life expectancy is taken as the best single indicator of national health levels, the quality of health care has obviously increased. Today the average life expectancy in all less developed countries is approximately 60 years, compared to an average of 73 years for the industrialized countries. Life expectancy in these same countries was only 40 years in 1960. In the newly industrialized countries, the life expectancy increased from 53 years in 1960 to 65 in 1989. The increase in literacy rate has also been significant, with the proportion of adults in less industrialized countries who are literate increasing from 40 percent in 1960 to 60 percent in 1989. There are major disparities between the individual countries within the less industrial world and the NICs, but the increases in health standards and literacy are important indicators that the less industrial world is slowly improving its well-being. In spite of these advances, for the poor members of the least industrial-

Calorie deficit: The total amount of food as measured in calories that would need to be added to provide minimum nutritional needs for the malnourished of the world.

ized realm, the issues of today are the need for long-term, sustained growth in per capita income and the need to distribute this income equitably to provide economic advancement for their populations. If the trends of the last two decades continue until the end of the century, people in the poorest countries of the world will be separated from those of the wealthy industrialized countries by an ever-widening gap. Advances in economic development in all but the NICs and capital-surplus, oil-exporting countries have been insufficient to close the existing gap. In real terms the quality of life for the majority of the world's population is becoming worse when compared to that of the wealthy minority population in the industrialized world.

Life in the industrialized realm. The wealthy industrial countries of the world have a life-style that is only dreamed of by the residents of the Third World. Most people in North America, Western Europe, Japan, and Australia/New Zealand enjoy freedom from want, a high degree of leisure time, and technology to free them from repetitive labor. The high degree of technology of these societies dictates a disproportionate demand for energy and other world resources. The United States, the epitome of the industrialized, technological world, consumes more than one third of the world's energy to support its life-style. Canada and the United States, with only 5 percent of the world's population, consume nearly 40 percent of the world's resources to maintain their quality of life.

Occupations in the wealthy industrialized world are different from those of the majority of the world's population. Very small numbers of individuals produce food for the rest (Figure 1.7), and the proportion of the labor force engaged in such activities is continually declining. The majority of the population in all industrialized countries lives in urban places, and the total rural population has been declining since 1900. The population characteristics in the industrialized countries are distinctive and provide useful indices to development. Birthrates are low, typically ranging from 12 to 18 births per thousand population per year. Death rates are also low with 8 to 14 deaths per thousand per year, resulting in a stable or slowly increasing population for most industrialized countries. Rarely does the growth rate in a fully industrialized country exceed 1 percent per year.

The predominant economic activity in the industrialized countries is manufacturing or distribution of manufactured items and provision of services such as banking, education, retail sales, or medical care. The countries that comprise the industrialized world normally rely heavily upon imported raw materials to supply their manufacturing sector. Then their highly skilled labor force with the necessary professional,

scientific, and engineering background transforms these raw materials into manufactured end products, which can be reexported at a profit. Some countries, such as the United States, Canada, New Zealand, and Australia, also export basic foodstuffs like grain or meat as well as manufactured goods. The advantage that the industrialized countries have in trade ensures that they will maintain their high proportion of the world's wealth. Only their requirements for energy or other raw materials that are in limited supply provide a mechanism for spreading their wealth to other countries. To date, only the oil-exporting countries have been able to obtain a part of the great wealth of the industrialized countries in this way.

As a result of the accumulation of wealth in the industrialized countries, their populations generally are free from hunger, reside in homes that would provide housing for several families in the less industrial world, and directly or indirectly consume enough food to support three persons in the less developed realm for each one in the developed region. Their literacy rates are essentially 100 percent; only individuals with learning disabilities or isolation from schools lack the opportunity to learn to read. Medical care in

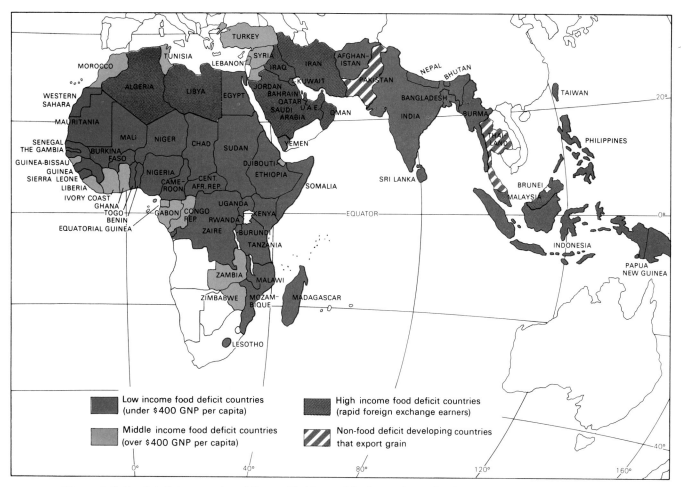

Figure 1.6 Food-deficit countries of the world. Many of the less industrialized countries of the world produce insufficient food to meet the needs of their population.

most industrialized countries (a major exception being the United States) is guaranteed for all members of society, and access to jobs is normally ensured by government programs. Certain minority populations in industrial countries (noteworthy being the United States) do not fully share in their country's wealth, but in all industrialized countries social programs designed to assist the poor protect them from the abject deprivation that the poor in the less industrial world view as their daily lot.

Governments in the industrialized realm are stable and actively engaged in a variety of social programs to ensure a high standard of living. Minimum education is compulsory and paid for by the government, housing is subsidized for lower socioeconomic groups, and government programs to train individuals for changing job requirements in a rapidly changing technological society are the rule rather than the exception. Personal incomes are high, with sufficient disposable income (what is left after meeting basic needs) to allow the majority to surround themselves with material goods. In most industrialized countries, owning a car is common; televisions, radios, refrigerators, and other consumer items are accepted as needs rather than luxuries; and the general life-style is one of affluence when compared to the less industrialized regions of the world.

This brief sketch of life-styles in the industrialized world, of course, ignores the complexity of life in these countries and the distinct characteristics of individual countries. But these internal differences have also been minimized in the industrialized societies by the increase in technology and the sophisticated level of transportation and communication systems. There is a high degree of interchange of products, which facilitates interaction with, and dependence on, other areas (Table 1.6). This interaction fosters greater uniformity in quality of life and life-style within countries of this group. Each person and area within the industrialized regions is highly dependent upon other individuals and places. There is considerable interchange of goods between places and also considerable

THE HUMAN FACTOR

TABLE 1.6 Interconnection Among Countries of the Industrial World: Building the Ford Escort Automobile

Country	Components
Austria	Tires, radiator and heater hoses
Belgium	Tires, tubes, seat pads, brakes, trim
Canada	Glass, radio
Denmark	Fan belt
Fed. Rep. of West Germany	Locks, pistons, exhaust, ignition, switches, front disc, distributor, weatherstrips, rocker arm, speedometer, fuel tank, cylinder bolt, cylinder head gasket, front wheel knuckles, rear wheel spindle, transmission cases, clutch cases, clutch, steering column, battery, glass
France	Alternator, cylinder head, master cylinder, brakes, underbody coating, weatherstrips, clutch release bearings, steering shaft and joints, seat pads and frames, transmission cases, clutch cases, tires, suspension bushes, ventilation units, heater, hose clamps, sealers, hardware
Italy	Cylinder head, carburetor, glass, lamps, defroster grills
Japan	Starter, alternator, cone and roller bearings, windscreen washer pump
Netherlands	Tires, paints, hardware
Norway	Exhaust flanges, tires
Spain	Wiring harness, radiator and heater hoses, fork clutch release, air filter, battery, mirrors
Sweden	Hose clamps, cylinder bolt, exhaust down pipes, pressings, hardware
Switzerland	Underbody coating, speedometer gears
United Kingdom	Carburetor, rocker arm, clutch, ignition, exhaust, oil pump, distributor, cylinder bolt, cylinder head, flywheel ring gear, heater, speedometer, battery, rear wheel spindle, intake manifold, fuel tank, switches, lamps, front disc, steering wheel, steering column, glass, weatherstrips, locks
United States	EGR valves, wheel nuts, hydraulic tappet, glass

SOURCE: *World Development Report 1987.*

mobility locally, nationally, and internationally among the residents of these countries. Many think little of commuting two hours to work and are willing to travel long distances for vacations and other leisure activities. The tastes and habits of the residents of the industrialized countries have been spread widely as a result of their travels. Coca-Cola has become a household word worldwide. Holiday Inns dot landscapes far from the shores of the United States. German beer gardens have spread through Africa and Asia in response to the far-flung travel habits of the West Germans, and Kentucky Fried Chicken and McDonald's restaurants are even found in less industrialized countries. At a more important level, architectural styles of commercial centers of the Western industrialized world are increasingly similar and are spreading to the capitals of even the less industrialized regions of the world.

Service industry is a major focus of the economic activity of industrialized countries, with the provision of goods and services for leisure activity comprising one of the largest sectors of the economy. The service industries include a wide range of activities, from airlines to golf courses, banks to schools, or hamburger stands to shopping centers. Leisure-time pursuits ranging from golfing to indoor tennis courts are increasingly viewed as a right rather than an option for the residents of these countries.

The irony of these life-styles is that the very wealth and affluence that characterize these regions ensures sufficient leisure time to allow the residents to contemplate their lives and question the utility or futility that they find. Have the wealth and life-style of some of the developed countries resulted in overdevelopment? Some observers maintain that the United States, Japan, and parts of Western Europe are excessively

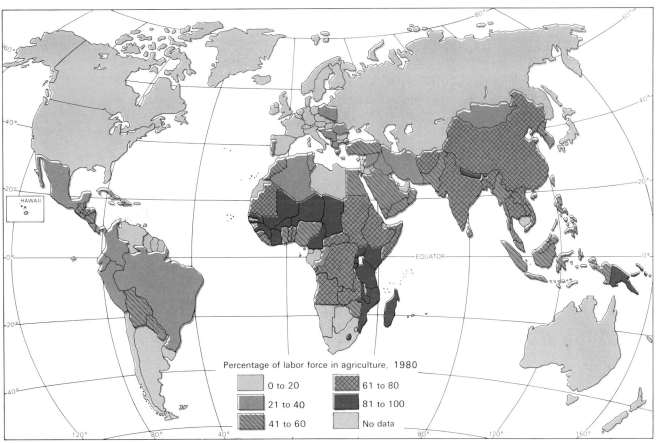

Figure 1.7 World patterns of agricultural employment reflect the degree of industrial development in the countries of the world.

industrialized, excessively urbanized, and under-humanized. Such observers see energy shortages, natural resource exhaustion, environmental pollution, congestion of population, and individual alienation as the signs of overdevelopment.

The NICs and capital-surplus, oil-exporting countries. The countries that comprise the newly industrialized and the oil-producing, capital-surplus are regions in transition. A conscious program to adopt industrialization as a tool of development or the availability of massive capital from exploiting oil resources has allowed these countries to begin the process of industrialization. Their economies are both industrial and nonindustrial in terms of occupations and quality of life. In both groups of countries a segment of the population has access to wealth, education, housing, and the other opportunities characteristic of the industrial world, while the majority of the population retains the general characteristics of the least industrialized world.

One part of the society has opportunities to participate in the emergent industrialization or control of an important segment of the wealth from the oil surplus. The life-style of these people is often more materially oriented than that of the typical resident of the wealthy, industrialized countries. For the rulers of the oil-rich countries of Saudi Arabia, Kuwait, or the United Arab Emirates, wealth is so abundant that it is difficult to spend it as rapidly as it accumulates (Figure 1.8).

People in the NICs who have jobs in the industrial sector, the emerging middle class, have higher incomes than the majority of the population of their countries, and their increased income provides opportunity for better training, better housing, education, access to medical care, and consumer items such as radios and televisions. But their standard of living is lower than that of most people in industrialized countries.

The other, far larger portion of the population of an NIC, and to a lesser extent the majority of the population in capital-surplus oil-exporting countries, live lives that are only marginally different from those of the residents of the least industrial world. The transition that is industrializing these countries is so recent that it has not yet included all their peoples.

The lives of the majority of the people involve a

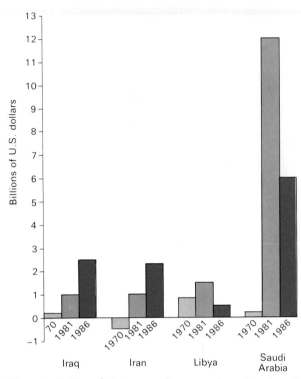

Figure 1.8 Growth in income from petroleum in selected surplus-capital oil-producing nations illustrates the infusion of capital into these countries in the last decade.

level of poverty seldom experienced in the industrialized countries. The bulk of labor is still accomplished by human or animal energy, although the oil-producing countries are rapidly constructing industries to utilize their energy resources. The majority of the population in both sets of countries is engaged in agricultural or other nonindustrial pursuits and lives in rural areas or in farming villages, where the economy is labor-intensive and subsistence-oriented. The farming sector produces little agricultural surplus to support large urban populations. Farms are small and often consist of several parcels of one acre or less in size.

In the few—often only—urban centers live those engaged in industry or other economic activity that has provided the wealthy elite with their high standard of living. If modernization continues at its present rate, there is promise that by the end of the twentieth century, the quality of life for the average individual in these countries may be substantially improved. Meanwhile, unless or until this transition is completed, these countries will remain areas of uneasy contrasts between rich and poor. The wealth of the elite in the NICs and the surplus-capital oil exporters will still be in stark contrast to the life of the majority of the population, and the conspicuous consumption of the wealthy class will continue to foster resentment and fuel political instability.

The population characteristics of the NICs and

Contrasts in housing between the huts of the poor and new apartment buildings in São Paulo, Brazil, illustrate the changes occurring in newly industrialized countries.

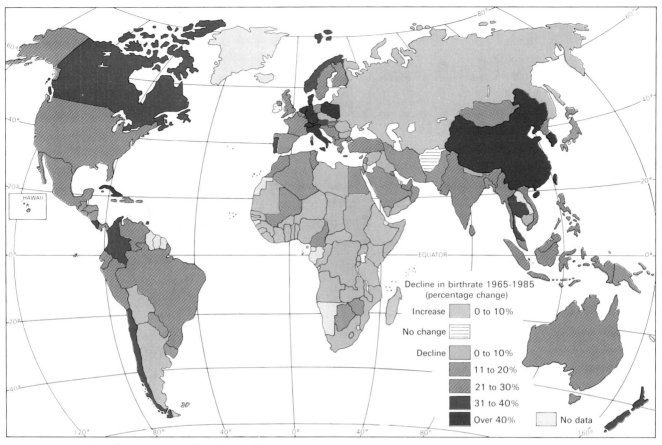

Figure 1.9 Changes in world birthrates illustrate that birthrates are declining in most countries of the world, but the most rapid declines are generally associated with the wealthier countries.

capital-surplus, oil-exporting countries are also in a state of transition. The birth rates are typically as high as in other nonindustrialized countries, but some of these countries have succeeded in lowering their birth rates and slowing population growth (Figure 1.9). Population growth rates have not declined as dramatically as the birth rates because improved medical care has resulted in an even greater decline in the death rates. The population of most of these countries is increasing rapidly, putting more pressure on the land and the economic system. Countries with a centrally planned economy have been more successful in reducing population growth than others in the NIC and oil-exporting group. Their successes are in large part the result of their ability to mobilize their population and channel their efforts into improving quality of life in many areas.

The nonagricultural sector of the economy in the NICs and oil-exporting countries depends on the export of a few staple products and some manufactured goods. In the oil-exporting countries, that one export (oil) is normally the primary basis of the economy. This provides massive infusions of capital now because of world demand for energy, but technological changes could abruptly end this income. The NICs are not as reliant on a single export item, but even these countries rely on exports of raw materials or food products. For example, Brazil is a major exporter of iron ore; Colombia is a world source of coffee; and Mexico exports petroleum and other raw materials for a significant portion of its export earnings. Much of the infant industrial sector of the NICs is heavily oriented toward extractive industries such as mining, petroleum production, or commercial agriculture.

The emerging manufacturing sector is based on relatively low-skill, repetitive, labor-intensive activities like the assembly of electronic components, manufacture of children's toys, production of textiles, and the manufacture of clothing. Management-level personnel and highly skilled technicians may be temporarily brought from more developed countries. The NICs thus exist basically at the discretion of the industrialized countries. Just as a change in technology could eliminate the demand for petroleum from the

Agricultural practices in the newly industrialized countries are little changed from those practiced by these farmers' ancestors.

oil-exporting countries, so could such a change eliminate the need for the labor-intensive industries in the NICs. More significantly, since the NICs realize a major proportion of their national income from export of the manufactured goods that they assemble, any change in the laws of industrialized countries that would discriminate against imports from the NICs could drastically hinder their economic growth. Tentatively entering the world of the industrialized, wealthy realm of Northern and Western Europe, these NICs are constantly subject to pressure from the truly industrialized countries that could effectively destroy the industrialization they have accomplished. Singapore, Taiwan, South Korea, Hong Kong, and Mexico are prime examples of this pattern.

The majority of the population of these countries resides in inadequate housing with few or no services such as water, electricity, or sewers. The cities with their industries and wealth are magnets for the poor from the countryside who perceive an opportunity for advancement in the cities. Lacking skills, education, and capital, the poor surplus population of the rural areas ends up in large ghettos of makeshift housing, crowded together in the undesirable locations of the large cities. They include the barrios of Mexico, the favelas of Brazil, and the squatters' shacks around Singapore.

Both the NICs and the oil-exporting countries are attempting to improve the standard of living of their populations through modernization and industrialization by using technology imported from the industrialized countries, but the combination of limited education and large populations makes the transition to an industrialized economy difficult. Even in the oil-exporting countries, with their surplus capital, the change to industrialization is proceeding slowly because of lack of trained personnel. Relying on foreign experts, the NICs and oil-exporting countries face a transition that will require as much as an entire generation to complete. If these countries can adequately educate the present younger generation and provide them with the skills necessary to manage and operate an industrial society, they will be able to complete the transition to industrialization. This hope for the future is a fundamental part of the rising expectations of their residents. The reality for the residents of the NICs with large populations is that the transition will not result in overwhelming wealth but will make life easier and provide a more reliable and comfortable standard of living.

The least industrialized countries. The least industrialized countries are those that have yet to begin major industrialization; per capita gross national product is low, and the quality of life is the lowest in the world. These countries comprise the majority. The per capita gross national product ranges from $1500 to less than $100 per year. The term *underdeveloped*, or *least industrialized*, should not be viewed as a value judgment about the style of life in these countries, but only a description of such factors as income and literacy and infant mortality rates. Such a designation reflects the underlying assumption that such improvements as a higher standard of living, higher literacy rates, and lower infant mortality rates result in a better quality of life but is not intended to indicate a negative assessment of the people or culture or political systems of the countries so designated.

Within this broad realm there is great variation between the truly impoverished countries like Burkina Faso in Africa with a per capita income of $150, and countries like Peru in South America with per capita gross national product of $1200. Although there are wide variations in income, the general characteristics of countries in this broad category of least industrialized areas are quite similar. In these countries live the poorest of the poor in the world, with limited hope for major improvement in the short-term future. They endure poverty, disease, hunger, and death as a way of life. Most live in villages or on small farms and practice subsistence agriculture. Subsistence agriculture emphasizes production of most of the food that the farming family consumes, with only a small proportion of goods left over for sale. Commercial agriculture is pursued primarily on plantations or the landholdings of the wealthy and emphasizes production of luxury items like bananas, coconuts, sugar, coffee, and other nonbasic foods for export to the wealthy industrialized world. These countries normally rely on a single food or other raw material for as much as 90 percent of export earnings.

The population growth in these countries is among the highest in the world. Birth rates average more than 35 per thousand and death rates hover near 15 per thousand. In some countries the birth rate exceeds 50 per thousand and the death rate is less than 20. The resulting population growth rate exceeds 3 percent per year, creating a population-doubling time of approximately 23 years.

Health and living standards in these countries are extremely low. The incidence of preventable disease is the highest in the world, and infant mortality rates may exceed 150 per thousand during the first year of life. Literacy rates rarely exceed 50 percent, and medical care is available to only a small porportion of the population. A tiny fraction of the population lives in luxury and engages in conspicuous consumption, while the great majority is subject to malnutrition, disease, and human suffering on a scale unknown in the in-

Taiwanese women assembling computers for export are one example of the emergent industrial base of newly industrializing countries.

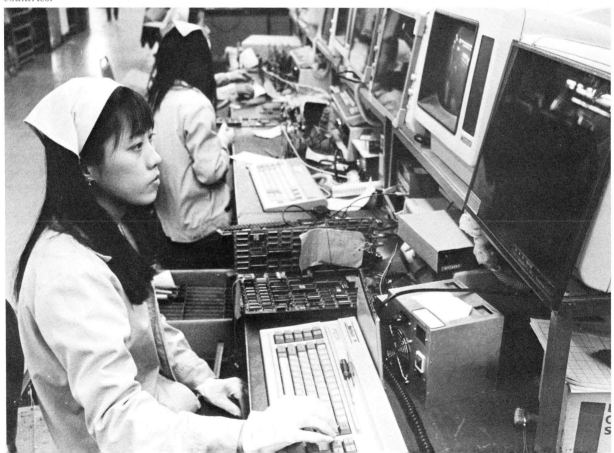

THE HUMAN DIMENSION

Paco Diaz of Mexico

Paco Diaz farms three well-watered acres of rolling countryside, raising a small quantity of corn, some wheat, potatoes, and pigs. In the process, he supports a family of eight.

He is up before dawn daily and hard at work in the fields almost before his roosters are awake and crowing. During the day, two of his sons join him in the fields, while his wife, Maria, and their two older daughters bring out breakfast and lunch on those days that Paco is especially busy.

Like many of his fellow farmers in the rich agricultural region around Toluca, west of Mexico City, Paco lives a rather precarious existence—made just a little easier by the introduction of special wheat and corn seed, and by a government-sponsored irrigation system that generally keeps land in the Toluca Valley well watered.

In a sense, both the seed, developed at the nearby International Wheat and Maize Improvement Center (CIMMYT, from its Spanish initials), and the irrigation effort are the keys to Mexico's expanding food production in recent years.

Chronic production shortages in the 1950s and 1960s forced Mexico in those decades to import as much as 40 percent of its food needs. Part of the problem was an antiquated agricultural system in which subsistence farming was the norm in many rural sectors. More than half, perhaps as many as 70 percent, of the 35 million Mexicans who live off the land are subsistence farmers and can barely feed themselves. Many of the rest eke out a paltry cash income working unviable small plots that resulted from the breakup of the big estates after the Revolution of 1910.

The newer landholding patterns may have given millions of peasants their own parcels of land, but they have done little to boost production and, in fact, have probably harmed farm output.

Moreover, all this took place as Mexico's population was soaring at a 3.5 percent a year clip, doubling every 20 years or so, with the resultant requirement for more food.

But farm production simply did not keep up.

Throughout the 1970s and early 1980s, however, there were changes for the better—increased government spending in the agricultural sector, Mexican and foreign-supported credits to small farmers, the work of CIMMYT in developing both high-yield and disease-resistant wheat and corn seed.

Mexico had gradually moved in the 1970s toward agricultural self-sufficiency—and the Paco Diazes of Mexico were in part responsible. Moreover, Paco and hundreds of thousands of small farmers like him began to experience change in their own livelihoods.

As his wheat yields grew from about half a ton an acre (about 1,200 kilograms per hectare) to the 1986 total of over 2.8 tons per acre (7,000 kilograms per hectare) Paco enjoyed increasing income. From his 1970 income of approximately $800 per year, he now receives the equivalent of almost $4,000 on the farm that he partly owns and partly rents. That's a very good income but typical of the money now being earned by Mexican farmers using the improved seed from CIMMYT experiments on well-watered land.

Paco recently purchased a 16-year-old car. "It isn't much," he smiles, "but I can drive Maria to Toluca to market and the children to the mountains for outings."

Maria, for her part, has a second-hand refrigerator, using electricity that came to their part of the Valley of Toluca in 1972. They have lights in their three-room house where five years ago they used candles.

All this has meant a significant change in the life-style of Paco and his family, as it is beginning to do for other farmers elsewhere in this cornucopia-shaped land.

The importance to the individual farmer is important and cannot be overlooked. But for Mexico as a whole, so beset with the population spiral, the change is even more important. Although self-sufficiency remains elusive, there is a humming prosperity that has bought Mexico time in its race with the twin threats of population growth and inflation.

Source: Adapted *from The Christian Science Monitor,* Jan. 12, 1979, and authors' fieldwork.

---

dustrialized world. Homes are often only shacks; electricity, running water, and sewage disposal systems are largely unknown for the majority of the population. Many of the people are fatalistic about their living conditions and have no hope of improving their life-style.

The plight of people in the least industrialized countries has received international attention. The United Nations and other international organizations are sponsoring programs to help these countries to provide housing, health, education, and food for their people. The magnitude of the population involved in the least industrialized countries (approximately 2.5 billion or 50 percent of the world's population) is so large that it is difficult to make any substantial progress. Even when foreign aid is provided, it may not have the beneficial effect that the donors anticipate. The complexities of development and the subtly interwoven connections between changes in one area of the society or economy of a developing country and all others are so poorly understood that efforts to assist these countries must be carefully planned and analyzed if they are not to create additional damage to the already fragile economies they are intended to help. This problem will be discussed in more detail in later chapters.

The majority of the residents of the least developed countries rely on subsistence agriculture for a livelihood. A woman of Burkina Faso (formerly Upper Volta) tills her field while tending her child.

GEOGRAPHIC FACTORS AFFECTING DEVELOPMENT

In the preceding pages we have introduced the vast differences between the wealthy countries of Northern and Western Europe, Anglo-America, Japan, Australia, and New Zealand and the majority of the world's residents. It should be apparent that each of the divisions of the world has its own problems. Residents of the industrialized nations may question their quality of life but rarely recognize the great gulf that exists between their standard of living and that of the majority of the world's population. The increasing awareness in the less industrialized regions of the quality of life enjoyed by the wealthy minority of the world poses the potential for international conflict.

The perception that industrialized countries consume an inordinate share of the world's resources and control an inordinate share of the world's wealth prompts hostility toward them. This hostility may result in overt action against the embassies, corporations, manufacturing plants, or other representatives of industrialized countries in the less industrialized realm. More often it takes the form of pressures on local governments by the residents of the less industrialized realm as they attempt to transform their lifestyles to reflect more nearly the potential quality of life they envision. The stresses that result from the rising expectations of the people of the less industrial world make their political system susceptible to promises by political revolutionaries who promise a better life for the poor of the world. An understanding of the factors that affect the relative prosperity of an individual country is important if we of the industrialized world are to be able to assist the less industrial world.

The complex interrelated factors that affect underdevelopment are only partially understood. The problems of maldistribution of wealth around the world are of concern to all of us and have attracted the attention of scholars from a variety of disciplines. The broad differences in levels of development are particularly useful for examining the geography of the world because of the geographers' interest in the earth as the home of the human population. Geographers are interested in the why of where. Patterns of development in the world pose such questions as:

1. Why are certain countries impoverished while others have an overabundance of wealth?
2. Why have some countries been able to begin the transition to development while others remain as impoverished as in the past?
3. Why did industrial development first occur in the Northern Hemisphere?

All these questions and others, too, are critical to an understanding of the geography of the world. Central to them all is the role of people in creating the distinct character of places that forms the mosaic of human culture on the earth's surface. The geographer is concerned with the interdependence and interactions of the earth and its human occupants. Understanding these relationships allows description of the world in which we live. The answers are rarely readily apparent and are often interrelated with other issues, preventing simple cause-and-effect statements. Each fact that a geographer examines is not the result of a single cause but of a series of causes and interdependent associations. Even a seemingly simple set of unrelated facts may be quite complex and interrelated. The goal of geography is to understand the interrelationships between facts that help to account for the patterns that characterize the earth.

Geography provides for such understanding, for it encompasses all physical, social, economic, behavioral, and political facts that have a bearing on the human occupants of a specific place in the world. Concern for the entirety of actions that affect a given place is referred to as a *holistic* approach to understanding the world in which we live. Of necessity, complete understanding of the world makes geography concerned with absolutely everything, but several broad concerns, among them the physical environment, the human occupants, and the spatial relationships that exist on the face of the globe, typify geographic studies. These central themes of geography are all related to the broad divisions of the world into rich and poor, and the patterns that emerge as the quality of life is examined are critical issues that concern modern geographers. The geographic concepts that explain these patterns of wealth and poverty are related to the physical world, the actions of the human population, and the resulting social, political, and economic organizations that exist in the world.

LOCATION: KEY TO PROSPERITY

One of the key factors that seem to affect the relative degree of economic development of a specific region, as will be apparent in later chapters, is its location relative to Western Europe. The location of places on earth is of special concern to geographers, since location is one of the central elements that contribute to the uniqueness of place. The most obvious aspect is **absolute location**—where is it? But the degree to which an area is accessible to ideas, goods, and material from other areas plays an even greater role in determining the extent of development. The relationship or **spatial interaction** between a place and the rest of the world depends on its **relative location,** its distance from other places, its accessibility or isolation, and its potential for contact at a local, regional, or global scale. This interaction may provide opportunity for assessment of local environments, which may result in increasing the pace of development. Certain locations are conductive to contacts with other areas, while others tend to isolate their occupants. Countries with a favorable relative location are accessible for trade, participate in the world exchange of goods and ideas, and are more likely to share in the higher standard of living of the wealthy countries. Countries that have a poor relative location are isolated by either physical or cultural phenomena, have more difficulty in acquiring ideas and goods from the rest of the world, and are likely to have a lower standard of living.

The island location of Great Britain and Japan has facilitated interaction by ship, which was the primary means of global transportation until the twentieth century. Other locations, including those isolated by mountains, deserts, or cultural phenomena such as language, have failed to benefit from the technological advances taking place in other areas of the world.

Among the poorest countries in the world, Chad, Rwanda, and Burundi are all located in the interior of Africa, separated from contact with industrial Europe by physical distance, climate, and landforms and by cultural barriers. Even today they lack adequate transportation facilities to integrate them fully into the world transport system, and they remain isolated. Many other members of the less industrialized world also have problems of physical location and resultant geographic relationships that have handicapped them and prevented them from acquiring necessary technology and expertise for development. Many less industrial-

Absolute location: The geographic location defined by latitude and longitude and specific internal geographic characteristics of a place. Used interchangeably with *site.*

Relative location: Refers to location of a place or region with respect to other places or regions. Used interchangeably with *situation.*

Spatial interaction: The relative amount of interaction between a place or region and other places or regions.

ized countries are also handicapped by their own internal physical characteristics. For example, Chad in Africa is mostly desert, which hinders attempts to expand agriculture, since there is inadequate water for irrigation. In other less industrialized countries rugged landscapes with little level land or other resources hinder economic growth.

The characteristics of the immediate location of a town or village are known as its *site*, while the location of a community relative to other places is its *situation*. The combination of site and situation historically has played a major role in affecting the degree to which a specific country or community developed. There are a number of other geographic concepts related to location that seem to affect the level of development, as can be illustrated by an examination of two cities.

LOCATION OF URBAN CENTERS: CONTRASTS BETWEEN THE INDUSTRIALIZED AND LESS INDUSTRIALIZED WORLD

The cities of Rotterdam, the Netherlands, and Calcutta, India, exemplify the importance of site/situation relationships and associated relative accessibility and interaction. The site of Rotterdam was conducive to location of a port city, since it was on the Rhine River. The Rhine is now, as it has been for centuries, the major transport link of Western Europe. The countries and states that today comprise Germany, Switzerland, France, Luxembourg, Belgium, and the Netherlands all utilized the Rhine River for transport in the preindustrial era with its reliance on water-transport systems. Located at the boundary between ocean transport systems and the transport systems using smaller craft on the Rhine River, the location of Rotterdam was one of several logical places to break the large bulk cargo of oceangoing vessels into smaller units for distribution on the Rhine and associated overland transport systems. This **break-in-bulk** function, in turn, fostered a demand for labor to handle the transfer of cargoes, specialists to process raw materials shipped from the farmlands of the interior of Western Europe and the imported goods from the rest of the world, and the development of merchants and governments to facilitate the transfers. The Rotterdam location was accessible to western European lands, and it became a central location for assembly of raw materials and subsequent manufacturing processes after the Industrial Revolution in the nineteenth century.

The Rhine River allowed Rotterdam to provide urban services for a broad area extending inland from the city proper. The area that a city serves is known as its **hinterland,** and Rotterdam's hinterland became the western European countries of Germany, France, Switzerland, the Netherlands, Belgium, and Luxembourg. In portions of all these countries Rotterdam was the focus of goods being shipped to market and the source of manufactured items or products imported from other lands. Rotterdam's growth to one of the largest cities in the world, and its present status as the world's busiest port in terms of tonnage of goods handled, reflects its advantageous site/situation relationships, its accessibility, and its access to a large hinterland (Figure 1.10).

The site/situation features that fostered Rotterdam's growth were multiplied in their effect because Rotterdam early developed **complementarity** with the hinterland it served. Goods were drawn from Western Europe for export or consumption in Rotterdam, but imported items were also provided to the residents of the towns and villages of the area the city served. This complementary relationship between Rotterdam and its hinterland is an important characteristic of major cities in the developed countries of the world. Such cities are known as **generative cities.** The city becomes a central place in which the urban functions of manufacturing, trade, and government administration are carried out for the benefit of the residents of both the city and hinterland it serves. By comparison, cities in less industrialized regions with favorable site/situation relationships have thus far failed to develop to the size and complexity of industrial cities like Rotterdam because they have a noncomplementary or *parasitic* relationship to their hinterlands.

Cities of the less industrialized world were located with respect to the same basic geographical principles but with an underlying goal that differed from that of cities in the industrialized countries. In the presently less developed countries the largest cities were typically established by the industrial powers as a focus

---

Break-in-bulk: Location where large cargoes are broken down into smaller units.

Hinterland: The area a city serves.
Complementarity: Production of goods or services by two or more places in a mutually beneficial fashion.
Generative city: A city with complementary relations with its hinterland.

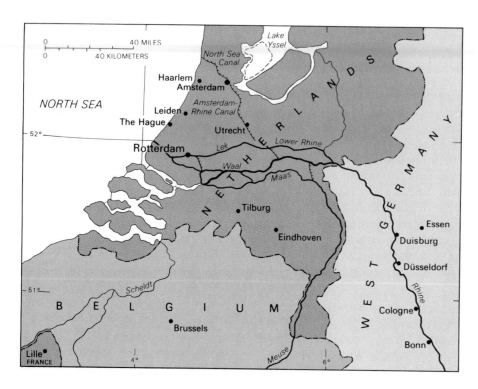

Figure 1.10 Situation relationships of Rotterdam and the Lower Rhine River. Rotterdam's location on the river downstream from other west European countries facilitated its emergence as the busiest port in Europe.

for economic activities designed to benefit the industrialized country that established the city. As a consequence the relationship that developed was based on extracting raw materials from the less industrialized country and exporting them to the industrialized world for manufacture and distribution. The cities of the less developed realm reflect this developmental pattern by focusing on accumulation of goods for export rather than provisions of services to the hinterland from which the materials are drawn.

The **parasitic cities** of the less industrialized world do not serve the same role as cities of the developed countries. The site/situation relationships that led to the original establishment of Rotterdam are similar to those that led to the establishment of colonial cities such as Calcutta; today's Kinshasa in Zaire, Africa; or Cape Town, South Africa. The process by which these cities developed resulted in major differences in their present function and importance. Whereas the cities of today's industrial world built complementary relationships with their hinterland, the cities of the less industrialized world established by colonial powers from the developed world maintained a parasitic relationship with their hinterland. Consequently the flow of goods and ideas has been essentially one way: from the less developed world to the developed region.

Parasitic city: A city that does not have a complementary relationship with its hinterland.

This traditional process has greatly contributed to the lack of development in the Third World today.

Calcutta, India, illustrates the role of site/situation relationships in the development of cities in the less industrialized world. Calcutta was established as a colonial city by the British in 1651 to acquire silk and saltpeter (for gunpowder) from the Indians. Calcutta is located on the Hooghly River, one of the distributaries of the Ganges River. The site is low and swampy, but river boats provided access to the Ganges River and the interior of India. Calcutta became the most important British colonial port in India as railroads were constructed to extend its hinterland to India's tea-producing regions. The level plains of the Ganges delta became the focus of production of jute, a fiber crop used to make burlap, fostering growth of processing industries in Calcutta. The growth of Calcutta under British rule continued until the independence of India in 1947. Population growth and development of industry and commerce were oriented to exploitation of India by the British, and the relationship between Calcutta and its hinterland never developed complementary. This asymmetrical relationship was typical of parasitic colonial cities. Since independence, Calcutta has become the focus of important industrialization efforts designed to assist all of India, but the city has yet to become a generative center for the country.

The natural site/situation relationships that originally distinguished individual sites of the world may

THE HUMAN DIMENSION

Life in the Poorest Countries

At least four major diseases afflict Wayen [a village in Burkina Faso in Africa]: Onchorerciasis, or river blindness, caused by the tiny worm transmitted by the black fly; a large parasite, known as the Guinea worm, which enters the human body by way of the foot; and leprosy and malaria. Intestinal parasites are also common.... All these diseases produce suffering and general debility. While I was amazed at the vigor and resilience of the villagers, I did in fact see many, especially children, knocked flat by sickness. It was always astonishing to see a person in seeming good health, although I saw many such people, too. But I doubt there is a person in Wayen who does not suffer from one disease or another. Many, such as the blind leper, Bayoure, have at least two plainly visible afflictions.

The closest public health clinic is too far from Wayen for convenience. Public transportation is rare, as uncommon as the cash needed to use it. The people of Wayen are left only traditional medicine to treat their ills. One sees little rags tied around bony ankles, one hears about magic incantations. But judging by the prevailing state of health and the high rate of mortality, it seems safe to say that Wayen cannot cure itself.

Source: Ray Withen, Action UNDP, May–June, 1977.

---

be less important than the actions of the groups of people who affect the use of these locales. The decisions of colonial powers from what is now the industrial world played a major role in the rate of development in the less developed world in the past three centuries. The changes in human activities and goals also affected the development within the present industrial world. Rotterdam was not the largest port in the Netherlands during the sixteenth, seventeenth, eighteenth, and nineteenth centuries, as Amsterdam, which also had suitable site/situation relationships, maintained this position. Both Amsterdam and Rotterdam occupied distributaries of the Rhine River. (Distributaries are channels into which a large river divides as it meets the coastal plain and the rate of flow slows down.)

Rotterdam's role as the world's major port developed only as events occurred in and around the Netherlands, reflecting changing forces and goals in the world. The Industrial Revolution of the eighteenth and nineteenth centuries and the political unification of Germany in 1871 led to the emergence of the German nation as a major industrial power. German industrial might was concentrated along the Ruhr River, a tributary to the Rhine. The Rhine thus became the major industrial throughway for Western Europe, and demands for improvements of the waterway led to the construction of the New Waterway in 1872, taking the major access to the Rhine through Rotterdam. In the twentieth century the increasing industrialization of Western Europe, with its concomitant demand for petroleum for energy, has increased the importance of Rotterdam as a port. Drawing petroleum resources from throughout the world, large ships bring the energy necessary to sustain the industrial might of the western European countries.

The continued growth of urban functions in Rotterdam has caused expansion and absorption of neighboring towns and cities. Urban growth has created a great urban conglomeration focused on Rotterdam called the Randstadt, a supercity combining Amsterdam, Delft, The Hague, and Leiden. This development of Rotterdam as a supercity (sometimes referred to as a **megalopolis**) represents the epitome of capitalizing on site/situation relationships in a changing world. The discussion of the growth of Rotterdam as compared to the cities of the less industrialized world illustrates that the realities of geographic location may present an opportunity for interaction and development, but the decisions and technology of the residents of that location may play a greater role in determining whether development occurs, even in seemingly ideal geographic locations.

GEOGRAPHY AND THE HOW AND WHY

The complexity of the relationships between the physical environment, location, resources, and human decisions in the pattern of wealth and poverty in the

---

Megalopolis: A term originally applied to the urban region from Boston to Washington, it refers to a group of large cities that has effectively become one large urban area through their growth.

world precludes simplistic explanations of cause-and-effect relationships. It should be apparent that an understanding of the geography of the world and its central issues of poverty and level of economic development involve not only an understanding of *where* the regions of development and underdevelopment are located, but factors that contribute to their relative standard of living. It is simple to say that Calcutta, India, is a city of more than 10 million people (including suburbs) in which the majority live lives of abject misery when compared to the residents of New York City. Such a statement ignores the differing values and goals of the residents of these two areas, the diverse cultures involved, and the processes that have led to industrial development in the United States and relative lack of industrial development in India. The simple questions of what and where, easy to answer, are essential to an understanding of the world. Without an understanding of what life is like in less developed regions, it is difficult to comprehend the problems and attitudes facing the residents of these regions. Without an understanding of where both developed and less developed regions are, it is impossible to grasp the magnitude of the challenge to the wealthy of the world if they are to assist in raising the standards of living of the world's poor.

Transcending both questions of what and where are the questions of *how* and *why* some countries are industrialized, wealthy, pleasant places in which to live, while others have persisted in or entered a world of poverty, ignorance, and disease. The purpose of this book is to provide both a basic knowledge of the *what* and *where* of the world in which we live and to suggest some explanations of *why* the problems persist. The discussion of location as a factor in development illustrates the complexity of attempting to answer this question. The following chapters will deal with other elements that seem to affect development, specifically the resources available in a specific location that have been important historically, and the specific experience of regions of the world in development and the resulting sense of place that exists within these regions.

The range of issues that affects the world is broad and interrelated in complex ways. The complexity of issues that face the world range from overpopulation to conspicuous consumption of resources; from infant mortality to the cancer and heart disease of the residents of the wealthy world; and from the cost of a new tractor in Western Europe to the difficulty of having a hand sickle repaired in Africa. The enviromental concerns of the residents of North America, the specter of disease and death that daily accompanies the residents of most African countries, and the perennial problem presented by life in harsh environments such as the great African desert, the Sahara, or the far north are all problems of the relation of the human population to the area in which it lives. These relationships are largely an outgrowth of the type and success of the economic, social, and political organizations that a group of people develops. The degree to which a specific area is industrialized or economically developed is often an indicator of the quality of life and the type of relationships that exist between the local occupants and their geographic setting. For this reason the theme of economic development will be used to unify this book. Not all issues of the world are directly attributable to underdevelopment or overdevelopment, but the relative degree of wealth or poverty in a specific region is closely related to the myriad problems and issues confronting its occupants.

ORGANIZATION OF THE TEXT

The factors that affect economic development vary in importance, and the degree to which they contribute to the quality of life in a specific area is debatable. Even a casual observer can see that certain elements do have a direct impact to the well-being of the residents of a place. Access to drinking water, land for producing food, climatic characteristics, and resource availability are all related to the nature and quality of life in a given place. The type and degree of social organization, political structure, and economic system similarly affect life and well-being. The characteristics of the population of an area are also related to the livelihood of the people. The reamining chapters in Part 1 analyze the physical and cultural factors that affect major issues of the world and the pattern of development and underdevelopment of the earth. Chapter 2 discusses in detail those aspects of the physical world that directly affect the potential for development of the world regions. Chapter 3 examines the population of the world and the problems that the less developed world faces because of its particular characteristics. Chapter 4 analyzes cultural organizations and linkages and suggests the effect of such forces on potential for development.

The information in Part 1 is essential background to understanding individual countries and their relative level of development. The answer to why underdevelopment exists in countries like Rwandi involves an understanding of more than simple location, more than the harsh environment they occupy, and more than the historical processes of socioeconomic and political forces that have affected this country, but these factors provide the starting point for explaining lack of development and potential problems such

Applied geography: what geographers do

The issues and problems presented by the world's geography are reflected in the diversity of professional careers of those who choose to study geography. Geographers are engaged in a wide range of employment relating to their training in Geography. Their expertise is used in activities ranging from analysis of international relations to selecting sites for retail outlets of fast-food restaurants; from planning land use for individual communities to regional planners dealing with development projects for entire countries; from elementary school teachers to employment in explaining and minimizing adjustment to natural hazards. The geographer who is involved in business-related employment may work for a large firm in helping them to determine the location of stores or factories, or they may be engaged in determining the best markets for a product.

Geographers map regions showing market areas, locate optimal sites for both factories and retail outlets, and analyze the opportunity for profit or potential for loss in differing locations and regions. Rarely will these people be called geographers. Geography is sometimes associated with only elementary geography classes focusing on memorization of places. Consequently, the geographer working in business is known as a facilities planner, a market researcher, a research analyst, a locational specialist, a traffic manager, a cartographer (map maker) or geoinformation processor. While the titles vary, each reflects the need for an individual trained in the ability to deal with real world problems that involve culture, economics, and environment.

Another large area of employment is associated with activities of governments of all levels. At the local level, geographers are engaged in land use and city planning, transportation planning, real estate, housing, economic development, or cartography. While there is a wide range of activities, all are involved with the general areas of research, policy formulation, program development and administration, and data collection and analysis relating to humans and their use of the environment. Geographers involved in city planning provide guidance for daily decision making concerning approval of subdivision plats and enforcement of zoning codes to prevent inappropriate land uses; are engaged in long-range planning to ensure that the future development of the city will result in a desirable environment in which to live; plan the routes and flows of traffic; and provide guidelines for housing and business for individual communities and regions. Geographers also provide guidance to local, state, or federal agencies as they attempt to minimize destructive natural hazards.

Understanding how people view hazards has resulted in federal programs to guide the location of housing and business and industry to minimize the damage from floods, earthquakes, or hurricanes. Geographers are actively involved in such activities under a variety of titles such as hazard mitigation specialist, natural-hazards planner, or cartographer. Geographers also continue to play a major role in teaching at all levels from the elementary school to the university.

---

countries face as they attempt to transform themselves and join the wealthy, industrialized world.

Part 2 examines the industrialized world with its affluence and conspicuous consumption. The bulk of wealth and economic development in the world is focused on the relatively tiny region of Europe. The level of development in the rest of the world is a function of accessibility to this European birthplace of industrialization through the past three centuries. Access to economic developments in western Europe came via migrants, traders, political dominance, or religious missionaries. Those areas of the world that entered into two-way trade with Europe on a reciprocal basis repeated the European industrial experience. Those areas that became the focus of parasitic European extractive trade have remained less developed. This book will examine the industrial world to show how development occurs. Although the European model has not been the basis of all successful attempts to revolutionize economies of the world, it has resulted in the highest individual standard of living. Canada and the United States (Chapter 7), Australia and New Zealand (Chapter 8), all adopted complementary trade relationships with Europe and adopted the technological and economic advances that fostered industrial development. Japan (Chapter 10) represents a unique case of development because it resisted European colonization attempts and prevented development of a parasitic relationship with Europe, but it adopted European economic and technological expertise. The resulting high standard of living in Japan represents the earliest example of a non-European society achieving a high level of income and technological development in modern times.

Other regions of the world that have achieved industrial and economic development have utilized different systems from the capitalistic European model. The Soviet Union, the subject of Chapter 9, is an example of a centrally planned economy that has managed to enhance greatly the standard of living of its residents. The satellite countries of the Soviet Union in Eastern Europe have utilized a similar model to transform their economies from agrarian, less developed systems to developing industrialized countries.

The balance of the world including Asia, Africa, the Pacific world, and Latin America can be typified as a region of lesser economic development and lower standards of living and is examined in Part 3. The countries in these regions have remained less developed for a variety of reasons. Many have attempted to revolutionize their economies using either the capitalistic model, the example of the centrally planned economies, or some combination or variant of the two. Chapters 12 and 13 of this book will examine the experience of the world's most populous countries, India and China, and contrast their developmental methods and resulting issues and advance. Chapter 14 (Southeast Asia and the Pacific) illustrates the problem of a region characterized by political and cultural fragmentation. Chapter 15 deals with the Middle East where the Islamic religion seemingly provides the potential for unity in development, a unity which to date remains unrealized.

Africa (Chapter 16) represents the most isolated area of the world, and development in Africa is at a lower level than other world regions. Latin America (Chapter 17) represents a region that was brought into the European sphere of influence prior to the Industrial Revolution, but its long history of association with European countries has not resulted in a uniformly high level of development. The Latin American experience illustrates the nature and direction of changes that may revolutionize the world distribution of wealth. In Part 4, Chapter 18, we will look at some of the issues on a world scale.

The division of the world on a continental basis obscures many important variations within regions. In Asia there are countries that are destitute, whose standards of living and technological level have been described as moving rapidly from the thirteenth century into the fourteenth; and at the same time there are countries of industrialization and affluence to rival any in the world. The variations and efforts to transform the economies of Asian countries range from democratic, individualistic, capitalistic entrepreneurship to the centrally planned, socialist economies of the people's republics of China and Vietnam. The contrasts in Latin America are nearly as great as those in Asia.

In spite of this complexity of developmental level within continents, the broad grouping of Europe, North America, and Australia and New Zealand as the industrialized world corresponds closely with the actual per capita income and quality of life index. The designation of Asia, Africa, Latin America, and the Pacific world as less developed countries includes a range of standards of living and levels of technological development, but a degree of distinctiveness when compared to the truly industrialized countries of the industrialized world justifies their inclusion in this broad region of underdevelopment. The geographic factors affecting each of these regions are partially responsible for the lack of development there. The nature of the climate, the fertility of the soils, and the availablity of resources differ widely from continent to continent and are in part responsible for the world pattern of wealth and poverty discussed earlier. The next chapter examines this physical world.

FURTHER READINGS

ADLER, R. et al. *Spatial Organization: The Geographer's View of the World.* Englewood Cliffs, N.J.: Prentice-Hall, 1971.

———. *Human Geography in a Shrinking World.* North Scituate, Mass.: Duxbury Press, 1975.

AMEDEO, D. and GOLLEDGE, R. *An Introduction to Scientific Reasoning in Geography.* New York: Wiley, 1975.

BROWN, LESTER R. et al. *State of the World, 1988.* Washington, D.C.: World Institute, 1988.

"Changing Patterns of Energy Use." *The Economist* as reprinted in *World Press Review*, February 1982.

CHORLEY, R. and HAGGETT, P., eds. *Models in Geography.* London: Methuen, 1967.

COLE, J. P. *The Development Gap: A Spatial Analysis of World Poverty and Inequality.* Chichester: Wiley, 1981.

———. *Development and Underdevelopment: A Profile of the Third World.* London and New York: Methuen, 1987.

CROW, BEN & THOMAS, ALAN, eds. *Third World Atlas.* Philadelphia: Taylor & Francis, 1984.

CUTTER, SUSAN L. et al. *Exploitation, Conservation, Preservation: A Geographic Perspective on Natural Resource Use.* Totowa, N.J.: Rowman & Allanheld, 1985.

DAVID, WILFRED L. *Conflicting Paradigms in the Economics of Developing Nations.* New York: Praeger, 1986.

DE BLIJ, HARM J. & MULLER, PETER O. *Human Geography: Culture, Society, and Space*, 3d ed. New York: Wiley, 1986.

DE SOUZA, A. R. and PORTER, P. W. *The Underdevelopment and Modernization of the Third World.* Washington: Association of American Geographers, 1974.

DUBASHI, JAY. "Must the Poor Get Poorer?" *India Today* as reprinted in *World Press Review*, July 1983.

"Economic Development." *Scientific American* 243, September 1980.

Geography and International Knowledge (Report). Washington: A.A.G., Committee on Geography and International Studies, 1982.

GEORGE, SUSAN. *How the Other Half Dies*. Montclair, N.J.: Allanheld, Osmun, 1977.

GLASSNER, M. and DE BLIJ, H. J. *Systematic Political Geography*. 4th ed. New York: Wiley, 1988.

GOULD, PETER R. *The Geographer at Work*. London: Routledge & Kegan Paul, 1985.

GRANT, LINDSEY. "The Cornucopian Fallacies." *The Futurist*, August 1983.

HART, J. F. "The Highest Form of the Geographer's Art." *Annals of the Association of American Geographers* 72, 1982, pp. 1–29.

JACKSON, ROBERT. *Global Issues 86/87*. Guilford, Conn.: Dushkin Publishing, 1986.

JAMES, P. E. and MARTIN, G. *All Possible Worlds: A History of Geographical Ideas*. 2d ed. New York: Wiley, 1981.

KAKWANI, NANAK. "Welfare Measures: An International Comparison," *Journal of Development* 8, 1981, pp. 21–45.

LANEGRAN, D. and PALM, R., eds. *An Invitation to Geography*, 2d ed. New York: McGraw-Hill, 1978.

LARKIN, R. and PETERS, G., eds. *Dictionary of Concepts in Human Geography*. Westport, Conn.: Greenwood Press, 1983.

LEY, D. and SAMUELS, M., eds. *Humanistic Geography: Prospects and Problems*. Chicago: Maaroufa Press, 1978.

MORRIS, M. D. *Measuring the Conditions of the World's Poor*. New York: Pergamon Press, 1979.

PATTISON, W. "The Four Traditions of Geography." *Journal of Geography* 63, 1964, pp. 211–216.

POLSGROVE, CAROL. "Strategic Minerals: Reality and Ruse." *Sierra*, July/August 1982.

ROBINSON, J. "A New Look at the Four Traditions of Geography." *Journal of Geography* 75, 1976, pp. 520–530.

SACHS, IGNACY. *The Discovery of the Third World*. Cambridge, Mass.: MIT Press, 1976.

SALTER, C., ed. *The Cultural Landscape*. Belmont, Calif.: Wadsworth, 1971.

SIMON, JULIAN L. "Life on Earth Is Getting Better, Not Worse." *The Futurist*, August 1983.

SIVARD, RUTH L. *World Military and Social Expenditures*. 12th ed., Washington D.C.: World Priorities, 1987.

SUNQUIST, FIONA. "Cut and Burn in a Nepal Park: All in the Plan." *Smithsonian*, March 1984.

SYMANSKI, R. and NEWMAN, J. "Formal, Functional, and Nodal Regions: Three Fallacies." *Professional Geographer* 25, 1973, pp. 350–352.

TATA, ROBERT J. and SCHULTZ, RONALD R. "World Variation in Human Welfare: A New Index of Development Status." *Annals of the Association of American Geographers* 78(4), 1988, pp. 580–593.

WHITTLESEY, D. et al. "The Regional Concept and the Regional Method." In *American Geography: Inventory and Prospect*, pp. 19–68. Edited by P. E. James and C. F. Jones. Syracuse, N.Y.: Syracuse University Press, 1954.

World Bank. *World Development Report 1987*. New York: Oxford University Press, 1987.

World Resources Institute. *World Resources 1986: An Assessment of the Resource Base That Supports the Global Economy*. New York: Basic Books, 1986.

Sichuán, China.

The Environmental Factor resources and development

MAJOR CONCEPTS

Club of Rome / Environmental perception
Tectonic forces / Gradational forces
Climatic types / Continentality
Precipitation types / Soil classifications
Environmental determinism / Ethnocentricity
Localized resources / Ubiquitous resources
Topography / Landforms / Dew Point
Cyclonic / Orographic / Rainshadow
Convectional / Biomass / Coniferous
Deciduous / Humus / Leaching / Alluvial
Seventh Approximation

IMPORTANT ISSUES

The limits of the earth
Relationship between environment and occupants
Global resource consumption
Access to adequate food by the world's poor
Global wealth transfers

Global Environmental Characteristics

1. Resources can be defined as naturally occurring elements that are perceived to be useful by a specific group. The utility of a resource is a function of geographic aspects of location, quality, and size of deposit, accessibility, and the kind of technology of the group.
2. Environmental factors such as landforms, climates, soils, and vegetation are integral parts of the character of each place.
3. The combination of environmental factors in any place affects the utility of that place for its residents.
4. Environmental factors do not determine human actions nor guarantee wealth or poverty.
5. The natural resources of most importance in a region are related to the type of economic activity found there.

The complex mosaic of wealth and poverty that characterizes the geography of the world is the result of a series of intricately interrelated factors. The primary factors leading to poverty seem to be human, social, and political problems, unfavorable natural conditions, insufficient access to financial and technical means of development, and general isolation from world interaction. The extent to which each of these factors is responsible for poverty or wealth is highly variable and difficult to assess adequately. It is unclear whether the lack of development in a specific place is primarily the result of human decisions or natural conditions. What is clear is that in order for development to occur, many of the factors must be favorable.

Among the obstacles to development are, first, unfavorable natural conditions such as harsh climate, or lack of natural resources (both underground mineral resources and on-the-ground resources of fertile soils, forest, or water). Second, there are human, social, and political problems such as social and political organizations that hinder the developmental process, the persistence of archaic customs and social habits, a low level of agricultural techniques and know-how, or the relatively recent establishment of educational systems. Third, places may lack access to financial and technical means of development, including capital, technology, and the world's wealthy markets. Fourth, isolation from the interactions that characterize the modern world, thus minimizing the possibility for capital accumulation, technology transfers, and modern agricultural methods may impede development.

This chapter examines the natural conditions of the earth that seem to be related either directly or indirectly to the development process. Although the presence or absence of resources does not determine that a specific region will be either industrialized or less industrialized, relatively abundant resources in a region, in many cases, seems to be related to ease and rapidity of development. Understanding the problems posed by environments in less developed regions is essential if the industrial, developed countries are to assist the impoverished world adequately in raising its standard of living. Examination of Figure 1.5 indicates that the majority of the less industrialized countries are found in areas with harsh environmental settings. Excessively hot, wet, or dry climates pose problems for development, which the countries of Western Europe and North America did not face.

THE LIMITS OF THE EARTH

There will be more than 5.3 billion people occupying the earth as of 1990. The resources necessary for survival of the earth's inhabitants, whether produced by nature or by people, are a primary factor in explaining the well-being of these billions of people. Some portions of the earth's surface are less suitable for occupation than others because of climatic or other geographical factors. Basic requirements for food and shelter cause the human population to be clustered in more favorable sites. The concentrations of population shown in Figure 2.1 are the result of past human decisions about obtaining food, shelter, and other needs. Densely inhabited areas, such as the Ganges River plain of India, historically were easier to occupy because of the low level of technology required in a warm, humid environment and flat land. By contrast, the Sahara of North Africa has never had more than isolated settlements because of the difficulties posed by a severely restricted supply of water. These simple examples of the role of the physical environment in population distribution mask important causal relationships. The lack of water in the Sahara does not *prohibit* people from living there, but it does present problems that can only be overcome through application of higher technology or at great cost.

The environment and the resources that confront the human occupants of the earth are a geographical reality, but a reality that also depends on the residents' assessment of the utility of the set of resources provided at a specific site. Most people view the Arctic lands of Canada and Alaska as less desirable locales, but the Eskimo population developed a suitable technology for utilizing the Arctic's resources. The Bedouin of the Sahara are able to live in a region that most Europeans and Americans would perceive as unsuitable for human occupation. Such examples of resource use in environmental settings with only sparse population illustrate the extent to which resources are a cultural and technological phenomenon. The resources that the Australian aborigines relied upon included rodents, small animals, wild plants, and a host of other naturally occurring foods. The majority Anglo population of Australia today does not recognize any of them as a basic resource. The contrast in assessment of resources between aboriginal and Anglo populations of Australia illustrates the changing relationship between the physical environment and the human use of the world.

The relationship between the human occupants and the physical environment of the earth is complex and difficult to categorize adequately. For some groups with only simple technology, the environment with its naturally occurring resources is a primary factor in affecting the quality of life enjoyed by the inhabitants. For residents of the wealthy industrialized world, the limits presented by the earth's physical attributes are more indirect. Residents of North America or Western Europe are largely insulated from the vagaries of day-to-day climatic phenomena, are insulated from hazards of drought or other food-threatening occurrences by efficient social organizations; but ultimately rely upon the earth because of their almost insatiable demand for mineral resources. The industrialized world contains 30 percent of the earth's population, but it consumes 87 percent of all energy resources each year. Consumption of some mineral resources such as aluminum or copper is almost entirely within the industrialized countries, which consume an average of 95 percent of world production.

The disproportionate share of resources that the industrialized world consumes epitomizes the complex role of the physical environment. Resources that might benefit the residents of a less industrialized country are too often exploited by industrial countries for use beyond the boundaries of the less industrialized region. The overwhelming concentration of resource consumption in industrialized regions suggests that in order to increase the standard of living of residents of less developed areas, residents of developed areas must be willing to curtail their consumption of the earth's finite resources.

THE CLUB OF ROME AND THE CRUEL CHOICE FOR THE WORLD

The widening gap between the residents of the wealthy industrialized countries and those of the less developed world has attracted the attention of many concerned world citizens. About 70 individuals from 26 countries have formed the Club of Rome. These people are not political leaders but are all successful residents of their respective countries who are united by their concern for the world's future. One of the accomplishments of the Club of Rome was the preparation and publication of volumes dealing with possible alternative futures of the world. The most famous of these volumes, *The Limits of Growth*, was first published in 1972. The data represent the analytical work of a group of scientists using a computer simulation of the interrelationship of factors affecting the world's future. The computer programs revealed just how finite are the world's resources.

It has long been unpopular in geography to maintain that the physical environment determines the actions of the human population, but the Club of Rome's findings suggest that there are absolute limits to the human use of the earth, given the present level of technology and present demands for material goods. The significance of the findings is that increasing demands for resources by an ever-growing population, with the associated pollution from an industrialized society, limit the extent to which all peoples can increase their standard of living (Figure 2.2). Increasing demands for natural resources ultimately lead to a decline in the quality of life and in total numbers of

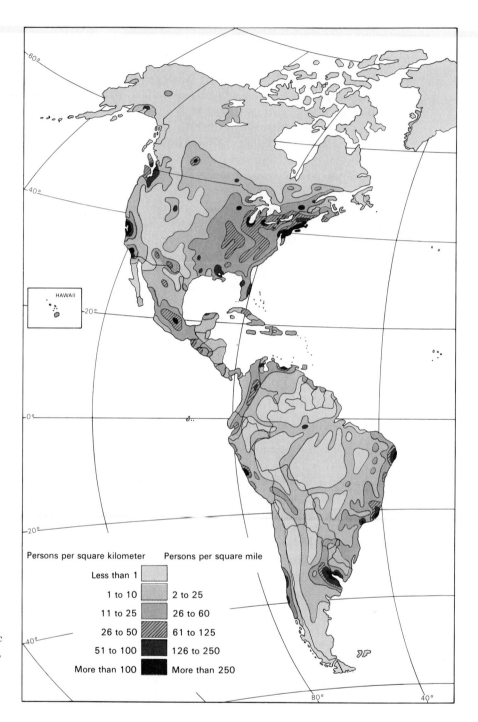

Figure 2.1 World population densities. The density of population in a specific location is related to physical geographic factors such as climate, vegetation, soils, and landforms as well as accessibility level of economic development, and historic development.

Persons per square kilometer / Persons per square mile
- Less than 1 / 2 to 25
- 1 to 10
- 11 to 25 / 26 to 60
- 26 to 50 / 61 to 125
- 51 to 100 / 126 to 250
- More than 100 / More than 250

people. The specific dates that the Club of Rome findings suggest for these events are less important than their prediction that scarcity of resources will ultimately lead to a decline in the quality of life. Since it is impossible for all the growing world population to experience an increase in wealth when the resource base is finite, the cruel choice becomes one of concern to residents of the industrialized countries.

The Club of Rome findings are supported by projections that show the gap between wealth and poverty in the world increasing (Figure 2.3). The concentration of capital, wealth, and power in the industrialized countries means that in the ever-increasing competition for resources, the industrialized world is in a position of dominance. If the world is to experience a general increase in the standard of living of all of its residents, those of the industrialized parts of the world must consider means of sharing their wealth with those in the impoverished less industrialized countries. Proponents of this view argue that the

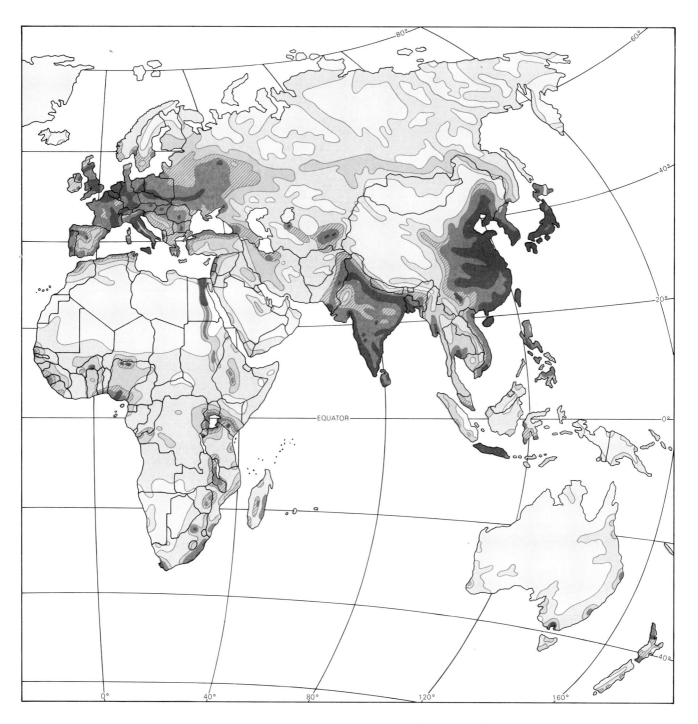

cruel choice is between voluntarily curtailing ever-increasing wealth and standard of living in the industrialized countries to allow the less developed world to begin to industrialize and increase their standard of living, and the present continued accumulation of wealth simply for the sake of wealth in the industrialized countries, ultimately leading to catastrophe for the entire world. The continued reliance on physical growth and wealth accumulation in the industrialized countries seems destined to create global poverty.

Obviously many observers do not accept the pessimistic view of the Club of Rome. Others maintain that the changing demands and the shortages of resources will result in new technological developments, which will expand the resource base. Proponents of this view argue that the increasing scarcity of petroleum, iron ore, or other basic resources simply reflects the existing technological level. They point out that during the seventeenth and early eighteenth centuries major cities of the United States, particu-

THE ENVIRONMENTAL FACTOR

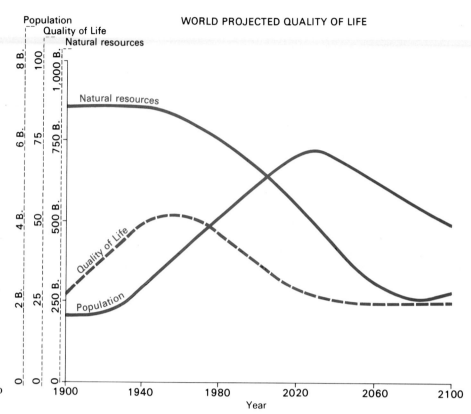

Figure 2.2 Club of Rome projections of future world quality of life. The Club of Rome's conclusions suggest that world population will rise rapidly as natural resources decline causing an associated decline in quality of life for the earth's population. They project world population peaking at 6 billion (left-hand scale) in the twenty-first century and then declining in response to the decrease in resources.

larly New York City, suffered annual energy crises as insufficient firewood was available for home heating and cooking. Ever-increasing prices for wood during the winter season ultimately brought about conversion to the use of abundant coal resources found in the Appalachians. The same process has been repeated throughout history. As one resource becomes scarce, another is substituted, either directly or indirectly, through technological advancement.

The contrast between these two views has many implications for the potential for the world's poor. The advocates of the Club of Rome's position view future industrialization of the less developed world as only a short-term solution to a long-range problem that can only be changed by a complete revolution of human expectations and lifestyles. The proponents of industrialization and technological change argue that all people of the earth can and will enjoy a higher standard of living through increasing world industrialization and economic development. Most people are neither as optimistic as the technology enthusiasts, nor as pessimistic as the Club of Rome members. The majority view among observers of the world would be described as one of guarded optimism.

The resources of the earth are limited, but the history of technological and economic progress suggests that increasing industrialization and technological and economic advancement will encourage a general increase in the standard of living on a world scale. This view assumes that technology will be introduced into the less industrialized countries to allow them to utilize their resources more efficiently to provide economic surpluses with which to improve their living standards. Such a guardedly optimistic view acknowledges that in order for the standard of living of the world's poorest majority to be increased, wise management is required within the individual less industrialized countries, and changes in relationships between rich and poor countries must be made.

Without adequate recognition by residents of wealthy countries of their obligation to the residents of poor countries, this optimism is replaced by pessimism. Some argue that unless the residents of the industrialized countries recognize the desirability and necessity of changing their life-styles to allow the underdeveloped countries to progress, the future of the world appears as bleak as the Club of Rome's report suggests. Development and relative standard of living are largely the result of human actions and decisions. The environment provides the resources that are utilized by people, but those resources are useful and valuable only as people value them.

THE EARTH AS A RESOURCE

The nature of the physical environment in each place on the surface of the earth affects the ability of humans to live there. The population distribution of the world shows that there are severe limitations on human use of the physical environment in some locations. High-latitude areas are of limited utility because of cold climates. The growing season is too short for most common agricultural crops upon which the human population relies. Provision of shelter is difficult in winter because of the need for supplemental heat for survival and because of the difficulties of providing water or maintaining transportation and communication systems. Occupance of such regions is difficult because of the unusual rigor of the environment. Population density is low and usually confined to exploiting mineral resources and furbearing animals.

Regions of the world in which there is inadequate water for crops, animals, and humans are also sparsely occupied. Although it is possible to import food into such locations, it is usually more economical to inhabit those regions of the earth's surface that have less extreme physical characteristics. But the adaptability of the human race allows its members to live essentially anywhere. Some areas may require extensive life-support systems, including inputs of energy, food, or other resources from outside of the area, but if people decide to inhabit a spot, they are able to do so. In this sense the earth poses no absolute limits to the human population, but the limitations of certain environments make their occupance costly.

The degree of difficulty that a specific environment presents is a function of the technology of its people and the interrelationship of the major geographical elements affecting each place on the globe. There are two major classifications of resources that affect the human inhabitants of the earth. The first group can be called **ubiquitous elements** because they occur everywhere. These ubiquitous elements include climate, soils, the nature of the land surface, and vegetation type. The second major type of resource is *nonubiquitous*—it occurs sporadically. Nonubiquitous resources include mineral resources, which, in economically recoverable quantities, are characteristically *localized*, abundant water resources, or fertile lands. The particular combination of ubiquitous and nonubiquitous resources in a given place is an important variable affecting the level of development and

---

Ubiquitous elements: Those elements that are found everywhere.

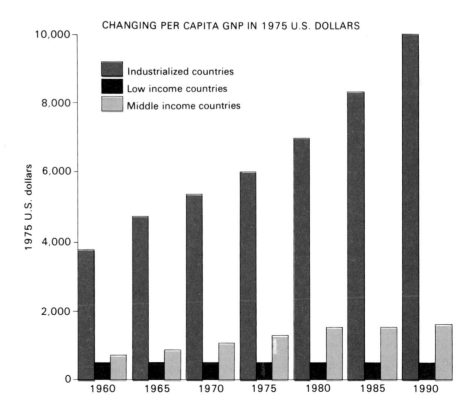

Figure 2.3 Changes in per capita Gross National Product projected to 1990. Note not only the existing gap between the industrialized and less industrialized countries but the slower rate of increase in per capita GNP in less industrialized countries that widens the gap through time.

THE ENVIRONMENTAL FACTOR

quality of life found there. The absence or presence of a specific resource does not determine that an area will be either industrialized or less industrialized but it does affect the cost and pace of development.

THE SURFACE OF THE EARTH: AN ENVIRONMENTAL CONSTANT

The surface features of a specific place are referred to as *landforms*. The characteristics and combinations of landforms at a specific location give character to the place and affect the type of activities found there. The Alps of Switzerland are an integral part of the mental image that the name of the country conveys. The Serengheti Plain of Kenya, the veldt of South Africa, or the Andes Mountains of Peru are equally important in determining the character of those countries. There is a distinctive sense of place associated with the cities of Switzerland's Alps, the farms of Kenya's Serengheti Plain, the mining camps of the veldt of South Africa, or the subsistence farms of the Andes of Peru. The combination of landforms in a particular place is referred to as its *topography*. The seemingly infinite variety of the topography of the earth's surface is an important variable that contributes to the uniqueness of each locale.

Landforms not only provide character to a place, but they also play an important role in the daily activities of the human residents of that place. The huge plains of the interior of the United States have facilitated modern mechanized agriculture, extension of railroad and highway transport links, and ease of communication with the rest of the world. The Himalaya Mountains of northern India and Tibet have combined to handicap the development of Tibet by isolating this mountain kingdom from the rest of the world. Mountains, deserts, swamps, or other landforms, by isolating certain regions, affect economic development and resulting standard of life. The existence of the Sahara served as an effective barrier to African contact with Europeans until late in the nineteenth century. The difficulty of providing adequate transportation and communication within the Appalachian Mountain region of the United States handicapped the integration of the residents of Appalachia into the mainstream of the wealthy American life-style.

The absence of landform barriers facilitates the exchange of information and technology and fosters economic development. The presence of large areas

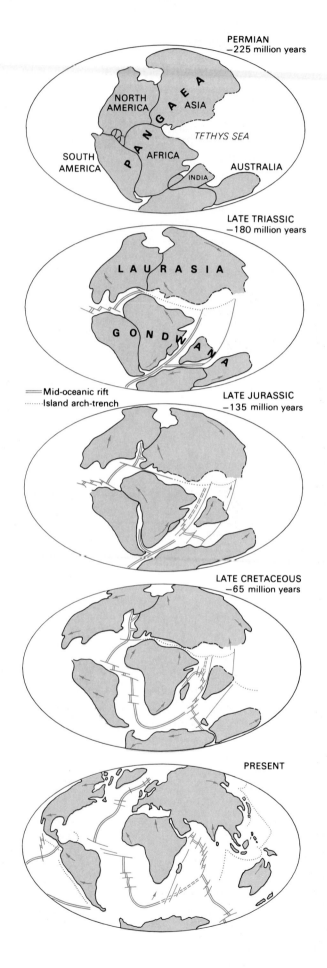

Figure 2.4 The proposed sequence of changes in the distribution of the earth's land mass from the time of the single continental land mass of Pangaea. The slow movement of the continental plates is still a basic cause of vulcanism.

Forces for changing the earth's surface

Two major forces are involved in changing the face of the earth, *tectonic forces* and *gradational forces*. Tectonic forces are those that lift, or build, the land surface, while gradational forces are those that wear it down. The tectonic forces are basically caused by pressures beneath the earth's crust. These pressures are a result of a number of processes, one of the most important being the movement of the continents. The continents were at one time united on one land mass that scientists call Pangaea (Figure 2.4). Because the interior of the earth consists of molten rock, or magma, the continents are slowly drifting apart, floating, as it were, on a sea of magma. The earth consists of a series of distinct plates that make up the individual continents and the ocean basins. Where these plates come into contact, the edge of the advancing plate is folded, creating mountains. The almost continuous ring of mountains extending from the tip of South America through North America and around the Pacific Ocean basin is a prime example. The folding of the earth associated with this mountain building is the cause of earthquakes and volcanoes. The great mountain arc around the Pacific Ocean is sometimes referred to as the Pacific ring of fire because of the volcanoes in it.

Gradational forces, which wear down the surface of the earth, include wind, water, ice, gravity, and the sea. These gradational forces create much of the physical character of the earth's landforms. Mountain chains like the Alps or the Himalayas have been severely eroded in the millennia since tectonic forces lifted them up. The materials that are eroded by wind, running water, or gravity are carried away from the original site of gradation. The process of removal of eroded materials and their subsequent deposit in other places by wind or running water creates other important landforms. The material deposited by rivers (*silt*) is often particularly suitable for cultivation of crops. Material deposited by wind (*loess*) in many areas such as northern China or portions of the Soviet Union are likewise extremely suitable for cultivation. Another major gradational force is ice. Large portions of the globe, particularly north of 40 degrees north latitude and south of 40 degrees south latitude have been subjected to the impact of ice during periods of continental glaciation. Glaciation has also occurred and is occurring in high mountains at the present time on a smaller scale.

Glaciation creates characteristic surface features. The advancing continental glaciers removed surface material and carried it in the advancing ice. As the ice melted, this fine material was transported by the wind and deposited, forming important loess landforms. Glaciation also changed the drainage pattern of areas it covered, resulting in poor drainage in much of Canada, northern Europe, the northern Soviet Union, and other areas of continental glaciation. The fjords of Norway were created when glaciers deepened river channels, resulting in flooding by the sea. Other landforms associated with glaciation include *moraines*, which are the ridges, hills, or plains created by the material pushed ahead of a glacier, or carried within it.

---

of relatively level land suitable for cultivation minimizes costs for production of crops to support a greater population than is normally found in mountainous areas. The importance of the landforms of the earth in affecting the distribution of world population and as a resource affecting livelihood is evident to any who have flown across the United States. The plains region of the interior United States is a region of agriculture. The mountains of the West and the Appalachians of the East have only isolated farming areas and large expanses of sparsely populated land. The pattern of agriculture and population distribution in the United States reflects more than landforms, of course. It is possible to have uninhabited plains when there are other environmental limits, as in the Sahara.

The landforms do more than characterize a place or provide potential for agriculture. The surface of the earth is a dynamic, ever-changing system. These changes may take place suddenly with catastrophic results, as when an earthquake occurs or a volcano erupts, or they may be a slow process like erosion in the Great Plains of the United States. The nature and pace of such changes is another factor affecting the utilization of any location on the globe.

The significance of the changing nature of the earth's surface is apparent. Catastrophic events like earthquakes or floods result in loss of life and property. Erosion of topsoil diminishes the ability of a specific site or a whole region to produce crops. The existence of noteworthy erosional remnants or other unique landscape features such as the Alps or the Grand Canyon become a local resource as tourist attractions. The catastrophic effect of the changing form of the earth is only one example of the impact of the changing earth on the activities of the people.

CAUSES OF CLIMATIC VARIATION

Another ubiquitous resource affecting the population of the world is climate. Climate affects the sense of place that characterizes each individual spot and each region on the earth's surface and the actions of the

An earthquake in the Republic of Armenia in the Soviet Union destroyed hundreds of apartment buildings and killed thousands of people in December 1988.

human residents of that place. Climate is a constant environmental factor that residents of any site must consider. Whether it be an Eskimo's concern for temperatures and resultant snow quality, the ski resort operator's concern for quality and quantity of winter snowpack, or the farmer's interest in rainfall for crops, climate is involved in all activities of the human population. It affects food production, housing types, incidence of disease, vegetation, soil quality, water supply, and through erosion, the landforms themselves.

The basic causes of climate are complex interrelationships of the amount of heat, precipitation, landforms, prevailing winds, and land-water relationships. Ultimately the climate of any place is caused by the amount of energy from the sun received at that location. The extent of solar energy is a function of the intensity of the sun's rays and the duration of time each day that the rays are striking the earth's surface. Generally, areas near the equator, where the sun's rays are concentrated, receive greater solar energy and consequently have climates with warm temperatures year-round. In the middle latitudes (from 25 to 55 degrees north and south of the equator) the amount of solar energy received varies with the seasons, resulting in distinct winter and summer. The high latitudes, north and south of 55 degrees latitude, receive only limited solar energy and have cooler temperatures throughout the year.

There are other influences besides latitude affecting climate, including surface features of the earth, prevailing wind systems, and the amount of precipitation. Land masses heat more rapidly in summer and cool more rapidly in winter than does water. This characteristic is called *continentality*. The climatic contrast between locations near large bodies of water and those in the center of continents result in part from these different characteristics of land and water. Coastal areas tend to have narrower ranges of seasonal temperatures, whereas the interiors of continents in the middle and high latitudes are characterized by great variation in winter and summer temperatures.

The impact of the differing nature of land and water on climate of the earth is intensified by the earth's prevailing wind systems. From 30 to 60 degrees north and south latitude the prevailing wind system blows from west to east (the Westerlies). It carries air from the oceans eastward over land; the oceans on the west coasts thus affect and moderate the western margin of the continents. The same Westerlies cause the greater extremes of the interior of the continents to affect the eastern margins, creating greater seasonal extremes than the west coasts experience.

A final climatic variable is the amount, duration, and seasonal distribution of precipitation. Precipitation results from air that contains water vapor being cooled to the temperature at which condensation occurs (*dew point*). The flat bottoms of clouds in the summer sky represent the dew point for that particular air mass. Air is cooled when an air mass rises to a higher altitude, where temperatures are cooler. Whether or not clouds and rain result from a rising air mass depends on the temperature and water vapor content of the air mass before it rises.

The actual amount of precipitation that a place receives depends on the air masses that affect that location and the amount of water vapor they contain. Certain areas of the earth have extremely heavy precipitation while other large areas receive insufficient precipitation for most common human activities. The great deserts of the Sahara, the Gobi, the Atacama, or Death Valley represent regions in which human occupancy is handicapped by the high cost of obtaining adequate water supplies. Conversely, the Amazon

Earth-sun relationships

The primary relationships between the earth and the sun result from the fixed orbit of the earth around the sun. The earth's axis is tilted with respect to the plane of its orbit, causing the area of the earth that receives the most direct rays of the sun to shift north and south of the equator during the year (Figures 2.5 and 2.6). The changing temperatures associated with the earth's rotation around the sun in the Northern and Southern Hemispheres provides the seasonal changes associated with the middle- and high-latitude areas. The variation in the directness with which the sun's rays strike a spot creates the seasons of spring, summer, fall, and winter. The periods when the sun's rays are vertical at the equator are known as the equinox and occur in spring and fall. "Equinox" means equal length of night and day at the equator; spring begins with the vernal equinox (March 21), while fall begins with the autumnal equinox (September 22). In winter and summer the sun's rays are vertical north or south of the equator. In the Northern Hemisphere the summer solstice occurs on June 22, when the sun's rays are vertical 23½ degrees north of the equator and summer begins. The winter solstice occurs on December 22, the first day of winter, when the sun's rays are vertical at 23½ degrees south of the equator. Were it not for the angle between the axis of the earth's plane of orbit around the sun, the earth would be divided into an equatorial zone of almost uninhabitable heat and an ice-covered polar zone. The only inhabitable area would be a transition zone north and south of the equator. The variation in solar energy that the earth experiences as it revolves around the sun makes much more of the surface of the earth inhabitable than is the case with other planets.

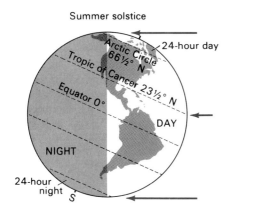
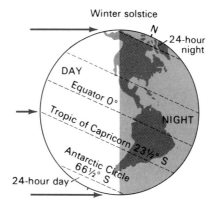

Figure 2.5 Relationship of the earth's inclination to the sun's rays.

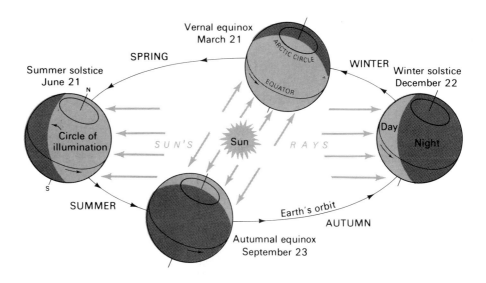

Figure 2.6 The rotation of the earth around the sun results in seasonal climatic change because of the tilt of the earth's axis.

THE ENVIRONMENTAL FACTOR

Water, water vapor, and precipitation

The amount of precipitation that a place receives is a critical variable affecting its usefulness. Regions that have either too much or too little precipitation present problems for the occupants. Basically precipitation in a given area results from one of three general causes, associated with the need to cool air to the dew point and continue to cool it to produce precipitation. The primary cause of such cooling is lifting air masses to higher elevations where temperatures are lower. The simplest mechanism for lifting air masses occurs in equatorial regions where air is heated by the warm earth. As air is heated, it expands and becomes less dense and is lifted as it is replaced by cooler, denser air. As this air is lifted, if it has sufficient moisture, the dew point is reached and precipitation can occur. Precipitation produced this way is known as *convectional precipitation* (Figure 2.7). Convectional precipitation occurs in tropical regions during most of the year and in the middle latitudes during the summer.

Another mechanism for lifting air occurs when an air mass is forced to rise over a mountain or other physical barrier. Such precipitation is referred to as *orographic precipitation* (Figure 2.7). As an air mass is forced to rise over a mountain barrier, the air is cooled beyond the dew point and precipitation occurs. The side of the mountain facing the prevailing wind is known as the *windward side* and has greater precipitation. The side away from the prevailing wind lies in a *rain shadow* because the air that has been lifted over the mountain becomes a drying wind as it descends the mountain range after being wrung of its moisture. Death Valley is an example of a rain shadow.

The third mechanism causing precipitation is the interaction between air masses of different temperature and water vapor content. Where such different air masses come into contact, they generally maintain their own characteristics, creating a line, or front, where they meet and cause *cyclonic precipitation*. The rotation of the earth causes these air masses to move around the earth, creating a series of storms in the middle latitudes, particularly in the winter, when there is greater contrast in temperature and pressure between the air masses. Frontal storms are responsible for the winter snowstorms in the Midwest and eastern United States, the gentle but persistent rains of England, and similar weather in the middle latitudes.

---

region of Brazil receives so much rain that lush vegetation and diseases have hindered development.

THE CLIMATES OF THE EARTH

The interaction of temperature, precipitation, winds, and air pressure creates a complex mosaic of climates on the surface of the earth. The simplest classification categorizes these climates from the equator north and south to the poles. Equatorial areas have hot, humid climates while polar regions have cold, dry climates. Between these two extremes are found the humid climates of the middle latitudes, the desert climates of the world, and all the varieties and combinations in between. The most common classification system used to describe the earth's climate is known as the Köppen system. The Köppen system recognizes five major climatic regions. The A climates are the humid climates near the equator that have no winter season. The B climates are the dry climates. The C and D climates are those found in the middle latitudes, with C climates having a milder winter than the D climates. The E climates are those of polar regions. The five climatic regions are subdivided into eleven distinct climatic types. Such a division obscures important local differences in climate, and the resulting map (Figure 2.8) of the world is misleading in showing a distinct line between regions of different climate. The change from one climatic zone to another is one of transition, with the line representing the approximate boundary of temperature, precipitation, or other variable.

CLIMATE AND VEGETATION

Although other factors affect the type of vegetation, the climate is important enough that on a global scale, patterns of vegetation broadly conform to the climatic regions. Within these broad zones there is great variation in vegetation when examined at a small scale, just as there is variation in climatic specifics from place to place within a climatic zone. But the general pattern of forested, grassland, and arid regions of the world follows the general pattern of precipitation. Forests require a humid climate, and grassland or the scattered brush and grass of deserts is found in drier regions. But the broad division into forests, grassland, and intermittent vegetation masks important vegetative differences and resultant utility for human occupants.

Forests of the world. The forests of the world can be divided into three broad categories, each with distinctive characteristics that affect their utility. The

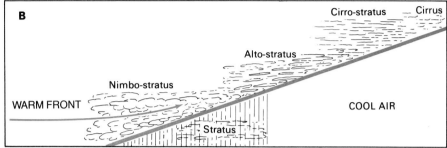
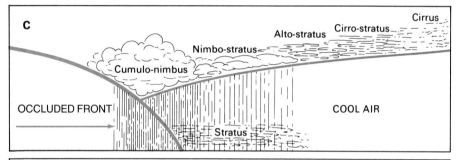
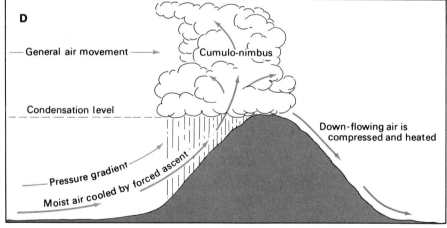

Figure 2.7 Mechanisms for lifting air, causing cooling and precipitation. Diagram A indicates a typical convective system. Diagrams B and C illustrate cyclonic (frontal) precipitation, with B showing a cross-section of the warm front sector and C, a cross-section of the cold front. Diagram D illustrates orographic precipitation associated with physical barriers.

tropical forests grow in areas with tropical rain-forest climates. The variety of species is enormous, and the thickness of the overhead branches and leaves (the canopy) limits the amount of solar energy that reaches the surface of the soil in the forest. The tropical forest has no winter season, so trees grow year-round, but the tremendous variety in species, and consequent scattering of any one type, have handicapped economic utilization of the abundant wood. Because of the high temperatures and continuous growing season, tropical rain forests create the greatest amount of *biomass* (vegetative matter) of any region on the earth's surface, but the same high temperatures and humid conditions foster the growth of disease organisms and parasites and generally result in health hazards for occupants of such regions.

A second large area of forests is found in regions of subarctic climates and some humid continental cli-

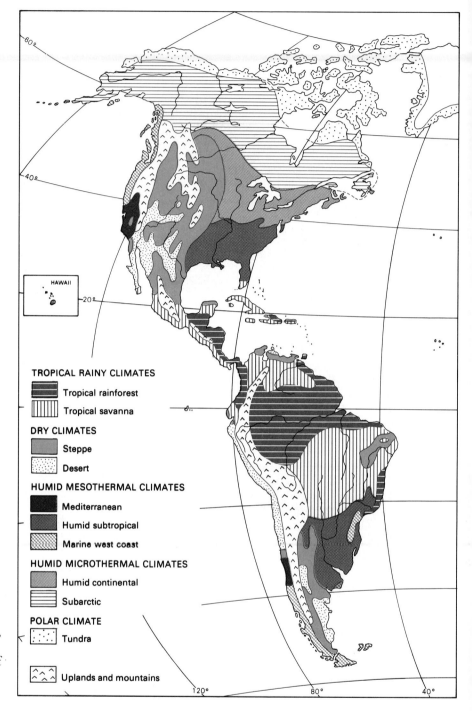

Figure 2.8 World climatic distribution based on the Köppen system. Although the specific climatic regions are shown as an abrupt line, the borders of climatic regions are actually zones of transition from one climate to another.

mates. Most of these forests are *coniferous* (cone-bearing). Coniferous trees have needle-like leaves that they generally do not shed in winter. The trees do not create a continuous canopy, and a single species may dominate large areas. Major types include the pine, fir, spruce, and larch. Because they tend to occur in single-species stands, some of the major forest resources of the world are coniferous forests of the marine west coast and subarctic climates of North America and the Soviet Union.

The third general type of forest is the *deciduous* forest. Deciduous trees, such as oak, maple, chestnut, elm, or walnut, lose their leaves during the winter. These trees are highly useful, as many of them have a dense wood (hardwood), used in making furniture or other items of high value.

PEOPLE, PLACES, AND POVERTY

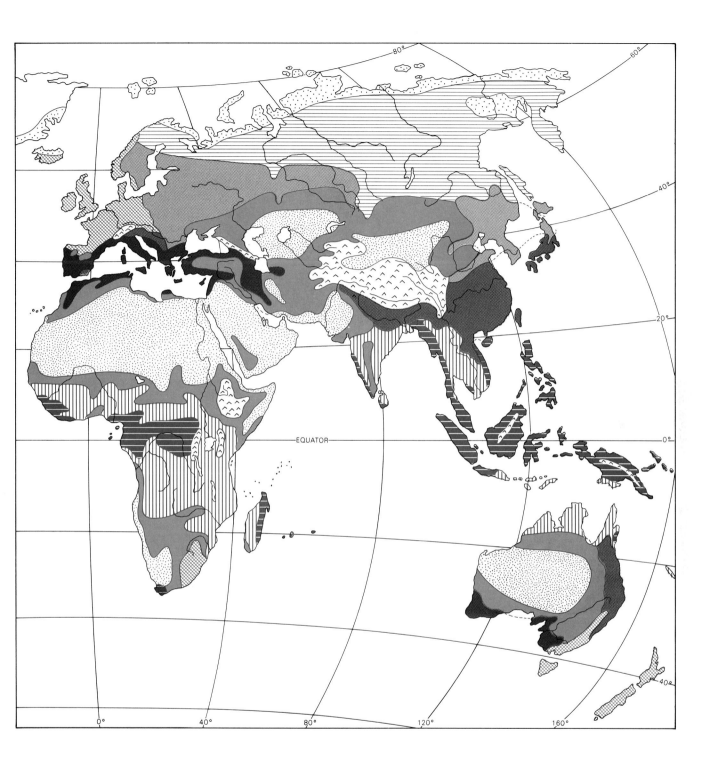

Grasslands of the world. In areas where there is insufficient moisture to support trees, grasslands become the dominant vegetation type. Grasslands include steppes, savannas, and prairies. Steppes represent a transition between the humid and arid climates; the typical vegetation is short grass, which becomes sparser as one travels from the humid to the arid region. The grasses of steppe lands are highly nutritious for grazing animals, and soils in steppe lands are quite fertile.

The grasslands of the tropical savanna are quite different from those of the steppe. Savannas have tall grasses, which are less suitable for most domestic animals. Prairie regions originally had deciduous forest vegetation, but repeated burnings by native populations allowed grass to become the dominant vegeta-

THE ENVIRONMENTAL FACTOR

47

Köppen's climatic types

1. *The tropical rain-forest climate* (Af). The tropical rain-forest climate is characterized by a hot, humid climate all year round. The climate has no month which averages below 64.4°F. (18°C), and no month in which rainfall is less than 2.4 inches (60 millimeters). Rainfall is heavy in the tropical rain forest, in some areas exceeding 400 inches (1,000 millimeters) and is convectional in nature.

2. *The tropical savanna* (Aw). The tropical savanna climate differs from the tropical rain forest in that it has a distinct dry season during the low-sun period. Rainfall is less than 2.4 inches (60 millimeters) per month during the dry period.

3. *The desert climates* (Bw). In the desert climates evaporation exceeds precipitation. Although there is no specific cutoff point, rainfall in desert regions normally is less than 10 inches (250 millimeters) per year.

4. *The steppe or semiarid climates* (Bs). The steppe climate has more precipitation than the desert climate. Steppe climates normally receive between 10 and 16 inches (250 and 400 millimeters) of rainfall but have greater vegetation than does the true desert.

5. *The humid subtropical climates* (Ca). The humid subtropical climates are located on the southeastern margins of continents in the middle latitudes. As their name implies, they receive abundant rainfall, normally between 25 and 50 inches (600 and 1,300 millimeters) per year. Summer seasons are hot, with the warmest month averaging over 71.6°F. (22°C). Winters are mild, with the coldest month averaging above 32°F. (0°C) but less than 64.6°F. (18°C).

6. *The marine west coast climates* (Cb and Cc). The marine west coast climate occupies the west coasts of continents pole-ward of approximately 45°. These climates are moderated by the ocean currents, so that they have a mild summer and winter. The warmest month averages below 71.6°F. (22°C) and the coldest month normally averages above 32°F. (0°C). The marine west coast climatic regions receive rain at all seasons of the year with no distinct dry season. Temperatures are rarely extreme.

7. *The Mediterranean climates* (Cs). The Mediterranean climates occur on the western margins of continents between 35° and 45° latitude. Their major distinguishing characteristic is a summer drought with associated high temperatures. Precipitation occurs in the winter season and is normally in the range of 20 to 30 inches (500 to 750 millimeters). Temperatures are extreme in the summer, with maximums reaching over 100°F. (38°C), but winter temperatures rarely are as low as 32°F. (0°C).

8. *The humid continental climates* (Da and Db). The humid continental climates occupy the central and eastern portions of large land masses in the mid-latitudes. As their name implies, they are humid, and their temperature regime reflects the impact of their continental location. They are recognizable by their colder winter climates, with the coldest month averaging below 32°F. (0°C). They can be further subdivided into a humid continental—warm summer (Da) in which the warmest month averages over 71.6°F. (22°C), and a humid continental—cool summer (Db) in which the warmest month averages below 71.6°F. (22°C). Precipitation is abundant throughout the year with totals of 30 to 60 inches (750 to 1,500 millimeters). These climates are found in the latitudes of 40 to 55°.

9. *The subarctic climatic type* (Dc and Dd). The subarctic climate occupies the continents between 55° and 80° north and south of the equator. Its major characteristic is cold and snow. The warmest month is below 71.6°F. (22°C), and it has less than four months above 50°F. (10°C). The coldest subarctic climates are the Dd climates in which the coldest month averages below −36.4°F. (−38°C). The subarctic climates experience this extreme cold because they receive only limited amounts of solar energy, and the large landmass cools dramatically during the winter.

10. *The tundra climates* (Et). The tundra climates are characterized by cold summers with the warmest month averaging under 50°F. (10°C). Precipitation may total less than 10 inches (250 millimeters), but low temperatures result in a low evaporation rate, so that precipitation is sufficient for growth of low-growing shrubs, mosses, and grass. The limiting factor is, therefore, not water but temperature.

11. *The ice cap climates* (Ef). The distinguishing characteristic of the ice cap climate is that no month has an average temperature exceeding 32°F. (0°C). Thus, its name *ice cap* is descriptive of the physical characteristics of regions with this kind of climate.

---

tion. Prairies were found in the United States in Ohio, Indiana, Illinois, and eastern Nebraska and Kansas before these areas were cultivated. The pampas of Argentina is also a great prairie region.

Vegetation of dry lands. In areas where there is insufficient moisture for grass to grow, vegetation adapted to drought predominates. Such **xerophytic vegetation** has thick bark, deep roots, water-storing capabilities, or spines to prevent damage to the surface of the plant, all combining to allow growth in arid

Xerophytic vegetation: Vegetation that is tolerant of arid conditions.

regions. Such vegetation is found in the deserts and in portions of the Mediterranean climatic region. Xerophytic vegetation is sparse; depending on the amount of precipitation, it may range from scattered plants in the Sahara to the rather luxuriant shrub vegetation in areas of Mediterranean climate.

The tundra. In the tundra climatic zone there is sufficient precipitation for plant growth, but temperatures are the limiting factor. Tundra vegetation consists of low shrubs, willows, mosses and lichens, and short grass. During the short summer season these plants grow rapidly, and with the onset of the early fall they are frozen and covered by snow.

SOILS AND THEIR UTILITY

One of the geographical factors that plays a major role in the level of development, quality of life, and sense of place that characterizes each spot on the globe is the fertility of the soil. Soil is a product of the interaction of the underlying rock (bedrock), climate, and vegetation over thousands or millions of years. Soils affect the well-being of individuals by affecting crop yields, the nature and quantity of plant nutrients in specific crops, and the ability of a given area to support population. Soil is consequently one of the more important resources affecting development.

Soil fertility is a result of the complex association of minerals, gases, weathered rock, organic material, and microorganisms within the soil itself. *Humus*, the dead and decaying organic material on the top of the ground and in the top few layers of soil, is of critical importance to soil fertility. Rock fragments and pure sand without any humus are not soil. Humus is a source of nutrients in soil as water percolating through the humus dissolves its minerals, making them available to plant and animal life. Humus also affects the structure of the soil. An abundance of humus provides air spaces, allows circulation of moisture and root development, and increases the soils ability to store water.

The amount of humus in a soil reflects the local climate. Grasslands tend to have the greatest amounts of humus, followed by the deciduous forests of middle latitudes. Coniferous forests of the high latitudes have a thick layer of humus on the surface, but because of the cold and the high proportion of precipitation that falls as snow, the humus does not break down rapidly to become available to the plant and animal life. The needles of coniferous trees are slightly acid, and combined with percolating water, create a weak acidic solution that handicaps plant life and dissolves other minerals and nutrients in the soil.

Tropical forests produce the greatest amount of humus, but little of it is available for plant use even though leaves fall all year-round. The constant heat in the tropics fosters dense populations of microorganisms that consume the humus for their own livelihood, leaving little of the minerals and nutrients for plant use. Since the plants in the tropical forest remain green throughout the year with no dormant season, there is no accumulation of leaf material as in the deciduous forests of the middle latitudes. Humus is even more limited in the desert, of course, because of limited vegetation.

The actual amount of humus produced by the vegetation in a specific area is not a direct indicator of soil fertility. In areas with high precipitation nutrients of the humus are dissolved and carried away. The process by which water moving through the soil dissolves minerals and nutrients and removes them from the zone that can be reached by plant roots is known as *leaching*. Plants rely for nourishment on the absorption of minerals and nutrients in solution. If there is inadequate water in a soil, plants will die because there is no way to absorb plant foods from the soil. When there is a surplus of moisture, the plant nutrients are carried away from the plant roots by the movement of ground water. Leaching reduces soil fertility in tropical areas, coniferous forests of the subarctic and humid continental climates, marine west coast climates, and the subtropical climatic areas. Grasslands have greater accumulations of humus, less leaching, and more fertility. Unfortunately, the lack of precipitation may mean that they are also less productive for crops than more humid areas because the nutrients are unavailable to plants. Use of irrigation often makes such drier areas into some of the most agriculturally fertile lands in the world.

Generally, those areas of the earth with greater natural fertility are the locations of dense populations. Most regions of high soil fertility are located in the margins between the humid continental climates and the steppes, in river valleys, and on certain types of bedrock. Soils formed from limestone or certain volcanic materials tend to have higher fertility than other soils in areas of similar climate. Certain volcanic materials, particularly in warm climates, weather into highly fertile soils because of their mineral content. For this reason the islands of Indonesia, where soils would normally be leached and infertile, can support a dense population. Windblown material (*loess*) forms a particularly fertile soil as a result of its extremely fine particles. These fine particles provide more surface area in the soil itself to which water-dissolved plant nutrients can become attached. Materials deposited by river or stream (*alluvial*) action tend to be finer, are level, and have a rich humus content as a result of periodic flooding. Unfortunately, these areas, so suitable for cultivation, may also be subject to disastrous flooding.

THE LIFE-OR-DEATH ISSUE: FOOD RESOURCES OF THE WORLD

The patterns created by the landforms, climates, vegetation, and soils of the earth are significant because of their effects on people's lives and activities. Examination of the distribution of industrialized versus less industrialized countries indicates that most of the less industrialized countries are located in regions of harsh climates and little soil fertility. The tropics, the deserts, and the savanna climates of the world have tended to be locations of low industrialization. Although the causes of low levels of industrialization are only partly related to these physical features, they are important because of the relative ease or difficulty of development that they provide. Where climates are severe, vegetation sparse, or soils infertile, it is difficult, costly, or impossible for humans to exploit the environment.

At the most fundamental level, the critical issue facing the world's population is that of food production, and climate and soil fertility directly affect crop yields. Food production per capita in developing countries grew at an annual rate of 1.3 percent between 1975 and 1985, and in Africa it actually fell by 1.7 percent per year. The least developed countries imported food in 1981 costing over $28 billion, but by 1990 this had increased to over $100 billion. During the period from 1970–1988 per capita food production in developed regions increased at a rate of more than 2 percent per year.

Meanwhile, the population of the less developed world increased at a rate of more than 2 percent per year during the 1970 to 1988 period, changing many of the less developed countries from food exporters to food importers (Figure 2.9). At present (1990), an estimated one quarter of the earth's inhabitants are persistently undernourished. The availability of fertile soils, suitable climate, and adequate water is a critical question, on which hinges survival in much of the less industrialized world. People in less developed countries must often face the reality that the environment does play a critical role in their well-being. Drought in the margins of the Sahara region during the early 1980s, flooding in Bangladesh, or lower crop yields in India as a result of lack of precipitation present immediate issues for survival to people in these countries. The ever-growing population of the world, particularly in the less developed countries, dictates that the issue of food production receive central attention in the coming decades.

THE LAND RESOURCE BASE

Land for production of the food necessary to support tomorrow's children and to increase the standard of living for today's residents is a limited resource. The land area of the globe (excluding the ice caps of Greenland and Antarctica) totals 32.1 billion acres (13 billion hectares). Climate, landforms, and soil characteristics limit the usefulness of most of this land for cultivation. Approximately 6.4 billion acres (2.6 billion hectares) are limited because freezing temperatures persist for nine or more months per year. An additional 4.7 billion acres (1.9 billion hectares) receive inadequate precipitation for nine or more months of the year. The

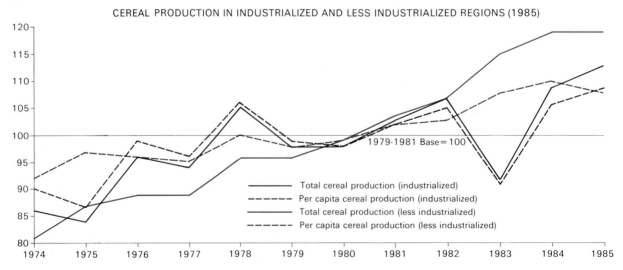

Figure 2.9 Cereal production in the industrialized and less industrialized regions (1985). While total cereal production has increased more rapidly in the less industrialized countries, per capita production has grown more slowly due to population growth.

TABLE 2.1 Estimated Cultivated and Potentially Cultivatable Land in the World in Millions of Hectares (Acres)

	Cropped Area Without Irrigation		Cropped Area Added by Irrigation		Total Potential Gross Cropped Area	
Africa	705	(1,742.13)	290	(716.62)	995	(2,458.75)
Asia	625	(1,544.38)	475	(1,173.77)	1,100	(2,718.15)
Australia and New Zealand	123	(303.93)	2	(4.94)	125	(308.87)
Europe	205	(506.56)	40	(98.84)	245	(605.40)
North America	535	(1,321.99)	60	(395.38)	595	(1,717.35)
South America	635	(1,569.09)	80	(197.69)	715	(1,766.78)
U.S.S.R.	325	(803.08)	30	(74.13)	355	(877.21)
Total	3,155	(7,796.01)	1,077	(2,661.37)	4,232	(10,457.38)

SOURCE: Adapted from Roger Revelle, "The Resources Available for Agriculture," *Scientific American* 235, No. 3, Sept. 1976: 168.

remaining 21 billion acres (8.5 billion hectares) receives sufficient moisture and heat for cultivation, but large areas have other limitations restricting their use.

Mountainous and arid regions covered with stony or shallow soils comprise 6.4 billion acres (2.6 billion hectares). There are 4.2 billion acres (1.7 billion hectares) of desert, and there are 1.7 billion acres (0.7 billion hectares) of unusable sandy soils. All 1.24 billion acres (0.5 billion hectares) of the tundra are unsuited for crops. Soils of tropical areas include 3.46 billion acres (1.4 billion hectares), much of which is heavily leached, waterlogged, or otherwise unsuitable for cultivation.

It is possible to live and to support human life in areas of unproductive soils, but it is difficult to support a dense population in such locations. Sheep and goats can graze on the sparse vegetation of deserts, but such economic pursuits support only a limited number of people. Tropical areas with infertile soils can be used by farmers who cultivate the land for only one or two years, releasing the limited fertility of the soil by burning the forest and allowing their plants to absorb the minerals and nutrients in the ash. Then after one or two years these farmers move on to new forest. Obviously such *slash-and-burn* farming techniques support only limited numbers of people.

About 24 percent of the earth's surface (excluding ice caps) is capable of producing crops without irrigation, a total of 7.8 billion acres (3.15 billion hectares) (Table 2.1). Addition of lands with water for irrigation increases this to 32 percent or 10.45 billion acres (4.2 billion hectares). This figure represents the presently accepted maximum possible limits of cultivation and would require large investments of capital. The human population today relies on only 3.36 billion acres (1.36 billion hectares) of land to produce the world's food. Expansion to the maximum extent will occur only when demand for food increases to such an extent that the additional costs of expanding the cultivated land become less important than production of the food.

Economic Realities: Food Allocation Between Rich and Poor. World food production is sufficient to feed the total population in abundance, if it were distributed uniformly. The reality is that the wealthy industrialized world consumes a disproportionate share. The United States, for example, is the world's leading producer of basic food grains of wheat, corn, and soybeans, and more than 75 percent of these grains are used to produce meat for the wealthy American population. At the same time residents of African, Asian, and Latin American countries have inadequate grain for proper nutrition. One of the issues that must be resolved if the less developed world is to achieve the goal of adequate food for all its population is the worldwide distribution of the earth's food resources.

Variations in food habits. The contrast between the diets of residents of the wealthy industrial countries and those of the poorer countries illustrates the disposition of the food produced in the world. The typical resident of a less industrialized country consumes primarily grains or other staple products such as potatoes or other *tubers* (root crops). The diet of the resident of the less industrialized world lacks animal protein such as meat, milk, eggs, fish, or poultry. In some countries, cultural taboos preclude eating meat, but in most it reflects poverty. Even when the resident of the less developed country receives adequate grains or other vegetable products, these foods lack the protein necessary to ensure good health (Figure 1.6). By

Soil classification

Soils are the result of the interaction of climate and vegetation on the parent material (underlying rocks) over time. The resulting pattern of soils is much more complex than the climatic patterns since there are more variables affecting the soil in a specific place. The generally accepted system for classifying the world's soils is known as the Comprehensive Soil Classification System, or "Seventh Approximation," to distinguish it from previous attempts at comprehensive soil classification. The seventh approximation contains ten orders—broad categories that can be divided into a total of 10,000 specific soils.

Soils of the tropics are primarily *oxisols*, but also include the related *ultisols* and *vertisols*. These soils are highly leached of nutrients and have a red or yellow color due to high concentrations of iron and aluminum oxides. These soils are infertile, and the luxuriant vegetation survives by absorbing nutrients directly from decaying plant matter rather than from the soil. Many farmers in the less developed countries must rely on oxisol soils, thereby limiting agricultural productivity. Ultisols are generally less leached and more fertile than the oxisols.

Aridisols refer to the soils found in dry climates. These soils receive inadequate precipitation. Lack of leaching allows salts to accumulate in such soils, but with irrigation and adequate drainage these soils can be highly productive. The western United States, southern South America, northern Africa, Australia, and Central Asia have aridisol soils.

High-latitude locations with cool climates have *inceptisols*. These soils are found in the tundra and northern subarctic where short growing seasons and limited vegetation are associated with infertile soils. The *entisols* are similar to the inceptisols in that they are generally infertile. Entisols are recent soils, commonly associated with sandy areas or the river floodplains where erosion and deposition of the surface has prevented development of a mature soil. Entisols are concentrated in Africa and Australia. *Histosols* refer to soils that are found in cool, damp climates with inadequate drainage. Referred to as bog soils, these soils are associated with the peat bogs of northern Canada and Europe, and are normally unsuitable for agriculture.

Spodosols, *mollisols*, and *alfisols* are commonly associated with humid mid-latitude climates. Spodosols are the least fertile, occupying the subarctic and northern humid continental climates. The cool, damp climate supports dense forest, but soils have a gray, ashlike appearance and only limited fertility for agriculture without application of fertilizer. Mollisols are the most fertile soils, and occupy the drier margins of the humid mid-latitudes where leaching has not been severe. The most fertile of the mollisol soils are called *chernozems*, a Russian term meaning black earth, reflecting the abundance of humus in the soils. The mollisols developed under mid-latitude grasslands and today are the productive grain-producing regions of North America, South America, Europe, Asia, and Australia. Alfisols are not as fertile as mollisols, but have more fertility than spodosols. They are associated with the humid continental climates of Europe and North America, but are also found in more tropical areas where local conditions modify the leaching process. Alfisols are very important for agriculture in Europe, India, Australia, Africa, and Brazil.

---

contrast, many residents of the wealthy industrial countries tend to be overconsumers of both total calories and animal protein. Some of the industrialized countries, such as Canada and the United States, or Australia and New Zealand, are in the enviable position of having a large, fertile land resource base and a relatively small population. Their overconsumption of food directly effects the ability of residents of the less developed world to obtain adequate food supplies.

Five hundred pounds of grain will provide sufficient calories for an adult for a year. In the United States it takes 1,000 pounds of grain to fatten the cattle for the average 80 pounds of red meat consumed by each man, woman, and child in the United States each year. Each American also consumes additional grain in the form of alcohol, pork, poultry, and milk products. It is estimated that a resident of the United States consumes every day twice the amount of protein that the body can utilize, and as much as one-third more calories than needed. The quantities of grain required to produce milk, meat, eggs, and other high-value foodstuffs for a wealthy resident of the industrialized world would support four residents of the less developed world.

Overconsumption by some residents of the industrialized world directly affects the residents of the less industrialized world because the vast majority of the world's population relies either directly or indirectly on only a few basic crops for the majority of their food needs (Figure 2.10). Even in the consumption of four basic foods, wheat, rice, maize (corn) and potatoes, there is regional contrast between the developed and less developed worlds. Most residents of the industrial, developed world prefer wheat. Residents of the less developed world rely much more heavily upon rice, maize, and to a lesser extent potatoes to provide their minimum caloric requirements. The major sources of animal protein in the world are pork and beef. People in less industrialized countries eat more pork than beef because pork can be produced on a small scale by feeding agricultural waste products to the pigs.

Residents of less industrialized areas have a much more restricted variety of foods, a lower quantity of food, and a diet that often lacks basic minerals and nutrients. The limited variety of foods that have traditionally been important indicates the need either to revolutionize the food production patterns of the world or to distribute the foods that are produced more equitably. Unless this issue is addressed by the residents of the industrialized countries, the world can confidently expect crisis situations in food availability in less industrialized regions in the near future.

RESOURCES FOR INDUSTRIALIZATION: CONTRIBUTORS TO WEALTH

All human populations require food if they are to survive. The availability of either fertile lands for production of crops or wild plants and animals upon which to subsist is essential. The variety of food crops and the variety of animals used for meat indicate that not all people recognize the same items as desirable or necessary to their survival. The same is true of the mineral resources that industrialized countries have come to depend on. For Western industrialized societies, the presence or absence of petroleum is critical, but for a subsistence hunter/gatherer in the Congo River basin of Africa, such a resource is marginal, if it is useful at all. The definition of a resource depends on the physical presence or absence of the specific substance as well as the needs and views of the people who use it. Resources, then, are both a cultural and a physical variable. Resources are usually defined in terms of their utility to the human population.

There are certain products of the physical environment to which a value can be ascribed, such as iron ore, gold, diamonds, or petroleum. For the industrialized, technical societies of the West these are indeed resources. What the value of wilderness may be is less apparent; yet, for many people it constitutes a resource. The controversy in the United States over whether it is more important to build a dam to prevent floods and generate electricity or to allow a free-flowing stream to ensure the continued existence of specific plants or animals exemplifies the difficulty in defining a resource. While recognizing that resource appraisal and use involve cultural values, it is nonetheless important to point out the availability of the common and commonly perceived resources that seem important in economic development. For the purposes of clarity, we will define *resources* as all those things that are used or viewed as potentially useful by humans. The physical resources of the earth can be categorized in a number of ways. Resources are often divided into nonrenewable and renewable resources. Nonrenewable resources include coal and petroleum—once they are used, they ae gone. Renewable resources, like forests or water, can be replenished and maintained. Such a simple dichotomy obscures important differences, and it is essential to recognize that within these two broad categories a number of subdivisions can be usefully recognized. Some nonrenewable resources, such as copper, have a high potential for reuse, while others, such as petroleum, are essentially consumed upon use. Renewable resources can be divided into two broad groups: those that are essentially unaffected by human activities, such as the space of the earth itself, which we cannot substantially expand or contract; and those that can be modified by human activities, such as water or air. The importance of these subdivisions is apparent. Resources that can be reused will last longer than those that are consumed. Resources that the human population can contaminate or destroy will hopefully attract more attention than those we cannot.

Exploiting these resources has implications far beyond using up the resource itself. Resource use by the industrialized countries has resulted in pollution of air and water, destruction of farm and forest, and isolation of the population from the physical environment. In the less industrialized world, pollution of water from untreated sewage, soil erosion from inadequate agricultural techniques, and destruction of forest to provide fuel for cooking and heating present immediate challenges. The impact of the industrialized wealthy world on the world's resources has been more dramatic because of the voracious appetite for industrial minerals, but the environmental problems created by the residents of the less industrialized world are no less critical, because they threaten the very livelihood of the majority of the world's people. Continued erosion of the soil, overgrazing of grasslands, and overexploitation of forest resources may create an irreversible decline in the ability of the world to support its population. (See accompanying box.)

Resources for industrialization. The contrasts in demand for resources between the developed and the less developed world illustrate the importance of certain resources in industrialization. Certain mineral resources have traditionally played a major role in this process. Among the most important are the fuels: coal, petroleum, natural gas, and to a lesser extent, ura-

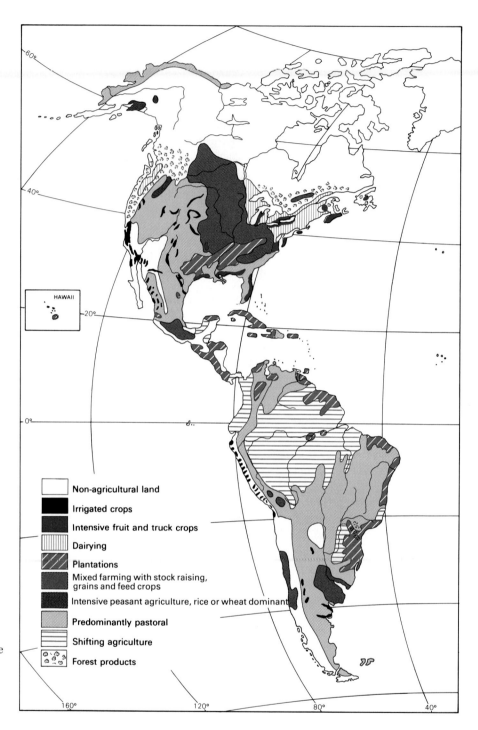

Figure 2.10 Major agricultural regions. Shifting agricultural and pastoral activities dominate the majority of the surface of the earth used for agriculture, but the areas designated grain or mixed farming produce the majority of the world's foodstuffs.

nium. The presence or absence of coal seems to have been one of the critical factors affecting the degree and rapidity of industrialization and accumulation of wealth. In the present decade the presence or absence of petroleum may be of even more critical importance in allowing capital accumulation necessary for industrialization. Mineral resources by nature are highly localized, with relatively few areas having large quantities (Figure 2.11).

The concentration of coal in North America, Western Europe, and the Soviet Union contributed to industrialization. The concentration of petroleum was less favorable with regard to the industrialized countries, but because of the economic power they had already acquired, they were able to obtain the limited resource from the less industrialized countries. At the beginning of the last decade of the twentieth century, the less industrialized countries with

PEOPLE, PLACES, AND POVERTY

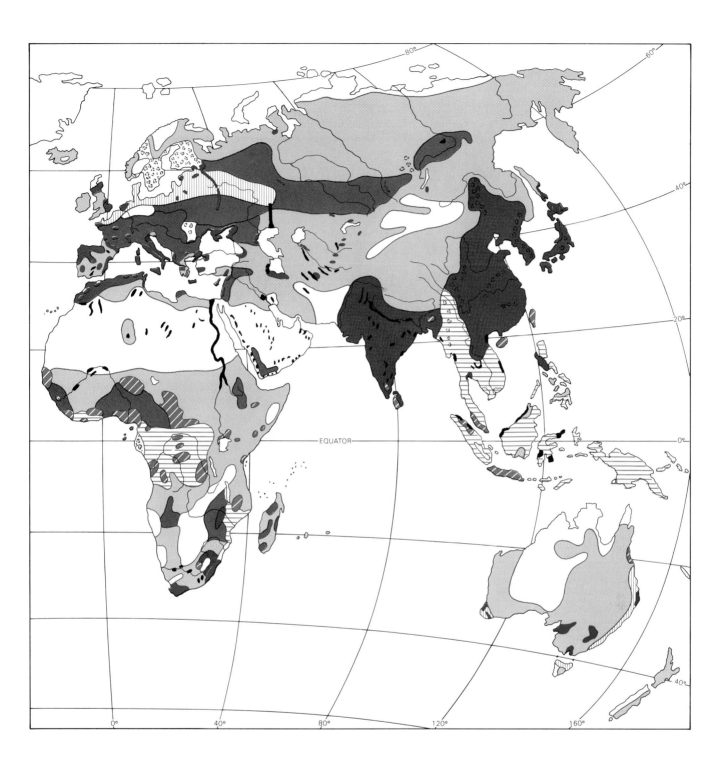

large petroleum deposits have gained control of their own resources, and the changing pattern of wealth distribution created by the purchase of petroleum by industrialized countries have affected the pattern of wealth and poverty in the world. The changing pattern of fuel resource exploitation, and associated changes in distribution of world wealth, illustrate the constantly changing role of resources in explaining the map of wealth and poverty in the world.

THE HUMAN FACTOR: DETERMINANT OF RESOURCE USE AND DEVELOPMENT

The physical environment provides resources of agricultural land, climate, water, minerals, or other useful items, but the relationship between the occupants of the earth and their physical world determines their use. Individuals and groups directly affect the physical environment through their utilization of resources, as

THE ENVIRONMENTAL FACTOR

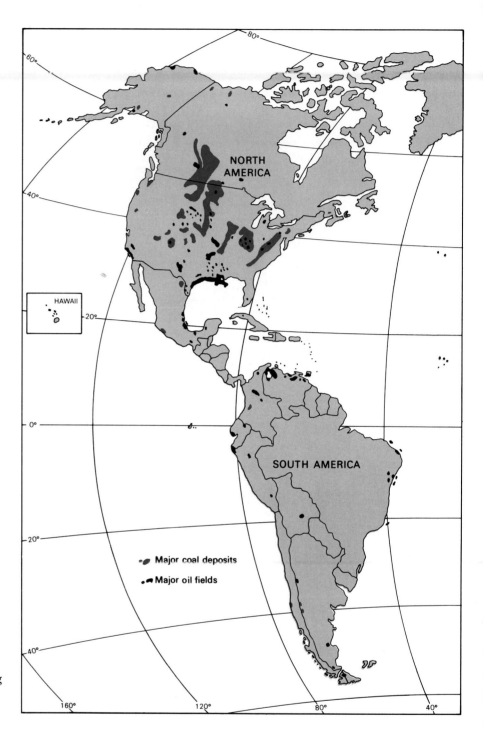

Figure 2.11 Coal and petroleum resources are unequally distributed throughout the world. North America, Europe, and Asia have the greatest reserves of coal, whereas Africa and Latin America have relatively minor coal reserves. Distribution of these and other resources is an important variable affecting the level of economic development in the world.

well as through their perception of what constitutes a resource. The environment we occupy has constraints; yet, these constraints are relative to our capabilities to manage our wants and demands within the scope of available technologies. Physical constraints of finite quantities of fertile land, petroleum, coal, or water do not mean absolute limits; rather, they are relative to human wisdom and ingenuity. Whether 1 billion people enjoy wealth while more than 4 billion suffer poverty, or all the earth's inhabitants experience a relatively high standard of living depends on the technology and management of agricultural and other resources necessary for the livelihood of the human population.

A specific area can support from ten to a thousand or more people per square mile, depending on the use and management of the available resources. A desert is not impossible to occupy; it is only difficult.

Arid lands can be irrigated, alternate resources adopted, new technologies invented, but all such changes reflect the wants, goals, and abilities of the human occupants of the earth. That industrialized countries exploit mineral and fuel resources in the Arctic or desert regions disproves the view of the environment as a determinant of human action. The cost of maintaining cities in the Siberian regions of the Soviet Union, in the deserts of Saudi Arabia, or on the north slope of Alaska are great, but where people perceive an advantage from so doing, they are not constrained.

In the past the lack of development was blamed on the physical environment, with European observers concluding that the climates of the less developed regions of the world in some way prevented their occupants from making economic progress. Proponents of this view argued that the residents of the humid tropics were lazy because the environment was hot, humid, and uncomfortable, thus making people reluctant to work. This view is labeled **environmental determinism.** European observers were accurate in their observation that people in warm lands had less stamina for work and moved more slowly than Western Europeans. The Europeans recognized a symptom—but their diagnosis was inaccurate. People in hot, humid climates have little stamina or strength because of persistent malnutrition from starchy diets that are low in proteins. The high incidence of disease and parasitic infections resulted in physical debilitation and incapability to work as hard or as long as many Western Europeans. And even healthy people with good diets "pace" themselves in tune with the heat. The physical environment does affect economic development. Certain human diseases that thrive in hot climates hinder advancement. Lack of fertile soils or other necessary resources precludes development if capital to import technology is lacking. But these problems can be overcome.

More recent observers of the location of the less industrial world have concluded that the cause of underdevelopment is the cultural values of the residents. These observers maintain that culture determines the level of development, and that societies can be classified in terms of their social, political, and technological organization. This view concludes that Western Europe developed first because of its Protestant religion, which emphasizes work and individual achievement, whereas other cultures have failed to develop because they lacked the individual ethic and were more cooperative in pursuing community goals. Such a view is as erroneous as the environmental determinist view. To assume that a culture is static and devoid of understanding of the method of development is a prime example of **ethnocentricity.** The cultural values of a group do affect its way of life but do not preclude development and an increase in the standard of living.

The relationship between the human population and the physical environment is more complex than either of these simplistic views allows. People's decisions are based on a combination of physical and cultural variables. The relationship between environment and people is based on the perceptions of the residents of a place. The way in which people perceive

---

Environmental determinism: The theory that the physical environment controls certain aspects of human action.
Ethnocentricity: The idea that one's own ethnic group, race, or other group is in some way inherently superior to others.

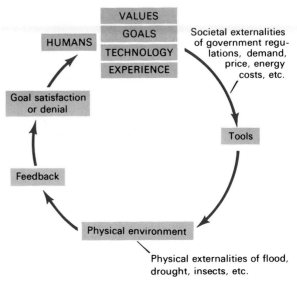

Figure 2.12 The model of the perception of the environment illustrates the complexity of the interaction between people and the world around them.

the environment is in turn affected by their technology, their goals, their cultural background, and external forces that affect the group (Figure 2.12). We all view the world through a filter—a filter made up of experience, goals, technology, and cultural values. These filters affect the decisions of an individual or human group. The culture may restrict certain types of economic activity, as the traditional ethic of African tribal groups that specify certain economic activities, such as trade, as being suitable only for women.

The technology available to an individual or a group affects their use of the environment by enabling new types of activity. The Navajo of the Arizona–New Mexico–Utah area did not use the great coal deposits that underlay their reservation. The demands for electricity from southern California prompted large power companies to import the necessary large excavating equipment, which has made possible the strip mining of large areas of the reservation—with resulting land and air pollution. The goals of an individual or a group affect their relationship to the environment by affecting their attitudes. The Navajo believes in living in harmony with the environment, while the typical Western European–North American has a long history of environmental exploitation. On a world scale, the American with goals of maximization of individual material wealth has a much different impact on the environment than the resident of India with a goal of simply maintaining life. In either case external events can affect this relationship to the environment, as in

THE HUMAN DIMENSION

The Human Modification of the Environment

The human species evolved late, very late, in the history of our planet; but during their short time on Earth, humans have wrought pervasive changes—in air, water and soils, in other living creatures, in the whole tightly connected system of interacting parts of which the environment consists—and this in but the last instant of geological time. The history of man's changing relationship with his habitat can be divided into four distinct but overlapping ecological phases: the *Primeval Phase*, the *Early Farming Phase*, the *Early Urban Phase* and, finally, the *Modern Industrial Phase*.

During the **Primeval Phase** human populations probably differed little from those of other omnivorous mammals in the extent and nature of their interaction with the ecosystems of which they were a part. Then, probably about half a million years ago, in the volcanic lands of the East African rift valleys man began to use fire, deliberately and regularly, both for protection and as a means of obtaining and cooking food.

Forest and grass fires became much more frequent events, bringing about important changes in the plant cover of certain areas of the Earth's surface. Humans had already become a significant ecological force, and human groups no longer fitted into their local ecosystems in quite the same way as other animals.

The **Early Farming Phase** began about 10,000 years ago. With the domestication of animals and plants, for the first time in the story of the Earth, a living creature came, quite consciously and deliberately, to manipulate the biological processes of nature for his own purposes. As domestication spread and farming techniques advanced, the total impact of human society on the environment increased significantly. The face of the Earth was beginning to be changed by the hand of man, to the extent that, in the northern hemisphere, vast areas of forest land disappeared, to be replaced in some instances by farming land and in others by less luxuriant and more open vegetation.

The **Early Urban Phase** started about 5,000 years ago, with the establishment of the first cities in Mesopotamia and, not long after, in China and India. These early cities ushered in the third ecological phase of human existence, which was marked by a series of fundamental changes affecting the organization of human society and the life experience of its members. The most obvious changes were simply the huge increase in the number of people gathered together in one place and the fact that the majority of their inhabitants were not directly involved in such subsistence activities as farming, fishing, and hunting. Until very late in this period, few cities contained many more than 100,000 inhabitants. However, although cities have until very recently contained only a very small percentage of the total human population, they have had an ecological impact out of all proportion to the number of people living in them.

The **Modern Industrial** or **Technological Phase** began with the so-called industrial revolution. Though it began only seven or eight generations ago, this phase has had an impact on the Earth and its workings quite disproportionate to its duration. Its ecological characteristics are not shared with any of the preceding phases of human existence.

One outstanding feature of the modern industrial phase, and one which determines many of its other characteristics, has been the massive introduction of machines and manufacturing processes powered by inanimate energy. In highly industrialized countries like the United States, the amount of power used on a per capita basis is now 30 times greater than it was prior to the industrial transition. Globally, the amount of energy flowing through human society is doubling about twice as fast as the population.

A significant change of a qualitative nature has been the synthesis, often in very large amounts, of thousands of new chemical compounds, many of which have potent effects in biotic systems. Vast quantities of these synthetic chemical substances are finding their way into the oceans, the soil and the atmosphere and, of course, into living organisms.

Complexity and unpredictability are two other characteristics of the present phase of human existence. Modern technology and the economic web of interdependence of countries are making problems of land use and resource management much more complex than they were even a generation ago. Machines are now available, for example, for clearing large areas of forest in a very short space of time. With modern communications technology, a decision taken now in a financial center in a temperate zone can lead to the felling, within a week, of a tropical forest tree 10,000 kilometers away which has taken 100 years to grow.

If certain widely accepted predictions about future increases in the level of energy use turn out to be correct, by the year 2050 humankind will be using about as much energy as is used by all other animals and plants put together. One does not have to be a specialist to see that this kind of growth in energy use, and the concomitant increase in the use of resources and the outpouring of wastes, cannot go on indefinitely.

Adapted from Stephen Boyden and Malcolm Hadley, "The Hand of Man,", Unesco Courier, July 1986, p. 35–37.

---

the case of the Navajo and the exploitation of their coal by southern California interests. The Navajo's long tradition of harmony with the environment resulted in relatively minor impact except for the grazing of their livestock. The impact of the external force of power demand from southern California has resulted in greater environmental destruction than 1000 years of Navajo occupance.

World food crops

Wheat is the single most important crop produced on the earth. It has specific requirements for optimal production, but can be grown in most middle-latitude climates. It does not do well in hot, tropical climates because of a disease called smut, which affects it under damp conditions. In practice, wheat tends to be forced from the best lands because it will produce well in less optimal conditions than crops like corn and soybeans. Wheat is concentrated in the steppe climates of the world and to a lesser extent in the humid continental, marine west coast, and Mediterranean climatic zones. As a crop, wheat has unique attributes that make it suitable for human use. It matures relatively quickly, and a crop of wheat can be grown in any region with at least 60 days growing season. A longer growing season will result in higher yields, but wheat will yield even under shorter growing seasons and under conditions of limited moisture. Wheat stores well after harvest and is preferred as a foodstuff because it contains gluten, which allows bread dough to trap the gases generated by yeast, resulting in a lighter bread. Wheat is the only major cereal with this characteristic. It is also useful as an animal feed. Much of the wheat produced in the world is consumed by animals.

Rice needs a longer growing season than wheat and is concentrated in the Mediterranean, subtropical, and tropical environments. The primary advantage of rice as a food grain is that under optimal conditions it will yield more calories per acre than any other crop. It is the common food throughout most of the less industrialized regions of Asia, Africa, and portions of Latin America.

The potato is an important crop because it will grow in cool, damp climates. A tuber, it is unique among root crops in having a fairly high protein content. It yields a very high tonnage per acre, but much of the potato is simply water and carbohydrates. It is difficult to store, and must be kept cool and dry to prevent decay.

Corn (maize) will produce under a wide variety of climatic settings. It does best in a hot, humid environment with fertile soil, and highest yields are obtained in the humid continental climates. Unlike the other major food crops, the majority of corn produced in developed regions is consumed by livestock. It is used in poor areas for human consumption because it stores well and can be kept for a long period of time without significant deterioration.

The other major food crops of barley, sweet potatoes, manioc (cassava), soybeans, millet, and yams have varying qualities. Barley is produced primarily in climates characterized by drought. It will mature rapidly, so that it is grown in the margins of steppe lands, Mediterranean lands, or in climates where the growing season is short. It is widely grown in the industrialized regions of the world where it is used as livestock feed. In the less industrialized regions barley is produced as a survival food in the poorest regions, such as Ethiopia in Africa.

Manioc, or cassava, is a tropical root crop that will grow in poor soils, and it is concentrated in the tropics and savanna region. Manioc yields a high volume per land unit, but consists mainly of carbohydrates with little protein or mineral nutrients. It does not store well when removed from the ground but does provide food over a long period, as it can be left in the ground without deteriorating. Manioc is primarily consumed by the farmers who grow it rather than entering the worldwide market exchange system. Manioc is characteristic of the poor countries of the world.

Oats are similar to barley in that they will grow in climates with a short growing season; they do best in cool, damp climates. Oats are primarily used for animal feed, and the producing areas are concentrated in the higher latitudes in the developed countries. Sorghum and millet are grown for cereal grain, and some species of sorghum are produced for their sugar content. The grain sorghums and millet serve as emergency food sources in the savanna lands and the margins of desert regions of the underdeveloped countries. Some varieties of millet are referred to as "hungry millet" because, while yielding a very low volume, they will produce some yield even when there is persistent drought.

Soybeans are an important crop because of their high protein content, and they have much the same growth requirements as corn. Major producing areas are the United States, Brazil, and China, but much of the high-protein soybeans produced are consumed by animals or exported to the wealthy industrial countries. Processed for their high oil content, soybeans form the basis for cooking oils, margarines, and other industrial items in the industrial countries.

The production of beans and peas represents a relatively small proportion of total food in the world, but they are of great importance. Beans, peas, lentils, and certain other crops of this family have a high protein content. In less developed countries these small bean crops are an important source of essential proteins not found in the staple foodstuffs of rice, potatoes, cassava, or manioc. Beans and peas store well after harvest and can be used in a variety of dishes. Certain varieties produce not only the bean but also leaves for green salads; some can be eaten before they are mature and are suitable for production of food year-round in areas of long growing seasons.

Grounded in the particular values, goals, technology, and experience of an individual or group, people make an impact upon the environment. It may range from the overwhelmingly destructive practices of strip mining in West Virginia and Kentucky in the first three quarters of the twentieth century to the seasonal planting of crops over millennia in India. In either case the tools and technology available affect the relationship between resident and environment. Once a specific action is taken affecting the environment, the local population receives some type of feedback, which either does or does not satisfy their goals. This feedback, in turn, affects future decisions.

This model of the way in which men and women perceive their environment, interact with it, and satisfy their goals helps to explain the great variety of levels of livelihood and standards of development in the world. The variables that have affected wealth and poverty are a combination of cultural, technological, attitudinal, and physical constraints. Such a model indicates the complexity of changes that will be required if all individuals in the world are to achieve the standard of living of the industrialized countries. The dilemma facing the concerned world citizen centers on whether such a transformation would, in fact, be beneficial. The pollution, waste, and general environmental degradation of the wealthy industrial countries may represent too high a price for the residents of the less industrialized world. It has been suggested that one resident of the industrialized world affects the environmental and resource base as much as 40 residents of India or other less industrialized countries. The consumption of resources in the industrial countries suggests that the present environmental impact of residents of the industrial countries cannot be sustained indefinitely. This view argues that the environment poses absolute limits. Proponents of this view argue that the world's nonrenewable resources are finite, and the world has at best 50 to 100 years in which to make fundamental changes in lifestyles. As indicated earlier in the discussion of the Club of Rome, these observers argue that the only fair model for development is one based on sharing the known nonrenewable resources on a global scale. Any development that does not increase the standard of living of the poorest will only maximize the environmental impact of the calamity in the future.

Other observers of the model of human-environmental interaction argue that the resources of the world are unlimited. Such optimists maintain that the important resources are technology, attitude, and educated people. They conclude that the greater these resources of human nature, the greater the potential for even more. This optimistic view concludes that there are no meaningful limits to growth in sight and that if any long-term limits of a finite earth really exist, they can be overcome by technology, which will allow the human population to exploit the universe. The basic premise of this view is that necessity fosters invention and that as the need arises, alternative resources of technologies will be adopted.

The impact on the world of adoption of one or the other view is profound. The adoption of the view of a limited resource base of the earth suggests that the residents of the presently wealthy countries must immediately begin to transfer technology, resources, and capital to the less developed world to assist those countries to raise their standard of living. As a part of this transfer, the lifestyle of the industrialized countries may experience fundamental changes.

Adoption of the optimistic view of infinite resources denies the need of such income transfers and argues that the industrialized countries should continue to accumulate wealth but encourage the less industrialized world countries to industrialize at the same time. Proponents of the two views are not equally vociferous, with those arguing that the resources of the earth are finite being most vocal. It is not the purpose of this text to answer the question of finite versus infinite resources, only to suggest that the present distribution of wealth and poverty is largely the result of human decisions concerning the use and development of the environment. The human population is the critical factor: critical in deciding what constitutes a resource, in deciding what form development should take, and through sheer numbers deciding what standard of living an area will enjoy. Regardless of the view one adopts concerning the limits of the earth, it is apparent that in some countries rapid population growth is defeating any attempts to overcome the poverty and underdevelopment that they presently experience.

FURTHER READINGS

BAILEY, MARTIN. "The Secret Uranium Trade." *South* as reprinted in *World Press Review*, April 1983.

BEGLEY, SHARON. "On the Trail of Acid Rain," *National Wildlife*, February/March 1987.

BOTKIN, DANIEL B. and KELLER, EDWARD A. *Environmental Studies*. Columbus, Ohio: Merrill, 1982.

BRADY, NYLE C. *The Environment: Managing Natural Resources for Sustainable Development*. Washington, D.C.: U.S. Agency for International Development Office of Publications, 1987.

BROWN, LESTER R. *Our Demographically Divided World*. World Watch paper 74, December 1986.

———. *State of the World, 1987*. New York: Norton, 1987.
———. *State of the World, 1988*. New York: Norton, 1988.
BROWN, LESTER R., FLAVIN, CHRISTOPHER, and WOLF, EDWARD C. "Earth's Vital Signs." *The Futurist*, July/August 1988.
BRYSON, R. A. and MURRAY, T. J. *Climates of Hunger: Mankind and the World's Changing Weather*. Madison: University of Wisconsin Press, 1977.
BYRON, WILLIAM, ed. *The Causes of World Hunger*. New York: Paulist Press, 1982.
CHANDLER, WILLIAM U. "Converting Garbage to Gold: Recycling Our Materials." *The Futurist*, February 1984.
CLARK, AUDREY N., compiler. *Longman Dictionary of Geography: Human and Physical*. White Plains, N.Y.: Longmans, 1985.
COOK, E. *Man, Energy, Society*. San Francisco: Freeman, 1976.
COURTENAY, P. P. *Geographical Studies of Development*, London: Longmans, 1985.
CUTTER, SUSAN L. et al. *Exploitation, Conservation, Preservation: A Geographic Perspective on Natural Resource Use*. Totowa, N.J.: Rowman & Allanheld, 1985.
DEUDNEY, DANIEL and FLAVIN, CHRISTOPHER. "Shapes of a Renewable Society," *The Humanist*, May/June 1983.
DICKENSON, J. et al. *A Geography of the Third World*. London: New York: Methuen, 1984.
"The Dynamic Earth," *Scientific American*, Special Issue, September, 1983.
EL-HINNAWI, ESSAM and MANZUR-UL-HASHMI, eds., *Global Environmental Issues*. Dublin: Tycooly International Publishing, Ltd., United Nations Environment Program, 1982.
"Forecasting the Weather," *The Economist*, 11 June 1983.
GANN, L. H. and DUIGNAN, PETER. "Nationalization as a Trade Culprit," *The Wall Street Journal*, May 24, 1984.
GILBERT, SUSAN. "America Washing Away," *Science Digest*, August 1986.
GOLDEMBERG, J., JOHANSSON, T. B.; REDDY, A.K.N., and WILLIAMS, R. H. *Energy For a Sustainable World*. New York: Wiley, 1988.
GOUDIE, ANDREW. *The Human Impact: Man's Role in Environmental Change*. Cambridge: MIT Press, 1986.
GREGG, D. *The World Food Problem 1950–1980*. Oxford: Basil Blackwell, 1985.
GRIBBIN, J. *Climate and Mankind*. London: Earthscan, 1979.
GWATKIN, D. and BRANDEL, S. "Life Expectancy and Population Growth in the Third World," *Scientific American*, May 1982, pp. 57–65.
HARLAN, JACK R. *Crops and Man*. Champaign, Ill.: American Society of Agronomy, 1975.
HOLE, FRANCIS D. and CAMPBELL, JAMES B. *Soil and Landscape Analysis*. Totowa, N.J.: Rowman & Allanheld, 1985.
HOLLING, CRAWFORD S. "Predicting the Unpredictable," *Unesco Courier*, August/September, 1982.
HOSMER, ELLEN. "Paradise Lost: The Ravaged Rainforest," *Multinational Monitor*, June 1987.

JACKSON, ROBERT. *Global Issues 86/87*. Guilford, Conn.: Dushkin Publishing, 1986.
JOHNSON, ARTHUR H. "Acid Deposition: Trends, Relationships, and Effects," *Environment*, May 1986.
JOHNSTON, RON J. and TAYLOR, PETER J., eds. *A World of Crisis? Geographical Perspectives*. New York: Basil Blackwell, 1986.
JUMPER, S. et al. *Economic Growth and Disparities: A World View*. Englewood Cliffs, N.J.: Prentice-Hall, 1980.
KATES, ROBERT W. "The Human Environment: The Road Not Taken, The Road Still Beckoning," *Annals of the Association of American Geographers*, 77(4), 1987, pp. 525–534.
KAYA, YOUCHI, KONDO, SHUNSUKE, KOBAYASHI, HIKARU, SUZUKI, YUTAKA, TONAKA, TSUTOMO, and MUROTA, YASUHIRO. "Management of Global Environmental Issues," *World Futures*, Vol. 19, Gordon and Breach, Science Publishers, 1984.
KELLOGG, WILLIAM W. and SCHWARE, ROBERT. *Climate Change and Society: Consequences of Increasing Atmospheric Carbon Dioxide*. Boulder, Colo.: Westview Press, 1981.
KNAPP, BRIAN. *Practical Foundations of Physical Geography*. London: Allen and Unwin, 1981.
LAZZLO, E. *Goals for Mankind: A Report to the Club of Rome on the New Horizons of Global Community*. New York: Dutton, 1977.
"Living in a Global Greenhouse." *The Economist*, November 29, 1986.
MCKNIGHT, TOM L. *Physical Geography—A Landscape Appreciation*. 2d ed. Englewood Cliffs, N.J.: Prentice-Hall, 1987.
MATHER, JOHN R. *Water Resources: Distribution, Use, and Management*. New York: Wiley, 1984.
MEADOWS, DONELLA. *The Limits to Growth. A Report for the Club of Rome's Project on the Predicament of Mankind*. New York: Universe Books, 1972.
MURDOCK, WILLIAM W. *Environment: Resources, Pollution and Society*. Sunderland, Mass.: Sinauer Associates, 1975.
MYERS, NORMAN. "The Conversion of Tropical Forests." *Environment*, July/August 1980.
O'HARE, GREG and TIVEY, JOY. *Human Impact on the Ecosystem*. Edinburgh and New York: Oliver and Boyd, 1981.
OHLENDORF, PAT. "The Coming Water Crisis." *Maclean's* as reprinted in *World Press Review*, February 1983.
OVCHINNIKOV, YURI. "Seeds of Plenty: The Promise of Biotechnology," *Unesco Courier*, April 1984.
PEASE, ROBERT W. "The Average Surface Temperature of the Earth: An Energy Budget Approach," *Annals of the Association of American Geographers* 77, 1987, pp. 450–61.
Report on Natural Resources for Food and Agriculture in Latin America and the Caribbean. Rome: Food and Agriculture Organization of the United Nations, 1986.
REVELLE, PENELOPE and REVELLE, CHARLES. *The Environment: Issues and Choices For Society*. Boston: Jones and Bartlett, 1988.

Rostow, W. *The Stages of Economic Growth*, 2d. ed. London: Cambridge University Press, 1971.

Simmons, I. G. *The Ecology of Natural Resources*. New York: Halsted Press, 1974.

Smil, Vaclav. *Energy, Food, Environment: Realities, Myths, Options*. London: Oxford University Press, 1987.

Southgate, Douglas D. and Disinge, John F., eds. *Sustainable Resource Development in the Third World*. Boulder: Westview Press, 1987.

Stolz, Joelle. "TV Captivates Algeria," *Le Monde* as reprinted in *World Press Review*, January 1983.

Strahler, Arthur N. and Strahler, Alan H. *Elements of Physical Geography*, 3d ed. New York: Wiley, 1984.

Trewartha, Glenn T. *The Earth's Problem Climates*. Madison: University of Wisconsin Press, 1981.

Winder, David. "The World's Shrinking Forests." *The Christian Science Monitor*, January 10, 1984.

World Resources Institute. *World Resources 1986: An Assessment of the Resource Base That Supports the Global Economy*. New York: Basic Books, 1986.

Srinagar, Kashmir.

3

The Human Equation the central issue

MAJOR CONCEPTS

Arithmetic growth / Exponential growth / The Malthusian dilemma / Neolithic revolution / Labor specialization / Emergence of cities / Population doubling time / Demographic transition / Lifeboat earth / Triage / Small planet / Rural-urban migration / Industrial Revolution / Population explosion

IMPORTANT ISSUES

Population growth and resource allocation in less industrialized countries

Changing technology, population growth, and resource exploitation

Foreign aid: gift or requirement from wealthy countries?

Foreign aid: handicap or asset to less industrialized countries?

Urbanization of the world's population

Global Population Characteristics

1. Global population growth and related environmental change has been slow throughout most of time.
2. Technological changes, especially those of the Neolithic and Industrial Revolutions, were associated with rapid growth in global population.
3. Beginning with the theory of Malthus during the Industrial Revolution, observers have warned of the consequences of unchecked population growth.
4. Population growth rates are slowing in all major geographic regions at the present.
5. The less industrialized countries of the world typically have the highest growth rates, and Africa has the highest rates of any region.

An old riddle asks when a lily pond will be covered by the leaves of the lilies. If the lily pond has only one leaf on the first day and the number of leaves doubles each day, by the end of the third day there will be four leaves on the pond. If the pond is full on the thirtieth day, at what point is it half full? The answer is the twenty-ninth day. Many observers maintain that world population is similar to the lilies of the pond. Just as on the twenty-ninth day the lily pond is a pleasant combination of blue water and green leaves, for most people the earth in 1990 is basically a desirable place in which to live. On the thirtieth day the lily pond is covered with leaves, the water is obscured, and a process of deterioration begins in the pond itself.

The human population of 5.3 billion in 1990 may well be like the lily pond on the twenty-ninth day. If so, the present pattern of development and underdevelopment will only become more extreme as the thirtieth day arrives. If the earth is presently half full, the next generation, which is the equivalent of the next day in the lily pond, could completely consume all of the resources that the earth can provide. Many observers are accepting the idea that resources for development are presently nearing their full utilization. The basic raw materials that humans depend upon include croplands, fisheries, grasslands, forests, and mineral and fuel resources. Assessment of each indicates that the relationship between population and resources is nearing the critical point.

Most of the 5.3 billion human beings presently alive have rising expectations. The poorest of the less industrialized world are no different from those of the wealthy world: Both desire a better life tomorrow from what they experience today. The evidence suggests that the demands of the human population are exerting such pressure on the earth's resources that the long-term ability of the earth to support population is actually declining. The productivity of oceanic fisheries is falling as world catch exceeds reproductive ability. Forests are disappearing on a worldwide scale as the growing populations of the less developed world destroy them for more cropland, building materials, or fuel for cooking and heating. The pressure of the ever-growing human population is central to any efforts to overcome the world imbalance of wealth and poverty. It is particularly critical to recognize that overcoming the lack of economic development will require balancing the human equation. When populations of countries such as India continue to grow, catching up with the additional needs of existing residents plus the new multiplying generation is nearly impossible.

The population issue is a little like a person with a savings account of $100,000. If the account earns 10 percent interest per year, the owner has a disposable income of $10,000. If the individual spends $20,000 per year, the savings will be depleted within eight years. The analogy to population and development is that when the natural ability of the earth's major life-support systems is exceeded, the actual ability of the earth to support people declines. Unless this issue is addressed, not only will the less industrialized countries remain underdeveloped, but they and the industrialized world can expect to find their "income" declining as their savings account balance is utilized for present consumption.

The critical equation results from the limits of the earth compared to the number of people and their consumption rates. So long as one group has consumption rates that demand a disproportionate share of the income from the "savings" (in this case the world's available resources), the balance of the earth's population must live on an ever-declining income. The plight of the residents of the less industrialized world is compounded by their ever-growing populations, which demand a portion of the already small income of these countries. Population growth in the wealthy industrialized countries of the world has a compound effect on the less developed world because each new resident in North America, Western Europe, Japan, or Australia and New Zealand will absorb another increment of world income, further depriving the less industrialized world. If the world is indeed in the twenty-ninth day, the population issue represents a worldwide crisis that must be resolved, not only to allow an increase in the standard of living of the world's poor, but to prevent the disintegration of the world economic order as we know it.

The full significance of the population crisis can be partially understood by looking at the simple arithmetic of population growth. In 1990, 148 million children were born. Fifty-three million individuals died from old age, disease, accidents, or malnutrition. The estimated net growth in world population of 95 million in 1990 is nearly equal to the population of Mexico and greater than the population of any individual European country. Two years' addition would exceed the population of Japan, and three years' would be greater than the population of the United States. The daily net increase in world population yields over 260,000 additional mouths to feed. Each week a new city the size of Chicago is created by the additional population growth. In the time it has taken you to read the numbers in this paragraph, 180 new people have been added to the earth's population: 180 per minute, three per second, three additional world residents every time your heart beats!

What this means for development of the world centers on the need for food, clothing, shelter, medical care, and jobs for this ever-growing population. In order to provide each new resident of the earth with a single glass of milk and a single one-pound loaf of bread per day would require adding the equivalent of 125 dairy cows and an additional 15 acres of land to raise the wheat. On a yearly basis more than 60,000 cows and 6200 acres of wheat would need to be added simply to provide this minimal food ration. Simple numbers do not indicate the long-term significance of this population growth, as the majority of these children are born in the less industrialized world where they not only are likely to receive inadequate food and other basic needs but may contribute to a lower standard of living for the existing residents.

THE HISTORY OF THE POPULATION PROBLEM

A look at the growth of world population reinforces the analogy of the lily pond (Figure 3.1). Although estimates for world population before A.D. 1650 are at best vague reconstructions, they illustrate the slow rate at which population has grown for most of human history. Generally accepted estimates suggest that human beings have occupied the earth in one form or another for the past 1.5 to 2 million years. One million years ago there were an estimated 125,000 human inhabitants. In the next 700,000 years, this population may have increased to 1 million. The growth rate during the first 2 million years of human occupance of the earth was slow because the means of livelihood available were limited to hunting and gathering, and availability of foodstuffs in the natural environment limited population growth.

THE NEOLITHIC REVOLUTION

The first great change affecting the world population occurred approximately 8000 years ago with the **neolithic revolution**. The neolithic, or agriculture, revolution involved the change from dependence on gathering wild plants and hunting wild animals to plant and animal domestication and sedentary farming. The neolithic revolution allowed people to establish permanent residence in places suitable for farming and ultimately fostered emergence of civilization as we know it. Technical advances dating from the postneolithic period included the use of the written alphabet, the wheel, the sailing ship, and the emergence of cities and political empires. The neolithic revolution probably began in either southeastern Asia or the

---

Neolithic revolution: The change from dependence on gathering wild plants and hunting wild animals to plant and animal domestication and sedentary farming. Also called the first agricultural revolution.

WORLD POPULATION GROWTH

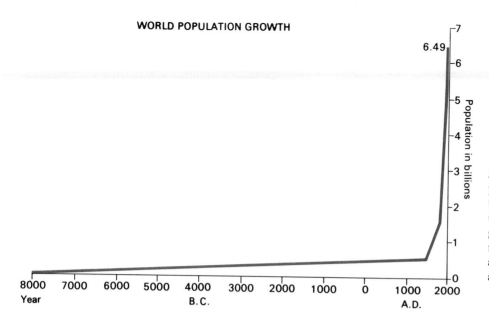

Figure 3.1 World population growth reflects the exponential increase of the past 500 years. Both growth rate and total numbers increased rapidly as technology, medicine, and health improved.

mountains of the present-day Middle Eastern countries of Israel, Jordan, Syria, Iran, and Iraq. Archaeological evidence suggests that the wide variety of environments at different elevations in the uplands of the Middle East was the site of domestication of wheat, sheep, goats, and dogs. This theory concludes that the variety of plants and animals in a relatively small area ultimately fostered cultivation of preferred species. Another view suggests that the initial idea for domestication occurred in southeast Asia where land and water meet and where gathering societies could easily accumulate a food surplus. This food surplus allowed semipermanent residence and eventually the breeding of plants to obtain poisons for stunning fish. Breeding led to cultivation of plants with particularly desirable attributes, and from here the idea of domestication spread to the Middle East (Figure 3.2). Whichever view is accepted, the significance of the neolithic revolution on world population was immediate and profound (Table 3.1). Rapid adoption of plant and animal cultivation increased the ability of favored sites to support population, and total numbers of humans increased ten to twenty fold. Within 2000 to 3000 years the earth's population had increased twenty-five fold from an estimated 5.3 million to more than 86 million.

Just as important as the massive increase in population were the changes associated with livelihoods based on domestication. People moved from highlands to the floodplains of the great rivers of the world where spring flooding of the rivers brought a new layer of fertile soil. Invention of the plow and the use of domestic animals to pull it were the basis for a great change in the *character* of the human population. The creation of a reliable agricultural base, with its resulting food surplus, freed some individuals from the constant need to acquire food. The emergence of a segment of population that was not actively engaged in agriculture was the beginning of today's **labor specialization** and urban life. Labor specialization refers to development of an economy in which an individual or group performs only one activity. Instead of each individual producing food, building a home, and making tools, with labor specialization some individuals are farmers, some carpenters, some artisans, and so forth. Associated with labor specialization was the concentration of population in villages where the products and labor could be exchanged. By 400 B.C. villages had become a well-established form of human occupance.

The valleys of the Tigris and Euphrates rivers of the Middle East, the Nile Valley of northern Africa, the Ganges and Indus of the Indian subcontinent, and the Chang Jiang (Yangtze) and Hwang He rivers of China became the focus of agricultural population and developed early cities. The cities served as cultural crossroads where peoples of different cultures met. The early towns in all these river valleys were very similar. A main cereal crop (wheat, barley, or rice) was the primary food source. The development of bronze and other metal tools and transportation advances resulting from the sail and the wheel accompanied and facilitated urban growth. Although cities before the time of Christ remained small, rarely ex-

---

Labor specialization: The emergence of specific occupational categories that replaced general labor requirements of the early neolithic period.

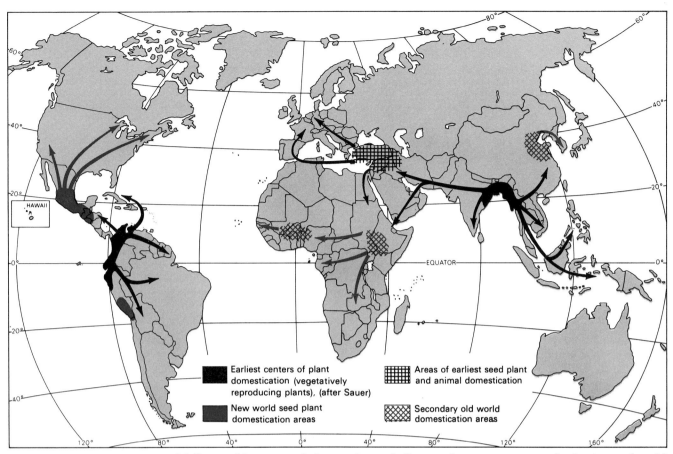

Figure 3.2 Origins and diffusion of domesticated plants and animals illustrate that present centers of technology and wealth have not always dominated the world.

TABLE 3.1 Population Growth of the World

Date	Area Populated	Assumed Density per km^2	($mile^2$)	Total Population (Millions)
Years Ago				
1,000,000	Africa	0.00425	(0.011)	0.125
300,000	Africa and Eurasia	0.12	(0.3)	1.0
23,000	Africa and Eurasia	0.04	(0.10)	3.34
10,000	All continents	0.04	(2.59)	5.32
6,000	All continents	1.04	(2.7)	86.5
30 B.C.	All continents	1.0	(2.59)	133.0
A.D. 1650	All continents	3.7	(9.6)	545.0
1750	All continents	4.9	(12.7)	728.0
1800	All continents	6.2	(16.1)	906.0
1900	All continents	11.0	(28.49)	1,610.0
1950	All continents	16.4	(42.48)	2,400.0
1990	All continents	35.4	(91.62)	5,300

SOURCE: E. S. Deevey, "The Human Population," *Scientific American* 203, No. 3 (1960): 195–204, and authors' calculations.

THE HUMAN EQUATION

TABLE 3.2 Approximate Population (millions) of the World—8000 B.C. to A.D. 1600

Year	World	Europe	Asiatic Russia	Southwest Asia	India	China Major	Japan	Southeast Asia, Oceania	Africa	The Americas
8000 B.C. A.D. 1				Data not available						
1000	275	42	5	32	48	70	4	11	50	13
1100	306	48	6	33	50	79	5	12	55	17
1200	348	61	7	34	51	89	8	14	61	23
1300	384	73	8	33	50	99	11	15	67	28
1400	373	45	9	27	46	112	14	16	74	30
1500	446	69	11	29	54	125	16	19	82	41
1600	486	89	13	30	68	140	20	21	90	15

SOURCE: Annabelle Desmond, "How Many People Have Ever Lived on Earth?" *The Population Crisis*, ed. Larry K. Y. Ng and Stuart Mudd (Bloomington, Ind.: Indiana University Press, 1965), p. 30.

ceeding 10,000 people, these small cities began the process of change that ultimately culminated in the Industrial Revolution and modern urbanization. They were the focus of social, economic, and political changes including development of complex irrigation systems, the calendar, written language, mathematics, astronomy, and social and political organization.

Early cities developed along major transportation routes through which new ideas and innovations flowed. The large concentration of specialists, crafts, and industries encouraged innovation and stimulated desire for philosophical and scientific investigation. As innovations found fertile ground in the city, technological advance continued and in turn facilitated increased growth in population (Table 3.2). Advanced technology depended, in turn, on an increasingly complex division of labor, and the labor specialization increased population potential. At the same time, a more reliable food source and incipient sanitation lengthened the individual life span. Although only a tiny minority of the population resided in cities after the neolithic revolution, the increased flow of trade and ideas between cities allowed technological advances to spread throughout the world.

After the neolithic revolution, and concomitant with the advent of the city, social and political changes led to the great empires of the world such as the Babylonian, Assyrian, Persian, Roman, Chinese, and Indian civilizations. These empires incorporated a highly organized city life based on surplus food from rural areas.

Advances in technology and political organization, however, were not matched by advances in medical care, and high infant mortality rates and a limited life span combined to keep population growth rates well below 1 percent per year. It is estimated that 2000 years ago the largest population of the earth was on the Indian subcontinent with 100 to 140 million people. China contained as many as 70 million individuals, and imperial Rome had approximately 54 million people. The average life expectancy in Roman times was probably about 30 years, and the general average for the rest of the world was likely about the same. The longevity rates remained relatively constant until the advent of scientific medicine in western Europe in recent times.

RECENT POPULATION GROWTH

The population of the world reached an estimated 400 million by A.D. 1500. Greatest concentrations were in Asia, with Europe a distant second. Africa and Latin America had much smaller populations. The growth of population from 5 million at the beginning of the neolithic revolution, to 400 million at the beginning of the European age of exploration in A.D. 1500 represented an abrupt increase unmatched by any preceding period.

The period following A.D. 1500 is known as the age of discovery and colonization. In the short period from 1450 to 1600, Europeans sailed over essentially all of the globe and made contact with inhabitants of most populated regions. Improvements in shipbuilding by the Portuguese and refinement of navigation methods through use of charts and maps enabled European nations to form empires throughout the world, particularly in the Americas. The growth of colonies increased trade and commercialization (Figure 3.3).

This age of discovery was paralleled by migration and population movement on a scale previously un-

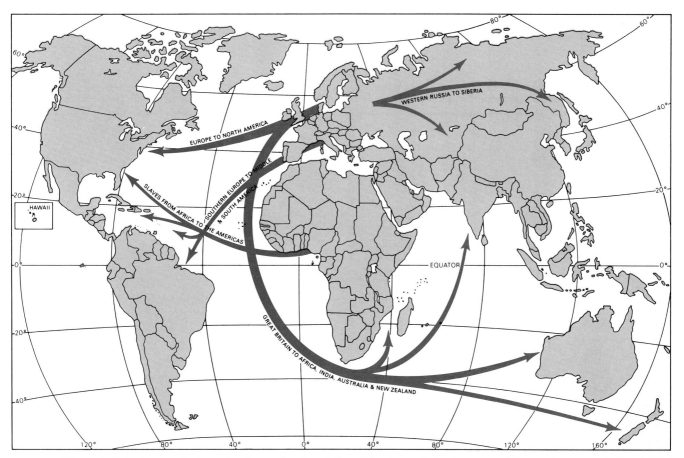

Figure 3.3 Major population migrations related to industrialization and development of Europe between 1600 and 1900 illustrate the influence of Europe on the rest of the world.

known. Among the most important migrations were those: (1) from Europe to North America, (2) from Mediterranean countries of Europe to Middle and South America, (3) from Great Britain to Austrialia, New Zealand, India, and Africa, (4) of slaves, from Africa to America and to Arab and Turkish areas, and (5) eastward, of Russians into what is now Siberia. European colonists diffused to North America, Australia and New Zealand, parts of Africa, and parts of Latin America in the period from 1500 to 1800. African natives were an unwilling part of this great population movement as approximately 20 million blacks were forced from their homeland and dispersed throughout the world as slaves. The organizational and technological abilities of the European cultures enabled them to dominate the native cultures that they contacted; even where Europeans did not actually colonize, they affected the population by their economic and political dominance. The availability of new lands provided an outlet for population increase in Western Europe; an increase that was the beginning of the tremendous population growth that continues to this day.

The industrial revolution: stimulus of population growth. The expansion of European colonization after A.D. 1500 was a result of changing technology in Western Europe. Improvements in agriculture resulted in greater food availability, greater food availability fostered both larger populations and a higher standard of living, and technological changes made possible greater productivity of material goods. The rate of population growth in Western Europe increased dramatically after 1700 with social changes of the **industrial,** technological, and agricultural revolutions. Industrialization facilitated a host of related improvements in the productivity of labor, new methods of transportation, the discovery and use of new forms of energy, and an explosion of inventions, giving humans

---

Industrial revolution: The Industrial Revolution involved the substitution of machine power for muscle power sources, allowing production increases and a growth in demand for resources.

THE HUMAN EQUATION

freedom and opportunities on a scale never before dreamed possible. Associated advances in science and technology increased awareness of the cause of disease and prevention of illness. Simple advances in hygiene and sanitation, including waste disposal and cleanliness, greatly expanded life expectancy of Europeans. The impact of the declining death rate brought about by medical and hygiene advances is illustrated by the Netherlands. In 1840 one fourth of all infants born in the Netherlands were dead by age 2 1/2. A century later, in 1940, one fourth of infants born would not be dead until age 62 1/2!

The spread of European culture increased the population growth rate throughout the world. World population reached 1 billion by 1825, after nearly 2 million years of the human occupance of the earth. With the continual increase in rate of growth per year, the second billion were added in only 100 years (Table 3.3).

With ever-increasing numbers of people and ever-greater chances of survival to adulthood, the time required to double the world's population continues to diminish. Where the doubling of the world's population from 3 million to 6 million took 20,000 years, each subsequent doubling has required less time. The frightening aspect of present population growth is that the doubling rates for *billions* of people is becoming smaller than the doubling rates for *millions* in past times. The third billion population took slightly more than 30 years, the fourth billion only 15 years, and the fifth billion took only 12 years. The literal explosion of population that results from the decline in death rates and longer life spans associated with the Industrial Revolution and related medical advances represents one of the most critical issues facing the entire world.

The demographic transition. The rate of population growth in some countries, particularly industrialized ones, has followed a pattern of increase and then decline. The experience of the major industrialized countries in Western Europe and North America in a period of rapid population growth followed by declining growth rates is referred to as the **demographic transition** (Figure 3.4). The process can be divided into four stages. The first stage represents the situation in the majority of the world before 1700. Birthrates were high, averaging more than 40 per thousand, but death rates also remained high, resulting in slow rates of growth. Good harvest years might result in increased birthrates and lower death rates, allowing an increase in population. But the earth's population

TABLE 3.3 World Population Growth

	Year	Average Annual Growth (%)	Doubling Time (Years)
1st billion	1825		
2nd billion	1927	0.68	102
3rd billion	1960	1.23	57
4th billion	1975	1.92	36
5th billion	1985	2.23	31

SOURCE: William W. Murdoch, *Environment, Resources, Pollution, and Society*, (Sunderland, Mass.: Sinauer Associates, 1975), p. 49, and calculations by authors.

Demographic transition: Model of population change that suggests countries move from a slow population growth stage with both high birth- and high death rates, to a stage of rapid population growth when death rates drop, to a stage of slow population growth as birthrates also fall, to a potential negative growth rate.

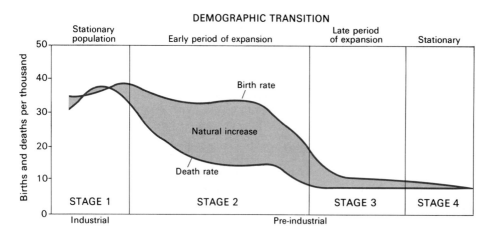

Figure 3.4 Demographic transition. Countries with either a high birth and death rate or a low birth and death rate will grow slowly. The World population "explosion" is taking place in countries with a high birthrate and a low death rate.

Contrasts in lifestyles within and between the less industrialized countries are illustrated by these photographs of children involved in agricultural activities in Brazil (harvesting sugar cane), Egypt (plowing), Morocco (transporting harvested grain), Philippines (spinning thread).

remained small because disease and limited capacity for food production resulted in high infant mortality rates, if not conscious strategies to minimize births. Furthermore, the average life expectancy was much lower than it is today, causing populations to remain relatively slow-growing even though birthrates were very high.

The second stage of the demographic transition is marked by high birthrates but low death rates. Birthrates continue to average 40 or more per thousand, but death rates rapidly decline to 20 or less per thousand. The difference between the high birthrate and the low death rate explains the high growth rate experienced in the world since the Industrial Revolution. Western Europe was in stage 2 from 1700 to 1900, and its population increased dramatically. The countries of the less industrialized world have entered the second stage since 1800, and their populations continue to increase. In stage 2 adoption of better hygiene and availability of medical care ensures that more infants will live to maturity, and more children will consequently be born.

Some countries of the world have moved beyond stage 2 with its characteristic population explosion into a period of declining birthrates and slower growth. Stage 3 is marked by a rapidly declining birthrate and continued falling death rate. The industrialized countries of the world, with their ready access to medical care, abundant and nutritious food, and long life expectancies, are in this stage. For example, Europeans can anticipate an average life of 74 years, while people of the less industrialized world average only 58 years.

The most mature industrialized countries of the world enter stage 4, reaching a level of stability between birth- and death rates. Death rates are low, averaging only about 10 per thousand, but the birthrates are also low, in some instances below the death

THE HUMAN EQUATION

rates. In most industrialized countries the birthrates are about 12 to 14 per thousand, and the population grows slowly or not at all. In some countries in the fourth stage negative growth has occurred, as in Hungary and Denmark.

The stage of the demographic transition for each country of the world is closely related to its level of economic development. Those countries that have entered stage 3 or 4 typically are industrialized, whereas those in stage 1 or 2 typically possess the characteristics of underdevelopment discussed earlier.

Will all countries ultimately reach stage 3? Analysis of the forces that seem to have affected the European transition indicates some of the problems facing the less industrialized countries. The decline in birthrate in western Europe seems closely related to the actual and perceived value and benefit of children as opposed to their costs. The impact of a child born in a family in the industrialized world is much different from what it is in less developed countries, as is the relative cost and utility of each additional child. The benefits of children in various societies may include their labor, security for parents in old age, personal satisfaction, or such miscellaneous benefits as tax deductions or government subsidies. The disadvantages or costs of children include the child-rearing expenses of education, food, clothing, and shelter; the loss of the mother's labor during part of the pregnancy and early life of the child; restricted options for the parents because of the responsibility of caring for children and providing for their needs; and a variety of minor disadvantages such as mental stress on the parents.

The relative advantages and costs of children vary between industrial and less industrialized countries. In less industrialized countries a child may be the only form of security for old age that is available to an average family. Governments rarely provide old-age assistance programs (social security) like those in the industrialized world. Because the economies of the less developed countries are predominantly agricultural, children become at least marginally economically useful at an early age. Children under six can weed the garden, tend poultry, collect firewood, or perform other tasks. As a source of personal satisfaction in less industrialized countries, children often are the only area of distinct achievement that a woman or couple achieves. Social status associated with large families, particularly with male children, acts as an incentive to bear additional children (Table 3.4).

The cost of additional children in less developed countries is less than in industrialized countries. Lack of educational opportunities limits the cost of providing education, and the majority of children in less industrialized countries are provided with only minimal quantities of food, clothing, or other consumer items. In a less industrialized country the mother can more quickly return to work, since she can take her new infant to the fields or to the kitchen with her, unlike industrialized countries where people have traditionally not been able to bring their young children to, say, the office or factory. In the less industrialized countries parents' options are less restricted by additional children because the primary restriction on options is poverty, and additional children are not perceived as adding materially to that restriction.

In industrialized countries the benefits and costs of children are directly the opposite of those in the less industrialized countries. The government provides care for the elderly, there are various ways in which an individual or couple may derive self-satisfaction other than through bearing more children, and social and legal restrictions prevent children from working until maturity. The costs in the industrial society are perceived as much higher for each additional child because of the expectation that children will attend school for ten to sixteen years, be provided with a variety of consumer items including televisions, clothing, and even automobiles, and the restrictions on actions of the parents are even greater because of the nuclear family organization in industrial countries that often precludes leaving children with grandparents or other relatives while parents engage in their activities. In industrialized societies children lose much of their economic significance, and having fewer children becomes a benefit. Here, the smaller the family, the greater the freedom of choice for either material things or personal goals. By limiting the number of children, a family may purchase an additional car, buy a larger home, or engage in travel or other personal hobbies. In a less industrial society children can help earn the living and support the parents when they are old, but in an industrial society children are an economic liability.

The perceived value of children in industrial societies leads to a decline in the birthrate and ultimate stability of the population. Most industrial regions such as Western Europe, North America, and Japan have reached this point. The contrast between countries in stage 4 and those in stage 2 of the demographic transition is readily apparent from the age structure of countries in each stage (Figure 3.5).

The *population pyramid* is simply a bar graph showing the percentage of the total male and female population in each age group. Depending on the scale, each bar normally represents a five- or ten-year interval. The population pyramids of countries in stage

TABLE 3.4 Attitudes About Family Size in African Nations

Country and year	Percent of currently married Women with children who:			Desired/ideal family size Percent of currently married women who:		
	Do not want more children	Are undecided about having more children	Want more children	Gave a nonnumerical response	Have 9+ surviving children	Desire family size of 9+
West Africa:						
Benin 1981–82	8	18	74	NA	NA	NA
Ghana 1979–80	12	10	78	11	2	15
Ivory Coast 1980–81	4	5	90	26	2	49
Nigeria 1981–82	5	11	84	32	2	40
Senegal 1978	8	5	87	29	2	54
Central Africa:						
Cameroon 1978	3	18	79	33	2	45
East Africa:						
Kenya 1977–78	17	15	68	19	5	26
Southern Africa:						
Lesotho 1977	15	1	84	2	0	15
Arab Africa:						
North Sudan 1979	17	6	77	18	4	22

SOURCE: Odile Frank, "The Demand for Fertility Control in Sub-Saharan Africa," *Studies in Family Planning*, 18 (1987), p. 181–201.

4 of the demographic transition show a relatively uniform number of residents in all age groups until late middle age. Population pyramids of countries in the stage 2 have a true pyramidal shape with each lower age group comprising a larger percentage of the total population. The preponderance of children under age 15 is a major problem for less developed countries; in many of them as many as 45 percent of the population is 15 years old or younger. Meanwhile, people in the industrialized world are concerned about the ever-increasing proportion of elderly in their populations as increasing life spans since the 1950s have created a new nonproductive sector, the elderly retired.

POPULATION GROWTH AND WORLD IMPLICATIONS

The implications of the relative rate of population growth in a country are of worldwide significance. In the early years of the Industrial Revolution an amateur English economist by the name of Thomas Robert Malthus observed that the population of the world

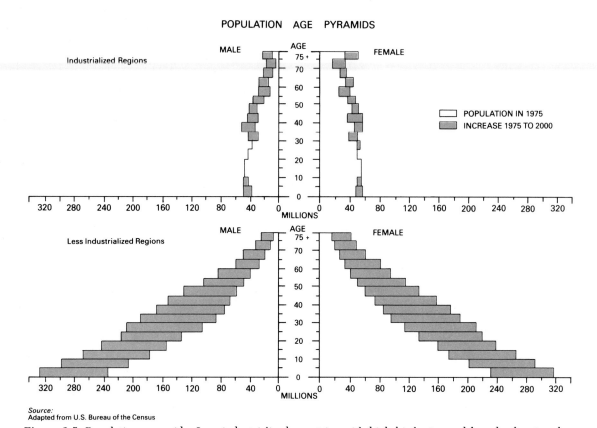

Figure 3.5 Population pyramids. Less industrialized countries with high birthrates and low death rates show the typical pyramidal shape caused by the high percentage of their population in the lower age groups.

would tend to increase more rapidly than the food supply that was needed to sustain it. In 1798 he published an essay in which he explained what subsequently would come to be called the *Malthusian equation.* The basic premises of Malthus' theory were that population always increases **exponentially** while food supply increases only **arithmetically.** Thus 2 people become 4, 4 become 16, and 16 eventually becomes 256. In the same period 2 tons of food become 3, 3 tons increase to 4, and 4 increase only to 5. According to Malthus the two chief human desires, food and procreation, are powerful, uncontrollable, and unchanging, and ultimately lead to misery for the majority of the human race. Malthus maintained that the only checks to prevent population exceeding food supply are moral restraint or wars, famines, and pestilence (disease). The ever-growing population will not respond to moral restraints because each individual is either unwilling or unable to see the significance of his or her actions on a global scale. We are all concerned with maximizing our own pleasure, and the result is a decreasing potential for the world to support its burgeoning population. According to Malthus, ever-increasing populations with increased pressures for resources result in malnutrition and disease, famine when food supplies are limited by drought or other natural disasters, and wars to gain control of the limited resources available. Famine, disease, and war, then, halt rapid population growth and bring periods of at least temporary population stability.

At the time Malthus wrote, Western Europe was in stage 2 of the demographic transition, and his gloomy projection did not materialize. Europe's colonization and settlement of North America and Australia and New Zealand provided relief from the pressures of population growth, and "moral restraints" associated with changing values of children in an industrial society moved the European countries into stage 3 of the demographic transition. In recent years, some observers have maintained that Malthus was correct; he was just ahead of his time. Others argue that Malthus

---

Exponential growth: Referring to the increase associated with increasing any number by some exponent or power. Malthus maintained that populations tended to double, thus the exponent is 2.

Arithmetic growth: Refers to simple addition as a means of increase.

was completely in error, and that human ability to solve problems will prevent the cycle of war, famine, and disease, with its associated human misery, from occurring. Whatever the ultimate outcome, the Malthusian equation points to the problems of population growth in less developed countries.

Writers have responded to the rapid population growth in less developed countries—and what should be done about it—in various ways. Three of the most widely discussed views are the *lifeboat earth* concept, *triage*, and the *small planet* concept.

The lifeboat earth view of world population argues that there is only a finite quantity of resources, just as a lifeboat from a sinking ship has only limited quantities of water and food. If a lifeboat is surrounded by other passengers who are swimming in the ocean, pulling more survivors into the lifeboat is not in the best interests of those already occupying the boat. Adding survivors will only limit further the amount of time the water and food in the lifeboat will last, and if enough survivors are pulled into the boat, may only ensure that no one will reach land. This view argues that the industrialized, developed countries should concentrate on ensuring their well-being and allow the less industrialized countries, which represent the passengers who are not in the lifeboat, to sink or swim on their own.

The term *triage* comes from modern warfare. Following any battle, there are always more wounded than the limited available medical personnel can handle. The challenge for the medical personnel is how best to utilize the limited facilities to assist the survivors. The military medical person's answer to the problem is to divide the casualties into three groups: those who are slightly wounded and will recover without any additional medical care; the seriously wounded whose prospects for survival will be greatly improved by medical attention; and finally the very gravely wounded, for whom a doctor can do little or nothing, or whose wounds would require a disproportionately large amount of medical attention. If a choice must be made under battle conditions, the first and third groups are ignored, and all attention is given to the second group. Those who epouse this view in response to world population growth argue that only those less developed countries with smaller and slower-growing populations should receive aid from the industrial countries. Those with overwhelming population problems should be categorized as unsavable; for them, aid would only delay ultimate disaster while not aiding the general world population. Those with potential to become developed and to overcome their population problems should receive aid from industrialized countries since they would benefit most from such assistance.

Proponents of the small planet view argue that both the lifeboat and triage are not only wrong but are immoral and represent only the rationalizations of the wealthy of the world as they attempt to justify their continued dominance of resources and wealth. The small earth view maintains that everybody lives on the same earth, and all are equally entitled to its resources. The population problem that must be solved if the inequity in distribution of wealth is to be overcome is a world problem, not a regional problem. The solution is not the arbitrary abandonment of large segments of the earth's population to a life of misery and human degradation, but the need for international cooperation to harmonize the demands for resources with the population in each region. Therefore, residents of the wealthy industrialized countries must recognize their disproportionate drain on the world's resources and be willing to adjust their life-styles so that people of the less industrialized world may gain access to a higher standard of living.

WORLD DISTRIBUTION OF POPULATION

The estimated population of the world in 1990 is about 5.3 billion. If they were evenly distributed over the land area of the earth (including Antarctica), there would be approximately 92 persons per square mile (35.5 per square kilometer) of land area. But physical, technological, social, political, and economic factors combine to prevent a uniform distribution of the earth's population. Instead, the pattern is highly complex because of interaction of these factors over the past 2 million years (Figure 2.1). The greatest number of people is found in Asia, where more than one half of the world's population lives. Within Asia, the countries of China, India, Indonesia, and Japan have the greatest numbers.

Europe has a greater population *density* than Asia, even though it has a smaller proportion of the world's population, because of its smaller land area. Overall, Europe has a population density approximately one-third greater than does Asia overall. The density for some individual countries in Europe rivals that of the most densely settled areas of the world in Bangladesh in south Asia (Figure 2.1). The greatest population density in Europe occurs in the Netherlands and industrialized portions of the British Isles, France, Germany, and other Western European countries.

North America has a center of dense settlement in the northeastern industrialized area, but most of the continent has very low population densities. Latin America, Africa, and Australia are characterized by low population densities with a few denser clusters

TABLE 3.5 Population Density by Region

Region	Area (sq. miles)	Population (per sq. km.)	Population (per sq. mile)
Africa	11,692,673	20.46	53
Asia	11,425,952	108	280
Europe	1,881,870	101.9	264
Latin America	7,930,816	20.85	54
North America	8,308,538	12.74	33
Oceania	3,287,265	3.09	8
Soviet Union	8,649,496	12.74	33

along the coast and water routes or other especially favorable locations. The low population densities of Latin America, Africa, and Australia are in marked contrast to the high densities of Western Europe and Asia. Availability of large expanses of land with limited population offers a potential for population expansion as world numbers continue to increase (Table 3.5).

POPULATION GROWTH RATES

The rate of change in population, like the distribution, is uneven (Figure 3.6). The general trend for slow growth rates in industrialized countries and more rapid growth rates in the less industrialized world masks important regional variations (Table 3.6). The rate in an individual country is a critical consideration, since it determines part of the increase in demand for the basics of life that the country must supply. Those countries with an annual rate of population increase of 2 percent double their population in only 35 years. Countries with a growth rate of 3 percent double their numbers in only 23 years. The importance of these figures is easily overlooked. A growth rate of 2 or 3 percent on money invested in a bank seems like a very small return, but because population growth is exponential, the impact of even a moderate growth rate can be overwhelming if maintained over an extended period or if applied to a large population base (see accompanying box).

The regions with the lowest growth rates are those of northern, western, and eastern Europe, which have the greatest wealth to support additional inhabitants.

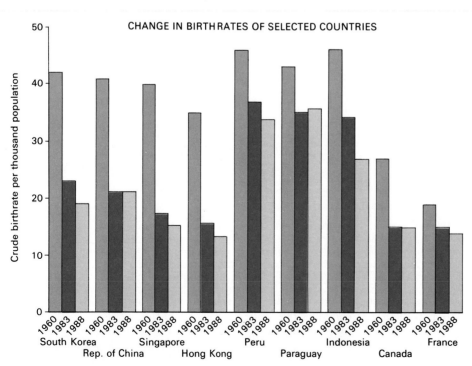

Figure 3.6 Rate of population growth. Nearly every country has had a declining birth rate in the last few decades.

PEOPLE, PLACES, AND POVERTY

TABLE 3.6 World Population by Continent

Region	1988 Population (in Millions)	Rate of Natural Increase	Projected by 2,020 (millions)
Africa	623	2.9	1,497
Northern Africa	138	2.8	279
Western Africa	194	2.9	479
Eastern Africa	186	3.3	512
Middle Africa	64	2.7	152
Southern Africa	40	2.3	76
Asia	2,995	1.8	4,629
S.W. Asia	124	2.8	257
Middle South Asia	1,137	2.1	1,987
Southeast Asia	433	2.2	720
East Asia	1,302	1.3	1,665
North America	272	0.7	327
Latin America	429	2.2	711
Middle America	111	2.5	197
Caribbean	33	1.8	47
Tropical South America	238	2.2	401
Nontropical South America	48	1.5	67
Europe	497	0.2	499
Northern Europe	84	0.2	84
Western Europe	156	0.2	157
Eastern Europe	113	0.4	122
Southern Europe	144	0.3	145
U.S.S.R.	286	1.0	354
Oceania	26	1.2	36

SOURCE: World Population Data Sheet (Washington, D.C.: Population Reference Bureau, 1988).

The regions with the highest growth rates are Africa, parts of Asia, and Latin America with their prevailing poverty and underdevelopment. Although birthrates are falling rapidly in newly industrializing countries, Latin America and Africa continue to have the highest birthrates in the world (Figure 3.7).

Less industrialized countries have the potential to increase their growth rate even more as a result of modern medical care and worldwide programs to provide better nutrition for children. Improvements in health services today reach poorer areas relatively quickly, allowing more people to live through the childbearing years, in turn producing more children, who also live through the childbearing years. The concern for the health of people of the less industrialized world, manifest in Western-funded programs that have eradicated smallpox and greatly minimized other childhood diseases, results in ever greater populations in the less industrialized world.

The implications of increasing populations in the less industrialized world without parallel increases in productivity and wealth comprise one of the central issues that affect all other issues facing today's world. It is incongruous that the major concerns of wealthy Americans and Europeans are inflation, energy shortages, or acquiring more material wealth when the continued growth of world population almost certainly dictates that inflation will continue to increase, energy shortages will become more acute, and the material standard of living in the industrialized countries will decline relative to the past. The vociferous demands of the various groups in the industrialized countries will do little to resolve the issues facing the industrialized world unless the fundamental world problem of population is addressed.

The daily economic pressures of millions of individuals attempting to provide a decent life for themselves may lead to the ultimate destruction of our

The Population Explosion

The term *population explosion* has sometimes been applied to the phenomenal increase in population in the less industrialized world. The figures are indeed alarming when applied to specific countries. For example, China had a population estimated at 583 million in 1953. Thirty years later it was nearly double at over 1 billion. Growth rates of 1990 if continued will give China over 2 billion people by the year 2030, the equivalent of more than 40 percent of today's world population. Rapid growth in African countries is even more alarming. Africa has 660 million people today, but 30 years ago there were only 265 million people in the continent. If present rates continue, there will be more than 1 billion people in Africa by 2005. Individual African countries are growing at an even more dramatic rate. Kenya has some 25 million people today, more than double the number in the country in 1965. There will be 46 million by the year 2005, 90 million by 2022, and 180 million by 2040 if present rates continue.

In Asia small countries have become large over the last few decades as a result of exponential growth. Pakistan and Bangladesh each had only about 50 million people in 1965. Today they each have more than 110 million, and by the year 2015 each will have as many people as the United States does today if present growth rates continue. Such continued high growth rates are analogous to an explosion both in terms of relative sudenness and in destructive impact on the environment.

At today's rate there will be more than 9 billion people on the earth in 30 years. Nearly 8 billion will be in the less industrialized world as compared to 4 billion today. Only 1.4 billion will be in today's industrial world, an increase from the 1.2 billion of today. The ever-increasing numbers and proportion of the world's population living in the less industrialized world poses one of the most critical challenges facing the world.

environment. The economic advancement of less industrialized world countries often is impossible with their rate of population growth. Rapidly increasing populations make it difficult to provide the additional schools, utilities, and capital necessary to maintain or improve the present standard of living. A country with a rapid growth rate must spend a disproportionate share of its income supporting the consumer needs of its population. Little is left to spend on improvements, and the growing population places additional pressure on the limited land resources. For example, demand for wood for fuel in Africa results in denuding thousands of acres yearly, and subsequent erosion makes reforestation difficult if not impossible. Erosion fouls waters in streams, rivers, and lakes—water sources that are often used directly by the African people for drinking and cooking purposes. The polluted water causes intestinal problems and further weakens the people's health.

The impact of ever-growing populations in the less industrialized world is not confined to the countryside or rural areas. Surplus population of farms and villages often migrates to cities; the most rapidly growing cities in the world are the large cities of the less industrialized world. Although the large cities of the less industrialized countries contain a much smaller proportion of the total population of their respective countries than do large cities of industrial countries, these cities are focal points of those countries' poverty and human misery. The migrants from rural areas to the cities perceive opportunities for education, jobs, medical care, and cultural benefits not found in the countryside. But the reality is that in the city, unemployment is high, particularly among the newly arrived migrants with their low level of skills. The standard of living for new migrants from the rural areas is low, and the life-style of many new arrivals is lower than what they had in the country. Nevertheless, in spite of the realities of life in the city in the less industrialized countries, many people continue to leave the land in the hope of obtaining a better life for themselves or their children.

The challenge facing the government of the less industrialized countries is integration of the millions of rural migrants into a meaningful economic order in the cities. The development of cities in the industrialized world occurred during a period of rapid expansion of world trade, industrial productivity, and capital accumulation. The changing economic order in Western Europe and North America created a new social class of urban residents. In the less industrialized world this process is only partially occurring, and the rural migrant remains just that: an individual from a rural area attempting to make a living in an urban setting. So long as population growth rates remain high, additional migrants will move to the cities and remain a sizable minority of impoverished rural people there. The explosive potential of these populations is great since they are poor, desperate, and susceptible to promises by political or social organizations opposed to existing governments.

The problems of the rural migrants in the less developed world are compounded by the growth of population among the new migrants in the cities. The

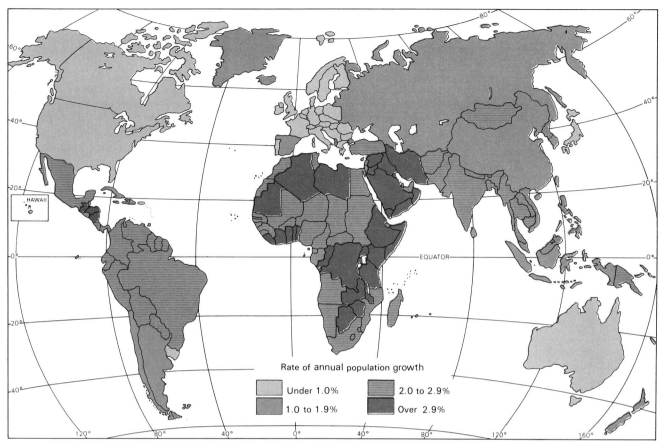

Figure 3.7 Change in birthrates of selected countries. Some countries in stage two of the demographic transition have lowered their birthrate by more than 50 percent in the past decade.

high birthrates typical of less industrialized countries in general prevail among the poor of their cities where urban medical care causes the death rate to decline further, resulting in an even greater urban population growth in less developed cities. The problem of providing housing alone is staggering, not to mention roads, sewage systems, water supplies, or education. The less developed countries of Africa, Asia, and Latin America will account for 85 percent of world population growth in the next 30 years, and the additional stress created by population growth in the cities may create problems that could develop into political and social situations that might result in world tragedy.

SLOWING THE GROWTH RATES

Because the rate of population growth is so fundamental to all of the other aspects of a country's well-being, the population issue is a critical one in our world. Unless the less industrialized countries can limit their population growth rates, they will never be able to increase the standard of living of their people materially.

Recognition of the critical nature of the population issue has prompted many less industrialized countries to undertake population control programs. The traditional methods for controlling population have been the war—famine—disease alternative of Malthus, or social restraints like later marriage, control of conception, or abortion. The majority of countries in the less industrialized world have focused on a combination of programs designed to encourage late marriage and to prevent conception. Educational programs may show that the small family is happier because there are more resources for each member. Other programs attempt to instill in the public an awareness of the magnitude of the population problem facing a country, with its implications for economic growth and well-being. In conjunction with such educational programs, governments may provide public assistance for sterilization or distribution of birth control information and materials.

Such programs are often handicapped by cultural values including religion, the status of male children, or rules of community status, which encourage large families. Overcoming these kinds of barriers is essen-

World Urbanization: Opportunity or Disaster

The population growth in the less industrial world (about 2.1 percent annually) is fueling a phenomenal urban growth. In 1950 only five cities of the world had more than 5 million inhabitants. In 1985, 42 cities had more than 5 million, and by the year 2025 an estimated 135 cities will be that large.

In 1975, six of the largest metropolitan areas of the world were in the wealthy industrial world. By 1990 it is estimated that only three metropolitan areas in the industrial world will be in the top 10. By the year 2025, 114 of the 135 metropolitan areas with more than 5 million people will be in less developed countries (see Table 3.7).

While populations of less industrialized countries double in two to three decades, population in developing world cities is doubling every 10 to 15 years. Already overcrowded cities must deal with millions more, millions who generally add to the population of the poorest residents of the city. The causes of urban growth are twofold: a high rate of population growth among existing residents, and a high rate of rural-to-urban migration. Much of the growth occurs in the form of slums with few of the urban amenities typically associated with cities: New migrants are generally squatters who establish some type of shack on vacant land.

The ever-growing slum areas that surround urban centers in the developing world are focal points for social, economic, and political unrest. Whether pushed to the city by exploding rural populations or attracted by the perception of a better life, many migrants are disillusioned by the reality of slum life. These slums are fertile ground for extremist groups advocating revolution of the existing system. Divorced from traditional familial or village support systems, slum residents are presented with the contrast between their poverty and the conspicuous consumption of the wealthy elite in cities in developing countries.

Pessimistic observers argue that the cities of less developed countries will be the nuclei of a new geography, a geography of terrorism, anarchy, and human suffering, as continued population growth overwhelms the capacity of these cities to govern or provide services. Optimistic observers argue that the burgeoning urban populations provide the opportunity to expand the economy by capitalizing on cheap labor and to increase the standard of living by providing education, medical, and sanitation services. For less developed countries the central issue is whether they can improve their economies and expand services rapidly enough to prevent the outcome foreseen by pessimists.

TABLE 3.7 Population of World's Largest (Metropolitan) Areas (in millions)

1975		1985		2000	
New York	19.8	Mexico City	18.1	Mexico City	26.3
Tokyo–Yokohama	17.7	Tokyo/Yokohama	17.2	Sao Paulo	24.0
Mexico City	11.9	Sao Paulo	15.9	Tokyo/Yokohama	17.1
Shanghai	11.6	New York	15.3	Calcutta	16.6
L.A.–Long Beach	10.8	Shanghai	11.8	Bombay	16.0
Sao Paulo	10.7	Calcutta	11.0	New York	15.5
London	10.4	Buenos Aires	10.9	Seoul	13.5
Buenos Aires	9.3	Rio de Janeiro	10.4	Shanghai	13.5
Rhine–Ruhr	9.3	Seoul	10.2	Rio de Janeiro	13.3
Paris	9.2	Bombay	10.1	Delhi	13.3
Rio de Janeiro	8.9	L.A.–Long Beach	10.0	Buenos Aires	13.2
Peking	8.7	London	9.8	Cairo/Giza/Imbaba	13.2

tial if less industrialized countries are to improve the standard of living of their people.

CONCLUSION: POPULATION IN RETROSPECT

The central place of population dynamics in explaining not only poverty versus wealth on a global scale, but a host of related issues that merit attention in the 1990s should be apparent. Growing populations in less industrialized countries handicap attempts to increase the standard of living of existing populations, and growing populations in industrialized countries can use an inordinate percentage of the earth's resources, thus consigning the world to ever greater contrasts in affluence and misery.

Posters such as this one from Singapore illustrate the types of educational programs being used by less industrialized countries to slow population growth rates.

Not all observers accept that population is *the* central issue, and some less industrialized countries attempt to raise the standard of living through lowering population growth rates by indirect methods. A popular school of thought in the 1970s maintained that simply industrializing a country's economy would cause the same decline in birthrates that had occurred in industrial countries of northern and western Europe. The weakness of this view lies in its failure to consider that the process of industrialization takes time to accomplish. During the time of transition to an industrialized society the populations of most less industrialized countries would at least double, offsetting any gains made by advances in industry. For example, in India efforts at industrialization have not been matched by investment in limiting population growth. Although India has become one of the world's major industrial producers in several specific areas, the pop-

ulation has increased so rapidly that even the increased industrialization is insufficient to meet the needs of existing residents.

Unless industrial advance is accompanied by declines in population growth, materially improving the lot of the average person in less industrialized countries is impossible. Prevention of the birth of one additional child in India provides in disposable capital the difference between what an individual Indian would consume during a lifetime and what he or she would produce. Multiplied by millions, limiting births in less industrialized countries would provide additional capital to facilitate the improvement of quality of life around the world and accelerate the process of industrialization. It may well be in the best interests of the existing industrialized countries to aid this transformation, as failure to do so will ultimately generate worldwide conflict and diminish the standard of living for western industrialized countries, too.

The irony of the relationships between the rich and poor countries of the world as we approach the end of the twentieth century is that the economies of the industrialized countries rely heavily upon resources, labor, and food from the less industrialized world; but industrial countries are generally reluctant to aid the people of the less industrialized world. The leaders of the less developed countries have suggested

TABLE 3.9 Change in Birthrates of Selected Countries

Country	Crude Birthrate per Thousand Population	
	1960	1988
Canada	28	15
China, Rep. of	40	21
France	18	14
Hong Kong	35	13
Indonesia	47	27
South Korea	41	19
Paraguay	43	36
Peru	47	34
Singapore	39	15
United States	24	16

SOURCE: World Population Data Sheet, 1988 (New York, The World Bank, 1988)

TABLE 3.8 Percent of GNP Given as Official Development Assistance, 1986

Country	% of GNP
Canada	.49
France	.78
Germany, Fed. Rep.	.47
Iraq	−.08
Italy	40
Japan	28
Kuwait	3.16
Libya	.59
Netherlands	1.00
New Zealand	.27
Nigeria	.06
Qatar	0.03
Saudi Arabia	2.88
Sweden	.88
Switzerland	.30
United Kingdom	.33
United States	.23

SOURCE: World Development Report (New York: The World Bank, 1988).

that the industrialized countries contribute one half of 1 percent of their yearly gross national product to the less developed to assist in overcoming population growth and related problems hindering development. To date many of the wealthiest countries of the world have been reluctant to do so (Table 3.8). The Scandinavian nations presently give a greater percentage of their gross national product in foreign aid to less industrialized countries, but the United States is overwhelmingly the greatest contributor in total amount.

The aid that has been given to certain countries has been partially successful in achieving the goals of population limitation with industrial development (Table 3.9). The birthrate is falling in many parts of the world, and the majority of such countries have been recipients of significant quantities of foreign aid. Countries like Taiwan and South Korea, focus of American foreign aid, have joined the newly industrializing segment of the world. Additional aid from the Western and industrialized countries will be of central importance in transforming the economies of the less industrialized world.

The process of development requires both population control and capital for investment. Accomplishing both of these goals will be a major step in the process of moving toward industrialization and an acceptable standard of living for all the inhabitants of the earth. Other factors are necessary before a complete transformation of the economies of the world can occur, but overcoming the population problem will make other issues easier to resolve.

FURTHER READINGS

Agenda. Washington, D.C.: Agency for International Development 1979–1982.

"The Aging World." *Unesco Courier*, October 1982.

ANDRAE, BERND. Translated by Howard F. Gregor. *Farming Development and Space: A World Agricultural Geography*. Berlin and New York: de Gruyter, 1981.

BAYLES, MICHAEL D. *Morality and Population Policy*. Tuscaloosa: University of Alabama Press, 1980.

BENNETT, D. GORDON. *World Population Problems: An Introduction to Population Geography*. Champaign, Ill: Park Press, 1984.

BLAND, JOHN. "The Bright Lights Beckon." *World Health*, July 1983.

BOUVIER, LEON F. "Planet Earth 1984–2034: A Demographic Vision." *Population Bulletin* 39. Washington, D.C.: Population Reference Bureau, 1984.

———. "Human Waves." *Natural History*, August 1983.

BROWN, LESTER et al. *Twenty-Two Dimensions of the Population Problem*. Washington: Worldwatch Institute, 1975.

———. *Our Demographically Divided World*. Washington: Worldwatch paper 74, December 1986.

———. *State of the World, 1987*. New York: Norton, 1987.

———. *State of the World, 1988*. New York: Norton, 1988.

CARTER, GEORGE F. *Man and the Land*. New York: Holt, Rinehart and Winston, 1975.

CLARKE, JOHN I. *Population Geography*. New York: Pergamon Press, 1975.

———. *Population Geography and the Developing Countries*. Oxford: Pergamon Press, 1971.

———. *Geography and Population: Approaches and Applications*. Elmsford, N.Y.: Pergamon Press, 1984.

DOGAN, MATTEI and KASARDA, JOHN D., eds. *The Metropolis Era: A World of Great Cities*. Newbury Park: Sage Publications, 1988.

DRAKAKIS-SMITH, DAVID. *The Third World City*. London: Methuen, 1987.

FABIAN, RAINER. "Mexico's Urban Wilderness." *Stern* as reprinted in *World Press Review*, July 1982.

GOULD, W.T.S., and LAWTON, R., eds. *Planning for Population Change*. Totowa, N.J.: Barnes and Noble, 1986.

HAMIL, RALPH E. "The Arrival of the 5-Billionth Human." *The Futurist*, July/August 1987.

HEWETT, JENNI, MCDONALD, HAMISH, GUPTA, RAJAN, and DUNCAN, WARREN. "The Population Bomb Ticks On." *The Sydney Morning Herald* as reprinted in *World Press Review*, April 1984.

HORNBY, WILLIAM F. and JONES, MELVYN. *An Introduction to Population Geography*. New York, London: Cambridge University Press, 1980.

JONES, HUW R. *A Population Geography*. New York: Harper & Row, 1981.

KLIENMAN, DAVID. *Human Adaptation and Population Growth*. Montclair, N.J.: Allanheld, Osmun, 1980.

MERRIK, T. "World Population in Transition," *Population Bulletin*, 41(2) April 1986. Washington, D.C.: Population Reference Bureau.

———. "Population Pressures in Latin America." *Population Bulletin*, 41 (1986). Washington, D.C.: Population Reference Bureau.

MYERS, NORMAN. "The Disappearing Tribes." *The Guardian* as reprinted in *World Press Review*, January 1983.

NEWMAN, J. and MATZKE, G. *Population: Patterns, Dynamics, and Prospects*. Englewood Cliffs, N.J.: Prentice-Hall, 1984.

PETERS, GARY L. and LARKIN, ROBERT P. *Population Geography: Problems, Concepts, and Prospects*. Dubuque, Iowa: Kendall-Hunt, 1983.

ROBINSON, HARRY. *Population and Resources*. New York: St. Martin's Press, 1982.

SELIM, ROBERT. "The 1980s: A Decade of Hunger?" *The Futurist*, April 1980.

STANDFORD, O. N. *The World's Population: Problems of Growth*. New York: Oxford University Press, 1972.

TEITELBAUM, M. S. "Relevance of Demographic Transition Theory for Developing Countries." *Science* 188, 1975, pp. 420–425.

"The Population Debate." *World Press Review*, October 1984.

VAN DE KAA, DIRK J. "Europe's Second Demographic Transition." *Population Bulletin* 42, 1987. Washington, D.C.: Population Reference Bureau.

VINING, DANIEL R., Jr. "The Growth of Core Regions in the Third World." *Scientific American*, April 1985.

War on Hunger. Washington, D.C.: Agency for International Development, 1975–1978.

YATES, WILSON. *Family Planning on a Crowded Planet*. Westport, Conn.: Greenwood Press, 1983.

ZELINSKY, W.; KOSINSKI, L. A.; and PROTHERO, R. M., eds. *Geography and a Crowding World*. New York: Oxford University Press, 1970.

ZIMOLZAK, E. and STANSFIELD, C., Jr. *The Human Landscape: Geography and Culture*. Columbus, Ohio: Merrill, 1979.

San Juan, Puerto Rico.

Poverty and Wealth variation in economic decisions

MAJOR CONCEPTS

Capitalism / Mercantilism / Labor exploitation / Socialism / Communism Peasants / Extended families / Dual societies Subsistence systems / Slash-and-burn agriculture / Acculturation / Complementary resources / Relative deprivation / Periodic markets / Neocolonialism / Institutional framework / Clan / Ethnocentrism Schistostomiasis

IMPORTANT ISSUES

The dilemma of conflicting economic systems with their attendant values

Population pressure and the dilemma of primitive societies faced with demands to alter life-style

The impact on developing and underdeveloped countries of foreign-owned industry within their borders

The responsibility of the industrialized world for the persistence of poverty in the less developed countries

Problems in the harsh environments of many nonindustrialized countries

The challenge to poor countries of ever-growing youthful populations.

Global Economic Characteristics

1. Three general economic systems dominate the world's economic geography: capitalism, socialism, and communism.
2. Within the three broad economic systems, economic activity occurs in market (cash) exchange, peasant, and subsistence economies.
3. The market exchange economies of the industrial world are highly interconnected with one another, whereas peasant economies have fewer international linkages.
4. Harsh environmental conditions are often associated with less industrialized countries.
5. Social and historical characteristics associated with rigid class structure, political instability, and rapid population growth are more common in the less industrialized world.

One of the major patterns apparent on the earth's surface is areas of wealth and poverty. The broad division discussed in Chapter 1 between the industrialized, the newly industrialized, and the least industrialized sections of the world masks important regional and intraregional variations in wealth and standard of living. At one extreme are the industrialized nations of Western Europe whose incomes average over $10,000 per capita. These countries are characterized by a high standard of living for most of their population. The people are engaged in manufacturing, commerce, and government service and tertiary activities to an overwhelming extent, and less than 10 percent are normally engaged in agriculture or other primary economic activities. In these Western European nations the income for individuals even in the lowest-income groups exceeds the average per capita income in impoverished countries.

At the opposite extreme are the countries like Bangladesh, Bhutan, or Chad. Incomes in 1990 would be below $250 per year per capita if the total gross national product were divided equally. Even this abysmally low figure does not accurately reflect the poverty in these countries. There is a vast gap between the few at the top of the economic ladder, who control a majority of the wealth, and the masses at the bottom, who control a minimum of the country's wealth and resources. In such countries the upper income is usually controlled by a small minority of the population who receive from 50 to 70 percent of the total gross national product (Table 4.1). The poorest half of the population normally receives 20 percent or less of the total gross national product. The majority of the population is engaged in agriculture or other primary economic pursuits. Health services are limited, and health problems are compounded by poor hygienic practices, low standards of sanitation, and contaminated water supplies. In the poorest areas of these countries, as many as five out of every ten children born may die before age five. Not only are these poorest countries characterized by abject poverty, but the incomes of the poor are actually declining in real purchasing power as a result of inflation.

In the middle group of new industrializing countries, like Korea, Singapore, and Taiwan, variations in relative access to wealth exist, but there is also an emerging middle class. The wealthy elite still control the bulk of the income, and the poor majority still has extremely low standards of living, but the emerging middle class provides a goal to which the poor can aspire. In these countries there is a range of income levels, and there are increasing job opportunities in secondary and tertiary areas. Governments in these countries are providing education and health services, literacy is increasing, the infant mortality rate is declining, and life expectancy is longer.

In some rich oil-producing countries the per capita gross national product is higher than that for the wealthy industrial countries of Europe and North America. But the figures are misleading, since in these countries the wealth is not as evenly distributed as it is in the industrialized Western countries. A few enjoy

TABLE 4.1 Proportional Distribution of Income in Selected Countries

Country	Poorest 40%	Richest 20%
Brazil	7.0	66.6
Chile	13.4	51.4
Finland	18.4	37.6
India	16.2	49.4
Peru	7.0	61.0
Philippines	14.1	52.5
United States	17.2	38.0
Venezuela	10.3	54.0

SOURCE: World Development Report (New York, The World Bank, 1987).

tremendous wealth while the majority struggle with the same basic problems found in other less industrial countries. Some oil-producing countries are using their wealth to increase industrialization and provide jobs, education, and other services for their population, but they remain essentially poor countries with pockets of wealth. At the other extreme are China, the Soviet Union, Cuba, and other avowed Communist countries with central government planning. The political regime in these countries is officially dedicated to improving the standard of living of all citizens. Even though average per capita gross national product is normally only marginally better than that of the poor countries, the actual standard of living as measured in physical terms is much higher. In Cuba, for example, life expectancy averages 74 years and literacy is almost universal; yet per capita gross national product is under $2000 yearly. A system of subsidies and rationing is designed to guarantee a minimum level of food, clothing, and shelter to everyone.

Similar success has been achieved by countries with low per capita incomes through welfare intervention in a basically democratic society. Sri Lanka has a per capita gross national product of less than $350 per year, but life expectancy averages 70 years, and approximately 85 percent of the population is literate. Sri Lanka accomplished these standards by providing universal access to education and health programs and by a food subsidy. Comparison of Cuba and Sri Lanka indicates the variety of forms that attempts to overcome world poverty can take. In examining such attempts, their relative success, and the issues and implications they present, it is imperative to understand the economic and government infrastructure they reflect.

INSTITUTIONAL FRAMEWORK: CAPITALISM, SOCIALISM, AND COMMUNISM

The distribution of wealth and poverty in the world is affected by the institutions through which the economies of the world function. The formal **institutional framework** that determines the type of economy in an area reflects the history, tradition, and values of a particular area. The economic system in an area reflects the ideas of those who have power within that country or region. Those who control the wealth usually control the economic system, and that system is designed to promote or maintain the interests of the wealthy. In most countries there are several levels of economic activity, but in the following pages we will discuss the types of economy practiced by the majority of people within a given country or region. It is imperative to understand how these systems work, since the basic institutional frameworks determine how the scarce resources of land, labor, and capital (the factors of wealth) are allocated. As a result the institutional framework plays a critical role in the relative level of poverty and wealth. The role of the institutional framework is not a simple one, as there are great variations between countries seemingly using the same framework. India's democratic institutions and economy are much different from those of the United States, for example.

Would any of these institutional frameworks provide the means for assuring everyone a fair share of world wealth through redistribution of land, labor, and capital at local, national, regional, and global scales? This question implies that a higher standard of living as measured by income and wealth in the Western world is a positive goal that countries would achieve, given the opportunity. This view of the development process does not imply that the cultural relationships within the poorer countries need to be changed, only that better allocation of the world's wealth will result in minimum standards that all people can hope to achieve. The goal of development thus defined is ensuring that people of the poorer countries have opportunities for secure jobs, adequate nutrition for themselves and their children, health and health care, adequate and safe water supplies, housing, and education. The extent to which these items are inadequate

---

Institutional framework: Refers to the customary or legal processes that set the parameters within which the economy of a country functions.

or lacking in the poor nations of the world partially reflects the present framework for allocation of wealth. Three major institutional frameworks affecting the world's economic geography are recognizable.

THE CAPITALIST MARKET SYSTEM

The capitalist system is based on the concept of private ownership of property and individual profit motive. It is associated with those countries that have achieved a high standard of living in northwestern Europe and in regions directly colonized by migrants from that region. By their overwhelming wealth and technological dominance in the past the capitalist countries have extended their influence to much of the rest of the world. The less industrialized countries are tied to the capitalist, industrialized countries through either former colonial control or economic dependence. Using capitalism as a basis, the world can be divided into those countries that dominate (including non–Communist Europe, North America, Australia, Japan, and other members of the industrialized world), and those dominated (the less industrialized countries of Latin America, Africa, and Asia, with their traditional economies).

Capitalism emerged in Europe in the seventeenth and eighteenth centuries as a replacement of the mercantile system. *Mercantilism* emphasized the role of the government in directing the economy, and the main goal was to accumulate wealth for that government, in effect, the monarch. Government intervention limited free trade; guilds, which limited entry and practice, carried on manufacturing, and the government chartered large trading companies designed to bring wealth to the monarch's coffers. Capitalism, by contrast, emphasized freedom of the individual to engage in economic activity and to receive the benefits of that economic activity. Adam Smith's *Wealth of Nations*, published in 1776, extolled the virtues of capitalism and argued that it would result in greater per capita income and a higher standard of living.

The basis of capitalism, as Smith expressed it, was that there should be no governmental control and that production of goods should respond only to the laws of supply and demand. Smith's argument maintained that the demand for a particular good would lead to sufficient production to ensure that the demand could be met at a price people were willing to pay. A secondary point of Smith's thesis was that the ability of a nation to accumulate wealth as a group of individuals was a function of the skill of the labor force and the proportion of the labor force engaged in wealth-producing (productive) labor. As part of this thesis he emphasized the need for division of labor to increase productivity. A cobbler under the guild system could produce only a few pairs of shoes per day, but by dividing the labor among specialists who produced soles, heels, the leather uppers, and sewed a particular portion of the components together, productivity would be greatly increased.

According to Smith the increased productivity would result in greater capital accumulation, which would enable construction of more plants, more machinery, and greater employment in the productive sector. In this fashion the market was viewed as a self-governing mechanism, which would automatically ensure that workers received an adequate wage, while providing more goods and services, greater employment, and a higher income for all. Capitalism was adopted by Western European countries (particularly Britain) and the United States in the eighteenth, nineteenth, and early twentieth centuries. It was used as justification for the expansion of Western interests into the less industrialized world, and fostered the industrial development of Western Europe.

More recently, economists have suggested modifications of the free-enterprise capitalistic system. Chief among these modifications are the views of John Maynard Keynes, who argued that the government needed to control the level of economic activity in the national interest to ensure that there would be adequate employment. Otherwise, he still advocated that the economy should generally be left to self-governing sources of supply and demand. His views enjoyed their greatest popularity in the Great Depression of the 1930s.

Over the years capitalist countries have modified the ideas of Smith and Keynes to a greater or lesser extent to provide greater governmental control of the economy, but in the industrial world, capitalism and the individual profit motive remain the primary mechanisms affecting the economy. As an institutional framework, capitalism thus remains generally identical to its early origins in the seventeenth and eighteenth centuries, but with a greater emphasis on state control of the level of economic activity.

KARL MARX AND THE SOCIO-COMMUNIST PHILOSOPHIES

In the twentieth century there have emerged two major systems that contrast with capitalistic economic institutions. Both owe their fundamental doctrines to the work of the economist Karl Marx, a German who lived in London from 1849 until his death in 1883. Marx argued that the social effects of the free enterprise system of capitalism with its emphasis on labor specialization and large factories were detrimental to the well-being of people. Marx witnessed the social situation in England during the Industrial Revolution of the late nineteenth century in which workers labored 10 to 12 hours per day for low wages, resided in inadequate housing, and generally suffered intense poverty. By contrast, the owners of factories or land accumulated massive fortunes and engaged in flagrant conspicuous consumption.

Marx argued that the basis of wealth could only be labor and that the wealth of the rich was accumulated at the expense of the laboring masses who had produced the wealth. Marx maintained that the value of goods produced in England's industrial economy primarily represented labor inputs into the raw materials. Workers received only a fraction of the value of their labor, and the riches amassed by the factory owners represented exploitation of the laboring class. Marx further maintained that in their drive to accumulate ever greater fortunes the capitalist countries were forced to expand into the less industrialized countries where they could find more markets for the goods that they produced. Marx maintained that the final result of the capitalistic system would be the creation of ever greater wealth concentrated in the hands of fewer and fewer companies and a monopoly system in which supply and demand did not regulate price, but price would be artificially set by the monopolist. Marx predicted that in their drive for markets and wealth the capitalist countries would ultimately come into conflict over control of the nonindustrial regions of the world. He explained the colonial expansion of European powers as a part of this drive.

In Marx's view the end of this evil of capitalism would come when workers revolted against their oppression and oppressors and took over the factors of production for themselves. This stage of economic institution he called *socialism*; in it the society, as represented by the workers, controlled the factors of production and the wealth derived therefrom. The fundamental tenet of socialism is that the workers who produce the labor, which is the basis of wealth, should be the ones to benefit from the wealth that they have created. These ideas, which were first espoused by Marx, have been adopted in varying degrees. The Scandinavian countries are said to have moderate socialism with government ownership of selected enterprises such as the railroads, coal mines, or other critical industries, and government attempts to ensure a minimum standard of living while preventing the great wealth accumulation that was such a conspicuous part of Western European capitalism in the eighteenth and nineteenth centuries. In spite of state ownership of selected economic activities, these nations are better classified as **state-oriented capitalist** societies.

State-oriented capitalism: When the government regulates selected aspects of the economy, but lets other aspects function under laws of supply and demand.

Marx's ideas were carried farthest in the Communist countries, where all the factors of production were nominally controlled by the laboring class. In reality the central government controlled the total economy; the state owned the factors of production in the name of the workers. Private ownership was for all intents and purposes nonexistent, and competition was unacceptable. In theory, in a true communist system each worker would receive payment in terms of need rather than actual labor. Committed members of a Communist country would perform to the maximum of their ability for the benefit of the entire society. In practice, Communist states usually have a single political party, a centrally planned economy, and lower standards of living than the older capitalistic countries, but several (including China and the Soviet Union), began encouraging more capitalistic practices in the 1980s. Countries of western and northern Europe are sometimes called socialist because of state ownership of selected industries, and the Soviet Union, the Eastern European countries, China, Cuba, and a few others are Communist. The primary differences are in their degree of private ownership, central planning, and single-party government.

THREE APPROACHES: MARKET, PEASANT, AND SUBSISTENCE SYSTEMS

The institutional frameworks of capitalism, communism, and socialism can be implemented at any level, from the national to the individual farmer or other entrepreneur. The manifestation of the institutional frameworks is found in the day-to-day activities of the people of a country or region. The particular institution that determines the economic framework also affects the life-styles and cultural values of the people who participate in that system by affecting the availability and costs of goods and services, and the extent and nature of individual, government, and commercial entities.

THE MARKET SYSTEM

A market exchange economy is characteristic of wealthy industrial countries. The system functions on a *cash* (including checks and credit cards) exchange basis and is characterized by impersonal relationships between buyer, seller, and producer of goods and services. Society is organized in nuclear households consisting of a parent or parents and their immediate juvenile offspring. Emphasis is on individual property ownership and wealth accumulation. A market system is

The southern part of Manhattan Island is the heart of New York's business district. The concentration of financial and other economic activities in this area is rivaled by few other major cities of the industrialized world.

often part of a state capitalist system or moderate socialist system of northern Europe, or even the Communist systems like the Soviet Union or China. The particular institutional framework is less important than the emphasis on cash as a medium of exchange for all goods and services.

Market exchange economies usually are associated with the relatively high standard of living typical of industrialized countries. These markets have a high degree of interaction with the rest of the world in terms of trade, tourism, and ideas. Educational and literacy levels are high, as are health standards and nutrition levels. Countries where the market exchange economy predominates are in stage three or stage four of the demographic transition or rapidly moving into these states. Related families rarely live near one another, as the forces affecting the market exchange economy dictate that employees go where jobs are to be found.

Market exchange economies have highly developed communication and transportation networks and social systems designed to provide minimal standards for the poor, the elderly, the invalid or handicapped, or for others who are unable to participate fully in society. Finally, in market exchange systems particular roles in the society are not inheritable. That is, a person's job or role in the society is not necessarily the same as that of his or her parents, nor need it remain the same throughout his or her life. The market exchange system in its modern and highly developed form is closely associated with the industrialized countries and involves a minority of the world's population.

PEASANT SOCIETIES

The term **peasant society**, or peasant economy, as used to discuss economic development has multiple meanings. It refers to a society in which roles in the market are highly specific and are often a function of birth or of living in a specific village. In a peasant market system nonfarming members of one village may specialize in production of a specific item such as pots, or service such as carpentry. Individuals from these villages participate in these roles as a result of birth rather than choice. Roles for men and women are highly specific, with each having designated duties. The extended family is more commonplace than in modern market economies. It may consist of the grandparents, sons and daughters-in-law, and grandchildren as well as cousins or other family relatives. Figure 4.1 shows a diagram of family relationships in peasant societies.

Peasants: Rural farmers in a stratified society. Peasants may or may not own land; the applied term refers to a way of life.

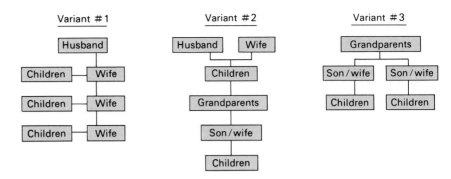

Figure 4.1 Family relationships in peasant societies.

Most people who live in countries with peasant systems are engaged in primary activities, particularly agriculture. There is always need for additional labor during planting or harvest time. Since the peasant world is characterized by limited mechanization, the extended family provides a means for ensuring adequate labor at specified times during the year. The extended family also provides continuity from generation to generation of roles in the society and security in old age. Peasant societies rarely have government services of social security, welfare, or universal health care. The existence of a large family is insurance that the needs of all members can be met. The extended family also serves as a substitute for the friendship relations that exist in modern market societies, and it can be a means of accumulating wealth. Any excess of able-bodied members of the extended family can be sent to work in emerging industries or for neighboring families that lack workers.

Another characteristic of the peasant economy is in the medium of exchange, where exchange of products often replaces cash purchase. The farmer may simply exchange his surplus of wheat for a carpenter's services in building a new cart or plow, or for crops not produced on his or her farm. Markets are characterized by not only specialization of product by town, but by the multiple roles played by the participants. Since the majority of the population in peasant societies are agriculturalists, they are both sellers and buyers in the market. The limited cash exchange that

A periodic market in Peru. Markets such as this are typical of the type of exchange system utilized in many less industrialized countries.

takes place is not used for capital accumulation but is immediately invested in other goods.

The significance of the peasant market system is that residents of these countries are often engaged in several different economic activities on an individual basis. While the farmer may both grow and sell wheat, more often the farmer grows grain and other crops while his wife and children engage in the market function. This specialization of roles by sex or age is typical of peasant systems. The market itself may be one of a number of types, ranging from sales to a middleman or woman who visits country areas and accumulates goods for resale in the cities, to a barter system in which surplus goods from one individual are traded to another for goods or services, or it may be a cash exchange system. The markets may be held daily, or they may be **periodic markets** that meet only at a specified time and place. Whatever form they take, peasant marketing systems provide not only an opportunity for exchange of goods and services, but also an important social function for the exchange of ideas and social interaction.

This simple description of peasant market systems obscures many important qualitative and quantitative differences among them. Peasant markets may range from sophisticated systems involving loans from a middleman to finance production of a crop, to a quasi-subsistence economy in which only a small amount of the product is sold. Whatever the variations, the peasant market system hinders capital accumulation, slows industrial development because the participants are also producers for whom individual maximization of profit is not necessarily the sole motive, and maintains traditional roles and methods of exchange and livelihood.

The majority of the world's population lives in countries that are part of the peasant world. The significance of this for modernization, development, and increasing standard of living is profound. Yet, within these systems there are seeds of change. Some have described peasant systems as **dual societies** because modern market exchange systems exist alongside peasant market exchange systems. The upper classes

---

Periodic markets: Markets that operate on some regular interval (as every Saturday for example) rather than daily. Each village in a region will have its own market day so that farmers can bring their products for sale or exchange.

Dual societies: Societies in which both market and peasant exchange systems operate.

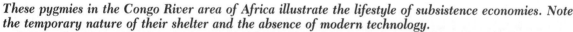
These pygmies in the Congo River area of Africa illustrate the lifestyle of subsistence economies. Note the temporary nature of their shelter and the absence of modern technology.

and emerging middle class are engaged in a cash exchange economy, and cities are the bastions of these modern capitalistic systems. The rate at which the modern market systems spread from the urbanized areas to the masses of the rural regions is a major factor in effecting the change from less developed, agrarian status to developed, industrialized status.

SUBSISTENCE ECONOMIC SYSTEMS

Subsistence economic systems are those in which the primary economic activity is provision of the foodstuffs for consumption by the immediate family. Isolated peoples such as the !Kung San (Bushmen) of the Kalahari Desert of Africa, the pygmies of the Congo rain forest, and certain tribal groups in Africa, Latin America, and Asia have subsistence economies. Within the peasant market system, part of the produce may be produced for subsistence, but a true subsistence economy operates essentially without cash.

Subsistence societies are normally bound to a specific area and have only limited ties to the broader world. Roles are segregated by sex, and the culture is characterized by an existence at the margins of necessity. Agricultural products may include livestock, particularly cattle and goats, and small fields of vegetables and grains and may be supplemented by hunting and gathering. The majority of the population is engaged in food-producing or food-procuring activities. Normally there are no cities in subsistence cultures, and most of the social activity takes place in the extended family and **clan.** Goods and services that are necessary but are not produced by the subsistence system are obtained through bartering with other subsistence or peasant societies.

The standard of livelihood is typically very low, and most subsistence societies are in stage one or two of the demographic transition. Education levels are low, literacy rates are low, industry and technology and related activities in transport and communication are low, and in general the subsistence societies epitomize life on the margins of existence. But such a life is not necessarily unappealing to those in the society. These societies have endured generations with little change, roles are specified, and the only threats to continued existence are natural hazards or expanding populations in the peasant market sectors. True subsistence economies usually constitute a minority system within a country dominated by peasants.

The greatest threat to the subsistence agriculturalists and pastoralists in the less industrialized countries is the expanding peasant population. Most peasant economies are in stage two of the demographic transition, and their expansion leads to pressure on the lands of the subsistence societies. Subsistence populations rarely have any legal title to their lands, and expanding peasant populations push them from their traditional territory. For those engaged in hunting and gathering, the loss of territory leads to a severe diminution of resources for food. Subsistence farmers often practice **slash-and-burn,** or **shifting,** agriculture (Figure 4.2). Slash-and-burn agriculture is particularly common in tropical and savanna climates where clearing the forests with their limited underbrush is easier than removing the second growth, composed of dense brush and saplings covering the entire land surface.

Slash-and-burn agriculture requires a fairly large acreage in comparison to sedentary peasant farming, and when the two systems come into contact, the former is the loser. Subsistence economies are normally found in conjunction with a dominant peasant economy and are the focus of government attempts to resolve the conflict between peasant and subsistence farmers. The government response to pressures on subsistence systems is to encourage them to locate in permanent villages and to adopt the culture and life-styles of the peasant system.

The pressures on subsistence economies are part of the ongoing process by which all areas of the world are being brought more closely together and made more similar. Attitudes and values of the industrialized world are spreading to the rest of the world. The wealth associated with the modern capitalist market societies is an attractive alternative to the poverty experienced in much of the world. There is an increasing attempt by the peasant market systems and their governments to adopt the characteristics and technology of the modern market economy countries. In addition, the expansion of the capitalistic market economies of northern and western Europe into the rest of the world in the seventeenth, eighteenth, and nineteenth centuries led to the domination of the poor countries of the world by the rich. Although most of the areas that were at one time political colonies of

---

Clan: The clan is a small group of extended families.

Slash and burn (also known as *shifting agriculture*): Consists of clearing and burning the vegetation, planting crops for one or two years until new vegetation begins to overrun the fields and then moving on to a virgin area to repeat the process.

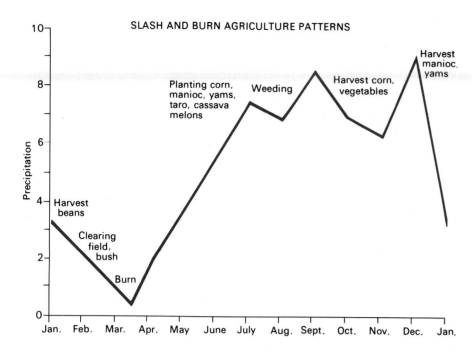

Figure 4.2 The succession of activities during the year in a slash-and-burn agricultural system reflects precipitation, with clearing and burning in the dry season.

the market economies have gained their independence since World War II, they are still effectively dominated by the economies of the western market economy countries, a situation known as **neocolonialism**. Marxists and others maintain that this dominance results in handicapping development of the less developed world, in maintenance of their poverty, and in the enrichment of the already wealthy countries of the West. Other observers maintain that the reliance of the less industrial world upon the industrial world for technology and manufactured items is balanced by the industrial world's reliance upon the less developed world for raw materials, resources, and specialized foodstuffs. Whichever view is adopted, obviously the modern world is made up of interconnections, and the peasant and subsistence economies are affected by events beyond their communities to a greater or lesser degree.

THE INTERCONNECTED WORLD OF THE TWENTIETH CENTURY

LINKAGES BETWEEN INDUSTRIALIZED COUNTRIES

The connections between the countries of the world are complex, numerous, inequitable in their impact, and variable in strength and importance from country to country. The connections between the industrialized wealthy countries are most numerous and of greatest strength. Industrialization has created a significant amount of integration of production and marketing among countries with modern market systems. These connections take many forms: communication networks, transportation nets (Figures 4.3 and 4.4), multinational business firms, cultural values, and even political and quasi-political arrangements such as NATO (North Atlantic Treaty Organization). Even the surface connections of land and water transportation and communication fail to indicate the extent of interconnections between developed countries. Twentieth-century technology allows the use of satellites to relay communications. Sharing of television programs between the United States and Britain and other industrialized countries is the rule rather than the exception. Industrialized countries are connected through trade organizations of many types designed to facilitate the movement of either raw materials or finished goods. Western European nations are tied in an economic union known as the European Economic Community (Common Market), designed to facilitate trade between member nations while protecting them from competition from outside.

Multinational organizations may have their headquarters in any industrialized country and have plants in both industrialized and nonindustrialized countries. For example, General Motors of the United States, Volvo of Sweden, the Nestlé Corporation of Switzerland, and a host of other multinational firms are found throughout the industrialized as well as the

---

Neocolonialism: Control of former colonies by colonial powers, especially by economic means.

PEOPLE, PLACES, AND POVERTY

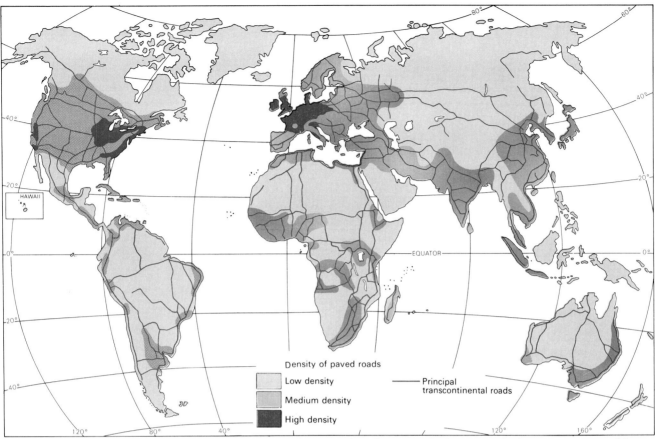

Figure 4.3 Level of interconnections as indicated by paved roads. The lack of transport facilities is a major problem in the less industrial countries of the world.

less industrialized world. The industrialized countries are culturally similar; sharing common views on dress styles, arts, and life-styles. Throughout the industrialized countries there is an emphasis on materialism and consumption of goods to maintain industrial productivity.

LINKAGES BETWEEN INDUSTRIALIZED AND LESS INDUSTRIALIZED COUNTRIES

The industrialized countries are tied to each other through a complex system of linkages, but are also tied to the less industrialized countries, and to a lesser extent, the Communist countries. The less industrialized world is a primary source of raw materials for the industrialized countries, particularly fuels such as petroleum and mineral resources such as iron ore. These goods make up the bulk of the oceanic trade in terms of tonnage, and flow overwhelmingly from the less industrial countries to the industrialized ones. The less industrialized countries also provide a host of foodstuffs to the industrial world, particularly foods of tropical origin such as coffee, cocoa (cacao), sugar, and tropical fruits. Return flows from industrialized countries include principally manufactured goods. Less industrial countries today are developing subsidiary manufacturing operations in which assembly of components, or in certain types of industries the entire manufacturing process, is attracted to the cheap labor of the less industrialized world (Figure 4.5). Textiles, boots and shoes, electronic assembly, clothing factories, toy manufacturing, and a host of other industries for which labor is a predominant input have been prominent in the movement.

In addition to the two-way exchange of raw materials and goods, the less industrialized world is tied to the industrial world through the dissemination to the less industrialized world of Western styles and values. Trade ties include organizations to move raw materials from the less industrial world to the industrial world. They also include such things as the World Bank, which provides funding for development projects in less industrial countries, a plethora of programs of the United Nations designed to improve the standard of living in less industrialized countries, and a

THE HUMAN EQUATION

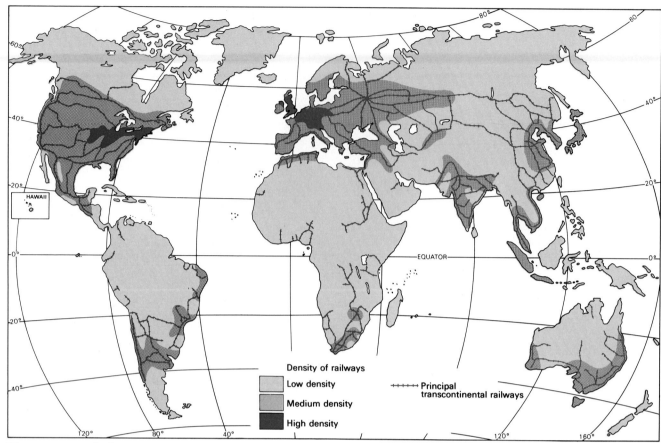

Figure 4.4 Level of interconnections as indicated by railroads. The absence of railroads over the majority of the less industrial world is a major handicap to their economic development.

host of private charitable organizations involved in similar projects.

IMPLICATIONS OF AN INTERCONNECTED WORLD

The significance of the interconnections of the world is profound. At the most superficial level a harsh winter in Brazil may result in higher coffee prices in Europe and United States. Brazilian attempts to gain self-sufficiency in petroleum, involving the use of sugar cane to make alcohol to mix with gasoline, have resulted in a shortage of molasses, which is important for a number of industries of Western countries. Destruction of tropical forests may affect climate, sea level, and economic activity in the entire world.

Influences in the other direction also have an impact. Less industrial countries rely on the industrial world for technology and markets, forcing them to pay for high-value manufactured goods with generally low-value raw materials (the exception being petroleum). Consequently, even though the less industrialized countries may export a large quantity of materials, it is difficult for them to earn sufficient capital to improve the lot of their impoverished people, since the money goes to pay for imported manufactured items.

The impact on the local economy of foreign firms established to utilize cheap labor, has also been criticized. In Haiti, American firms now produce essentially all the baseballs used in the United States. Wages for the women sewing the covers on the baseballs average about $4.00 per day, which is less than one-fifth the labor cost in the United States. Critics maintain that such low wages are basically an exploitative system that maintains the poverty of less industrialized countries. Other observers point out that paying wages that are excessively high by local standards would cause rampant inflation and thereby cause even greater damage to most of the citizens of the host country.

At an even more serious level, the interconnected nature of the modern world may cause not only inflation in less industrialized countries but detrimental cultural changes. In the 1970s, for example, Western firms, faced with declining markets for baby food as the industrialized countries moved fully into stage 3 of the demographic transition, began to sell and advertise these products in less industrialized countries. Health officials have been extremely critical of massive

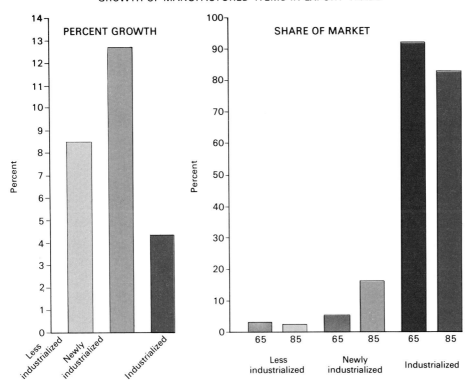

Figure 4.5 Growth of manufactured items in export trade. Both the rate of growth and the market share of manufactured goods has increased more rapidly in the less industrialized world than in the industrialized world in the last two decades, but nearly 90 percent of exported manufactured products still come from industrialized countries.

advertising campaigns that encourage giving up breast-feeding for infant formula. Critics point out that the mothers in these countries could feed their infants at almost no cost until the advertising campaign convinced countless millions that instant infant formula was better for their children. This has resulted in further impoverishing the families and has exacerbated health problems among infants in the less industrialized countries. Most women in poor countries have inadequate facilities for sterilizing bottles and nipples, and consequently disease has increased. Because of the cost of the baby food, many women dilute the formula beyond the recommended quantity, resulting in a diet that is deficient in nutrients required for normal growth.

Some observers maintain that the interconnected system maintains less industrialized countries in an impoverished condition. Proponents of this thesis argue that these countries are at the mercy of economic situations beyond their borders because of their role as purveyor of raw materials to the industrial countries, a problem exacerbated by the tendency of most less industrialized countries to produce only one or two staple items (Table 4.2). Thus Bolivia is a major exporter of tin, and tin comprises the majority of its export earnings. When world tin prices fall, the economy of Bolivia is affected to a greater extent than that of industrial countries. In similar fashion, when the Brazilian coffee trees were damaged by frost, the entire economy of Brazil was severely affected.

Leaders of less industrialized countries argue that they should be allowed to have a set price for their goods or at least have some type of organization that allows them to affect production and thus prices. World organizations such as the OPEC (Organization of Petroleum Exporting Countries) were successful in increasing the price of their raw materials for a time because there was a greater demand than supply. OPEC has been less successful in keeping prices high because of overproduction by some of its members. Other organizations, such as the Organization of Coffee Exporting Nations, have been less successful because they are not dealing with a finite quantity and a limited number of producers. Attempts to regulate the amount of sugar cane produced have failed because countless numbers of small farmers produce sugar cane and rely on it for their major cash income.

Achieving what the producing countries feel is an adequate price has been impossible because countries protect their producers against competition from elsewhere. For example, in the United States sugar beet farmers in the American West produce sugar at a cost roughly triple that to produce cane sugar in the Caribbean. In order to protect the American sugar beet growers, the federal government places quotas on incoming sugar to guarantee a higher price for American

THE HUMAN EQUATION

TABLE 4.2 Proportion of Total Exports Represented by Major Staples

Country	Product	Percentage of Total Exports Represented
Chad	Cotton	91.9
Chile	Copper	48.6
Iraq	Petroleum	98.6
Ivory Coast	Coffee, Cacao	47.2
Mauritania	Iron ore	49.7
Nigeria	Petroleum	94.2
Reunion	Sugar	75
Rwanda	Coffee	81.8
Somalia	Live animals	72.8
Sudan	Cotton, oil seed	77.4
Togo	Fertilizer	50.1
Uganda	Coffee	91.8
Zaire	Copper	42.4

SOURCE: *Europa Yearbook*, 1988.

beet growers. This discriminates against the sugar cane growers of the impoverished countries. Such inequities in the world economic order are pervasive and difficult to rectify because of the conflicting interests of the importing versus the exporting countries.

But the interconnected world order has benefits as well as problems. Industrial countries provide a significant amount of aid funds, food supplies, and technical expertise to assist the emerging countries in eradicating poverty, illiteracy, and disease (Figure 4.6). Some programs have been more successful than others; for example, the worldwide health programs, which have been highly successful in eradicating diseases that in the past decimated the infant and adult populations of the impoverished countries. The World Health Organization has eliminated smallpox as a major threat in the world.

We cannot categorize the interconnected world system as either all good or all bad. Certain aspects of it function to the advantage of the industrialized countries, certain aspects function to the advantage of the less industrialized countries, and certain aspects neither benefit nor hinder either. It is hoped that through the efforts of the United Nations, world and regional trade organizations, and an educated and informed world population, the inequities associated with the interconnected world system can be overcome, while the benefits can be maximized. Central to attempts to accomplish this goal is the drive to increase the standard of living of residents of the less developed world.

CAUSES OF WEALTH AND POVERTY

The fundamental question or problem that needs to be resolved before this goal can be reached centers on the causes of underdevelopment and poverty. Most observers agree on the character of underdevelopment. The great contrast between the world of wealth

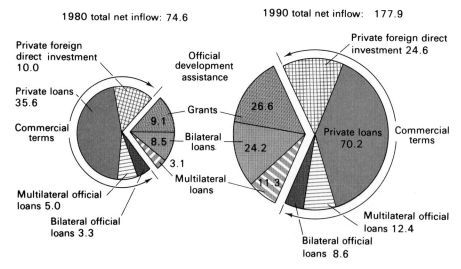

Figure 4.6 Changes in types and amounts of aid to developing countries projected to 1990. The amounts indicated as official development assistance are outright grants or loans given at preferential rates. The sector indicated as "commercial terms" provides capital for developing countries but requires repayment as market-established interest rates. The increasing debt burden of the less industrial countries is an important issue in international relations. (Source: *Finance and Development*, September, 1981, p. 17, [World Bank, New York].)

and the world of poverty is recognized by anyone who has traveled beyond the borders of the wealthy world. One observer has graphically described what life in the world of poverty is:

Everyone knows an underdeveloped country when he sees one. It is a country characterized by poverty, with beggars in the cities and villagers eking out a bare subsistence in the rural areas. It is a country lacking in factories of its own, usually with inadequate supplies of power and light. It usually has insufficient roads and railroads, insufficient government services, and poor communications. It has few hospitals and few institutions of higher learning. Most of its people cannot read or write. In spite of the generally prevailing poverty of the people, it may have isolated islands of wealth, with a few persons living in luxury. Its banking system is poor; small loans have to be obtained through moneylenders who are often little better than extortionists. Another striking characteristic of an underdeveloped country is that its exports to other countries usually consist almost entirely of raw materials, ores, or fruits or some staple product, with possibly a small admixture of luxury handicrafts. Often the extraction or cultivation of these raw materials/exports is in the hands of foreign countries.

Some of these underdeveloped countries are new. Other of the underdeveloped countries, oddly enough, are among the oldest known to history and were the seats of refined and elaborate cultures while Europe was in a state of barbarism and America was yet undiscovered. In some of these ancient centers of culture, such as India and China, there is reason to believe that the life of the average man is worse than it was centuries ago. Today in these countries population has increased to such an extent that the ancient methods of agriculture and handicraft can no longer provide adequate food and goods to meet even basic needs.[1]

Even a casual observer is struck by the difference between the wealthy and the impoverished countries of the world. The characteristics of the two are obvious, but what is less obvious is why one has emerged as a center of wealth, while others have remained impoverished or have actually declined in standard of living. Although few researchers agree on the causes of limited industrialization or the methods necessary to eliminate it, there is a general group of factors commonly recognized as contributing to the world of poverty and its problems.

These general factors affecting the lack of development can be divided into a number of types. These include problems of physical geography, problems of isolation and accessibility, the social structure of the individual country, and historical development and relationships with other countries. The degree to which each of these contributes to economic development varies from country to country, but all are a part of the problem.

HARSH LANDS: THE PHYSICAL GEOGRAPHY OF THE WORLD OF POVERTY

As pointed out earlier, the less industrialized countries are overwhelmingly concentrated in the tropical and subtropical regions of the world. Consequently, in the past, explanations of economic development have begun from the premise that the condition was related to climate. Such theories reflected the environmental determinist views of nineteenth- and early twentieth-century geographers and other scholars. This view argued that the tropical and subtropical areas were centers of poverty and limited technological development because the climate was warm all year round, plants produced food throughout the year, and little effort was required for human existence. Since adequate food was readily available, there was no incentive for people to be creative.

European observers looked at the indigenous population, which was malnourished, diseased, and generally capable of only limited physical activity, and argued that the heat of these climates caused the natives to be lazy and indolent. Consequently people in these areas had never been forced to exert themselves or to develop their intellectual capabilities and remained childlike with a simple mentality and doomed to a life of subservience to the perceived intellectually advanced Europeans. European explorers maintained that people in such countries did not have the mental capability to make technical advances necessary for development.

Behind this deterministic view was the **ethnocentrism** of the Europeans visiting the tropical and subtropical countries. Seemingly incapable of grasping the nuances of cultures and values that differed from their own, they interpreted all life-style characteristics of non-European civilizations as inferior. They used this simplistic explanation for the lack of development as justification for their conquest of poorer areas of the world in the seventeenth, eighteenth, and nineteenth centuries and continued rule of much of the less developed world until after World World II.

---

[1]Paul Hoffman, *One Hundred Countries, One and One-Quarter Billion People* (Washington, D.C.: Committee for International Economic Growth, 1960), p. 14.

Ethnocentrism: Belief in the superiority of one's own ethnic group.

Floods in Bangladesh in 1988 killed thousands and left hundreds homeless. Even in Dhaka, as indicated in the photo, damage from the flooding affected homes and stores.

Although simplistic, this explanation of the role of climate in development should not be completely discarded. Climate does play a role in the lives of people in tropical and subtropical areas, although the environmental determinists reached erroneous conclusions concerning this role. Excessive heat in the northern Sahara, where temperatures reach 54.4°C. (130°F), creates problems for preservation of food, difficulty of labor, and impact on life. All the tropics are not so hot, although some areas have high humidity, which makes the temperatures seem uncomfortable. The problem is that no adequate research has yet been done on the effect of the hot, humid climate alone. Its impact is so tied to related factors of disease that the actual effect of the climate is impossible to determine.

Another part of climate is precipitation. For much of the less developed world a basic problem for agricultural development is too much or too little precipitation or its timing and distribution. In July of 1972, 4.5 meters (160 inches) of precipitation fell on the island of Luzon in the Philippines. In this most densely settled area in the Philippines, the deluge destroyed 1 million acres of rice and 30 percent of the sugar crop, which is the country's main export.

In the Sahel of Africa, recurrent drought in the early 1970s and again in the 1980s led to the loss of hundreds of thousands of animals, the death of an estimated half million people, and destruction of the vegetation through overgrazing (Figure 4.7). The people of the Sahel migrated southward, searching for water for their surviving animals, and their economy and life-style were totally destroyed. Since the primary economic activity in all the less industrialized world is agriculture, precipitation or its lack has much greater importance than in industrialized countries.

Temperatures in these regions also affect agriculture disproportionately. The absolute maximum temperature is less important than the lack of freezing temperatures, or a dormant season. As a result of the lack of a dormant season, plant and insect growth is continuous. Rapid evolution can occur; when a new crop is introduced into the tropical or subtropical environments, weed and pest mutants rapidly evolve and compete with the desired crop. Competition by undesirable plants lowers the yield, as available nutrients must be spread among crops and weeds. As much as one third of the total agricultural production may be lost to insects, birds, and disease prior to the harvest (Figure 4.8).

DISEASE: THE PLAGUE OF THE LESS INDUSTRIALIZED WORLD

The impact of the high temperatures is not restricted to plant diseases and parasites. The people of the less industrialized world also suffer as a result of the high temperatures throughout the year. For example, the World Health Organization estimates that in Africa each person suffers a minimum of two infections at a time, and one half of the population is affected by bilharzia (*schistostomiasis*), a disease carried by parasitic liver flukes. Depending on the degree of severity, bilharzia reduces the person's strength from 15 to 80 percent. The resulting apathy toward work (or any other exertion) is part of the reason why Europeans assumed that residents of tropical areas were lazy. There are an estimated 750 million victims of bilharzia in Latin America, Africa, Asia, and the Middle East. There is no effective cure for the disease; yet, it is spreading because of contaminated water sources and the increasing mobility of residents of less industrialized countries.

Malaria is another plague of tropical and subtropical regions of the world. Spread by mosquitoes, it was largely eradicated as a result of the use of DDT during World War II and thereafter. But as a result of mutants in the mosquito carrier, malaria is now making a reappearance. The World Health Organization estimates that some 1 billion people are presently under grave risk from malaria. In Sri Lanka alone the past elimination of malaria has been reversed, and there are now over 2 million new victims as a result of new mosquito mutants that are resistant to DDT.

Intestinal worms are even more widespread than bilharzia or malaria, constituting a pandemic affliction.

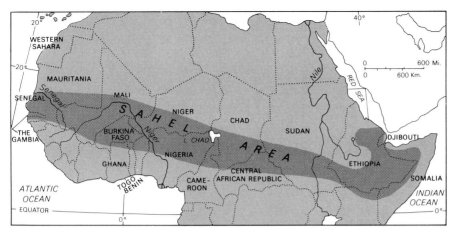

Figure 4.7 The Sahel area on the southern margins of the Sahara has been the focus of world attention because of the increasing damage caused by repeated droughts. Increasing populations of both humans and their herd animals result in even greater losses when drought occurs.

More than 1 billion people in the less developed world are estimated to suffer from one or more parasitic invaders. Hookworms are the primary parasite, with an estimated 600 million victims. Hookworms cause anemia, mental and physical retardation in children, and apathy toward work or other physical exertion. Filariasis is another disease that is spread by mosquitoes, causing blindness in its most severe form. Elephantiasis is a variety of filariasis with 100 million victims, primarily in a mild form. Intestinal and other parasitic worms are a persistent problem in the less developed world because of contaminated water sources, inadequate waste treatment, and improper hygiene among the majority of the population.

Other diseases associated with the hot tropical and subtropical climates include yellow fever, leprosy, and a host of others. Most medical research concentrates on heart disease, cancer, or other afflictions of the wealthy. Even in less industrialized countries, most doctors treat the middle and upper classes and their diseases. The poor masses with limited funds do not have access to proper medical care, and there are few doctors willing to devote their lives to research into tropical diseases when there is no money to support such research. Less than 10 percent of the amount spent on cancer research yearly in the United States is spent worldwide on tropical disease research.

Just as parasites and diseases affect the human population, there are many diseases that plague livestock in the less indusrialized countries. Intestinal parasites in domestic animals mean less milk from dairy animals, less meat from beef animals, and less ability

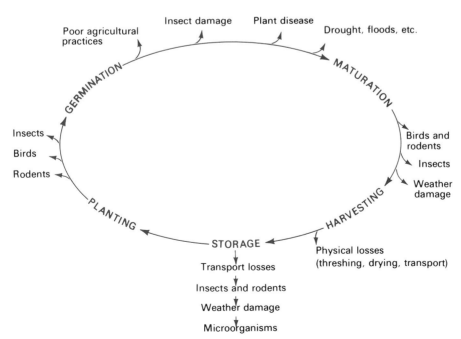

Figure 4.8 Harvest losses in less industrialized nations. The problem of providing adequate food is compounded by losses to crops during or after harvest. Industrial nations have lowered loss rates through better cultivation, harvest, and storage techniques.

THE HUMAN EQUATION

Refugees fleeing from the drought in the Sudan in 1985 carry their meager belongings across the parched fields of the Tigre region.

to perform work from draft animals. Production of wool is also handicapped by diseases, and morbidity rates among animals are high, handicapping efforts of farmers in the less industrialized world to increase their flocks and herds. The tsetse fly of Africa spreads fatal disease to livestock and handicaps use of much of the African savanna zones.

SOILS OF THE THIRD WORLD

The problems of livestock are but a fraction of the many difficulties facing farmers in the tropics, subtropics, and savannas of the world. The soils associated with these climates are excessively leached of nutrients. The impact of these soils on agriculture is obvious: Crop yields are lower, nutrients in the crops produced are lower, and the high rainfall in the tropics makes it difficult to counteract the infertility of the soils (Figure 4.9). Artificial fertilizers developed for the middle latitudes are inadequate for tropical soils because they are designed to release their nutrients over several weeks or months. High precipitation in the tropics results in leaching the fertilizer from the soils before it can be used by the plants. Even if adequate amounts of fertilizer were available, modifications would be necessary for use in these regions of the less developed world.

RESOURCE AVAILABILITY

The poor quality of the soil resources of most less industrialized countries is matched by the availability of certain important mineral resources. The presence or absence of elements that have played a role in the development of other countries may well be one of the factors handicapping the less industrialized world.

For example, Africa has only 2 percent of the world's known coal resource, Latin America has only 0.5 percent, and Asia outside of China has only 3 percent. By contrast, the United States, Europe, and the Soviet Union have over three-quarters of all known coal reserves.

The availability of petroleum is likewise limited to a small portion of the less industrialized countries. Some have an abundance of petroleum, iron ore, tin, or other resources, but these riches have not led to development. For example, Latin America has both large areas of fertile lands and extensive resources of iron ore, petroleum, and tin, as well as important petroleum deposits in Venezuela and Mexico. Yet Latin America remains primarily a less industrialized region. Indonesia has both fertile soils and an abundance of resources; yet it remains an underdeveloped area. The critical element seems to be less the total absence of resources in the less industrialized countries than a lack of *complementary resources*. Resources are said to be complementary when the usefulness of each is enhanced by presence of the other. Both coal and iron ore are necessary for steel production, for example. In addition, the resources that the Third World countries do have were developed by foreign owners for use outside the less industrialized world for the benefit of the wealthy industrial World.

ACCESSIBILITY OR ISOLATION

The role of climate in certain less industrialized areas is apparent, and the absolute lack of resources in certain others severely handicaps them. Nevertheless, these things alone do not determine whether a country will be rich or poor. Even where soils are not leached, as in the Altiplano of Bolivia or in Afghani-

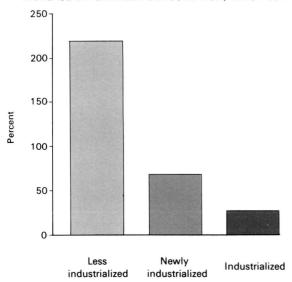

Figure 4.9 The rapid increase in fertilizer consumption in the less industrialized world represents efforts to increase agricultural productivity to raise the standard of living of farmers and to meet growing demand for food as populations increase.

stan, low levels of economic development still persists. One of the key determinants affecting the development of the world has been the access to information. The concept of accessibility versus isolation seems to be of paramount importance in explaining the lack of economic development.

If we examine the physical geography of the less industrialized countries, we find that there are some common factors that have worked to isolate much of it. Latin America has few navigable rivers that flow through areas of high population density. Africa consists of a plateau over which rivers descend in a series of falls and rapids near the coastline, preventing access to the interior by boat. Much of Asia is mountainous and difficult to reach. Both Latin America and Africa have large areas of mountains, deserts, tropical rain forests, or other barriers to free movement of people, goods, and ideas. Even in the late twentieth century, much of the less industrialized world has limited internal transportation networks. Lack of contact with areas beyond their borders has been one of the prime problems hindering economic development in the less industrialized world.

Examination of the relative location of the less industrialized countries shows this problem. Isolated spatially from trade routes, industrialized countries, and new technological advances, they have been unable to develop with the industrialized world. Lacking the initial impetus for development, they have been constantly trying to catch up. Advances in communication and transport have come late to these areas because of their relative location, making the gap between the industrialized and less industrialized even larger.

SOCIAL ORGANIZATION

The rate at which development has occurred in the less industrialized world is a function of more than just physical environment and relative location. The social organization has likewise played a role in rate of development. Society in less industrialized countries is characterized not only by a maldistribution of wealth but by a class stratification unknown in industrial societies. Each individual is aware of his or her particular role and level in society, and there is very little upward mobility. The family and its organization determine the role of the individual. Societal organization is weak, with tribal groups predominating in Africa, and village allegiance greater than allegiance to the nation elsewhere. As a result many less industrialized countries are plagued by instability in government. Some countries average a new government each year, for example Bolivia, which had 150 different governments in its first 146 years of independence.

Differences in social standing are often reflected in dress and in the region of residence. In Latin American countries, the difference between those of European ancestry, the mestizos (Indian and European mixture), and the Indians represents a gulf in values, outlook, education, and life-styles that is almost impossible to bridge. In India, social stratification is intensified by the caste system of the Hindu religion, which specifies individuals' value by their occupations. At the lowest level over 100 million untouchables live in separate villages, have their own vocabulary, and are generally distinguishable from the other members of the society. Such social stratification, in which roles are not interchangeable, handicaps advancement and the emergence of a middle class.

Partly as a consequence of social organization, the business community in the less industrialized world is much different from that in industrialized countries. Between 60 and 80 percent of the population is normally engaged in peasant agriculture, and business exchanges tend to be trading in bazaars rather than the impersonal exchange economy of the industrial world. Individual traders are the rule, capital formation is minimal, and consequently, large companies with economies of scale are slow to develop.

The role of the government in the daily lives of citizens of the less industrialized countries differs from country to country, but in general there is a high proportion of government employees compared to industry. In countries like India, government employees tend to make up a class in themselves, one in which nepotism and graft are the rule rather than the

Explanation in Geography: Models of Development

Examination of the regional patterns of global wealth and poverty reveal a basic north-south contrast. Countries of the north are industrialized, wealthy, and powerful, while countries of the south are poorer, and have less political and economic power. This basic division is reflected in terms such as *developed world*, *underdeveloped world*, *First World*, and *Third World*. Each term includes recognition of the gulf in wealth and poverty that separates the two broad regions, and related contrasts in life-styles, human welfare, society, economy, and political systems.

Geographers and other social scientists have attempted to develop models that explain why and how "development" occurs. Central to each model is an implicit acceptance of a definition of development that assumes economic growth will be associated with improvements in society, which will allow the realization of the potential of human ability. This definition does not preclude development of alternatives that are not fashioned on the industrial model of the north, but it does assume that development occurs only as poverty levels, unemployment, and inequality in distribution of wealth in a country decline. To date, the industrial model has been most successful in attaining these declines.

Models of how and why development occurs take several forms. One of the oldest is the environmental determinism model that tried to explain economic development in terms of the natural environment. As discussed earlier, this model in its extreme form concluded that human economic behavior was dictated by the environment and that tropical areas were unsuited for human activity. While the environmental determinism theory is no longer accepted, the environment is still important in affecting the ease and cost of development. Some geographers have adopted the term *tyranny of geography* to refer to unusually harsh conditions that are difficult to overcome. Periodic drought, earthquakes, or other natural disasters over which the earth's residents have no control are examples of the tyranny of geography. Accessibility, general climatic characteristics, and resource availability are also part of the tyranny of geography. Even where the natural environment is harsh, development can occur if the residents are willing to pay the cost.

Another early model of development was developed by the historian Arnold Toynbee. Toynbee examined 5000 years of world history and tried to explain the rise (development) and fall of civilizations. Toynbee's theory implicitly accepts the idea of the tyranny of geography, as he concluded that achievement was essentially a response to the challenge presented by environmental situations of increased challenge or opportunity. In Toynbee's model, great civilizations arose when a challenge was great enough to bring to bear a group's creative abilities, but not so great that it overwhelmed them. Thus the Egyptian civilization according to Toynbee developed when climatic change forced people from the Sahara as it changed from a well-watered grassland into a desert. Settling in the Nile River valley these people developed into one of the world's major civilizations. Like environmental determinism, Toynbee's model is criticized as too simplistic, but the challenge of the environment is still an issue the residents of the less economically developed countries must face.

In 1983 the European historian E. L. Jones concluded from an examination of Asia and Europe (with special emphasis on A.D. 1400–1800), that it was less environment than social conditions that were necessary for economic development. Professor Jones concluded that birthrates and resultant population growth were much lower in Europe than in Asia, creating lower population densities and associated resource availability in Europe. Moreover, social organization in Europe after 1400 was changing, allowing more social mobility, intellectual inquiry, and adoption of new technology and economic systems.

Perhaps the most famous model that attempts to explain development was formulated by Walter Rostow in 1960. Rostow likened economic development to the process of an airplane setting off down a runway. After a long "pre-industrial" taxiing process the economy reaches sufficient speed to "takeoff". Rostow's theory of economic development recognized five stages in economic growth, representing the level of technological development in an individual country. The first stage Rostow defined as the *traditional society*; it is characterized by limited local economic activity, low levels of technology, and a rigid social structure highly resistant to change. The second stage is one of *preconditions for takeoff*, in which birthrates fall, and an elite group undertakes programs designed to encourage political unity and loosen the rigid social structure. The economy diversifies as

exception. Government functions usually operate through involvement of numerous agencies and bureaus. Obtaining clearance for a business permit may involve visiting dozens of offices to obtain stamps, signatures, and permits. Since the wage structure is low, each of these offices may necessitate a small bribe to facilitate the transaction.

The society of the less industrialized world countries is usually made up of a mass of the working class, who are primarily peasant cultivators; a small and po-

employment in the manufacturing and service sectors increase, transportation and communication networks are increased and some of the hostile elements of the physical environment are brought under control through better management techniques.

Rostow's third stage is one of economic *takeoff*, which he defines as the industrial revolution stage of development. This stage is characterized by mass production, technological advances, the slowing of population growth, urbanization, and continued expansion of manufacturing. The fourth stage is the *drive to maturity*, a stage that occurs in industrialized countries after the takeoff stage has continued for several decades. This stage is characterized by diffusion of industrialization and technological advances to all of the country, increased national and international trade and interconnections, and growing industrial specialization. The fifth stage is one of *high mass consumption* with high per capita incomes, an abundance of material possessions, the dominance of consumer products, and employment concentrated in occupations of the service economy. The fifth stage is epitomized by the United States, with its high per capita incomes, sophisticated technology, and materialistic society.

Another theory of development is based on the idea that development begins in urban centers that serve as growth poles in a region, and that development then "trickles down" or "spreads out" to the rest of a region or country. Central to this model of economic growth is diffusion from a core to peripheral areas. The *core-periphery* model hypothesizes that primary growth centers are catalysts from which industrial growth spreads out to other smaller centers, eventually leading to development of an entire country. On a global scale the core-periphery model describes the relationship between the urban-industrialized core of Europe and North America and the rest of the world. At the local scale the core-periphery model describes the relationship of the dominant commercial-manufacturing center of a country and peripheral rural areas.

Another model of economic development is called the *structural change* model. This model is based on the changes that occur in the structure of the economy of a country as it reaches the level of industrialization. Proponents of this model point out that in the least industrialized economy the level of savings and investment, expenditures on education and public utilities, government revenues, and value of manufactured goods is less than half that found in newly industrializing countries. These structural changes in the economy are often associated with changes in land ownership that provide more farmers land and reduce the concentration of wealth in the hands of a small elite. The structural change model concludes that development occurs as countries reach a minimum level of per capita GNP that allows greater savings and investment, greater government revenues and expenditures, and more consumer demand to support manufacturing establishments.

A final model of economic development can be described as the *revolutionary change* model. This model hypothesizes that a revolution is required in order to reach Rostow's takeoff level, or to create the necessary structural changes in an economy to allow development. Often associated with Marxists, the revolutionary change model hypothesizes that a revolution in distribution of wealth (especially land), political power, economic control, and society is necessary to move societies from what Rostow would call the traditional stage to industrialization.

None of the models have been completely successful in explaining how less industrialized countries should develop. The growth-pole and trickle-down theories of the core-periphery model have been the basis of development planning in many countries, while revolutions with varying degrees of success have been undertaken in a variety of countries attempting to make structural changes. The underlying goal of development in each case is one of increasing the quality of life of the residents by providing at least a minimal level of food, shelter, education, employment, medical care, and equality for the residents of the country. The reality of the world at the end of the twentieth century illustrates that none of the models of development have been successful in providing the tools necessary to transform the lives of the poor of the world. Continued population growth in the less industrialized world, in combination with other problems they face, continues to maintain the global contrast between rich and poor. The poorest 20 percent of the world's population share only 1.6 percent of the world's GNP, and the poorest 60 percent combined share only 7.3 percent of global GNP. By comparison, the richest 20 percent of the earth's inhabitants share 74.2 percent of the world's GNP.

litically weak middle class; and a tiny, wealthy upper class with political power. The wealth of the upper class is often tied up in land or in foreign investments, and the elite are reluctant to invest it in domestic business and industrial functions, so that even what little wealth these countries do produce is not effectively utilized in the development process.

The rapidly growing population with its youthful component is a critical handicap to development. Children under fifteen years of age are usually consumers rather than producers. Although it can be argued that these youth are marginally productive in agriculture, they consume more than they produce. The productive capacity of countries with rapidly growing populations must be devoted to increasing the basic necessities for the survival of this large youthful population rather than in investments for development. In order to provide adequate education and training for the exploding population, less industrialized countries would need to invest essentially all their capital in schools. Since they are unable to do so, they wind up with an ever larger population of poorly trained individuals. With little training, no capital, and no opportunity for upward mobility, the vast masses of the less industrialized world remain peasants in upbringing and outlook. Job opportunities are minimal, and the majority of the population survives at only subsistence levels.

HISTORY AND THE EMERGENCE OF INDUSTRIALIZED AND LESS INDUSTRIALIZED COUNTRIES

Large populations living at high densities do not alone explain development. The Netherlands has a large, dense population; yet it is an area of wealth. A final class of factors that affect the distribution of wealth in the world is related to the historic process by which the nations of Western Europe gained control of much of the rest of the world by the twentieth century. Because of their central location and the emergence of more modern systems of transport, trade, communication, and business, these nations could expand their political power to encompass much of the world. In the process they gained control of areas that are now part of the less industrialized world. Those who argue that this history effectively hindered development of what we now call the less industrialized world point out that in the process of expanding their economic control, the wealthy nations of Western Europe gained control of the government and economies of the less industrialized world and directed their activities in a fashion that would benefit them. The dominated areas were maintained as suppliers of raw materials and wealth for the industrialized nations. A favorite theme of Marxists and others, this view is used as a justification for the condemnation of capitalist nations.

Whether or not the industrialized nation's dominance of the less industrialized world was intentional, it is apparent that the process of economic development in these countries has been handicapped by their subservience to the industrial nations. The monopoly of the wealthy world on technology has effectively prevented the less industrialized countries from modernizing their economy. To what extent lack of development is a product of Western domination is impossible to say. What role this domination played seems to depend largely on the politics of the observer, with Marxists arguing that Western domination is the primary determinant of lack of development and capitalists claiming that it had little direct effect and that colonial powers helped development by building railroads, improving food supplies, building schools and hospitals, and providing a stable government and markets for local products.

DEVELOPMENT IN RETROSPECT

The forces affecting the rate of development of a country and its concomitant wealth appear to be as variable as the countries of the less industrialized world themselves. It is important to remember that there are great regional and national differences within the less industrialized world. For some countries isolation has been a major cause of lack of development. In countries such as Rwanda and Burundi in the tropical rain forest of Africa, the climate appears to have been a primary factor, but it is compounded by location. In India and Vietnam, colonial domination appears to have been a major factor in the lack of development. In all the countries the preexisting or imposed characteristics of society have handicapped development of a higher standard of living.

A critical question in the discussion of distribution of wealth in the world and economic development is the need and desirability of development. Some observers maintain that Western nations have become so wealthy that materialism and related social problems represent a greater cost than the benefits of industrialization and economic development. Observers argue that the extended family with its emphasis on mutual support and concern for the individual creates a much better quality of life than the nuclear family typical of wealthy countries. Proponents of this view state that the wealthy world has been dehumanized by the necessity to engage the majority of the work force in repetitious jobs as on the assembly line. The values of independence and sharing associated with rural farm families in the peasant world are lost in wealthy industrial countries.

Others maintain that this bucolic view of peasant societies is completely erroneous. The disease, poverty, illiteracy, and general hopelessness associated with the world of poverty are too great to allow normal human existence. Proponents of this view claim that those who insist that there is something inherently good in a rural life-style are myopic and emotional rather than dispassionate observers. These conflicting views illustrate the importance of understanding the processes and problems involved in economic development if the problems facing the world are to be solved. Even for people of the less industrialized world the process and anticipation of economic development holds both favorable and unfavorable aspects. The existing peasant system is the one they know, one in which their roles are specified, and one that has existed for centuries. The prospect of change promises not only a higher standard of living, but a potential disruption of the entire fabric of society. Nonetheless, for the majority of the people of the less industrialized world this disruption is preferable if it means that families can be assured an adequate standard of living, opportunity for meaningful work, and security from famine and disease.

What is less apparent is the institutional framework under which development should take place. Some countries are opting for the Western capitalistic model, while others are following a path of Communist economies. In other countries, Iran for example, the drive for economic development is facing opposition from religious or other conservative forces.

It seems apparent that the demands by people of the less industrialized world for an adequate standard of living will result in continued economic convergence in the world. People of the less industrialized world are aware of the higher standards of living enjoyed in industrialized countries, and they have rising expectations for their lives. Even in countries that are partially industrialized the concept of **relative deprivation** raises the demand for ever greater standards of living.

Assuming that development will occur, the question becomes what is necessary for it to occur at a rapid rate? Radical observers argue that development can only occur through revolution. This revolution can take many forms but is necessary to change the societal, political, and economic organizations of the individual country to allow development. These observers argue that the elite members of the less industrialized societies oppose rapid development because it would diminish both their power and their wealth. Consequently, if development is to occur, those forces that maintain the existing society and power structures must be overcome. Other observers argue that a revolution is necessary, but it is a revolution in access to technology and the world's resources. These observers maintain that the less industrialized countries cannot develop until the technology of the industrialized countries is more available to them, along with the resources necessary to utilize that technology. They argue that the existing economic system, which favors the wealthy countries at the expense of the poor, must be changed to provide the emerging countries greater access to markets.

It is difficult to determine what type of revolution needs to occur, but it is apparent that there must be a change in the societies of individual countries before development that will involve all members of the society can occur. The following chapters will examine the experience of the industrialized countries, Communist nations, and important emerging countries in their attempts to transform their societies and economies. Examination of their experiences provides insights into how the wealthy countries have become wealthy and the specific factors in other individual countries that prevent them from raising their standards of living. Through such examination it is possible to see the extent to which revolutionary change has either benefited or hindered the world's population, and to understand the regions that make up the geography of the world.

Relative deprivation: Referring to relative levels of poverty rather than absolute poverty. A millionaire may feel poor compared to a billionaire, but both are wealthy by absolute standards.

FURTHER READINGS

BECKFORD, G. L. *Persistent Poverty: Underdevelopment in Plantation Economies of the Third World.* New York: Oxford University Press, 1972.

BELSHAW, S. *Traditional Exchange and Modern Markets.* Englewood Cliffs, N.J.: Prentice-Hall, 1965.

BERRY, B. J. L. "Hierarchical Diffusion: The Basis of Development Filtering and Spread in a System of Cities." In *Growth Centers in Regional Economic Development.* Edited by N. M. Hansen. New York: Free Press, 1972.

———. CONKLING, E. C. and RAY, D. M. *The Geography of Economic Systems.* Englewood Cliffs, N.J.: Prentice-Hall, 1981.

BLACK, C. E. *The Dynamics of Modernization: A Study in Comparative History.* New York: Harper & Row, 1966.

BOSERUP, E. *The Conditions of Agricultural Growth: The Economics of Agrarian Change Under Population Pressure.* Chicago: Aldine, 1965.

———. *Population and Technological Change: A Study of Long-term Trends.* Chicago: University of Chicago Press, 1981.

BROOKFIELD, H. C. *Interdependent Development.* London: Methuen, 1975.

BROWN, L. R. *The Global Economic Prospect: New Sources*

of Economic Stress (Worldwatch Paper No. 20). Washington D.C.: Worldwatch Institute, 1978.

———. *State of the World, 1988.* New York: Worldwatch Institute, 1988.

CHAMBERS, ROBERT, LONGHURST, RICHARD, and PACEY, ARNOLD, eds. *Seasonal Dimensions to Rural Poverty.* Totowa, N.J.: Allanheld, Osmun Publishers, 1981.

CHISHOLM, M. "The Wealth of Nations." *Transactions of the Institute of British Geographers* 5, 1980, pp. 255–276.

———. *Modern World Development: A Geographical Perspective.* Totowa, N.J.: Barnes and Noble, 1982.

CLARKE, J. I. *Population Geography.* New York: Pergamon Press, 1975.

COLE, J. P. *The Development Gap: A Spatial Analysis of World Poverty and Inequality.* New York: Wiley, 1981.

———. *Geography of World Affairs.* Norwich: Butterworth, 1983.

DEGREGORI, R. and PI-SUNYER, ORIOL. *Economic Development: The Cultural Context.* New York: Wiley, 1969.

DE SOUZA, A. R. and PORTER, P. *The Underdevelopment and Modernization of the Third World.* Washington, D.C.: Association of American Geographers, 1974.

DICKENSON, J. P., CLARKE, C. G., GOULD, W. T. S., PROTHERO, R. M., and HODGKISS, A. G. *A Geography of the Third World.* New York: Methuen, 1983.

DOVE, MICHAEL R. *Swidden Agriculture in Indonesia: The Subsistence Strategies of the Kalimantan Kantu.* Amsterdam: Mouton, 1985.

DUMONT, RENE. *Types of Rural Economy.* Oxford: Parker and Sons, 1969.

EL-SHAKHS, S. "Development, Primacy, and Systems of Cities." *Journal of Developing Areas* 7, 1972, pp. 11–36.

FOUST, J. G. and DE SOUZA, A. R. *The Economic Landscape: A Theoretical Introduction.* Columbus, Ohio: Merrill, 1978.

FRIEDMANN, J. and WEAVER, C. *Territory and Function: The Evolution of Regional Planning.* London: Edward Arnold, 1979.

GALLON, GARY, "The Aid Fix: Pushers and Addicts." *International Perspectives*, November/December 1983.

GESHEKTER, CHARLES L. "Society, Culture, and Ecology: The Somali Nomads of Northeast Africa." *The University Journal*, California State University, Chico, Fall 1981.

GILBERT, A., ed. *Development Planning and Spatial Structure.* New York: Wiley, 1976.

GILBERT, ALAN and GUGLER, JOSEF. *Cities, Poverty and Development: Urbanization in the Third World.* Oxford: Oxford University Press, 1982.

GINSBURG, N. "From Colonialism to National Development Geographical Perspectives on Patterns and Policies." *Annals of the Association of American Geographers* 63, 1973, pp. 1–21.

———. ed. *Essays on Geography and Economic Development.* Chicago: University of Chicago Press, 1960.

GRIGG, DAVID. *Agricultural Systems of the World.* New York: Cambridge University Press, 1974.

———. *An Introduction to Agricultural Geography.* London: Hutchinson Education, 1984.

GROSSMAN, L. "The Cultural Ecology of Economic Development." *Annals of the Association of American Geographers* 71, 1981, pp. 220–236.

HECHT, S. B., ANDERSON, A. B., and MAY, P. "The Subsidy From Nature: Shifting Cultivation, Successional Palm Forests, and Rural Development." *Human Organization* 47, Spring 1988, pp. 255–35.

KAMEREK, A. M. *Tropics and Economic Development.* Baltimore, Md.: Johns Hopkins University Press, 1976.

KELLY, MICHAEL G. "Grappling with International Debt." *International Perspectives*, March/April 1964.

LETICHE, JOHN M., ed. *International Economic Policies and Their Theoretical Foundations: A Source Book.* New York: Academic Press, 1982.

LOSCH, A. *The Economics of Location.* Translated by W. H. Woglorn and W. F. Stolper. New Haven, Conn.: Yale University Press, 1954.

MOHANTI, PRAFFULLA. "A Village Called Nanpur." *Unesco Courier*, June 1983.

MYRDAL, G. *Economic Theory and Underdeveloped Regions.* London: Methuen, 1957.

———. *Rich Lands and Poor.* New York: Harper & Row, 1957.

NAIR, KUSUM. *In Defense of the Irrational Peasant.* Chicago: University of Chicago Press, 1979.

PETERS, F. L. and ANDERSON, B. L. "Industrial Landscapes: Past Views and Stages of Recognition." *Professional Geographer* 28, 1976, pp. 341–348.

ROSTOW, W. W. *The Stages of Economic Growth: A Non-Communist Manifesto.* 2d ed. Cambridge: Cambridge University Press, 1971.

RUTHENBERG, HANS. *Farming Systems in the Tropics.* Oxford: Clarendon Press, 1971.

SMITH, M. et al. *Asia's New Industrial World.* London: Methuen, 1985.

STOHR, W. B. and TAYLOR, D. R. F., eds. *Development from Above or Below? The Dialects of Regional Planning in Developing Countries.* New York: Wiley, 1981.

STONE, P. B. "Development at a Crossroads." *World Press Dialogue* as reprinted in *World Press Review*, March 1983.

TAAFFE, E. J., MORRILL, R. L., and GOULD, P. R. "Transport Expansion in Underdeveloped Countries: A Comparative Analysis." *Geographical Review* 53, 1963, pp. 503–529.

Towards a World Economy That Works. United Nations, 1970, pp. 69–70.

TURNER, B. L., II and BRUSH, STEPHEN B., eds. *Comparative Farming Systems.* New York: Guilford, 1987.

WILKINSON, R. G. *Poverty and Progress: An Ecological Model of Economic Development.* London: Methuen, 1973.

WOLF, R. *Peasants.* Englewood Cliffs, N.J.: Prentice-Hall, 1966.

"A World of Communications Wonders." *U.S. News and World Report*, April 9, 1984.

Part Two

THE WORLD OF WEALTH
regions and problems

Every Society In The World Is Being Profoundly Affected By The Scientific And Technological Revolution That Started In Western Societies Approximately 500 Years Ago. In A Very Real Sense, [The World] Is Being Transformed By This Scientific And Technological Revolution. Where We Are Heading And Where We Will Wind Up Is Very Uncertain.

JACK D. DOUGLAS, THE TECHNOLOGICAL THREAT

The Only Real Trouble With Our Time Is The Future Is Not What It Used To Be.

PAUL VALERY (1891–1945)

Bonn, West Germany.

EUROPE PROFILE (1990)

Population (in millions)	499.6	Energy Consumption Per Capita	
Growth Rate (%)	.3	(kilograms)	4858
Years to Double Population	266	Calories (daily)	3408
Life Expectancy (years)	74	Literacy Rate (%)	99
Infant Mortality Rate		Eligible Children in Primary School	
(per thousand)	13	%	99
Per Capita GNP	$8170	Percent Urban	75
Percent of Labor Force Employed in:		Percent Annual Urban Growth	
Agriculture	12	1980–1985	1.6
Industry	40		
Service	48		

5

Europe an industrialized market economy

MAJOR CONCEPTS

Point settlements / Colonialism
Neocolonialism / Head Link
Interconnections / North Atlantic Drift
Agricultural revolution / Democratic
revolution / Industrial revolution
Scientific revolutions / Regionalism
Devolution / Polders / Nation State
Shatter belt / Centrifugal forces / Nation
Nationalism / Cultural convergence
Lingua franca / Fjords / North Atlantic
Drift / State / Tariffs / Centripetal forces

IMPORTANT ISSUES

Europe's impact on the developing and less developed regions of the world: beneficial or detrimental?

National unity versus regionalism in Europe

Economic and social confrontations in industrial countries

Obsolescence, resource exhaustion, and environmental quality questions in industrialized countries

Separatism, racial and ethnic tensions, and terrorism in Europe

European Geographical Characteristics

1. Europe is the home of revolutions (democratic, industrial, and scientific) that have affected the entire world.
2. Europe is densely inhabited and yet has the largest number of countries in the world with a high standard of living.
3. European countries are in either stage 3 or 4 of the demographic revolution, and face challenges of slow or negative population growth.
4. Europe is the world's leading importer and exporter, epitomizing the industrialized world.
5. The emergence of the economic union represented by the Common Market creates an economic power rivaled only by the United States.

Europe is that part of the Eurasian landmass located west of the Soviet Union and Turkey, bounded by the Atlantic Ocean on the north and west, the Mediterranean on the south, and the Eurasian landmass to the east. (Figure 5.1). Its size is belied by its importance; with an area of only 1,542,017 square miles (3,993,824 kilometers), Europe nevertheless had a population of nearly 500 million in 1990. With a population density of some 330 per square mile (125 per square kilometer), it is one of the most densely occupied regions of the world. Individual countries, for example, the Netherlands, rank among the most densely populated areas on the surface of the earth (Figure 5.2). Within large European cities are found even greater population densities.

Even the sheer numbers in such a small area provide but an indication of Europe's world importance. Europe epitomizes the industrialized, wealthy, consumption-oriented industrial world. Its residents by world standards have high per capita incomes, are predominantly engaged in secondary or tertiary economic activities, have high literacy rates, long life expectancy, and a general surplus of material goods and leisure time. Switzerland, Sweden, Norway, and Luxembourg are among the world leaders in per capita income. The high standards of living in this region rely heavily on interconnections with the less industrial world, interconnections that often benefit Europe more than the less industrial countries. Resources and foods or low-cost labor (directly or in the form of manufactured items) are imported into Europe. High-value manufactured or processed materials are exported to the less industrialized countries.

The people of Europe are among the most highly trained and skilled work forces found anywhere and represent the mature economies of the industrial world. European industries produce a significant proportion of the major industrial products upon which the world relies. With 10 percent of world population, Europe produces more than one quarter of the world's industrial products (Figure 5.3). Although the individual countries rarely rank as the world's leading producer, taken as a group, the industrial production of Europe is of worldwide significance. The export of industrial items provides many of the less industrialized countries with access to modern technology (Figure 5.4). Europe's high standard of living makes the region one of the world's major importers, particularly of raw materials. Agricultural products from Australia, New Zealand, Argentina, and North America; industrial raw materials from the Middle East, Latin America, Africa, Australia, and Asia; and low-cost manufactured goods from less industrialized countries dominate imports into the European region (Figure 5.5).

Even the role of Europe as a center of population, a producer of exports, and a market for imports does not fully explain its importance in the world. The modern world has been influenced and shaped by events in Europe so numerous, so subtle, and so interwoven into the fabric of individual non-European countries that it is almost impossible to appreciate fully the continent's worldwide impact. Europe was the source of migrants who colonized Canada, the United States, Australia, New Zealand, and large portions of Latin America and who established important concentrations of population in South Africa. The lan-

Figure 5.1 Location, political divisions, and major cities of Europe.

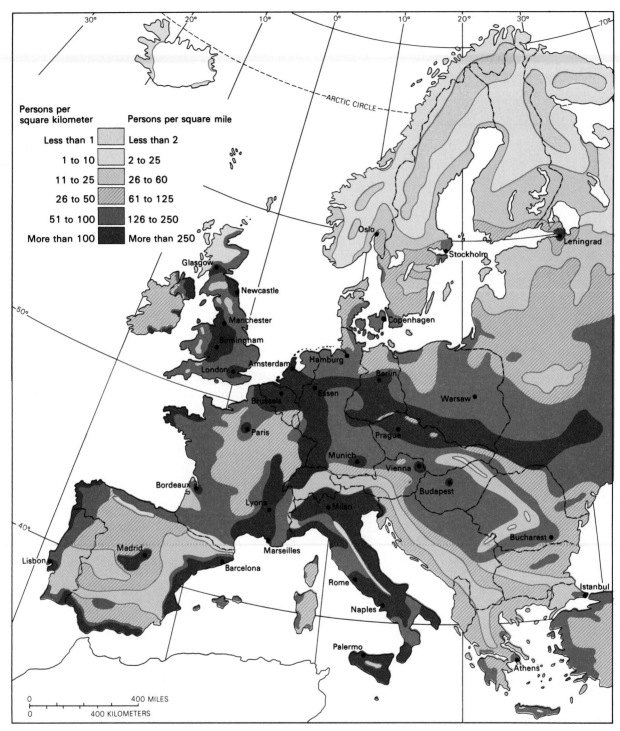

Figure 5.2 Population distributor of Europe.

guage, cultural values, and attitudes toward the environment of these colonized regions reflect their European heritage. Language alone gives them a tie to Europe that is overwhelmingly stronger than ties to the rest of the world. Cultural traits of religion and family organization, business and economic traditions and practices, and legal and institutional forms are closely related to those of Europe, creating a European cultural realm with its distinctive languages, values, and landscapes.

Even in areas of the world where they did not occupy the land, Europeans have played a role far greater than the population of Western Europe today would indicate. Colonization led to adoption or im-

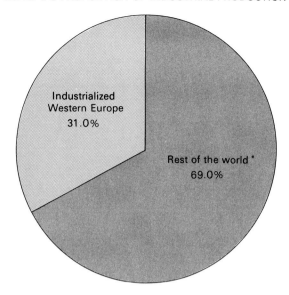

*Excluding centrally planned economies

Figure 5.3 Europe's proportion of industrial production, based on value added by manufacturing, 1986.

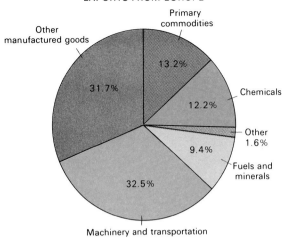

Figure 5.4 Composition of European exports, 1986.

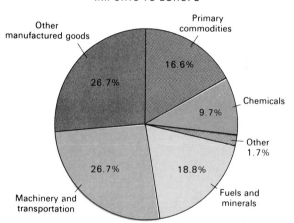

Figure 5.5 Composition of European imports, 1986.

plantation of European cultural traits in Asia and Africa (Figure 5.6). As a consequence of the European dispersal through much of the world in the past 300 years, the European languages serve as a **lingua franca** for use in trade, diplomacy, and tourism. In areas of former colonial rule a European language is often more frequently used for trade and diplomacy than the tongue of the local inhabitants. French in Southeast Asia and English in India and the Middle East are both examples of the dominance of the European language of the former colonial powers. The governments and civil services of former colonies also are modeled after those of Europe.

The past and present role of Europe makes it significant to all other areas of the world. In today's interconnected world it is impossible for most countries to progress and proceed without being affected by external events. This is particularly true of Europe. As the world's greatest net importer of materials and as one of the world's greatest exporters, events that affect the economy of Europe are translated into events that affect the economy of the rest of the world. When the economies of the European countries are strong and healthy, their almost insatiable demand for materials has a significant impact upon the economies of less industrialized countries. Governments of less industrial countries recognize this fact and are attempting to encourage European countries to be more active in sharing their technological expertise and assisting the less industrialized countries to transform their economies and standards of livelihood. The concentration of wealth, technology, industrial productivity, and diplomatic power in Western Europe make it central to events that will transpire in the world in the coming decades. The development of the less industrialized areas, the harmonization of Communist and non–Communist countries, and the general increase in world standards of living will be in large part affected by the active or passive role assumed by the European countries. Whether consciously or unconsciously, Europe affects the destinies of many people in the world and will serve as a model for good or ill

Lingua franca: Refers to use of a second language that can be spoken and understood by many peoples to overcome diversity of language in an area.

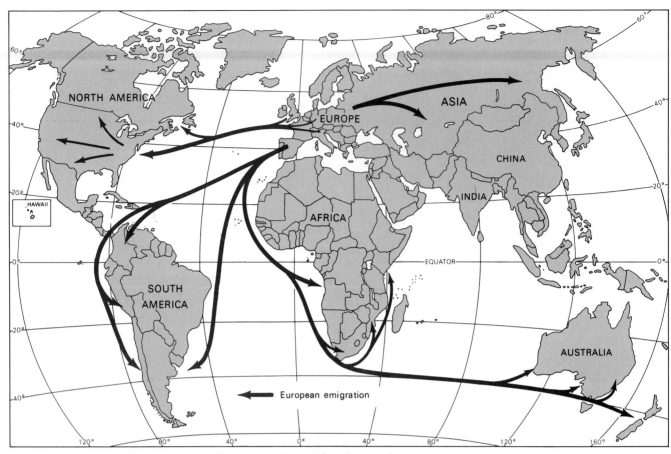

Figure 5.6 Migration of Europeans to other parts of the world in the past four centuries.

for those countries attempting to industrialize and to transform their economies.

THE ENVIRONMENTAL SETTING

What factors have allowed Europe to become such an important area? An examination of its physical base partially explains its rise to dominance in the late seventeenth and eighteenth centuries. It also indicates the problems that Europe faces today.

First, the continent's high-latitude location is of paramount importance. Its southernmost points in southern Spain and Greece lie in a latitude similar to that of Kentucky or Virginia. Major cities such as Madrid (which lies at approximately the same latitude as New York), or Paris (which lies at the same latitude as Seattle), or London (which lies at approximately the latitude of Hudson's Bay) are even farther north. Copenhagen is at the same latitude as central Canada (Figure 5.7). The high latitude affects climate, soils, crop production, energy requirements, and life-styles.

A second major characteristic of Europe is the peninsular nature of the continent. Europe occupies a major peninsula and several smaller ones extending seaward from the Eurasian landmass. The Scandinavian peninsula (Norway, Sweden) the Jutland peninsula (Denmark), the Iberian (Spain, Portugal), the Italian, and Balkan peninsulas (Greece, Albania, Yugoslavia, Bulgaria) extend outward from the main European peninsula. In addition, there are important islands: the British Isles and those that are part of Denmark, France, and Italy. The effect of this interfingering of land and water cannot be overstated. It affects the climate, it affects commerce, it has affected European development, and it continues to affect Europe's role as a major trading power.

The sea is important for another reason. Because of the relationship of land and water, the climates of Europe are warmer than would be expected for such high latitudes. They are predominantly maritime as a result of the proximity of water and the prevailing winds.

THE WORLD OF WEALTH

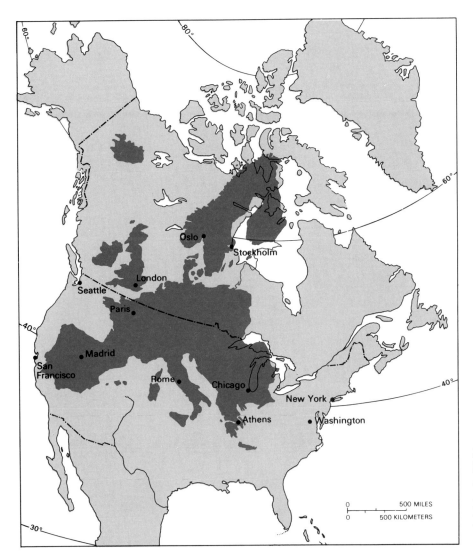

Figure 5.7 Relative location and size of Europe compared to North America. Much of Europe lies north of the latitude of the border between the United States and Canada.

The Swiss Alps include the highest land forms in Europe, and their scenic beauty is a major tourist attraction of Switzerland.

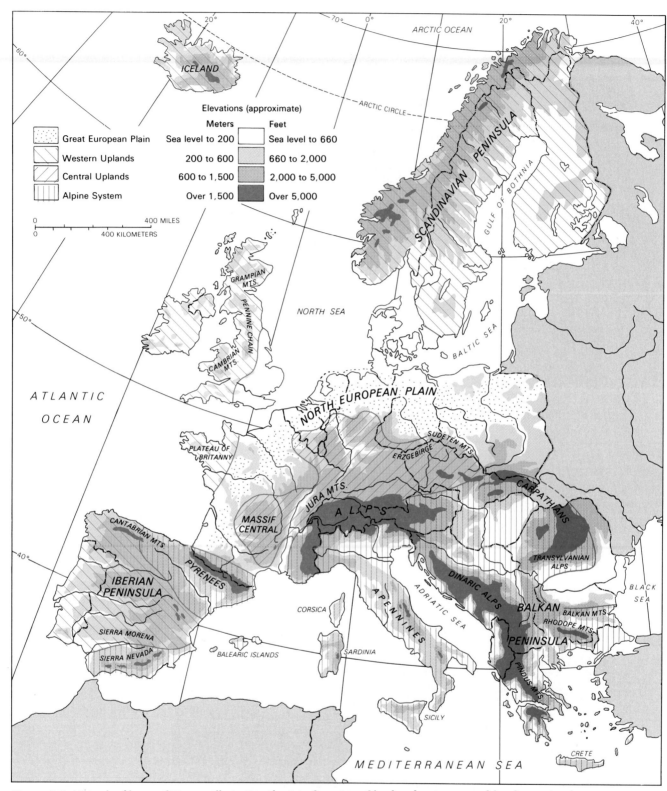

Figure 5.8 Major landforms of Europe illustrating the interfingering of land and water created by the peninsular nature of Europe.

The polder lands of the Netherlands have been reclaimed by diking and draining shallow coastal seas.

LANDFORMS OF EUROPE

The important landforms of Europe are a northern mountainous area, a northern and western plains region, a central mountain system, and a southern mountain complex. To the north the Scandinavian peninsula is dominated by a core of relatively low mountains where elevations range from 1000 feet (300 meters) to a maximum of 8104 feet (2500 meters) in the south of Norway at Mount Glittertinden. These are old mountains that were covered by the huge continental glaciers, and they are less rugged than the Alpine mountain systems of central Europe. The high-latitude location combined with the altitude of the northern mountains makes them some of the most sparsely settled areas of Europe, essentially uninhabited except for logging activites, mining, hunting, or other activites associated with **point settlements** (isolated settlements).

The western portion of the European plain is the most densely occupied area of Europe (Figure 5.8). It is an area of low, gently rolling topography. On its western margins it is slightly below sea level in the reclaimed lands (*polders*) of the Netherlands and Germany, and nowhere does it exceed 500 feet (150 meters) above sea level. This is the major agricultural and industrial region of Europe and has been important historically for migrations of peoples and movements of armies (including German armies in World War II). It remains the focus of confrontation between the Soviet Union and the Western powers. Important rivers crossing this plain, including the Rhine, make this region even more accessible for trade. The European plain is widest in Poland and extends into the Soviet Union; to the west, it narrows and fragments through the Netherlands, Belgium and western France. Eastern England is also part of the lowland plains areas of Europe, while Scotland and Ireland are similar to the Scandinavian peninsula with low, rounded, glaciated mountains.

To the south the European plain is bounded by the Pyrenees between France and Spain. The Pyrenees are young mountains, rising to an elevation of more than 11,000 feet (3,400 meters). South of the Pyrenees, the Spanish plateau lies at an elevation of between 1000 and 4000 feet (300 and 1000 meters) with mountain ranges like the Sierra Nevada rising to more than 11,000 feet (3,500 meters). To the east, the European plain is bordered on the south by the Swiss and Austrian Alps where elevations reach more than 15,000 feet (4,500 meters). The Alpine system continues in the Carpathian mountain range, which stretches across Czechoslovakia and Romania to separate the western European plain from central and southern Europe. Farther south, the landforms become even more complex with a series of mountains in Italy, Yugoslavia, and Greece. Major ranges include the Apennines, which form a rugged backbone of the Italian peninsula, the Dinaric Alps of Yugoslavia, and the Pindus of Greece. These mountains of southern Europe are young and are still geologically active. Consequently, this area is subject to earthquakes. Within the complex mountains of southern Europe is a series of important lowlands. The Po Valley of northern Italy and the lowlands along the Danube River are important agricultural regions of Europe.

Point settlements: Point settlements are located in harsh environmental settings to capitalize on unique site features such as resources. They are generally extractive in nature with little or no hinterland.

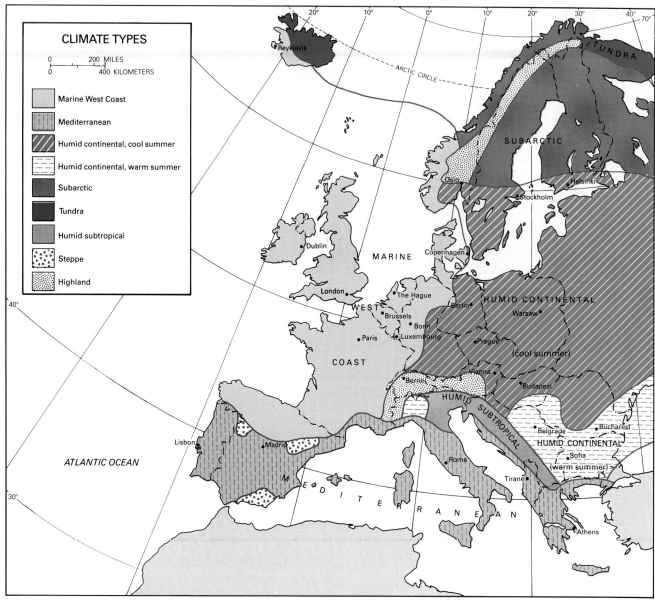

Figure 5.9 Climatic types of Europe.

The landforms of the four major areas of Europe are quite different from one another. The mountains of northern Europe are rounded with subdued relief as a result of the great glacial (Pleistocene) epoch. Glaciers extended down the slopes of these mountains, deepening river valleys to create the famous *fjords* of the western coast of Norway. Fjords are simply drowned river mouths and channels, which make excellent harbors. The glaciers spread beyond these highlands of northern Europe to the European plain as far south as the Rhine River distributaries of the Netherlands and the Carpathian mountains. This plains area has some important areas of fertile loess soils, as in southern Poland north of the Carpathians. In southern Sweden, northern Poland, Denmark, and Finland the topography has been made more level by glacial action.

CLIMATES OF EUROPE

The interaction of the climatic controls of latitude, land-water relationships, and landforms results in a complex pattern of climates in Europe, with eight climatic types represented (Figure 5.9). The dominant type in terms of areal extent is the marine west coast. It extends from the Arctic Circle in western Norway south along the coast and through southern Norway and Sweden, and inland through central Poland and as far south as western Czechoslovakia before being blocked by the central mountain ranges. In additon,

THE WORLD OF WEALTH

all of Denmark, the British Isles, the Netherlands, Belgium, Luxembourg, and much of France, Germany, and even a narrow strip of the northern coastal area of Spain and Portugal have a marine west coast climate. This climatic type extends inland and essentially follows the European plain in its distribution. To the north, distance from the sea in Poland moderates the air masses sufficiently to eliminate the marine influence.

The influence of the sea on the climate of Western Europe causes winter temperatures in southern England (at a latitude north of the United States northern border) to be similar to those found in Virginia in winter. Summer temperatures are cooler than those in Virginia, resulting in a climate that is moderate throughout the year, with rain during most seasons. The region receives between 20 and 40 inches (500 and 1000 millimeters) of precipitation in most areas, with higher precipitation totals as a result of orographic lifting in Ireland, Scotland, and other highland regions. On the continent, the marine influence is modified by the land; winter temperatures in eastern France and in Germany are lower than those in the British Isles, and summer temperatures are higher. The moderating effect of the Atlantic Ocean is the result of the *North Atlantic Drift* (Gulf Stream). Benjamin Franklin observed 200 years ago that the Gulf of Mexico was a source for a stream of water he named the Gulf Stream, which flowed northeastward and warmed the shores of the British Isles and Western Europe. More recent research indicates that the North Atlantic Drift actually comes from the central portion of the Atlantic and is deflected northeastward toward Europe by the rotation of the earth. The source for the warmer water that moderates Western Europe's climate is the equatorial regions with their great heat accumulations.

Inland from the coastal areas, the climate of Europe is much different. In northern Scandinavia there is tundra climate across the tip of Norway. Northern Sweden, Norway, and Finland have a humid continental, cool-summer climate with winter temperatures below 0°F (-20°C) and fewer than four months of temperatures warmer than 50°F (10°C). Southern Sweden and southern Finland, eastern Poland and portions of Hungary and Romania have a humid continental, warm-summer climate. They have a hot, humid summer with maximum temperatures reaching the eighties during the daytime, and winters with subzero temperatures and permanent snow cover. Along the lower Danube in Romania and northern Bulgaria is a warm, humid continental climate similar to that found in the Midwest of the United States. Winter snow cover rarely persists for more than a few weeks, and summer temperatures are hot and humid.

There is a small zone of humid subtropical climate in the area east of the Alps and north of the Mediterranean countries. The Po Valley of Italy and portions of the Danube region of southern Hungary and northern Yugoslavia have such a climate. Covering a larger area is the Mediterranean climate around the Mediterranean Sea. Precipitation totals range from 18 to 25 inches (400 to 500 mm), nearly all of it coming in winter, and a summer drought is characteristic of the area. Much of Portugal, Spain, Italy, Yugoslavia, and Greece have Mediterranean climate.

SOILS AND VEGETATION IN EUROPE

The soils and vegetation of Europe reflect the climate, but they are highly modified by centuries of human occupation and exploitation. The soils in their natural state are generally acidic and of lower fertility (alfisols, inceptisols, histosols, spodosols, and entisols). In Scandinavia, Poland, and Germany, lime and other bases have been added to offset this acidic tendency. Where the soils are treated in such a manner, they tend to be moderately suitable for crop production. There are some areas of extremely favorable soils in Europe, including the loess deposits of central France and southern Poland and mollisols in the Danubian Plains.

To the south in the Mediterranean lands, the soils were also originally quite fertile. They produce well when adequate precipitation or irrigation is available. The soils of the Great Hungarian Plain and the Wallachian Plain are mollisols and are among the most fertile in Europe. Acidic soil areas are found in Scotland, Ireland, and much of Scandinavia.

Vegetation in Europe is, again, highly modified. Since Scandinavia is one of the few areas in Europe not extensively cleared for agricultural purposes, its coniferous forests remain an important vegetation type. In the past, Germany and Poland also had large coniferous forests, but in both countries extensive clearing has reduced their extent. Only in areas of more rugged topography, less suited for agriculture, do the natural forests remain. Forests are heavily utilized and are also planted as tree farms and cared for as a crop in less rugged areas. In southern England and France, the eastern European countries of Czechoslovakia, Hungary, Romania, and Bulgaria the predominant vegetation was in the past a mixed deciduous forest. These areas have been largely cleared for building materials, medieval smelting of iron, and the provision of more agricultural land. The Mediterranean countries at one time had extensive oak forests in the higher elevations and typical Mediterranean xerophytic vegetation in the lower elevations. Today most of the oak forests have been cleared, and only the xerophytic *maquis* (brush) of the Mediterranean remains. Reforestation efforts in many of the Mediterranean areas are trying to restore the oak and other forests that are more useful.

EUROPEAN RESOURCES

There are a number of physical resources that are of paramount importance. Among these are the river systems that complement Europe's maritime location. Three general river regimes can be recognized: those of Northern Europe, those of Central and Western Europe, and those of Southern Europe. The rivers of Northern Europe flow from the Scandinavian mountain area and tend to be swift in the summer and frozen in winter. Because of their steep gradients they are important for generation of electricity but are of limited use for navigation.

The rivers of Southern Europe include some that are very long, such as the Tejos (Tagus) of Portugal and Spain, but because of the summer drought of the Mediterranean climate, they are of limited usefulness. The Tejos, the Guadalquivir, Ebro, and Douro are all used for some navigation, especially in the late winter and early spring when water flows are higher. During late summer, at the peak of the Mediterranean drought, these rivers have low flows as a result of limited precipitation and diversion for irrigation purposes.

The rivers of Central and Western Europe are much more useful for navigation, and the entire West and North European plain is crisscrossed by a series of canals connecting the major waterways. The most important river of western and central Europe is the Rhine. The Rhine carries a greater volume of freight than any river system in the world (Figure 5.10). It has its origin in the Alpine mountain systems of central Europe and flows northward, then westward through Switzerland, and north through Germany, France, and the Netherlands. It is navigable to Basel, Switzerland, providing that landlocked country with water access to the oceans. Tributaries like the Ruhr provide access to the major industrial areas of Europe and the distributaries are the location for some of the major ports in Europe, including Rotterdam, which is the world's number one port in terms of volume.

The Danube is the longest river of Europe but is handicapped by its site-situation characteristics. There are rapids in the middle Danube at the so-called iron gate that hinder navigation, and the river flows through a region that has historically been more agriculturally than industrially oriented. Communist domination of the countries through which it flows has handicapped free trade along its extent. A final problem with the Danube is its terminus in the Black Sea, which is far removed from the major trade routes of Europe.

Other important rivers and the cities that rely upon them include the Thames (London), the Rhône

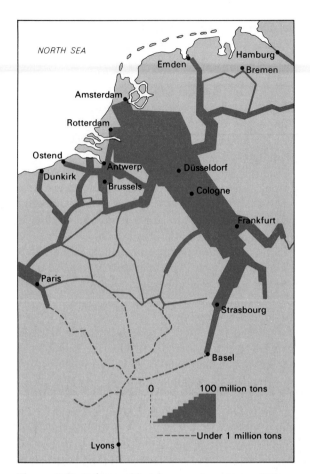

Figure 5.10 Volume of freight moved on rivers of Western Europe. Dashed lines indicate proposed improvements to extend the inland waterways of the region.

(Marseille), the Seine (Paris), the Schelde (Antwerp), and the Elbe (Hamburg). Even this partial listing of the major rivers is sufficient to indicate the importance of these navigable streams of the European plain. Traffic on the river systems of Europe and their interlocking canals constitutes one of the main forms of transportation in the region. Present plans of Poland and Czechoslovakia to connect the Danube drainage to the streams flowing northward into the North Sea and the Baltic will further expand the transportation potential of Europe's rivers.

Europe is also endowed with some important mineral resources. It is not the purpose of this brief introduction to list them all, but there are some that have been particularly important in Europe's past rise to power, and there are recent discoveries that will affect its continued role in the world. Important coal deposits affected the early emergence of the industrial powers of England and Germany. On the mainland major coal deposits are found in the Ruhr of West Germany, Silesia in Poland and Czechoslovakia, the Saar in Germany, and the Sambre-Meuse and Cam-

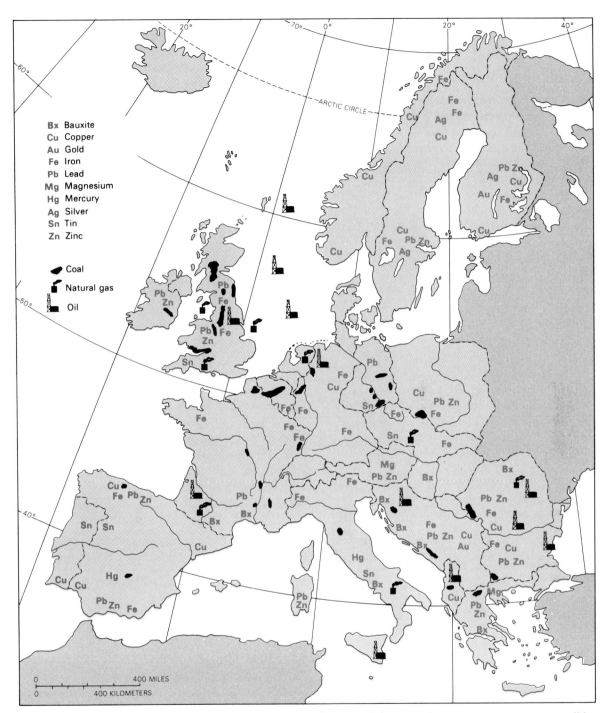

Figure 5.11 Mineral resources of Europe. The majority of the mineral deposits of European countries are small by world standards.

pine in the Netherlands and Belgium. There are several major coal fields in the British Isles, the most important at the present being the Yorkshire field (Figure 5.11).

There are also major deposits of iron ore that played an important role in Europe's development, beginning in the north at Kiruna, Sweden, and extending through the Ore Mountains (Erzgebirge) of western Czechoslovakia to Alsace-Lorraine of France. There is iron ore in the Midlands region of the British Isles, as well as along the northern coast of Spain and in scattered and minor deposits in Austria, Romania, and Yugoslavia (Figure 5.11). In combination, the iron ore and coal were the basis for Europe's steel industry.

Deposits of zinc, lead, copper, and bauxite are found in Spain and southern France. Bauxite is par-

EUROPE

ticularly important in the south of France and in Yugoslavia. Petroleum, the critical energy resource in modern times, has been found only in limited quantities in the mainland of Europe. There are important petroleum resources in the eastern Carpathians of Romania, and natural gas and petroleum have been produced in northern West Germany, Albania, and Hungary. More recently, major resources of petroleum and natural gas have been discovered beneath the North Sea. Norway's share of these resources has allowed it to become a petroleum exporter, and the United Kingdom's share created an oil boom that greatly changed Scotland (which is closest to the oil), and freed England from imported oil for at least the next few decades.

There are also fishing grounds in the North Atlantic around Iceland and in the North Sea, focused on the Dogger Banks. Both these areas and the Arctic Ocean remain a source of protein for Europe's population and provide high-value exports. The Mediterranean is also an important fishery, providing for the protein needs of the peoples of these lands.

A final resource that should be mentioned is the scenery of the continent. The spectacular Alps of Switzerland and Austria, the beaches with their mild winters in Mediterranean lands, and the cultural landscape of the diverse European countryside and cityscape attract millions of tourists each year. Tourist travel to Europe is greater than to any other region of the world (Table 5.1) and provides a major contribution to the European economy.

In combination the resources available to the occupants of Europe are not the most abundant, but they are adequate to allow the residents of Europe to have developed a highly viable economy. The physical environment, while not providing large expanses of extremely fertile land, nevertheless has areas of sufficient fertility to allow production of major crops such as wheat, sugar beets, fruits and vegetables. Europe is, however, the world's largest importer of food. Major hazards such as hurricanes or tornadoes are rare, and only in the Alpine regions are there hazards from avalanches. The mountains around the Mediterranean are subject to earthquakes, but for most of Europe's population, the physical environment remains rather benign.

CULTURAL DIVERSITY IN THE PENINSULA

The people who utilize Europe's environment are a highly diverse group. Europe has been occupied for thousands of years, and during that time there have been repeated movements of population from east to west and from north to south, resulting in a complex cultural mosaic. Some of the more important movements that have affected the present cultures of Europe began with the expansion of the Roman Empire north from Italy to incorporate the British Isles, France, part of Germany, Spain, Italy, Switzerland, Austria, Yugoslavia, Albania, Greece, and Romania, as well as what are now the Benelux countries of Netherlands, Belgium, and Luxembourg. To the north and east, the Roman Empire came into contact with Germanic, Slavic, and Scandinavian cultural groups. With the collapse of the empire the boundary between the two broad areas was destroyed, but many areas to which Roman power extended still share a common language base. Since the time of the Romans there have been movements by Germanic peoples westward and eastward, movements by Asian peoples westward into what

TABLE 5.1 World Tourism Patterns

Region	Arrivals (Millions)	Percent	Departures (Millions)	Percent
Europe	224.5	67.4	232.6	69.86
North America	32.5	9.75	30.9	9.28
Latin America and Caribbean	20.3	6.98	20.8	6.25
Middle East	7.1	2.13	6.82	2.0
Africa	9.0	2.7	5.3	1.6
East Asia and Pacific	37.0	11.1	31.1	9.3
South Asia	2.54	.76	5.4	1.62

SOURCE: *Yearbook of Tourism Statistics* (Madrid: World Tourism Organization, 1986).

is now Europe, as well as persistence of groups that had been in the area prior to Roman expansion.

Broadly speaking, there are four cultural regions based on language (Figure 5.12). Three are part of the Indo-European language family: Romance, Germanic, and Slavic. The Latin, or Romance, language derivatives of Southern Europe are found in France, southern Belgium (Walloon), Portugal, Spain, and Italy. To the north, Romania is an outlier of the old Roman Empire that managed to retain its Romance language in the face of invasions of other groups. The Germanic language group consists of German, English, Danish, Norwegian, Swedish, Dutch, and Flemish (spoken in part of Belgium). A third major group consists of the Slavic languages of Yugoslavia, Czechoslovakia, Poland, and Bulgaria (and the Soviet Union). There are many Slavic tongues, including Polish, Czech, Slovak, Slovene, Croat, and Serb. The pattern of Slavic languages in the Balkan peninsula is highly complex as a result of their position on the crossroads between east and west. The Slavs of the Balkans have been invaded by Turks, Mongols, Greeks, Romans, and Germanic peoples, further complicating the language pattern in the region.

In addition to these three major language groups, several other languages in combination form a fourth region. To the west are the Celtic languages, which represent an early group who have since been pushed to the margins of Europe by subsequent invaders. The Bretons of the Brittany peninsula in France, the Welsh, the Scots, and the Irish all speak relic tongues of Celtic origin. To the north and east the Finns and the people of Hungary speak Uralic languages that are not related to languages of Western Europe. The Hungarian language is the Magyar, and the Hungarians refer to their country as the Magyar Republic. Greece has an ancient language of its own. Finally, Albanian is an Illyrian language (Indo-European) modified by interaction with the invaders of the Balkan peninsula. The language distribution is broadly comparable to the three most important ancient occupying powers of Europe: the Roman Empire, the Germanic empire (first called the Holy Roman Empire), and the Byzantine Empire, each of which occupied large areas of Europe at one time or another.

It is also possible to recognize three major cultural divisions of Europe based on religious systems. The south and Mediterranean lands are dominated by the Roman Catholic church. The north and northwest have a greater concentration of Protestant churches, especially the Church of England and Lutheran churches. To the east are the Eastern Orthodox churches, originally Greek Orthodox; now each individual country has its own Orthodox church. The Roman Catholic church is also dominant in Poland and East Germany, but communism has greatly affected the role of religion in the Eastern European countries. Ironically in spite of Communist governments Pope John Paul of the Catholic church is from Poland, and his visits to Poland illustrate the unique combination of Catholicism and communism in that country. This threefold division of Europe on the basis of religion masks important regional contrasts in religion, as between the Roman Catholics and Protestants of Ireland. The dominance of Christian religion in Europe as a unifying cultural feature has not been sufficient to overcome the **centrifugal** forces of **nationalism** and economic well-being of individual states.

ECONOMIC DIVISIONS

A division of Europe based on economy is even more evident. In general terms there are three broad regions: Northern and Western Europe, Southern Europe, and Eastern Europe. Northern and Western Europe are characterized by wealth, high literacy rates, a highly urbanized population, and a preponderance of industrial as opposed to agrarian economic activity. Southern Europe is less industrialized than Northern Europe, with an average per capita income of one-half that of Northern and Western Europe. Industrialization in Southern Europe is concentrated in the capital cities such as Madrid, Rome, and Lisbon or in large centers such as Barcelona or Marseilles. The cities of Milan and Turin in the Po Valley are an exception as the Po Valley of northern Italy has emerged as an industrial area more similar to Western and Northern Europe than to Southern Europe. Southern Europe has a high proportion of its labor force engaged in agriculture and tends to be somewhat similar to less developed areas with extremely small and fragmented farms, a high reliance on animate power, and more limited transportation and other connective links to tie the region together for maximum economic efficiency. The countries of Suthern Europe are presently experiencing economic growth rates that are higher than Western and Northern Europe. Eastern Europe has traditionally been the most economically and culturally backward of Europe's regions. In this area feudal estates persisted for a long time, and on the eve of World War I much of the region was dominated by the Austrian Hapsburg Empire. Industrialization has been increased as a result of post–World War II activities by Communist governments. The

---

Centrifugal forces: Forces that tend to weaken or destroy a country's unity.

Nationalism: The feeling of pride and/or ethnocentrism focused on an individual's home territory.

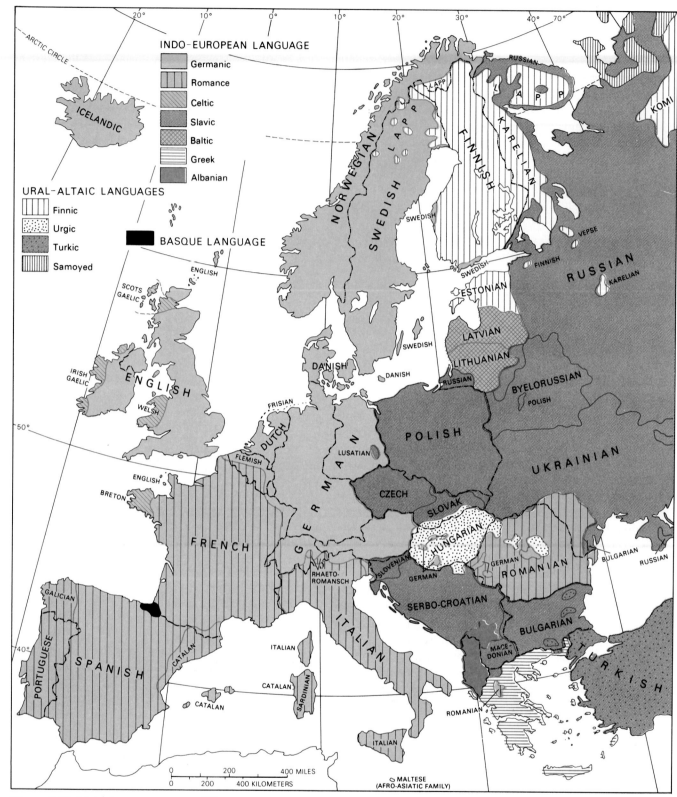

Figure 5.12 Major language groups of Europe.

influence of communism in these states is another fact that separates them culturally from the rest of Europe.

POLITICAL FRAGMENTATION IN EUROPE

This division of Europe in terms of language, religion, and economy illustrates the fragmentation of the entire area. Although there are several countries that have been in existence for a long period of time, most of the countries (**states**) of Europe reflect rather recent events. The complexity of cultural groups means that few countries are occupied by a single distinctive group with a common language, religion, or other unifying characteristic. Conflicts over boundaries and territory have been frequent as a result of the cultural complexity. Political changes were usually associated with conflicts among the countries of Europe, and the map of European countries has been revised numerous times.

The most extreme example of these frequent changes has been the Balkan Peninsula. The Balkan Peninsula has the greatest multiplicity of languages and ethnic backgrounds in Europe. Serbs, Croats, Slovaks, Slovenes, Bulgars, Albanians, and Magyars have repeatedly fought one another and forces from outside the region. The individual peoples and countries have been part of various empires, and countless boundary changes have occurred. Even outside powers have been unable to overcome the regional differences of language, religion, and associated nationalism. Boundary and political changes have repeatedly separated people with a shared culture (a **nation**) and united those with no similarities. The degree of political fragmentation and instability has made the region synonymous with territorial dispute and struggle, and the term *balkanize* (or balkanization) is today used to refer to the concept of a **shatter belt** that has been repeatedly conquered, occupied, subjugated, and reestablished through time as have the Balkan countries.

By contrast, France was constituted as a separate political unit in A.D. 843 under Charlemagne and has retained its identity as a separate state and nation ever since. Not until the 1500s did the other states of Europe begin to emerge in present form. The United Kingdom was united under one rule after this time; Germany was not united until the late 1800s; and the other European states, particularly in the Balkans, owe their existence to the post–World War I period. This long process of formation of states has not led to a complete harmonization of nation and state. Switzerland has four official languages, for example, and stamps and money are printed in Latin to save space. German predominates (70 percent), followed by French (20 percent), and Italian (9 percent), with a small group using Romansh language (a Romance tongue). Nevertheless, the Swiss have a strong national identity based on nearly a millennium of peace and prosperity.

The countries of Eastern Europe owe their existence to the post–World War I period. Czechoslovakia consists of two major regions, the western Czech portion characterized by industrialization and urbanization, and an eastern Slovak portion, which has remained rural and less economically developed. Yugoslavia remains an uneasy amalgam of four major and a host of minor language and cultural groups.

Only 40 years ago the great boundary changes of World War II again changed the political map of Europe. Relics of this conflict remain in the division between the German Democratic Republic (East) and the German Federal Republic (West Germany), East and West Berlin, and in Communist control of Eastern Europe. Nevertheless, in the 45 years since the end of World War II, a generally improving economy in all of Europe has tended to stabilize the present boundaries. It is questionable whether there will be any serious attempts to harmonize cultural and political boundaries in Europe in the short term. Future events may well result in creation of new states in places like Yugoslavia, the United Kingdom, or Spain, but the general trend appears to be for peaceful evolution rather than political revolution.

EUROPE: HOME OF REVOLUTION

Everyone is familiar with the importance of Europe as the home of world conflicts in the twentieth century. More important, Europe has been the site of important changes that have had such a dramatic affect on the geography of the world that they are called "revolutions" even though they occurred over an extended period of time. Several occurred concurrently, and together they comprise the greatest changes that have affected the geography of the world. The effect of these changes is imprinted in the present pattern of industrial versus less industrial countries. The degree to which these revolutions have occurred in individual countries is closely related to the level of

---

State: The formal name for the political units we commonly call countries.

Nation: Refers to a group of people with a distinct culture that may or may not coincide with political boundaries.

Shatter belt: Region located between stronger countries (or cultural-political forces) that is recurrently invaded and/or fragmented by aggressive neighbors.

development and standard of living their people enjoy.

Four major revolutions can be recognized: the *Democratic Revolution*, which took place between the fifteenth and eighteenth centuries, the *Agricultural Revolution* of the seventeenth century, the *Industrial-Technological Revolution* of the seventeenth, eighteenth, and nineteenth centuries, and the *Scientific Revolution* of the nineteenth and twentieth centuries. The first of these revolutions is the Democratic Revolution. Historically, and even today, most peoples have not had a voice in their government. But the Democratic Revolution involved more than just free and open elections. The important features of the Democratic Revolution are sometimes described as the five freedoms: freedom from taxation without representation; freedom from arbitrary acts of those in power; freedom of knowledge; freedom of public assembly and speech; and freedom to vote your conscience through use of a secret ballot. The importance of these freedoms in the development of Europe can be debated, but they seem to have provided for greater flow of information between people, greater personal initiative, and greater desire to accumulate knowledge and to increase the standard of living. Freedom of speech, assembly, and access to knowledge from whatever source allowed countries like England to profit from knowledge developed in other areas. In addition, the democratic process and the guarantee of equal treatment before authorities allowed the development of a group of people who felt free to experiment, to challenge traditions, and to attempt new and innovative life-styles.

Concomitant with the transition we refer to as the Democratic Revolution was the so-called Agricultural Revolution of Western Europe. The predominant agricultural system of Europe in 1500 was little changed from what had been introduced 2000 years before. Full-grown cattle, for example, weighed an average of only 600 pounds, while full-grown sheep averaged only 30 to 40 pounds. Grains were scattered broadcast, often in combination with one another (wheat, oats, rye, or barley), and harvested with hand tools. Fields were defined as *in-field* and *out-field*, with the in-field being cultivated yearly because it was near the settlement while the out-field was cultivated only intermittently, with alternating fallow periods.

The later Agricultural Revolution had its origin in the Netherlands, Belgium, and England, and consisted of selection of certain species of plants and animals to give a higher yield or greater size and productivity. Crop rotation was developed, rotating timothy (a grass) with a root crop such as turnips, and grains. These changes may seem small, but their impact was profound. Provision of timothy and turnips

Looking east into East Berlin, the Berlin Wall stands as a vivid reminder of the continued political division that represents one of the most obvious barriers to truly unifying Europe.

as regular crops allowed farmers to have a stored source of animal feed for use during the winter. With a reliable source of feed, animals were no longer turned loose during the winter to survive on their own. Not only did larger livestock result, but farmers and urban dwellers had a more reliable food supply. Storage and distribution systems became more complex and efficient. With a surplus food supply, population increased, particularly in towns.

As a result of the increased food supply and increasing trade, there developed the Technological Revolution, which culminated in the Industrial Revolution. The impetus for the Technological Revolution began with Gutenberg's invention of metal type cast in molds, which allowed widespread use of printing and the printing press. The printed Bible of 1456 was a momentous breakthrough in the preservation and distribution of knowledge, and by 1482 the first pocketbooks were printed in Italy by Manutius. Voyages of exploration led to the invention of the compass, one of the first achievements of western Europeans not borrowed from the Greek, Roman, or Byzantine civilizations. Advances in the sailing ship that allowed it to tack into the wind, the invention of saws operated by waterpower, mills using wind or waterpower, and machines for harvesting crops were products of the Technological Revolution that ultimately led to the Industrial Revolution. By the end of the seventeenth century the Renaissance, which had originated in northern Italy, was replaced by technological and scientific progress centered on the English Channel in England, France, and the Netherlands.

The interest in technology and science led at the end of the eighteenth century to advances that are sometimes referred to as the Industrial Revolution. The Industrial Revolution was the process by which inanimate power sources were substituted for animate sources. The origin of the Industrial Revolution lay in the saws and mills that utilized water or wind power for operation. The quickening pace of change is evident from the fact that of some 26,000 patents taken out in England during the 1700s, one half were in the period from 1785 to 1800.

One of the first activities to be affected by the Industrial Revolution was in the textile industry, beginning with the spinning jenny invented by James Hargreaves. The spinning jenny spun thread at such a rate that it took twelve weavers to keep up with the output from just one. In 1769 a spinning frame powered by water was patented by Richard Arkwright, symbolizing the new machine age with its greatly increased productivity. Shortly thereafter, spinning and weaving were changed to steam power through use of the steam engine based on James Watt's work, which was patented in 1769.

By the end of the 1700s, steam engines were being used in mines for lifting and drawing cars along a track to a stationary source, and were being seen as a major power source. In 1829, George Stephenson's first steam locomotive moved a 13-ton load at 29 miles per hour over a set of rails: the epitome of the Industrial Revolution with its emphasis on inanimate power sources. Prior to this time, the speed of travel had been restricted to the pace of a galloping horse or a ship driven by strong winds. Suddenly railroads made it possible to move large quantities of materials at rapid speeds. For example, in 1750 it took 48 hours to travel from London to Birmingham by horse or coach; in 1850 it took only 4 hours by rail. The importance of railroads was widely recognized, and they spread rapidly across western Europe, increasing from less than 15,000 total miles in England, France, and Germany in 1850 to more than 75,000 miles by 1900. During the same period, major advances in the technology of making iron and steel led to enormously greater productivity to supply those railroads. Coal, made into coke, replaced charcoal for smelting iron ore and making iron as early as 1709, but the use of coke was not widely adopted until the mid-1700s. Then the use of coke greatly expanded the quantity of steel that could be produced, but the steel was inferior and brittle. In the late 1700s an English watchmaker developed crucible steel, which was made in small batches, using coke to heat it for an extended period to separate the impurities (slag). This process was slow and produced only limited quantities of higher-quality steel. The invention of the Bessemer converter in 1856 and the open hearth furnace in 1857 provided a means of making large quantities of high-quality steel.

The impact of the Industrial Revolution on England, and then on the European mainland and areas colonized by Europeans, marks the distinction between the wealthy industrialized world and the less industrialized, impoverished world. Of major importance was the tremendous increase in productivity. On a daily basis one miner eating at most 3500 calories of food could produce a minimum of 600 pounds of coal with a pick and a shovel. Used for fuel in Watt's engine, this coal produced energy equal to four horsepower, a 500-fold increase in the energy provided initially by the miner. This tremendous increase in energy allowed greater productivity and the development of a society based on mass production and consumption. The factory system, which developed from the use of inanimate power sources, allowed concentration of production in one location (factory system). The result was to attract large numbers of people to the cities to provide labor in the newly emerging industries, thereby increasing the number and size of urban areas.

Prior to the Industrial Revolution cities were very small. London had only 100,000 inhabitants in 1600, and the second largest community in England, Bristol, had only about 20,000. The majority of towns had populations of less than 10,000 people, and in London the majority of the population lived in abject poverty. With the Industrial Revolution, communities grew at a rapid rate. The population of London was 865,000 in 1800, and in 1850, 2,362,000. There were 1 million in Paris by 1850, and between 1850 and 1900, seven more cities in the world reached 1 million population. This pace has increased during the twentieth century with 30 additional cities having populations of more than 1 million by 1930. An additional 50 cities exceeded 1 million population by 1955, and by 1990 there were more than 200 cities of 1 million or greater population.

Coupled with increasing city size and industrial productivity has come a massive increase in circulation and movement of goods and peoples. In less indus-

This hand-powered spinning jenny of the late 1700s enabled the single female operator to produce as much thread as 16 women using traditional spinning wheels.

Awkright's spinning frame substituted water power for the labor of the female operator of the spinning jenny, greatly multiplying productive capacity.

trialized countries today, and in industrialized countries before the Industrial Revolution, travel required large amounts of time or money or both. Since the Industrial Revolution with its expanding transportation network the population of the earth has become mobile.

The Industrial Revolution's impact on the environment has been equally profound. Increased consumption of raw materials, and associated pollution and destruction of landscapes, are major problems in industrialized countries. But the impact on human life as observed from the twentieth century has been beneficial in terms of convenience and standard of living. During the Industrial Revolution in Western Europe, however, social conditions for those working in factory and mine were often intolerable. It was common to employ children 8 or 10 years of age to work 12 to 14 hours per day, six days per week in spinning mills, factories, and mines. Public concern with such excesses fostered the Factory Acts of the 1830s and 1840s in the United Kingdom, establishing a minimum age for employment. The negative impacts of the Industrial Revolution on the working class prompted the theories of Karl Marx. Viewing the Industrial Revolution in England in the late 1800s, he predicted that there would be a revolution to ensure that those who performed the labor adequately shared in the resulting wealth.

Because of the Industrial Revolution, England's productivity increased dramatically. Exports increased two and one-half times during the seventeenth century alone. The resulting wealth fueled greater industrialization in England and financed expansion of its colonial empire. In spite of the social and other problems, other European countries attempted to adopt the developments of the Industrial Revolution. England's efforts to maintain a monopoly on the Industrial Revolution failed, and in the 1800s, Western Europe developed rapidly. After 1850 France and Germany embarked on industrialization on the scale found in England, and by the end of the nineteenth century, the combination of France, Germany,

THE WORLD OF WEALTH

and the United Kingdom produced about 60 percent of total world manufactures. The regional concentrations of industries had already been established in England by the beginning of the nineteenth century, and by the end of that century, similar concentrations existed in France and Germany. In Germany the Ruhr coalfields were the center for the heavy iron and steel industry. In the north of France heavy industry developed in the Alsace-Lorraine iron ore fields and the Sambre-Meuse coalfields. In Belgium and the Netherlands, the Campine industrial concentrations emerged. In England, there were industrial concentrations on each of the major coalfields, as well as in the Midlands.

The cities that became the *head link* for each country developed rapidly. Head-link cities are those that link the country with the rest of the world. Goods destined for export are transported to the head-link city, and imported materials are dispersed from it. London and Liverpool became the trading ports for imports and exports, Amsterdam and Rotterdam developed as the access to the Ruhr, and Paris grew as a focus of transport routes in France. The dominance of these large centers antedates the Industrial Revolution. Major European cities such as London, Paris, Berlin, or Copenhagen all owe their origins to medieval trade and trade fairs, which were concentrated where natural routes or a source of protection existed. Site characteristics of a good harbor (Rotterdam), an easy river crossing (London), a defensible position as an island in the river (Paris), or a junction of routes (Berlin), provided impetus for growth before the Industrial Revolution. The site-situation relationships that had affected each center's initial development and subsequent regional dominance ensured that these places remained important after the Industrial Revolution.

The site-situation relationship affecting development of cities like Düsseldorf, Manchester, Leeds, Liège, and others reflected need for raw materials and illustrates how site-situation relationships are related to technology. The early industrial centers that relied on waterpower often located where waterfalls or rapids could be harnessed to drive the machinery. Sources of raw materials (wool, iron, coal) dictated the establishment of other centers.

The size and relative importance of individual towns and cities of Europe are functions of their hinterland. An individual industrial city may have only a limited hinterland, while London has as a hinterland all of the British Isles, and, for selected activities such as financial services, much of the world. The actual population of cities reflect the demand of their hinterland for service. In the hierarchial order—from city to town to village—the smallest villages may provide only a grocery store, pub, post office, and church; there are thousands of such places in any country. At the other end of the spectrum are cities like London, Rotterdam, Paris, or Copenhagen, which provide all the services of smaller centers plus government, financial, medical, educational, trade, and governmental services to a national or even international hinterland.

The process by which the Industrial Revolution spread from England to Europe is analogous to the way in which it later spread to southern Europe, eastern Europe, the Soviet Union, and Japan. More recently, the emerging countries of the less industrialized world are looking at the technology of Europe and other industrialized centers to aid them in transforming their economies. As they attempt to import the expertise necessary to industrialize, they face problems that Western Europe did not. Many of the emerging countries do not have a complete array of resources that complement one another. The iron ore and coal of England made possible its Industrial Revolution. Other countries are less fortunate in their access to these important elements. Furthermore, the increased demand for energy in the industrialized countries and the increased price in the 1990s will handicap industrialization in less industrialized countries. It can be argued that the events that led to the Industrial Revolution in Europe may not be replicated beyond the existing industrial powers. Alternate energy sources may change this pessimistic view, but until they are readily available, newly emerging countries are faced with an enormous challenge if they wish to emulate Western Europe. Future economic development in the less industrial countries may not be based on the European model of basic heavy industries of iron and steel. Although some countries like Brazil or South Korea are developing an iron and steel industry, others like Singapore are industrializing in areas that do not rely on heavy industry.

INTERNATIONAL INTERDEPENDENCE AND COMPETITION: RESULTS OF THE INDUSTRIAL REVOLUTION

One of the major effects of the Industrial Revolution has been the increase in interconnections of a global nature. The increased productivity that resulted from the improvements in industry in Western Europe in the nineteenth and twentieth centuries led to an increased demand for raw materials. The growth of population that could be supported by industrial societies with their great demand for labor was not met by a concomitant increase in food productivity in these countries. Although the scientific advances that accompanied the Industrial Revolution led to productivity increases of two or three times in England, Ger-

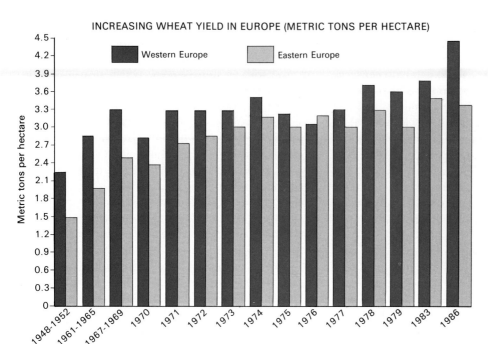

Figure 5.13 Increasing yields of grain in Europe, 1948–1983.

many, and other industrialized countries, there was still inadequate food to feed the burgeoning populations (Figure 5.13). But meanwhile, the application of the steam engine to ocean transport was revolutionizing the speed and cost of ocean shipping. The result was a reliance by European countries on nonindustrialized areas for food and other raw materials. Grain from Canada, the United States, and Australia; meat from New Zealand, Australia, Argentina, and Uruguay; wool from Australia, Uruguay, and New Zealand; tropical products like coffee, cocoa (cacao), bananas, and peanuts from Africa and Latin America; tea from Ceylon, India, and China all reflected the burgeoning markets in the industrialized countries. In return the industrialized countries exported manufactured items to the nonindustrialized countries: transport equipment, agricultural equipment, woven cloth, dresses—the list is almost endless. The economy that resulted from the Industrial Revolution of Western Europe led to an interconnection between the industrial and nonindustrial countries upon which both relied. The less industrial countries were relegated to production of raw materials, luxury items, or tropical foodstuffs. The advent of the internal combustion engine and ever-increasing technological advances has only strengthened this connection.

In modern times this pattern is beginning to change as the technology of the industrialized countries is diffused to the less industrialized countries. Increasing wage costs in Europe have resulted in the movement of the earliest industries of the industrial revolution to the developing countries. Textiles, boots and shoes, even cameras and electronic equipment are increasingly manufactured at least partially in developing countries. Instead of the simple interconnection in which the less industrialized countries provided raw materials and the industrialized countries provided the manufactured goods, we see in the late twentieth century a system in which the industrialized countries concentrate on high-technology items like computers, sophisticated transport systems such as aircraft, and specialty items, while the developing countries concentrate on those elements that require a maximum of labor combined with a lower level of technology. Thus South Korea, Hong Kong, Mexico, India, and Brazil are increasingly becoming the source area for textiles, shoes, and similar items. The changing nature of the international interconnection of the twentieth century represents one of the more important geographical changes now taking place. In only 50 years the pattern of imperialistic domination by European powers, which developed in the late nineteenth and early twentieth centuries as European countries obtained colonial possessions for markets and raw materials, has been almost completely transformed.

In the latter half of the nineteenth century, European **colonialism** brought essentially all the less economically developed lands of the world under the control of a European power. Most of the former col-

---

Colonialism: The political and economic control of a country by a foreign power.

onies of Great Britain had gained their independence by this time, but the increasing pace of the Industrial Revolution prompted the nations of Europe to confront one another over the remaining less industrialized areas of the world. Although colonialism was nominally justified on the basis of need for markets and raw materials, these countries were poor and unable to absorb a large share of the production of the industrial powers. They became symbols of the political power and prestige of the ruling nations. Competition for colonies had its ugly times as Germany confronted France, France confronted England, and all confronted one another in such far-flung areas as Africa, Asia, and the Pacific. Latin American countries remained nominally independent but became economic colonies inextricably tied to the industrial nations of Europe and the United States. In India the United Kingdom had already consolidated its position. In Asia and Africa European powers controlled colonies greatly exceeding the size of the ruling nations (Figure 5.14).

The impact of this colonial era on the less industrialized countries of today cannot be overstated. The motives that led to European expansion and domination of these areas were a combination of self-interest, aggrandizement, and altruism. The results of the domination remain in place-names, political organization, laws, language, religion, and commercial activity. Of more critical concern to less industrialized countries is the real or perceived economic domination of the European nations. The continued reliance on technology from Europe and America and the control of manufacturing and other economic activity within less industrialized countries by European or other industrialized nations is viewed as a form of colonialism called *neocolonialism*.

As the colonial empires began to gain independence following World War II, these former colonies looked to their former rulers for support in the transition to a viable independent economy. In the case of some European powers the transition was relatively peaceful with support from the original colonial power (for example, India under the United Kingdom). In other cases independence came only after a protracted conflict that destroyed the economy of the area (Algeria and Vietnam under the French).

Regardless of one's viewpoint on the colonial effort of the European powers in less industrialized

Figure 5.14 The colonial possessions of European nations at the start of World War II reveal the vast amount of territory controlled by the countries of Western Europe.

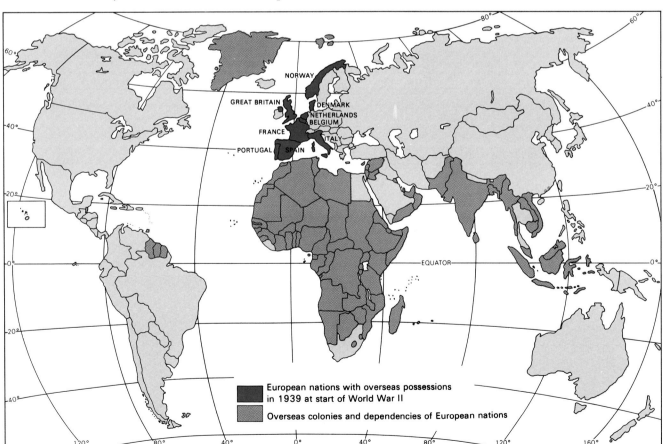

countries in the late nineteenth and early twentieth centuries, it obviously played a major role in the present status of the less industrial countries. As such it is an important example of the spread of European influence throughout the world. The development of the image of Europe as the epitome of education standards, artistic standards, cultural values, and scientific advancement has not only caused irreparable losses to less industrialized countries by emigration of the skilled elite to the West (the "brain drain"), but has also led to **cultural convergence** in architecture, values, dress, and political systems. An example of the magnitude of this impact can be seen from the figures for the loss of educated and trained specialists to Western areas between 1960 and 1985. During this period it is estimated that over 600,000 skilled people migrated to the industrialized West from the less industrial world, nearly 75,000 physicians and surgeons, more than 150,000 engineers and scientists, and about 175,000 technically qualified people among them. The estimated cost to the economies of the less developed countries was $46 billion for the decade 1961 to 1972 alone, an amount as large as the official foreign aid flows from the industrial countries to the less industrialized countries during this time.

SOCIAL REVOLUTIONS IN EUROPE

The migration of skilled people from the less developed to the industrial world in the twentieth century is an interesting reversal of the historic trend of European population. Europe was the world's major source area for migrants from 1500 to the post–World War II era. European migration resulted in peopling large areas of the world, including Canada and the United States, Australia, New Zealand, and many Latin American countries, as well as South Africa and Zimbabwe. The migration of Europeans reflects changing social conditions in Europe that resulted from the Industrial and Technological revolutions, from industrialization and changing population components. Prior to the Industrial Revolution, Europe lost one-half of its population in epidemics of bubonic plague from 1347 to 1402. Following this loss of population, Europe's population grew slowly as a result of the high infant mortality rate of stage 1 of the demographic transition. With the advent of the scientific advances of the Industrial Revolution, Europe entered stage 2 of the demographic transition. Birthrates remained high throughout much of Europe, but death rates for infants began to fall, resulting in a higher net increase.

The causes of the declining death rate were advances in food quantity and quality and the development of better medical care, including the discovery of the germ theory of disease. The result was a population increase similar to that of the less industrialized countries that today are in stage 2 of the demographic transition. Large numbers of people were unable to find work, particularly in countries of Europe that were just beginning the Industrial Revolution but to which the medical and scientific advances of the industrial countries had spread, resulting in a higher growth rate. These numbers of surplus people formed the basis for migration during each period of the Industrial Revolution. During the initial stage they came from Great Britain, Germany, and Scandinavia. As these countries moved into stage 3 of the demographic transition, the countries of eastern and southern Europe became the source areas for migrants (Table 5.2).

Coupled with population growth were political and economic events that forced people from Europe. The potato blight in the mid-1800s resulted in a loss of nearly two-thirds of Ireland's population. Nearly 1 million died of famine and 2 million emigrated. In similar fashion, technological changes in France led to unemployment for large numbers of weavers, who migrated to the United States. Political upheaval in eastern Europe, with pogroms against Jews, resulted in migration of Ukranians and Polish people to the United States and Canada. An examination of migrants to the United States (Table 5.2) illustrates the changing nature of the source region for European migrants. Migration to the rest of the world has increased Europe's world importance. For North and South America, Australia and New Zealand, and South Africa, Europe remains the ancestral homeland. As all European countries have moved into stage 3 or 4 of the demograhic transition, it has become increasingly less important as a source area for migrants (Figure 5.15). The causes for Europe's transition to stage 3 and 4 with its concomitant decline as a source of migrants are explained by its ever-increasing industrialization, urbanization, and social changes.

SOCIAL CHANGES IN EUROPE

There have been a number of important changes in the society of Europe that reflect the Industrial and Democratic revolutions. Foremost among these is the emergence of a nonagricultural society. Industrialization with its demands for labor removed the necessity for all individuals to provide for their own subsistence through farming. The resulting concentration

---

Cultural convergence: The tendency of world cultures to become more alike as European culture and/or technology is dispersed, particularly to urban areas of less industrialized countries.

TABLE 5.2 Origin and Volume of Selected Foreign Migration to the United States, 1820–1979 (in thousands)

Date	Total Immigrants	United Kingdom	S. Ireland	Scandinavia	Germany	E. Europe	Mediterranean
1820–1830	128	26	52	0.3	6	0.1	3
1830–1840	538	74	171	2	125	0.3	5
1840–1850	1,427	219	656	13	385	0.6	4
1850–1860	2,815	445	1,028	25	976	0.5	19
1860–1870	2,081	533	427	96	724	2	19
1870–1880	2,742	578	422	208	752	36	67
1880–1890	5,249	811	674	672	1,445	190	290
1890–1900	3,694	329	406	391	579	460	652
1900–1910	8,208	469	345	488	329	1,656	2,166
1910–1920	6,347	372	166	238	174	1,219	1,570
1920–1930	4,296	338	207	178	387	175	693
1930–1940	499	54	36	17	119	16	108
1940–1950	857	125	22	22	119	7	71
1950–1960	2,499	184	56	57	576	8	261
1960–1970	3,321	210	37	32	190	6	441
1970–1979	7,292	108	13	18	59	97	353
1980–1986	3,403	92	16	11	50	92	117

SOURCE: United States Bureau of Census, Department of Commerce, *Historical Statistics of the United States, Colonial Time to 1970*, and supplements (Washington, D.C.: U.S. Government Printing Office, 1975), and 1988–1989 Statistical Abstract of the United States.

of population in urban areas led to the development of an extremely poor laboring class in the early stages of the Industrial Revolution. The laboring class in Europe was like that in less industrialized countries in that it was landless, but there was a major difference, since workers had a source of income. As the Industrial Revolution progressed, this working class in most European countries shared increasingly in the benefits of the Industrial Revolution through unions and government programs providing for the workers. The result was the emergence of a middle class. Today this middle class is dominated by white-collar workers in management, sales, banking and finance, and other tertiary activities. The blue-collar worker, who is directly engaged in the manufacturing process, shares many of the same cultural characteristics as the middle-class workers in white-collar jobs.

The emergence of the middle class in Europe marked the first time in world history that there had been anything other than the rich elite and the poor farming class. For the world economy the middle class of Europe represents a great market for goods, services, and food products, affecting people of almost all countries. In their own countries the middle-class citizens represent a stabilizing force. This force is most evident in the countries of Western and Northern Europe, where the Industrial Revolution has had the greatest and longest effect. Here the middle class with its emphasis on work and high standard of living aids in maintaining political and economic stability. In Southern and Eastern Europe, where the Industrial Revolution is still in progress and where the Democratic Revolution has been only incidental, the middle class represents a smaller proportion of the population. Moreover, because the middle class in these countries has not had a perceived equal share in the wealth, it has tended to be less of a stabilizing force.

In the Southern and Eastern European countries, where the wealth from industrialization is less equally divided, another social revolution has been of great importance. The development of communism as a political force in Italy and in all the Eastern European countries partially reflects the lack of an emergent middle-class majority who share in the wealth of their country. The emergence of communism and other leftist political groups is part of the process of the emergence of the middle class. In these countries, recognition (or perceived recognition) that the wealth is inadequately shared or that the needs of the poor are inadequately met has prompted the emergence of

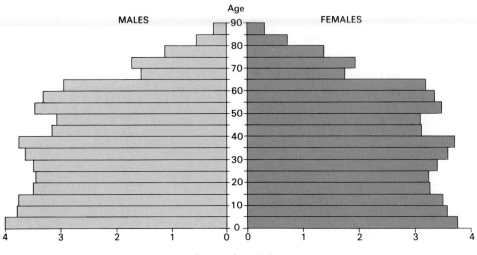
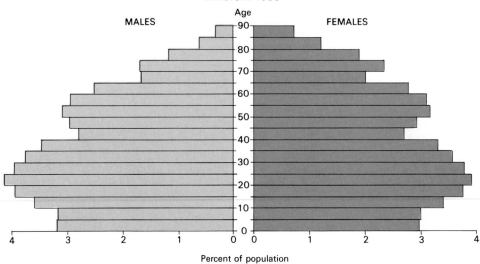
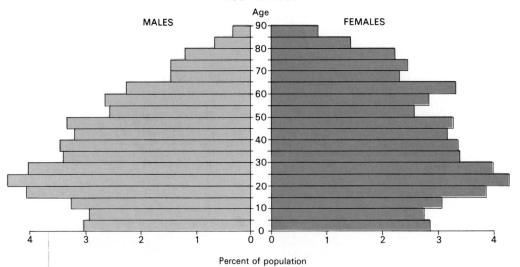

Figure 5.15 Representative population pyramids for European countries, 1980.

THE WORLD OF WEALTH

communistic political parties. In simplistic terms, all are dedicated to the gaining of greater control of private industry to ensure that the productive capability is used for the benefit of the people. This process has its greatest impact in the Eastern European countries where Communist governments dominate, but here the original Communist governments came to power as a result of Soviet intervention after World War II.

The process of government intervention in industry and private enterprise (as opposed to the total central control of governments in the Communist lands) varies from country to country but is an important issue in all. The concept of government or public ownership of basic industries such as rail lines, coal mines, steel industry, and assorted others (socialism) is most highly developed in Scandinavia and the United Kingdom, but is prevalent in the other countries of Europe. Even here, however, private ownership far exceeds public ownership, and in recent years the United Kingdom and other Western European countries have been returning government-owned industries to private ownership. Socialism in part represents an attempt to spread the benefits of an industrial society more widely and to insure a minimum standard of living for all people. In part, it also results from the problems that face the industrialized societies in the late twentieth century.

PROBLEMS AND CONFLICTS IN THE INDUSTRIAL WORLD

OBSOLESCENCE, RESOURCE EXHAUSTION, AND POLLUTION

The trio of problems in the industries that made the Industrial Revolution of Europe possible is partially responsible for the continuing importance of socialism and communism in Europe. Obsolescence has greatly affected those industries that formed the basis of the Industrial Revolution. Europe's iron and steel factories, early textile and chemical plants, and similar industries suffer not only because their equipment normally dates from the end of World War II or earlier, but because plants were located inland at the site of power and raw materials. It is difficult for these industries to compete with those in Japan or in newly industrializing countries that use later technology and are located at break-in-bulk points where they can minimize transport costs. Originally located rationally with respect to the raw materials, many of Europe's industries have exhausted the higher-quality ores upon which they were based and must turn to imported raw materials.

Many of the older plants were built at a time when environmental pollution was not an issue, and the old industrial areas of the Midlands of England, the Ruhr of Germany, and portions of France and Italy suffer from severe environmental degradation. As countries attempt to improve their air and water quality, these industries are faced with the high cost of modernizing their plants to minimize pollution. Faced with increased costs of raw materials, pollution, and obsolescent equipment, many of the factories would close. In the face of worker opposition to job loss, however, governments have nationalized some of these industries. The resulting nationalized industries are often not competitive economically with those in other areas of the world, and in order to protect them, governments must institute protective policies. In consequence of these problems, coupled with increased competition from emerging countries, Europe's industrial powers face severe challenges at the close of the twentieth century. At issue are the questions of how to ensure a high standard of living, adequate jobs, and a clean environment. At the same time there is a need to make the transition that will allow European industry to abandon uneconomical production sites and types of industry and shift to the high-skill and high-technology areas in which they can successfully compete. In the 1980s several European countries (most notably the United Kingdom) have reduced the level of government ownership of industry in a process called **privatization.**

Traditionally, Europe has been a center for production of metals, textiles, transportation equipment, and chemicals. As emerging countries make inroads into textile and metal industries, it becomes imperative that Europeans focus their industrial activities on production of electronics, high-technology transportation equipment, and similar products. If they do not, we may well see development of protectionism to a degree unseen in the past. In order to maintain uneconomical industries, Europe will have to discriminate against imports from poorer countries to an even greater extent than it does now. In a world based on an interconnected economy, such a course would result in increased direct and indirect costs to the entire world. One way that the newly industrializing coun-

Privatization: The process of transferring ownership and operation of manufacturing, service, or other activities from the government to private ownership and operation.

tries can progress to the point that they can share in the wealth and standard of living experience in Western Europe and other industrialized countries is by trade, and protective **tariffs** are a threat to this trade.

Just as the obsolescent industries of Europe affect areas beyond its boundaries, so pollution affects European countries beyond the heart of the industrialized regions. So-called acid rains, which are caused by sulfur in the atmosphere reacting with the hydrogen and oxygen in the air to create weak acidic solutions, are damaging the forests of Northern Europe, but the major sources of the sulfur are Germany and England. Pollutants in the Rhine from Germany affect the Netherlands significantly. Chemical pollution in Spain, Italy, Germany, and France bring injury and death to plants and animals. The Western European countries are making progress in overcoming pollution, and the Rhine and Thames rivers are actually cleaner than they were a decade ago. The less industrialized countries of Southern and Eastern Europe have not yet approached the issues of pollution to the same extent. The countries surrounding the Mediterranean Sea are now embarking on a program to limit the flow of raw sewage into the Mediterranean in order to prevent further deterioration of water quality and of the beaches frequented by tourists. The pollution problem remains in Europe as in all industrialized countries and will persist until such time as countries adequately fund programs to overcome it.

The pollution, resource exhaustion, obsolescence, and foreign competition that affect industry in Europe should not obscure the fact that Europe is still one of the major industrial producers in the world. The United States, Europe, the Soviet Union, and Japan dominate the productive capacity of the world when all industrial productivity is considered. In spite of the difficulties they presently face we can expect that European countries will not only continue to maintain their position of importance in the world economy, but may well expand their role to one of dominance as a result of events now occurring.

ECONOMIC UNIFICATION: THE EUROPEAN DREAM

One of the most important geographical events to occur in the world in the past quarter century is the European attempt to unify economies. Since Europe was united under Charlemagne in A.D. 800 the idea of reunification has persisted. Attempts to implement

---

Tariff: A tax placed on imports or exports into a country. Often used on imports to protect an economy from lower cost imported goods.

a united Europe took the form of armed conflict on numerous occasions. With the end of the latest effort by Hitler under the German fascists, Europe was in a state of ruin. Eastern Europe was occupied by Soviet troops who proceeded to implant Communist governments. Other portions of Europe not occupied by the forces of the Soviet Union, including Yugoslavia and Greece, were faced with internal conflict that resulted in the emergence of a Communist-oriented government in Yugoslavia and fears of a Communist-dominated government in Greece.

Faced with the problems of Western Europe, the United States in 1947 instituted the Marshall Plan. The Marshall Plan provided European countries the opportunity to receive aid from the United States to redevelop their economies. All the countries of Europe, with the exception of Soviet-occupied East Germany, Poland, Czechoslovakia, Romania, Hungary, Bulgaria, Yugoslavia, and Albania, agreed to participate in the Marshall program. To distribute the funds provided by the United States, these nations formed the Organization for European Economic Cooperation (OEEC), which was simply a vehicle for dispersing American aid to the countries of greatest need.

In 1950, as the Marshall Plan resulted in success in rebuilding the ravaged economies of war-torn Europe, the Schumann plan emerged. Robert Schumann, the foreign minister of France, noted that Germany had coal needed for manufacture of iron and steel, and France had iron ore at Alsace-Lorraine, but neither effectively had both. He suggested that the nations of Europe cooperate in manufacturing iron and steel to maximize their productive capabilities while minimizing their costs. The Schumann plan called for removing tariff barriers on imports and exports of the raw materials and finished products of the iron and steel industry.

All nations of Europe were once again invited to attend and participate. The Eastern European countries, again, were prevented from joining by Soviet wishes. Sweden did not join because of its commitment to neutrality. Finland was reluctant to join Germany because of fear of Soviet reaction. Norway and Denmark were not producers of iron and steel, but when the British Isles determined not to join because of historic conflicts with Germany and fear for its own iron and steel industry based on local coal and iron ore, Norway and Denmark opted to remain out because of their trade relationships with the British Isles, the largest food importer in Europe. Switzerland as a neutral country refused to join, and Austria was partially occupied by the Soviet Union, which refused to allow Austria to join Germany in any type of trade or economic union. The totalitarian governments of Spain, Portugal, and Greece were excluded because of their political systems. Consequently, in 1952 only Germany and France, the two major beneficiaries, with the Netherlands, Belgium, and Luxembourg, which traded heavily with them, joined with Italy, which had extensive trade relationships with Germany and France, to form the European Coal and Steel Community (ECSC).

The impact of the ECSC on the economies of the involved countries was profound. The productive capability of the iron and steel industries increased dramatically. Marginal producers, hitherto able to exist because of protective tariffs, were forced from business. Unemployed workers from such establishments were given the opportunity of transferring to new locations where jobs were provided by the governments in newer, expanded facilities. The ECSC was so successful that the six countries involved suggested in 1955 that Europe consider a common economic community. Once again, all nations of Europe were invited to join. Still, the Scandinavian and Alpine countries declined, for the same reasons. The United Kingdom remained reluctant to join because of its relationship with its former colonies, known as the Commonwealth countries. Canada, Australia, New Zealand, India, and other former colonies of the British Commonwealth were the major sources of food and raw materials for the British economy. The United Kingdom was reluctant to join a European economic organization that would have discriminatory tariffs to protect French and German agriculture, since this would increase the costs of food in the United Kingdom and affect the economies of the former colonies. The result was that the six nations of the ECSC were the ones that formally established the European Economic Community (EEC), or "Common Market," in 1955 (Figure 5.16 and Table 5.3).

In the 1960s the economies of the nations in the Common Market generally grew more rapidly than those of the balance of Europe, particularly the United Kingdom. The result was repeated applications by the United Kingdom to join the Common Market during the 1960s. In 1972 the United Kingdom, Ireland, Norway, and Denmark held a national vote (plebiscite) on joining the Common Market. As a result of these plebiscites, all except Norway joined. The United Kingdom joined because it needed to trade with the other manufacturing nations of Europe, and Ireland and Denmark relied extensively on their trade with the United Kingdom. These nations had found in the decades since the formation of the Common Market that their economy suffered by being outside the protective tariff of the Common Market. Norway also traded extensively with the United Kingdom, and it was assumed that it would join, but when the plebiscite was held, membership was defeated. The pri-

Figure 5.16 Evolution of the European Community.

mary reason was its oil in the North Sea and a campaign by the Norwegian fishing industry, which maintained that as part of the Common Market, it would lose its exclusive fishing rights to the area within Norwegian waters (Figure 5.17). As of 1990, Greece, Spain, and Portugal have become full members, and Austria, Sweden, Switzerland, and Finland have special trading relationships with the Common Market to prevent their exclusion by the tariff barrier.

For trade and economic purposes, there are standard tariffs on all goods imported into the EEC. There are standard programs for agriculture and agricultural support prices necessary because of an overproduction of certain items and a prevailing high cost for goods produced by west European farmers. As a result of membership in the Common Market, less industrialized nations such as Ireland, Spain, and Portugal, where labor has traditionally been cheap, have become the focus of development by other European firms. The resulting increase in economic activity has increased the standard of living in these poorer nations. At the individual level, residents of the Common Market countries have an identity card allowing them to move freely between member countries. Since 1985 most member countries have issued the distinctive "European Community" passport for travel outside the

TABLE 5.3 Milestones Toward a Unified Europe

1946
Winston Churchill calls for a "United States of Europe."

1950
French Minister of Foreign Affairs, Robert Schumann, proposes the first European Community (dealing with coal and steel) on May 9, celebrated as the official "birthday" of the Communities.

1951
Paris Treaty of April 18 sets up the European Coal and Steel Community (ECSC).

1957
The Treaties of Rome, signed on March 25, establish the Common Market and Euratom, creating three "European Communities," the EEC, ECSC, and Euratom.

1958
The Councils and Commissions to manage the Common Market and Euratom are set up.

1961
Greece signs an association agreement.

1962
The common agricultural policy is born in January, based on the principle of a single market and common prices for most agricultural products.

1963
Turkey becomes an associate member.

1964
A common agricultural market is set up, with supporting marketing organizations, and uniform prices for cereals, to go into effect in 1967. The European Agricultural Guidance and Guarantee Fund begins operations.

1967
The Treaty establishing a single European Commission and one Council comes into force.

1968
The customs union is completed, abolishing all duties on trade between members. A common external tariff is set up. In practice formalities still hinder trade within the EEC.

1973
Denmark, Ireland, and the United Kingdom become members, enlarging the Community to nine members.

A free-trade agreement comes into force with some members of the European Free Trade Association that are not members of the EEC.

1974
The European Council is born, to supplant earlier irregular summit meetings. It is decided to hold elections to the European Parliament by direct universal vote.

1975
The Community signs the Lome Convention with 46 countries of Africa, the Caribbean, and the Pacific (ACP) to foster commercial cooperation by giving the ACP countries free access to the Community market, and guaranteeing stable earnings for 36 commodities from those countries.

1979
The Council, meeting in Paris on March 9–10, brings the European Monetary System into operation. The EMS comprises the European currency unit (ECU), an exchange and information mechanism, credit facilities, and transfer arrangements.

First direct elections to the European Parliament take place.

A new Lome Convention is signed (Lome II), this time with 58 countries in Africa, the Caribbean, and the Pacific.

1981
Greece becomes the tenth member of the Community.

Greenland, an "autonomous region" of Denmark, opts out of the Community but is granted the same associate status as overseas countries and territories governed by member nations.

1984
Lome III is signed, this time with 65 partners in Africa, the Caribbean, and the Pacific regions.

1986
Portugal and Spain become members.

1992
Removal of trade barriers between members of the EC.

EEC. Certain roads are designated "E" roads, and international travel by Common Market members is facilitated by minimizing barriers at individual country borders.

At the regional scale there are changes taking place that indicate a possible trend toward even greater unity among the Common Market countries. A common port for Europe has been constructed down-

The Galeries Lafayette department store in Paris exemplifies the increasing interconnections between the countries of the European Community. Removal of most remaining tariff barriers between EC members will prompt increasing homogeneity among the European markets.

stream on the Rhine from Rotterdam. This new Europort serves as a focus for imports and exports for the entire Common Market. A Common Market parliament was established when the EEC was first organized. Membership in the parliament is based on the size of the participating countries, and members were appointed by the countries until 1979. Since 1979, free community-wide elections were held for the Common Market parliament that meets in Strasbourg, France.

The Common Market parliament makes regulations affecting the entire community, particularly in economic matters; levies certain taxes and collects fees; has an independent budget that it administers; and in general, acts as a quasi-governmental body. There is also a Common Market Supreme Court that rules on community-wide appeals. The members of the Common Market are presently attempting to develop a common currency based on the so-called Eurodollar. All these activities are indicative of a general move toward unity among the European nations, particularly economic unity. The EEC parliament is treated as an independent government by other countries, who have ambassadors representing them at the EEC. In the 1980s the ECC became the EC (European Community), reflecting the goal of European unity. In 1992 all internal trade barriers within the EEC are to be removed, creating the true common market envisioned by its founders.

Whether this signifies a movement toward the age-old dream of a united Europe is debatable. The major **centripetal** (unifying) forces are those associated with the advantages of economic interaction. Major centrifugal forces are the individual cultures, heritages, and national identities of the nations of Europe. Residents of the British Isles have always viewed themselves as distinct from those on the continent, but the Eurotunnel under the English channel will connect them to France and the continent after 1993 if the tunnel is completed on schedule. The great variation in standard of living, education, and outlook between Northern and Western European countries and those of Southern Europe make true political unification doubtful. The centuries of national independence with associated national heroes and symbols effectively prevent the short-term realization of the goal of a united Europe. It is possible that over the next century, these centrifugal forces may be overcome by processes now in action. Unless and until the centrifugal forces are replaced by a loyalty to Europe as a whole or other centripetal forces, it is questionable whether the regionalization of Europe will give way to the Common Market unification. The potential of a united Europe is one of the most significant

---

Centripetal forces: Refers to forces that tend to unify a country or area.

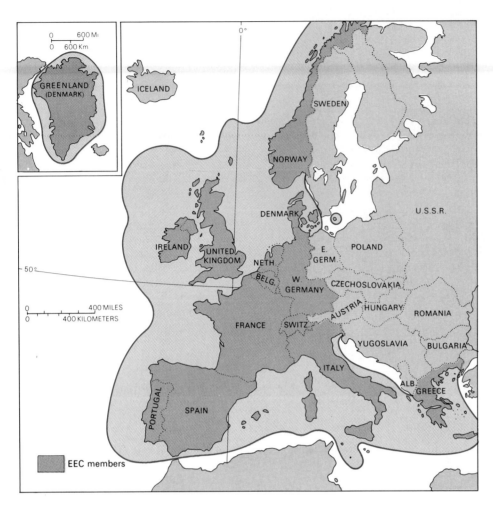

Figure 5.17 Territorial claims of EEC members.

prospects in the geography of the world. If European nations are able to unify their economies and progress toward political unification, Western Europe could well become the most powerful nation economically in the world (Figure 5.18).

ALIENATION AND REGIONALIZATION IN EUROPE

Weighed against the promise of European unification are problems of persistent regionalism and separatism in Europe. The historic self-identification of Scandinavians, the British, Mediterranean Europe, or Germans is compounded by separatist movements within individual countries. We commonly assume that the nations of Europe represent homogeneous entities. In reality, each individual country is composed of an intricate mosaic of groups, each of which has a unique history. Just as in the United States it is possible to recognize distinct regions such as the Northeast, the South, the Mormon culture region, and even a southern California car culture, so in Europe a host of subregions comprise each country. This regional mosaic is even more complex than the United States because of Europe's long history, with its repeated conflicts, migrations, and settlement attempts. In the time since the Roman Empire, political units have been formed, dissolved, reformed, realigned, relocated, and dissolved over and over again. The result is that countries like France, Switzerland, Germany, or even the United Kingdom are not homogeneous entities, but a series of groups. Each group has unique histories, cultures, customs, landscapes, religion, and even dress.

To the casual tourist the differences in such minor things as house style, field patterns, or family relationships may not seem significant. To the resident of a specific locale, however, these are the very fabric of life and are central to the identity that characterizes each subregion of Europe. Where such cultural differences are reinforced by language differences, economic differences, or a sense of persecution or lack of equal treatment by the government, demands for separatism or more regional autonomy may flourish.

Significant cultural differences may not always dictate a demand for separatism. In nations such as Switzerland where the government is a form of con-

THE WORLD OF WEALTH

THE HUMAN DIMENSION

The Evolution of the Common Market

In the last 40 years, the tentative beginnings of economic cooperation among European nations has blossomed into the economic power of the Common Market and the renewed stirring of a possible political unification of the continent. Some 36 years after the birth of the European Community, and 26 years after the establishment of the EEC with its 6 members, the Community expanded to 12 members on January 1, 1986, with the addition of Spain and Portugal. Today it is the largest trading group in the world, and 1992 promises to make it even more powerful (Figure 5-18).

The 30-year-old Common Market is supposed to make good on its name in 1992. If the 12 members of the European Economic Community do actually manage to tear down all intra-EEC trade barriers, goods and people will then move as easily from Portugal to Germany as they now do from Iowa to Michigan. The result will be the industrialized world's single largest market: 320 million increasingly prosperous consumers. Hopes are that it will nudge the Continent toward new political unity as well.

European technocrats have already (1989) pushed 70 out of 300 barrier-breaking rules through the Community Council. Since January 1988, member nations have been using a common customs form instead of the code of many colors that traditionally snarled border traffic and trade. The coming spate of deregulation will eliminate separate national standards for everything from food additives to noise levels of machinery.

The agencies involved in removing trade barriers are the EEC "government," consisting of executive, legislative, and judicial branches.

Executive: The Brussels-based Commission of the European Communities. This is like the executive branch in the United States. Members are appointed by governments of the 12 Community nations for four-year terms. The president and vice-presidents serve for two-year renewable terms. A sprawling bureaucracy comes under the jurisdiction of the commission. It is headed by a secretary-general and has 22 departments, ranging from foreign relations to fisheries.

Legislative: The Council of Ministers and European Parliament.

The *Council of Ministers* resembles the U.S. Senate when governors appointed senators. It is composed of ministers from each of the member states, delegated according to what subject (foreign affairs, economics, agriculture, etc.) will be discussed. They meet once a month. The presidency changes each six months.

Voting is either by simple majority, qualified majority, or unanimity, depending on the subject. Qualified majority voting gives large nations such as France, Germany, Italy, and Britain more votes than smaller ones. Government leaders from the 12 member countries meet in a summit twice a year at the capital of the nation with current presidency to oversee development of the EC.

Like the U.S. House of Representatives, the *European Parliament* is elected directly by Europeans (last elections were in June 1989). It has 518 members and is generally an advisory body. The Parliament is based in Luxembourg, meets annually in Strasbourg, and holds committee sessions in Brussels. The multiple locations of its activities serve to hamper the effectiveness of the Parliament.

Judicial: The Court of Justice, based in Luxembourg, is like a combination of the U.S. Supreme Court and the Justice Department. It has 13 judges and 6 advocates general, all appointed for renewable six-year terms by governments of member states. This gives the judiciary substantial independence. The Court of Justice rules on cases involving EC treaties.

The dream for 1992 is of a European region in which goods and people can travel as freely as they do between the states of the United States. If it actually occurs on schedule, the true common market will be the major milestone on the journey to a unified Europe dreamed of by numerous citizens and politicians of Europe (Table 5.3).

Adapted from Finance & Development, September 1986, p. 30.

---

federation with a great deal of local autonomy, individual regional differences are expressed at the local level. In such cases, the cultural groups have a voice in government affairs and seem to feel no great motivation for separatism. In countries with a strong central government that dictates policies and practices with relatively little local input, the drive for separatism is much greater. The United Kingdom, France, Belgium, and Spain have been most influenced by demands for regional autonomy, and potentially are most apt to have major changes in their political structure. In these countries, the issue of *devolution* is extremely strong. Devolution is the process by which a part of a country becomes completely or partially independent from the nation. In the United Kingdom, both Scotland and Wales are demanding establishment of separate parliaments to deal with local matters, and separatist parties (the Scottish Nationalist Party and the Plaid Cymru in Wales) are strong. In both countries, there is a rising demand for independence from the United Kingdom, and significant changes in their relationship to the central government in London will probably occur. France, as the first nation of Europe to establish a strong central government that

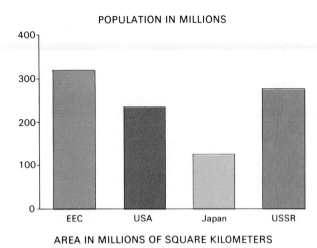
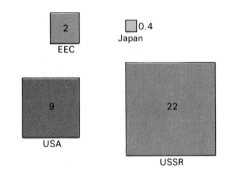
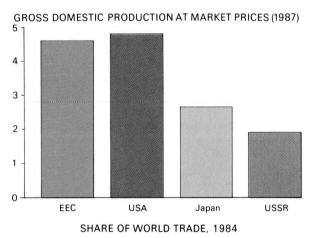
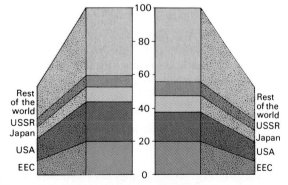

Figure 5.18 EEC compared to the other industrial powers.

unified the divergent cultural groups that comprise modern France, faces challenges to the Paris government in several areas. These are particularly strong in Brittany, Alsace, Savoy, and Corsica. In Belgium, which was one of the later nations to become formally independent, a major cultural division exists. The northern part of Belgium speaks the Flemish (Dutch) language, while southern Belgium speaks Walloon (French). Contrasts in culture and economic and political power cause persistent tension and raise the potential for complete separation.

Spain is the epitome of the regional fragmentation that threatens the fabric of a united Europe and indicates why the Common Market will probably never be more than a customs (trading) union. One of the most vociferous separatist movements in Europe is that of the Basques of northern Spain and southern France. Located in and near the Pyrenees Mountains, the Basques have a long tradition of ethnic separation and declared themselves independent in the Spanish Civil War in the 1930s. Their language is different from that of Spain, and their dress, diet, religion, and economic standards differ from the rest of the country. The result has been a violent movement for separatism resulting in death and injury. Less violent but no less important to Spain have been regional demands for autonomy from the province of Catalonia, located around Barcelona, or in Galicia in northwestern Spain. The Spanish parliament granted limited local autonomy to Catalans by reestablishing their local parliament in Barcelona in 1977 and extended similar powers to the Basques in 1979. (See "The Human Dimension".)

The geographical significance of these regional demands for separatism are obvious. The conditions that lead to success or failure by separatist movements also reflect geographical conditions. Where a region is isolated from the rest of the country by geographic barriers or simple distance, traditional cultural values persist, reinforcing the sense of separation caused by location. In addition, when regions are isolated from the central portion of their respective country, they tend to be less integrated into the economic framework of their nation and may have a lower standard of living. The existence of these two conditions in a country that insists on subjugating local control to central governmental control may result in separatist movements.

CULTURE, BIAS, AND HOSTILITY

One of the major factors that affects separatist movements is the actual or perceived discrimination that a distinct cultural group suffers from the majority population of its country. Such discrimination is not re-

THE HUMAN DIMENSION

Preserving Basque Identity in Spain

Adjacent to France's very southwest corner are three Spanish provinces (Vizcaya, Guipuzcoa, and Alava), which have agitated for independence as a Basque republic. The city of San Sebastian near the French border is a stronghold for Basque separatists since in the past they could flee into France if pressure in Spain became too great.

The Basque separatist group ETA (Euzkadi ta Azkatasuna or Basque Homeland and Liberty) has a record of violence dating from the mid-1960s. Its attacks are sporadic, and are almost always directed against representatives of the central government in Madrid—soldiers, policemen, politicians.

A surface sense of normality reigns in the region. Since a foreign visitor fears little physical harm, he can spend his time sunbathing on San Sebastian's huge, crescent-shaped beach. But as soon as he decides to take a stroll on the boardwalk, he sees signs of the strain. Almost every wall is marked with nationalist graffiti.

In the 1980s the ETA periodically assassinated those they felt represented Madrid, killing more than 25 people in 1985. Locals are unwilling to discuss the ETA, and a strong sense of fear prevails. Even innocuous questions about San Sebastian's history go unanswered in order to "avoid trouble."

In light of Basque history, the reaction is understandable. Though speculation holds that they are the last remnant of the Iberians, no one knows their origin for sure. Surviving successive waves of conquerors, the Basques remained different from their Spanish and French neighbors. For example, their complex language bears no relation to any Romance language, or in fact, any other European tongue. The population of the entire Basque region is about 2.5 million.

Guarding such a distinct identity takes toughness, and the Basques are a tough people. Their architecture is black and white with none of the romantic flashes of the Spanish. Their Roman Catholicism is so solid as to resemble Calvinism. Their favorite sports are pelota and jai alai, the fastest of all ball games.

Ever since Philip II made Madrid the capital in the sixteenth century, Basques have felt stifled and exploited by the central Castile region. Conversation after conversation reveals that the Basques feel they provide an unfair proportion of Spain's brawn and brainpower. During the nineteenth century, the Basques developed much of the country's industry, and the region remains more industrialized than the rest of Spain. Drive down the coast from the French border and smoky plants overwhelm once quaint villages.

As the economy boomed, modern Basque nationalism exploded. In 1976, the Basques lost all of their remaining autonomy. Dissatisfaction with the corrupt, autocratic governments in Madrid increased, and calls for an independent Basque nation were heard.

Interestingly, nationalism did not develop among French Basques. Madrid and Paris agreed to cut the Basque country in two in 1660, and to this day, the 200,000 Basques north of the border remain primarily rural and poor. No industry mars their picturesque farms and ports. The French Basques also never have felt exploited by an autocratic government. Indeed, they are proud that their distinctive beret has become a symbol of Frenchness.

In 1979, the Spanish Basques were granted a large degree of autonomy, including powers over taxation, television, education, and police. The Basque Nationalist Party (PNV), a moderate party which disavows violence, was elected in 1980 to run the autonomous region. Of course, no amount of autonomy will satisfy the ETA hardliners—or the legally represented political party that supports them, the Herri Batasuna. The party says it wants an independent state and polls about 150,000 votes every election.

A majority of Basques, though, realize that an independent Basque state would not be workable. They vote for the PNV or nonnationalist parties. A recent opinion poll showed that more than 75 percent of the Basques oppose the ETA's violent campaign to gain independence. While the Basques have retained a distinctive character, the majority seem to feel that this character has become inseparably intertwined with the Spanish.

---

stricted to the longtime residents of Europe who are culturally different. It is greatest and finds its greatest expression in areas where non–Europeans have migrated to Europe in large numbers. Problems of racism and racial strife are particularly acute in England and France, where an influx of residents of former colonies has resulted in cultural conflict. In England, Indian and Pakistani people have become the focus of blatant racial hostility. Blacks from Jamaica have also been discriminated against in England (Table 5.4). In France, migration of Arab peoples from formerly French Algeria has resulted in racial confrontations. To a lesser extent, the migration of southern Europeans, particularly Yugoslavs and Turks to Germany and other industrialized European nations to meet the demand for cheap, unskilled labor in menial or unpleasant occupations, has resulted in strife and persecution. Eight percent of the French and German population and 6 percent of England's were foreign as of 1990. Ethnic conflict increases with perceived competition for jobs or other economic benefits. Normally, actual violent confrontations are sporadic, but persistent and subtle persecution and prejudice against those who are different exists constantly throughout Europe and rep-

TABLE 5.4 Origin and Volume of Immigration to the United Kingdom 1970–1984

Country	Immigrants (in thousands)
Commonwealth Citizens	1985.7
Commonwealth Countries	1356.0
Australia	314.9
Canada	118.4
New Zealand	108.8
African Countries	270.1
India, Pakistan, Sri Lanka	50.8
West Indies	62.8
Others	337.8
Foreign Countries	798.7
South Africa	165.3
Latin America	35.5
U.S.A.	107.7
Western Europe	215.5
Others	192.2
Aliens	904.7
Commonwealth Countries	31.6
Foreign Countries	873.2
U.S.A.	274.9
Western Europe	248.2
Others	400.5
Total	8,049.3

SOURCE: *Europa Yearbook* (London: Europa Publications Limited, 1974, 1977, 1980, 1984, 1987).

resents a major challenge for European society. If such ethnocentricity cannot be overcome, there is little hope that a unified Europe will ever be possible.

Another form of conflict in Europe partially related to the ethnic minorities is that of radical terrorism. Terrorism in Europe takes several forms, ranging from the nationalism of the Basque party (Euzkadi) of Spain, to the broad revolutionary goals of the Red Brigades of Italy, and to simple criminals interested only in economic profit. Nationalistic movements that have spawned terrorist groups include the Euzkadi (the Basques) and the Irish Republican Army (IRA). The Red Brigades of Italy and the Red Army faction of West Germany (Baader-Meinhof Gang) are based on the premise that capitalistic societies need to be overthrown. Some, like the West German group, are composed of university graduates who are highly literate but who are frustrated by the perceived lack of success of the affluent society of West Germany in dealing with social problems that the radicals view as important. Poverty, lack of adequate housing, or other social goals have prompted such groups to violence as an alternative to the democratic process, which they believe has not adequately changed the capitalistic industrial society of the West. By comparison, the Italian terrorists resent the inability of their political system to deal with the social problems of Italy, but also resent the seeming ineptitude of the governments of Italy themselves. The end result of the terrorist activities, whatever their motivation, is violence and destruction. The issues that have fostered them may well not be resolvable. For the separatist groups, provision of local autonomy or outright independence can lead to an end to the terrorism. For the radical groups of Germany and Italy, no such solutions are apparent. A growing terrorist problem in Europe is associated with immigrants and guest workers from Third World countries. Actions of groups associated with Palestinian, Iranian, Iraqi or other groups include attacks on embassies and transport facilities.

EUROPE IN RETROSPECT

The emergence of Europe as the first and preeminent industrial power in the eighteenth and nineteenth centuries marks the beginning of the change that is reflected in the present world patterns of wealth and poverty. The European experience of industrialization with its middle-class majority enjoying a relatively high standard of living, with leisure time, freedom from famine, and freedom to express opinions and vote for political representation, is one of the major events affecting our globe. As Europe has developed and emerged as a center of wealth, it has become an example of wealth; a market; a supplier; and a focus of resentment on the part of some who do not share the wealth. European society of today represents the end product of a series of changes over 300 years. Unlike less industrialized areas, Europe's countries were able to develop *before* they had an overwhelming population and without the dual burden of colonial administration and exploitation. The European experience seems to indicate to some the importance of revolution in development. What it does not indicate is the precise nature of the revolution or revolutions that are required. The Democratic Revolution may or may not be essential, but to proponents of this view, a revolution of some type is necessary to overcome the inertia of existing economic systems in developing countries.

As European nations wrestle with the problems of a mature industrial society, they may face new revolutions of their own. Changes within Europe may provide a model for other areas of the world. The transitions of the twentieth century that have seen the maturation of the industrial society have prompted a new set of social and political problems, problems that differ in their impact in different regions in Europe. The general geography, both physical and human, of Europe partially explains the regional variation in Europe, but masks important contrasts that make the individual nations and regions of Europe distinctive.

FURTHER READINGS

ASHWORTH, G. J., BATEMAN, M., and BURTENSHAW, D. *The City in Western Europe.* New York: Wiley, 1981.

BAECHLER, JEAN, HALL, JOHN A.; MANN, MICHAEL, eds. *Europe and the Rise of Capitalism.* Oxford: Basil Blackwell. 1988.

BAMFORD, COLIN G. and ROBINSON, H. *Geography of the EEC: A Systematic Economic Approach.* Plymouth, U.K.: MacDonald & Evans, 1983.

BAYLIS, JOHN. "NATO Strategy: The Case for a New Strategic Concept." *International Affairs*, Vol. 64., No. 1. Winter 1987/88, pp. 45–59.

BERENTSEN, W. H. "Regional Policy and Industrial Overspecialization in Lagging Regions." *Growth and Change* 9, 1978, pp. 9–13.

BERG, V. DEN L. et al. *Urban Europe: A Study of Growth and Decline.* Oxford: Pergamon Press, 1982.

BRAUN, AUREL. *Small-State Security in the Balkans.* Totowa, N.J.: Barnes and Noble, 1983.

BROWNE, G. S., ed. *Atlas of Europe: A Profile of Western Europe.* New York: Scribner's, 1974.

BURTENSHAW, D. et al. *The City in West Europe.* New York: Wiley, 1981.

CLOUT, H., ed. *Regional Development in Western Europe.* 2d ed. New York: Wiley, 1981.

———. *Western Europe: Geographical Perspectives.* White Plains, N.Y.: Longman, 1985.

DAHRENDORF, RALF. "The End of the 'Labor Society.'" *Die Zeit* as reprinted in *World Press Review*, March 1983.

DE SOUZA, ANTHONY R. "To Have and to Have Not: Colonialism and Core-Periphery Relations." *Focus*, Fall 1986.

DIEM, AUBREY. "Pollution on the Roof of Europe: Are the Alps Dying?" *The Geographical Magazine*, Vol. 60., No. 6., June 1988, pp. 2–8.

DIEM, A. *Western Europe: A Geographical Analysis.* New York: Wiley, 1979.

FOSTER, C. R., ed. *Nations Without a State: Ethnic Minorities in Western Europe.* New York: Praeger, 1980.

GLEBE, GUNTHER and O'LOUGHLIN, JOHN, eds. *Foreign Minorities in Continental European Cities.* Wiesbaden, West Germany: Franz Steiner Verlag, 1987.

GOTTMAN, J. A. *Geography of Europe.* New York: Holt, Rinehart and Winston, 1969.

JONES, E. L. *The European Miracle: Environments, Economies, and Geopolitics in the History of Europe and East Asia.* Cambridge: Cambridge University Press, 1981.

JORDAN, T. J. *The European Culture Area.* 3d ed. New York: Harper & Row, 1988.

HALL, P., ed. *Europe 2000.* London: Duckworth, 1977.

HALL, P. and HAY, D. *Growth Centres in the European Urban System.* Berkeley: University of California Press, 1980.

HUDSON, R. and LEWIS, J. R., eds. *Regional Planning in Europe.* London: Pion [London Papers in Regional Science II], 1982.

IBERY, B. *Western Europe: A Systematic Human Geography.* New York: Oxford University Press, 1986.

LANSTON, J. *Geographical Change and Industrial Revolution.* Cambridge: Cambridge University Press, 1982.

LIKENS, G. et al. "Acid Rain." *Scientific American* 241, October 1979, pp. 43–51.

KNOX, PAUL L. *The Geography of Western Europe: A Socio-Economic Survey.* Totowa, N.J.: Barnes and Noble, 1984.

MALMSTROM, V. H. *Geography of Europe.* Englewood Cliffs, N.J.: Prentice-Hall, 1971.

NELSON, DANIEL N., ed. *Communism and the Politics of Inequalities.* Lexington, Mass.: D.C. Heath, 1983.

NYSTROM, J. W. and HOFFMAN, G. *The Common Market.* 2d ed. New York: D. Van Nostrand, 1976.

ODELL, P. R. "The Energy Economy of Western Europe: Return to the Use of Indigenous Resources." *Geografisch Tijdschrift* 8, 1979, pp. 1–14.

Organization for Economic Cooperation and Development. *The State of the Environment in OECD Member Countries.* Paris: OECD, 1979.

PARKER, G. *An Economic Geography of the Common Market.* New York: Praeger, 1969.

———. *The Logic of Unity: A Geography of the European Economic Community.* 2d ed. (Longman Geography Paperbacks). New York: Longman, 1981.

PARKER, GEOFFREY. *A Political Geography of Community Europe.* London: Butterworths, 1983.

PAWLICK, THOMAS. *The Killing Rain: The Global Threat of Acid Rain.* San Francisco: Sierra Club Books, 1984.

ROSENCRANZ, ARMIN. "The Problem of Transboundary Pollution." *Environment*, June 1980.

ROSTOW, W. W. *How It All Began: Origins of the Modern Economy.* New York: McGraw-Hill, 1975.

SEERS, DUDLEY and OSTROM, KJELL, eds. *The Crises of the European Regions.* New York: St. Martin's Press, 1983.

SOMMERS, L. "Cities of Western Europe." In *Cities of the World: World Regional Urban Development.* Edited by S. Brunn and J. Williams. New York: Harper & Row, 1983, pp. 84–121.

SULLIVAN, S. "The Decline of Europe." *Newsweek*, April 9, 1984, pp. 44–56.

THERBORN, GORAN. "Migration and Western Europe: The Old World Turning New." *Science*, September 4, 1987.

WHEELER, J. and MULLER, P. *Economic Geography.* New York: Wiley, 1981.

WHITE, P. *The West European City: A Social Geography.* New York: Longman, 1984.

West Germany.

EUROPE REGIONAL PROFILES: (1990)

Population (millions)	499.6	Percent of Labor Force Employed in Industry		Eligible Children in Primary School (%)	
Northern	27.1				
Western	213.5	Northern	35	Northern	99
Southern	121	Western	42	Western	100
Eastern	137	Southern	38	Southern	100
Growth Rate (%)		Eastern	44	Eastern	100
Northern	0.2	Percent of Labor Force Employed in Service		Percent Urban	
Western	0.2			Northern	85
Southern	0.3	Northern	57	Western	84
Eastern	0.4	Western	52	Southern	68
Years to Double Population		Southern	41	Eastern	63
Northern	373	Eastern	39	Percent Annual Urban Growth 1980–1985	
Western	393	Energy Consumption Per Capita (kilograms)			
Southern	219			Northern	.7
Eastern	190	Northern	5998	Western	0.90
Life Expectancy (years)		Western	5114	Southern	1.97
Northern	75	Southern	2387	Eastern	1.8
Western	75	Eastern	4487	Infant Mortality (per thousand)	
Southern	74	Calories (daily)		Northern	9
Eastern	71	Northern	3267	Western	8
Percent of Labor Force Employed in Agriculture		Western	3527	Southern	16
		Southern	3455	Eastern	18
Northern	8	Eastern	3389	Per Capita GNP	
Western	6	Literacy Rate (%)		Northern	$9950
Southern	21	Northern	99	Western	11380
Eastern	18	Western	99	Southern	5660
		Southern	94	Eastern	5104
		Eastern	99		

6

Regional Variations in Development in Europe

MAJOR CONCEPTS

Invisible exports / Economies of scale
Cooperative farming / Shatter belt
Balkanization / Cultural fragmentation
Nationalization / Buffer zone / Irish curve
Irredentism / Islam / Insular

IMPORTANT ISSUES

Client-dependent relations of Eastern Europe and the USSR
Technical obsolescence
Resource exhaustion
Environmental problems in industrialized nations
Unequal development

Geographical Characteristics of Europe's Regions

1. Western and Northern Europe have higher standards of living and greater industrialization than Eastern and Southern Europe.
2. England, France, and the Federal Republic of Germany dominate the geography and economy of Western Europe.
3. Italy dominates the countries of Southern Europe economically.
4. Economic growth in Southern Europe associated with membership in the EEC may raise substantially the Mediterranean region's standard of living.
5. Eastern Europe is distinct from the rest of Europe because of centrally planned economies and communist governments.

The emergence of Europe as the core of the industrialized, wealthy, developed world in the nineteenth century and its continued development in the twentieth century has created an uneven pattern of development in Europe. Countries of Europe that were first to industrialize, particularly the United Kingdom, remain major industrial powers, but their individual standard of living has been surpassed by others, such as some Scandinavian nations and Switzerland, which enjoy the highest per capita incomes in Europe. At the other end of the spectrum are some of the Mediterranean countries, which have low per capita incomes. The uneven pattern of economic development in Western Europe reflects each individual country's unique combination of geography, social and political organization, resources, and accessibility.

It is difficult to divide Europe into homogeneous regions on the basis of economic development, because of the complex variation in economies within individual nations. Regionalizing Europe in terms of culture presents problems, because each independent country is largely distinct from surrounding countries in language, history, and values. Division simply on the basis of geography separates nations that are similar and combines nations with important differences. Because of the complexity of the European physical, cultural, and economic map, it is common to treat each individual nation of Europe separately. An approach such as this provides details on each nation but often ignores the unifying forces that have affected the entire area.

Overcoming the **cultural fragmentation** that has resulted from the long history of human activity in Europe is almost impossible, but for the purposes of illustrating the process of development, we will discuss four broad regions: the Scandinavian North; Western Europe; the Mediterranean lands of Southern Europe; and the Eastern European nations (Figure 6.1). This division of the thirty nations of Europe (including the microstates) is not as arbitrary as it may appear at first glance. The Scandinavian nations have important climatic and cultural similarities that transcend their national boundaries. They have a high level of industrialization, enjoy high per capita incomes, literacy, and life expectancy and have a very low infant mortality, minority group membership, and poverty.

The nations of Western Europe include the United Kingdom, Ireland, France, West Germany, Belgium, Netherlands, Luxembourg, Austria, and Switzerland. Geographically this region is extremely complex, ranging from the **insular** (island) location of the United Kingdom and Ireland to the mountainous areas of Switzerland and Austria. Culturally it is just as complex, with a dozen major languages and dialects. The justification for regarding this group of countries as a region is the combination of climate, economic development, and transportation networks that together provide a degree of uniformity. Grouping together nations with differences as broad as those between Austria and the United Kingdom is offset by the similarities within the group. All nations of Western Europe generally enjoy the characteristics of industrialized regions. As the Industrial Revolution developed in Great Britain, Germany, Belgium and Netherlands, and France, it spread to the neighboring nations of Scandinavia and to Austria and Switzerland.

Cultural fragmentation: The presence in one region or/country of a host of cultural traits as opposed to dominance of one trait over a large area.

Insular: Referring to an island or locations having island-like characteristics.

THE WORLD OF WEALTH

Figure 6.1 Major regional groupings of Europe based on relative level of economic development. Northern and western European countries are generally more industrialized than those of eastern and southern Europe.

The Industrial Revolution has had less impact on the economies of the nations of Southern Europe, and Eastern Europe was essentially typical of less industrialized nations until after World War II. The nations of southern Europe (Portugal, Spain, Italy, and Greece) are separated from the rest of Europe by landform barriers. The climate is distinctive and further sets the region apart. The Industrial Revolution began later in this region, and rural areas still have many of the characteristics of less industrialized countries. Literacy rates are the lowest in Europe, the percentage of population engaged in secondary and tertiary occupations is lower than in the industrial Northern and Western European nations, and per capita incomes are markedly lower.

The nations of Eastern Europe comprise a distinct region based on their adoption of a centrally planned, state-controlled model of development patterned after that of the Soviet Union. Prior to World War II this region had only a few industrialized areas, while the majority of the population was engaged in agriculture. The level of economic development in Eastern Europe was similar to that found in some less industrialized areas today. Since the end of World War II important development has occurred in Eastern Europe, with particularly notable advances in the areas of medical care, education, and provision of basic human needs. Today the literacy levels, population growth rates, and life expectancy rival those of Western and Northern Europe. However, per capita incomes are significantly lower in Eastern Europe, and a high proportion of the population is still engaged in agriculture.

THE NATIONS OF WESTERN EUROPE

The Nations of Western Europe were the birthplace of the Industrial Revolution and the source for the diffusion of European technology to many areas of the world. The nations of Western Europe were the imperial powers that dominated much of the less industrial world, and the region typifies mature industrial economies. The major powers are the United Kingdom, the Federal Republic of Germany (West Germany) and France, followed by the Benelux (Belgium, Netherlands, Luxembourg) nations. Switzerland is a major economic center with a unique focus on banking and tourism, and Austria has developed less rapidly because of its political history.

THE BRITISH ISLES

The nations that comprise the British Isles (Ireland and the United Kingdom) are different from the other nations of Western Europe for a number of reasons, one of which is geographic location. Their insular location has provided major advantages to their development. The British Isles are within 21 miles of the mainland of Europe at the closest point, but this distance has been sufficient to minimize involvement of the islands in the conflicts on the mainland. (Although this isolation is decreasing because of membership in the EC and the Eurotunnel under the English Channel, which will connect England's transport system to that of France.) The island nature of these nations minimized the need for a large standing army to defend borders, prevented involvement in each minor dispute on the mainland, and provided opportunities for trade that were unrivaled in Western Europe. Even their political and social systems developed differently from those of the European mainland. Most important was the development of a democratic parliamentary government with a relatively powerless monarchy. The political map of the British Isles encompasses two countries and four nations of people. The largest political state is the United Kingdom, which includes England, Wales, Scotland, and Northern Ireland. The state of Ireland encompasses the bulk of the island of Ireland, and most of the Irish nation. The industrial development of the British Isles has historically been centered in the United Kingdom.

The landforms of the United Kingdom can be divided into the northern highlands of Scotland, Northern Ireland, and northern England, and the lowlands of southern England. The economic importance of the United Kingdom began with the rise of London as a major trading center for both the British Isles and the then known world, long before the Industrial Revolution. The Thames River provided access to the interior of the British lowlands, and London's location opposite the major river systems of Europe made it the major port area for shipping from the British Isles to Europe. The English Channel is shallower along the eastern shore near France, so that north-south shipping stays closer to England. London developed into the world's largest city at the beginning of the Industrial Revolution, and the emergence of industrial centers in the Midlands area of England produced industrial surpluses that could be exported through London to other areas.

The first industrial centers were located in today's Leeds-Bradford urban area, using waterpower from the Pennine mountain chain to produce woolens, and in Manchester, on the west side of the Pennines, where cotton textiles were produced (Figure 6.2). Coal and iron ore deposits fostered development of iron and steel manufacturing at Birmingham, Sheffield, Newcastle-upon-Tyne, and Middlesborough. The industrial area stretching from Birmingham north to the Bradford-Leeds area and west from Leeds to Liverpool is the second most important industrial region in Europe, exceeded only by the Ruhr in West Germany. City location reflects local (wool in the Leeds-Bradford area), or imported (cotton for Manchester), raw materials, or head-link functions (Liverpool).

The impact of the Industrial Revolution on the cities of the United Kingdom was profound. Cities such as Birmingham, located near raw materials or power sources, grew up in this era. With the exception of Liverpool and London major cities of the United Kingdom reflect location for industrial purposes (Figure 6.3). The cities themselves are specialized in their type of manufacturing, with Leeds remaining the major woolen manufacturer in the United Kingdom and Manchester still the dominant cotton textile center. Birmingham and its suburbs are an important producer of iron and steel for use in a variety of products including aircraft and other transport equipment. Sheffield is a focus for specialty steels, particularly stainless steel. On the east coast Newcastle-upon-Tyne is a center of shipbuilding and chemical industries.

London antedates the Industrial Revolution, and remains the dominant urban area of the United Kingdom today. Industrialization fostered more trade and finance, and these activities were concentrated in a small portion of London. In the past London was the center for the government and financial activities of not only the United Kingdom but for the world's largest colonial empire. Today it remains the focus of essentially all company headquarters in the United Kingdom and is a major world trade center and financial capital. The greater London metropolitan area, with a population of 13 million, contains nearly one fourth of the population of the United Kingdom.

Other areas of the United Kingdom are less industrialized than the area around the Pennines and southern England around London. Wales has coal, but has been isolated from England by mountains and cultural differences. Much of the Welsh landscape is still devoted to grazing and other agricultural activities, although the majority of the population lives in towns and cities. Scotland is mostly uplands unsuited for anything but grazing activities. The economic center of Scotland is the Scottish lowland where coal fields provided the basis for industrialization. Coupled with nearby iron ore, the Scottish coal was sufficient to provide Scotland a major manufacturing base focusing on iron and steel for shipbuilding, and textile manufactures. Glasgow and Edinburgh are the dominant centers of Scotland, but discovery and development of petroleum and natural gas in the North Sea have prompted development in the port of Aberdeen and in the Shetland and Orkney Islands (Figure 6.3).

The United Kingdom today has problems typical of early industrialized regions. The lead in industrialization was not sufficient for the country to maintain its role as the dominant industrial power of the world.

London is the heart of the United Kingdom, and the Houses of Parliament and Big Ben are symbolic of the emergence of this small country to a position of world influence.

Figure 6.2 Major cities and industrial regions of the United Kingdom.

First it was surpassed by Germany, then the United States, the Soviet Union, and Japan. Today its iron and steel production and other manufactures are challenged by other nations. The main problems facing the United Kingdom are obsolescence of industries, a declining resource base, and changing social organization. Location of the industrial cities on the coal beds was a rational geographic decision when iron and steel was the major industry and relied exclusively on local coal resources for energy. Today coal produces only one quarter of the total energy in the United Kingdom as petroleum and nuclear power are substituted. Petroleum discoveries in the North Sea have made the United Kingdom an important petroleum producer. Although the reserves of petroleum are much smaller than those of the Middle East, the United Kingdom was the world's seventh-largest producer in 1989. The factories that were built in the past 100 years are technologically obsolete and structurally inadequate. Industry is moving to port cities where imported raw materials of coal, iron ore, and petroleum are available. Many of the old industrial centers are suffering from economic stagnation and high unemployment. Union demands to maintain jobs in mines and factories that are not economically competitive lead to strikes and demands for government assist-

THE WORLD OF WEALTH

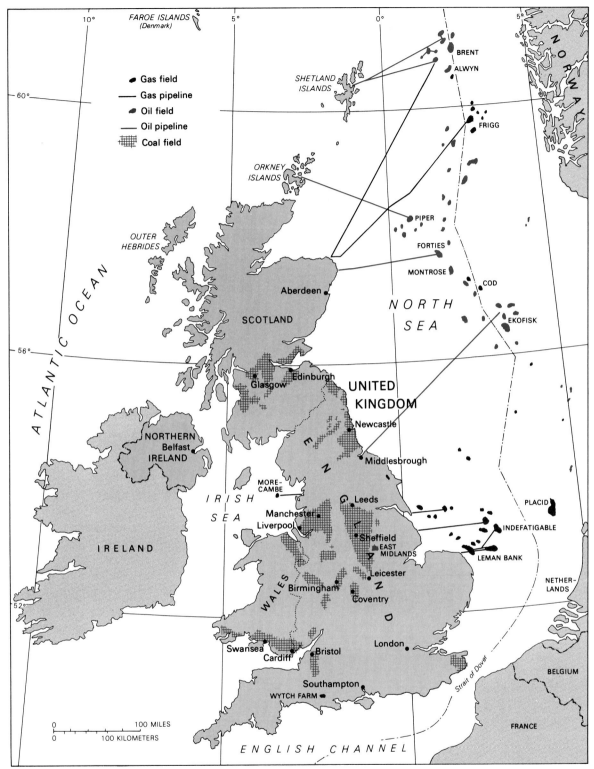

Figure 6.3 Energy resources of the British Isles, including those in the North Sea that the United Kingdom claims.

The old industrial cities and towns of Western Europe epitomize problems of the earliest industrial centers. Obsolescence and competition have created economic problems for towns such as Slaithwaite, Yorkshire, England.

ance. Between 1950 and the 1970s, the British government responded to these problems by **nationalizing** basic industries such as coal, iron and steel, railroads, and telephone and telegraph. In the 1980s in a major change, the United Kingdom began selling the state-owned industries to private companies, a process known as **privatization.** This is one of the most important changes now affecting the economic development of the mature industrial economy of the United Kingdom. The challenge for the United Kingdom is to utilize the country's greatest resource, its highly skilled, educated population, to develop areas of manufacturing where the skills of 200 years' experience with industrialization can provide the nation a competitive edge.

Industrialization also affected the agriculture of the United Kingdom. Although the nation produces only about 50 percent of its food, it does so with one of the most efficient agricultural systems in the world. Agriculture is highly mechanized, and less than 3 percent of the population is engaged in agriculture. Yields per acre are among the highest in the world, but the high yields are offset by the high population density of the island republic. Major crops include wheat, barley, oats, potatoes, and sugar beets. Much of the British Isles is suitable only for grazing because of the marine west coast climate and the topography. The climate handicaps production of many agricultural items, and tropical crops are notable among British imports.

The Republic of Ireland (Eire) came into existence as an independent country in 1921 as part of the British Commonwealth of Nations. The Irish are distinct in language and culture from the English, being predominantly Catholic compared to England's Protestant majority. Ireland was exploited as a colony until it gained independence, and is still closely tied to the United Kingdom by trade and other economic bonds. Ireland's history is somewhat analogous to that of other colonial possessions of Western European powers. Industrialization has been late in coming to the island because of its dependency relationship to the United Kingdom and the physical barrier of the island of Great Britain between it and Western Europe.

Nationalization: The process by which ownership of private property is assumed by the government.

Privatization: The process of transferring ownership of government operated activities to private ownership.

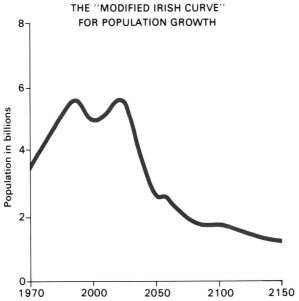

Figure 6.4 The "Irish Curve" of population growth reflects the experience of Ireland where population decreased rapidly as a result of famine and disease associated with potato blight. It is hypothesized that serious disasters cause temporary population decline, followed by long-term decline and subsequent stabilization at lower levels.

The problems of Ireland were compounded by rapid population growth in the nineteenth century as it became a single-crop economy based on the Irish potato. Then disease (potato blight) in the mid-nineteenth century caused famine and mass migration from the island, cutting the Irish population to less than half. Since that time Ireland's economy has been much lower than that of other Western European nations, with a per capita income at present only slightly more than half that of the United Kingdom. Ireland has managed to develop a slow population growth rate even though it has primarily been an agrarian, Catholic, and relatively poor state. The so-called Irish curve is hypothesized as one alternative outcome of present world population growth (Figure 6.4). The Irish curve shows rapid population growth followed by a precipitous drop in numbers and stabilization at a lower level. In Ireland's case, at first, adoption of the potato as a food crop allowed the population to double and triple in size. The potato blight effectively removed the resource base of the population, and subsequent generations have maintained the population at a low level. Some observers feel that unless worldwide population is controlled, it will ultimately decline rapidly as did the Irish population and then maintain equilibrium at a lower level.

Ireland continues to reflect its colonial heritage in that the Irish people are divided between Northern Ireland, part of the United Kingdom, and the Republic of Ireland. Approximately 60 percent of the population of Northern Ireland is of non-Irish ancestry, but the minority Irish maintain that the geographical realities of the island make unification of the Republic of Ireland with North Ireland sensible. Conflict over this issue is a common occurrence in Northern Ireland today.

THE CONTINENTAL NATIONS OF EUROPE

Separated from the United Kingdom and Ireland by the English Channel and the North Sea, the continental nations of Western Europe present a complex continuum of development. Belgium, the Netherlands, and what is now Federal Republic of Germany (West Germany) were the focus of early industrialization on the continent. The single most important industrial area of Europe today is based on the coal of the Ruhr region and from there extends through northern France and the Netherlands and Belgium (Figure 6.5).

The core of this industrial region is the Ruhr of Germany, a concentrated manufacturing region with more than 10 million people in over 50 cities. Major cities include Cologne, Dortmund, Essen, Düsseldorf, and Duisburg, all with populations between 500,000 and 1 million. This is the focal point for the German industrial economy with its concentration of iron and steel manufacturers, textile manufacturers, automobile manufacturers, and the range of industrial products found in modern industrial societies.

The industrial development of France is based on developments from the Middle Ages, especially silks and textiles, and subsequent iron and steel manufacture. France has in the past had heavy industry concentrated in the north near Nancy, based on iron ore, and Lille, based on coal resources. Paris was the number one industrial area for France historically, and remains so today. The industrial power of France has never equaled that of Germany, but France is much more important in terms of agricultural productivity than West Germany.

Western Europe extends from the European lowlands into the mountainous terrain of Switzerland and Austria, but the bulk of the population is concentrated in the European lowland. The largest nation in area is France with 213,000 square miles (551,670 km²), but West Germany has the largest population at 61.2 million. The other nations of Western Europe are very small in area—Luxembourg has only 990 square miles of area (2,564 km²) but nearly one-half million residents. Belgium has only 12,000 square miles (31,080 km²) but 9.9 million people. The Netherlands has only

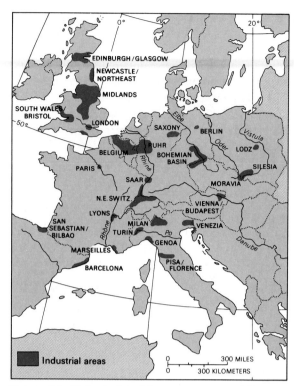

Figure 6.5 Major industrial concentrations of Europe indicating the dominance of Western Europe.

13,000 square miles (32,370 km²) but a population of 14.8 million. Switzerland has nearly 16,000 square miles (41,440 km²) of area but a population of 6.6 million, and Austria's 32,374 square mile area (83,848 km²) is home for 7.6 million people. Thus, the five smallest nations of continental Western Europe have only 39.3 million people, compared to the 120 million people in France and West Germany.

All these nations have high per capita incomes, with Switzerland leading the group at $18,000 per person per year. The lowest per capita income among the continental nations of Western Europe is in Belgium, but even here it is nearly $10,000. Literacy rates are near 100 percent in all these nations, and other characteristics of industrial nations are concentrated in this area. Industry is highly oriented toward production of metals, chemicals, transport equipment, textiles, and high-technology items.

The development of Germany and France illustrates the problems and potentials associated with geographical location. France is the oldest nation of Europe, and Germany is one of the most recent, having been formed in 1871. The two nations have different site-situation relationships, and at first glance it may appear that France has the better. France has ports on the Atlantic and the Mediterranean Sea, a much larger land area, and a climate more suited to agriculture than does West Germany. West Germany has a more accessible location, however, because of the navigability of the major rivers, particularly the Rhine and the Elbe River in the north with Hamburg as a major port. France has suffered historically from the nature of the rivers that drain the French countryside. French rivers are shallower and less suited for navigation than either the Rhine or the Elbe, and no port in France is as ideally suited as the port of Rotterdam in the Netherlands is for Germany. In terms of simple accessibility Germany had, and still has, the most accessible location.

Unfortunately, the very accessibility of Germany has contributed to major problems in the past. Germany is located in the European plain, and there are no natural boundaries within which the German peoples could form a state. The lack of boundaries led Germans to expand eastward into Poland, Czechoslovakia, and even the Soviet Union. Attempts to unify the German people were a significant causal element in both World War I and World War II as **irredentism** developed among the German people and government. The lack of definitive boundaries has been compounded by the arbitrary division of Germany into East and West at the end of World War II. East Germany (the German Democratic Republic) is much smaller and is dominated by the old capital of Berlin. The economy of East Germany historically was more agricultural than that of West Germany with its Ruhr area, and the standard of living in East Germany has never rivaled that of West Germany.

The largest city on the mainland of Europe is Paris with a population of 9 million (including suburbs). Nevertheless, France as a whole is less urbanized than West Germany. France has a much larger agricultural population (8 percent of the labor force versus 4 percent for West Germany). Paris became the focus of dominance because of its site-situation relationships. The location of the initial settlement at Paris was on the island in the Seine River known as the Ile de la Cité. Several tributaries join the Seine near Paris, providing access to major regions in the country. The area surrounding Paris is noted for its soil fertility, and the productive agricultural hinterland contributed to the emergence of Paris as a major continental city before the Industrial Revolution. The dominance of Paris was ensured by actions of the French political leaders who established it as the focus of activity in the French nation. Today Paris is the major manufac-

Irredentism: Claims on lands culturally or historically related to a nation that are now occupied by a foreign government.

Grape production on the south-facing slopes of the Rhine River. Note the number of vessels using the Rhine for transport also.

turing center in France, specializing in such items as transport equipment, perfumes and other toiletries, and other high-value products. Marseille, Lyon, and Lille, France's other major cities, barely reach 1 million population. By comparison West Germany has no individual city that rivals Paris, but many of 500,000 to 1 million.

The problems of Germany and France are similar in that both are small nations by world standards (though large by European standards) with mature economies. France is a major agricultural producer and exporter of such items as grapes and wine, potatoes, wheat, apples, and sugar beets. Wheat is the most important crop in France, and France normally ranks in the top five wheat-producing nations in the world. The production of wine has long been associated with France, and export of French wines is a major part of the country's international trade.

Germany by comparison does not produce sufficient food for its population and must import significant quantities of agricultural products. Major crops in West Germany include potatoes, sugar beets, small grains, and forage crops. Along south-facing slopes of the Rhine Valley grapes are an important crop and provide Germany with significant wine production. The production of livestock is the most valuable agricultural activity in Germany today. Agriculture is highly subsidized in all of the countries that are members of the EC. Food surpluses that result (especially cheese and dried milk) are a major source of contention among EC members.

In comparison to the dominant countries of Germany and France, the other nations of Western Europe are small but highly important. The Netherlands, Belgium, and Luxembourg illustrate that high population densities, a limited land area, and limited resource base do not determine a low standard of living. These nations have a per capita income rivaling that of the United States. This has been achieved through intensification of industry and agriculture. The Netherlands is synonymous with human ability to obtain resources for survival in marginal lands. The expansion of Dutch agricultural land through reclaiming lands from the sea (*polders*), scientific farming, and application of capital-intensive measures are world-renowned. The Dutch have added more than 500,000 acres to their farmland through reclaiming lands from the sea (Figure 6.6). The reclaimed land is very expensive, and a national land use plan has been developed to ensure that it will be used wisely. Most of the Dutch are urban dwellers concentrated in the megalopolis known as the Randstaad. It extends from Rotterdam and The Hague in the south to Amsterdam and Haarlem in the north. The Dutch have a high population growth rate for an industrialized nation (0.4 percent per year), which, coupled with the small area and the large population, makes problems of urban life a significant issue for the Dutch today. The need to provide open space for recreation and to maintain precious farmland in the face of ever-growing urban populations is one of their greatest challenges. Rotterdam and Amsterdam are the focus of the economy of the Netherlands, and Rotterdam is the busiest port in Western Europe. The industries associated with imports of raw materials, particularly petroleum and petrochemicals, provide much of the employment in this urban agglomeration.

Belgium has many of the same problems as the Netherlands, but does not have quite the population pressure. The problems of Belgium are compounded by a cultural division of the country into Flemish and Walloon. The Flemish are Dutch-speaking people in northern Belgium and comprise 60 percent of the population. The Walloons speak a distinctive form of French and live in southern Belgium. This cultural division is an important issue that has led to a plan to separate Belgium into three semiautonomous regions: Flemish north, the Walloon south, and the Brussels region as a separate entity because of its unique role in political, economic, and cultural activities in the nation.

None of Belgium's cities rivals Rotterdam in terms of trade, but the most important is Antwerp on the Schelde River. In the past Antwerp was the focus of

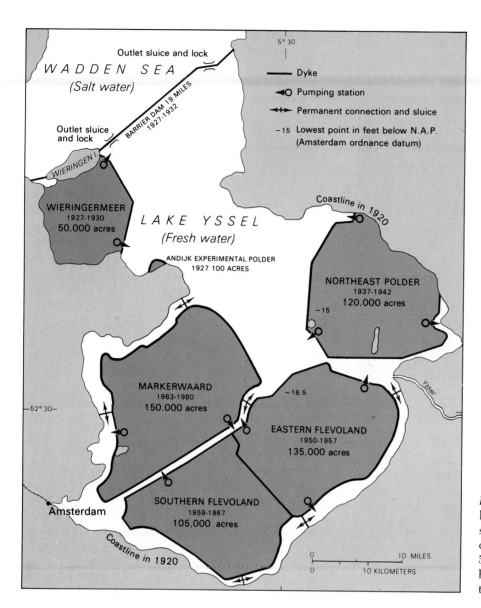

Figure 6.6 Reclamation of land from the former saltwater Zuider Zee has created approximately 560,000 acres (216,000 hectares) of farm land and the fresh water Lake Yssel.

Belgium's trading activities, but since the Schelde passes through Dutch territory, it suffered a period of isolation when the river was cut off by Dutch forces during conflicts between the two nations. Today Antwerp is the major port for Belgium but is insignificant compared to Rotterdam. Brussels is the capital of Belgium and is the headquarters of international organizations including the North Atlantic Treaty Organization and important functions of the European Common Market.

The location of the Low Countries, as Belgium and the Netherlands are called, has provided the opportunity for them to become "middlemen" for the industrialized nations of Europe. The Netherlands particularly has capitalized on this role in the port of Rotterdam and has become a focal point for importing raw materials for Western Europe and exporting manufactured items. The construction of the Europort downstream from Rotterdam signifies the importance of this city to the economies of Western Europe. The Netherlands and Belgium have capitalized on their location to import low-cost raw materials and export high-value manufactured products. As a simple example, the Dutch import cacao (from which chocolate is made) from Africa and reexport high-value chocolates.

Small size and generally mountainous terrain have not handicapped Switzerland's or Austria's emergence as members of the industrial world. Switzerland has capitalized on a long history of neutrality to become the banking center for much of the world. The industries of both Switzerland and Austria focus on high-quality manufactured products such as watches, electronics, cameras, and high-quality tools. Both countries have developed as important tourist centers, providing additional *invisible exports*. Invisible exports bring income into a country for "products" such as the technical expertise of the population, the scenery, or banking services that do not actually leave the country. Unlike exports of automobiles from Germany, wine from France, or cheese and chocolate from the Netherlands, invisible exports do not require supplies of raw materials. Invisible exports benefit a country by providing jobs and income while not demanding inputs other than the skills of the local population.

The issues and problems facing the industrialized nations of Western Europe are varied, ranging from problems of overcapacity in the iron and steel industry with resulting unemployment and nationalization to agriculture-related issues. The highly productive and technologically efficient European agriculture presents the dilemma of agricultural surpluses in a region that still relies heavily on agricultural imports. Western European farmers produce surpluses of milk, wine, and other products, but the high cost of production makes them noncompetitive in world markets. Major problems of the differing levels of efficiency and production in the member nations have faced the EC. Agricultural problems fostered establishment of a host of EC regulations, regulating such things as standards for green onions, support prices for milk, and the amount of sugar that wine producers can add to grapes.

The high standard of living in Western Europe has attracted migrant labor from Southern Europe, Turkey, Yugoslavia, and North Africa. These migrants provide cheap labor, but their treatment in Western Europe is a major problem for the region. In some countries they can become residents, whereas in others they are not even eligible for the social services that the country provides its own residents. Reliance upon imports poses a constant challenge to these nations, and the increasing cost of petroleum is a particular problem. Competition from newly industrialized areas threatens older industrial activities with less efficient or obsolete equipment. Government programs to ensure employment protected obsolete factories in the past, but in doing so, state-owned and operated industries and protective tariffs resulted in higher costs to Europeans for many manufactured

Many towns in Europe are located on rivers. River transportation is still important, as seen in this photo of the Moselle River in Germany.

THE HUMAN DIMENSION

A "Completely Normal Austrian"

Hedwig Bachler works in the Post Office. Postal workers are to be found all over the world, and it is not difficult to draw comparisons. Miss Hedwig Bachler is 25 years old and lives in the beautiful and historic town of Steyr, in Upper Austria, a thousand-year-old community with a population of 40,000 and an ancient center of iron and steel making. She is the eldest of three children from a working-class family. Her father works with the firm of GFM, making special machinery that is in demand all over the world. For example, if you happen to drive a Cadillac, a Peugeot, a Fiat, or a Volvo, the chances are that father Bachler had something to do with the making of its engine.

Hedwig is actually a trained nursery-school teacher, a "Kindergartnerin," and she worked at this job in Vienna initially. Then, when she came back to Steyr, all the available posts were occupied, so she applied to the Post Office for a job, and for four years now she has been practising her patience on customers at the counter instead of on children. Presently a nonestablished civil servant, she can expect to be upgraded and employed on a permanent basis early next year after passing a second-grade service examination. This future upgrade does not bring any immediate financial improvement, but the advantage is that, as an established civil servant, she will be practically secure from unemployment (if she does not commit any serious misdemeanor), and if she stays with the Austrian Post Office until the age of 60, she can expect to retire with a pension of 80 percent of her last salary scale.

That amount will naturally be somewhat higher than it is today, for Hedwig Bachler has to make do with a gross monthly salary of 8600 Austrian schillings (about $500), which is, however, paid fourteen times a year, as is usual in Austria. After the necessary tax deductions, her yearly income is about 90,500 schillings ($5,300), or 7500 schillings per calendar month ($420).

That amount also can and will change in the course of the years, when Hedwig moves up to the "gehobenen Fachdienst," the next rung up the ladder after the next examination. Her boss sets an excellent example here, for the head of the Steyr Post Office, one of the largest in Austria, is a woman who also began in the "Group 1c" like Hedwig Bachler. Today, Frau Danzcu "rules" over more than 160 women and men, and her personal secretary is a man.

But Hedwig has a bit to go before she reaches that stage. How does she manage on her income? Around 2000 schillings go for housing—for that she has the good fortune to be able to rent a "Garconniere" (one and a half rooms plus kitchen and bathroom) in a house belonging to Steyr City Council for only 1200 schillings, including heating—a blessing in oft chilly Steyr. There are plenty of new flats to rent in the town, but the rents are often as high as 4000 or 5000 schillings, and that is a lot in an industrial town that has been hard hit by recession.

Two thousand schillings go quickly through one's stomach, whereby the meal vouchers from the Post Office help to some extent. The rest is consumed by the usual items—clothing, cosmetics, and, not least, the little Ford car for which the young woman saved for so long. And something has to remain for the holidays, presently four weeks in the year, later rising to five.

One week of the holiday she will spend with her fiance on a skiing holiday in the Salzburg mountains, where the young man, training to be a teacher of economics in Linz, has a holiday job as a ski instructor. In summer there will be three weeks in not-too-expensive private accommodation on the Italian or Yugoslavian coast—for swimming, because Hedwig is athletic, does gymnastics in the Post Office Sporting Club, and in winter is an enthusiastic ice skater.

Source: Austria Today, April 1984, p. 50–51.

---

items. In consequence, all of the countries of Western Europe are experimenting with privatization.

THE COUNTRIES OF NORTHERN EUROPE

The northern countries of Europe include Iceland, Norway, Sweden, Finland, and Denmark. Culturally there is a strong similarity among these nations, since more than 80 percent of the people are members of the Lutheran church, and with the exception of Finland the languages are Germanic in orgin. Finnish is a Uralic language, but the Finns are still predominantly Lutheran. The Industrial Revolution came late to these countries. They all have a high-latitude location, small populations, and relatively large areas with few inhabitants. The only sizable areas of lowland are located in the south of Sweden, the south of Finland, and in Denmark. Norway is dominated by the highlands of northern Europe, and in all these nations landforms have been affected by glaciation. The economies of the individual nations differ in the degree of reliance on agriculture, fishing, or industry, but all have very high standards of living. Each individual country has an almost homogenous population with few ethnic minorities. Finland is an exception, as it

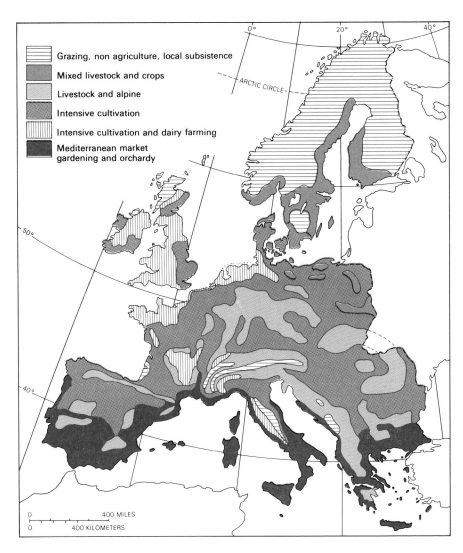

Figure 6.7 Major agricultural regions of Europe. Distinct regions in Europe specialize in a range of products from wheat to olive oil.

has a significant number of Swedes in the southwest. The populations of the countries are small, with Sweden's being the largest (8.4 million) and the 200,000 of Iceland being the smallest. The growth rates in all of the nations are under 0.2 percent per year, with the average being 0.15 percent per year. Literacy rates are nearly 100 percent, and per capita incomes rival those of all industrial nations.

Denmark is agriculturally the most important of this region, with three quarters of the total land area cultivated (Figure 6.7). Denmark exports high-quality agricultural items such as specialty meats, eggs, and vegetables to the other Common Market countries, but relies upon imports of low-cost grains from the United States, Canada, and Australia for feed. The Danes have achieved a highly efficient and productive agricultural system in spite of relatively small farms. The mechanism that contributed to the agricultural success of Denmark was the development of agricultural **cooperatives.** The advantages of cooperatives lie in providing *economies of scale*. A small farm of 20 acres or less could not justify the cost of large harvesting equipment or produce enough to guarantee a reliable supply for a specific market. By pooling their resources, the members of a cooperative can purchase one large harvester for a number of farmers, and by pooling their products, have a large and stable supply for market. The cooperatives provide technical advice on crops, accounting procedures, and quality control for the individual members. Cooperatives have been extremely successful in Denmark and have allowed the Danes to compete on a world scale with their agricultural products.

Cooperatives: In cooperatives a group of farmers pool their resources to purchase equipment and to market their products, but individuals farm their own land.

The majority of the people of Denmark are engaged in industry and reside in urban areas. Fully one fourth of the 5.1 million people of Denmark live in one city, Copenhagen. Copenhagen is the center of industrial activity in Denmark, and no other Danish city rivals it in size or importance.

Finland is another small Northern European country with a high standard of living based on extractive industries. Finnish industry is heavily oriented to production of pulp, paper, lumber, prefabricated housing, and other items that can utilize the large forests of the nation. Agricultural products, furs and hides, and other items are also important exports. The Finnish population is concentrated in the southern plains area of the nation, and the northern part of the country is essentially unoccupied. Finland's economy has been handicapped in the last decade by enforced neutrality, which prevents it from joining the Common Market, its logical trading partner. Treaties with the Soviet Union prohibit Finland from joining the rest of Western Europe in such custom unions, and in the future the trade of Finland may move increasingly to the Soviet Union.

Sweden is the largest Northern European country. It has important resources of high-grade iron ore (at Kiruna), manganese, copper, zinc, and lead. Southern Sweden is an important agricultural producer of dairy cattle, poultry, wheat, sugar beets, and forage crops for livestock. The industrial base is diversified with manufacture of such disparate items as automobiles and jet aircraft, surgical instruments and machine tools.

Norway is quite different from Sweden in that little of the nation's territory is suitable for cultivation. The coastal area of western Norway is dissected by fjords, which make excellent sites for harbors, but the lack of a hinterland to provide the economic base for harbors prevents the fjords from being important on a world scale. The bulk of the population of Norway is concentrated in the southeast where the land is less rugged and the wide valleys provide space for cities and industrial development. Norway has always relied more extensively on the ocean than other Scandinavian nations and remains a major fish producer for Europe and the world. Norway has also become a major producer of petroleum from finds in the North Sea. Provision of equipment for the recovery of the petroleum from the North Sea has prompted economic growth in the ports of southwestern Norway such as Stavanger. The emergence of Norway as an oil-surplus nation completely changes its relationship to the rest of Europe. Norway is also different from most of Europe in that its 4.2 million people are concentrated in a tiny fraction of the total land area, and the bulk of the country is uninhabited.

The important characteristic of the Northern European nations is that they have adopted the elements of the Industrial Revolution and combined them with their existing democratic principles to create one of the highest standards of living in the world. The Northern European nations have all adopted a strong state-controlled economy in which the basic needs of each individual are guaranteed. The extremes of wealth and poverty found in larger industrialized nations are absent. The major issues facing the Northern European countries relate to the high costs of such social programs and problems deriving from the industrial centers of the United Kingdom and the Ruhr. High levels of sulfur from the heavy industrial core of Europe are carried over Northern Europe by the prevailing wind systems, and precipitation falls as acid rain. (See box.) This acid rain handicaps the growth

Massive production facilities are necessary to exploit resources of the North Sea. This Norwegian platform is as high as a 60-story building and will house 200 production personnel.

Regional Patterns of Pollution: Acid Rain

Acid rain and its effects have been widely recognized in the past decade. Acid rain occurs as sulfur and nitrogen are emitted into the atmosphere by burning of coal, oil, or natural gas. The sulfur or nitrogen combines with moisture in the atmosphere to form weak solutions of sulfuric or nitric acid. Acidity is measured by the concentration of hydrogen ions, which gives a pH scale of 0 to 14. Distilled water has a pH of 7, vinegar 3, and most acid rain is between 4 and 5. The scale is logarithmic, so a pH value of 4 is 10 times more acidic than 5, and 100 times more than 6.

Acid rain causes damage to statues (especially marble or limestone), historic stone buildings, exposed steel (as in railroad tracks, bridges, electrical lines), water, vegetation, and soil. In water, the first effect of acid rain is a fish kill brought about by aluminum poisoning. Concentration of aluminum increases dramatically as pH falls below 4.5, apparently as the higher acidity leaches it from rocks and soil. Heavy metal concentration (lead, zinc, manganese and cadmium) increase, and although not high enough to threaten humans at present, their long-term accumulation may do so. Corrosion of pipes increases and the growth of forests is slowed.

In the past the solution to the acid-rain pollution in the highly industrialized regions of the United Kingdom and Western Europe was to build higher smokestacks so that the gases and particulate matter created by burning would catch upper air currents and be dispersed over a wider area. Unfortunately, the upper-air circulation pattern that prevails over Europe moves in a north–northeast direction, focusing on Scandinavia. The geographic pattern of damage associated with acid rain is thus extended from the industrial regions to the less industrial regions of Europe. Sweden, the Scandinavian country with the greastest concentration of heavy industry, produces an estimated 300,000 tons of sulfur pollutants per year, 100,000 tons of which fall within the country, while 200,000 are carried on to Finland or the Soviet Union. But Sweden also receives an estimated 400,000 tons of sulfur pollutants from other West European countries. The situation is even worse for Norway, Denmark, and Finland countries that produce relatively little domestic pollutants.

The Long-Range Transboundary Air Pollution Convention of the United Nations theoretically provides a legal framework to allow countries receiving pollutants from others to receive reimbursement and require air-quality improvement. The technology to minimize nitrogen and sulfur emissions is available but costly. Since the most critical environmental impact affects the forest, recreation, and tourism of the less industrial countries of Scandinavia, rather than the producers of the pollutants themselves, progress toward environmental improvement has been slow, and acid rain will remain a central issue in the geography of Europe for the foreseeable future.

---

of the forests that are a major resource of Northern Europe. The Northern European countries have demonstrated that it is possible to overcome the problems of a less-than-perfect environment through dedication and commmitment. The reservoir of highly skilled workers in the Northern European countries promises well for their future standard of living.

THE MEDITERRANEAN LANDS OF EUROPE

The Southern European countries of Spain, Portugal, Italy, and Greece have a total population of over 118 million. As a group these nations are distinct from either Western or Northern Europe. Climatically this region is dominated by Mediterranean climate with its hot, dry summers and mild winters. Economically these nations have been slower to industrialize, and per capita incomes are markedly lower than those in Northern or Western Europe. Italy has the highest per capita income, but it is only half that of Switzerland. Literacy rates range from 80 percent in Portugal to 98 percent in Spain, but birthrates, population growth rates, and infant mortality rates are higher than those in Northern or Western Europe.

Culturally Southern Europe was at one time the center of Western civilization. The civilizations of the Romans, the Greeks, Phoenicians, and other peoples have focused on this area. In the past this region developed major advances that made it a center of learning unrivaled in Europe. At the present time the poverty levels of Southern Europe are the highest of the regions of Europe discussed to this point. In some cases areas within these countries parallel less industrialized countries. The standard of living varies greatly within individual regions of countries, with contrasts between such areas as the Po Valley of Northern Italy and the southern region of the Italian peninsula being especially marked. Culturally, the Mediterranean region is dominated by Catholicism, either Roman Catholic centered in Rome or Greek Orthodox centered in Athens. Throughout this region agriculture is much more important in terms of employment than it is in the other regions of Europe. The percentage of population engaged in agriculture ranges from 35 percent in Greece to 10 percent in Italy. Major crops are wheat and barley planted in the fall to capitalize on the winter rainfall, citrus fruit in favorable coastal areas, the ubiquitous olive, tobacco, grapes, and vegetables. Livestock, particularly sheep and goats, are an important element of the rural economy. The agricultural resource base of the Mediterranean lands suffers from millennia of occupation, and much of the region is plagued with soil erosion caused by overgrazing. Tourism is an important and growing part of the economy of Mediterranean Europe.

The settlement pattern of the Mediterranean lands is analogous to that of less industrialized areas with one or two cities dominating each nation. The national capitals (Athens, Madrid, and Lisbon) and the few truly industrial cities (Milan, Barcelona, Turin) dominate their respective countries' industrial development.

The change of the Mediterranean countries from the center of the civilized world to a peripheral location reflects changing situation relationships. So long as the trade and cultural and handicraft centers of Europe were concentrated around the Mediterranean, these nations had a central location. With the Industrial Revolution in Western Europe and its subsequent diffusion to North America, Australia and New Zealand, and Japan, the Mediterranean became more isolated from the world's major trade routes. Thus isolation added to the problems caused by the millennia of occupation and resource exhaustion to lead to a relative decline in importance.

Within Southern Europe the most highly industrialized nation is Italy, which has one half of the population of the region. Italy's development corresponded with the development of trade and the emergence of Western Europe in the Middle Ages. Italy's Po Valley was the focus of this industrialization and remains the most industrialized portion of Southern Europe. The Po Valley has a modified humid subtropical climate, and additional moisture comes from streams flowing from the Alps to the north and the Apennine Mountains to the south. This area produces crops similar to those of Western Europe, including rice, corn, sugar beets, and a variety of fruits. Milan and Turin (7 million and 2 million, respectively) are the focus of manufacturing in Italy. Fiat automobiles, aircraft, motorcycles, and a host of other manufacturing activities are concentrated in this area. Accessibility to Western Europe's industrial core stimulated the industrial development of the Po Valley and today some 40 percent of Italy's population resides in the valley. The Po Valley enjoys the highest standard of living in Italy even though the government has attempted to diffuse development and industrialization throughout the peninsula.

South of the Po Valley, Italy is characterized by rugged topography and rural poverty interspersed with islands of industrialization around cities such as Naples, or Palermo on the island of Sicily. Agriculture is handicapped by the Mediterranean climate, and only the irrigated areas produce high yields per acre. The industrial revolution has affected this area through diffusion of such things as televisions, radios, motorcycles, and the other elements of the modern consumer society, but the relative standard of living remains low. Land is owned by large corporations or landlords, and the majority of the private farms are less than five acres in size.

The other nations of Southern Europe also reflect this dichotomy of urban industrial centers and rural poverty. The contrast between Southern Europe and Western Europe results in continual out-migration of southern Europeans to the industrial core of Western Europe. These migrants have provided a major source of income for those who remain behind, but many rely upon continued well-being of the economy of Western Europe. Other migrants from Southern Europe have settled in North America, Latin America, and Australia on a permanent basis in the past. The continued poverty of much of Southern Europe is the major challenge facing these nations, as it affects stability of governments, international relations, and internal relationships between regions in each nation. Membership in the EC is fostering rapid economic development and continued growth in the tourism industry, providing optimism about the future of the region.

THE NATIONS OF EASTERN EUROPE: ALTERNATIVE TO DEVELOPMENT

The eight nations of Eastern Europe are similar to those of Southern Europe in that in the past they have been much less industrialized than the core area of Europe. These nations are peripheral in location because of their isolation from the ocean trade routes of Western Europe. East Germany (the Democratic Republic of Germany), Poland, Czechoslovakia, Hungary, Romania, Bulgaria, Albania, and Yugoslavia are grouped together because of their shared traits of development and developmental types. Parts of this region industrialized at an early date similar to Western Europe, notably in what is now East Germany, the western portion of Czechoslovakia, and southern Poland. Important mineral resources in Poland and Czechoslovakia (the Silesian coal fields and the Erzgebirge—Ore Mountains—of Czechoslovakia) provided a basis for economic development. For most of this region, however, industrial development came very late, and at the beginning of World War II the majority of the population relied upon agriculture for a livelihood. The end of World War II marked the beginning of major changes in this region. All nations of Eastern Europe adopted a strong, centrally planned economy somewhat modeled on the Communist experience in the Soviet Union. Development in these nations has progressed unevenly as the changes made in Western Europe diffused over the cultural and geographic distances separating Eastern from Western Europe. The western portion of today's Czechoslovakia and today's East Germany were industrialized by the end of the nineteenth century. Poland and Hungary industrialized somewhat later, and Bulgaria, Romania, and Yugoslavia have been actively involved in industrialization primarily since the end of World War II. Albania is only now beginning a concerted effort of modernizing and industrializing its national economy.

Within Eastern Europe climates range from humid continental with cool summers in northern Poland to the Mediterranean climates of the Yugoslav and Albanian coast. The Eastern European countries are dominated by the humid continental, warm-summer climates of Bulgaria and Romania, and a transitional humid subtropical climate of portions of Hungary and Yugoslavia.

Resources are important in Eastern Europe, and the most important is the agricultural land of the Danubian Plains in Hungary, Romania, and Bulgaria (Figure 6.8). The Great Hungarian Plain and the Wallachian Plain are the focus of highly productive agricultural systems quite distinct from those of Western Europe. Crops include corn, fruits, sunflowers (for oil for margarine and cooking), and vegetables. Mineral resources include Silesia's important coal deposits, northern Romania's Ploesti oil fields, and lead and zinc scattered throughout the region but with the greatest concentrations in Poland, Yugoslavia, and Bulgaria. A host of other materials are found in small quantities throughout Eastern Europe, and nearly every country has some coal or petroleum plus other small mineral deposits. However, none has large deposits rivaling Western Europe except Poland, Czechoslovakia, and Romania. The coalfields of Silesia are roughly equivalent to the Ruhr in area and approach the Ruhr in terms of quality. The Ploesti oil fields were of primary importance in Europe before World War II, but more recent discoveries in the North Sea overshadow them.

The peoples of Eastern Europe are diverse, and in only a few of the eight nations is there a single dominant ethnic group. This region has become known as a *shatter belt* because of the repeated invasions by surrounding powers and subsequent destruction, recreation, and boundary changes. The Balkan peninsula is the focus of such changes and the term *balkanization* has come to refer to the process of creating many states or political instability in a region. The ethnic makeup of the area reflects the past role of the region as a *buffer zone* between Eastern and Western civilizations.

Yugoslavia represents the extreme example of the effects of balkanization and subsequent attempts to create a nation. The Yugoslav population consists of eight significant ethnic groups including Serbs (who comprise 40 percent of the population), Croats (22 percent), Muslim (8.4 percent), Slovenes (8.2 percent), Macedonians (5.8 percent), Albanians (6.4 percent), Montenegrins (2.5 percent), and Hungarians (2.3 percent), creating continued ethnic unrest. The other nations of Eastern Europe are not so culturally fragmented, but only Poland and Germany have a nearly homogeneous ethnic makeup. Bulgaria and Romania have approximately 85 percent of their population of the primary ethnic group, while Czechoslovakia is divided between Czechs (46 percent), Slovaks (50 percent), and Magyars (4 percent). Hungary is dominated by Magyars who comprise 92.4 percent of the population. In Hungary, Bulgaria, and Romania, gypsies comprise significant elements of the population, numbering over half a million in Bulgaria and Hungary alone.

The nations of this region are primarily Slavic, and Eastern Orthodoxy and Roman Catholicism dominated the region in the past with the exception of Albania, which had a strong **Islamic** (Muslim) majority but today is officially an atheistic state. Several other nations of Eastern Europe have significant numbers of Muslims, particularly Yugoslavia and Bulgaria.

Islam: Religion founded by the prophet Mohammed (Muhammad) in Saudi Arabia around A.D. 624. Islam is the name of the religion and means submission to the will of one God (Allah). Muslim or Moslem refers to a member, one who submits himself or herself to the will of Allah.

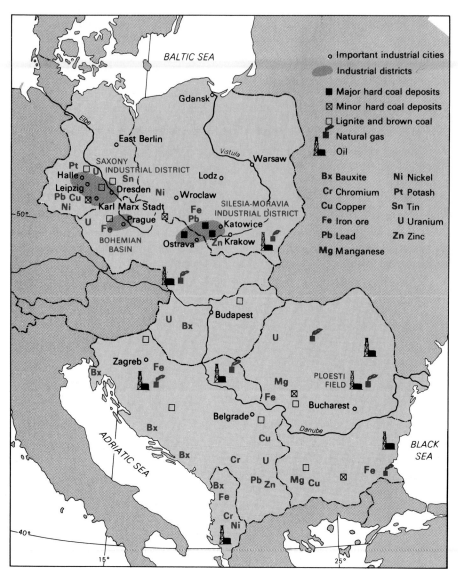

Figure 6.8 The major mineral resources and industrial centers of Eastern Europe illustrate the industrial dominance of Poland and Czechoslovakia. The coal resources of the Silesia-Morovia district and the Ploesti oil fields are particularly important in Eastern Europe.

Industrial development of the region as a whole generally began at the end of World War II. All these nations have a centrally planned economy utilizing five-year plans designed to expand industrialization and overcome the problems of underdeveloped areas. Albania and Yugoslavia have maintained complete independence from the Soviet Union, but the other six nations of Eastern Europe have been closely associated politically, militarily, and economically with the Soviet Union.

Albania at present is attempting to develop a Communist system in isolation from the rest of the world. Its economy is more agriculturally oriented than any other nation of Europe, with 61 percent of the population still engaged in agriculture. Literacy rates in Albania are the lowest in Europe at only 70 percent. In general the characteristics of Albania outside of the capital city are similar to those of less industrialized nations. Yugoslavia was able to maintain complete independence from the Soviet Union because of its isolation from the border areas of that power and the presence of a strong guerrilla movement at the end of World War II. The charismatic leadership of Marshal Tito, who had led the Yugoslavs in World War II, provided the basis for a stable government, which was able to overcome the centrifugal forces of the multiple ethnic groups within the country. Tito led Yugoslavia until his death in 1980 and left a governing system based on rotating the presidency among the various ethnic groups in the country. It is too early to draw conclusions about the success of this arrangement, but the economy of Yugoslavia has developed in a distinctly Yugoslavian pattern. Nearly one third of the population (30 percent) is still engaged

THE WORLD OF WEALTH

in agriculture, and the agriculture is typical of Southern Europe with small, fragmented farm holdings and a preponderance of hand labor or draft power. Industrial development has been concentrated in the major cities where the standard of living and employment patterns are different from the countryside. The centrally planned government and economy of Yugoslavia allows a great deal of private ownership of land and small businesses in addition to the centrally (state) owned factories and industrial activities. Yugoslavia was one of the least industrialized of the Eastern European nations at the end of World War II and had come into existence as a nation only at the end of World War I. Nevertheless, per capita income is now roughly equivalent to Portugal, the poorest of the Southern European nations.

The other nations of Eastern Europe vary in industrial development from northwest to southeast, with East Germany being the most highly developed and Albania the least. Germany, Czechoslovakia, and Hungary all have less than 20 percent of their work forced engaged in agriculture and are important producers of a range of industrial items. Poland, Bulgaria, and Romania have approximately one third of their population engaged in agriculture, but Poland has a parallel development of industry, making it one of the major industrial powers in Eastern Europe. Poland is a major manufacturer of ships for both Eastern Europe and the Soviet Union and Western European nations; East Germany shares in the high level technological development of East Germany and manufactures cameras, electronics, and other high-technology items for export to a variety of countries; Hungary is a major producer of transport equipment; and all the nations provide important industrial products for the Soviet Union. All these nations provide basic human needs of education, jobs, medical care, and housing for their population through their centrally planned economies.

Agricultural development in these nations has paralleled industrialization since World War II. Agriculture in Eastern Europe is a blend of state ownership and private holdings, with Poland and Yugoslavia characterized by private ownership of land and the balance of the Eastern European countries characterized by state ownership in which the land is divided into collective farms where the workers are paid a share of the production after state quotas and taxes are met.

In both industry and agriculture Communist governments have placed great emphasis on improving transportation networks to overcome the isolation of the region. The Eastern European nations are tied to Western Europe by rail lines, highways, and pipelines. Rail lines in Eastern Europe are the same gauge as those in Western Europe but narrower than those of the Soviet Union, which hinders movement of goods to the Soviet Union. Outside of Czechoslovakia, Ger-

Women bundling cut grain by hand illustrate the broad contrast in the level of mechanization of agriculture in Europe, and the dominance of women in manual labor in agriculture in the Communist bloc countries.

THE HUMAN DIMENSION

The Less Developed World in Europe

Immigrants from the less industrialized world are increasingly important in Europe. The United Kingdom counts 3.5 million, West Germany 4.8 million, and France 4.5 million among their populations at the present. More than 12 million "guest workers" came to Europe from the less industrialized world in the 1950s and 1960s in response to the labor demands of Europe's booming economy. Nearly all came from former colonies (Pakistanis, Indians, and Caribbean Islanders to England; Algerians and Moroccans to France or Indonesians to the Netherlands) or from Yugoslavia and Turkey. Viewed by the Europeans as temporary guest workers, many subsequently sent for wives and children. Today's "foreign workers" include millions more who were born of migrant parents in Europe and view European countries as their only home.

Immigrant workers and their families have the lowest standard of living, the highest birthrates, and the highest unemployment rates in Europe. Their geographic impact ranges widely. Culturally the migrants and their children tend to be Muslims, to be black or Asian, and have lower education levels. Mosques, social organizations, grocery stores, and restaurants catering to the migrants occupy the streets of the low-income neighborhoods where migrants are concentrated. Racial hostility is manifest in riots in England, extremist politicians' calls for an end to migration and forced removal of guest workers, attacks on ethnic or racial minorities by European youth, and *de facto* segregation of migrants in public housing projects. Public opposition to migrants is expressed in complaints that they cause unemployment, crime, and declining quality of public schools in areas with a high proportion of migrants. For the immigrant from a less industrialized country or area or for his/her family, however, life in Europe with public provision of housing, medical care, and education is generally preferable to returning home to face unemployment and an even lower standard of living. Continued high birthrates among migrants, continued illegal immigration, and unwillingness of migrants to leave Europe indicate that Europe's population from the less industrialized world will continue to be a major issue.

---

many, and Poland, major industrialization has been concentrated in the capital cities. In many Eastern European countries the urban pattern reflects that of less industrialized regions of the world: a few large cities and a host of smaller ones. Each of the Eastern European countries has a city that essentially dominates the country.

The Eastern European nations have had stable boundaries since the end of World War II, representing the longest period of peace they have known in this century. The political systems vary in terms of dictatorial powers, but the Soviet Union serves as the enforcer of peace in this region (outside of Yugoslavia and Albania). Soviet troops have invaded Hungary and Czechoslovakia to prevent a change to more liberal governments, and this client-dependent relationship reflects one of the major problems facing Eastern Europe. Faced with the reality of the dominance of the Soviet Union, each individual nation is attempting to transform its own economy utilizing a blend of Western managerial and worker incentive programs in conjunction with their centrally planned economies. The standard of living in the Eastern European countries varies from Czechoslovakia and the German Democratic Republic, which are characterized by a per capita gross national product of about one-half the Western European average, to Poland, Hungary, Bulgaria, and Romania at less than one-fifth that of the Western European average.

PROBLEMS IN THE INDUSTRIALIZED EUROPEAN CORE

The problems facing the European industrial core of the world are a result of the early emergence of industry in this area. Technical obsolescence has combined with increased standards of living to make many of the traditional European industries noncompetitive with new industrializing regions. Steel production peaked in Europe in 1974 and has suffered severe recessions since. As recently as 1948, 50 percent of world shipbuilding was in the United Kingdom. Today it produces less than 4 percent. Similar declines in textiles, electronics, and automobiles challenge the old industrial core of European countries. Their economy is further handicapped by resource exhaustion from decades of mining activity. Environmental degradation from industrialization has created landscapes of industrial waste. The mining areas of the British Isles around the midlands of the Pennines are sometimes called the "Black Country" because of the grimy conditions in the old towns. In West Germany the

Foreign workers in France. These North African migrants in Marseilles are Muslims and are facing Mecca to pray at noon.

heavy industrial cities of the Ruhr complex are constantly covered with a pall of smog. Deposition of industrial wastes over a century has completely changed large segments of the European landscape and caused massive air and water pollution. Rivers such as the Ruhr and the Rhine have been polluted with oil, chemical, and heavy metal wastes in significant quantities. The challenge to maintain a high standard of living and prevent further environmental degradation while reversing existing problems is a major issue facing Europe.

Another major problem is the complex ethnic makeup of the region as a whole. Contrasting cultural backgrounds, political systems, and standards of livelihood create friction between many countries of Eastern and Western Europe. Turks and Yugoslavs travel freely to Western Europe where they fill jobs unwanted by the wealthy industrial Western Europeans. The people of West Germany can visit their relatives in East Germany relatively easily, but East Germans are hedged in by a bureaucratic maze of restrictions in their attempts to visit family members in the West. Hungary has liberalized its border restrictions to allow Austrians to move relatively freely into and out of Hungary, and Hungarians are able to move relatively freely between Austria and Hungary. Poland has also liberalized travel requirements for its citizens. Even in members of the EC, there are tensions between nations such as Germany and the United Kingdom and Italy and other nations. Citizens of the Common Market countries are supposedly free to move freely between nations and to have their educational background recognized as equivalent, but some Western European nations are reluctant to grant full recognition to degrees taken at other nations' universities.

The complexity of the cultural development of Western Europe makes it difficult to overcome the centrifugal forces of the region's mosaic of culture. Within individual nations there are regional differences and resultant provincialism. The south of France is very different from Paris or the north of France. The Albanians of Yugoslavia view themselves as distinct from the Serb majority. The issues for Europe in the next decades will focus on the need to overcome the centrifugal forces of culture, economy, and politics in this complex region while developing a European economy that will allow the peoples of this area to maintain the high standard of living of the West and North while raising the standard of living in Southern and Eastern European nations.

FURTHER READINGS

BANDERA, V. N. and MELNYK, Z. L., eds. *The Soviet Economy in Regional Perspective.* New York: Praeger, 1973.

BEAUJEU-GERNIER, J. "Toward a New Equilibrium in France." *Annals of the Association of American Geographers* 64, 1974, pp. 113–125.

BECKINSALE, M. and BECKINSALE, R. *Southern Europe.* London: University of London Press, 1975.

BERENTSEN, W. H. "Regional Change in the German Democratic Republic." *Annals of the Association of American Geographers* 71, 1981, pp. 56–66.

BOAL, F. and DOUGLAS, J., eds. *Integration and Division: Geographical Perspectives on the Northern Ireland Problem.* New York: Academic Press, 1982.

CHAPMAN, K. *North Sea Oil and Gas.* London: David and Charles, 1976.

CHISHOLM, M. *Rural Settlement and Land Use: An Essay in Location.* 3d ed. London: Hutchinson University Library, 1979.

CHISHOLM, MICHAEL. "The Impact of the Channel Tunnel on the regions of Britain and Europe." *The Geographical Journal* 152, November 1986, pp. 314–334.

CLAVAL, P. "Contemporary Human Geography of France." *Geoforum* 7, 1976, pp. 253–292.

CLOUT, H. D., ed. *Regional Development in Western Europe.* New York: Wiley, 1981.

CLOUT, HUGH D. *Regional Variations in the European Community.* Cambridge: Cambridge University Press, 1986.

DEMKO, GEORGE J., ed. *Regional Development Problems and Policies in Eastern and Western Europe.* New York: St. Martin's Press, 1984.

"Despite the Union's Clamor, Mitterrand is Pushing His Steel Plan Ahead." *Business Week*, April 23, 1984.

DURY, G. H. *The British Isles.* 5th ed. New York: Barnes and Noble, 1973.

DUTT, A. K. and HEAL, S. "Delta Project Planning and Implementation in the Netherlands." *Journal of Geography* 78, 1979, pp. 131–141.

EANS, E. E. *The Personality of Ireland: Habitat, Heritage and History.* London: Cambridge University Press, 1973.

FULLERTON, B. and WILLIAMS, A. *Scandinavia: An Introductory Geography.* New York: Praeger, 1972.

GREEN, A. E. The North-South Divide in Great Britain: An Examination of the Evidence. *Institute of British Geographers. Transactions.* New series. Vol. 13, No. 2, 1988, pp. 179–198.

HOFFMAN, G. W., ed. *Eastern Europe Essays in Geographical Problems.* London: Methuen, 1971.

———, ed. *Federalism and Regional Development. Case Studies on the Experience in the United States and the Federal Republic of Germany.* Austin: University of Texas Press, 1981.

———, ed. *A Geography of Europe: Problems and Prospects.* 5th ed. New York: Wiley, 1983.

HOUSE, J. W., ed. *U.K. Space.* London: Weidenfeld and Nicolson, 1979.

HOUSTON, J. M. *The Western Mediterranean World: An Introduction to its Regional Landscapes.* New York: Praeger, 1967.

JACKSON, PETER. "Beneath the Headlines: Racism and Reaction in Contemporary Britain." *Geography*, Vol. 73, Part 3, No. 320, June 1988, pp. 202–207.

JOBSE, REIN B. and NEEDHAM, BARRIE. "The Economic Future of the Randstad, Holland." *Urban Studies*, Vol. 25, No. 4, August 1988, pp. 282–296.

JOHNSTON, R. and DOORNKAMP, J., eds. *The Changing Geography of the United Kingdom.* London: Methuen, 1983.

KING, RUSSELL L. *The Industrial Geography of Italy.* New York: St. Martin's Press, 1985.

LUCIANI, G. *The Mediterranean Region: Economic Independence and the Future of Security.* Beckenham, U.K.: Croom Helm, 1984.

MCINTYRE, ROBERT J. *Bulgaria: Politics, Economics and Society.* London: Pinter Publishers, 1988.

MACLENNAN, D. and PARR, J. B., eds. *Regional Policy: Past Experience and New Directions.* Oxford: Martin Robertson, 1979.

MANNERS, I. *North Sea Oil and Environmental Planning: The United Kingdom Experience.* Austin: University of Texas Press, 1982.

MELLOR, R. *The Two Germanies: A Modern Geography.* New York: Barnes and Noble, 1978.

MAIER, C.S., ed. *Changing Boundaries of the Political Map of Europe.* New York: Cambridge University Press, 1987.

ORME, ANTONY R. *Ireland.* 2d ed. White Plains, N.Y.: Longman, 1987.

SCARGILL, P. I. *Problem Regions of Europe.* London: Oxford University Press, 1973–1975. A series of small books.

SEERS, D. et al., eds. *Underdeveloped Europe: Studies in Core-Periphery Relations.* Atlantic Highlands, N.J.: Humanities Press, 1979.

SEGALA, GIAN PAOLO. "East Germany's Alienated Young." *Europeo* as reprinted in *World Press Review*, December 1982.

SHORT, JOHN R. and KIRBY, ANDREW M. *The Human Geography of Contemporary Britain.* New York: St. Martin's Press, 1984.

STAMP, L. D. and BEAVER, S. H. *The British Isles.* New York: St. Martin's Press, 1971.

TAMES, R. *Economy and Society in Nineteenth-Century Britain.* London: Allen and Unwin, 1972.

TUGENDHAT, CHRISTOPHER. *Making Sense of Europe.* New York: Columbia University Press. 1988.

WABE, J. W. "The Regional Impact of De-Industrialization in the European Community." *Regional Studies* 20, 1986, pp. 540–559.

WILLIAMS, COLIN H. "Ideology and the Interpretation of Minority Cultures." *Political Geography Quarterly* 3, 1984, pp. 105–125.

WOODELL, STANLEY R. J., ed. *The English Landscape: Past, Present and Future.* New York: Oxford University Press, 1985.

Houston.

ANGLO-AMERICAN PROFILE: (1990)

Population (in millions)	276		
Growth Rate (%)	0.7	Energy Consumption Per Capita—	
Years to Double Population	99	pounds (kilograms)	28,850
Life Expectancy (years)	75		(11,750)
Infant Mortality Rate		Calories (daily)	3,663
(per thousand)	10	Literacy rate (%)	99
Per Capita GNP	17,500	Eligible Children in Primary	
Percentage of Labor Forc		School (%)	98
Employed in:		Percent Urban	76
Agriculture	3	Percent Annual Urban Growth,	
Industry	31	1970–1984	1.5
Service	66	1980–1985	2.3

7

The Anglo-American Realm

MAJOR CONCEPTS

Cultural realm / Emergent landforms
Submergent landforms / Estuaries / Prairie
wedge / Cultural pluralism / Fall line
Cultural conflict / Piedmont / Assimilation
Urban morphology / Relative deprivation
Megalopolis / Ghettos / Culture of poverty
Seigneur / Acculturation / Postindustrial
society

IMPORTANT ISSUES

Resource depletion and environmental
  degradation
High per capita consumption of resources
Minority groups in industrialized nations
Loss of agricultural land
Urban sprawl and the role of cities

Geographical Characteristics of Anglo-America

1. Anglo-America has many of the same characteristics as Western Europe, the hearth for the bulk of the peoples and ideas now found in the United States and Canada.
2. The physical geography of Anglo-America is characterized by large size; climatic and resource variability, abundance and complementarity; and external isolation.
3. Canada's high-latitude location makes much of its area unsuited for dense settlement.
4. Anglo-America represents the world's largest industrial producer and largest and richest market.
5. Anglo-America's population is the world's most urban, mobile and wealthy group.

The North American continent extends from Central America to the North Pole and consists of ten independent countries (Figure 7.1). The continent exhibits the stark contrasts that distinguish the industrial world from the less industrialized world. Nowhere in the world is there such a marked break between wealth and poverty as at the border between the United States and Mexico. Nowhere is there such a contrast between less industrialized and industrialized with no transition zone between. South of the United States border the countries have the characteristics typical of the less industrialized world, while Canada and the United States, the topic of this chapter, are the epitome of the industrial world.

The Mexico-United States border is more than a political or economic division. It is a cultural boundary between the Latin American **culture realm,** with its legacy from southern Europe, and the Anglo-American culture realm. In Anglo-America the cultural legacy is largely from Western Europe. The two countries of Anglo-America were both colonized by western Europeans and have participated in the revolutions that have made Europe a center of wealth.

Alike in general ethnic origins of their majority populations, the United States and Canada are very different in climate, resources, and size and importance of population. The United States has 250 million people and plays a role in the world as a counterbalance to the Soviet Union. Canada has only one tenth as many people as the United States, but is an important part of the world's peacekeeping forces because of the world perception that it has no colonial ambitions or desire to manipulate other countries. The two countries are similar in that they represent the end product of the Industrial Revolution: wealthy, mobile, highly educated societies with an abundance of leisure time and consumer goods. They are unusual in the world because the overwhelmingly dominant middle class exists in a society based on social mobility rather than social stratification. The commitment of the Anglo-American people to equality regardless of race, social position, birth, or ethnic background is an important example for the world.

Although Switzerland has a per capita gross national product that exceeds that of Canada and the United States, these two nations combined represent the world's largest concentration of middle-class people. But on a world scale "middle class" is misleading, since the average person in either the United States or Canada enjoys a life of luxury compared to at least 70 percent of the world's peoples. They are middle class only because there exists an even wealthier elite in each country. These two countries collectively enjoy a standard of living that is only a dream for the world's hungry, illiterate, ill-housed, ill-fed, and underemployed majority. Although all may not desire the specific culture of the United States and Canada, the poor in most parts of the world would willingly accept the designation of "middle class" if it enabled them to have the wealth and material well-being of the citizens of Anglo-America.

The wealth of Anglo-Americans is the result of an economic system that utilizes a significant portion of world resources. With one-sixteenth of the world's

---

Culture realm: A group of countries sharing related cultural traits.

THE WORLD OF WEALTH

Figure 7.1 The Anglo-American Realm.

total population, the two countries consume approximately one third of all resources used yearly in the world. The massive consumption of resources and associated wealth and material well-being of Anglo-America are not equally distributed. Although the majority of Anglo-Americans share in the high standard of living, there are areas and groups that have largely been excluded. Blacks, Hispanics, American Indians, and residents of economically depressed areas with a white majority, such as Appalachia, have standards of living below the average for the two countries. Such relative poverty should not be confused with the absolute impoverishment found in less industrialized countries, since government income transfer programs generally prevent the widespread absolute poverty like that in less industrialized countries.

The large size of the Anglo-American population makes it difficult to eradicate poverty completely, but government programs and policies are committed to providing the basics of minimal education, food, and shelter for all. Critical observers argue that the governments of the two countries could do more to increase the standard of living of the poorest one fifth of each country, but in total the Anglo-American population enjoys the highest standard of living of any large group on earth. Unlike people in the less industrialized countries, the poor in Anglo-America are rarely starving, and they represent the minority rather than the majority. The development of the Anglo-American industrial society with its affluence and material possessions has taken place concomitantly with the Industrial Revolution. Endowed with a large, fertile, and resource-rich area, Canada and the United States were settled at a time when the Industrial Revolution was transforming the economies of Europe. Because it was a new land, the economic transformation of the Industrial Revolution was paralleled by a social transformation which makes the Anglo-American society unique in the world.

THE ANGLO-AMERICAN ENVIRONMENT

The geography of land and resources in which Anglo-American development has occurred can be characterized by a number of key elements. The general characteristics of Anglo-America's geographical relationships include a large size, a compact shape, a large latitudinal extent, isolation from other countries, internal accessibility, fertile level lands, and an abundant and complementary resource base.

SIZE AND DEVELOPMENT: THE ROLE OF AREA

The two countries of Anglo-America are individually among the world's largest in terms of land area. Canada is second only to the Soviet Union in size, with a land area of 3,852,000 square miles (9,971,500 square kilometers), while the United States, only slightly smaller, is the fourth largest nation with a total land area of 3,677,700 million square miles (9,525,240 square kilometers). The potential advantages of such a large area from a geographical standpoint include room for expansion, the potential for additional resources, and the basis for world power. The disadvantages are accessibility, transport and communication, political and cultural fragmentation, and the necessity of defending long boundaries.

The advantages have outweighed the disadvantages in the United States. Since its occupation by people from the United Kingdom, Spain, and France after the voyage of Columbus, the United States has been able to gain independence and national unity and develop one of the world's major economic and military forces. The large resource base of the continent, the ability to absorb large numbers of immigrants because of its large territorial expanse, and the variety of climatic types contributed to this development. The mere size of a country is insufficient to guarantee wealth and importance, however, as manifested by Canada. Although larger than the United States, Canada is much less important economically, politically, and militarily. Canada's high-latitude location limits occupation of much of its territory. The large majority of Canada's population is concentrated near its border with the United States, while the bulk of the country is only sparsely occupied (Figure 7.2). The large area of the two countries, combined with only sparse population in the north, would seem to create problems of defense and national unity. This has not occurred because Anglo-America developed in relative isolation from the rest of the world, and each country developed a common culture and economic system.

ISOLATION: A CRITICAL RELATIONSHIP

Unlike the states of Europe, Canada and the United States are isolated from other major countries by the great oceans of the world. The relatively short border between the United States and Mexico represents the only land access to Anglo-America from other regions. In the north, the United States is very close to the Soviet Union, but the physical geography and the climate in this area aid defense. Isolated from the wars

Figure 7.2 The population distribution of Anglo-America reflects both the physical geography and the historical development of the continent.

and revolutions of Europe and the rest of the world, Canada and the United States have been free to develop independently. The initial advantages of this isolation have been reinforced by the political and economic developments within Canada and the United States. Both have developed transportation systems to unify their peoples; both have political systems that emphasize democracy, and both have developed economies that provide wealth transfer payments to insure a minimal standard of living for all their people.

THE ANGLO-AMERICAN REALM

Although the relative isolation of North America from Europe and Asia has changed with technological advances in transportation, the continent is still effectively isolated militarily. The 3,000 miles (4,827 kilometers) separating the United States from Europe and the 5,000 miles (8,045 kilometers) separating the western United States core area in California from Japan serve as effective means for minimizing the potential for invasion by hostile forces. The isolation of Anglo-America in the past when transportation was less able to overcome the distance separating Anglo-America from the rest of the world allowed the two countries to develop their political and economic systems to the point that external threats are a relatively minor problem. The development of internal unity over the past 200 years of isolation also contributes to the safety of the two countries, since it minimizes the potential for internal revolution.

THE PHYSICAL BASE OF ANGLO-AMERICA

The emergence of a relatively homogeneous population in Canada and the United States with a high standard of living is the result of a number of factors, but an important one is the physical environment. Although technology can overcome many problems of harsh environmental settings, the ease of occupation of much of the United States and southern Canada contributed to the rapid development of the continent. The two countries have extensive areas with relatively level land and a variety of climates presenting few problems for habitation.

CLIMATIC CHARACTERISTICS OF ANGLO-AMERICA

The great latitudinal extent of the continent results in a wide variety of climatic types in Canada and the United States. The other factors contributing to the continent's climatic pattern are its size; the prevailing westerly winds; the warm, moist air from the Gulf of Mexico; and the land and water relationships in and around the continent. In general the climates reflect latitude, but they are modified by the mountain chains of the Sierra Nevada, Cascades, and Rocky Mountains (Figure 7.3).

Anglo-America has nine of the eleven major climatic types. In the north is an area of tundra that is essentially uninhabited. Subarctic climate occupies much of Alaska and Canada and helps explain the more limited economic development in Canada. An area of coniferous forests, it is sparsely inhabited except for point settlements for mining, trapping, or other extractive economic activities.

The marine west coast climate extends from 60° north latitude to approximately 40° north latitude on the west coast of Canada and the United States. This area has relatively intensive settlement, particularly south of the Canadian border. The bordering ocean gives this region relatively cool summer temperatures and mild winters for the latitude along with abundant precipitation to support coniferous forests. The moderate temperatures provide a long growing season. The region is less developed in Canada than in the United States because of rugged landforms in Canada.

Part of California has a Mediterranean climate with hot, dry summers and mild, wetter winters. Although precipitation totals between 16 to 25 inches (390 and 620 millimeters) per year, the Mediterranean climate has the greatest population concentration in California, which is the most populous state in the United States, with more people than all of Canada.

The mountainous western portion of the United States has desert and steppe climates. Some of the driest areas of the world are the interior deserts of California, Nevada, Arizona, and New Mexico. In Canada, cooler temperatures prevent deserts and minimize the extent of steppe regions.

The eastern portions of the United States and Canada have humid continental climates. Both warm- and cool-summer types are found, and are associated with some of the most fertile agricultural lands in the world.

Humid subtropical climate dominates the southeastern United States. Summers are hot and humid, and winters are typically moderate with long periods of subfreezing temperatures uncommon. The southern tip of Florida and the islands of the Keys have a savanna climate with a hot, humid summer and a warm drier winter, which has contributed to its development as a winter tourist center.

The essentially north-south orientation of the mountains in Anglo-America results in important modifications of the humid continental and humid subtropical climates of the United States. The absence of any physical barrier across the central plains allows cold Canadian and Arctic air to move farther south in the United States than in any other continent. Consequently the humid subtropical climate of the United States has freezing temperatures periodically, sometimes damaging citrus and vegetable crops.

THE LANDFORMS OF ANGLO-AMERICA

The landforms of Anglo-America have also been conducive to human use. The greater part of Canada consists of an old igneous rock mass that has been highly glaciated, important because of its diverse and abundant mineral resources. Characterized by gently roll-

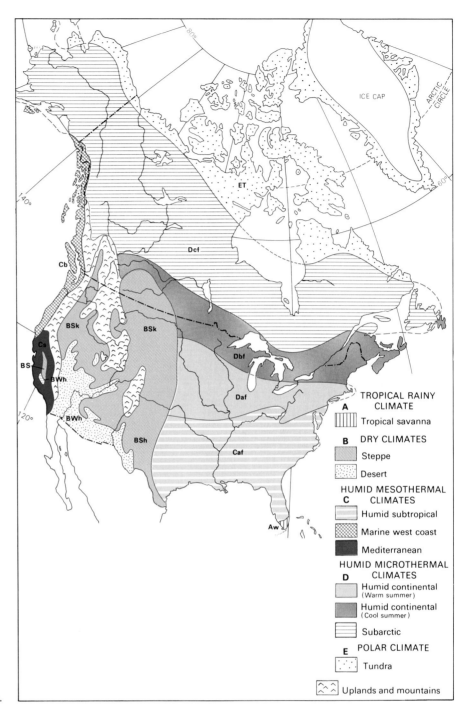

Figure 7.3 Climatic patterns of Anglo-America.

ing to nearly level landforms, it was the center of the glaciation in Anglo-America that smoothed the land and caused numerous lakes, marshes, and ponds (Figure 7.4). The other landforms, which are common to both the United States and Canada, consist of the coastal plain, the piedmont, the Appalachians, the central lowlands, the high plains area, the western mountains, and the Pacific coastal plain.

The Atlantic coastal plain is a broad, easily accessible region. There are two recognizable sections. North of Cape Fear in North Carolina is an area of *submergent* coast where the coast has sunk in past times, flooding the river mouths. South of this point to the Mexican border the region is generally an emergent coastal plain where the land is slowly rising. The submergent portion is the location of the flooded, or drowned, river mouths of the St. Lawrence, Hudson, Delaware, Chesapeake, and Susquehanna rivers. The drowned river mouths are known as *estuaries* and provide excellent harbors allowing oceangoing ships

THE ANGLO-AMERICAN REALM

183

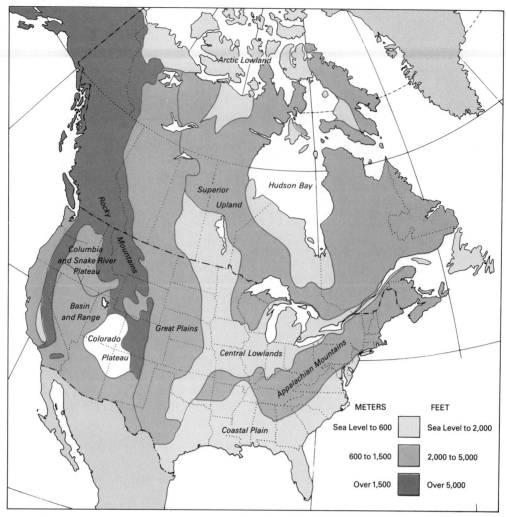

Figure 7.4 Landforms of Anglo-America. The large amount of relatively level land in the continent has facilitated occupance of the continent.

to penetrate into the land. The coastal plain is narrower in the area of submergent coast, but the existence of the drowned river mouths allows access farther inland. The emergent coastal plain in the southeastern United States has poorly drained lands that have only recently emerged from under the sea. Rivers are shallow and winding, and the sites for ports are few and less suitable. The west coast of the United States is different from the east coast, with the mountains rising abruptly from the seashore and limited coastal plain areas.

Between the Atlantic coastal plain and the Appalachian Mountains is an area of *piedmont*, the transition zone between mountains and plains. It is separated from the coastal plain by the *fall line*, the point at which streams descend from the high piedmont to the lower coastal plain. Rarely is the fall line an actual waterfall here; it is normally a series of rapids.

The Appalachians are a series of parallel mountians whose elevations rarely exceed 5000 feet (1500 meters), but because of their north-south orientation, they have historically been a barrier to easy transportation from east to west in the United States. The Appalachians have extensive deposits of high-grade coal in their sedimentary rocks.

The central lowland formed by the Ohio-Mississippi-Missouri drainage constitutes the agricultural heart of the American continent. This area of fertile, well-drained soils is generally level and suitable for mechanized equipment. West of the central lowlands, the land rises in the Great Plains of the United States and Canada. Extending from San Antonio in the south to the lowlands along the Arctic Ocean, the plains region rises from 500 feet (150 meters) above sea level in the east to 5000 feet (1500 meters) above sea level in the west. The combined area of the Great Plains and central lowlands of the United States constitutes more than one-half of the total landmass. This broad expanse of level, fertile land has provided a unique opportunity for agricultural development.

The fall line on the James River at Richmond, Virginia. Although the falls are actually only rapids, they effectively blocked the movement of ships up the river.

The Great Plains merge into the north-south-trending Rocky Mountains complex with elevations rising to over 15,000 feet (5,000 meters). The mountains are important because of their impact on the climates of the continent, their resources, and the barrier they have provided to uniting the eastern and western portions of the continent. The Canadian portion of the great Rocky Mountain chain merges into the coastal range of mountains, but in the United States, the coastal range is separated from the Rocky Mountains by a series of plateaus, block mountains, and basins, collectively called the Basin and Rane region. The Basin and Range region of the United States is a region of arid lands with internal drainage such as the Great Salt Lake. The complex series of basins and plateaus comprising the Basin and Range is bordered on the west by the high mountains of the Sierra Nevada. West of the Sierra Nevada is the Central Valley of California, which is separated from the Pacific by the California coastal ranges. North of the Sierra Nevada, the Cascade Range of Oregon and Washington consists of volcanic mountains.

AMERICA'S RIVER SYSTEMS: ACCESSIBILITY FROM NATURE

Anglo-America is endowed with natural transportation arteries rivaled only in Western Europe. The river system of the continent in the past provided routes for expansion of settlement and fostered unification. Today they provide major transport arteries for industrial and agricultural development. The most important river systems are the St. Lawrence–Great Lakes system and the Mississippi and its tributaries (Table 7.1). The St. Lawrence River provides access to the Great Lakes, which can then be navigated halfway across the continent. Initially the St. Lawrence River provided access to shipping only as far inland as the fall line rapids near Quebec. Above Quebec the river was navigable once again to Lake Ontario, but it was difficult to get from one lake to another because of falls of as much as 356 feet (110 meters) between the lakes. At first these barriers were overcome by transporting goods overland around the rapids and falls. The Sault Sainte Marie canal and locks, constructed

TABLE 7.1 Shipping on Inland Waterways (Billions of Ton/Miles)

System	1960	1965	1970	1980	1985
Atlantic Coast	28.6	27.8	28.6	30.6	24.8
Gulf Coast	16.9	21.8	28.6	36.6	36.5
Pacific Coast	6.0	6.6	8.4	14.9	19.9
Mississippi River	69.3	96.6	138.5	228.9	224.7
Great Lakes	99.5	109.6	114.5	96.0	75.8

SOURCE: *Statistical Abstract of the United States* (Washington, D.C.: U.S. Department of Commerce, Bureau of Census, 1988).

The Mississippi River carries more tonnage than any other American river. Coal, wheat, iron, petroleum, and other bulk items constitute the majority of the volume of shipping on the river.

in the nineteenth century, allowed ships to proceed past those rapids without the necessity of off-loading their cargo. Construction of the St. Lawrence Seaway project by the United States and Canada in the 1950s allowed oceangoing ships to travel inland as far as Chicago on Lake Michigan or Duluth or Thunder Bay on Lake Superior. It has provided relatively low-cost shipment of bulk items such as grain, coal, steel, and other items but is handicapped by its northern location, which limits the navigable season.

The second major river system in the United States is the Mississippi and its tributaries, including the Arkansas, Ohio, and Missouri rivers. This system drains the entire central area of the continent. The Mississippi is second only to the Rhine in importance for river transportation. It is navigable by locks as far upstream as Pittsburgh on the Ohio and Minneapolis–St. Paul on the Mississippi, and it is tied to Lake Michigan by the Illinois-Michigan Barge Canal. In the past, the Missouri River was navigated by steamboats as far upstream as central Montana, but today navigation on the Missouri is primarily from Omaha to its junction with the Mississippi. The Arkansas River is also navigable for barge traffic, and a massive canal-building project has made Tulsa, Oklahoma, a seaport. The Mississippi system provides a low-cost means of hauling bulk items of petroleum, steel, coal, grain, and chemicals. The movement of bulk cargo on the Mississippi makes New Orleans the largest port in the United States in terms of total tonnage.

In the western United States two main river systems provide access to the interior. The Columbia River is navigable for barge traffic as far inland as Idaho. The Sacramento River is navigable today as far inland as Sacramento, California. Other western rivers, including the Colorado, are not important for transportation, but provide water to the arid region. The importance of the water of the Colorado for agriculture, industry, and residential use in the arid American West makes it as significant to the inhabitants of the lands through which it passes as the Mississippi is to the central lowland.

AMERICA'S RESOURCE ENDOWMENT: OPPORTUNITY FOR WEALTH

The abundance of water in most of the continent masks the importance of water as a resource. The river systems of Anglo-America have been utilized for both transportation and water for powering electric generators, cooling thermal and nuclear generators, for irrigation, and for disposal of wastes. The growth of urban concentrations such as that stretching from Washington and Philadelphia to Boston with its population of more than 20 million has created water shortages even in the humid areas of Anglo-America with resulting competition between cities to tap water sources ever farther from their boundaries. The water resources of Anglo-America, however, are generally abundant and are more than adequate for the needs of the present and projected population, if used wisely. As with many other resources of the continent, water resources have been abused and only an expensive program of pollution control will rectify the

The fertile lands of the central United States are among the world's most productive. The large scale and mechanization allow one family to farm hundreds of acres, as in this scene from Minnesota.

shortage of clean water. Only in the 1970s did Americans become aware of the finite nature of water and other resources.

Another major resource of Anglo-America that has been abused is the land itself. The expanses of level land in the central lowlands, the Great Plains, the Central Valley of California, and the coastal plain and piedmont provide some of the most suitable areas for agriculture in the world. As a result of the interaction of the warm, humid air from the Gulf of Mexico and the cooler, drier air from Canada, there exists in the Midwest of the United States an area known as the *prairie wedge*, consisting of Missouri, Iowa, Ohio, Illinois, and Indiana. This prairie wedge constitutes the largest, most naturally fertile area of soil in the world. Soils in this area are mollisols, with chernozem soils in the western margins and prairie soils in the east. The lands farther west in the Great Plains border are drier, but soils have a high level of natural fertility and produce good yields of grain with available precipitation. When irrigated, they produce even higher yields. The level lands with their complementary climate and soil fertility have facilitated the emergence of the United States and Canada as the great food-surplus region of the world. These two countries combine to produce nearly one-half of the world's grain, corn, and soybean exports (Table 7.2).

As with the water resources, however, the very abundance of Anglo-America's lands has contributed to inadequate conservation measures until the past few decades. Much of the land in the Piedmont has been eroded and is today suitable only for forest crops.

THE ANGLO-AMERICAN REALM

TABLE 7.2 World Wheat Exports (1985)

Area	Total (1,000 metric tons)	Percent
WORLD	105,221	100.00
Argentina	9,716	9.23
Australia	15,781	15.00
Canada	17,363	16.50
Europe (Western)	29,872	28.39
Europe (Eastern)	3,052	2.9
United States	26,141	24.84
U.S.S.R.	1,557	1.48
Other	1,739	1.66

SOURCE: FAO Trade Yearbook; 1985.

Portions of the Great Plains have also suffered erosion as a result of poor agricultural techniques. Today the land as a resource is the focus of attention of private and government agencies that are attempting to maintain and/or increase the quality of the land we have left. America's land is also threatened by the spread of the ubiquitous American suburbs, which effectively eliminate an estimated 1 to 3 million acres (404,700 to 1,214,100 hectares) of high-quality cropland each year. Approximately 1 million acres (404,700 hectares) are actually transformed into housing or industrial or transportation uses, but as many as 2 million more acres (809,400 hectares) are lost to agriculture by urban sprawl. These lands near urban areas are purchased by developers who anticipate subdividing them. Soil fertility is not maintained, and often the land is allowed to lie idle. The loss of high-quality agricultural land is a major issue, as there exist only an estimated 400 million acres (161,880,000 hectares) of prime agricultural land in the United States (Table 7.3). The protection of the high-quality lands is one of the critical problems facing Americans.

MINERAL RESOURCES OF ANGLO-AMERICA

Another important geographical factor that has affected the North American continent is the abundance and variety of its mineral resources. The United States and Canada have large deposits of petroleum, natural gas, and coal. Natural gas is found in Texas, Louisiana, and Alberta (Figure 7.5). The United States produces much more natural gas than does Canada, and natural gas is found scattered over a broad area of the continent. Petroleum follows the same general pattern. Petroleum is particularly important since in the last quarter century it has provided approximately half of all energy used in the two countries. The resources are unequally distributed, with the United States having an estimated 7 percent of the world's total reserves and Canada having only 2 percent.

The major oil-producing areas are the Gulf Coast of Texas and Louisiana, interior Texas, Oklahoma and Alaska. California has also been a major producer, and the western region of Colorado, New Mexico, Wyoming, and Utah is becoming increasingly important. Canada's petroleum is concentrated in Alberta and Saskatchewan, but recent discoveries have been made in the Atlantic off the maritime provinces. Major fields are Le Duc and Pembina near Edmonton, Alberta. The United States was the world's largest producer of petroleum until the 1970s. It's known reserves are estimated to last for approximately nine years at present consumption, assuming no new discoveries. Because of past exploitation, the United States relies increasingly upon other countries for petroleum. Both

TABLE 7.3 Urban-related Land Conversion in the United States in Acres (hectares), 1967–1975

	Converted to Urban Areas	Converted to Water Areas	Total Converted	
			Areas	Percent
Prime farmland	6,200,000 (2,509,105)	1,191,000 (481,991)	7,391,000 (2,991,096)	32.0
Non-prime farmland	6,128,000 (2,479,968)	1,862,000 (753,541)	7,990,000 (3,233,509)	34.5
Non-farmland	4,044,000 (1,636,584)	3,637,000 (1,471,873)	7,681,000 (3,108,457)	33.5
TOTALS	16,372,000 (6,625,657)	6,682,000 (2,707,405)	23,055,000 (9,333,062)	100.0

SOURCE: Linda K. Lee, "A Perspective on Cropland Availability" (Agricultural Economic Report No. 406, Washington, D.C.: U.S. Department of Agriculture, 1978), tables 8, 9, pp. 14, 15.

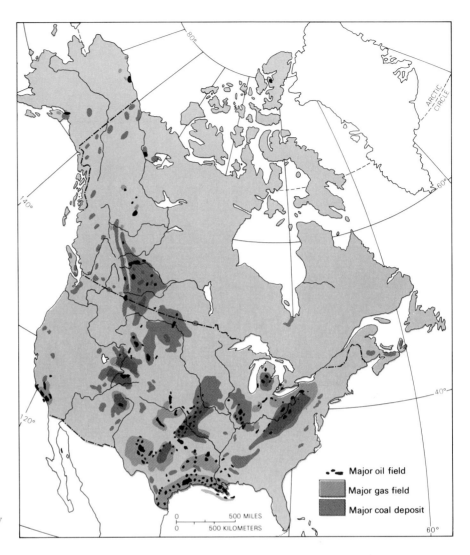

Figure 7.5 Fossil fuel energy resources of Anglo-America.

the United States and Canada have large reserves of energy in their *oil shales* and *tar sands*, but the cost of recovering them is greater than present cost of imported petroleum. It is estimated that the oil shales in Colorado, Utah, and Wyoming alone are sufficient to equal the total known world reserves of petroleum, but the environmental problems associated with exploitation may prevent their use for several decades.

Coal. The United States and Canada also have major deposits of coal. The United States has the better reserves of high-grade coal with known reserves sufficient for at least the next two centuries. The coal reserves of the United States are rivaled only by those of the Soviet Union. Canada has much less coal than the United States, but does have some in the Rocky Mountain and prairie regions of western Canada. Coal is divided into three classes. *Lignite* is a soft brown coal used primarily for power generation and production of coal tar for chemicals. *Bituminous* is the soft black coal used in making coke, used in manufacturing iron and steel, and in power generating plants. *Anthracite* coal, often called hard coal, exists in much smaller quantities, mainly in eastern Pennsylvania, and has been used primarily for heating purposes in the past.

The vast coal resources of the United States were a major factor in the industrial development of the nation before the twentieth century. The use of coal for fuel in trains, ships, and factories overshadowed all other fuel sources until after the development of the internal combustion engine and the use of the automobile. Coal production declined with the adoption of petroleum products, which are cleaner, easier to store, and until recently were cheaper (Figure 7.6). The increasing costs and limited nature of petroleum reserves in the United States are providing renewed impetus to the production of coal.

There are three major coalfields: the Appalachian field, the interior field, and the Rocky Mountain fields (Figure 7.5). The Appalachian fields have been exploited intensively in the past. They have been easily

THE ANGLO-AMERICAN REALM

Soil erosion caused by lack of proper conservation techniques resulted in major destruction to American cropland in the past, as illustrated in this 1950 photo.

accessible and were near the populated and industrial centers of the nation. The Appalachian fields contain an estimated 20 percent of total known reserves. The eastern and western interior fields combined have approximately 10 percent of the nation's coal. Much of it is exploited through open-pit mining, and as in the Appalachian field, it is primarily bituminous with a relatively high content of sulfur, which forms the pollutant SO_2 (sulfur dioxide) when it is burned. Strip mining consists of using large mechanized scoop shovels to remove the surface materials and reveal the coal seams. Strip mining is cost-efficient but can cause severe environmental degradation in the absence of strong controls. The Rocky Mountain field contains more than 25 percent of all coal in the United States. Much of the high-grade bituminous coal in the Rocky Mountains must be mined underground because of the complex folding and faulting of the Rockies. Large quantities of lignite suitable for use in power generation are found in the Powder River basin of Wyoming and Montana, and can be mined by strip methods. Both Rocky Mountain and Powder River coal have a low sulfur content and are presently the focus of intense developmental pressures to provide coal that can be burned in thermal generating plants without polluting the environment.

Canada's coal is concentrated in the Rocky Mountains of Alberta and in Saskatchewan in an extension

Figure 7.6 The relative importance of a particular energy source reflects the complex relationship between physical characteristics of a specific source, its availability, and socioeconomic factors.

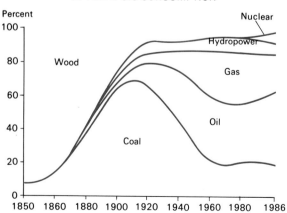

THE WORLD OF WEALTH

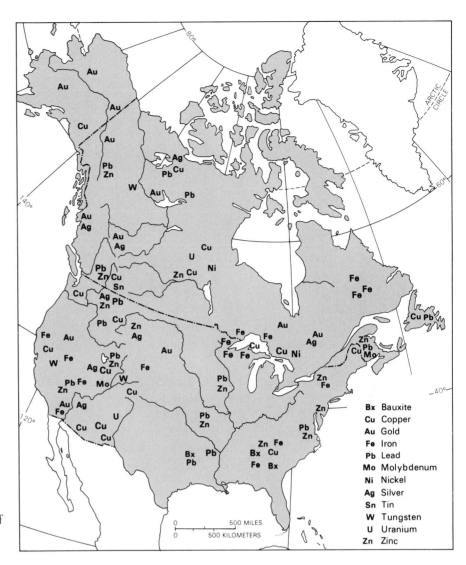

Figure 7.7 Distribution of major metallic minerals in Anglo-America.

of the U.S. Rocky Mountain fields. There is also bituminous coal in New Brunswick and Nova Scotia, but reserves are small and at great depths.

METALLIC MINERALS

The United States and Canada also have abundant reserves of the major metallic minerals. Major iron ore deposits are located around the Great Lakes in the United States and Canada, and in the Quebec-Labrador area of Canada (Figure 7.7). Ores of the Great Lakes mines have been exploited for nearly 100 years, and the high-grade ore is largely exhausted. Today most of the iron must have its iron content increased through a process known as beneficiation in order to make it usable in steel mills. The great Labrador-Quebec field of Canada is an increasingly important area of production. The United States produces much more iron and steel than Canada and consequently imports one-fourth of its total iron ore needs annually, much of it from the Canadian fields. Nearly one-half of the United States iron ore imports come from Venezuela or less industrialized countries of West Africa.

The United States and Canada also have large reserves of copper, lead, zinc, nickel, gold, and silver, but are notably lacking in tin and bauxite. The United States is one of the world's leaders in aluminum production but relies on imports for approximately 97 percent of the bauxite (aluminum ore) it uses. The United States and Canada combined have large productions of nonmetallic minerals and are major world producers of sulfur, phosphate, and potash. All these materials are important for use in chemicals and chemical fertilizer. Both the United States and Canada are important producers of uranium, with Canada normally being the world's largest single producer.

THE ANGLO-AMERICAN REALM

Strip-mining in the winter wheat belt of North Dakota. The wheat lands on the left are in stark contrast to the spoil heaps of strip mining on the right. (Mining is proceeding from right to left in this photo.)

ENVIRONMENTAL ISSUES IN INDUSTRIALIZED COUNTRIES

The major environmental issues in America center on resource depletion and environmental degradation. The United States and Canada have been utilizing their resources for the past hundred years in relatively intensive industrial activity. Resource demand from ever-greater production has depleted much of the high-grade ore, and the great demands of the large market of the United States results in ever-greater reliance on imported raw materials (Figure 7.8). Anglo-American imports of raw materials may benefit less industrialized countries through provision of capital and jobs by transferring income to the developing countries. Some observers maintain that the benefits to less industrialized countries do not equal the loss of resources that they might use in the future. The resources are purchased at low prices while the manufactured end products from the United States and Canada represent the high cost of labor in industrial countries. A frequent comment is that less industrialized countries must process and manufacture goods from their resources before export if they are to improve their economy.

The second major issue is the impact on the environment of resource utilization. Air pollution from increased use of coal to replace dwindling petroleum reserves poses important social and economic costs. Coal can be burned as cleanly as petroleum, but the costs of cleaning the emissions from thermal plants and home furnaces will be great. The large areas that will need to be strip mined in order to utilize the western coal and oil shales will cause major environmental degradation in the West. The issue is complicated by the high rate of energy consumption in the American economy. The United States and Canada consume more energy per capita than any other country, with typical consumption rates being two to three times as great as other industrial countries.

Of critical concern to many is the thesis that the world is now moving from a period of surplus to a period of scarcity. This argument maintains that the easily accessible, high-grade, relatively cheap resources of the world are now nearing exhaustion. The great wealth and materialistic society epitomized by Anglo-America is the result of the age of surplus, and it cannot be maintained as we enter an era of scarcity. The proportion of resources consumed by Anglo-Americans will decline as resources become less available and as developing countries demand a greater share. The present generation will be faced with making the transition from the wasteful habits of the resource-surplus era to the conservation ethic mandated by the evergrowing scarcities of the era of shortages.

DEPENDENCY OF THE UNITED STATES ON IMPORTED RESOURCES (1986)

Minerals and metals		Major foreign sources (1982-1985)
Columbium	100%	Brazil, Canada, Thailand, Nigeria
Graphite	100%	Mexico, China, Brazil, Madagascar
Manganese	100%	South Africa, Africa, France, Brazil, Gabon
Mica (Sheet)	100%	India, Belgium, France, Japan
Strontium	100%	Mexico, Spain
Platinum group	98%	South Africa, United Kingdom, U.S.S.R.
Bauxite & Alumina	97%	Australia, Guinea, Jamaica, Suriname
Cobalt	92%	Zaire, Zambia, Canada, Norway
Diamond (Industrial)	92%	South Africa, United Kingdom, Ireland, Belg.-Lux.
Tantalum	91%	Thailand, Brazil, Australia, Malaysia
Fluorspar	88%	Mexico, South Africa, China, Italy
Chromium	82%	South Africa, Zimbabwe, Turkey, Yugoslavia
Nickel	78%	Canada, Australia, Norway, Botswana
Potash	78%	Canada, Israel, East Germany, U.S.S.R.
Tin	77%	Thailand, Brazil, Indonesia, Bolivia
Zinc	74%	Canada, Mexico, Peru, Australia
Cadmium	69%	Canada, Australia, Mexico, West Germany
Silver	69%	Canada, Mexico, United Kingdom, Peru
Barite	66%	China, Morocco, India, Chile
Tungsten	62%	Canada, China, Bolivia, Portugal
Asbestos	61%	Canada, South Africa
Iron ore	37%	Canada, Liberia, Brazil, Venezuela
Gypsum	36%	Canada, Mexico, Spain
Silicon	35%	Brazil, Canada, Norway, Venezuela
Copper	27%	Chile, Canada, Peru, Mexico
Aluminum	26%	Canada, Japan, Ghana, Venezuela
Gold	21%	Canada, Uruguay, Switzerland
Iron & Steel	21%	EEC, Japan, Canada, South Korea
Lead	20%	Canada, Mexico, Peru, Australia, Honduras
Nitrogen	17%	Canada, U.S.S.R., Trinidad & Tobago, Mexico

Figure 7.8 Dependency of the United States on imported minerals. Canada is a principal source for over two thirds of the individual minerals indicated in the chart.

Higher prices for petroleum are mirrored in increasing costs of other limited resources. Gold increased from $36 an ounce (28 grams) in 1970 to $700 an ounce in 1980, and was at about $400 in 1989. Silver increased from $2 an ounce to as much as $55 an ounce in the same period. Petroleum prices have increased from less than $2 per barrel at the beginning of the 1970s to as high as $50 a barrel in the late 1970s before falling to $16 in 1989. Increasing costs of these resources, it is argued, are indicative of the issues that the era of scarcity will present. Increasing shortages of resources may well dictate a change in life-style for North Americans.

Others maintain that increasing prices will prompt development of substitutes for the limited and costly resources. Proponents of this view claim that the greatest resource of Anglo-America is to be found in the skills and training of the region's residents.

AMERICAN PEOPLES: UNITY OR DIVERSITY

The population of Anglo-America totals nearly 276 million people. The nearly 250 million in the United States and the more than 26 million in Canada are often described as relatively homogeneous societies. The homogeneity reflects attitudes and cultural traits that have developed in these two countries over the last 200 years. These traits and values are those of occupying peoples rather than those of the original inhabitants. They are largely derived from western Europe and reflect the attitudes of the industrial and democratic revolutions emphasizing the work ethic, individual freedom, materialism, democracy, private ownership, and competition.

The work ethic and individualism are highly developed in the United States and Canada. The development of the two countries from the time of European colonization has emphasized the accomplishments of the individual. The emphasis upon individual rights has fostered the development of the private enterprise system to a degree found in few other countries. The abundance of free land and resources has facilitated individualism and the work ethic, as most people in the two countries could obtain either land or business or a job through diligence and hard work. This emphasis on individualism and the work ethic creates problems in the United States as the society moves into the era of scarcity, since the government plays an increasing role in providing basic social needs. Traditional hostility to government makes the transformation of the society to greater government control controversial. Canada has been more willing to accept direct government management of the economy, including operation of one of its two railroad companies.

The emphasis on individualism, hard work, and private enterprise led to the concept that "more is better," a dominant attitude in the two countries for the past 200 years. This emphasis on materialism helps to explain the constant demand for more and bigger houses, cars, recreation equipment, clothing, and other items of material wealth. As the world begins its transition to the era of scarcity, this cultural trait may diminish in importance.

The American ethics of individualism, competition, private ownership, and materialism are in stark contrast to the traditional values of the less industrialized world. Societies such as those of India have existed for millennia with little change because their societies accept submission of individual desires to the needs of the village or family group. Recognition of the finite nature of the resource base in the cultures of the less industrialized world have minimized materialistic demands. Consequently, it has been said that one American is the equivalent of 100 Asian Indians in impact on the environment. One Asian Indian requires less grain and other foods for an entire year than a typical American consumes directly or indirectly in one month. The typical Indian requires less than one one-hundredth as much energy as a typical American. The environmental impact of Americans leads some critics to maintain that America, not India, is overpopulated.

THE DEMOGRAPHIC CHARACTERISTICS OF THE AMERICAN PEOPLE

The concept of an overpopulated America is only possible on the grounds of the consumption and environmental impact of the continent's inhabitants. The actual population is small compared to the land area, with the United States averaging only 69 people per square mile (27 per square kilometer), and Canada averaging 7 (2 per square kilometer). Both countries are in stage three of the demographic transition, with crude death rates of 7 to 9 per thousand and crude birthrates of 15 to 16. The birthrate declined in both countries during the 1970s as increasing costs and shortages forced more women into employment outside the home. This trend to lower birthrates has been emphasized by the increasingly urban nature of the population of the two countries. Only 2 percent of the U.S. labor force and 5 percent of Canada's is engaged in agricultural activities. Approximately two-thirds of the labor force in each country are employed in tertiary occupations. Less than 5 percent are in primary activities; with the balance in the secondary sector. Secondary and tertiary occupations are concentrated in urban areas, and nearly 80 percent of the people in each country lives in urban places.

THE ETHNIC COMPOSITION OF ANGLO-AMERICA

In Anglo-America there are four major racial groups: Indian, Black, Caucasian, and Asian. The first is the original Indian population, who today represent a minority group (Table 7.4). The second is the descendants of European colonists who emigrated to the two countries before the end of the nineteenth century. These majority populations normally speak English, are highly educated, and for the most part are culturally homogeneous in broad cultural values. A third group is made up of blacks, primarily descended from slaves brought to Anglo-America before 1860. The final group are Asians, the earliest of whom arrived in 1849 in California.

The original inhabitants. The North American Indian population suffered dramatically as a result of European exploration and colonization from the time of Columbus. European contacts before formal colonization of the continent led to a tremendous decline in the Indian population as epidemics of smallpox and other European diseases to which the Indians had no resistance swept through the population. Varying estimates place the Indian population of Anglo-America at the time of the settlement of Jamestown as between 2 and 4 million, only one-third as many as 200 years before. The impact of the Europeans after the initial

TABLE 7.4 Population by Selected Ethnic Origin[a]

Group	1990 U.S. Total (1,000)	1990 Canada Total (1,000)
English	54,153	16,514
French	3,329	6,919
German	18,746	578
Italian	6,675	585
Polish	3,821	141
Russian	3,038	74
Spanish	19,887	78
(Black)	31,148	34
(Indian)	1,628	415

[a] Estimates based on *Statistical Abstract of the United States*, 1988 and *Canada Yearbook*, 1988. The majority of U.S. population is not designated by specific ethnic group, so total does not equal total U.S. population.

epidemics was primarily one of exploitation, subjugation, and cruelty.

The European settlers had cultural values that differed markedly from those of the Indian population. European values of private ownership and individual motivation for personal gain were not highly developed among the Indians. Small in number and lacking a central political organization and the sophisticated technology of the Europeans, the Indians of both the United States and Canada were ultimately overwhelmed by the European settlers. After a long history of greater or lesser warfare, the "ultimate solution" to the problems of cultural conflict between the Indian and European settlers in both countries was a form of segregation on Indian reservations. In the United States, as European immigration continued and population expanded, the Indians were pushed into ever-smaller or less desirable areas. The Indian was made a ward of the government and placed on reservations and was not granted citizenship in the United States until the twentieth century.

The Canadian experience with the Indians is similar, though less severe. In Canada there were not the mass expulsions found in the United States, but in both countries the Indian population continued to decline in numbers until after World War II (Figure 7.9). There were an estimated 1,628,000 Indians in the United States and 415,000 in Canada in 1990. They tend to be underenumerated because of the difficulty of reaching them on the reservation, problems of language, and the Indians' general reluctance to deal with the government. In both the United States and Canada they are concentrated in the West, with California having the largest number in the United States, followed by Oklahoma and Arizona.

In general the Indians in the United States live in poverty. It is estimated that nearly one-third of them are functionally illiterate. The average Indian income is only one-fourth of the average per capita income in the United States, and the average life expectancy for the American Indian is 20 years less than that for the rest of the nation. Although Canada and the United States have established programs to alleviate some of their problems, the Indian minorities remain the poorest segment of both countries' populations. Separated spatially on reservations, restricted to areas that often have fewer resources than other areas, and isolated by culture, the Indians of the United States and Canada present a major challenge. As the region of the world with the largest numbers of people enjoying a high standard of living, a challenge for Anglo-America in the 1990s is extending that standard of living to Indians and other minority groups.

The European migrants. The first European occupants of North America were the Spanish in Florida and what are now New Mexico and Texas. Settlements by the French date from 1608 in Canada and the British from 1607 in the United States. Thereafter, migrants to North America increased slowly over the next 200 years. The initial French occupation of Canada was exploitative in nature, focusing on furs. The French encouraged settlement, but they granted land in large tracts to feudal lords known as *seigneurs*. It was difficult to attract migrants from France to New France (Canada) to live under a feudal system different from France only in that it was a new land with all the risks and hazards inherent in occupying a new land. As a result the French population of Canada, numbering less than 80,000 by 1750, was concentrated along the St. Lawrence River, where the land division created a unique landscape. Because of the limited numbers of people and the emphasis on fur trapping, the rivers became the chief transport arteries. Land was divided in narrow strips called *long lots* with a narrow end on the river and extending far back from the shoreline. The landscape is typical of French settlements in Canada and along the Mississippi River in the United States.

Settlement of what is now the United States proceeded much more rapidly. The Pilgrims arrived in Plymouth, Massachusetts, in 1620, followed by the Puritans in Massachusetts Bay in 1629. By 1640 an estimated 25,000 British migrants had moved to Massachusetts. The primary motivation of these initial groups was economic and religious. Both were interested in freedom of worship, but after establishing their villages in the New World, they did not extend this religious freedom to others. Subsequent migrants came because of the opportunity of free land and financial gain. In the southern portion of the United States the Jamestown colony, settled in 1607, ulti-

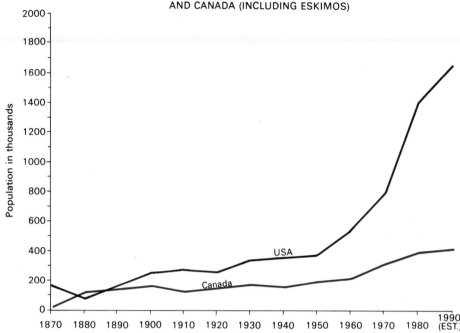

Figure 7.9 The population of American Indians in both the United States and Canada is increasing much more rapidly than the general population. The 1980 Indian population in Canada and the United States is still probably lower than it was in 1492.

mately began to produce tobacco, which provided a cash crop. The need for labor for the tobacco plantations and for production of rice and indigo for European markets exceeded the number of migrants, and as a result, the importation of slaves developed in this region.

The population of the middle colonies of New York, Pennsylvania, Maryland, and New Jersey also grew rapidly. Pennsylvania became a haven for those who disagreed with the religious beliefs of the Puritans and the Pilgrims. The availability of rich, fertile, low-cost land attracted Europeans from Germany, Britain, Switzerland, and other areas of western Europe. The population of Europeans and blacks in North America increased slowly during the first 150 years of settlement, and in 1776 there were only 2 million in what was to become the United States.

The major migrations of people to the New World came as a result of the Industrial Revolution and political conflicts in Europe in the nineteenth century. The numbers of migrants during any given period depended on political and economic conditions at that time in Europe. In the early 1800s large numbers of Scottish people emigrated as a result of loss of lands to landlords interested in grazing sheep for the emerging textile mills of England. During the mid-1800s large numbers of Irish migrated, many as a result of the potato famine. The increased population of Europe resulting from stage two of the demographic transition in the nineteenth century forced large numbers of British, Germans, Scandinavians, and Swiss to the new lands. Five main periods of migration to North America can be recognized. From 1650 to 1800 the emigrants were primarily from Britain, Germany, and Ireland. From 1830 to 1860 they came from the Celtic fringes of Britain (Scotland, Wales, and Ireland) and the Rhine River area of Germany. From 1860 to 1890

This Navajo hogan in the Monument Valley area of Utah is still home for an American Indian family. Isolation from schools, medical care, and employment severely handicaps many American Indians.

THE WORLD OF WEALTH

The distinctive long lot pattern of land division produced by the French seigneurial system is evident in this photo.

the migrants came primarily from the British Isles, Germany, and Scandinavia. From 1900 to 1914 they were primarily eastern and southern Europeans. From 1965 to the present, immigrants are overwhelmingly from Latin America, the Caribbean, and Asia.

As a result of the migrants of the nineteenth century, the United States and Canada have sometimes been described as a melting pot. The new migrants to the United States were expected to learn English, as the public school system made no provisions for ethnic groups to maintain their own languages except in the case of the French in Quebec. Thrust into a new social environment where individuals were judged on their own willingness to work and participate, most European migrants encouraged their children to adopt the traits of the dominant culture.

The twentieth century witnessed a change in origins of migrants to the United States and Canada. From 1900 until World War I, large numbers of peoples from southern and eastern Europe, especially Italy, the Ukraine, and Poland, came to the United States as a result of wars, political unrest, and economic problems. At the same time numbers of Chinese were brought to California and Hawaii to work in agriculture. Because the cultures of these people were recognizably distinct from those of the western Europeans, the United States adopted a restrictive immigration system in the 1920s. This quota system was based on the numbers of a specific ethnic group in the United States as of the 1890 census, thus very effectively discriminating against Asians and southern and eastern Europeans. Consequently, large numbers of eastern Europeans from the Ukraine, Poland, and Russia immigrated to Canada in the period between World War I and World War II. In 1965 the United States changed its immigration laws, and once again the origin of migrants has changed. Since the 1960s the immigrants to the United States have increasingly represented Asian and Latin American rather than European origins (Figure 7.10).

CULTURAL PLURALISM AND ETHNIC MINORITIES

Most of the migrants to the United States and Canada have been integrated into one superficially homoge-

Immigration: an American ethic

Between 1600 and 1985 over 75,000,000 immigrants abandoned their homelands with the specific purpose of establishing a new life in a new country. More than 50 million of these migrated to the United States and more than 7 million to Canada. Over the past four centuries nearly three-fourths of all the world's migrants have come to Anglo-America, over two-thirds to the United States. This situation continues today, as the United States accepts more migrants than all the rest of the world combined.

No one knows exactly how many migrants enter the United States yearly. The law specifies that a total of 290,000 migrants will be admitted to the United States yearly, plus refugees. Refugees are defined as people who cannot return to their homes "because of persecution or a well-founded fear of persecution on account of race, religion, nationality, membership in particular social groups, or political opinion." The number of refugees fluctuates and exceeded the number for general migrants in the late 1970s and 1980s as Cuban, Vietnamese, Laotian, Iranians, Russians, and others who met the definition of refugee came to the United States. In addition, an unknown number of people entered the United States illegally, pushing the average annual number of immigrants to an estimated 600,000 to 800,000 per year in the last decade.

The geographical image of immigration is far-reaching. Historically New York City was the gateway for migrants, with over 80 percent of migrants during the peak period of 1880–1924 passing through New York. Immigration since 1965 is at a level that will rival the 1880–1924 period if continued, but today's migration enters primarily through California (especially Los Angeles), Arizona, New Mexico, Texas, and Florida. The impact of millions of refugees and immigrants in the last 20 years has been dramatic. In Los Angeles County in 1960, only 1 percent of the population was Asian, while 10 percent was Hispanic. In 1985 one-third was Hispanic, 10 percent Asian, and Los Angeles ranked second to Mexico City as home for Mexicans. The influx of Cubans, Central Americans, and Caribbean Islanders to Florida has led to classification of part of the city of Miami as "Little Havana," and intensified racial and ethnic conflict.

The issues raised by increased immigration are numerous and will undoubtedly increase as there does not seem any acceptable way of curbing illegal immigration from the less industrialized countries of Central and South America. The United States Immigration Service captures some 1 million illegal immigrants yearly and returns them home, but an untold number avoid capture and as many as 1 million annually enter the United States, some of whom become permanent residents. Ironically the economic and political conditions that motivate the illegal migrant may be as great as those of the legal refugee, but the United States does not recognize those fleeing the poverty of Central America, South America, or the Caribbean Islands as refugees. The geographic realities of the United States border which is both long and largely unguarded presents the opportunity for illegal entry, while the nation's perceived social and economic opportunities provide a strong magnet to the ever growing populations of poorer neighbors. Faced with a life of poverty at home, millions of people are determined to try to become part of the "American Dream," a dream of a lifestyle characterized by freedom, social mobility, and economic opportunity. The dream has changed little in the last three centuries; only the geographic, ethnic, and racial origin of the dreamers change.

neous whole. With a few minor exceptions most migrants have adopted English as their primary language, basic dress styles are similar for all groups except during ethnic festivals, and the values of individualism, competition, and materialism have come to dominate nearly all residents. Nevertheless, there are important ethnic groups that have their own cultural characteristics.

Ethnic groups in America. There are major minority groups in both Canada and the United States but they are different. In Canada one minority group, the French, constitutes 30 percent of the country's population. With slightly more than 7 million members, the French-speaking minority, located largely in Quebec, poses important problems for the country. The French are predominantly Catholic while the rest of Canada is predominantly Protestant, and they are separated geographically from the rest of the country. Election in 1976 of a party dedicated to separation of Quebec from the rest of Canada led to passage of legislation specifying that in Quebec only those children whose mother or father attended English elementary schools will be allowed to attend English-speaking schools. By the mid-1980s all businesses in Quebec were required to have certificates certifying that they are promoting the use of French and the employment of French-speaking people. French is to be used as the language of public administration in government, schools, and social services. The avowed goal of the separatist party of Quebec is withdrawal from the confederation of Canada and creation of a new, independent French nation of Quebec. In 1980 the residents of Quebec voted on whether or not they

CHANGING PATTERNS OF IMMIGRATION INTO THE UNITED STATES

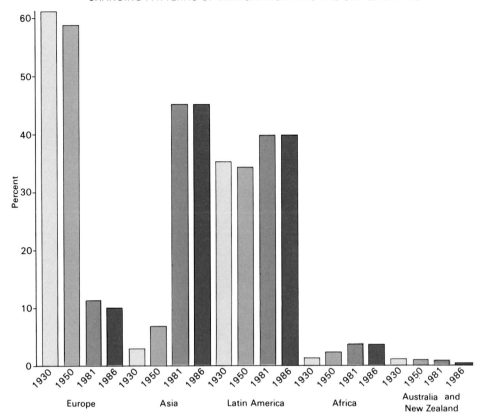

Figure 7.10 In the past decades the percentage of immigrants of European origin has decreased dramatically while the arrivals from Asia have experienced a marked increase. Change in origins and numbers arriving from the less industrialized world have prompted attempts to change the immigration laws.

should separate from Canada. By a narrow majority the Quebec voters defeated the separatist proposal, but the French language and separatist issue remains a major centrifugal force in Canada.

The United States has a more varied ethnic minority pattern, without the dominance of one minority group in a specific geographical area. The largest ethnic group in the United States is the blacks, totaling an estimated 31 million in 1990, or 12 percent of the population. Unlike the French, the black population of the United States is not culturally and geographically isolated in one contiguous area. Slightly more than half of American blacks live in the South. The black American speaks English, is as apt to share the characteristics of competition, materialism, and individualism with other citizens, and has no distinctive religion. Blacks in the United States have not been treated equally in the past or at present, even though legally no official discrimination is allowed. Over 57 percent of American blacks live in central city slum areas while only 27 percent of whites live in the central city (Table 7.5).

The black population of the United States is overwhelmingly concentrated in areas that are effectively segregated from the balance of society. Known as **ghettos,** these areas comprise older sections of large cities that have been abandoned to blacks or other low-income groups. They are characterized by abandoned buildings, abandoned stores, dilapidated housing, inadequate police and fire protection, the highest crime rates in American cities, and a general sense of deterioration. The residents of the ghetto, whether black, Spanish, or other minorities, have been said to have a *culture of poverty*. The culture of poverty,

TABLE 7.5 Black Population in Central Cities

Year	Total (1,000)	Percentage
1960	9,702	51.4
1970	13,145	58.2
1980	15,028	57.8
ª1990	17,680	57.0

SOURCE: *Statistical Abstract of the United States*, 1983 and 1988.
ªProjected from *Statistical Abstract*, 1988.

Ghetto: A distinct section of a city characterized by a particular ethnic composition. In American cities ghettos often comprise the old central city with its high concentration of poor, elderly, or ethnic groups unable to move to more desirable locations.

Illegal aliens crossing the border between the United States and Mexico near Tiajuana, Mexico, illustrate the gulf in economic opportunity between the two countries.

some maintain, is distinct from the dominant U.S. culture composed of middle-class, future-oriented families. People with a culture of poverty tend to be fatalistic and present-oriented. A high proportion of families is headed by the mother (for the black population 59 percent of all families with children have only a female parent). These people do not share the common outlook of optimism of a majority of the residents of the United States. Because interactions with authority figures normally take the form of confrontations with police, welfare workers, school officials, or landlords who are seen as threatening, this group tends to be hostile to all figures of authority. The result is the widespread feeling that their poverty is the result of conscious action on the part of the white majority population. There is constant potential for violence against authority figures such as police, landlords, or even fire departments. It has taken its greatest toll in the form of riots in places like Watts in Los Angeles, in Chicago, in Detroit, in Newark, and most recently in Miami. The death and property damage from such riots, however, is overshadowed by the persistent loss of human resources because of the existence of the ghetto and the destructive attitudes it fosters. The persistence of the culture of poverty is an important element of the geography of the United States. The concentration of the poor through economic discrimination creates distinctive landscapes in the United States, whether in rural locations or the ghetto. Appalachia, inner-city slums, and Indian reservations comprise distinctive regions recognizable largely by their poverty.

The Spanish-speaking minority. The Spanish-speaking minority (Hispanic) consists of three major groups: residents of Puerto Rico and the eastern seaboard cities, Mexican-Americans concentrated in the southwestern border states, and Cubans concentrated in Florida. There were 14.6 million Hispanic citizens of the United States according to the 1980 census, and an estimated 20 million in 1990. Hispanic Americans are underrepresented in the census because many of those from Mexico and Latin American countries are illegal aliens who have come to the United States to work. These "undocumented residents" are afraid to respond to census enumerators and are estimated at between 5 and 8 million. Actual numbers of Hispanics approach the total black population and should make them the largest American ethnic group by 2000.

Nearly three quarters of the Mexican-Americans are concentrated in Arizona, California, Colorado, New Mexico, and Texas. They represent the most rapidly increasing population group in the United States because of a high birthrate and a high immigration rate. As a group they are culturally more distinct from the typical American than the blacks. They are predominantly Catholic, are fatalistic, and have less education than even the blacks. Incomes vary for the different Spanish-speaking groups, but only the American Indian has a lower standard of living than the Puerto Ricans, with over one quarter of the Hispanic population receiving less than the official poverty income of $11,500 for a family of four. This income is high by world standards, but because of **relative deprivation,** it is insufficient to allow its recipients to live in a manner similar to the mass of the American population.

The Spanish-speaking minority in the United States is reluctant to adopt the values of the dominant cultural group. There is increasingly a demand for bilingual education to allow Spanish-speaking children to utilize Spanish in their educational programs. Although such an action would possibly facilitate the acquisition of academic information during the first years of school, it would handicap the Spanish-speaking children in their attempts to gain economic and cultural equality. The existence of a large and growing minority population such as the Spanish-speaking Americans, who are increasingly committed to maintaining their own cultural traits, is one of the issues facing Anglo-America in the future. Seventeen states had adopted English as their official language by 1989 in response to ethnocentric fears of increased numbers of Spanish leading to bilingualism. The old concept of a melting pot is being replaced by the concept of a plural society. The problems this creates in terms of communication, cultural integration, and education are obvious.

ASIAN MINORITY GROUPS

Other minority groups in the United States are smaller than the Spanish-speaking or black populations and have been less vocal in their demands for equality and cultural recognition. Major groups include the Japanese, the Chinese, and the Vietnamese. The Japanese have historically constituted the largest Asian minority group although none came to the United States until 1868. Unlike other groups, they tend to equal or exceed average per capita incomes and literacy rates and are not part of the culture of poverty.

The Chinese in North America came in large numbers to serve as laborers in California after the gold rush of 1840. Because of white persecution the

Urban blight strikes even buildings that are structurally sound, as in these apartment buildings in Yonkers, New York. Ghetto areas of the United States metropolitan centers contain many neighborhoods where blight is rapidly destroying the housing of low income families.

Chinese have historically been forced into ghettos (Chinatowns) in the major cities, allowing them to maintain themselves as ethnically distinct. Few fall below the poverty level, and they have a higher education level and average family income than the white majority population. They are concentrated in Hawaii, especially in the Honolulu area, the Boston-New York region, and in the San Francisco-Los Angeles region in the United States. Vancouver is a major center for the Chinese in Canada.

The Vietnamese have come since the Vietnam War. Large numbers have come to the United States as a result of political problems in Southeast Asia; they number over one-half million at the present time. They are spread widely across the United States but are concentrated in large cities, especially in California.

Such an analysis of the complex American ethnic groups is at best only superficial. Within each group the cultural nuances are so subtle and interwoven with the personalities of the group members that they are difficult to isolate. This brief overview of the ethnic groups in the United States and Canada is simply designed to indicate the complexity of the American population. Although commonly viewed as one homogeneous whole, both Canada and the United States are made up of numerous ethnic groups of greater or lesser uniqueness. The issues that these ethnic groups pose center on the questions of **assimilation** versus

---

Relative deprivation: Referring to relative levels of poverty rather than absolute poverty. A millionaire may feel poor compared to a billionaire, but both are wealthy by absolute standards.

Assimilation: The process of integration of one culture into another so that the former loses its distinctiveness.

acculturation and resulting **cultural pluralism.** Assimilation of cultural groups through mandatory schooling using only English enables them to move fully into the economic and social mainstream of Canada or the United States. Cultural pluralism is a centrifugal force as the case of the French Canadians of Quebec shows. The issue is whether the centripetal forces of cultural assimilation are of sufficient value to outweigh the cultural loss associated with enforced assimilation. In the United States, a secondary issue is that of racial and ethnic prejudice. Coupled with this is the broad gulf between the incomes of those in the culture of poverty and the majority of the American population. The necessity for overcoming racial and ethnic prejudice and the need to develop an adequate system to assure that all of America's peoples share in its wealth are some of the most pressing issues facing North America.

THE GEOGRAPHIC MANIFESTATION OF ETHNIC GROUPS

There are many geographic manifestations of ethnic groups and the culture of poverty in the United States, beginning with the concentration of ethnic minorities in the central cities. Ironically, in spite of the federal efforts to end segregation and discrimination, the proportion of blacks forced into the poorest neighborhoods and least desirable areas of America has increased. It is sometimes erroneously argued that the black ghetto persists because blacks or other ethnic minority groups chose to live there. While it is true that new migrants from the rural South or from foreign areas find the ghettos a home for a period of transition to life in the large city, the majority of blacks and other groups want to leave the central cities with their high crime rates and other social problems as much as anyone else. Overt and covert restrictions have effectively prevented them from moving to the predominantly white suburbs.

Another example of the geographic impact of minority groups is in the routes of freeways. They are apt to go through the ghetto areas of cities rather than through the wealthier white areas because of the political power of the white population. In addition, the location of new hospitals, clinics, schools, and other urban services tends to reflect the political power of the white majority population.

Another geographic manifestation of ethnic groups is the Indian reservations. Tradition, poverty, lack of skills, administrative convenience, and personal choice have combined to restrict many Indians to the reservations. It has been easier for the public agencies charged with administering programs to act when the Indians are isolated from the majority population. Few people visit the Indian reservations, and there has not been public outrage over the impoverished conditions. It is also cheaper for the government to operate a program when the recipients of the services are concentrated in one place. For the Indian the reservation provides the security of a familiar culture, even though the standard of living may be lower in other locations. Even when Indians move to the city, they are still discriminated against in housing, jobs, and general welfare because their background, education, and training have not adequately prepared them to move into the mainstream of America's economic life.

CITIES: HOME OF AMERICANS

The population of the United States and Canada is concentrated in urban places. Canada's population is clustered near the border of the United States, with over three fourths living within 200 miles of the United States border. The major centers of population in Canada are in Ontario and Quebec, where 60 percent of the Canadian people reside. Major centers include Montreal, Quebec, Ottawa, and the western edge of Lake Ontario from Toronto south. In addition, there are three major clusters of population in western Canada at Vancouver, Calgary, and Edmonton. In the United States the population is much more widely distributed over the entire country. In 1990 the center of population was just south and west of St. Louis. The population of the four major regions of the United States is shown in Figure 7.2.

The present distribution of the American population masks important changes that will affect the future of the two countries. The people of Canada and the United States represent one of the most mobile populations in the world. It is estimated that nearly 1 in 5 citizens of the United States changes residence yearly. The great mobility of the American population can be classified into several distinct trends:

1. The historic trend of movement from rural to urban areas.
2. Movement from the central city to the suburbs within urban areas.

Acculturation: Changes occurring in a culture through intercultural borrowing. Changes may include technology, language, values, and so forth.

Cultural pluralism: Variety of cultures living in a country, but not necessarily in distinct territories.

3. Movement from the Northeast to the southern and western states.

The mobility of the American population is characteristic of an industrialized market economy. Social and geographical location are highly flexible, and an individual's location is a function of personal preference and career goals rather than birth and place in society. The mobility of the American population has also been fostered by cultural homogeneity. Lack of language and legal barriers has allowed residents to move to areas perceived to have a higher quality of life rather than being restricted as a result of dialect, tradition, or social stratification as is common in less industrialized countries of the less developed world. In the past, the movement of population in the United States was from rural areas to cities with their perceived higher quality of life and access to educational, cultural, and job opportunities. This trend is changing as modern transportation and communication provide even rural areas with such amenities, while urban areas are perceived as less desirable because of problems of obsolescence, crime, and environmental pollution. Many Americans seem to view the central urban area as an undesirable place to live (Table 7.6). Suburbs, smaller towns and cities, and rural areas are attracting migrants today. The perception of rural areas as more desirable is heightened by the movement of industry to these same areas to take advantage of lower land prices, lower taxes, and lower wages of nonunion labor forces. (See Box.)

THE DEVELOPMENT OF URBAN AMERICA

In spite of the changing nature of internal migration, over three fourths of the Anglo-American population lives in towns and cities. The concentration of population in cities is a phenomenon of the twentieth century and the result of the intensification of the industrial and technological revolutions. Prior to the twentieth century, Anglo-America was similar to other less developed countries in that most of its population was engaged in agricultural activities. The development of the cities of North America reflects the technological changes associated with the Industrial Revolution. Cities grew and developed with changes in transportation.

The first period of urban development in North America was similar to the emergence of incipient urban centers anywhere transportation was by animal or wind power. Ships were the basic movers of bulk goods, and break-in-bulk points grew up along coastal areas and fords across rivers. The most important cities in the United States during the sail and wagon era from 1790 to 1830 were New York, Boston, Philadelphia, and Baltimore. In Canada, Quebec and Montreal grew during this period. In addition, inland cities developed where land and water transportation met, as at Pittsburgh, Cincinnati, and Louisville on the Ohio River. Where land routes met water routes, bulk cargoes of ships were broken down into smaller lots for transport by wagon. Such locations are known as break-in-bulk points, and most of the largest cities in Anglo-America and the world serve a break-in-bulk function.

The next major era of city growth in the United States came with the technological innovation of canal boats. Canals allowed movement of bulk goods overland, and cities like Albany, Buffalo, and Rochester on the Erie Canal emerged as break-in-bulk points during this period. At the same time, New York City experienced a period of rapid growth as its hinterland expanded across the Appalachian Mountains into the Ohio Valley and central lowland via the canals.

The canal era had just begun when the steam engine was invented, and steamboats became an important form of transportation on America's inland waterways. The steamboat allowed movement of greater volumes of goods. It quickly overshadowed the canals as a means of transportation. Cities that developed rapidly during this time included New Orleans and St. Louis on the Mississippi and Cincinnati, Pittsburgh, and Louisville on the Ohio, all of which benefited from the increasing traffic along the Mississippi.

The steamboat era lasted from 1830 to 1860, but as the steamboats were reaching their peak in numbers, they began to be challenged by the steam railroad. In 1850 there were only 9,000 miles (14,500 kilometers) of railroad track in the United States, but by 1870 there were 53,000 miles (85,300 kilometers) including a line spanning the continent from the east to the west coast. The cities that developed most rapidly were initially the water break-in-bulk cities, as the railroads were first viewed as simply a means of extending water transportation. Cities such as New York, Baltimore, and Philadelphia all grew rapidly as a result. Other cities based on the railroad then grew up, the most important being Chicago which grew from 4,479 in 1840 to 109,000 in 1860 as a result of

TABLE 7.6 Population of United States in 1986 (in Thousands)

Large metropolitan	115,300
Other metropolitan	69,455
Nonmetropolitan	56,323
Total	241,078

SOURCE: *Statistical Abstract of the United States,* 1988.

becoming a railroad hub. Omaha, Nebraska; Albuquerque, New Mexico; and Cheyenne, Wyoming, developed as points to provide service for the railroad itself. The increased demand for rails and rolling stock affected the iron and steel industry, leading to growth of cities such as Pittsburgh; Birmingham, Alabama; Bethlehem, Pennsylvania; and Gary, Indiana, to provide greater quantities of steel.

The emergence of the internal combustion engine in the twentieth century affected cities less in terms of new foundations than in changing the **morphology** of existing cities. Prior to the widespread adoption of the automobile urban functions of manufacturing, commercial, and residential uses reflected transportation by ship, wagon, or railroad. Economic activities were located along the transport route, either at the harbor or the rail terminal. A central business district normally developed near the manufacturing core, and residential areas for workers were near the factories. Wealthier families developed higher-cost housing in suburbs accessible by rail and later by urban trolley lines. Difficulty of transportation spawned neighborhood stores throughout the city.

The advent of the automobile freed Americans from the constraints of the railroad, as it freed urban residents from the need to live close to their work or to a mass transportation system. The result is the sprawl city, typified by Los Angeles, which has led to the highest per capita gasoline consumption in the world and filled the air with smog. All American cities share the urban sprawl of Los Angeles to some degree. Suburbs and freeways have proliferated, the downtown business district faces increasing competition from suburban shopping malls accessible only by auto, and manufacturing activities are moving to suburban locations. Small neighborhood stores have largely disappeared as rapid transport by auto obviates their necessity. The air age has simply contributed to the existing pattern that was set during the initial stages of transportation technology and intensified by changing technology and human decisions.

The importance of transportation in understanding city development is not only the speed of transportation but the frequency. The large cities of Anglo-America are tied together by an intricate network of rail, road, water, and air linkages. The net effect of these multiple linkages is to minimize the travel time between large centers. The development of multiple transportation linkages reflects the importance and size of the cities involved. Major centers such as Los Angeles and San Francisco or Washington, D.C. and New York City have more interconnections than places such as New York City and Ogdensburg, New York. The effect of such linkages is to make it possible to travel from New York City to Chicago more rapidly than from New York City to Ogdensburg, although Chicago is three times as far away.

MEGALOPOLIS

The largest and most important urbanized region in Anglo-America is the Northeast, from Boston to Montreal, Montreal to Milwaukee, Milwaukee to St. Louis, St. Louis to Washington, and Washington to Boston. This region includes over half the population of Canada and United States combined. It is the most intensively developed area and has historically dominated the continent. Within this major urbanized area of North America are found most of the largest cities of both countries (Figure 7.11).

The focal point of this urbanized region stretches from Boston to Washington, D.C., and is known as **Megalopolis.** Approximately 50 million people, one-fifth of the population of the United States, as well as one-quarter of the value added by manufacturing, and one-third of the wholesale trade of the country are found here. It is a major producer of all types of manufactured goods, and yet much of the region has only limited resources. The largest cities of Megalopolis almost completely lack local fuel sources. Local farms produce only a fraction of the food needed by the area. The most urbanized portion of the region has limited space for expansion and recreation and faces severe problems of adequate water and waste disposal. The dominance of North America by Megalopolis is a result of its situation, which has led to its becoming a focal point for transport routes, both nationally and internationally. These transport routes allow the cities of Megalopolis to tap much of Anglo-America for food, fuel, and markets, creating a core area unrivaled elsewhere on the continent.

The cities within Megalopolis are important collectively and individually. New York City is the largest in North America and is a world center for manufacture of men's and women's clothing, for banking and finance, and leads in cultural trends. Washington is the national capital and is increasingly the focus of business as a result of the role of the government in the emergent central state capitalist system. All major cities of Megalopolis—Washington, Baltimore, Philadelphia, New York, and Boston—are old break-in bulk points that have maintained their importance in

---

Urban morphology: The pattern and relationship of land use, economic activity, and population distribution in urban areas.

Megalopolis: Term used to refer to "supercities" formed by coalescing of large cities as in the Boston to Washington area.

THE WORLD OF WEALTH

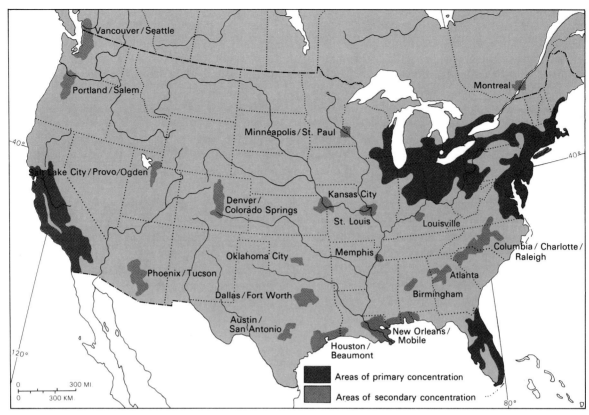

Figure 7.11 Present and projected areas of metropolitan concentration of population in Anglo-America.

spite of changing transportation technology. New York City's location on the Hudson River and Manhattan Island created the initial favorable site-situation relationships to allow it to develop, and the emergence of the Hudson-Mohawk routeway of the Erie Canal as the major route across the Appalachians insured its continued dominance. Subsequent development of railroads followed the same general route, ensuring that New York would remain the dominant eastern city. Philadelphia and Baltimore, although at one time larger than New York, were never able to expand their hinterlands across the Appalachians as New York City did.

NONCONTIGUOUS METROPOLITAN AREAS

There is no other area in North America with such a concentration of urban activities as New York and Megalopolis. There do exist a number of urban areas that are not contiguous to Megalopolis but are of major national significance.

Older major cities of Anglo-America outside Megalopolis include Montreal and Toronto in Canada and Detroit, Chicago, and other industrial cities in the United States. Toronto's role in Canada is similar to New York's in the United States. It is the largest single industrial city within Canada in terms of value of production. It is a focus of land, rail, and air transportation systems. Toronto is a break-in-bulk point for goods moving from the fertile Ontario Peninsula to Lake Ontario and the Saint Lawrence Seaway. The western end of Lake Ontario from Oshawa (east of Toronto) to Niagara Falls is known as the *Golden Horseshoe* and is the center for production of iron and steel, automobiles, and other manufactured goods.

Chicago has had a role analogous to New York City for the central part of the United States. It serves a manufacturing, banking, government, cultural, and port function for the interior. Cities such as Buffalo, Cleveland, and Pittsburgh were all established as break-in-bulk points and today are important manufacturing centers providing the full range of urban functions. Thunder Bay on the western shore of Lake Superior is the major break-in-bulk point for western Canada.

TWENTIETH-CENTURY METROPOLITAN CENTERS

The movement of population in the United States and Canada in the last two decades has fostered the growth of several large urban centers in western Canada and the United States and in the southern United States (Table 7.7). Atlanta has emerged as a regional center for the southeastern United States. It has been im-

THE ANGLO-AMERICAN REALM

portant because of its site-situation relationships since the nineteenth century and in the twentieth century has became a hub of air routes that extends into the Caribbean and Latin America.

Houston and Dallas are among the most rapidly growing in Anglo-America. Their modern development reflects government spending on the space program and the increasing importance of the Gulf of Mexico's fuel resources. The importance of the two cities as banking, manufacturing, research, educational, and cultural centers for the south central United States makes them rivals of Chicago as the focus of urban activities in the United States interior.

The growth of western metropolitan centers in both the United States and Canada reflects the increasing importance of mineral resources of the West and government defense spending. Cities like Denver or Seattle in the United States have experienced rapid population growth as military spending provided a catalyst to the local economy. In Canada, Calgary and Edmonton are centers for Canada's petroleum resources and industry. The petroleum dollars have attracted banking and manufacturing to the two cities, which provide the same general services for the Canadian interior as Dallas, Houston, and Denver do for the interior United States. It is doubtful if the twentieth-century metropolitan areas will equal the New York–Boston–Washington megalopolis in size or importance, but they are indicative of the increasing size and importance of the urban areas of North America. Increasing costs of fuel may change the role of the metropolitan areas as automobile-generated sprawl is slowed or reversed. Changes in the economy from technological and social changes may foster renewed population growth in older metropolitan areas.

THE ECONOMY OF NORTH AMERICA

The economy of North America epitomizes that of the developed, industrialized, interconnected world. The productivity of American industry and agriculture is the basis not only for the wealth of its inhabitants but for a variety of exports that affect much of the rest of the world. Decisions affecting American interests also affect both the less industrialized world and the balance of the industrial world. American tariffs to protect sugar beet farmers from the cheaper cane sugar effectively reduce the market for the major export crop of much of the Caribbean and contribute to the poverty of many farmers of that area. In similar fashion, American unions' demands for high wages have prompted American firms to locate assembly plants in Mexico, Haiti, Hong Kong, Taiwan, and in other emerging countries with low wages. The wealth of the American population and its constant demand for food and manufactured goods either directly or indirectly affects the life and livelihood of a significant portion of the world's population. The demand for petroleum by the auto-oriented American culture is a major factor in determining world price and availability of petroleum. The United States and Canada together constitute the premier food-surplus region of the world, and the very livelihood of much of the world is dependent upon grain exports from these two countries.

AMERICAN AGRICULTURE: CORNUCOPIA OR DISASTER

Agriculture in the United States and Canada employs a small minority of the labor force. Two percent of the population of the United States and approximately 5 percent of the population of Canada is engaged in farming, and yet the United States and Canada combined are the world's leaders in production of corn, soybeans, meat, milk, eggs, and a host of other products (Table 7.8). The United States produces more total grains than either China or India, but it does so with a fraction of the workers. The great productivity

TABLE 7.7 Twenty Largest Metropolitan Areas in North America (1986)

		Population (Thousands)
1.	New York, New York	8,473
2.	Los Angeles, California	8,296
3.	Chicago, Illinois	6,188
4.	Philadelphia, Pennsylvania	4,826
5.	Detroit, Michigan	4,335
6.	Dallas–Fort Worth, Texas	3,655
7.	Washington, D.C.	3,563
8.	San Francisco–Oakland, California	3,522
9.	Toronto, Ontario	3,427
10.	Houston, Texas	3,231
11.	Montreal, Quebec	2,921
12.	Boston, Massachusetts	2,824
13.	Nassau–Suffolk, New York	2,635
14.	Atlanta, Georgia	2,469
15.	St. Louis, Missouri–Illinois	2,438
16.	Minneapolis–St. Paul, Minnesota	2,295
17.	Baltimore, Maryland	2,280
18.	Pittsburgh, Pennsylvania	2,123
19.	Newark, New Jersey	1,889
20.	Cleveland, Ohio	1,850

SOURCE: *Statistical Abstract of the United States,* 1988, and *Canada Year Book,* 1988.

TABLE 7.8 Agricultural Production of the United States Compared to World (1986)

Commodity	Percent of World
Corn	45.9
Meat	15.7
Sugar	6.01
Rice	1.3
Wheat	12.9

SOURCE: *Statistical Abstract of the United States*, 1988.

of North American agriculture is the result of several factors, including application of industrial technology to agriculture, favorable environmental conditions, and an affluent society and large market. The combination of these three factors results in great total productivity, but the yield per acre is lower than that of Japan or of other countries practicing intensive garden-type agriculture.

American farms are large-scale, extensive operations. The average farm size in the United States is 475 acres (192 hectares); in Canada it is 460 acres (186 hectares). The large Canadian and United States farms are possible because of application of industrial technology through mechanization, which also allowed the farm population to decline from 25 percent to less than 3 percent of the total population between 1930 and 1990. Farms have become larger, fewer in number, and are increasingly agribusinesses rather than farms. The largest farms in the United States and Canada constitute a minority of the total but produce a majority of total crops. More than 60 percent of farm sales come from less than 10 percent of all the farms.

The primary market for American farm products is the United States and Canada, one of the wealthiest and best-fed areas of the world. Each American consumes more than 100 pounds (45 kilograms) of beef, 60 pounds (27 kilograms) of pork, and 70 pounds (32 kilograms) of poultry yearly. It takes 10 pounds (4.5 kilograms) of grain to produce a pound (1/2 kilogram) of beef, 5 pounds (2 kilograms) of grain to produce a pound of pork, and 3 pounds (1 1/2 kilograms) to produce a pound of poultry. Each American annually consumes approximately two-thirds of a ton of grain in meat products alone. Tremendous quantities of grain are also used for alcohol, bread and pastries, and feed for dairy animals.

American farming is capital intensive and relies on world market conditions for profitability. In addition to the large internal market, the United States and Canada rely heavily on foreign markets for agricultural sales. The production of one-third of the farmland in the United States is exported. The United States exports two-thirds of its rice production, half of the wheat and soybeans, one-third of the cotton, and one-fourth of the corn it produces. The United States is the source of one-half of the world's exports in coarse grains (barley, rye, and oats) and soybeans, yet it produces only one-third of the world's total coarse grains. It accounts for 30 percent of the world trade in cotton and wheat, though it produces only 16 percent of total world wheat. The United States and Canada account for nearly one-half of total wheat exports, which is only 12 percent of the world's total production of wheat but the margin between survival and famine for many countries in the world.

Agricultural regions of North America. The United States and Canada are unequally endowed with agricultural land. The majority of Canada's land area is unsuitable for agriculture because of climate. Agriculture is concentrated along the St. Lawrence from Montreal to the southwestern end of the Ontario Peninsula and in the western part of Canada from Winnipeg to the Rocky Mountains. The United States has agriculture in most areas east of the Rockies with the exception of rugged or poorly drained regions. Broad agricultural regions in the two countries are based on the primary crop-livestock combination (Figure 7.12). In the northeastern portion of the United States and the bulk of Canada's agricultural areas, general farming emphasizes grains and livestock, particularly dairying. The typical agriculture in this area is known as mixed farming because no one crop dominates and because the farmers rely on both crops and livestock for their income. The primary reason for specialization in livestock and dairy products in this region is the large urban market of the Northeast. There are important regions of vegetable and fruit production within this region as well in response to economic and environmental factors.

The southeastern United States is commonly called the Cotton Belt. But cotton has declined dramatically in importance in this area as a result of loss of soil nutrients, competition from irrigated cotton lands farther west, and development of alternative crops such as soybeans and peanuts. The Southeast is still a distinctive region because of its climate. It is bounded to the north by length of growing season, to the south by heavy rainfall during the cotton harvest season, and to the west by lack of precipitation. South of the Cotton Belt the Gulf Coast and Florida are important areas for production of subtropical crops of sugar cane, rice, citrus fruits, and other subtropical crops that require a long growing season.

The Midwest of the United States from the Appalachians through Illinois, Indiana, Iowa, Missouri,

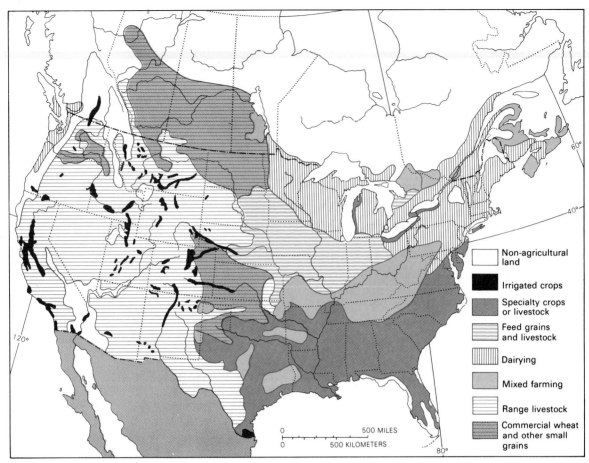

Figure 7.12 The agricultural regions of Anglo-America exhibit a regional pattern that broadly reflects the landforms and climate of the continent.

Minnesota, Wisconsin, Michigan, and Ohio is known as the Corn Belt. This region has fertile soils and a warm, humid growing season particularly suitable to the production of corn and has the world's largest concentration of corn production. Soybeans are nearly as important as corn in total production, and the corn and soybeans are used to support an intensive production of beef and pork for the American market. West of the Corn Belt, where the climate becomes more arid, wheat is the dominant crop in both the United States and Canada. In the winter wheat belt the seed is planted in the fall, while in the spring wheat belt it is planted in the spring.

West of the wheat belts is less favorable area for agriculture as a result of more arid climates. The Rocky Mountains and Basin and Range have patches of irrigated agriculture concentrating on a variety of crops from sugar beets to alfalfa, from dairy to turkey production. In the Central Valley of California is another agricultural region of national and international significance. Much of the Central Valley has a climate that is conducive to production of early vegetables in the spring, and the southern portion produces winter vegetables. This agricultural region is a major producer of fruits, nuts, and vegetables. Major fruit crops include grapes—used as fresh fruit and for raisins and wine—apricots, and a host of berries such as strawberries. It also produces a wide variety of vegetables, ranging from artichokes to lettuce, from carrots to radishes.

Two major issues are presented by the abundant American agricultural base: (1) maintenance of the resource base, and (2) the social and economic problems of American agriculture nationally and worldwide. The first issue relates to the loss of agricultural land to erosion, overgrazing, and conversion to urban and industrial use. This absolute loss of farmland is also increased by problems of environmental pollution from use of artificial fertilizers. The United States uses more fertilizer per acre than any other country, and the process of leaching moves much of it to lakes and rivers, causing pollution problems.

The second issue centers on the role of the United States as a food-surplus region of the world. The world's greatest stores of grain are held by the United States and Canada. Because of periodic grain shortages caused by fluctuations of production in other countries it has been suggested that the United States and Canada be major participants in establishing a world grain storage system. This idea raises important questions concerning who would control such stocks, on what continent they would be located, and how their release would affect the economics of countries like the United States that rely heavily on grain exports. A secondary issue relates to the great productivity of the American farm sector. Because of a highly efficient but energy-consumptive system, American farmers have traditionally produced surpluses of most agricultural commodities. The federal government has historically guaranteed a set price for these products if the farmer participates in government programs. The result is production of large quantities of food for which there are no adequate markets. Much of the grain is made available to less industrialized countries at favorable rates, but provision of such food at a subsidized cost may effectively attempts by such food-deficit countries to become self-sufficient.

Another issue growing out of the American agricultural economy is government reclamation attempts. Throughout the arid western regions, the government has been involved in reclamation projects centering on building dams to provide additional irrigation water. The farmers cannot afford the full cost of such water; consequently, the new land is heavily subsidized by the general taxpayer. A similar process is taking place in regions of higher water tables or swamps and marshes along the Mississippi, Missouri, and Ohio river systems. Reclamation projects reclaim new land, which then contributes to the agricultural surplus of the nation. In both cases the farmer is subsidized in two ways, in the cost of developing the land for agriculture and in the actual price received for the product. The large-scale reclamation projects also create environmental disruptions in areas affected.

A final values question focuses on the small family farms of the United States and Canada. Increasing cost for machinery, fertilizer, seed, and other factors of production make it less and less economical for the small farmers to compete. An important question is whether the governments of the United States and Canada should subsidize small farmers and enact legislation to limit the growth of large corporate farms. Those who favor such a course of action argue that the small farmers represent the individualism, independence, and freedom that have characterized American development and are essential to the countries.

AMERICAN INDUSTRY

At least one-fourth of the labor force of Canada and the United States is engaged in manufacturing, and manufacturing accounts for about an equal amount of the national income in the two countries. Both countries have a diverse and technologically advanced industrial economy. In the United States, the leading industries in terms of value and labor force include electrical equipment, transportation, nonelectrical machinery, food and related products, and metals and metal products. Canada has a similar manufacturing composition as measured by number of employees, with foods and beverages first, transportation equipment second, and electronics and electrical equipment, clothing and primary metals rounding out the top five.

Manufacturing is found in all areas of Anglo-America, but there are important regional concentrations. In Canada, industry is highly localized with one-half of the manufacturing labor force found in Ontario and an additional one-third in Quebec. Ontario is a center for heavy industry including iron and steel, auto manufacture, and other metal fabrication industries (Figure 7.13). Quebec has more production of clothing and agricultural products. In the United States, the most important industrial region is in the Northeast. Within this region New England is known for a highly diversified industrial base producing electrical equipment and electronics, instruments, leather, and some apparel and textiles. The Middle Atlantic region of New York, Pennsylvania, and New Jersey is a major area for production of instruments, apparel, metals, printing and publishing, and chemicals. The north central region of Ohio, Illinois, Indiana, and Michigan produces primary metals with an emphasis on iron and steel, fabricated metals, and transportation equipment, especially automobiles. The Southeast region concentrates on fabrics and fabric production, including textile weaving, carpets, and apparel manufacturing. The South region concentrates on chemicals, petroleum equipment and refining, textiles, and lumber products. The West Coast region of the United States is a diversified manufacturing area with transportation equipment, particularly aircraft, being the dominant industry.

The industrial regions of Canada and the United States reflect the level of interconnection in the industrialized, developed world. These interconnections integrate specific regions and facilitate the process by which components of a single item are assembled into a final product. The concentration of auto-related industries in Michigan, Illinois, Ohio, and Indiana provides an excellent example of these linkages. Tires may be manufactured in one city, engine blocks in another, and generators and other electrical parts in still others. Ultimately, the components converge at still other plants where they are assembled into an automobile.

The American industrial regions are also connected to the rest of Anglo-America and the world.

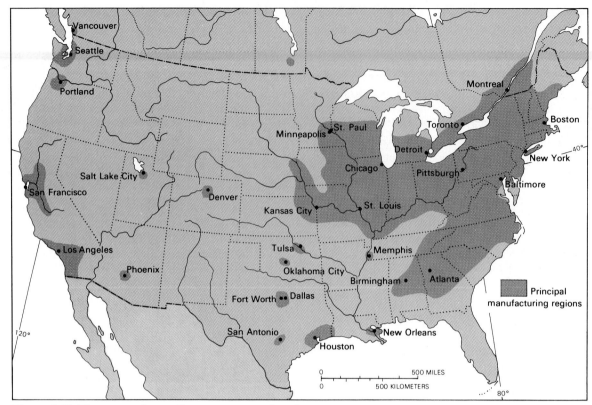

Figure 7.13 The principal manufacturing regions of Anglo-America.

The complex infrastructure provides access to raw materials and markets upon which the industrial economy depends. Coal and iron ore flow into the steel mills around Chicago, and assembled cars move out of Detroit and other cities to the towns and cities of the United States. The high level of interconnection and the complementary relationships within and between the manufacturing regions of Anglo-America and the emphasis on high-technology products typify the economy of the industrialized world.

The productive American industrial base raises environmental and social issues. Resource extraction and industrial manufacturing have led to air and water pollution. The problems of air pollution, water pollution, and destruction of land and water by resource extraction have prompted Canada and the United States to take government action to protect the environment. In the United States, the National Environmental Protection Act of 1970 set guidelines for air and water quality. At the beginning of the 1990s, some progress had been made in cleaning the rivers and lakes of Anglo-America, particularly Lake Erie, which is surrounded by industry of both the United States and Canada. Air pollution has been more difficult to overcome because of the central role of the automobile in its production. The necessity of improving air quality remains a critical issue facing manufacturing in the United States and Canada.

An important social issue grows out of the industrial productivity of Anglo-America and the tremendous demand for resources that this industrial monolith generates. Consumption of resources increasingly becomes a world issue when less industrialized countries attempt to transform their economies from poverty to affluence. In order to make this transformation, they will need to share more equally in the resources of the world. Since Anglo-Americans have the wealth, they can always outbid the countries of the less industrialized world, thus effectively prohibiting them from catching up.

THE AMERICAN EXPERIENCE IN RETROSPECT: EXAMPLE OR LESSON

Many people of the less industrialized world look to Anglo-America in their attempts to increase their standard of living. In retrospect the American experience raises certain issues. Americans have the largest numbers of people enjoying the highest standard of living in the world. But they also have significant minority groups who do not adequately share that wealth. Anglo-Americans are world-renowned for their medical expertise; yet they have the highest infant mortality rate of all the industrialized countries. In an industrial world in which literacy is essential, the United States has a higher illiteracy rate than most other industrial-

The Postindustrial Society

Changes in the economic base of Anglo-America have given rise to the concept of a *postindustrial society* that differs from the industrial society. Although different in emphasis, it represents only one more in a whole series of changes that have affected the economic geography of the industrialized world since industrialization began.

Changes focus on what are popularly called the "high-tech" activities but are reflected in many aspects of the region's geography. High-technology industries focus on production of computers, electronic industrial robots, and related products that capitalize on the high educational and training opportunities of industrialized, developed countries. Unable to compete with the labor-surplus countries of the less industrial world, Anglo-America emphasizes production of items that are not labor intensive, or on techniques to minimize labor requirements in traditional industries to compete with ever-increasing imports of steel, textiles, shoes, and other goods from the less industrialized world.

The underlying concept of the designation "postindustrial society" is a society that has gone beyond the industrial model. The postindustrial model is characterized by increasing importance of technical- and professional-level workers rather than manufacturing workers, and ever greater emphasis on education and training. The geographic implications of the postindustrial model are numerous. "High-tech" industries that rely on human knowledge rather than natural resources are much less restricted in terms of location. Access to major universities and sources of funding rather than to iron ore and coal dominate as locational factors. The regional impact of the changing economy is manifest in the growth of high-tech industries in the "Silicon Valley" of the San Francisco Bay area capitalizing on the universities in the area, on the climate in the Phoenix area, on universities in the Boston area, and so forth. At the same time employment is growing in the high-tech industries it has declined drastically in the traditional heavy industries of iron, steel, and transportation of the old industrial Northeast. Unfortunately, the new postindustrial society relies on relatively few scientists, engineers, and skilled production workers and instead on large numbers of automated industrial robots. The loss of jobs in the traditional industries is not equaled by new jobs in the high-technology industries, raising questions of the future standard of living for the majority of today's working population in industrial countries. New jobs are increasingly found in service activities that require low skill and have low wage rates. Some observers fear that the postindustrial society will be characterized by a large lower class of low-skill, low-paid service workers rather than the large middle class that typifies industrial societies.

ized countries. The United States, with its emphasis on free-enterprise medicine, has the largest number of discretionary surgeries of any country in the world. Meanwhile, the poor are not receiving adequate medical care, resulting in high infant mortality rates for children of black, Indian, and Spanish-speaking minorities. Important segments of minority groups suffer from malnutrition with some of the characteristics found in less industrialized nations. Hundreds of thousands of Americans are homeless, in contrast to the general level of wealth found in the United States. In spite of these seeming contradictions in a continent of affluence, Anglo-America remains the most affluent region of the world and one envied by much of the world's population. The issues for Americans focus on sharing that wealth adequately for their own residents and accepting their natural role as a leader in assisting the peoples of less industrialized countries and areas to emerge from poverty into security.

AMERICANS AND THE FUTURE: THE WINDS OF CHANGE

The issue of increasing world demand for a share in the global wealth, coupled with growing numbers of people, is viewed by some as leading to an era of scarcity. The American economy has been based on an era of surpluses: surpluses in agricultural and industrial raw materials that have fueled great industrial growth. This surplus of raw materials was compounded by a surplus of cheap labor as a result of high birth and high immigration rates during the nineteenth and early twentieth centuries. The great industrial complexes of United States Steel, Standard Oil Company, and others resulted from a combination of cheap labor and abundant resources. The actual process of development followed the same pattern as that of Western Europe, with labor exploitation predominating until unions gained sufficient power to secure a fair share of the value of their work. Lower birthrates and organized labor's attitudes toward immigrants as a threat to wages has effectively curtailed the labor surplus. Coupled with the increasing scarcities in industrial raw materials, this scarcity presages a new era characterized by labor shortages and technical obsolescence. Obsolescence and locational anomalies are making it harder for American industry to compete effectively on a worldwide scale. With increasing reliance on imported resources, the natural location for American industry is near the coastal areas, but in the past, the heavy industry has been located in the north central states to take advantage of local raw materials. The exhaustion of these raw materials and the age of the industrial complexes associated with them make the cost of such industrial production excessively high. Anglo-American industry is building new plants in less industrialized countries to capitalize on resources and cheap labor.

In the past, Anglo-Americans have been guilty of conspicuous overconsumption, but the era of scarcity makes the continuance of such a trend unlikely. Faced with ever-greater demands from the poor of the world, the Anglo-American people need to recognize their obligation to modify their life-styles. If they are unwilling to do so, we may expect greater conflict over acquisition of resources and maintenance or improvement of life-styles.

FURTHER READINGS

ABLER, R. and ADAMS, J. S. *Comparative Atlas of America's Great Cities*. Washington, D.C.: Association of American Geographers; Minneapolis: University of Minnesota Press, 1976.

ADAMS, J., ed. *Contemporary Metropolitan American*. 4 vols. Cambridge, Mass.: Ballinger, 1976.

Atlas of North America: Space Age Portrait of a Continent. Washington: National Geographic Society, 1985.

BAERWALD, T. "Urban Transportation in North America." *Focus*, November–December 1983.

BAUMANN, D. and DWORIN, D. *Water Resources for Our Cities*. Washington, D.C.: Association of American Geographers, 1978.

BENNETT, C. F., Jr. *Conservation and Management of Natural Resources in the United States*. New York: Wiley, 1983.

BERNARD, R. and RICE, B., eds. *Sunbelt Cities: Politics and Growth Since World War II*. Austin: University of Texas Press, 1984.

BERRY, B. J. L. "The Decline of the Aging Metropolis: Cultural Bases and Social Process." In *Post-Industrial America: Metropolitan Decline and Inter-Regional Job Shifts*, pp. 175–185. Edited by G. Sternlieb, and J. Hughes. New Brunswick, N.J.: Center for Urban Policy Research, Rutgers University, 1975.

———. "The Geography of the United States in the Year 2000." *Transactions of the Institute of British Geographers* 51, 1970, pp. 21–53.

———. "Inner City Futures: An American Dilemma Revisited." *Transactions of the Institute of British Geographers, New Series* 5, 1980, pp. 1–28.

BEYERS, W. B. "Contemporary Trends in the Regional Economic Development of the United States." *Professional Geographer* 31, 1979, pp. 34–44.

Beyond the Urban Fringe: Land-Use Issues of Nonmetropolitan America. Minneapolis: University of Minnesota Press, 1983.

BIRDSALL, S. and FLORIN, J. *Regional Landscapes of the United States and Canada*. 3d ed. New York: Wiley, 1985.

BLACK, MERLE and REED, JOHN SHELTON, eds. *Perspectives on the American South*. New York: Gordon and Breach, Science Publishers, 1981.

BORCHERT, JOHN R. *America's Northern Heartland*. Minneapolis: University of Minnesota Press, 1987.

BORGESE, ELISABETH MANN. "The Law of the Sea." *Scientific American*, March 1983.

BOURNE, L., ed. *Internal Structure of the City: Readings on Urban Form, Growth, and Policy*. 2d ed. New York: Oxford University Press, 1982.

BROWNING, C. *Population and Organized Area Growth in Megalopolis, 1950–1970*. Chapel Hill, N.C.: University of North Carolina, 1974.

BRUNN, S. and WHEELER, J., eds. *The American Metropolitan System: Present and Future*. New York: Halsted Press/V. H. Winston, 1980.

CHRISTIAN, C., and HARPER, R., eds. *Modern Metropolitan Systems*. Columbus, Ohio: Merrill, 1982.

CHUDACOFF, H. *The Evolution of American Urban Society*. 2d ed. Englewood Cliffs, N.J.: Prentice-Hall, 1981.

CUFF, D. and YOUNG, W. *The United States Energy Atlas*. New York: Free Press, 1980.

DESBARATS, JACQUELINE. "Indochinese Resettlement in the United States." *Annals of the Association of American Geographers* 75, October 1985, pp. 522–538.

DICKEN, P. and LLOYD, P. E. *Modern Western Society: A Geographical Perspective on Work, Home, and well-Being*. New York: Harper & Row, 1981.

EHRLICH, PAUL R. "North America After the War." *Natural History*, March 1984.

"Energy: A Special Report in the Public Interest." *National Geographic*, February 1981, pp. 1–114.

ESTALL, R. *A Modern Geography of the United States*. Harmondsworth, England: Penguin, 1976.

GARREAU, J. *The Nine Nations of North America*. Boston: Houghton Mifflin, 1981.

GASTIL, R. *Cultural Regions of the United States*. Seattle: University of Washington Press, 1975.

GOLDBERG, MICHAEL A. and MERCER, JOHN. *The Myth of the North American City: Continentalism Challenged*. Vancouver: University of British Columbia Press, 1986.

GUINESS, PAUL and BRADSHAW, MICHAEL. *North America: A Human Geography*, Totowa, N.J.: Barnes and Noble, 1985.

HARP, J. and LOFLEY, J. R., eds. *Poverty in Canada*. Scarborough, Ontario, Canada: Prentice-Hall, 1971.

HART, J. F. "Cropland Concentrations in the South." *Annals of the Association of American Geographers* 68, 1978, pp. 505–517.

KEMPE, FREDERICK. "Violent Tactics: Terrorist Attacks Grow But Groups Are Smaller and of Narrower Focus." *The Wall Street Journal*, April 19, 1983.

LEWIS, P. F., TUAN, YI-FU, and LOWENTHAL, D. *Visual Blight in America*. Washington, D.C.: Association of American Geographers, 1973.

LOUV, R. *America II*. Los Angeles: Jeremy Tarcher/Houghton Mifflin, 1983.

MCCANN, L., ed. *Heartland and Hinterland: A Geography of Canada*. Scarborough, Ontario, Canada: Prentice-Hall, 1982.

MAYER, H. M. "Geography in City and Regional Planning." In *Applied Geography: Selected Perspectives*. Edited by J. Frazier. Englewood Cliffs, N.J.: Prentice-Hall, 1982.

MAYER, H. M. and WADE, R. C. *Chicago: Growth of a Metropolis*. Chicago: University of Chicago Press, 1969.

MEINIG, D. "Symbolic Landscapes: Some Idealizations of American Communities." In *The Interpretation of Ordinary Landscapes: Geographical Essays*, pp. 164–192. Edited by D. Meinig. New York: Oxford University Press, 1979.

MEINIG, D. W. *Southwest: Three Peoples in Geographical Change*. New York: Oxford University Press, 1971.

MEINIG, DONALD W. *The Shaping of America: A Geographical Perspective on 500 Years of History, Volume 1, Atlantic America, 1492–1800*. New Haven: Yale University Press, 1986.

MILLER, CRANE S. and HYSLOP, RICHARD S. *California: The Geography of Diversity*. Palo Alto, Calif.: Mayfield Publishing, 1983.

MORRILL, R. L. and WOHLENBERG, E. H. *The Geography of Poverty in the United States*. New York: McGraw-Hill, 1971.

MORTLAND, CAROL A. and LEDGERWOOD, JUDY. "Secondary Migration Among Southeast Asian Refugees in the United States." *Urban Anthropology* [Brockport, NY], Vol. 16., Nos. 3–4, Fall–Winter 1987, pp. 291–326.

MULLER, P. *Contemporary Suburban America*. Englewood Cliffs, N.J.: Prentice-Hall, 1981.

NOYELLE, T. and STANBACK, T. *The Economic Transformation of American Cities*. Totowa, N.J.: Rowman and Allanheld, 1984.

Oxford Regional Economic Atlas of the United States and Canada. 2d ed. New York: Oxford University Press, 1975.

PALM, R. *The Geography of American Cities*. New York: Oxford University Press, 1981.

PATERSON, J. *North America: A Geography of Canada and the United States*. 7th ed. New York: Oxford University Press, 1984.

PHILLIPS, P. and BRUNN, S. "Slow Growth: A New Epoch of American Metropolitan Evolution." *Geographical Review* 68, 1978, pp. 274–292.

PIERCE, NEAL R., and HAGSTROM, JERRY. *The Book of America: Inside the 50 States Today*. New York: Norton, 1983.

PUTNAM, D. and PUTNAM, R. *Canada: A Regional Analysis*. 2d ed. Toronto: Dent, 1979.

REISNER, MARC. *Cadillac Desert: The American West and Its Disappearing Water*. New York: Viking, 1986.

ROBINSON, J. LEWIS. *Concepts and Themes in the Regional Geography of Canada*. Vancouver, B. C., Canada: Talon Books, 1983.

ROONEY, JOHN F., Jr., ZELINSKY, WILBUR, and LOUDER, DEAN R., gen. eds. *This Remarkable Continent: An Atlas of United States and Canadian Society and Culture*. College Station, Tex.: Texas A & M University Press for the Society for North American Cultural Survey, 1982.

ROSS, THOMAS E. and MOORE, TYREL G., eds. *A Cultural Geography of North American Indians*. Boulder, Colo.: Westview Press, 1987.

SAXINEN, A. "The Urban Contradictions of Silicon Valley." In *Sunbelt/Snowbelt: Urban Development and Regional Restructuring*, pp. 163–197. Edited by L. Sowers and W. Tabb. New York: Oxford University Press, 1984.

SCHERTZ, L. et al. *Another Revolution in U.S. Farming?* Agricultural Economic Report 441. Washington, D.C.: U.S. Department of Agriculture, 1979.

SCHILLER, HERBERT. "Information: America's New Global Empire." *Channels*, September/October 1982.

SMITH, D. M. *The Geography of Social Well-Being in the United States*. New York: McGraw-Hill, 1973.

STODDARD, ELLWYN, R. *Patterns of Poverty Along the U.S. Mexico Border*. El Paso: University of Texas at El Paso, Center for InterAmerican Studies, 1978.

[United States] *Atlas of American History*/Robert H. Ferrell, Richard Natkiel. New York: Facts on File, c1987.

WARD, D., ed. *Geographic Perspectives on America's Past*. New York: Oxford University Press, 1979.

WATSON, J. W. *Social Geography of the United States*. New York: Longman, 1979.

WATSON, J. WREFORD. *The United States: Habitation of Hope*. London: Longman, 1982.

WATSON, J. W. and O'RIORDAN, T., eds. *The American Environment: Perceptions and Policies*. New York: Wiley, 1976.

WEYR, THOMAS. *Hispanic U.S.A.: Breaking the Melting Pot*. New York: Harper & Row, 1988.

WHITE, C. L., FOSCUE, E. J., and MCKNIGHT, T. L. *Regional Geography of Anglo-America*. Englewood Cliffs, N.J.: Prentice-Hall, 1973.

YEATES, M. *North American Urban Patterns*. New York: Halsted Press/V. H. Winston, 1980.

YEATES, M. H. and GARNER, B. *The North American City*. 3d ed. San Francisco: Harper & Row, 1980.

ZELINSKY, W. *The Cultural Geography of the United States*. Englewood Cliffs, N.J.: Prentice-Hall, 1973.

———. "North America's Vernacular Regions." *Annals of the Association of American Geographers* 70, 1980.

Sydney.

AUSTRALIA AND NEW ZEALAND PROFILE

Population (in millions)		Energy Consumption Per	
Australia	16.75	Capita—pounds	(14,427)
New Zealand	3.35	(kilograms)	(6558)
Growth Rate (%)	0.8	Calories (daily)	3380
Years to Double Population	88	Literacy Rate (%)	99
Life Expectancy (years)	76	Eligible Children in Primary	
Infant Mortality Rate		School (%)	100
(per thousand)	10	Percent Urban	85
Per Capita GNP		Percent Annual Urban	
Australia	$12,000	Growth, 1970–1981	1.9
New Zealand	$7,500	1980–1985	1.2
Percentage of Labor Force			
Employed in:			
Agriculture	8		
Industry	35		
Service	57		

8

Australia and New Zealand Outposts of the industrialized world

MAJOR CONCEPTS

Generative centers / Repatriation / Physical remoteness / Migration / Replicative cities Outback / Import substitution

IMPORTANT ISSUES

Development of the Outback and North
Trade relationships with other developed economies
Foreign ownership and exploitation of resources
Protective tariffs
Immigration pressures

Geographical Characteristics of Australia and New Zealand

1. Australia and New Zealand represent the most isolated part of the periphery of the industrialized world.
2. Australia is nearly as large as the United States excluding Hawaii and Alaska, but has fewer than 7 percent as many inhabitants.
3. Australia has harsh climates and extremely low population densities except in the southeastern and southwestern margins of the continent.
4. New Zealand, only a fraction the size of Australia, consists of two major islands dominated by a marine west coast climate.
5. Both Australia and New Zealand rely heavily on exports of agricultural and mineral products to Japan and the industrialized countries of Europe.

Australia and New Zealand are outposts of the industrialized core of the world. (Figure 8.1). Their development makes them unique in several ways. They are the only representatives of the fully industrialized world in the Southern Hemisphere. The economies of both Australia and New Zealand are characterized by high per capita incomes, and the standard of living of the European majority of each country is similar to that of residents of the United States or European industrialized core. The development of these two nations roughly paralleled that of North America in time, but European interest was based on a somewhat different perception.

Australia and New Zealand offer contrasts in physical geography, with one large and the other small; one mineral-rich and the other with few mineral resources; one an island nation, the other the only nation occupying a continent; and one reliant upon agricultural exports almost exclusively, the other reliant upon both mineral and agricultural exports. In spite of these contrasts the two nations are similar because of their developmental process. Isolated from the European cultural hearth, they are examples of **replicative societies,** maintaining many of the characteristics of the United Kingdom, the homeland of their first European settlers. The development of towns, governments, and societal values consciously followed the model of England initially. Since World War II, towns and cities have spread to suburbs as high-rise buildings similar to those of the United States have come to dominate the downtown skyline. Both nations have developed economies that share the characteristics of high literacy levels, high incomes, large amounts of leisure time, and individualism and materialism found in other regions developed on the European model by European colonists. The economies of both nations are highly reliant upon their interconnections to the balance of the industrialized world, but both nations also have important connections to the less industrialized world.

The similarities of the two nations based on economic development and cultural background mask important cultural variations. The residents of Australia occupy a landmass equivalent in size to the United States excluding Alaska, nearly 3 million square miles (7.77 million square kilometers), while those of New Zealand have only slightly more than 100,000 square miles (259,000 square kilometers), nearly identical to the state of Oregon or Colorado. New Zealand has 3.35 million people compared to Australia's 16.75 million.

Replicative societies: societies that result from conscious or unconscious efforts to model themselves on another society.

THE AUSTRALIAN EXPERIENCE

Australia is unusual in the world for a number of reasons. The typical stereotypes of Australia—kangaroos, koala bears, and sheep ranches—represent but a tiny fraction of its diversity and uniqueness. The colonization of Australia by Europeans reflected their conclusion that the continent was unsuited for exploita-

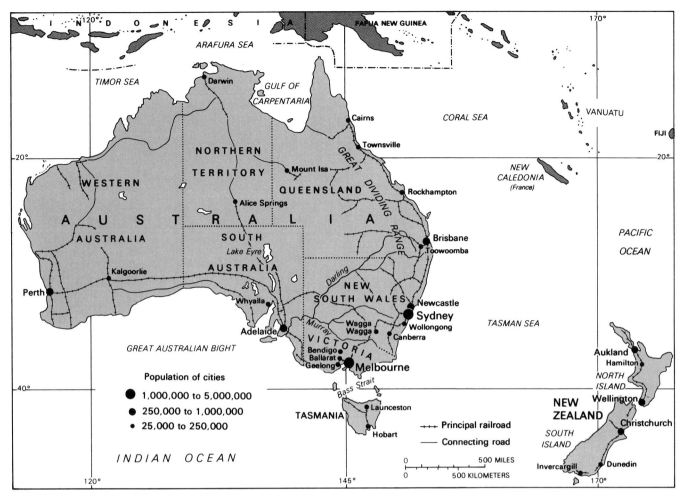

Figure 8.1 Australia and New Zealand.

tion and that its only resource was its isolation. In the days of sailing ships the 13,000 miles (20,900 kilometers) of ocean between Europe and Australia represented a journey of six months. To the British, this isolation became a resource, and after the American Revolution, Australia became a dumping ground for social undesirables. Debtors, shoplifters, and other convicted criminals were sent to Australia to ease prison crowding in England. The first penal colonies were started in 1788, and Australia's development into the next century reflected this role as a remote prison. The cities of Sydney, Brisbane, and Hobart were originally penal colonies, while other cities combined penal colonies with the need for garrison cities to protect trade routes or to exploit the resources of the continent. Ultimately five major cities (Sydney, Melbourne, Perth, Adelaide, and Brisbane) came to dominate the continent. For more than half a century settlements in Australia were subsidized by Britain as grain, manufactured items and administrators were sent out to prevent failure of the colony. In time each main port city became more self-sufficient and developed into a generative center, stimulating growth in the surrounding region. These centers attempted to reproduce some British forms in their development and close ties were maintained with England. Ultimately each of the major cities became the focus of a self-governing colony under general British control. This earlier lack of a central government caused each colony to develop its own transport system. Railroads in the various colonies differed in the distance between the rails (gauge), effectively preventing a unified continental transportation network.

National government was established in 1901 when the six colonies united as the Commonwealth of Australia with its six states and two territories (Table 8.1). The form of government combines elements of both the U.S. division of power between state and federal levels and the British parliamentary system. The long history of development as a British colony with autonomous regions poses only minor problems for the individual development of the states. Not until 1970

TABLE 8.1 Australian States and Continental Territories: Area and Population (1990)

State	Area in Square Miles	(sq. km.)	Population (thousands)	Capital City
New South Wales	309,433	(801,432)	5,800	Sydney
Victoria	87,884	(227,620)	4,300	Melbourne
Queensland	667,000	(1,727,530)	2,700	Brisbane
South Australia	380,070	(984,381)	1,420	Adelaide
Western Australia	975,920	(2,527,633)	1,500	Perth
Tasmania	26,383	(68,332)	465	Hobart
Northern Territory	520,280	(1,347,525)	155	Darwin
Australian Capital Territory	939	(2,432)	375	Canberra
Total	2,967,909	(7,684,452)	16,715	

SOURCE: *The Europa Yearbook 1988* (London: Europa Publications, 1988, and 1990 estimates).

did standard gauge links allow movement from Sydney to Perth without changing trains at least once. The fact that each colony (and later, state) had a unique rail gauge reinforced the control of each capital over its territory, leading to the capital cities containing nearly two-thirds of Australia's population in 1990. The earlier competition between the dominant cities of each state, which prompted development of a capital territory and city at Canberra, is still evident, but these are minor centrifugal forces and do not threaten the Australian nation.

THE PHYSICAL ENVIRONMENT

The importance of transportation in a nation as large as the 48 contiguous states of the United States is obvious. The problem is compounded by the relative unsuitability for settlement of much of the Australian continent. Most of Australia has arid (desert or steppe) climates, which are not suitable for agricultural activities other than extensive grazing of sheep and cattle.

Australia has five general climatic regions. Amount and distribution of precipitation is of critical importance in the climatic regions (Figure 8.2). The eastern coastal area from Brisbane south to Melbourne enjoys abundant precipitation year around. The climatic types range from marine west coast in the south around Melbourne and Canberra to humid subtropical in the Brisbane area. The major highland of Australia extends along this eastern coast in a belt 100 to 250 miles (160 to 400 kilometers) wide. These rugged but low mountains rarely exceed 3000 feet (900 meters) elevation. The highest point, Mount Kosciusko, is only 7316 feet (2200 meters), the highest elevation in Australia. The rest of Australia is predominantly smooth to low rolling hills with a few weathered low mountain

Kangaroos have been protected for the past few years after their previous slaughter by hunters and ranchers. At present, their numbers are so large that they are a nuisance to farmers, ranchers, and even urban lawns in the Outback.

Sheep products have traditionally been one of Australia's premier export items, but their relative importance is declining with the growth of mineral exports.

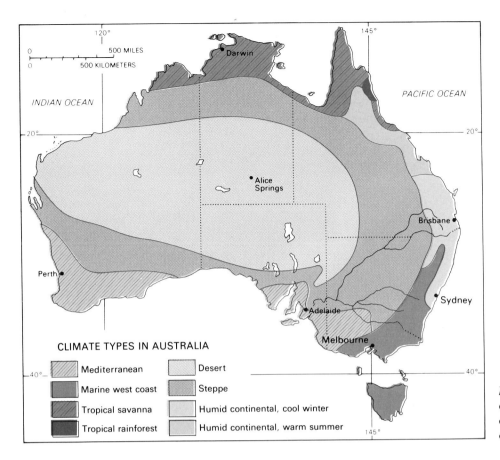

Figure 8.2 Australia is dominated by dry climates except in the coastal areas.

ranges. The relatively narrow and isolated coastal lowland of this eastern and southeastern portion of Australia is the center of population. Here the cities of Sydney and Melbourne alone account for 40 percent of the total population of the nation. Sydney is the largest city with more than 3 million people, but is followed very closely by Melbourne. The other centers include Brisbane (1 million), Adelaide (1 million), and Perth (800,000).

The southwestern and southern portions of Australia have a Mediterranean type of climate with hot, dry summers and mild, moist winters. In the Southern Hemisphere the summer dry season is from October to April and the winter wet season is from June to September. The Mediterranean climate of southwestern Australia is a second center of more intense population, outside of the heavily populated southeast and east coast area (Figure 8.2). The northern coastal region of Australia has a savanna climate with rainy summers and dry winters. Precipitation exceeds 20 inches (508 millimeters) through most of this northern region, but a dry season and high temperatures handicap agriculture.

About two-thirds of Australia is characterized by arid and semiarid climates. The central portion receives less than 10 inches (254 millimeters) of rainfall per year and is surrounded by a steppe land receiving 10 to 20 inches (254 to 508 millimeters). This great, dry interior the Australians call the Outback.

THE SETTLEMENT AND DEVELOPMENT OF AUSTRALIA

The physical characteristics of Australia have limited the possibilities for European-type agriculture to the east and southeast coastal regions and the southwestern corner of Australia, where the population is concentrated (Figure 8.3). These clusters of concentrated population developed after Australia's original role as a prison was transformed to one of an agricultural hinterland for the United Kingdom and Western Europe. Improvements in ocean transport, especially refrigerated ships (1881) made it possible to export agricultural products to feed Europe's growing nonfarm population.

Wool for the textile mills of Europe was the first major export from Australia, and production of sheep for both wool and meat remained the major source of export earnings until after World War II. Today sheep production is concentrated in the area west of the eastern highlands and in the southeastern and southwestern portions of the continent (Figure 8.4). Rising

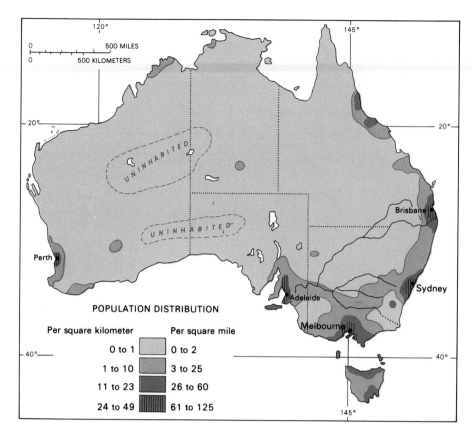

Figure 8.3 Population distribution in Australia is closely related to the precipitation received by a region.

prices for wool made it the largest source of income from exports in 1988 and 1989. The driest central portion of Australia is unsuitable even for grazing sheep. The cattle industry is important in the semiarid and savanna lands of northern and northwestern Australia. More than 50 percent of the cattle production is exported, primarily to the United States for use in manufacturing meat patties for fast-food restaurants. Australian cattle are range-fed rather than grain-fed as in the United States, and Australian animals tend to be leaner and less suitable for steaks and higher quality meats. Australia's exports of mutton (meat from sheep) used to go primarily to the British Isles, but British membership in the EEC has reduced this market.

Ayers Rock, sacred to the Aboriginal population of Central Australia, rises above the flat plains of Australia's Outback.

THE WORLD OF WEALTH

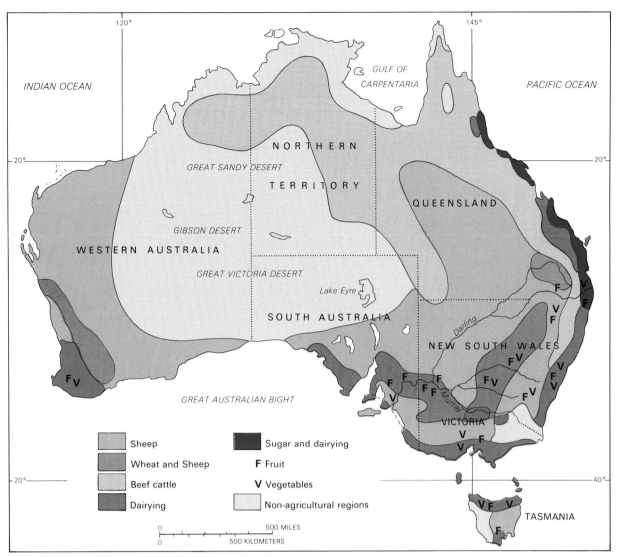

Figure 8.4 Major crop and livestock patterns in Australia.

Today lamb and mutton are exported to Japan, the United States, Persian Gulf countries (after slaughter according to the Islamic ritual), and the EEC.

Exports of beef, mutton, and wool are atypical for an industrialized nation, but the large area of the continent combines with the small population to give the country an agricultural surplus. Australia's trade relationships with Japan, United States, and Western Europe illustrate some of the same problems that less industrialized countries face. The production of beef in Australia for export to the United States is at the mercy of American politicians who set quotas on the amount of beef that can be imported. Quotas affect the price received by Australian ranchers and farmers. The market for wool, the major export item from Australia in the first 150 years of its occupance by Western Europeans, reflects the other impact of industrialized economies on the rest of the world. Increasing use of synthetic textiles cut the market for wool dramatically and sheep numbers declined proportionately until a resurgence in the 1980s as world demand increased.

Another important agricultural crop is wheat, which accounts for 10 percent of Australia's exports by value. The production of wheat in Australia is highly localized in southeastern and southwestern Australia, where it usually occurs in conjunction with sheep production. Australia ranks third in wheat exports, after the United States and Canada. Major markets are the People's Republic of China, which is the main customer for wheat, Egypt, Pakistan, Japan, and the Soviet Union. As in the United States and Canada, the production of wheat is highly mechanized, and the farms may exceed 5000 acres (2024 hectares) in the more arid regions of western Australia.

AUSTRALIA AND NEW ZEALAND

The entire range of other agricultural products demanded by industrialized societies is produced in Australia because of the climatic diversity of the continent. Apples, pineapples, citrus, grapes, and bananas are grown. Much of the fruit is exported, particularly apples, since they are available in Europe during the off season for apples in the northern latitudes.

Australia is also an important producer of sugar cane, and represents one of the few areas where sugar cane is produced on family farms by the landowner. Production is concentrated on the coastal plain of Queensland where protective tariffs and production quotas established by the national government make it cost-competitive. The use of tariffs to protect the sugar industry of Queensland illustrates the effect of industrialized economies on the Third World. By a tariff on incoming sugar, Australian sugar-cane farmers are protected from the competition from lands with low labor costs. Without the protective tariff, sugar cane from Asia or Latin America or the Caribbean islands would undersell domestic sugar and force Australian sugar-cane growers to convert to other crops.

MODERN DEVELOPMENT IN AUSTRALIA

The emergence of Australia as an agricultural hinterland to Europe's industrial core in the eighteenth and nineteenth centuries did not lead to the underdevelopment found in other colonies of European states. Australian settlement by the British ensured that the emerging country had access to the technology of western Europe and prevented the racial discrimination and economic exploitation experienced by other regions that were colonial possessions but not settled by Europeans. After the creation of the commonwealth in 1901, Australia adopted a "white only" immigration policy and the Negrito laborers from the Melanesian islands who had been imported to work in the sugar-cane plantations of Queensland before confederation were **repatriated** (returned to their home islands). Some Chinese who came during the Australian gold rush remained to form a small Asian minority. Until 1956, Negroid and Asian peoples were excluded from immigration, but since that time non-Caucasians with certain skills have been admitted and are allowed to apply for citizenship after 5 years. Australia remains overwhelmingly European in ethnic makeup. Even after World War II, when the origin of many migrants changed to South Europe from West Europe, the British Isles still sent more migrants. The values of the people reflect their European background and development in a region with space for expansion.

Industrial development began with early settlement, increased greatly during World War I, and has made major advances since World War II. Australia is fortunate in having an abundant mineral resource base for industrial use (Figure 8.5). It produces 70 percent of its petroleum needs, primarily from the Bass Straits field between mainland Australia and Tasmania. Petroleum production uses offshore drilling technology borrowed from the United States. Coal, concentrated in the area around Sydney, and inland and north from Brisbane in the Bowen Basin, is abundant, and Australia is a major exporter of coking coal to Japan and South Korea. Australia is the world's second largest producer of iron ore and is a large exporter to Japan, South Korea, Taiwan, and other newly industrializing countries in Asia. Iron mining is concentrated in the west at Mount Newman, Mount Whaleback, Mount Tom Price, and other mines in the Hamersley and Opthalmia ranges. New railroads connect the region to harbors at Port Hedland and Dampier.

Australia is the world's largest producer of bauxite, the ore from which aluminum is made. It is also the world's largest producer of alumina (processed bauxite used in refining aluminum), although little aluminum is produced in Australia. A host of other minerals is also produced, ranging from gold to uranium, including nickel, lead, copper, zinc, and manganese. Australia now accounts for about one-fifth of world coal exports, one-third of bauxite, and one-fifth of alumina.

The mineral wealth of Australia is the basis for export trade with Japan, the European Common Market, the United States, and other nations (Table 8.2). The iron ore of the Hamersley and adjacent districts is carried to ports in automated trains carrying only iron ore. There automated dock facilities load the ore into large ships for export to Japan and other markets. The mineral resources have also been the basis for development of a diversified economy in Australia, but these *import substitution* industries function only because of protective tariffs. There are, for example, five automobile manufacturing companies in Australia (the three largest having United States ties—General Motors, Ford, and Chrysler), and they far exceed the market capacity of the small population. These companies exist only because protective tariffs make the price of imported automobiles double that of domestically produced autos. Industrial development is concentrated in the five largest cities in Australia, each

---

Repatriation: The process of returning people to the country of their birth or where they hold citizenship.

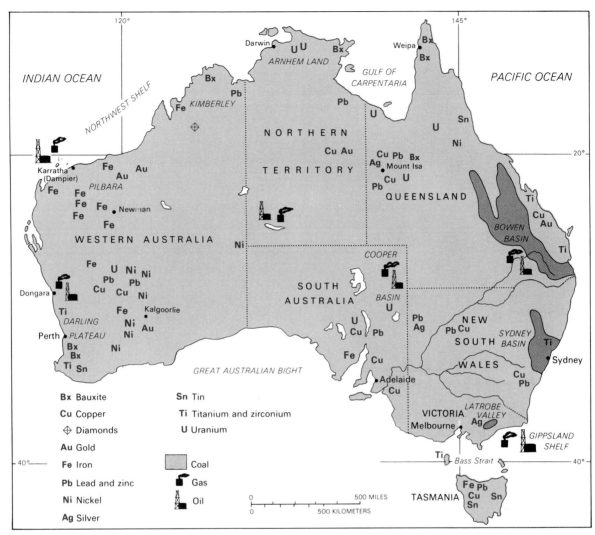

Figure 8.5 The distribution of mineral resources in Australia. Note the regional concentration.

TABLE 8.2 Australia's Trading Partners

Country	Percentage of Total Exports	Percentage of Total Imports
Japan	28.5	24
New Zealand	4.6	4.2
United States	10.0	21.0
Other industrialized countries	7.9	29.0
Developing countries	42.0	19.8
Countries with centrally planned economies	4.0	1
Capital-surplus, oil-exporting countries	3.0	1

SOURCE: *Europa Yearbook* 1986 and *World Development Report* 1987.

of which is a seaport. The large cities are similar to cities in other industrial nations, but they have relatively less pollution because their Industrial Revolution has been more recent and the major cities lack the heavy industries such as iron and steel. Steelmaking is located in smaller coastal cities north and south of Sydney. The urban axis of Australia is the two cities of Melbourne and Sydney, with Melbourne the financial center and Sydney the center of commerce and communications.

Melbourne continues to challenge the dominance of Sydney, but Sydney's spacious harbor and central

AUSTRALIA AND NEW ZEALAND

Melbourne epitomizes the replicative nature of Australian culture, with its parks, gardens, architecture, and trolley lines reflecting the European origin of most of the inhabitants.

The Aborigines of Australia live on the fringe of Australian society. These structures near Darwin are homes that the government constructed for Aborigines.

location with respect to the populated east coast makes the prospect that Melbourne will overtake Sydney unlikely.

ISSUES AND PROBLEMS IN AUSTRALIA

The problems that face Australia are similar to those of other industrial nations: rising costs for labor, concern for the environment, and the need to maintain a competitive position in world markets. Australia also has some problems that are shared by some industrial nations but not others. The original inhabitants of Australia, the aborigines, suffered much of the same treatment as the aboriginal population of North America. Today there are fewer than 50,000 aborigines of pure ancestry and another 150,000 to 200,000 of mixed ancestry left in the nation. They are similar to the Indians of the United States in having the lowest standard of living in the country, being confined to menial jobs, and facing the loss of their culture. The aborigines operate at the margin of the industrial society or attempt to maintain their traditional life-styles in regions that are undesirable to the European society.

Another problem facing the Australian nation is the foreign ownership of companies engaged in resource exploitation. As a single example, the Utah Development Company (completely owned by the General Electric Company of the United States) is a coal- and iron-ore–mining firm headquartered in Queensland. It is estimated to realize a 55 percent per year return on its investment in Australia, and pays over $100 million to American stockholders yearly. More than 80 percent of corporate profits are paid to American stockholders. Many Australians would like to see an excess profits tax to ensure that more of this profit stays in Australia, but foreign companies maintain that the high profits are necessary to encourage the risk involved in developing of mineral resources. They also point out that the Utah Development Company pays more than $200 million in taxes yearly to the Australian government and that additional taxation would harm the economy of Australia. The issue of foreign ownership and exploitation is one that will continue to challenge Australia, especially since growing exports to Japan have made Australia the third largest recipient of Japanese foreign investment (after the United States and Indonesia).

A final problem that faces Australia is its isolated location. Australia is a European enclave that is relatively close to the heavily populated Asian landmass. Maintenance of cultural ties to Europe has been fostered by trade relationships and immigration policies, but the continually growing population of the less industrialized countries of Asia at Australia's doorstep and the growing role of Australia in the trade of Asia, may ultimately pose an immense challenge to the Australian people. With a land that is only sparsely populated by Asian standards and a standard of living rivaling that of European nations, Australia may eventually be forced to admit more Asian immigrants. (At present, Australia admits about 100,000 immigrants per year, but only 10 percent are Asian). The concern that such a program would ultimately lead to cultural conflict and loss of European domination continues to affect Australian attitudes toward immigration and poses an important challenge to Australia's tomorrow.

THE WORLD OF WEALTH

NEW ZEALAND

The islands of New Zealand are home for 3.3 million people of European descent. The nation consists of two large islands (Figure 8.6), North Island, which contains nearly three-fourths of the population, the larger South Island, and a number of other smaller ones. New Zealand is characterized by high mountain ranges, particularly the Southern Alps of South Island, which reach 12,349 feet (3,764 meters) at Mount Cook. The Canterbury Plain of the east central portion of South Island and the coastal plains and the lower slopes of the uplands are the centers of population. The four largest cities contain over half the population, with one-fourth living in the Auckland urban area alone.

New Zealand's history is similar to that of other areas colonized by European industrial nations. The first colonists from the United Kingdom confronted the Maori peoples when they began to occupy the two

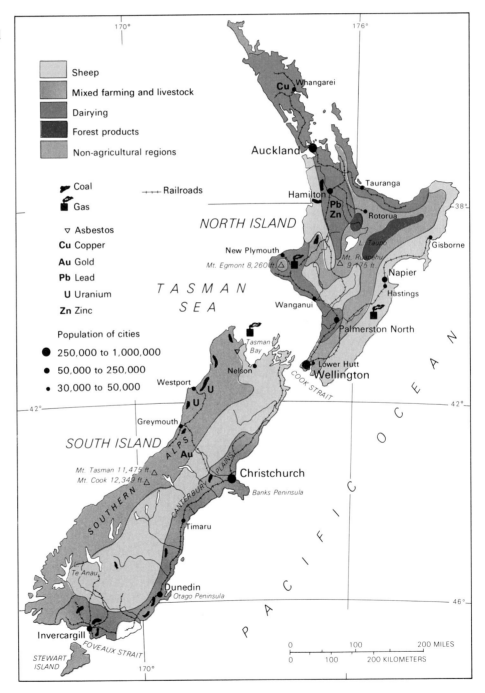

Figure 8.6 Cities, minerals, and agricultural regions in New Zealand.

AUSTRALIA AND NEW ZEALAND

Lake Hayes on South Island. New Zealand's landscapes are generally rural, and agricultural products dominate exports.

islands. The Maori were much more technically advanced than the aborigines of Australia, but repeated contacts and conflicts between the two cultures in New Zealand decimated the Maori. Like the American Indian, the Maori population declined rapidly as European settlement progressed, and at the turn of the century reached a low point of fewer than 40,000. Since that time the New Zealand government has attempted to change attitudes toward the Maori, and the population has grown to 325,000. The Maori have full citizenship but are still a minority group in their own land, and in the last few years conflict between them and other New Zealanders over Maori claims to lands based on traditional use has increased.

European occupation of New Zealand parallels that of Australia. Settlers have been almost exclusively from the United Kingdom, and the cultural landscape has been modeled after that which the settlers perceived existed in England. Cities like Christchurch (325,000) are replica cities, modeled after those of England. The rural agricultural economy was based on the perception of what England was like for the upper classes and reflects a conscious attempt by the settlers to recreate their British homeland. The economy of New Zealand is heavily oriented toward the export of raw materials, with meat, dairy products, and wool making up two-thirds of exports. The climate of New Zealand is marine west coast, and fully one-half of the nation is suitable for intensive grazing. The production of wool and mutton for export to Europe, especially England, has been the major economic activity from the time of the first European settlement. Perishable meat and dairy products joined wool in the export economy after 1881 when refrigerated shipping started. Since the United Kingdom joined the EEC, New Zealand's markets for wool and mutton have changed. Loss of English markets for dairy products is leading to an economic restructuring, and national (and per capita) incomes have declined.

Industrial development has been based on small deposits of coal, iron ore, and a number of minor minerals, but the country is still essentially oriented toward exporting foodstuffs to the industrial core areas. As in the case of Australia, most manufactured products are import substitutes and rely on high tariffs for protection. Only the great distance from the industrial core of Europe and the nationalist feelings of New Zealand justify the industrial base that exists. It would be cheaper to create a truly symbiotic relationship with the industrial core and focus attention on producing food products. Perception that such a relationship would result in exploitation of the nonindustrial partner (New Zealand) has prompted New Zealand's efforts to industrialize and be self-sufficient. Unlike today's less industrialized countries New Zealand has had the technical expertise and the cultural and political connections to accomplish independence and economic development. New Zealand's importance is primarily as an example of the dispersion of the European industrial and technological revolutions and their impact on native peoples. The attempts to recreate elements of the English landscape in New Zealand typify the role of European settlers in the development of the periphery of the industrial world. Borrowing of elements of European culture, especially British, continues today. Like the United Kingdom, New Zealand is beginning to privatize elements of its economy, including New Zealand Air.

THE AUSTRALIA–NEW ZEALAND PERIPHERY IN RETROSPECT

The Australian and New Zealand experiences are unlike those of European sojourners in Asia and Africa. Large-scale European settlement failed in Asia and Africa because of the large indigenous population in comparison with the limited number of European colonists. In Australia and New Zealand the small indigenous populations were no match for the invading European forces, and the two nations have developed as a political, military, and economic peripheral area of the industrial world. Both exhibit all the characteristics of industrial areas, except that they rely heavily on exports of raw materials. Their economies are highly affected by decisions made beyond their borders, as with the formation of the EEC or U.S. tariffs.

FURTHER READINGS

BARRETT, REES D. and FORD, ROSLYN A. *Patterns in the Human Geography of Australia.* South Melbourne: Macmillan of Australia, 1987.

BOLTON, G. *Spoils and Spoilers: Australians Make Their Environment, 1788–1980.* Boston: Allen and Unwin, 1981.

BURNLEY, IAN H. and FORREST, JAMES, eds. *Living in Cities: Urbanism and Society in Metropolitan Australia.* Winchester, Mass.: Allen and Unwin, 1985.

CAMERON, R. *Australia: History and Horizons.* New York: Columbia University Press, 1971.

COCHRANE, P. *Industrialization and Dependence: Australia's Road to Economic Development, 1870–1939.* St. Lucia, Australia: Queensland University Press, 1980.

CRABB, P. "A New Australian Railroad—Then North to Darwin." *Geographical Review* 72, 1982, pp. 90–93.

CUMBERLAND, K. B. and WHITELAW, J. S. *New Zealand.* London: Longmans Group, 1970.

DALY, M. *Sydney Boom, Sydney Bust.* Sydney, Australia: Allen and Unwin, 1982.

DAVIS, PETER. "Health Patterns in New Zealand: Class, Ethnicity and the Impact of Economic Development." *Social Science & Medicine* 18, No. 11, 1984, pp. 919–925.

GALE, F. "A Social Geography of Aboriginal Australia." In *Australia: A Geography.* Edited by D. N. Jeans. New York: St. Martin's Press, 1978.

GILPIN, A. *Environmental Policy in Australia.* Brisbane, Australia: University of Queensland Press, 1980.

———. "Problems, Progress, and Policy: An Overview of the Australian Environment." *Environment* 25, 1983, pp. 7–38.

HEATHCOTE, R. L. *Australia.* New York: Longman, 1976.

JEANS, D. N. ed. *Australia: A Geography.* New York: St. Martin's Press, 1978.

LINGE, G.J.R. and MCKAY, J., eds. *Structural Change in Australia: Some Spatial and Organizational Responses.* Canberra: Australian National University, 1981.

LIVINGSTON, W. S. and LOUIS, W. F., eds. *Australia, New Zealand, and the Pacific Islands Since the First World War.* Austin: University of Texas Press, 1979.

MORAN, W. and TAYLOR, M. eds. *Auckland and the Central North Island.* Auckland, New Zealand: Longman Paul, 1979.

NICHOLLS, J. L. "The Past and Present Extent of New Zealand's Indigenous Forests." *Environmental Conservation* 7, 1980, pp. 309–310.

OLIVER, W. H., with WILLIAMS, B. R., eds. *The Oxford History of New Zealand.* Oxford: Clarendon Press, 1981.

RICHARD, JOHN. *Australia: A Cultural History.* London: Longman, 1988.

RIMMER, P. J. "Transport." In *Australia: A Geography.* Edited by D. N. Jeans. New York: St. Martin's Press, 1978.

ROBINSON, K. W. *Australia, New Zealand, and the Southwest Pacific.* London: University of London Press, 1974.

SADDLER, HUGH. *Energy in Australia: Politics and Economics.* Sydney, Australia: George Allen, 1981.

SPATE, O.H.K. *Australia.* New York: Praeger, 1968.

STOKES, EVELYN. "Maori Geography or Geography of Maoris." *New Zealand Geographer,* Vol. 43, No. 3, December 1987, pp. 118–123.

The Pacific in Transition: Geographical Perspectives on Adaptation and Change. Cambridge, England: W. Heffer and Sons, 1973.

WHITE, R. *Inventing Australia: Images and Identity, 1688–1980.* London: Allen and Unwin, 1981.

Moscow.

SOVIET UNION PROFILE: (1990)

Population (in millions)	292	Energy Consumption Per Capita—pounds (kilograms)	13,488 (6,131)
Population Growth Rate (%)	1.0		
Years to Double Population	68		
Life Expectancy (years)	69		
Infant Mortality Rate (per thousand)	25	Calories (daily)	3,440
		Literacy Rate (%)	99
Per Capita GNP	$7,400	Eligible Children in Primary School (%)	100
Percentage of Labor Force Employed in:		Percent Urban	66
Agriculture	14	Percent Annual Urban Growth, 1970–1981	1.8
Industry	45		
Service	41	1980–1985	1.6

The Soviet Union
The communist alternative

MAJOR CONCEPTS

BAM / Marxism / Mir / Expropriation
Collectivization / Sovkhoz / Kolkhoz
Incipient industrial base / Exotic stream
Permafrost / Taiga / Virgin Lands Project
Full Employment

IMPORTANT ISSUES

Centralized state planning as a development strategy
Development in a harsh and dispersed land
Rigidity of agricultural planning
Underallocation of agricultural sector
Low productivity of the worker
Overemphasis on the military budget
Population policy in nations in stage 3 of demographic transition

For most of the world the Soviet Union looms as a land of mystery. The geographic character of the country tends to be either unknown or grossly distorted in the minds of Westerners. The Communist government is equally misperceived, as viewers may categorize it in extremes of merciless dictatorship or the only real people's democracy, or various levels between. As the world's largest and first Communist nation the Soviet Union is a reality in today's world, a reality that cannot be ignored. The land with its abundant resources will become increasingly important as the world moves from an era of surplus to an era of scarcity. The success or failure of the political system in continuing to develop the country will significantly affect the less industrialized countries as they search for alternatives to develop their economies. Whatever the decisions made in Moscow, they will have an impact far beyond the boundaries of the Soviet Union even though the country is only partially integrated into the interconnected system of today's world. The Soviet model of economic development is distinct from that of Western Europe and has been a model for many other countries. Although it has a lower level of economic development than other industrialized nations, the Soviet Union is an industrialized country with high levels of literacy, medical care,

THE WORLD OF WEALTH

Figure 9.1 Political subdivisions and major cities of the Union of Soviet Socialist Republics. Although there are fifteen republics, the Russian Republic dominates the country areally and politically, as the shading indicates. (Non-Russian republics shown in darker shades.)

and other basic needs. Its political system and economic organization are markedly different from that of the European core of development.

THE COMMUNIST EXPERIENCE

The Union of Soviet Socialist Republics (U.S.S.R.) is the official name of what most people still refer to inaccurately as Russia. (Figure 9.1) Russia was the name of the vast territory in the centuries it was ruled by the czars. As with all totalitarian governments, the nature of the czarist rule varied from period to period depending upon the character and qualifications of the czar who happened to be in control.

In general the situation in Russia was similar to that in many less industrial countries today. The vast majority of the populace was engaged in agriculture, living in poverty, illiterate, and barely subsisting. At the other end of the spectrum there were a few living in wealth and engaging in conspicuous consumption and ostentatious display of their riches. Between these two extremes was a relatively small group of what might be called middle class: merchants, prosperous farmers, and professional people. Industry existed,

> **Geographical Characteristics of the Soviet Union**
>
> 1. The world's largest country, the Soviet Union is more than 2 times the size of Canada (the second largest country), and 41 times the size of France, the next largest European country.
> 2. The bulk of the territory of the Soviet Union is in high latitudes with harsh climates.
> 3. The population, agriculture, and industry of the Soviet Union is concentrated in a small area of the western part of the country.
> 4. The Soviet Union represents the longest experience with communism and a planned economy in the world, and has been a model for other centrally planned economies dominated by the communist party.
> 5. The Soviet Union is now in the process of reevaluating its centrally planned economy to encourage greater individual initiative.

but as in most less industrialized countries it was largely controlled by outsiders who used Russian labor and/or resources. Wages were low, the workweek was long, and dissatisfaction among the laboring class had been increasing throughout the early years of the twentieth century.

In agriculture the peasants faced similar conditions. Not until the 1860s did the czars abolish serfdom, but the land given the newly freed serfs was inadequate to maintain anything but a subsistence economy. By the turn of the century a few of the former serfs had emerged as larger landowners (kulaks) who hired landless peasants to farm their land. In 1905 the czar sponsored a new reform aimed at solving some of the problems of the agricultural sector, but the primary result was simply consolidation of the individual peasant's scattered plots. The bulk of the population remained in abject poverty, whether in the rural areas or in the towns.

As in most developing countries the workday in industry was long, often more than 12 hours a day for six days a week. As in England and in other Western nations at the beginning of the Industrial Revolution, women and children constituted an important part of the work force, and children were sent out to labor for 12-hour days before age 12. The factors that caused wages to be higher in developing nations such as the United States were absent and allowed continuation of low pay. The tremendous poverty of the peasantry ensured a continuous flow of surplus population to the emerging cities, where workers were forced to labor for whatever they could get. As a result the average pay in the Moscow area in 1914 was only $7.00 per month. Compared to the peasant, however, these people were well-off, as the peasant averaged only about $2.00 a month cash income, although the peasants did grow their own food. Because of the dismal working conditions and low wages, strikes plagued czarist Russia in the first decades of the twentieth century. Between 1912 and 1914 more than 2000 strikes occurred each year involving more than 20 percent of the labor force.

As industry developed in the Soviet Union, cities also grew as they are doing in the less industrial countries today. By 1910 Saint Petersburg (now Leningrad) had 2 million people and Moscow 1.5 million. By 1917, 20 percent of the population of the Soviet Union was living in urban areas.

The growth of cities was also affected by the land reforms of 1905. Prior to this the peasants had lived in villages in which the land was owned and cultivated communally in what was known as the *mir*. In 1905 the land was changed to private ownership, leaving many with no land or too little for subsistence. The landless and malcontents flocked to the cities and added to the existing numbers of dissatisfied workers. Their cries for reform ultimately toppled the czarist regimes as the peasantry united with urban residents in the revolt that brought down the czars in 1917.

THE COMMUNIST REVOLUTION

The 1917 revolution ended with the Bolsheviks in power. From this revolution the Communist party ultimately emerged as the dominant force and created the new country of the Union of Soviet Socialist Republics. The basis for the Communist party ideology was the work of Karl Marx as interpreted and modified by Vladimir Ilyich Lenin. Marx observed the capitalist societies in the 1800s while living in England. He postulated that as industrialization developed under capitalist doctrine, it would lead to continued accumulation of wealth for the factory owners and constant poverty for the workers, since industrialists would never pay more than enough to sustain life in the working class. Ultimately conflict would occur between nations as they attempted to expand their markets and gain ever-greater wealth for the industrialists. The solution to this problem, Marx maintained, was a revolution to shift power from the wealthy industrialists to the working class. Marx maintained that labor was the only true source of wealth and that the rich had gained their wealth through exploitation of the laboring class. He argued that those who performed the labor should benefit from the wealth it produced. The theories of Marx are commonly referred to as *Marxism*. The theoretical basis of the government that developed in the Soviet Union on marxist principles maintains that the factors of production (labor, land, capital) are controlled by the masses rather than by a privileged elite. In theory the result is a classless society in which all men and women are equal, with all receiving according to their labor.

To facilitate the transformation of czarist Russia with its privileged classes and impoverished masses into the classless society, certain methods were used. Initially, one was **expropriation** by the state of all foreign-owned industries. These industries were then operated by the workers with government direction (which was sometimes more repressive than that of the old absentee owners). The estates of the kulaks were occupied by the peasants and the land expropriated and divided up among the landless and the small farmers under the concept of land for those who worked it.

THE IMPACT OF THE COMMUNIST REVOLUTION

The immediate result of the revolution, other than overthrowing the czar, was to create political turmoil. Into this turmoil the exiled Bolshevik leaders, Lenin, Stalin, and Trotsky, returned. The Bolsheviks gained power after the revolution as a more democratic coalition was overthrown in an armed coup d'etat. The Communists consolidated their power between 1917 and 1921 as they fought what was known as the Bolshevik Revolution against other forces unwilling to accept the Bolshevik leadership. During this period important territories were lost as Finland, Poland, and today's Soviet Republics of Estonia, Latvia, and Lithuania became independent. At the same time parts of today's republics of Belorussia, the Ukraine, and Moldavia were lost to neighboring countries (Figure 9.2). Approximately 1 1/4 million square miles (3 1/4 million square kilometers) of land and more than 60 million people were taken from the old czarist territory during this period of conflict. By 1921 the Bolsheviks had gained complete power but were left with a country whose economy had been totally disrupted by World War I and the ensuing conflicts associated with the revolution. Disease, famine, and war had taken the lives of approximately 9 million people. Urban areas were in disarray as the food supplies from the country had been disrupted. People were concerned about survival rather than type of government, and there were riots in the cities.

Riots for food in the cities and peasant concentration on production for their own consumption led Lenin to abandon the Communist ideal in 1921. At this time the NEP (New Economic Policy) was announced. Under the NEP the peasants were encouraged to produce agricultural surplus and to trade it freely on the urban market. In essence private enterprise was reinstated. This same system was adopted in the factories to allow the economic base of the country to be restored. By 1925, 90 percent of the factories in the new Soviet Union were operating as a part of a private enterprise market economy, and agricultural and industrial production had regained their prerevolutionary level. Under Lenin's direction the NEP restored conditions for the average person to the pre-World War I level.

Lenin died in 1924, but the NEP continued until 1929. From 1924 to 1929 Stalin consolidated his position in the Communist hierarchy and emerged as the primary power in the Soviet Union. In 1928 the ever-present specter of famine once again faced the Soviet Union. The 1928 grain harvest was below the previous year's harvest, and food for the urban dwellers was once again in short supply and high priced.

Partially in response to these problems the government instigated the first five-year plan in 1929. The plan ended the NEP and changed essentially all agriculture into large communally operated, state-owned farms known as *collectives*. Industries were returned to state control with production quotas based on the five-year plan. The avowed aim of the plan for industry was to guarantee the continuation of the revolution and the development of the Soviet Union into one of the world's industrial powers by following Marxist economic theory. After a decade of retreat to a free

---

Expropriation: The process by which a country seizes the property of private citizens, businesses, or corporations. Ownership is then maintained by the government either directly or indirectly.

Portraits of statues of Lenin are as frequent in the landscape of the Soviet Union as billboards for advertisements are in the United States. Here a woman passerby removes snow from a relief of Lenin in Moscow.

enterprise system, the Communists were ready to embark on their great social experiment: a planned society.

Sixty years have passed since Stalin implemented the collectivization of agriculture and the first of the great five-year plans. The success of the Soviet Union is difficult to assess because strict comparisons to countries such as the United States reveal great disparities between the economies and standards of living. In 1990 the average Soviet citizen earned less than half as much as the average American, had much less in the way of consumer items, and in general was living at a level more typical of the lower middle class in the United States (Table 9.1). But when the Soviet Union is compared to what it was when the Bolsheviks seized power in 1917, the picture is much brighter. Prior to the revolution the nation was an agrarian,

Figure 9.2 Boundary changes of the Soviet Union since the 1917 communist revolution illustrate that much of the territory gained after World War II was simply the reclaiming of territory lost between 1917 and 1939.

THE WORLD OF WEALTH

TABLE 9.1 Amount of Work Time Required to Earn the Price of Items in Three Cities

Item	Washington Time	Moscow Time	Paris Time
White bread (1 lb.)	6 min.	17 min.	20 min.
Rice (kilo)	6 min.	49 min.	30 min.
Beef (kilo)	46 min.	111 min.	120 min.
Chicken (kilo)	18 min.	189 min.	31 min.
Sugar (kilo)	6 min.	52 min.	11 min.
Ice cream (liter)	13 min.	107 min.	21 min.
Butter (kilo)	40 min.	195 min.	63 min.
Milk (liter)	4 min.	20 min.	8 min.
Eggs (10)	5 min.	50 min.	17 min.
Potatoes (kilo)	9 min.	11 min.	9 min.
Cabbage (one head)	7 min.	7 min.	9 min.
Red beets (kilo)	20 min.	9 min.	22 min.
Peas (kilo)	9 min.	49 min.	19 min.
Dry beans (kilo)	9 min.	84 min.	29 min.
Apples (kilo)	18 min.	28 min.	16 min.
Oranges (kilo)	10 min.	112 min.	15 min.
Orange juice (liter)	7 min.	151 min.	18 min.
Cola (liter)	7 min.	58 min.	21 min.
Coffee (kilo)	67 min.	1115 min.	121 min.
Regular gasoline (gal.)	6.5 min.	64 min.	25 min.
One pair of jeans	4 hrs.	56 hrs.	10 hrs.
One raincoat (men's)	11 hrs.	98 hrs.	30 hrs.
2-piece suit (men's)	18 hrs.	118 hrs.	34 hrs.
24-inch color TV	30 hrs.	669 hrs.	106 hrs.
Monthly rent	78 hrs.	11 hrs.	71 hrs.
Small car	5 mos.	45 mos.	8 mos.

Source: Keith Bush, "Retail Prices in Moscow and Four Western Cities," *Radio Liberty Research Supplement*, 1986, and personal comparisons by authors, 1988.

semifeudalistic society with poverty in the villages and impoverished workers in the cities slaving for low wages and facing the future with no security. Under communism it has been transformed into a nation in which the majority of the people are urban dwellers, enjoying a standard of living that ranks them at least at the lower level of other industrial nations of the world, and where the fear of the future has been removed by the ever-present Communist party's programs for social welfare. This has been accomplished, it should be remembered, in the face of world opposition to communism, World War II with its disastrous impact, and the excesses of Stalinism until his death in 1953.

The Soviet experience and success or lack thereof can only be understood in the framework of the land with which both Communists and czars worked. The histories of the czarist Russia and the Soviet Union have largely been written as a result of the harsh geographic realities of this immense land. The Soviet Union's 8.6 million square miles (22.3 million square kilometers) of territory constitute the world's largest independent country. More than two and one half times as large as Canada, China, or the United States, the next largest countries, the Soviet Union sprawls nearly halfway around the world (Figure 9.3), through 170 degrees in longitude and covering eleven time zones. When people in Leningrad in the west are going to bed, those in Vladivostok in the east are ready to begin work. The immensity of this longitudinal extent can perhaps be visualized by imagining the western part of the Soviet Union placed on the western border of Alaska. The Soviet Union would extend across Alaska, Canada, the North Atlantic, and reach Norway.

The second important fact about the location of the Soviet Union is its northerly location. In latitude it extends from 35 degrees north to nearly 78 degrees north, but the vast majority of the area is north of the northern border of the 48 conterminous United States. This is the most densely populated land in the world

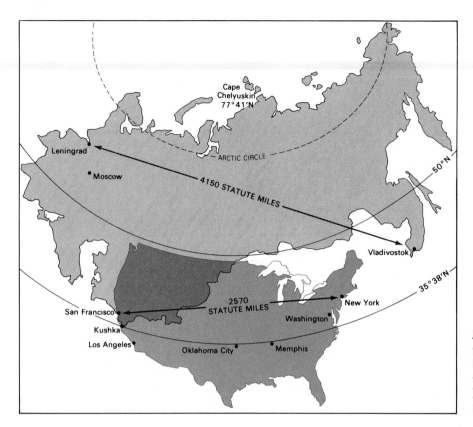

Figure 9.3 Comparison of the United States and the Soviet Union, illustrating the great land area and high-latitude location of the Soviet Union.

at such a high latitude. If you remember that Moscow lies north of the southern border of Alaska, you can see the significance of the high latitude. Leningrad is located at the same latitude as Stockholm, Sweden, and is north of the latitude of Juneau, Alaska.

THE LAND

The vast land of the Soviet Union is of paramount importance in understanding the character of the people of the Soviet Union. The climate is extreme and is plagued by the factors associated with continentality: drought, intense cold, and tremendous seasonal variations. Nevertheless, the land is home. Today's Soviet citizen is as emotionally tied to the land as the prerevolutionary peasant. The land with its great featureless plains plays a role in the consciousness of the citizens that cannot be overstated. When the government sought to punish dissident author Alexander Solzhenitsyn, it expelled him from the country. For the inhabitants of the Soviet Union the land is home, Mother Russia.

The Soviet Union occupies one-sixth of the habitable land area of the world. It is nearly 3 times the size of the United States and is 90 times the size of Great Britain. The country can be divided into the Russian Plain, the Ural Mountains, the West Siberian Lowland, the Central Siberian Uplands, and the eastern and southern mountain ranges (Figure 9.4).

The advantages of such a vast geographic area are partially offset by its problems. The Soviet Union has more than 10,500 miles (16,900 kilometers) of land frontiers and 26,700 miles (43,000 kilometers) of sea frontier to defend. The northerly location makes exploitation of mineral resources difficult and limits the land available for agriculture. The Soviet Union cultivates roughly one and one-half times the amount of land as the United States even though it is so much larger, but the cultivated land is generally not as suitable for farming. Geographically the land appears to be in a central position with respect to the world's landmass, but actually it is quite isolated by distance and cold. Access to the Atlantic Ocean is available only through the easily blocked narrows between Sweden and Denmark, and the Dardanelles (Turkey), or north through the port of Murmansk and the Arctic Ocean. Most of the northern coastal areas are frozen except during a short period in summer. Most of the southern border from the Caspian Sea to the Pacific lies in rough, mountainous country, which makes transport and communication difficult. Whether the advantages of the land outweigh the disadvantages is debatable, but it has certainly been a factor in the rate and success of development of the Communist nation.

THE RUSSIAN PLAIN

The Russian Plain stretches 1100 miles (1770 kilometers) from north to south and 1500 miles (2400 kilometers) from east to west, from the Polish border to the foothills of the Urals. This plains area is characterized by rolling topography interspersed with rivers and streams. The highland areas in the Russian Plain rarely rise more than 1000 feet (300 meters) above sea level. The Smolensk Ridge, from northeastern Poland to the Moscow area, is the drainage divide between the Black Sea and the Caspian to the south and the Baltic to the north. It has an elevation of only 200 to 700 feet (60 to 220 meters) but has long been an important transportation route because it is higher and drier than the surrounding land. Just north and west of Moscow the Lenin Heights rise to 1000 feet (300 meters) above sea level, and on the west bank of the Volga River the Volga Heights also rise to approximately 1000 feet (300 meters) elevation. In the Volga Heights at the great eastward bend of the Volga are the Zhiguli Hills. These are not high—only 400 to 600 feet, or 122 to 183 meters—but stand out as more mountainous topography than anything else around, as they are steep and drop off abruptly to the river. Otherwise, hills in the Russian Plain rarely rise more than a few hundred feet.

St. Basil's Cathedral in Red Square in Moscow is one of the premier architectural monuments of the Soviet Union. Today it is not used for religious purposes.

THE URALS

The Ural Mountains are a series of parallel ranges running approximately 1500 miles (2400 kilometers) north-south at about the sixtieth degree east longitude. They extend from the island of Novaya Zemlya in the north to their terminus north of the Caspian. Physiographically they are similar to the Appalachians of the United States and are not a great barrier to transportation. They reach a maximum of 6185 feet (1900 meters) at Mount Narodnaya, but most are between 750 and 2000 feet (230 and 610 meters) above sea level. They are 50 miles (81 kilometers) wide in the north and widen to 140 miles (225 kilometers) in the south where they are also lower. Their main importance is that because they are the "roots" of an old mountain system, they have major deposits of minerals around their flanks.

THE WEST SIBERIAN LOWLAND

East of the Urals is the broadest and the most level expanse of land in the world, the West Siberian Lowland. Even though it extends more than 1200 miles (1930 kilometers) from east to west and 1100 to 1600 miles (1170 to 2574 kilometers) from north to south, at no place does the elevation exceed 400 feet (122 meters) above sea level. The Ob is the main river flowing through it, but as would be expected, drainage is poor all year around, especially in the spring when the high water causes flooding over large areas. This area is increasingly important to the Soviet Union because of recent oil and gas discoveries. Presently the southern portion of the region is the prime wheat producer in the Soviet Union.

CENTRAL SIBERIAN UPLANDS

To the east of the West Siberian Lowland are the east Siberian uplands and plateaus. The Yenisey River marks the boundary between the two physiographic areas. The plateau region has an elevation of between 2000 and 6000 feet (610 to 1830 meters) but is highly dissected and constitutes a very rugged region that is an effective barrier to transportation. It is an important area for resources and lumber production but is sparsely populated. East of the Siberian plateaus is a complex of mountain ranges, rising to elevations over 10,000 feet (3,100 meters). This area is not only sparsely populated, but most parts of it are only superficially explored.

SOUTHERN MOUNTAINS

To the south the Soviet Union is ringed by a series of great mountainous areas. In the Tien Shan and Pamirs

THE SOVIET UNION

237

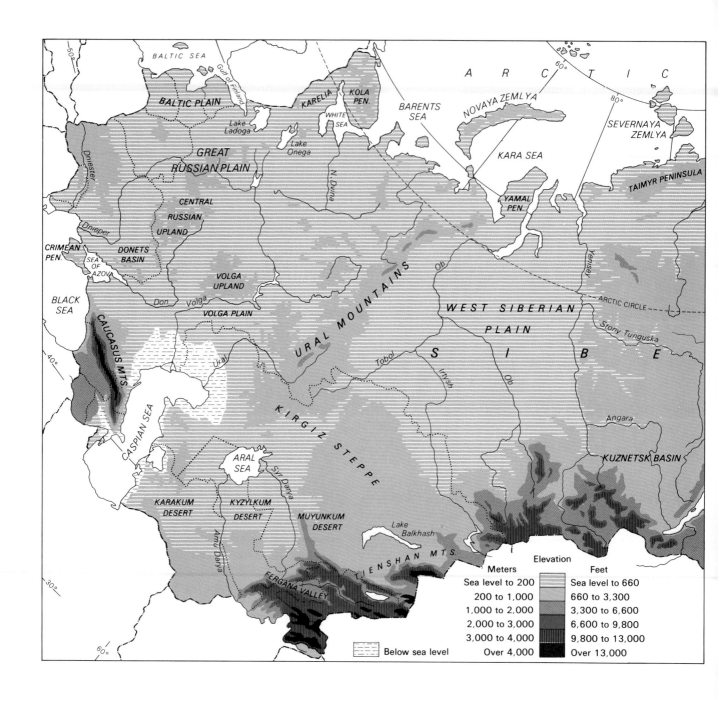

peaks are between 16,000 and 20,000 feet (5,000 and 6,200 meters). A host of mountains make up these ranges, but their importance lies in their function as a barrier and as a source of moisture to the adjacent deserts in Soviet Central Asia. Soviet Central Asia extends south of the latitude of the northern border of the 48 adjacent states of the United States, and the climate is warm enough for cotton and other crops that cannot be produced elsewhere in the country. The mountains trap moisture as a result of orographic lifting, and this moisture provides the irrigation water for cultivation of the desert lands.

RIVERS OF THE SOVIET UNION

The rivers of Russia historically were primary arteries of transportation (Figure 9.4). Even today the people of the Soviet Union use them extensively for transport, but the physical characteristics of the rivers handicap this use. The majority of the rivers flow north into the Arctic Ocean, and their use is consequently limited to the short frost-free period of summer. Moreover, since the movement of goods and people is primarily east to west, the north-south orientation of most rivers of the country further hinders their use.

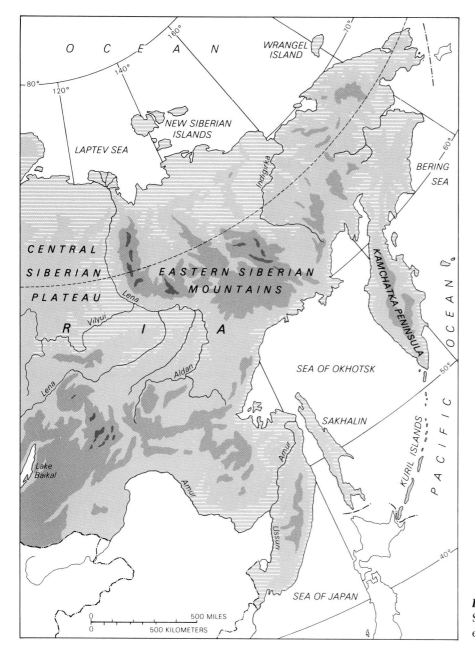

Figure 9.4 The landforms of the Soviet Union are dominated by large expanses of low-lying plains.

The rivers characteristically flow slowly and are shallow with sand bars; navigation requires periodic dredging.

The most important river system for transportation is the Volga and its tributaries. Like most Russian rivers it is frozen from late November to early April. The Volga flows south from a humid region into and through a dry region (an **exotic stream**) and in its natural state had low flow in late summer. Under the Soviet Union's planned economy the Volga has been transformed into a series of huge reservoirs, which provide navigation, hydroelectricity, and irrigation.

The Ob River with its tributaries is the longest river of the Soviet Union, but the Yenisey has a greater volume. With its tributaries the Ob is nearly 3500 miles (5630 kilometers) long, but it has a very low gradient, as it descends only 300 feet (90 meters) in the 1200 miles (1930 kilometers) of its lower course. The river flows so slowly that in places during the summertime the water becomes stagnant and fish die.

Exotic stream: A river that originates in a humid region and flows through a dry region.

THE SOVIET UNION

The lower Ob is navigable but is much less important than the rivers of the western part of the country. The Yenisey is three to four miles wide in its lower reaches, and large oceangoing vessels can navigate as far upstream as Igarka, some 450 miles (724 kilometers) from the ocean.

In the far east the dominant river is the Amur with its tributary the Ussuri, which form part of the boundary with China. The stream is navigable in its lower reaches and provides irrigation, electricity, and other benefits. In Soviet Central Asia the Amu Dar'ya and the Syr Dar'ya are highly significant for irrigation of the desert areas.

In addition to the major rivers of this huge land there are a number of lakes that are among the world's largest. The Caspian Sea covers more than 152,000 square miles (393,680 square kilometers), roughly five times the size of the largest of the Great Lakes in the United States. Lake Baikal is the deepest lake in the world. Four hundred miles (645 kilometers) from north to south and 50 miles (80 kilometers) wide, it is a natural wonder in its own right. Fish and fauna are found in Lake Baikal that are found no place else in the world. The lake contains 20 percent of the world's and 80 percent of the Soviet Union's fresh water, more than all of the Great Lakes of the United States combined. It is highly important in Russian folklore and is the site of a large recreational development.

PORTS AND SEAS

Although the Soviet Union has the longest coastline of any country in the world, it is plagued by a lack of ports. Most Soviet ports are icebound at least part of the winter. The north coast along the Arctic is open from east to west from late July to early September only. To the east, Vladivostok is normally kept open the year around by icebreakers, but in many years it is closed for two to three weeks. To the northwest Murmansk is an ice-free port because of an arm of the North Atlantic Drift that drifts around Scandinavia and keeps the water temperature above freezing. To the west Leningrad can be kept open with icebreakers for all but a few weeks out of the year, and to the south Odessa is normally open all year around, but even the Black Sea is subject to ice on occasion.

The Volga River is the most important river for transportation in the Soviet Union, as illustrated by this view of the Volga at Gorky.

THE WORLD OF WEALTH

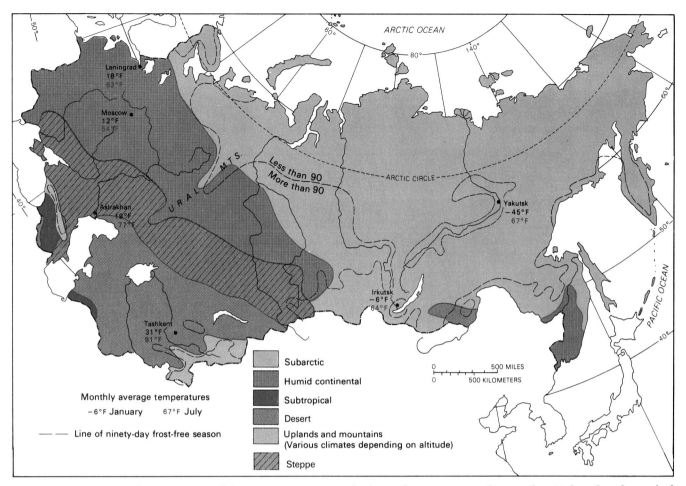

Figure 9.5 Climatic regions of the Soviet Union. Note the line indicating areas with more than 90 frost-free days, which is the minimum required for most crop agriculture.

THE HARSH LANDS

The climate of the Soviet Union is essentially harsh (Figure 9.5). The climatic controls include the large landmass, the high-latitude location, the lack of physical barriers across the vast expanse of the country, and the mountains to the south, which block warmer air from the lower latitudes from entering and cause a rain shadow in Soviet Central Asia. A final climatic control is the isolation of much of the landmass from sources of moisture. As a result of these controls the climate of the majority of the Soviet Union is continental, cold, and dry.

THE TUNDRA CLIMATE

The northern margin of the Soviet Union has tundra climate. Temperatures in this region drop to minus 50°F (−45°C) in the wintertime, and summer temperatures exceed 60°F (15°C). Vegetation consists of low-growing shrubs, willows, mosses, and lichens. The summer is short and cool, and the land is little used by people. Resource exploitation and reindeer herding are the primary economic activities in the tundra.

THE SUBARCTIC CLIMATE

The largest climatic region is the subarctic. Much of the subarctic and all of the tundra are underlain by permafrost. *Permafrost* is an area of permanently frozen subsoil. In the summers the top few inches melt, but because the soil below remains frozen, there is no place for the moisture to go. The cool climate retards evaporation and makes much of the tundra as well as portions of the subarctic marshy and boggy. Subarctic climate has extreme seasonal contrasts in temperature. Winter temperatures have reached −98°F (−73°C), and summer temperatures occasionally rise as high as 90°F (32°C). With nearly 200 degrees Fahrenheit (111°C) potential difference between summer and winter this climate is the epitome

Residents of a village near Archangelsk live in the wooden homes with triple-glazed windows and wear numerous layers of clothing to combat the intense cold of the subarctic winters.

of seasonal contrast. Winter temperatures regularly descend below −60°F (−51°C).

Vegetation in this area is coniferous forest, for which geographers have adopted the Russian word **taiga**. The Soviet taiga is the greatest forested area of the world. The Soviet Union has approximately one-fifth of the forested area of the earth, but because of the extreme cold and short growing season, it is slow-growing, scrubby, and deformed, especially in the north. Much of the taiga is unsuited for lumber.

Agriculture in the subarctice climate is limited. The growing season is short, ranging from 70 to 110 days. In this period there is a great deal of sunlight, and plants that grow quickly do well even north of the Arctic Circle. Cabbages, radishes, potatoes, and similar crops are grown of necessity, even though the yield is low, for the high cost of transporting foods to these isolated areas ensures that as much as possible will be grown. Soils are primarily spodosols, which are acidic and infertile. In the southern margins of the subarctic the soils become more suitable, and the growing season is longer, allowing production of wheat and other small grains. The Soviets have done a great deal of experimentation and have produced grain varieties that will yield even in such areas as Yakutsk. The yield per acre is extremely low, and the subarctic is not one of the major agricultural regions of the country nor is it self-sufficient in food.

THE HUMID CONTINENTAL CLIMATE

The humid continental climate of the Soviet Union occupies a narrow but important zone. Normally this area has ample precipitation all year, and the soils are adequate for agricultural purposes. Snow cover lasts from six months in the northern portions to three months in the southern portions, and winters are long, overcast, raw, and gray. Summers are humid and are normally not extremely warm. Temperatures in Moscow and Leningrad normally go no higher than 80°F (27°C) in the summer months. In August high temperatures in Leningrad may be as low as 60°F (16°C). Autumn is often the best season of the year with clear skies and mild temperatures.

Soil in the humid continental area range from true alfisols in the north to mollisols dominated by chernozems in the south where transition to the steppe climate occurs. Vegetation is a mixture of coniferous and deciduous trees changing to a mixed deciduous forest to the south. The area of humid continental climate in the Soviet Union is one of the most significant agricultural regions in the country.

STEPPE

The steppe climate of the Soviet Union is transitional between the humid continental and desert climatic types. The margin between the steppe and the humid continental area is a zone of very fertile black chernozem soils of the mollisols of the seventh approximation. These soils and the other steppe areas suffer from an unreliable climate, especially precipitation, but the region is the premier agricultural area of the country. It is the primary wheat area of the Soviet Union, but since rainfall is variable and erratic, there is always a potential for a shortfall in grain production. This area receives between 8 and 16 inches (203 to 406 millimeters) of precipitation annually, most of it in the form of heavy thunderstorms in summer. Win-

---

Taiga: The great northern coniferous forest, also referred to as the boreal forest.

THE HUMAN DIMENSION

Life in the Subarctic

Life in the subarctic climate is difficult. Unless you live on the southern margins, as most Soviets do, winter can be long and dreary. Verkhoyansk is an extreme example of the difficulty of the harsh environment. Located north of the Arctic Circle, this city of some 3000 people claims the distinction of being the coldest city on earth. No road or railroad connects it to the outside. The only connection in winter is via air to Yakutsk. In the summer when the Yana River thaws goods are also brought in by boat to supply the town for the winter. In the wintertime when the temperatures fall below $-58°$F ($-40°$C), even contact by air is cut off. The economic cost of maintaining a settlement at such a location is obviously high, but in the Soviet planned economy there is a need for the products that can be grown here, specifically furs for coats and decorations. In addition the original Jakuts tribesmen engage in reindeer herding and other activities. The clay yurts (huts) of the Jakuts mingle with the log houses of the early Russian colonizers and the two-storied stone school building to form a frontier community.

The difficulties in such a harsh region are obvious. Simple matters of water and sewage become monumental problems when the ground is permanently frozen. A central system heats water and then sends it out to the homes through a pipeline. If it is not immediately used, it is returned to the central heating plant to be reheated. Most families in Verkhoyansk still rely on ice for their water supply and simply have blocks of ice in front of the house, melting what they need in a pot over the fireplace. Sewage is simply allowed to freeze, and then the Yana River carries it out to sea in the summertime. Log houses have proved the most effective, but must be built on piles above the ground, so that heat seeping through the floor does not melt the permafrost and cause the house to sink. Automobiles and other internal combustion engines do not function at such cold temperatures unless they are kept inside or the oil is drained at night, heated, and placed in the engine at the time it is to be used. In Verkhoyansk there are no automobiles in winter. Since the roads are mere tracks, this is no great hardship. The community is growing as part of the Soviet effort to develop Siberia, and the residents continue to work to fill their quotas on the five-year plans.

Life in the small isolated communities such as Verkhoyansk is much harder than in other areas of the Soviet Union, and the government provides incentives for those willing to live in such locales. Wages may be as much as twice as high as in other areas, and workers in Siberia are entitled to a month's extra vacation in the more populated and inhabitable regions of the Soviet Union each year. But for the residents of communities like Verkhoyansk, there is a strong sense of commitment to developing Siberia. The limited goods in the stores, the difficulty of construction work at $-70°$F ($-56°$C), and the isolation are accepted as part of the costs of being pioneers. The government provides a department store that sells shoes, clothing, and other necessities, and a separate store sells food. Canned foods predominate, and in winter fresh vegetables are essentially unknown.

Life is possible at such extreme temperatures because the air is almost completely still. In Verkhoyansk the wind almost never blows because the intense cooling of the landmass produces a stable, cold, high pressure system. Thus there is no wind-chill factor, and the intense cold is at least tolerable if one is adequately dressed. For the people who live here the cold, isolation, and lack of culture and other amenities are acceptable for a variety of personal reasons.

Verkhoyansk, of course, is an extreme in that it is the coldest city in the country, but even in cities such as Yakutsk with approximately 200,000 people many of the same problems exist. At Yakutsk temperatures drop to a $-70°$F ($-57°$C), but construction continues. The Soviet Union has pioneered development of cold lands, and workers continue to pour concrete for their prefabricated buildings even when temperatures drop to $-50°$F ($-45°$C). The resulting buildings are less than elegant, often described as old before completion. Development of Siberia is part of national policy, and since this land occupies nearly 50 percent of the total area of the Soviet Union, it is easy to understand why the government expends such efforts in attracting and maintaining settlements in the great subarctic area.

---

ter snowfall is light, and the soil is subject to extreme frost action.

THE DESERT CLIMATES

The desert climates of the Soviet Union are important because they have hot summer temperatures, allowing production of specialty crops. Temperatures rise to over 100° (38°C) in the summertime, and in the wintertime they drop below freezing for short periods. Much of the desert area is subject to occasional cold spells, but where water is available, agriculture is practiced. Most of the desert area is devoted to grazing, primarily of sheep. Irrigated areas produce cotton, fruit, grains, and dairy products.

MINOR CLIMATIC TYPES

In addition to these broad zones, there exist in the Soviet Union some small areas of other types, which are extremely critical in spite of their small size. Heading the list is a small area of humid subtropical climate on the eastern margin of the Black Sea. Protected by the Caucasus Mountains to the north, this area has a climate warm and moist enough for citrus fruit and tea. It is also a summer resort area for the Soviet Union as tourists come to the beaches of the Black Sea.

Directly west of this humid subtropical region is a small region of Mediterranean climate on the Crimean Peninsula. The Yalta Mountains act to limit the impact of the colder air from the north and maintain mild winter temperatures. This small area of Mediterranean climate and the somewhat larger subtropical area are of great significance to the Soviets because of their tourist attractions and their suitability for certain crops that otherwise could not be produced in the country.

MINERAL RESOURCES IN THE SOVIET UNION

Although the climate of most of the Soviet Union is harsh and hinders agricultural development, the large area of the country makes it a potential storehouse for mineral resources. The Soviet Union has a vast array of resources of almost every conceivable type. It is the world's leading producer of iron ore, lead, zinc, and petroleum. It is usually number two in production of gold, chrome, copper, tin, and diamonds (Figure 9.6).

COAL

Traditionally coal was the most important resource to the country. Coal is found in numerous areas throughout the Soviet Union, but the most important producing area has long been the Donetsk Basin of the Ukraine. The Donetsk region as late as 1913 produced 90 percent of the coal mined in czarist Russia. Today only one-third of the total coal output of the country—but more than one-half of the coking coal—comes from here. The second major producer is the Kuznetsk area in central Siberia with the largest and best coal reserves in the entire nation. It is handicapped by its tremendous distance from industrial areas. Midway between Kuznetsk and the Urals another field, Karaganda, was developed to provide coal for iron and steel industries in the Urals. It has large quantities of

Figure 9.6 Principal resources of the Soviet Union. Many of the major deposits are located in regions with harsh climates that handicap their exploitation.

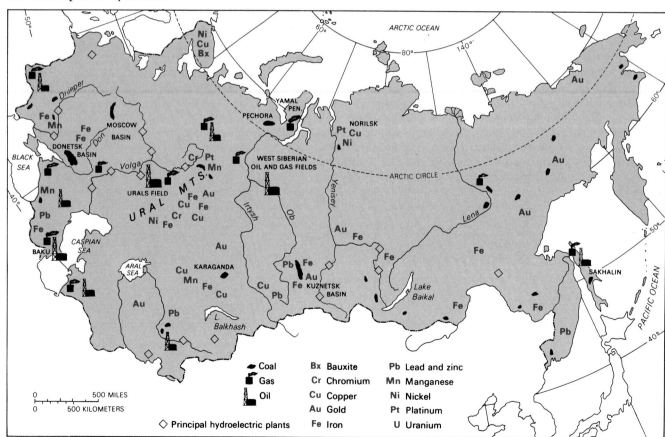

THE WORLD OF WEALTH

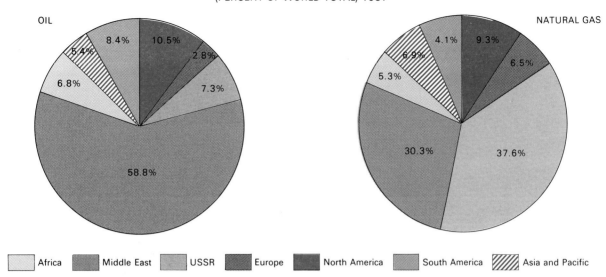

Figure 9.7 U.S.S.R. is one of the largest producers of oil and gas in the world.

coal, but it is of poor quality for coking with a high ash content and must be mixed with coal from the Kuznetsk.

Coal is produced in numerous other locations, including major lignite production in the Moscow area (Figure 9.6). The Soviet Union has large reserves of coal, claiming one-half of world reserves. Distance, accessibility, and deposits that are more accessible have prevented development of most of them, but mining takes place in the Transcaucasus region, in Siberia (for export to Japan), and elsewhere to meet local demands.

PETROLEUM

The Soviet Union has large deposits of petroleum and natural gas. The two most important producing areas at the present time are the Volga-Ural fields and the West Siberian Plain. Until the 1970s the Volga-Ural field was by far the leader, but it is now second after the Siberian fields, which have been developed in the last decade. Located north of Tyumen in the Ob Basin east of the Urals these large fields of petroleum now exceed the production from the Volga-Ural area, producing more than half of all petroleum in the Soviet Union.

In addition the Soviet Union historically produced a large amount of oil from the Caspian Sea near Baku. Although these fields have been surpassed by the Volga-Ural and Ob River areas, they are still important producing regions. With all its petroleum the Soviet Union must still import some from the Middle East to meet its needs. The cost of exploiting the petroleum in the northern portions of the country and commitments to export petroleum to east European countries have to date worked against total self-sufficiency.

The Soviet Union also has tremendous deposits of natural gas and is second to the United States in production. In the vicinity of the mouth of the Ob River alone, the Soviet Union has developed natural gas fields whose capacity is measured in the hundreds of trillions of cubic feet of reserves (Figure 9.7). These have now been developed, and two pipelines are transporting this gas westward to Moscow and the other populated regions, but still only a fraction of the reserve is being utilized. A large pipeline to export gas to Western Europe was completed in 1984, which provides up to one-third of the energy requirements of West Germany and France. The Soviet Union has adequate natural gas and petroleum for its own needs, but the difficulty of developing the infrastructure to exploit them fully in the harsh environment has hindered their development.

To meet its energy needs, the Soviet Union has also built large dams for power generation on most suitable rivers. Massive dams such as the one at Bratsk, or along the Volga River, have been insufficient to meet national energy requirements, causing the Soviet Union to invest heavily in nuclear electricity generating plants. Nuclear energy provided 10 percent of Soviet electricity in 1985; but the tragedy at the Chernobyl plant on April 26, 1986, when the plant suffered a near meltdown and released deadly radiation, raised serious questions about the safety of nuclear generators. In spite of the death and evacuation in Chernobyl and the long-term impact on cancer rates in the Ukraine, Poland, Finland, and Sweden where the radioactive cloud dispersed, it is unlikely that the

THE SOVIET UNION

Soviet Union will slow its growing reliance on nuclear energy. Today about one-fifth of electricity comes from hydroelectric plants, 70 percent from thermal electric plants, and the balance from atomic plants.

IRON ORE

The Soviet Union is the world's largest producer of iron ore. Major areas include the Krivoy Rog region in the Ukraine, which has traditionally been the major producer. At Magnitogorsk in the Ural Mountains are additional large reserves, which have been partially depleted but are still important. Midway between the Black Sea and Moscow at Kursk large new deposits are being developed at the Kursk Magnetic Anamoly. These ores are lower grade but are adequate for production of iron and steel. Deposits sufficient for local needs are found in the Kuznetsk basin, in Transcaucasia, in the northwest near Leningrad, and in many other areas. The Soviet Union has adequate iron ore for its present needs.

OTHER MINERAL RESOURCES

The Soviet Union has a variety of other mineral resources. It is the world's major producer of manganese, an essential element in the iron and steel manufacturing process. The largest reserves are found at Nikopol and Chiatura. By fortuitous circumstances Nikopol is located between the coal and iron ore deposits of the Ukraine. It is the major exploited area of manganese because Chiatura is located in Transcaucasia and is less accessible to the major iron and steel areas.

The Soviet Union claims to have the world's largest deposits of copper. Copper is mined at Gay in the Urals, at Dzhezkazgan in the Soviet Central desert, and at Kounradskiy north of Lake Balkhash, to mention but three of the important areas. Gold and diamonds sufficient to rank the Soviet Union number two in world production are found in Siberia. The ores for aluminum are available in adequate supply in the Soviet Union. In addition to using bauxite, which is available in only limited quantities, the Soviet Union is using alunite and nephelite, two clays that are high in aluminum oxide. The ores are processed into aluminum at sites where the abundant electricity required for the process is available. Major centers include Bratsk where hydroelectric power is used and Novokuznetsk where thermal electric power is utilized. Tin is mined from alluvial deposits in the streams and rivers of Siberia, chrome is mined in the Urals, and most other minerals are found someplace in the vast land of the Soviet Union.

COMMUNISM AND ECONOMIC DEVELOPMENT

The land of the Soviet Union, "Mother Russia," is a land of harsh environments, abundant mineral wealth, and vast extent. What have the Soviet leaders accomplished in their 70 years in power as they have attempted to transform a traditional agrarian economy into a modern, industrialized nation? Three broad periods of economic development are recognizable in both industry and agriculture; the postrevolutionary economy under Lenin's direction discussed earlier, the era of strong central planning and control of production introduced by Stalin, and the recent changes under the guidance of Mikhail Gorbachev, the communist party leader who came to power in the mid-80s.

Leadership of the Soviet Union under Stalin believed that industrial development was the most critical problem facing the Communist country. The Soviet Union had been involved in World War I prior to the revolution. Since it had lost significant territory as a result of the revolution and World War I, Stalin was convinced that the country must develop its industrial and military might to prevent any further territorial loss and to ensure that the country became a world power. In accomplishing this industrial revolution, the actions taken in agriculture played an important role.

An agricultural revolution was viewed as a necessity if the growing industrial population of the urban area was to be guaranteed a source of food. Moreover, since the overwhelming majority of the population was still living and working in rural villages, the agricultural sector was viewed as a source of labor for the industrialization that Stalin envisioned. To ensure that food was forthcoming and to transform the agriculture so that surplus labor could be drained off to the urban industrial areas, the government adopted a program of collectivization. Collectivization consisted of returning the villagers to a communal form of agriculture not unlike that of the mir that had existed in czarist Russia. The major difference was that instead of the villagers' collectively determining the use of the land themselves, decisions concerning crops and markets were determined by the state. These new collective farms were called **kolkhoz**. The land

---

Kolkhoz: A collective farm in which farming activities are performed by farm members who share in the profits after meeting quotas required by the central government.

that had been operated under the New Economic Policy (NEP) under private enterprise was now to be operated as a state-operated collective in which workers would be paid according to the amount of labor that they contributed.

Peasant unrest over collectivization resulted in major disruption of agriculture. Distribution of land after the revolution had been one of Lenin's main promises, and the peasants were reluctant to give up their newfound independence. Peasants killed their cattle, pigs, and sheep, burned agricultural equipment, and generally resisted collectivization. From 1929 to 1933 over one-half of the horses, one-half of the cattle, two-thirds of all the sheep, and most of the pigs in the Soviet Union were slaughtered. Stalin insisted on collectivization, and by 1934 the 100 million peasants had been largely forced into the collective system. The suffering of the peasants during this period was severe, with varying estimates of 5 to 10 million losing their lives as a result of famine or through resistance to the enforced collectivization. By 1938 the peasants had been completely collectivized, as nearly 250,000 kolkhoz replaced the 25 million independent farm holdings that had existed under the old NEP program.

Stalin saw several advantages to collectivization, not the least of which was maintaining an adequate supply of grain for the urban dwellers. The view of Stalin and his planners was that the small, individual peasant farms were uneconomical for mechanization. The independent farmer was primarily concerned with his own interests rather than with those of the state, and it was difficult to control the flow of crops from the peasants when there were 25 million individual farmers making decisions. Collectivization allowed the government to consolidate the landholdings into units large enough to justify mechanization, thus freeing labor for industrialization. The 250,000 collective farms also simplified control by the government, so that it could ensure that the collectives met their quotas of food for the urban dwellers. Finally, by organizing farmers into collectives, the government felt that it would be able to provide the basic necessities of housing, electricity, education, and cultural development more cheaply and simply than it could do in the numerous dispersed villages.

By the mid-1930s the goal of collectivization had been reached, the major farming activities of plowing, planting, and harvesting of grain had been largely mechanized, and food production had increased to the level of pre–World War I Russia. The peasant collective farmer was required to work on the collective, but he was also allowed to cultivate a small plot of land for his own use. Each peasant was allotted approximately one-half acre (.2 hectare) on which he could grow anything he desired, and he could do as he pleased with the crops. These private plots have remained in existence to the present and provide a significant proportion of the vegetables, fruits, eggs, meat, and butter in the Soviet Union. It is estimated that the private plots provide about 25 percent of the total agricultural production in the Soviet Union, although this figure is slowly dropping. In production of certain items this figure is much higher. The private plots produce more than 40 percent of all eggs and 30 percent of vegetables and milk.

As collectivization proceeded, the government expanded agriculture into areas that had previously not been cultivated, particularly Siberia and Central Asia. In these regions the government adopted a different style of collective known as the **sovkhoz.** The sovkhoz is characterized by tremendous size, fewer workers than the kolkhoz, essentially complete mechanization, and wages and a workweek similar to that of a factory (Table 9.2). On the sovkhoz, workers do not share in the profits but are paid a salary regardless of the amount produced. In the period since initial collectivization the government has increasingly moved toward the creation of sovkhoz or state farm operations, and today there remain only approximately 26,000 collectives in the Soviet Union. The remaining kolkhoz are much larger than the original kolkhoz as the numbers have been reduced, largely through amalgamation of collective farms to make larger units.

As a model for transforming agriculture the Soviet experience is an interesting example. For the decade

The housing provided for members of this collective in the western portion of the Soviet Union is representative of most collective farms. Houses are generally identical in size, and each family is provided a garden plot. Note that cars are noticeably absent, but electricity, television, and other amenities are provided.

Sovkhoz: A farm owned by the state; farm members are paid a set wage regardless of the profitability of the farm.

TABLE 9.2 Comparison of average of Kolkhoz (Collective) and Sovkhoz (State Farms), 1984

	Collective	State
Agricultural land, acres (hectares)	15,814 (6,400)	40,030 (16,200)
Arable land, acres (hectares)	9,250 (3,743)	14,750 (5,969)
Area sown, acres (hectares)	8,648 (3,500)	12,355 (5,000)
Number of households	482	n.a.[a]
Tractors	43	57
Cattle	1,933	1,880
Milk cows	609	608
Swine	1,120	1,151
Sheep and goats	1,695	3,090

[a] n.a. = not available
SOURCE: Alan P. Polland, ed. *USSR Facts and Figures* (Gulf Breeze, FL: Academic International Press, 1988).

following the revolution of 1917, land was returned to private ownership and divided among the peasants, so that the farmers tilled their own land. Since 1929 the land has been in possession of the state. As a method of increasing the standard of living of the rural peasants the collective has seemingly been successful. On the large sovkhoz the government pays a salary, which approaches that received by industrial workers, eliminating the great difference in income between urban and rural dwellers found in many agrarian economies. The workers on the sovkhoz, as well as all citizens of the Soviet Union, receive housing and the basic necessities of life at government-controlled prices. This means, for example, that no more than 6 percent of a worker's income can be required for housing, but at the same time the centrally planned society limits the amount of space per person on a fairly uniform scale. (The national standard is 9 square meters, or a room about 10 by 10 feet, per person). Farm workers have many of the same educational opportunities that city residents do, and although the rural areas do not have the same cultural opportunities as the large urban centers, rural farm life is better than it has ever been in the Soviet Union or was in czarist Russia. The farms are essentially mechanized for those jobs that can be mechanized, and although there are perennial problems of maintenance of equipment and shortages of trucks, grain harvesters, and tractors at times they are most needed, conditions still represent a marked improvement over czarist Russia.

On the collective (kolkhoz) farms, the conditions are also much better than they were in the pre-Communist era. Farms are now mechanized to a large extent, and the old tractor stations that provided equipment for numerous collectives have been abolished as the collectives themselves own their farm equipment. The government has increased the wages of workers on the collective farms. The majority now receive a cash payment on a monthly basis rather than a yearly basis to increase the standard of living of the collective residents. Although the workers on the collective farms are perhaps less well paid than those on the sovkhoz, the government has raised the pay level several times to make them more competitive with the industrial sector of the economy. The collective farms have been successful in feeding the country. Although there remain problems of drought and the generally harsh environment, the increase in cultivated land and the increased production has limited their impact.

Gorbachev has introduced reforms in the economy of the Soviet Union that directly affect agriculture. These policies are: *glasnost* (openness and frankness in admitting past mistakes), *demokratizatsiya* (democratic involvement of the people in the government and economy), and *perestroika* (restructuring of the economy). Although it is still too early to predict how successful these policies will be, as proposed they will make a major change in the agricultural system of the country. According to new regulations approved by the State Agro-Industrial Committee, the land (all of which is state owned) will be leased to private individuals or groups for up to 50 years. The machines, now owned by the state and collective farms, will be sold or leased to the new private entrepreneurs. These new farm operators will be able to hire additional labor, something prohibited by the Marxist-Leninist doctrines of the Soviet Union for more than 60 years. The new private farmers will have to pay wages that are not lower than those on the remaining state and collective farms. Whether Gorbachev's application of perestroika to agriculture will work remains to be seen, but it represents a revolutionary break with the agricultural policies in place since the time of Stalin.

REGIONAL COMPONENTS OF AGRICULTURE

Today's collective farms in the Soviet Union tend to be specialized by area of the country. In the irrigated areas of Soviet Central Asia the emphasis is on the production of cotton, alfalfa, and fruits and vegetables that require warm weather. In the mixed farming of the humid continental zone livestock are combined with production of small grains, sugar beets, potatoes, cabbage, and deciduous fruits such as apples and pears (Figure 9.8).

The Ukrainian steppes traditionally produced wheat and sugar beets, but the majority of the wheat has now been moved east of the Urals, and the Ukraine is now the area of concentrated corn (maize) production. Corn is marginal for this area, as the best locations are at a latitude similar to Minnesota, but it can be grown; it is produced on approximately 30 percent of the land in the Ukraine. Other crops include sugar beets, a variety of small grains (including wheat), and vegetables of all types.

In the southern province of Moldavia grapes are an important crop, and wine production is a significant part of the economy. In the subtropical climate of the Caucasus the major products are tea, citrus, and corn. In the far eastern portion of the Soviet Union flax, rye, and other grain crops are produced. In the subarctic region crops are grown for subsistence purposes for the urban developments in those areas. In the northwest of the Soviet Union production of rye, flax, oats, and animal products are the dominant activities.

Today the Soviet Union is one of the world's major agricultural nations. It cultivates approximately one and one-half times as much land per year as does the United States. It leads the world in production of barley, rye, wheat, sugar beets, flax, sunflowers, potatoes, butter, milk, and cabbage, but Soviet agriculture still faces major problems. One has to do with

Figure 9.8 Major agricultural regions in the Soviet Union. Crop and livestock production is concentrated in the southwest of the country.

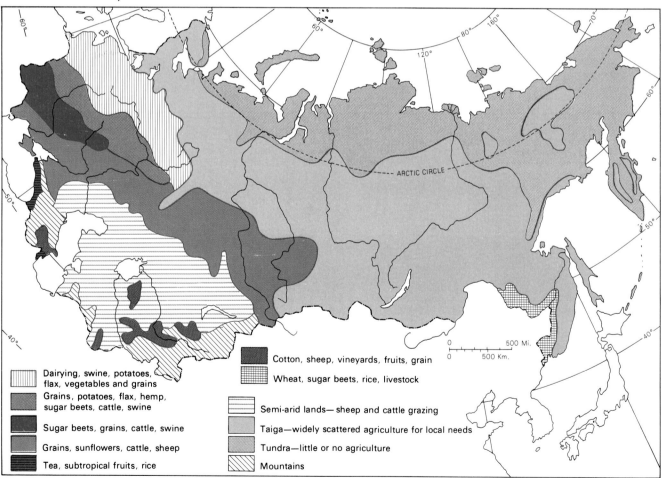

THE SOVIET UNION

249

central planning. Since the planning takes place for the entire nation with target production for each year of the five-year plan, the regional planning agencies must orient their planning to meet the goals of the five-year plans. This central planning affects the agricultural sector and makes responding to changes in demand, changes associated with poor yields, shortages of fertilizers, and other variables difficult. The centralized planning also affects agriculture through provision of machinery, seed, fertilizer, and other required elements. For example in 1978 the Soviet Union had the largest yield of wheat in the history of the country. Because there was inadequate equipment to harvest the grain, much of it was wasted. Because of the lack of trucks and drivers to carry the grain from the harvested areas to the storage areas and inadequate storage areas as well, much grain had to be stored outside, and some was destroyed by rain and insects. In 1984, environmental problems caused a shortage of grain.

Beyond the impact of shortages caused by the planning system agriculture in the Soviet Union suffers from a penchant of the planners to embark on massive experimentation programs with inadequate planning. The best example is the so-called Virgin Lands Project. The virgin lands is the belt of steppe and humid continental climate east of the Urals. In the past this land was devoted to grazing and pastoral activities. When Nikita Khrushchev visited the United States in the mid-1950s, he was greatly impressed by the productivity of the U.S. Corn Belt. Because meat and milk were perennially in short supply in the Soviet Union, Khrushchev and the planners embarked on a program to cultivate corn to provide feed for animals to ensure more meat and milk. The best land for production of corn was in the Ukraine, but this was also the best area for production of wheat. The planners examined the country and determined that the steppe region east of the Urals was suitable for producing wheat and other grains and could be cultivated if conditions were right.

More than 100 million acres (40,470,000 hectares) were plowed from grass and planted to grain in the 1950s and 1960s as the area was transformed into the major wheat-producing region of the Soviet Union. Unfortunately, this land has marginal precipitation, with yearly variation from 12 to 15 inches (305 to 381 millimeters) in the north to 10 inches (254 millimeters) in the south. The high variability of the precipitation results in an average of one year in five in which severe drought will affect the crop. As a result yields can vary enormously from year to year (Table 9.3). Grain yields have fluctuated nearly 100 percent from the low in 1966 to the high in 1978. This problem was not as severe in the Ukraine where corn has been planted, so that corn production tends to be more reliable than the grain production of the virgin lands. By plowing up an area equal to one-fourth the total cultivated area of the United States (equal to the entire cultivated area of France, Germany, and Great Britain combined), the Soviet Union has been able to increase greatly its production of meat, milk, and butter. The price it has had to pay has been the periodic shortage of grain as in 1972, 1975, and 1984. The country was able to maintain corn production and meat and milk were not completely absent from the markets although the quantities did decline as a result of the slaughtering of surplus animals during years of low grain production. The experiment with the virgin lands was successful in meat, milk, and corn production but left the country more vulnerable in production of wheat.

Another problem that affects the agricultural sector under the Communist system is the lack of incentive. Quotas have been set, and little recognition has been given to differences in quality of production. On the typical collective the workers were required to produce a quota of a crop, and then they shared in the balance. The lack of incentive led to the individual farmers' putting their best efforts into their own private plots where yields were much greater and the profit went entirely to the farmer. More recently the government has placed more emphasis on incentives. People who produce more than their basic quotas are rewarded with higher wages, awards, and other rec-

Propaganda posters are common in the Soviet Union. This photo from Moscow uses the portraits of Marx, Engels, and Lenin in conjunction with a banner contrasting Perestroika's reforms with the slow pace of progress in the past.

TABLE 9.3 Variations in Yield for Major Crops in the Soviet Union: 1960–1986

Year	Grains	Meat	Milk	Potatoes	Vegetables	Fruit	Sugar
			(million metric tons)				
1960	123	8.7	61.7	84.4	16.6	4.9	5.7
1965	114	10.0	72.6	88.7	17.6	8.1	9.2
1970	179	12.3	83.0	96.8	21.2	11.7	8.8
1971	174	13.3	83.2	92.7	20.8	12.4	9.0
1972	161	13.6	83.2	78.3	19.9	9.6	8.0
1973	214	13.5	88.3	108.2	25.9	13.4	8.1
1974	186	14.6	91.8	81.0	24.8	12.4	9.6
1975	134	15.0	90.8	88.7	23.4	14.2	7.7
1976	214	13.6	89.7	85.1	25.0	15.3	7.7
1977	188	14.7	94.9	83.6	24.1	15.3	7.4
1978	229	15.5	94.7	86.1	27.9	14.4	8.8
1979	174	15.3	93.3	91.0	27.2	16.3	9.3
1980	182	15.0	90.6	67.0	25.4	14.6	7.8
1981	165	15.2	88.5	72.0	25.6	14.5	7.2
1982	170	14.6	83.3	82.0	24.8	14.2	6.4
1983	192.2	16.4	96.5	82.9	25.1	14.5	13.5
1984	172.6	17.0	97.9	85.5	25.7	14.8	13.5
1985	191.7	17.1	98.6	73.0	24.5	14.3	12.8
1986	210.1	17.7	101.1	87.2	26.0	15.1	13.8

SOURCE: Lester Brown, *U.S. and Soviet Agriculture: the Shifting Balance of Power*, (Washington, World Watch Institute, 1982) p. 16 and Alan P. Pollard, *USSR Facts and Figures Annual* (Gulf Breeze, FL: Academic International Press, 1988), p. 143.

ognition designed to encourage and promote individual incentive under the title of "friendly socialist competition." Gorbachev's proposed changes acknowledge this weakness, and the new private farmer will keep any profit over the cost of leasing the land he or she farms. Perestroika's principles include financial rewards for those who work harder or who do a better job; and in theory the new private farmer will be well rewarded for leaving the security of the collective system for an untested quasi-capitalistic program.

A final problem of Soviet agriculture is that regardless of the organizational and institutional arrangements, the harsh environment persists. Droughts and other climatic impacts may well be the most important factors affecting agriculture in the Soviet Union today. The high latitude and continental nature of the U.S.S.R. subject agricultural activities to great variation from one year to another. It is clear that organizational and managerial strategies handicap Soviet agriculture, but regardless of the system, the environmental limitations of much of the Soviet Union play a major role in the ultimate success or failure of agriculture.

INDUSTRY IN THE SOVIET UNION

Industrial development in the Soviet Union was similar to that of most countries. In the feudal period industry was primarily handicraft-oriented or provided such services as blacksmithing for local farmers. As in Western Europe in the fifteenth and sixteenth centuries, there developed a fairly broad-based craft industry consisting of locksmiths, metal workers, leather workers, and linen manufacturers. Under the czars in the late seventeenth, eighteenth, and nineteenth centuries, foreigners were encouraged to invest capital in Russian industry. Under the impetus of outside investments Russia's industry developed rapidly, particularly iron manufacturing because of the abundant iron ore deposits of the country and the immense forests for making charcoal. Russia became a major producer, with total tonnage of three to four times that of England in 1750.

Technological innovation in the British Isles in the late 1700s and early 1800s left Russia behind. Under Peter the Great, a czar who desired to emulate

Western Europe, Russia turned westward and imported tremendous quantities of Western technology in the late 1700s. From 1800 to the 1850s industry advanced rapidly in Russia as foreign investors, capitalizing on the plight of landless peasants who were forced into the cities, established factories based on cheap labor. After 1861, when the serfs were freed in Russia, an even greater abundance of labor became available. As would be expected, textiles became the primary industry, as they were able to use unskilled labor. In 1890 cotton, linen, and woolen textiles represented one-third of all the manufactures in the country. In addition there were some machine works, the manufacture of locomotives, a beet-sugar industry, and a variety of others, but industry was small-scale and still producing primarily for an internal market.

The problems that prevented Russia from being a dominant industrial power in the 1800s are the same that have affected the country to the present. After Peter the Great's death the various czars had widely fluctuating policies toward the West, but in general the nation was isolated from technical advances of Western Europe. The vast area of the country was thinly populated, and lack of transportation prevented efficient concentration of the resources in central locations. Unlike Western Europe, particularly the United Kingdom, which was able to rely on the oceans to transport raw materials from its colonies, Russia was faced with the barrier of a large landmass with no cheap and rapid transportation system. The labor force was largely illiterate and poorly trained, further handicapping industry.

INDUSTRY ON THE EVE OF THE REVOLUTION

In spite of these limitations, Russia emerged as a significant industrial producer on the eve of the Communist revolution. The major spurt in growth took place from 1894 to 1917. Essential to this growth was the development of an internal transportation system, including the Trans-Siberian railway in the mid-1890s. Transportation, coupled with a favorable policy by the czars to foreign investment, led to tremendous investment by foreigners to capitalize on the cheap labor. By 1900 one-half of all the capital in Russian industries was foreign-owned, and this concentration increased by 1917. The huge coal deposits at Donetsk and the iron ore deposits of Krivoy Rog were developed by the Englishman John Hughes, and the Swedish firm of Nobel developed the oil at Baku. The mineral wealth of the southern Urals, including copper, gold, platinum, and the Magnitogorsk iron ore deposits, was developed by English, Belgian, and French firms in this period.

As a result, the iron industry in Russia increased three times as fast as did that of Germany and eleven times as fast as in England during the decades just prior to the revolution. Coal and textiles also increased rapidly, so that by 1914 Russia was the fourth largest industrial nation in the world after the United Kingdom, Germany, and the United States. It is important to remember that the foreign owners were primarily interested in maximizing their profits, and thus the wages paid in Russia were the lowest in Europe at the time, averaging only about $7.00 per month in the Moscow area in 1914. As a result of low wages and poor working conditions in the years just prior to World War I, the nation averaged over 2000 strikes per year involving more than 20 percent of the total labor force. These strikes were put down violently by the military and were a major factor affecting demands for revolution.

INDUSTRIALIZATION SINCE THE COMMUNIST REVOLUTION

Since the revolution the Soviet Union has emerged as the world's second largest industrial power as measured in total manufacturing (Table 9.4). In some areas the Soviet Union is the leader. After the revolution and the expropriation of all foreign-owned industry, productivity dropped markedly. Therefore, the New Economic Policy (NEP) essentially returned the industries to the owners' control, so that by 1928 industrial production once again returned to pre-1917 levels. After 1929 the industries were taken over and operated by the state as in the agricultural sector.

Prior to the revolution the annual growth rate in industry averaged 5 percent per year, compared to about 4 percent per year under the Communists. Lest this be viewed as proof that the Communist system is less effective, it should be remembered that two world wars and the Communist revolution disrupted industry and greatly handicapped development. The collectivization and nationalization of industry and agriculture after 1929 also greatly affected industrial growth. From 1929 to 1937 industrial production grew at a rate of 15 percent per year. It could be argued, therefore, that under Communist direction the rate of growth was triple that of the period under the czars.

Growth rates of such magnitude are not uncommon in nations as they begin rapid industrialization. The United States experienced similar or even greater growth rates between 1880 and 1917. Moreover, the Soviet Union had certain advantages as it attempted

TABLE 9.4 Production of Selected Items, U.S.S.R. and U.S.A., 1987

Product	Unit	U.S.S.R.	U.S.A.
Electricity	Billion KwH	1599	2652.7
Oil	Thousand Barrels per day	11754	8668
Natural gas	Billion cubic feet (meters)	24200 (691)	16,800 (480)
Coal (hard)	Million short tons (Metric tons)	564 (513)	785 (714)
Steel	Million tons (Metric tons)	177 (161)	122 (731)
Aluminum	Thousand tons (Metric tons)	3267 (2970)	3341 (3037)
Motor vehicles Passenger cars	Thousands	1300	7500
Tractors, large	Thousands	595	384
Television	Thousands	7935	22,585

SOURCE: Alan P. Pollard, ed. USSR *Facts and Figures Annual* (Gulf Breeze, FL: Academic International Press, 1988).

to industrialize after 1928. The **incipient industrial base** of the prerevolutionary period provided a basic reservoir of trained people. The industries developed under the czars and foreign investors at Donetsk, Krivoy Rog, Magnitogorsk, Baku, Moscow, and Leningrad provided the basis for additional development, so that the nation had a base to build upon, unlike many of the truly less industrialized countries of today. During the period of rapid growth from 1929 to 1937 much of the industrial world was suffering a severe depression and was eager to sell for export. Thus the Soviet Union imported technical knowledge, skills, and equipment at very low prices. Since the Soviet Union was late in industrializing, it was able to tap the accumulated technical knowledge of the previous two centuries from Western Europe without spending time to develop it themselves. Moreover, industrial development until after World War II was primarily in relatively unsophisticated areas of iron and steel, locomotives, tractors, and other low-technology items. Finally the market was tremendous, since there was a shortage of almost everything, and anything that could be produced could be used.

Incipient industrial base: The railroads, technology and skilled labor resulting from initial industrialization efforts by countries that speeds subsequent rapid industrialization.

As a result of these factors industry in the Soviet Union developed rapidly in the interwar years, but World War II dealt it a devastating blow. Essentially the industry of the entire European part of the Soviet Union, where 60 percent of all industry in the country had been located, was destroyed. Thus, after World War II there was once more a great need for industrial development, allowing rapid growth to take place.

THE PRESENT STATE OF INDUSTRY IN THE SOVIET UNION

Today the Soviet Union is the second ranking industrial power in the world, producing roughly 60 percent as much (measured by value) as the United States. The gross national product is approximately one-half that of the United States. The gap is even greater if specific items are examined, since in the Soviet Union production has been largely oriented toward heavy industrial, agricultural, and defense-related materials compared to the emphasis on consumer goods in the United States. Per capita production in the Soviet Union is only 40 percent of that in the United States. Return on capital invested in industry is much lower in the Soviet Union. It takes more than three times as much labor to produce a dollar's worth of gross national product and about the same amount of capital investment as in the United States. Industry in the Soviet Union has traditionally been viewed as a means

Siberia: Soviet Frontier

More than anything else Siberia epitomizes the impact of size on the Soviet Union's geographic relationships. Siberia stretches from the Urals to the mountains of the Pacific Coast, from the Arctic ocean to China, Mongolia, and the Kazakh Republic. Sprawling over nearly 4 million square miles it is a region of superlatives. It is a storehouse of mineral and fuel resources, with over one-half of the nation's production of petroleum and natural gas, and an even greater proportion of its reserves. The majority of the Soviet Union's coal, timber, gold, copper, diamonds, and hydroelectric potential are in Siberia. The rivers of Siberia could stretch around the world twenty-five times. The world's largest freshwater lake, the twenty-million-year-old Lake Baikal, is found here.

What Siberia lacks is people. Siberia comprises 40 percent of the territory of the Soviet Union, but fewer than 10 percent of its people live there in spite of the developmental efforts of the government. Isolation and a harsh climate have slowed economic development of Siberia and handicap attempts to attract more settlers. Albeit slower than the government might wish, Siberian development pushes forward. The Baikal-Amur Mainline (*BAM*) railroad was opened in 1984. The BAM is a 2000-mile, multibillion-dollar railway stretching across southern Siberia from Bratsk west of Lake Baikal to Sovetskaya-Gavan on the Pacific coast. Called the project of the century by Soviet planners, it is envisioned as the tool to develop many of the mineral resources of the region. (The strategic importance of the line—it is much farther from the Chinese border than the Trans-Siberian railway—is an obvious additional benefit.) The Soviet Union is committed to increasing the economic development of Siberia and new towns are springing up along the BAM to export resources. Pulp and paper plants, cellulose plants, copper mines, and coal mines are operating or under construction as the Siberia's storehouse of resources are tapped.

The cost of development is high. Most of the food, supplies, clothing, and manufactured goods must be shipped in to the new settlements of Siberia. Construction costs for housing, factories, roads, and services are higher in the harsh environment. Wage incentives that start at 10 percent and escalate to 50 percent in five years, free annual vacations to other parts of the Soviet Union after three years, and early retirement are designed to attract workers. In a free enterprise system where profit is the motivating force, Siberia would be even more sparsely settled and development even slower. Government subsidies of all elements of its development, however, were viewed as essential in the past. Whether Siberia will receive as much investment under the principles of perestroika remains to be seen.

---

to ensure the territorial integrity of the nation. Thus the emphasis has been on production of heavy equipment and armaments. Only in the past two decades has consumer production become a primary aim.

THE LOCATION OF INDUSTRIES IN THE SOVIET UNION

Since the Soviet Union has a centrally planned economy, the location of industries is based on planning principles established by the government planning agency (Gosplan). Because 80 percent of all industry was concentrated in the Moscow and Ukrainian areas in 1917, the Communist ideology stated that regarding future location of industry:

1. Industrialization should be rapid.
2. Economic activity should be evenly distributed.
3. Industry and economic development should be used to stimulate the development of backward areas.
4. Production should take place at markets or raw materials sites to minimize transport costs.
5. Specialized production should take place in regions possessing uniquely favorable conditions.

The dilemma posed to the planners of the Soviet Union is the conflicting nature of some of these goals. If industrialization was to be rapid, it was essential that the existing industrial base be expanded. Thus the uniform distribution of economic activity has never been achieved. The development of backward areas did not take place on a large scale until the German invasion in World War II, which forced relocation of industries to the Ural Mountains and east. After the war the industries that had been destroyed in the west were once more redeveloped, so that today the area from the Urals west is still the center of industrial activity while east of the Urals in Siberia industrial expansion is slow.

Several important regions of industrial activity are recognizable in the Soviet Union today, with the Moscow region being the most important (Figure 9.9). The Moscow region is a center for manufacture of consumer goods, textiles, automobiles, and synthetic rubber. Aircraft, river boats, subway cars, processed food, and furniture are all produced within Moscow proper. Ivanovo is a major area for manufacture of cotton textiles. Gorky manufactures aircraft, trucks, newsprint, and buses. Other communities could be mentioned, but this is indicative of the variety.

Figure 9.9 Major industrial districts and cities of the Soviet Union. Although the government has increased industrial production in the far east, there is no major industrial district rivalling those shown on this map.

The second largest manufacturing area is in the Ukraine, centered on the Donetsk–Krivoy Rog region. The dominant activity throughout this area is production of iron and steel and related products. The plants represent the largest iron and steel manufacturing production capability in the world. Specific plants are at Krivoy Rog, Donetsk, and Kerch. Krivoy Rog has the largest single plant in the area, presently having an annual capacity of 15 million tons (13.5 million metric tons), which is ultimately to be expanded to 22 million tons (20 million metric tons). (The largest U.S. plant has a capacity of 6 million tons, or 5 1/2 million metric tons). Iron and steel plants of the Ukraine produce 45 percent of pig iron, and 44 percent of steel in the country. In addition to producing iron and steel, this area produces agricultural machinery (Kharkov, Kiev, and Rostov-on-the-Don), supertankers (Kerch), and processes food (throughout the area).

The Leningrad area has historically been a major manufacturing center. Under Peter the Great the community of Saint Petersburg was built in the swamps near the mouth of the Neva River on the Gulf of Finland. The Neva delta was a marshy land, but here Peter built Saint Petersburg as his "window on the

THE SOVIET UNION

West," representing his desire to orient Russia westward and to adopt some of the ideas of the technological revolution of western Europe. After the communist revolution the city was renamed in honor of Lenin, and it remains a major industrial center. Leningrad developed as a center for innovations in industry, and today it manufactures electrical and electronic items. Hydroelectric turbines, machine tools, and shipbuilding are important industries. Cities around Leningrad are also important manufacturing centers.

The Ural Mountain area is traditionally a major manufacturing area similar to the Ukraine. Here is found heavy industry, including iron and steel and manufacture of agricultural equipment. Magnitogorsk is a major iron and steel center rivaling Krivoy Rog (15 million tons, or 13.5 million metric tons). On the eastern slopes of the Urals and extending into the Ob lowland are a number of larger important manufacturing cities including Sverdlovsk, which produces mining equipment at the Ural machine factory (Uralmash); Chelyabinsk where tractors, agricultural machinery, and iron and steel are manufacured; and Nizhniy Tagil for iron and steel and the manufacture of railway cars.

Another major area that is becoming significant as a center of growth is along the Volga River from Volgograd north to Kazan. Important centers in this region include Kuybyshev, the largest city along the Volga and a center for manufacture of petroleum-related equipment, roller bearings for a variety of uses, building materials, and consumer items. The Volga region has been a focus of industrial development in the post–World War II era in the Soviet Union and is a good example of the impact of the centrally planned economy on communities. At the present time the largest automobile plant in the country is located at Togliatti. Togliatti was named after a famous Italian Communist leader as part of the Soviet penchant for renaming cities. In the late 1950s Togliatti was a small town of about 20,000 people. During the 1960s it was designated as the center of the Soviet auto industry, and the Fiat company of Italy built the Zhiguli auto works there under contract with the Soviet Union. It now produces approximately one million automobiles a year. They are a modified version of the Fiat automobile and come in a 2- or 4-door model and a station wagon. In addition a few small pickups based on the same design are produced. Under the influence of this plan a huge plant stretching 5 miles was developed, and a planned community with a population of over 250,000 has been built to house the workers. The Togliatti auto works was one of the first industries to receive the benefits of perestroika. The plant's output is no longer measured only in quantity of autos produced, but in profit as measured by the difference between the costs of labor and parts to make the cars and their selling price. The plant keeps 47.5 percent of all profits since 1988, and allocates this money in workers' housing, plant reconstruction, and worker bonuses.

Volgograd (formerly Stalingrad) was the site of the first tractor factory constructed under the Soviets in the 1929–1932 period, and it remains a major pro-

A portion of the assembly line of the Likhachyou Motor works in Moscow showing the ZIL-130 truck body being attached to the frame.

ducer for agricultural equipment. Just north of Kazan on the Kama River at Naberezhnyye Chelny a new truck factory, consisting of six major facilities, represents the largest truck complex in the world. The plans call for production of 150,000 three-axle heavy trucks per year and an additional 250,000 diesel engines. (By comparison the United States produces approximately 100,000 three-axle trucks per year.) The entire region along the Volga is being developed as the major transportation-equipment area with Ulyanovsk, the home of Lenin, as the major producer of light trucks and army jeeps. Engels in the Volga area produces essentially all the trolley buses in the Soviet Union. The Volga is rapidly becoming a rival to the Ukraine and Leningrad as the second most important manufacturing region in the Soviet Union.

Textile manufacturing is important in Soviet Central Asia at such cities as Tashkent, and manufacturing for local needs is also concentrated in the valleys of the Caucasus Mountains at Tbilisi, around the Caspian at such places as Baku, and in other large cities. As in other industrial nations industry is found in most large centers, but a major difference is that in the Soviet Union large industrial centers are developed under government direction. The Volga auto plant at Togliatti and the Kama River truck plant are prime examples. Under Communist direction the Soviet Union has become one of the world's foremost industrial powers.

The problems of Soviet industry are similar to those faced by Soviet agriculture. In the past industrial plants were assigned a quota, and this quota was set in quantitative rather than qualitative terms. As a result much of the material produced was of low quality. In the past decade innovations that emphasize quality rather than quantity have been adopted. Nevertheless in the Soviet planned economy there is little room for variation of product, as the official view is that competition between items not only is unnecessary, but uses up scarce resources that could be better utilized elsewhere. Thus while there are four classes of automobiles available to the general public, the Zhiguli is designated as a car for the masses and is the one that is most readily available. The same holds true for other consumer items, in which there are several brands in differing price ranges, but little competition within a specific price range.

Another problem facing Soviet industry is the low productivity per worker. In a country in which **full employment** has been a national policy and wages have tended to be quite uniform within an industry, there has been little incentive to do more than the minimum. New incentive programs are beng adopted, but labor productivity remains a crucial problem and will remain so as long as the goal of full employment hinders investment in labor-saving technology in industry. Another problem is that the Soviet Union is somewhat isolated from technological advances because of government control of publications within the country. It is difficult for scientists to keep abreast of advances in other industrialized countries because of government censorship, and this factor has affected Soviet efforts to reach parity with the West in high-level technology. To overcome this problem the Soviet Union is increasingly turning to importation of technology from other technological nations, but the primary problem remains. With all these problems the Soviet Union is still progressing in industrial development, but it remains behind the United States in total production and behind all other industrial nations in level of technological development.

Perestroika, glasnost, and demokratizatsiya include attempts to deal with the problems of industry. Under Gorbachev's direction more industries are being given the opportunity to participate in deciding their own output, wages, investments, use of profits, and tying of rewards to merit and productivity rather than status and privilege. Although only 11 percent of all industrial output was from plants participating in perestroika in 1988, the Soviet Union plans to extend the same principles to all industrial activities by the end of 1990. Central to the success of perestroika in industry is the abolition of the notion of full employment, as firms will be able to fire unproductive workers or surplus workers. There is still a great deal of resistance to perestroika on the part of many mid-level communist party bureaucrats whose former role was to tell firms what they were to produce and who are no longer needed or powerful under perestroika. Whether all firms will in fact be operating under the principles of perestroika and glasnost in the 1990s is doubtful, but the beginnings of change in industry signal the recognition that some of the old ideas of the centrally planned economy need to be modified.

SOVIET PEOPLE: DIVERSITY IN THE COMMUNIST LAND

The Soviet Union has a large number of ethnic and nationality groups. Present estimates are that there are 292 million people living in the country. They are 99 percent literate, as a whole have received at least the equivalent of a high school education, and are

---

Full employment: The Soviet Union's official policy stating that it is the government's responsibility to provide jobs for all.

highly trained in technical skills. The society is highly organized as a result of the centrally planned economy.

There are more than 100 distinct ethnic groups recognized by the government with 22 groups having more than 1 million each, but they can be broken down into three dominant groups (Table 9.5). The largest population group is the Slavic people made up of the Great Russians, the Ukrainians, and the Belorussians (White Russians). The Great Russians (or simply Russians) numbered 150.4 million people or 51.5 percent of the country's total population. The next largest group is the Ukrainians with 46 million people or nearly 16 percent of the population. Thus, these two ethnic groups constitute nearly 70 percent of the total population, dominate all areas of the country, and represent a substantial minority population even in areas where other nationalities are more numerous. The 10 million Belorussians are the third of the great Slavic groups, making up about 3.5 percent of the population. Thus the Slavic groups represent 71 percent of the total population of the Soviet Union.

The second most important population group is the Turkic groups: the Uzbeks, the Tatars, the Kazakhs, the Kirghiz, the Turkmen and the Azerbaydzhanis. Together these constitute approximately 12 percent of the total population of the Soviet Union. A third general group is the Caucasians, a diverse group made up of such people as the Armenians and the Georgians. The area of the Caucasus contains a host of small language and nationality enclaves. "Others" include peoples ranging from the Estonians who speak a Uralic language similar to Finnish, to the Aleutian and Jakuts tribal groups.

TABLE 9.5 Groups in the Soviet Union, 1959–1990 (in millions)

Nationality	1959 Number	1959 Percent	1970 Number	1970 Percent	1979 Number	1979 Percent	1990* Number	1990* Percent
Great Russians	114.11	55.0%	129.01	53.0%	137.4	52.4%	150.4	51.5%
Ukrainians	37.25	18.0	40.75	17.0	42.35	16.1	46.0	15.75
Uzbeks	6.01	3.0	9.19	4.0	12.5	4.7	14.6	5.0
Belorussians	7.91	4.0	9.05	4.0	9.46	3.6	10.0	3.4
Tatars	4.97	2.0	5.93	2.0	6.3	2.4	6.8	2.3
Kazakhs	3.62	2.0	5.30	2.0	6.56	2.5	8.03	2.75
Azerbaydzhanis	2.94	1.0	4.38	2.0	5.5	2.1	6.4	2.2
Armenians	2.79	1.0	3.56	1.0	4.15	1.6	5.5	1.9
Georgians	2.69	1.0	3.24	1.0	3.57	1.4	5.1	1.75
Moldavians	2.21	1.0	2.70	1.0	3.0	1.1	3.5	1.2
Lithuanians	2.33	1.0	2.66	1.0	2.85	1.1	3.2	1.1
Jews	2.27	1.0	2.15	.8	1.8	.7	1.7	.7
Tadzhiks	1.40	.6	2.14	.8	2.9	1.1	3.5	1.2
Germans	1.62	.7	1.85	.7	1.94	.7	2.0	.7
Chuvash	1.47	.6	1.69	.6	1.75	.66	2.0	.7
Turkomen	1.00	.5	1.52	.6	2.0	.8	2.9	1.0
Kirghiz	0.97	.4	1.45	.5	1.9	.7	2.7	.9
Latvians	1.40	.6	1.43	.5	1.44	.55	1.46	.5
Mordovians	1.28	.6	1.26	.5				
Poles	1.38	.6	1.17	.5	1.5	.57	1.6	.55
Bashkirs	0.99	.4	1.24	.5	1.37	.52	1.6	.55
Estonians	0.99	.4	1.01	.4	1.02	.4	1.1	.3
All other groups	7.23	3.0	9.04	4.0	11.14	4.35	12.0	4.05
Total	208.83	100	241.72	100	262.4	100	292	100

SOURCE: *Narodnoe Khozyaistvo SSSR, 1970* (Moscow, 1971), pp. 15–16, 1984 Supplement to Paul Lydolph, *Geography of the U.S.S.R.* (Elkhart Lake, Wisconsin; Misty Valley Publishing, 1979).
*Estimates.

The diverse ethnic groups are the basis for the political subdivisions in the Soviet Union. The Soviet system recognizes 15 republics, which according to the constitution have a broad range of rights, including equality with the other 14, the right to coin their own money, and the option of withdrawing from the union whenever they desire. The republics do not exercise these rights, according to the Soviet government, because they accept the goals and progress of the Union of the Soviet Socialist Republics. The fifteen republics are designed to recognize the major ethnic groups in the country, with the Russian republic centered in Moscow occupying the vast bulk of the territory of the country. The other republics are along the borders and as initially constituted after the 1917 revolution were designed to attract the nationality groups in the border areas to ally with the new Communist state. At one point there was a sixteenth republic (Karelia) near the Finnish border. When the Karelians in Finland did not opt to join the Karelian republic, it was abolished, a process that would be like abolishing a state in the United States. Even though the independent republics do not actually enjoy the freedoms promised in the constitution, the Communist achievement in uniting the diverse peoples of their vast land must be recognized. To further recognize ethnic diversity, the Soviet Union grants 38 additional ethnic groups control of areas within the boundaries of the 15 republics.

The Soviet government has pursued a role of developing a so-called Soviet citizen through a standard program of education, teaching Russian as a second language in areas of other languages. At the same time it has allowed the retention of a great deal of national identity for each group. Thus local holidays, local heroes, local cultural traditions, and so on are encouraged to maintain ethnic consciousness and each nationality group is allowed to maintain its own first language if it so desires. The fact that the Communists have overcome the centrifugal effect of the numerous national groups as they developed the unified Soviet state is an achievement of major significance.

THE DEMOGRAPHIC CHARACTERISTICS OF THE SOVIET POPULATION

The family is an important unit to the Soviet people. Like most agricultural nations Russia under the czars was characterized by high birth and death rates and quite large families. Death rates in the prerevolutionary period were between 20 and 30 per thousand depending on the area, and the birthrate was 45 per thousand. Today the birthrate is 20 per thousand, and the death rate 10 per thousand. The decline in birthrates can be attributed to a number of causes. Paramount among these are increasing urbanization, a high proportion of women in the work force, and the cost of children in a technological society. As the population of the Soviet Union has become increasingly urbanized, there has been no need for child labor. The government guarantees the basic necessities of life, so children are no longer needed for social security purposes, and one or two children are usually sufficient to ensure the continuation of the family. As in other industrial countries, the fact that the majority of women in the Soviet Union are engaged in full-time employment means that relatively few women are able to stay home constantly to take care of children. The difficulty of obtaining adequate housing and the high cost of clothing and other consumer items also works against having large families.

Nevertheless, there are marked regional differences in family size. In the area west of the Urals in the Russian, Belorussian, Ukrainian, and Baltic republics a family of more than two children is unusual. In the more agrarian and Islam-dominated areas of Soviet Central Asia, the number of children in the family is more likely to be four or more. In Soviet Central Asia the birthrates are as high as would be expected in the less industrialized countries of the world, exceeding 29 per thousand. This disparity creates certain problems in the Soviet Union, as the dominant Slavic groups are decreasing as an absolute proportion of the total population because of their low birthrates. By the end of the 1990s the Russians may constitute less than a majority of the population for the first time in the history of the state. Government leaders also regard the low birthrate as a problem because it may mean an inadequate labor force in the coming years. As a result the government is constantly encouraging families to have more children. A supplement is paid to families with more than two children, but this subsidy is too low to affect the birthrate; the costs of children still far outweigh the minimal incentives. The government policy of free abortions, free birth-control materials, the general economy that requires mothers to work, and the difficulty of acquiring suitable housing keep families small. As a result, the population growth rates continue to decline.

URBANIZATION IN THE SOVIET UNION

An important factor in the declining population growth rate in the Soviet Union has been the urbanization of the country. At the time of the revolution the overwhelming majority of the Russian people lived in agricultural villages. Today approximately 66 percent of the population of the Soviet Union lives in cities. The

Baby Boom in the Soviet Union

Demographic changes in the Soviet Union in the past two decades raise fears of labor shortages as well as changing ethnic domination. During the 1976–1980 five-year plan the labor force increased by 11 million people, but in the present plan (1986–1991) it is less. In an economy in which increased productivity has resulted primarily from more workers rather than greater efficiency such figures are alarming. Labor concerns intensify ethnic concerns since the majority of the new workers in the present plan are non-Russians from Central Asian Republics. The official solution to the problem? Have more children. Subsidies for children were greatly increased in the late 1970s leading to an increase in the national birthrate, from 17.4 per 1000 population in 1970 to 20.1 per 1000 in 1989.

The increased birthrate represents the first increase in four decades in the Soviet Union and if continued, will have major implications for the country. The increase does not solve the dilemma of the Russian leadership of the Soviet Union since today's babies will not fully enter the work force for two decades and the Russian birthrate although up (17.6/1000 in 1989) is below the national average (20.1 in 1989), and compares poorly to non-Russian republics. The Uzbek Republic, for example had a birthrate of 35/1000 in 1989. Governmental incentives of time off for up to two years for new mothers, and financial, medical, and educational assistance for children are inadequate incentives to changing the birthrate among the Slavic western republics of the Soviet Union to equal the birthrate of the Central Asian Republics.

---

cause of this urbanization is obviously the increase in manufacturing. There are more than 200 cities in the Soviet Union with populations over 100,000 and 22 cities with populations over 1 million. The largest city is Moscow with 8.5 million people, followed by Leningrad (4.8 million), Kiev (2.4 million), and Tashkent (2 million). The government of the Soviet Union has planned for urban growth, and officially the largest cities are closed to future growth. Now the official goal of the government is to concentrate the people in intermediate size cities and to keep the larger cities livable with surrounding greenbelts of vegetation. This policy is meant to ease the problems caused by congestion in large cities: transportation to and from work, accumulation of goods to supply the city, and disposal of waste.

The morphology of the Russian city is unique. Since cities grow according to government plan, activities are widely dispersed. Housing takes the form of high-rise apartment buildings, with shopping facilities provided for each neighborhood. Since prices are controlled by the government and are uniform throughout the country, such planning minimizes the movement of people to the central city for shopping. The center of Soviet cities has a greater concentration of stores, but not to the same extent that a Western European or American city does. Since there is little advertising, the cities also present a rather drab appearance compared with Western cities. There are no large neon signs, none of the discotheques and boutiques of Western cities, and in general the functional provision of services of all types is emphasized. Transportation in cities of the Soviet Union is largely provided by public systems of trolleys or buses. The largest cities have subway systems. Fares on public transit are one kopek (approximately 6 cents). For a commuter from outside the city a three-month pass makes the ride even cheaper.

The distribution of cities in the Soviet Union is not uniform. (Figure 9.1). The European part of the Soviet Union contains two-thirds of the major cities in the country, concentrated in the major industrial regions. Beyond the Urals the population is even more concentrated in urban areas, primarily in the southern part of the region along or near the Trans-Siberian Railroad. In the Soviet Far East urban areas are found only at sites of mining or other resource exploitation.

There are two major types of cities in the Soviet Union. The first, administrative centers, is typical of the largest cities such as Moscow and Leningrad. These cities play a major role in administering the centrally planned economy but also provide a wide diversity of manufacturing and other employment opportunities. The second type of cities is associated with industrial development per se. Donetsk, Krivoy Rog, and Magnitogorsk all are examples in which manufacturing rather than administrative activities is the dominant activity.

EMPLOYMENT IN THE SOVIET UNION

As the Soviet Union has become more urbanized, the employment structure has changed. As late as 1959

The variety of ethnic groups in the Soviet Union presents a continuing challenge to the country's leaders. These photos show Russians in Moscow (photo 1); Tadzhik people in local costumes (photo 2); an Uzbek man and his daughter (photo 3); a Siberian native reindeer herder (photo 4); and two Turkish men in their traditional Karakul wool (Persian lamb) hats (photo 5).

THE SOVIET UNION

THE HUMAN DIMENSION

Portrait of a Family

Sergei Bayev, 34, works as a weaving-loom adjuster at the Shcherbakov Silk Mills in Moscow. He has a secondary technical education and is a member of the Communist Party. His wife Tatyana, 35, was trained as a paramedic. She now works as a laboratory assistant at the Research Institute of Biomedical Problems, also located in Moscow. The Bayevs have two sons, Misha, 10, and Seryosha, 8. The family lives in Skhodnya, 30 kilometers from Moscow.

Tatyana earns 140 rubles a month; Sergei earns 300. The family's total income of 440 rubles ($640) a month, is an average of 110 rubles per person. The Bayevs have a 2000-ruble nest egg. The money is being saved to buy furniture when the Bayevs receive their new two-bedroom apartment.

Tatyana and Sergei are both hard-working and like what they do. Though Tatyana doesn't see prospects for further promotions at her institute, she wouldn't want to stop working even if Sergei made a lot more than he does now. On the other hand, she would love to work only part-time. The Bayevs try not to economize on food. But they do avoid buying things that are above the average state prices, like expensive delicatessen items and sausage sold at cooperatives.

They buy most of their food in Moscow. Tatyana usually comes home from work with armloads of groceries. Sergei takes advantage of the food orders at his place of work. The essentials are available in Skhodnya—milk, sugar, butter, and bread. Other items, such as meat, fish, vegetables, and fruit, are easier to obtain in the capital.

Though everything in the Bayev household isn't new, their apartment is neat and clean. They have a small black-and-white television set, a 25-year-old refrigerator that Tatyana's parents gave them, all the necessary furniture, and a tape recorder (Sergei is a Beatles fan). The tape recorder was purchased with money the family got from selling Tatyana's winter coat, another present from her parents.

The Bayevs have a washing machine, four bicycles, and four pairs of skis, one pair of which Tatyana used as a child. Several years ago Tatyana bought herself a quilted down coat for 200 rubles—money Sergei earned working overtime. She bought it and an Alaskan jacket for her elder son from a private citizen, not in a store.

Sergei's commute into Moscow everyday takes at least an hour and 10 minutes. The journey is made up of three stages: the walk to the station, the train ride to the city, then the bus the rest of the way. Tatyana's commute is 40 minutes long. She walks to the station, takes the train into town and then, because the bus comes too infrequently, walks 20 more minutes to work. Still, Tatyana and Sergei believe that they are better off than many of their neighbors, whose commutes into Moscow are longer. Their children can walk to school, which is only a stone's throw away from home.

Having Tatyana's parents living close by is an enormous help. Tatyana doesn't have to worry at work, knowing that the boys will go home to their grandmother, who will feed them dinner. Then they can play outdoors. Skhodnya is a quiet town so there is nothing to worry about. When Tatyana gets home, the boys start their homework.

Sergei's work schedule, both the shifts and his days off, varies from week to week. Whenever the whole family is home on a Saturday or Sunday, they often go off together on skis or on bikes. Every year Tatyana has 24 working days off, and Sergei gets 18. They take their vacations at different times. One or the other of them goes with the children to visit Sergei's mother in another suburb of Moscow or to see other relatives.

Housing is a sensitive and complicated subject for the family. "We'll get a new apartment, and then everything else will fall into place," Sergei claims. At the mention of the apartment, Tatyana becomes nervous. The four Bayevs are registered in a one-bedroom apartment with 31 square meters (334 square feet) of floor space not counting the kitchen, bathroom, and hallways. Tatyana's 85-year-old grandmother, her sister, and her niece are also registered in the same apartment. But seven people couldn't possibly manage in such small quarters, so they rented an additional, one-bedroom apartment for 50 rubles a month.

When Tatyana's parents got their new apartment with 18 square meters (194 square feet) of floor space, they offered it to their daughter and her husband, who decided to move in. Conditions were still crowded, but at least only four of them were living together. Tatyana's parents took the other, larger apartment.

Having less than five square meters (54 square feet) per person, the Bayevs are on priority waiting lists at the local executive committee and at Sergei's work to receive new housing. Sergei has been promised a two-bedroom apartment. But if anything should happen to Tatyana's grandmother, the Bayevs will lose their "priority" status on the housing lists, in which case it could be years before they get new housing.

I ask a tactless question: "Why don't you buy a cooperative apartment?" Tatyana gives a sardonic laugh and answers, "We don't have rich parents who can pay the first installment for us."

Though the Bayevs work in Moscow and live close by, they don't consider themselves Muscovites. Tatyana isn't especially fond of her native Skhodnya, a former summer colony, which is now distinguished only by its beautiful countryside and two major enterprises. Still, she wouldn't want to move anywhere else.

"I've lived here for 12 years," says Sergei. "I'm used to it, but I don't think of Skhodnya as my home town. I grew up in Naro-Fominsk. My mother still lives there."

"So here we are, 35 years old. We've both worked most of our lives—Sergei since he was 15 and I since I got out of school. But, so far, what do we have to show for it? Not an apartment, not furniture, none of the good things in life," Tatyana sums up. "It's all right. Don't worry," Sergei reassures her, with characteristic optimism. "We'll get everything."

Adapted from Vitali Tretyakov, "Portrait of a Family," Soviet Life, October 1988, pp. 53–54. Used by permission.

A view of the central portion of Moscow across the Moscow River at the Rossiya Hotel.

manufacturing employed only 46.8 million workers as compared to 52.3 million workers in rural activities. By 1990 nearly three-fourths of all workers were involved in urban industrial activities with the balance engaged in farming or forestry. The dominant employment activities in the industrial sector are manufacturing, building and construction, and transport and communications, accounting for 61 percent of all employment. Trade and other service activities account for only 10 percent, but housing administration, education, government, and public health service add an additional 29 percent. Thus a very high proportion of the work force is engaged in the manufacturing sector. Women in the Soviet Union constitute nearly one-half of the total work force. In agriculture, 51 percent of all labor is performed by women, and more than 90 percent of all doctors are women. Yet, there are relatively few women in positions of power, namely the hierarchy of the Communist party. Wages vary somewhat according to occupation, but are generally within a rather narrow range of about $200 to $300 per month for the average worker (see Box.).

THE SOVIET UNION: AN ASSESSMENT

The increased standard of living in the modern Soviet Union is indicative of the changes under communism. As the first Communist nation of the world the Soviet Union is an example of a Communist state and provides an alternative form of development for the less industrialized realm. An analysis of the Communist experience in the Soviet Union reveals a number of major successes. The economy has been transformed from one similar to most less industrial countries in which the majority of the people are engaged in agriculture and live in rural villages to one in which the majority live in urban areas and are engaged in secondary and tertiary economic activities. From a primarily agrarian nation under the czars the country has emerged as the world's second greatest industrial power.

The social characteristics, which were typical of those in less industrialized countries, have been changed to those typical of industrial societies. Literacy rates are near 100 percent, education is universal, and abject poverty has been essentially eliminated. The great differences in wealth normally found in an industrialized country have largely been eliminated as the old landed class and aristocracy were removed and nominal egalitarianism substituted. The differences in wealth that normally exist between rural and urban dwellers have been minimized, as the urban dweller is only marginally better off in standard of living than the rural resident.

Under Soviet rule the people have developed a sense of unity and pride in the accomplishments of their country. Although there are dissidents who question the means by which ends are achieved, the great majority of the Soviet citizenry are proud of their country and of its accomplishments although only 17 million (6 percent) are members of the Communist party.

Communist accomplishments have been great, but they have come at a high cost in human suffering and personal freedoms. In order to reach the goals of the centrally planned economy during the formative years of the Communist experience, individuals were forced to give up many of those freedoms that Western nations regard as essential. In addition, the excesses of Stalinism with its purges and use of slave labor by prisoners to develop Siberia, industrial regions in the Urals, and elsewhere, and the development of the ever-present KGB (secret police) have extracted a tremendous toll in terms of individual loss of both freedom and life. It is estimated that there are more than 2 million prisoners in Soviet prisons and labor camps today, most in labor camps that still provide an important element in developing the harsh environment of the Soviet Union. Against these excesses the successes must be weighed.

When compared to Lenin's avowed goal of developing a classless communist society where each person worked according to his ability and received according to his needs without need for a formal government, the Soviet experience must be marked a failure. The old elite of the czars and aristocracy of Russia have been replaced by a new elite, the Communist party members and leaders. The centrally planned society tends to be highly ponderous in reacting to changing consumer demand and seems to stifle creativity and innovation in everyone, from farmers to industrial managers to artists and academia. An example is the present concern in the Soviet Union over Levi jeans. Levi jeans made in the United States sell on the black market in the Soviet Union for more than $100 a pair. Jeans are made in the Soviet Union, but the youth of the country complain that they look like work clothes rather than stylish jeans like those from the West. Although a seemingly trivial example, this problem has received a great deal of attention in the Soviet press as the director of the national Internal Affairs Ministry (which combats embezzlement and black marketeering) questions why industry must continue to produce what was fashionable yesterday instead of what is fashionable today. With a centrally planned economy based on five-year cycles in which each individual industry must meet quotas, it is difficult to provide for changing consumer tastes such as the demand for jeans. The directors of manufacturing concerns are conservative and unwilling to produce items not on their quota, as doing so would reflect negatively on them if these items subsequently did not sell well. As a provider of the basics the centrally planned economy in the Soviet Union has been successful, but it is difficult to build into it opportunities for rapid adjustment to changing demand. Is the Soviet experience, therefore, a valid model for less developed countries? The answer depends on whether or not one believes that provision of basics is the central issue in the truly impoverished less industrialized countries. Certainly for the leaders of the less developed the Soviet experience provides an important alternative development model.

THE SOVIET UNION AND THE FUTURE

With the increasing public discussion of the shortcomings of the planned society in providing for human material needs, what can be expected to take place in the Soviet Union in the future? There seems to be every indication that the Soviet Union will become more involved in the interconnected world in which we live. Although at the present its primary trade partners remain the Eastern European countries (more than 50 percent of all imports and exports to the Soviet Union come from this area), increasingly it is trading with the EEC, Latin America, Africa, and Asia. In recent years, it has begun to purchase many items from the United States and Canada as well. Internally the changes that are taking place seem to indicate

THE HUMAN DIMENSION

Soviet People; Western Veshch (Thing)

In Russian the term *Veshchism* is used to refer to the desire to have things, in this case material possessions such as a car, apartment, furniture, jewelry and so forth. More important the term signifies desire for Western "things," whether blue jeans, T-shirts, records, or even cars. In part, the desire for Western things reflects the desire for status that comes with having scarce, and thus expensive items. In part it may reflect the prevalent view that styles are set in the West. (Even the Russian word for jeans is Western, pronounced jeans-ee). But in large part the desire for Western things represents recognition that the centrally planned economy produces a plethora of low-quality goods that are also expensive.

The Soviet Union produces ten brands of refrigerator, for example. The Minsk is perceived as the best, and the Vega the worst. The result is that Vegas are unsold, and Minsks are unavailable. Continued production of items that are not in demand reflects the government commitment to full employment. Closing the Vega production plant would idle workers, so the Vega continued to be produced until after the 1986–1990 economic plan began.

Another example of the inadequacies of government planning, Soviet style is in school uniforms. The typical uniform for eighth- and ninth-grade girls was a dark brown dress with a black apron (white on holidays). In 1983 a new stylish uniform with blue pleated skirt, vest, and matching jacket was introduced. Unfortunately most were manufactured in large sizes, and jackets were rarely the same size as the skirt in an individual uniform.

The desire for Western goods that results from inadequate quality makes even goods with Western-sounding names popular. Denim clothing made in eastern European countries and sold in the Soviet Union under the brand name "Montana" is popular. Not only does the name Montana invoke images of the geography of America's Wild West, the quality is better than the Soviet-made goods. Perestroika is, in part, a response to the present inability of the centrally planned economy to meet the nation's needs.

---

increased emphasis on production of consumer items and increasing the standard of living, while maintaining strong central control by the Communist government. The goals of Gorbachev's principles of perestroika, glasnost, and demokratizatsiya do not include removing communist control, only revitalizing the economy. If successful, ongoing reforms will raise the standard of living of the people while allowing the nation's economy to begin closing the gap between the U.S.S.R. and other industrial countries.

Externally the Soviet Union seems less interested in influencing less industrialized countries, concentrating on its own economy. The Soviets have had only mixed success with their brand of diplomacy in the past, as their relationship with the so-called non-aligned countries tended to fluctuate widely from year to year. For example, investments in Egypt over a ten-year period led to the development of important Soviet influence in that country. In 1976 and 1977 this influence declined dramatically as Egypt turned completely from the Soviet Union and expelled Soviet technical and military advisers. In Afghanistan the Soviet Union supported a socialist government with troops and supplies from 1979 to 1989 before withdrawing and letting the various resistance groups fight for power. This same pattern has been repeated in numerous other areas. As a world force the Soviet Union will continue to take advantage of existing trouble spots in the world to attempt to gain influence with one side or the other. It is doubtful that it will engage in active conflict with the non-Communist world, but it may attempt to spread its influence at the expense of the non-Communist world provided it doesn't conflict with the goals of perestroika and glasnost. The continuing need to spend one-third of its gross national product for defense purposes has encouraged some agreement with the West to limit arms proliferation. The primary role of the Soviet Union in the future may well be one of demonstrating whether or not the theories of Marxism-Leninism can be modified to allow them to be competitive with the powerful free enterprise industrial countries. The continued integration of the Soviet Union into the interconnected world, in conjunction with a decrease in tension between the Soviet Union and other industrial nations, seems to indicate major changes will occur in the communist model of economic development.

FURTHER READINGS

ALLWORTH, E., ed. *Ethnic Russia in the U.S.S.R.: The Dilemma of Dominance.* Elmsford, N.Y.: Pergamon Press, 1980.

AMBLER, JOHN, ed. *Soviet and East European Transport Problems.* New York: St. Martin's Press, 1985.

ANDRUSZ, GREGORY D. *Housing and Urban Development in the U.S.S.R.* Albany, N.Y.: State University of New York Press, 1985.

BROWN, A. et al., eds. *Cambridge Encyclopedia of Russia and the Soviet Union.* Cambridge: Cambridge University Press, 1982.

CAMPBELL, R. W. *Trends in the Soviet Oil and Gas Industry.* Baltimore: John Hopkins University Press, 1976.

CLEM, R. "Regional Patterns of Population Change in the Soviet Union, 1959–1979," *Geographical Review* 70, 1980, pp. 137–156.

———. "Russians and Others: Ethnic Tensions in the Soviet Union." *Focus,* September-October 1980.

COLE, J. *Geography of the Soviet Union.* London: Butterworths, 1984.

CRITCHFIELD, H. *General Climatology.* 4th ed. Englewood Cliffs, N.J.: Prentice-Hall, 1983.

DEMKO, G. J. and FUCHS, R. J., eds. and trans. *Geographical Perspectives on the Soviet Union.* Columbus: Ohio State University Press, 1974.

DEUTSCH, ROBERT. *The Food Revolution in the Soviet Union and Eastern Europe.* Boulder, Colo.: Westview Press, 1986.

DEWDNEY, J. *A Geography of the Soviet Union.* 3rd ed. Elmsford, N.Y.: Pergamon Press, 1979.

———. *U.S.S.R. in Maps.* New York: Holmes and Meier, 1982.

DIENES, L. "Soviet Energy Resources and Prospects." *Current History* 71, 1976, pp. 114–118, 129–132.

DIENES, L. and SHABAD, T. *The Soviet Energy System.* New York: V. H. Winston, 1979.

DOSTAL, P. and KNIPPENBERG H. "The 'Russification' of Ethnic Minorities in the USSR." *Soviet Geography: Review and Translation* 20, 1979, pp. 197–219.

FRENCH, R. A. and HAMILTON, F.E.I., eds. *The Socialist City: Spatial Structure and Urban Policy.* Chichester, England: Wiley, 1979.

GEDZELMAN, S. *The Science and Wonders of the Atmosphere.* New York: Wiley, 1980.

GIBSON, J. *Imperial Russia in Frontier America.* New York: Oxford University Press, 1976.

GIESE, E. "Transformation of Islamic Cities in Soviet Middle Asia into Socialist Cities." In *The Socialist City: Spatial Structure and Urban Policy,* pp. 145–165. Edited by R. French and F. E. I. Hamilton. New York: Wiley, 1979.

GLASSNER, M. and DE BLIJ, H. J. *Systematic Political Geography.* 4th ed. New York: Wiley, 1989.

GOLDHAGEN, E. *Ethnic Minorities in the Soviet Union.* New York: Praeger, 1968.

GREGORY, J. S. *Russian Land, Soviet People: A Geographical Approach to the U.S.S.R.* London: Harrap, 1968.

GVOZDETSKY, N. A., ed. *Soviet Geography Today: Physical Geography.* Moscow: Progress Publishers, 1982.

HOFFMAN, G. W. "Rural Transformation in Eastern Europe Since World War II." In *The Process of Rural Transformation.* Edited by I. Volgyes et al. New York: Pergamon Press, 1980.

HOWE, G. M. *The Soviet Union: A Geographical Study,* 2d ed. Plymouth, England: MacDonald and Evans, 1983.

JACKSON, W.A.D., ed. *Soviet Resource Management and the Environment.* Columbus, Ohio: American Association for the Advancement of Slavic Studies, 1978.

JENSEN, R., SHABAD, T., and WRIGHT, A., eds. *Soviet Natural Resources in the World Economy.* Chicago: University of Chicago Press, 1983.

KALESNIK, S. V. and PALENKO, V. F., eds. *Soviet Union, A Geographical Survey.* Moscow: Progress Publishers, 1976.

KARKLINS, RASMA. *Ethnic Relations in the U.S.S.R.: The Perspective from Below.* Winchester, Mass.: Allen and Unwin, 1986.

KOSINSKI, L. A., ed. *Demographic Developments in Eastern Europe.* New York: Praeger, 1977.

LAIRD, R. D. et al., eds. *The Future of Agriculture in the Soviet Union and Eastern Europe.* Boulder, Colo.: Westview Press, 1977.

LINZ, SUSAN J., ed. *The Impact of World War II on the Soviet Union.* Totowa, N.J.: Rowman and Allanheld, 1985.

LYDOLPH, P. *The Climate of the Earth.* Totowa, N.J.: Rowman and Allanheld, 1984.

LYDOLPH, P. E. *Geography of the U.S.S.R.* New York: Wiley, 1977.

MATHIESON, R. S. *The Soviet Union: An Economic Geography.* 2d ed. New York: Barnes and Noble, 1975.

MELLOR, R. *The Soviet Union and Its Geographical Problems.* London: Macmillan, 1982.

PARKER, W. H. *An Historical Geography of the Soviet Union.* Chicago: Aldine-Atherton, 1969.

———. *The Soviet Union.* 2d ed. London: Longman (World's Landscapes Series), 1983.

PEPPER, DAVID and JENKINS, ALAN, eds. *The Geography of Peace and War.* New York: Basil Blackwell, 1985.

POND, ELIZABETH. *From the Yaroslavsky Station.* New York: Universe Books, 1988.

PREOBRAZHENSKY, V. S. and KRIVOSHEYEV, V. M. eds. *Recreation Geography of the U.S.S.R.* Moscow: Progress Publishers, 1982.

"Report." *Soviet Geography: Review and Translation* 20, 1979, pp. 567–586.

RUGG, D. S. *The Geography of Eastern Europe.* Lincoln, Neb.: Cliffs Notes, 1978.

SACKS, M. P. and PANKHURST, J. G. *Understanding Soviet Society.* Boston: Unwin Hyman, 1988.

SHABAD, T. *Basic Industrial Resources of the U.S.S.R.* New York: Columbia University Press, 1969.

SHIPLER, D. K. *Broken Idols, Solemn Dreams.* New York: Times Books, 1983.

SINGLETON, F., ed. *Environmental Misuse in the Soviet Union.* New York: Praeger, 1975.

SMITH, H. *The Russians.* 2d ed. New York: New York Times Publishing Company, 1986.

SYMONS, L. et al. *The Soviet Union: A Systematic Geography.* New York: Barnes and Noble, 1982.

TURNOCK, D. *Eastern Europe. Studies in Industrial Geography.* Boulder, Colo.: Westview Press, 1978.

ZUM BRUNNEN, CRAIG and OSLEEB, JEFFREY P. *The Soviet Iron and Steel Industry.* Totowa, N.J.: Rowman and Littlefield, 1986.

Tokyo.

JAPAN PROFILE (1990)

Population (millions)	123.7	Energy Consumption Per Capita—	
Growth Rate (%)	0.5	pounds (kilograms)	6,871
Years to Double Population	133		(3,116)
Life Expectancy (years)	78	Calories (daily)	2,856
Infant Mortality Rate		Literacy Rate (%)	99
(per thousand)	5.2	Eligible Children in Primary	
Per Capita GNP	$12,850	School (%)	100
Percentage of labor force		Percent Urban	78
employed in:		Percent Annual Urban	
Agriculture	*11*	Growth, 1970–1980	2.0
Industry	*34*	1980–1985	1.8
Service	*55*		

10

Japan
a unique pattern of development

MAJOR CONCEPTS

Job security / Worker-employee cooperation
Commitment to a nation / Double Cropping
Conurbation / Comparative advantage
Social revolution / Meiji Restoration
Samurai / Fallow / Clans and politics
Archipelago / Geopolitics

IMPORTANT ISSUES

Regional variation in development
Resource dependency for an industrial base
Environmental degradation versus economic growth
Values conflicts: urban versus rural values
Social changes in modern societies

Geographical Characteristics of Japan

1. Japan represents the only fully industrialized nation in Asia.
2. Japan's economic dominance of Asia today was paralleled by a period of military dominance before and during World War II.
3. Japan is a mountainous archipelago of islands off the east coast of Asia.
4. Unlike the United States and Canada, Japan has limited space and resources and high population densities.
5. Japan imports raw materials and exports manufactured goods to both industrialized and less industrialized nations.

Japan is unique among industrial nations. Culturally it represents the only non-European nation to have become a truly industrialized economy. Unlike the newly industrializing countries, Japan was never a European colony, but industrialization and modernization did represent awareness of the power of the United States and other industrialized countries. Japan's industrial revolution reflected the need to transform its economy to protect itself from Western powers, both politically and economically. Japan was isolated from the technological and social changes that fostered the Industrial Revolution in Western Europe, and it is surrounded by countries that in the past were not industrialized (Figure 10.1). Japan's emergence as a major industrial power does not reflect a large land area, rich resources, or central location with respect to the industrialized core area of the world. It represents a unique combination of oriental and occidental characteristics, one that is sometimes referred to as the economic miracle of Asia. This "miracle" is all the more impressive because it has been accomplished in the last four decades as Japan overcame the destruction of World War II to emerge as the third most productive industrial nation of the world after the Soviet Union and the United States. Some observers predict that Japan has the potential to surpass the United States and the Soviet Union to emerge as the dominant economic power in the world. As of 1990, Japan ranked either first or second in the world in total production of a host of industrial items including automobiles, cameras, bicycles, computers, televisions, and motorcycles. The intriguing question of Japan is what led to its emergence as an industrialized nation when the surrounding Asian countries continued as less industrialized colonial possessions of the industrialized Western European and North American powers?

THE JAPANESE EXPERIENCE

The label "made in Japan" has become so commonplace in the Western industrial nations that few Europeans or Americans pause to consider why Japan has become such a dominant industrial power. Japanese brand names like Sony, Kawasaki, Suzuki, Honda, Datsun (Nissan), Toyota, or Hitachi are common in Western industrial nations. Rarely do non-Asian observers ponder the transformation behind Japan's rise to industrial dominance.

THE JAPANESE PAST

The historical development of Japan led to the emergence of a distinctive Japanese culture, which affects the changes that have occurred and are occurring within the nation. The earliest occupants of Japan were the Ainu, who date from the last great glacial era. The major groups of migrants into the region came from China and Korea in the last 5000 years. The development of Japan during its first 4000 years started when separate **clans** in the south began practicing agriculture based on rice as a staple crop. In the northern Japanese islands the Ainu continued to live as hunters similar to the American Indians prior to European contact.

The great changes after A.D. 300 in Japan resulted from diffusion of migrants from China and the adoption of selected elements of the Chinese culture. Mi-

---

Japanese Clans: Clans consist of cultural groups based on extended family relationships. In Japan these clans expanded in size and power to control Japan's government and economy in a feudal system.

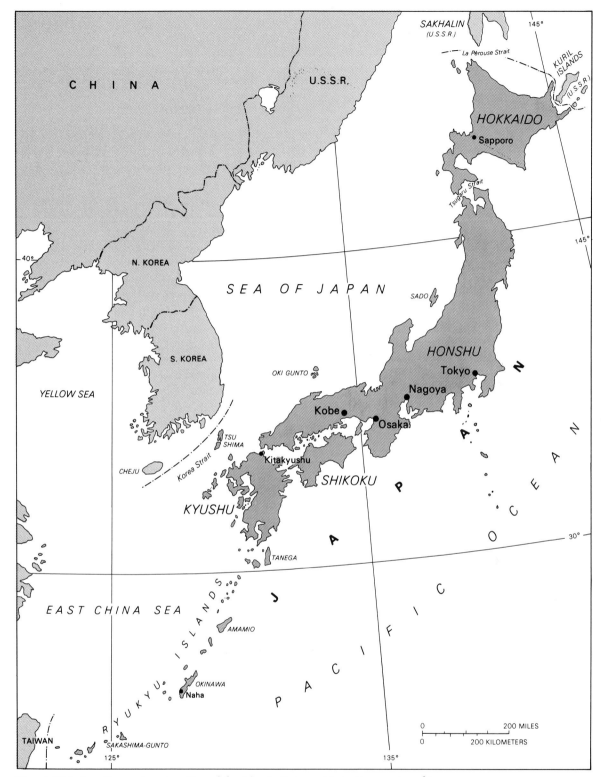

Figure 10.1 Location and major cities of the islands that constitute the country of Japan.

grants between A.D. 300 and 600 brought important Chinese cultural elements to the islands. Shortly after A.D. 600 Japanese leaders sent groups to the Chinese court to study the Chinese system and, upon their return, to help integrate the best elements of Chinese culture into a uniquely Japanese culture. The most important elements borrowed from China include a formal emperor over the clans, but unlike China, the

Japanese leaders maintained the clan and prohibited the widespread education and competency that developed in China. A written script was adapted from the Chinese system, creating a distinctive Japanese language using symbols borrowed from Chinese. In all the adoptions from China, the Japanese clan leaders never introduced any item that would dilute their power.

The result of the selective adoption of Chinese culture created a feudalistic system in Japan with rigid ordering of social strata based on the clans. Second only in power to the clan leadership were the **samurai**. The samurai were a professional soldier class that developed over the centuries as the Japanese extended their control northward and defeated the Ainu. The samurai were useful in maintaining the power of the feudal lords, and they had special privileges that enriched them economically through taxation of the peasantry. The samurai have been glamorized in Japanese history to represent the tradition of self-sacrifice, perseverance, discipline, and loyalty.

Below the samurai was the merchant class, townspeople who operated the economy of the nation, engaged in handicraft manufacture, and generally enjoyed a standard of living higher than the majority class, the peasants. The peasants in Japan were heavily taxed. As the largest single group they were the main food producers and provided the relative ease of life for the wealthy feudal lords, the samurai, and even the merchants. By A.D. 1600 there had developed in Japan a large (estimated 18 million) population with many of the characteristics of the Western European feudal societies during the Dark Ages. The peasants paid heavy taxes, had limited rights, and were subject to the whim of the samurai or the feudal lords.

After 1600 Japan was approached by European explorers from the Netherlands, Spain, and Portugal. Internal stress from peasant demands combined with the perceived threat of Western power and influence, particularly Christianity, led the feudal leadership to close Japan to outside influence. The Tokugawa clan became the dominant power in Japan for the next 250 years (1603–1867) and strictly enforced the edicts preventing contact with the West. Measures included limiting Japanese shipyards to building ships that could sail only in coastal waters, expulsion of foreign traders, both occidental and Asian, and outlawing of conversion by the Japanese to Christianity. The significance of the isolation of Japan from 1600 to 1850 is twofold. The two and a half centuries of strong military control by the Tokugawa military leaders allowed the Japanese to emerge effectively as a distinctive culture. The traits borrowed from China were modified to a distinctively Japanese form in art styles, architecture, political organization, and social relationships. Equally important, the isolation allowed Japan to escape the colonial dominance by Europe found elsewhere in Asia. Some claim that the exclusion of occidental influence was one of the most critical factors explaining the rapid industrial development of Japan after 1857.

THE MEIJI RESTORATION

In 1853 Commodore Matthew C. Perry of the United States Navy arrived in the harbor of Tokyo. The American steamships and the subsequent display of American technology that Perry presented impressed the ruling clans. They recognized the evident superiority of the military might of the West, and they began a deliberate program to transform their society from feudalism to a modern industrial state. By 1868 the Tokugawa clan had lost power as other clans encouraged the restoration of the emperor to a dominant role, in what is known as the **Meiji Restoration** because of the name given to the reign of the emperor Mutsuhito. The emperor in reality was still only a figurehead, as a relatively small group of people continued to run Japan in the same manner that the same clan leadership had controlled Japan for centuries, but the Meiji Restoration signaled the beginning of a rapid **social revolution.** All industrial nations have experienced a similar social revolution, which causes rapid and fundamental changes in the society. Of particular importance are the changes that destroy the fabric of the traditional society with its privileged elites who control the government, military, and economy in stark contrast with the majority poor of the country. Some observers maintain that in order for a modern industrial nation to develop, such a revolution must occur, and that the timing and nature of such social revolutions are the single most important factors affecting economic development.

In Japan several aspects of the social revolution associated with the Meiji Restoration had important consequences for modernization. The first was the abolition of landed feudalism and its related governmental system. Freeing the peasants from serfdom represented a societal change with broad and subtle ramifications. Direct rule of a bureaucracy with the emperor as the head and the establishment of new

---

Meiji Restoration: The calculated policy of modernization based on combining traditional Japanese values with western technology.

Social Revolution: A revolution in the organization, rules, and classes in a society. Some observers maintain that economic development cannot occur without social revolution to overthrow existing social systems that strive for stability of the system.

administrative regions transformed the political organization of the nation. The old political organization with tolls and other barriers to trade and free movement within Japan itself was abolished. The capital was moved from Kyoto to present-day Tokyo, and the feudal lords were replaced by governors. The military elite represented by the samurai had their power severely restricted as the military was transformed from a position of maintaining the rights of the elite in the traditional society into a force for propelling Japan into the industrial age. Ultimately many of the samurai groups became involved in the modernization of Japan, but those who refused to accept the permanency of the social and political changes of the Meiji Restoration were destroyed.

Following ancient tradition, students and observers were sent abroad to learn of the industrial nations. Often representing military clans and the samurai groups, they focused on developing a strong military to protect Japan from the colonialism of the West. Recognition of the need to develop industry to support such military expansion caused the development of heavy industry in Japan from 1870 to 1930. Japan's ability to adopt the Western industrial model in a short period of time reflected the previous development of the nation. Even before the Meiji Restoration the literacy rates for males were probably as high as 40 percent, providing a nucleus for developing a skilled industrial labor force.

While consciously adopting the technology of the industrial West, Japanese leaders carefully avoided cultural or social changes that conflicted with their perceptions of Japan's future. Eventually everything industrial was accepted, but Western ideals of democracy were avoided. By the end of the nineteenth century Japan had emerged as a major military force. Because of a rather restricted geographic base, the emerging Japanese military strength was used to expand Japanese control. The military expansion included claiming the Ryukyu Islands in the 1870s and military occupation of Korea and Taiwan in the 1890s. This military expansion was one side of the modernization of Japan, but it was paralleled by social developments that were to have even longer lasting influence. Compulsory education was introduced in Japan in 1872, less than three years after similar adoption in Great Britain. By 1900, 95 percent of the Japanese population could be classified as literate. This combination of an educated, skilled population and the demands for military expansion caused Japan to industrialize at a very rapid rate. Heavy industry such as iron and steel, railroad manufacturing, and armaments became the focus of investment efforts in Japan by 1900. Although textiles were the major industrial activity in terms of employment and value, the heavy industry necessary to support a strong military was the focus of the Japanese leaders' interests.

JAPANESE MILITARY EXPANSION

The demand of the growing Japanese industry for resources prompted the political leaders in Japan to utilize their military strength to gain control of large areas of Asia. In 1904 and 1905 Japan defeated the Russians in the Russo-Japanese War and acquired the southern half of Sakhalin Island, with its important coal and petroleum deposits, and part of Manchuria (northeast China) and Korea. Following World War I, the Japanese, who had fought with the Allied forces against Germany, were given control over important island possessions in the Pacific, including the Marianas (except Guam), the Marshalls, and Carolines. In 1932 Japan took control of all of Manchuria and began developing the coal and iron ore there to fuel its industrial and economic growth. In 1937 Japan invaded China and provoked a full-scale war. By 1940 Japan controlled an empire that stretched from the central Pacific to the Sakhalin Islands and gave it potentially the most productive empire in the Pacific (Figure 10.2).

In 1941, as the result of American restrictions on exports of strategic items to Japan, the Japanese attacked Pearl Harbor in Hawaii and attempted to gain complete control of the Pacific. The underlying assumption behind the attack at Pearl Harbor was that the American Pacific fleet could be destroyed and Japan could consolidate its control of the Pacific and East Asia before American naval power could be restored. Japan's leaders were unsuccessful in this goal, and the Americans terminated World War II by dropping the atomic bomb on the cities of Hiroshima and Nagasaki on August 6 and 9, 1945. The destruction from the bombs was catastrophic, but they effectively ended Japanese resistance in World War II and set the stage for modern development in Japan.

THE JAPANESE ENVIRONMENT

The physical geography of Japan is markedly different from that of most European industrial nations. Japan consists of a series of islands forming a 1,400-mile (2,250-kilometer) arc from northeast to southwest. The islands are the peaks of a volcanic chain that is part of the Pacific "Ring of Fire." Although these volcanic mountains are beautiful, the country periodically suffers destructive earthquakes (Table 10.1). The most recent severe earthquake in 1923 resulted in the deaths of an estimated 150,000 people. The hilly and moun-

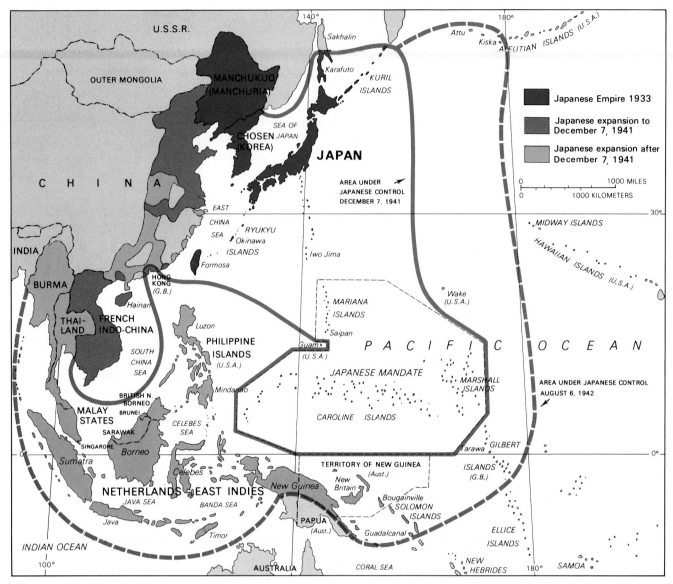

Figure 10.2 Expansion of the Japanese Empire during the World War II era.

tainous terrain of three quarters of Japan limits the area that is suitable for agriculture (Figure 10.3). Settlement is concentrated in the coastal plains, the level land in the small valleys between hills, and in the narrow river valleys. Rivers are short, and their steep gradients make them unsuited for navigation, but they are important for generation of electricity.

The Japanese **archipelago** consists of four main islands and more than 3000 smaller ones. The largest islands are Honshu, Hokkaido, Kyushu, and Shikoku (Table 10.2). Japan's total land area is only 143,706 square miles (372,200 square kilometers), 10 percent smaller than California, on which Japan maintains nearly five times the population of California. The nearly 124 million people who make up the Japanese population are concentrated on only 16 percent of the total land area. Although as a nation Japan does not rank first in population density, it is probably first in density of population per unit of cultivated land. Population densities approach 5000 people per square mile (1930 per square kilometer) in the cultivated areas.

The climate of Japan is an important element in understanding the high population density. Japan extends through a wide latitude if all the islands in the archipelago are included, but the main islands extend from just north of 30° north (the latitude of Jacksonville, Florida) to just north of 45° (the same as Halifax, Nova Scotia). Climate ranges from humid subtropical to humid continental. There is sufficient rainfall for agriculture on all the main islands, with no true dry

---

Archipelago: A group or chain of islands in close proximity to one another.

THE WORLD OF WEALTH

TABLE 10.1 Deaths from Major Earthquakes

Year	Location	Dead	Year	Location	Dead
1268	Silicia, Asia Minor	60,000	1935	Quetta, India (Pakistan)	60,000
1290	Hopeh Province, China	100,000	1939	Chillan, Chile	30,000
1293	Kamakura, Japan	30,000	1960	Agadir, Morocco	12,000
1531	Lisbon, Portugal	30,000	1962	Iran	10,000
1556	Shensi Province, China	830,000	1964	Alaska	131
1667	Shemaka, Russia	80,000	1968	Iran	11,588
1693	Catania, Italy	60,000	1970	Peru	66,794
1737	Calcutta, India	300,000	1972	Iran	5,374
1755	Northern Persia (Iran)	40,000	1972	Managua, Nicaragua	10,000
1755	Lisbon, Portugal	60,000	1974	Pakistan	5,200
1759	Baalbak, Lebanon	30,000	1976	Guatemala	22,778
1783	Calabria, Italy	50,000	1976	Hopeh Province, China	665,000
1797	Quito, Ecuador	41,000	1978	Iran	26,000
1828	Echigo, Japan	30,000	1980	Algeria	5,000
1906	San Francisco, Calif.	700	1980	Italy	3,000
1908	Messina, Italy	75,000	1982	North Yemen	2,800
1915	Avezzano, Italy	29,970	1983	Eastern Turkey	1,300
1920	Kansu Province, China	180,000	1985	Mexico	25,000
1923	Tokyo-Yokohama, Japan	142,807		Colombia	25,000
1932	Kansu Province, China	70,000	1988	USSR	40,000

SOURCE: *The World Almanac and Book of Facts 1988*, Boston Herald American, 1985.

areas. Precipitation totals range from 40 inches (1016 millimeters) in the north to 100 inches (2540 millimeters) in the south. Hokkaido and northern Honshu have cold winters, which limit agriculture to a single crop year (Figure 10.4). This northern area of Japan has a humid continental climate and commonly experiences snowfall in the winter season; in northern Hokkaido and in the higher elevations of the northern third of Honshu are many ski resorts. The southern two thirds of Honshu and the islands of Shikoku and Kyushu have humid subtropical climates. They are warm and humid in the summers, and moderate and wet in the winter season. Snow in the subtropical Honshu is limited to higher elevations. The level land of the coastal plains and interior valleys in this area is intensively cultivated, utilizing double cropping with irrigated rice in the summer and dry grains or other crops in the winter.

JAPANESE RESOURCES

Japan does not possess large deposits of mineral resources. The most significant one is coal, but Japan is limited in coking quality coals. The largest deposits of coal in Japan are found on Kyushu and Hokkaido, but they are generally unsuitable for use in iron and steel manufacture. The coal reserves of Japan are adequate only for the next fifty years at present rates of consumption and would be quickly exhausted if they were the only source of coal available to Japan.

Petroleum resources are essentially nonexistent. There are minor deposits in northwestern Honshu and Hokkaido, but they provide only a small proportion of Japan's petroleum needs. The short, swift rivers flowing from Japan's mountains provide cheap hydroelectric energy, but they are inadequate to meet Japan's energy needs.

There are widely scattered deposits of iron ore throughout the main Japanese islands, but they are rarely of sufficient quality to support large-scale industrial activity. The small size of the deposits and low quality of the ore make them of only minor importance to today's iron and steel industry. Japan does have relatively large deposits of copper but is unable to meet all its needs from local ores. There are also small deposits of lead, tin, zinc, graphite, and phosphates in Japan, but none are large enough to meet

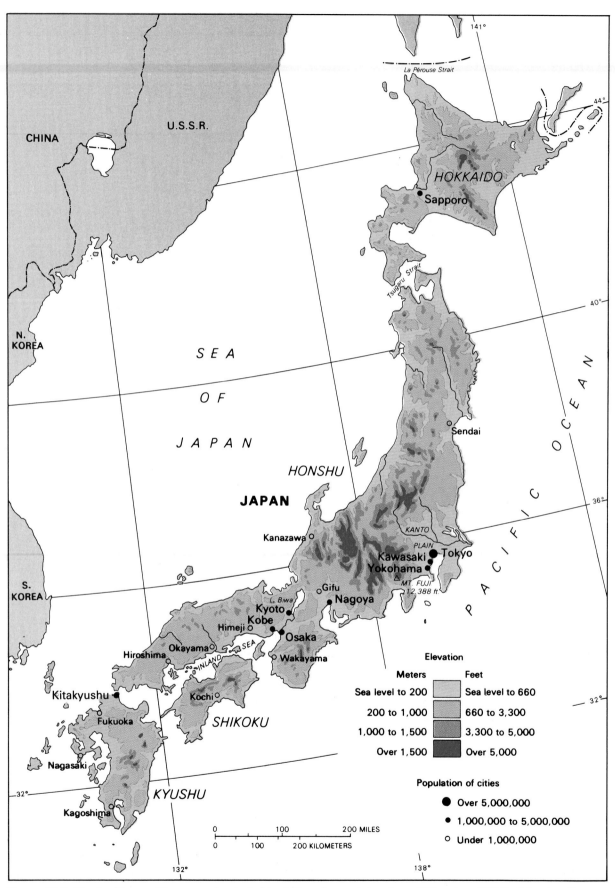

Figure 10.3 The landforms of Japan are dominated by hills and mountains above 200 meters (660 feet) in elevation.

THE WORLD OF WEALTH

TABLE 10.2 Population and Area of Main Islands of Japan

	Area		Population (1985)[a]	
Island	Square miles	(Square kilometers)	Percent	Number
Honshu	88,968	(230,427)	80	98,960,000
Hokkaido	30,334	(78,565)	5	6,185,000
Kyushu	15,756	(40,808)	11	13,970,000
Shikoku	7,280	(18,855)	3	3,711,000

[a]Figures rounded.
SOURCE: *Statistical Handbook of Japan*, 1988.

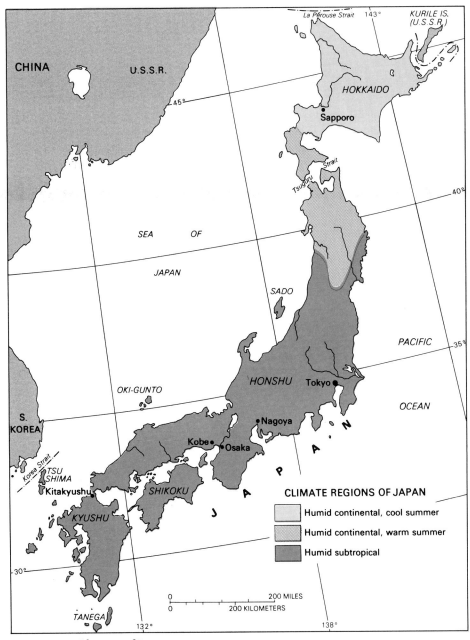

Figure 10.4 Climates of Japan.

A portion of Hiroshima nearly a mile from the center of the atomic blast. Destruction by the blast was compounded by uncontrollable fires as fire equipment, such as the trucks in this photo, were destroyed.

the total demand of the nation. Only in sulfur does Japan have a surplus capacity.

Japan's rather limited mineral resource base encouraged development of other resources. The extensive forests of the island nation played a major role in Japan's early development, as did the hydroelectric potential.

Of greater importance than these or the mineral resources were the human resources of the country. The emergence of a skilled population with a strong commitment to the development of the country was of greater importance than natural resources or their lack. In the initial stages of modernization Japan's peasants and workers suffered from the same problems of long hours, low wages, high taxes, and unsafe working conditions as those in European nations in the eighteenth and nineteenth centuries. Nevertheless, the Japanese people maintained their heritage of group cooperation, commitment to a nation, and deeply ingrained work ethic.

MODERN JAPAN: MIRACLE IN ASIA

World War II had a disastrous effect on the incipient modern industrial nation of Japan. Industrial production had focused on war material, and Japanese industry grew at a rapid pace to meet war demands. The decision to enter a full-scale war with the United States was taken by the small number of political and economic leaders in Japan. The manufacturing sector was controlled by a few large family or financial empires (combines). At the beginning of World War II the four largest combines—Mitsui, Mitsubishi, Sumitomo, and Yasuda—dominated every aspect of Japan's economy. These business combines had an inordinately large role in the political process in Japan and supported the militaristic goals of the ruling military party under Premier Tojo. The industrial development upon which Japan based its **geopolitical** aspirations was insufficient to match the tremendous productivity of a mobilized United States, and Japan's subsequent defeat left the economy and the nation in shambles. The destruction of Hiroshima and Nagasaki by atomic weapons was paralleled by destruction of over 60 percent of Tokyo and Osako by conventional weapons. Japan's factories, ports, roads, and residences were catastrophically damaged. The empire that had supported the expansion of industry with raw materials and markets was completely stripped from the nation, and the isolated Japanese were occupied by a foreign force: the Americans. The same nation that had first turned Japan to an industrial course less than 100 years previously now returned in a dramatic show of force to affect the destiny of Japan once again, directly and indirectly.

CHANGES UNDER AMERICAN OCCUPATION

The American occupying force was committed to a number of changes in Japan, but two were of greatest importance. First were steps to prevent Japan from having the ability or the desire to wage world war ever again. Second was to change the Japanese economy from a typically premodern industrial nation with wide variations in wealth and income to a democratic, middle-class society similar to that found in America. To accomplish the first goal the military forces of Japan were disbanded, and treaties that limited Japan's mil-

---

Geopolitics: The use of geographical relationships to justify a policy of national expansionism.

THE WORLD OF WEALTH

itary development to self-defense forces only and maintained American bases in Japan for the major protection of the Japanese islands were negotiated. The second goal was implemented through land reform, which eliminated the old land tenure system. Large landholdings were broken up, and the American occupying forces restricted landownership to 10 acres per family. To begin the process of transforming Japan to a middle-class nation, compulsory education was expanded to nine years, and the American forces introduced textbooks emphasizing democracy and downgrading militarism. The final significant change in Japan was the dissolution of the great trading combines that had dominated the economy prior to World War II. By 1948 the American occupying forces succeeded in dissolving many of these combines and creating a large number of individual companies from their components. Although some of the large trading companies continued on after the American occupation ended in 1952, the changes in land reform, education, and business organization had long-term importance for modern Japan society.

THE JAPANESE CORE: SPATIAL ARRANGEMENT OF ECONOMIC ACTIVITY

The effects of the destruction of World War II and the American occupation have caused major changes in Japan's economic development and the pattern of economic activity. The Japanese engaged in a program of rapid reconstruction after 1948 and the introduction of American changes, but the rebuilding recognized the importance of imports and exports to the small island nation. Industrialization increased so rapidly that the Japanese economy grew at a rate of 10 percent per year in the post–World War II decades. New industrial facilities were constructed at port locations to facilitate transforming raw materials into manufactured goods for reexport. A steel manufacturing complex contains a port for unloading coal, iron ore, and other raw materials; finished steel moves immediately into adjacent factories where it is manufactured into ships, cars, or other end products; and the end products are moved directly to the port for reexport or to the large urban areas for internal consumption. The efficiency of the new Japanese industrial complex is a significant factor in explaining its rapid rise to industrial preeminence in the world. Reconstruction centered on moving industrial plants from inland locations to coastal locations to maximize efficiency.

The concentration of industry in the coastal areas has increased the concentration of the population in a few areas of Japan. Based on population density and level of industrialization, three regions can be recognized in Japan: the industrial core, the industrial periphery, and the rural north (Figure 10.5). The majority of the industrial activity is concentrated in the Japanese core area from Tokyo to northern Kyushu. Within this zone of industrial and metropolitan landscapes are the four largest population and industrial

The ubiquitous American fast-food restaurants now found in many Japanese cities are but one example of the influence of western culture on Japanese society.

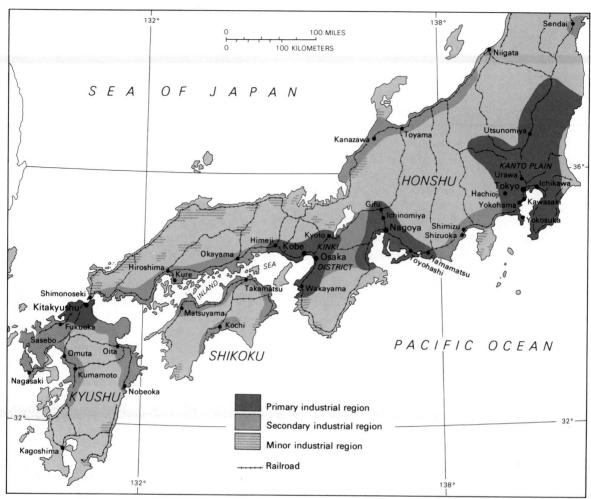

Figure 10.5 The industrial regions of Japan are concentrated in the low-lying plains of south and eastern Honshu and northern Kyushu.

centers in Japan: Tokyo, Nagoya, Kobe-Osaka, and Kitakyushu. The Tokyo **conurbation** is the primary manufacturing-industrial and population center of Japan. Manufacturing activities of all types are found here, from iron and steel plants, to motorcycle assembly, to petrochemicals, to shipbuilding. The Tokyo metropolitan area includes a number of large cities such as Yokohama, Kawasaki, and a host of lesser communities. The Tokyo metropolitan area is the largest in the world with a total population of 25 million. Although only slightly more than one-fifth of the population, the people of Tokyo produce a third of Japan's industrial output.

The second largest industrial center is concentrated in and around the cities of Kobe, Osaka, and the ancient inland capital of Kyoto. This part of Japan's core area focuses on the inland sea, which provides the most efficient transport linkage between the cities and industries of the Japanese coreland. The fragmentation of the Japanese national territory among numerous islands has always combined with the mountainous topography to make the ocean the major means of transport, and the sheltered inland sea has always played a dominant transport role. The cities of the Kobe-Osaka conurbation all rely heavily on the transport linkages provided by the inland sea. This region emphasizes shipbuilding, machine-tool manufacturing, petrochemicals, textiles, and oil refining. All the products of a mature industrial economy are manufactured in this region, which produces 20 percent of all of Japanese manufactures.

Between the two major urban and industrial centers of Kobe-Osaka and Tokyo is Nagoya. Nagoya produces less than 20 percent of Japan's manufactures and has a diversified industrial economy in which textile manufacturing is more pronounced than in other industrial centers in Japan. Textiles include both cot-

Conurbation: Extensive urban area formed by expansion of cities to form one continuous urbanized area. Used synonymously with megalopolis.

THE WORLD OF WEALTH

tons from locally produced and imported cotton, and wool and synthetic textiles based largely on imports.

The southernmost industrial center of Kitakyushu is based on local deposits of coal and iron ore and represents one of the old heavy industrial centers of Japan. Even here industry has been reoriented to the sea to facilitate imports of raw materials and exports of manufactured goods. The inland sea remains the focus of transport linkages and ties the Kitakyushu area to the balance of Japan's industrial and populated core. The entire northern Kyushu industrial region specializes in iron and steel, chemical fertilizers, and pulp and paper products. The old industrial centers of Kyushu are less suitably located for the emergent Japanese economy relying on imported raw materials and exported finished goods, but they are still the fourth most important manufacturing area in Japan. Kyushu, however, is the most slowly expanding part of the industrial core of Japan.

Within the Japanese core the level land is the scene of concentrated activity. Outside the large industrial centers land is farmed intensively. Population pressure necessitates use of hillsides where terraced fields cover suitable slopes. The rural landscape is one of small farms, rice fields, orchards, and small towns interspersed with growing industrial centers. Rice is the predominant crop, but citrus fruits, tea, and a variety of vegetables are produced. Expansion of urban centers, industrial activity, and transport systems creates tension between rural and urban life-styles. The limited amount of arable land dictates that new airports, highways, or railroads displace rural population. This has caused periodic confrontations between police and residents, such as that arising from the loss of land for construction of the Tokyo National Airport at Narita.

The industrial periphery is characterized by smaller communities, less industrialization, and physical settings that require greater labor and capital imputs. The west coast of Honshu has a colder winter climate than the protected coasts bordering the inland sea. The mountainous backbone of Honshu hinders linkages to the core area and increases transport costs. The coastal area produces rice, fruits, and vegetables, and some industry is found in the larger cities, but this area is peripheral to the major concentration of economic activity in Japan. The peripheral areas of Kyushu and Shikoku suffer from similar handicaps of accessibility. Agriculture is handicapped by the limited amount of level land, but the climate is more moderate than that of the west coast of Honshu. Throughout the periphery, handicraft or small-scale industrial activity employs a large part of the population.

The rural north is much less industrialized than the industrial core or its periphery. Northern Honshu is a hilly and mountainous area with limited arable land. Population is concentrated in the basins between the mountains, and agriculture employs nearly half of the labor force. The climate limits agriculture to a single crop per year, but small farms produce rice, sweet potatoes, mulberry trees, and small grains and vegetables. Little industry is located in northern Honshu, since it lacks raw materials or deep-water ports to facilitate imports of resources.

Hokkaido is even more distinct from the industrial core of Japan. Its isolation in the past from the main centers of settlement combined with its cool-summer, humid continental climate to delay its settlement. Today the landscape is distinct because land division during colonization utilized a grid system. The present economy is highly oriented toward agriculture, with crops reflecting both the climate and Japanese culture. Rice is the dominant crop, occupying 20 percent of the land, but sugar beets, potatoes, oats, and other small grains are sufficient. Hokkaido is the only area of Japan where production of beef and dairy cattle is significant.

FACTORS OF DEVELOPMENT

The emergence of Japan as the world's third greatest manufacturing nation in a 30-year span is the result of a number of factors. First among these is the historic development of a homogeneous Japanese population with a strong sense of national identity. Very limited numbers of remaining Ainu and Korean immigrants form the only permanent minority groups, and they are discriminated against by the highly ethnocentric Japanese. Although more than 800,000 Koreans remain in Japan from the 2 million brought as laborers before World War II, they are not allowed to become citizens. The Japanese people are committed to the principles glamorized in the samurai tradition of self-sacrifice, diligent labor, commitment to the nation rather than to the individual, and group benefit rather than individual achievement. The strong sense of *commitment to the nation* coupled with a willingness to labor long hours allowed Japan's industry to develop without conflict between workers and owners. In Japan the prevalent trend is for an individual to obtain employment with a company and to stay with that company throughout his or her life. Today an estimated one-third of Japanese employees have lifetime employment.

The second factor allowing Japan to emerge rapidly as an industrial giant was the existing skilled populace and the tradition of manufacturing. The American occupying force added another incentive by encouraging corporations owned by the public rather than by the individual family trading companies of prewar Japan. An estimated 20 percent of the Japa-

Terraced rice fields in Japan illustrate both the human effort required to cultivate steeply sloping land, and the difficulty of mechanizing the resulting small fields.

nese own stock in the publicly owned corporations today. Nevertheless, in spite of the dissolution of the trading companies under American pressure, today some twenty major trading companies once again control the majority of the large-scale manufacturing. Many of these companies are combines that include those that dominated Japan prior to World War II, a dominance that is of concern to some outside observers. The Japanese miracle is in part the result of the unique relationship between these trading companies and their employees. There is little of the adversary relationship found in the United States or Western European industrial companies where strikes are common and cooperation between workers and management handicapped. In Japan, workers tend to spend their entire working lives with the same company, the company provides vacations at resorts that it owns, and may provide subsidized lunches, recreation facilities, and social opportunities. The commitment to a unified Japan is absorbed by school-age children who are taught the basic geographic limitations posed by the Japanese environment and inculcated with a strong patriotic conviction of the need for Japan and its citizens to be efficient so that the nation can survive through world trade. For workers in Japan, successful achievement in increasing efficiency, productivity, or otherwise obtaining a competitive edge is a central value. Unlike the Western orientation to maximizing the individual benefit, the entire Japanese society advocates the commitment to the nation as a whole.

The uniqueness of the Japanese industrial development that has resulted from the unity between labor and management, the job security, and the industriousness of the society is matched by another unusual characteristic, a combination of large corporations and small manufacturing companies. Of a total of nearly 500,000 manufacturing establishments, only about 500 are large corporations with numerous plants and large numbers of employees. This small proportion of the manufacturing sector employs one-third of the Japanese work force and is the basis for most of the heavy manufacturing of iron, steel, automobiles, ships, cameras, and other items. Coexistent with the modern industrial economy are more than 400,000 establishments operating small-scale handicraft or home-type manufacturing. These groups may operate in a single home and involve only the members of an immediate or extended family in making such things as dolls, ceramics, or other handicrafts; or simple plastic objects, or consumer items such as linens, foodstuffs, or clothing. Some of the small establishments are sub assemblers of a larger manufacturing process. They may involve assembly of component parts for television, cameras, electronics, or even automobile parts. The unique combination of small- and large-scale manufacturing in Japan is derived from the high percentage of the labor force still engaged in such homecraft or workshop manufacturing processes.

The factories of Japan make Japan the leading exporter of a host of items, including automobiles, motorcycles, cameras, bicycles, televisions, and consumer electronic goods. Its steel- and shipbuilding industries are among the world's leaders, with Japan pioneering advancements in supertanker construction and the steel industry. Japan is a major exporter of goods to not only the United States, but all of the rest of the world. The miracle of Japan is its emergence as one of the leading industrial powers, and the leading exporter of manufactured goods in the world in spite of its seemingly unfavorable geographic relationships.

The dominant role of Japan in exporting manufactured products has caused some nations to erect restrictive tariffs. In response, the Japanese have built assembly plants in other nations (as auto plants in the United States), or entered into partnership with foreign firms such as General Motors. Today, Japan's role as a borrower of culture and technology has been reversed as other countries copy elements of its economic miracle.

MODERNIZING AGRICULTURE: A JAPANESE EXAMPLE

The transformation of Japan into a modern industrial nation has been partially paralleled by transformations in the agricultural sector. Rural areas are a unique blend of modern and traditional techniques, patterns, and life-styles. The long period of agricultural growth in Japan developed highly productive farms long before industrialization began. The same long development caused land division and fragmentation on a scale unknown in other industrialized nations. Seventy percent of Japan's farmers own less than 2.5 acres (1 hectare) of land, compared to the average farm size of more than 400 acres (162 hectares) in the United States. The small size of the farms reflects the limited amount of agricultural land in Japan, which has only 14 million acres (5.7 million hectares) available for cultivation. By comparison the United States annually cultivates approximately 400 million acres (162 million hectares) but could expand this at least 20 percent by simply cultivating land that is *fallow* (uncultivated).

The intense pressure of population on agricultural land can be seen by the ratio of people to cultivated land, which is equivalent to only slightly more than one-tenth of an acre (.04 hectare) per person. Of necessity the Japanese have had to develop a highly productive agricultural system to ensure adequate food for their population. The use of **double cropping** throughout most of the islands pushes the average amount of cultivated land available per capita to an equivalent of two-tenths of an acre (.08 hectare) per person per year. The fragmented nature of the Japanese farms was compounded by a landlord-dominated system prior to World War II. The American occupation forces engaged in land reform, which provided land to much of the landless rural population. In prewar Japan 54 percent of the work force owned land and 46 percent were tenants or landless laborers. The American-induced land reforms reduced the number of landless farm populace to only 8 percent. The provision of a near universal owner-operator farm class in Japan was the first of two major changes that have allowed Japan's agriculture to undertake its remarkable development process. The second was farm consolidation, which united the small parcels of each individual into a larger farm of 1 to 2 acres (.4 to .8 hectare) in size. Where the fragmented farm pattern of premodern Japan necessitated hand cultivation of all crops, the increase to several acres allowed adoption of mechanized equipment. But Japanese mechanization is unlike that found in other industrial nations, as it relies on small-scale implements similar to garden tractors used in the United States or other large Western nations.

Owner operators and farm consolidation and mechanization, combined with the availability of agricultural equipment from Japan's increasingly industrial economy, have vaulted Japanese farmers to a position of preeminence in agricultural productivity (Table 10.3). Japan's rice yields per acre exceed those of any other nation; they are nearly triple the average for Asia. Government-guaranteed prices for rice and other basic commodities encourage Japanese farmers to increase their efficiency, and Japan has had rice surpluses for the last decade. The increasing yields and guaranteed prices coupled with mechanization have increased the standard of living in Japanese rural areas to a level much closer to that found in the urban areas.

The Japanese government also has restrictive tariffs, which further protect farmers. In spite of the increases in the standard of living among the rural population, there has been a general movement from rural to urban areas as the increased mechanization of agriculture limits the need for farm labor. Although nearly 14 million Japanese were engaged in agricul-

---

Double Cropping: Planting two crops in succession in a field in the same year.

TABLE 10.3 Comparison of Rice Yields between Japan and the United States

	Japan		United States	
	Tons per Acre	(Metric Tons per Hectare)	Tons per Acre	(Metric Tons per Hectare)
1975	2.62	(5.95)	2.30	(5.23)
1976	2.33	(5.30)	2.30	(5.23)
1977	2.61	(5.94)	2.17	(4.94)
1978	2.72	(6.18)	2.20	(5.01)
1979	2.66	(6.06)	2.24	(5.08)
1980	2.25	(5.13)	2.18	(4.95)
1981	2.48	(5.65)	2.40	(5.46)
1982	2.50	(5.69)	2.32	(5.28)
1983	2.51	(5.70)	2.24	(5.14)
1984	2.26	(5.13)	2.44	(5.55)
1985	2.19	(4.98)	2.66	(6.05)
1986	2.22	(5.06)	2.78	(6.32)

SOURCE: Statistical Handbook of Japan 1988 and *Statistical Abstract of the United States, 1988.*

ture in 1960, by 1990 this figure had declined to less than 7 million. This rural-to-urban migration is dramatically changing the Japanese countryside, and the age of rural residents is higher than the average of the population as a whole.

The urban migration and its impact on rural areas is multiplied by the movement of industrial activities into rural areas. Increasingly the Japanese rural resident is a full-time employee of a factory and a farmer only part time. Nearly 80 percent of all Japanese farmers obtain at least part of their income from nonfarm sources. These changes in employment and income patterns allow the rural farm family in Japan to have most of the amenities that residents of urban areas in any industrial region enjoy. The tiny farms, small-scale agricultural equipment, and combination of modern and traditional buildings and appliances combine to make the Japanese rural area one of the most intriguing landscapes in the world. The field patterns are characteristic of less industrialized areas, but the mechanization and the standard of living are characteristic of the industrialized world. The rural Japanese countryside epitomizes the unique Japanese development process, combining both elements of Japan's heritage and the adoption of selected elements of the Western industrial capitalist model.

THE JAPANESE EXPERIENCE IN RETROSPECT

Japan's position as an industrial nation in the midst of less industrialized areas, as the only non-European nation to emerge as a fully industrialized state, and its combination of oriental and Western economies and life-styles poses important questions concerning its utility as a model for other developing countries. A number of distinctive characteristics found only in Japan contributed to the Japanese miracle. Japan's population is for all intents and purposes homogeneous. The tiny minority population of the Ainu or Koreans has never presented a significant centrifugal force in the post–World War II era. The early emphasis placed on education in Japan provided it with a highly literate work force, which today is the most literate society in the world. Essentially 100 percent of the Japanese people are literate, and education is a consuming and driving force among younger Japanese. The lack of internal problems of ethnic minorities greatly simplified Japanese development.

The early emergence of Japan reflected its insular location, which allowed it to borrow ideas from the Chinese on the mainland, but freed it from involvement in conflicts of the Chinese themselves. The long tradition of borrowing cultural traits that the Japanese perceived could be used advantageously aided their adoption of the Industrial Revolution. The Japanese character traits of industriousness, commitment to nation over self, loyalty, and commitment made possible the emergence of an industrial power based on the efforts of a homogeneous, skilled labor force.

The Japanese example would be difficult to incorporate in a majority of the less industrial countries because of the social, political, and economic fragmentation in such countries. The political development of Japan with a strong central government and a tradition of support of the emperor further aided its development. Japan's decision and ability to prevent European contact for 250 years spared it from colonialism with all its parasitic relationships that have hindered the development of so many other countries of the world. The Japanese miracle in Asia, based on distinctive Japanese history, ethic, and attitudes, may not be replicable on a world scale.

Japan's location on major Pacific sea routes provided it a comparative advantage, allowing it to dominate not only its own shipping but a large share of the trade of other Asian communities. The commercial and financial infrastructure that developed in conjunction with Japan's trade provided important foreign exchange for capital investment in industry. Japan's historic ability to control its population size prevented it from facing the problem of modernization in a labor-surplus economy. As an example, following World War II a baby boom resulted in a rapid increase in population, and government concern over the population issue culminated in the eugenics protection laws that legalized abortion and made birth control readily available. The drop in birthrate in Japan has been dramatic and is now identical to other industrial nations.

The site/situation relationships of Japan have also been important in its development. Japan's insular location only tens of miles from Korea and only a few hundred miles from China allowed Japan to benefit from the resources of the continent and to adopt cultural innovations that were perceived as benefiting the Japanese people while maintaining its territorial integrity.

JAPAN: ISSUES AND PROBLEMS

In assessing the Japanese experience, it should be noted that the same characteristics that allowed the Japanese miracle to occur also compounded the negative impact of industrialization. The concept of loyalty to company and commitment to ever-greater productivity to increase trade have handicapped efforts to minimize environmental degradation. Companies have been reluctant to undertake environmental controls, which add costs to the manufacturing process but do not increase productivity. Reluctance to criticize manufacturing concerns or others who were contributing to Japanese economic growth delayed corporate acceptance of environmental problems that they caused.

The most famous example is the mercury poisoning of fish in Japan. The companies around Minamata Bay dumped wastes heavily contaminated with mercury into waters of the bay. The mercury ultimately concentrated in the fish that lived in the bay. The local people relied upon the fish for a major part of their protein, and children born to women who had consumed the mercury-contaminated fish suffered serious physical deformities. Only in the last two decades have Japanese companies begun, reluctantly, to accept environmental responsibility.

The smog of Japanese urban areas rivals that of any industrial region in the world and obscures the beauty of important parts of the Japanese landscape. Mount Fuji, to most people the symbol of Japan, is rarely visible from downtown Tokyo. Tokyo policemen must often wear oxygen masks as they direct traffic, and the problem of pollution remains unresolved.

As critical as problems of pollution in Japan are the social problems that have resulted from Japanese economic development. Japan's goal of becoming a major world economic power concentrates capital investment in productive areas rather than in provision of services to the public. Japan's social service budget is much smaller than any other technically advanced nation, there are major shortages in housing, in urban transportation systems, and in higher education opportunities. The intense competition for land for homes has driven land prices in Tokyo and other metropolitan regions to levels rarely exceeded in other urbanized areas. Absence of housing for families dictates that only the upper middle class is able to own a home in the style of America's suburbia. Even these homes are small by comparison to Western homes and reflect the high cost of land in Japan. (See Box.)

Japanese emphasis on trade, coupled with protective taxes, presents additional problems for Japan in the modern world. Japanese attempts to expand trade relationships to the balance of Asia are often met with suspicion by the nations who remember the Japanese expansion prior to and during World War II. Protective taxes that enable Japanese industry to compete on a world scale are criticized by other Western nations who claim that Japan discriminates against imports of manufactured goods but expects other nations to reduce tariffs on manufactured items.

The relatively small proportion of land devoted to streets and highways in Japan as compared to other industrialized nations causes tremendous traffic jams in metropolitan areas with even a smaller number of automobiles (Table 10.4). The urban infrastructure is backward by comparison with other industrial areas. The Japanese have made amazing technological advances in some areas of transportation, including the famous bullet trains, which travel between the major industrial regions at speeds of up to 160 miles per hour. By comparison the subways and bus systems in these urban areas are inadequate, and the Japanese government hires people to push passengers into subways to maximize their capacity during rush hour. Equally serious is the lack of highways and railroads connecting the various islands to each other. Japan has completed a tunnel to provide rail linkage from

THE HUMAN DIMENSION

The Japanese middle class

One of the things that has set Japan apart in Asia is its large middle class, common to industrial societies. Like their counterparts in Europe or North America, the middle class are made up of salaried workers and small businessmen (businesswomen are uncommon). Families are small, with one or two children the norm. Nearly 90 percent of Japan's residents list themselves as part of the middle class in regular government surveys, yet their standard of living is much different from that of the American middle class.

Meet Masao Watanabe and his family (a wife and seven-year-old son), members of the Japanese middle class, with all of the hopes and fears of others in Japan. Mr. Watanabe is a government employee who resides in the Tokyo metropolitan area (but he could just as easily be a member of a trading firm or work for a factory as a mid-level manager). At 36, Mr. Watanabe earns $35,000 per year, but must budget nearly half of his $1700 per month take-home pay for rent and utilities for the three-room, 484-square-foot apartment he shares with family. In spite of the hot, humid summers in Tokyo, the Watanabes do not have an air conditioner since it would increase their utility bill by up to 30 percent. Masao spends three hours per day commuting to and from his work, but his wife works in a nearby department store and only spends an hour in commuting. Her clerk's salary pays for her clothes, transportation for both, and leaves about $150 per month for saving for a home.

The Watanabes face one of the most serious problems affecting the middle class of Japan—obtaining a home. The price of land in Tokyo city limits is so high that a lot covering just 1,100 square feet (one-fortieth of an acre) would cost $530,000. Construction of a two-story house with 1300 square feet of space would cost an additional $125,000. Annual payments would total nearly $52,000, more than most Japanese earn in a year. In 1988, 73 percent of Japanese households earned from $19,000 to $60,000, but most households are clustered between $20,000 and $30,000. Homes and other material possessions reflect the lower salaries. While the average American home has 1,585 square feet, the average Japanese home has only 925 square feet. The average American family has 2.2 cars, the average Japanese only .88. The average Japanese family spends more than 25 percent of their income on food, while the American family spends less than 15 percent.

To obtain a home the Watanabes will have to move farther out into the suburbs of Tokyo, adding to the commute time of both Mr. and Mrs. Watanabe. To get to work they will need to leave home before 7:00, take a short bus ride to the train station, and spend an hour to an hour-and-a-half standing on a crowded train. If they want to obtain a seat on the train, they will need to move even farther out near the end of the line and spend even more hours commuting. Most of the middle-class suburbanites of the Tokyo metropolis will not arrive home from work before 7:00 or 8:00 in the evening.

Unfortunately for the Watanabes, it will be at least a decade before they get a home of their own because they need to save the 20 percent down payment. Rising costs for both land and materials will continue to drive up the cost of their home, possibly making them use a new form of loan that has payments that will be passed on to their children, thus low-

---

Honshu and Hokkaido, but the sea remains a major barrier to inter-island transport. Rugged topography handicaps expansion of dense rail and road routes from the core to the periphery and north and provides an important challenge for the country.

The migration of people from the rural to urban areas is causing a breakdown in the old extended family system in Japan, laying the groundwork for increasing social problems in the future. In spite of these dilemmas caused by the Japanese economic experience, the Japanese have one of the highest standards of living in the world. Life expectancy for both men and women (75 and 79 respectively) is exceeded by only one or two Western European nations and is above that found in the United States. The Japanese people recognize the social costs that have resulted from their rapid development, but are amazingly confident that they can continue to maintain a high level of economic development and overcome the social problems of housing, transportation, and other urban infrastructure. The challenge for Japan in the future will be to maintain the existing agricultural base and its competitive position in international trade as the cost of imported resources continues to rise. These

TABLE 10.4 Automobile Registration in Selected Industrialized Nations

Nation	Population per Car
Australia	2.3
Canada	2.2
France	2.6
Japan	4.4
Switzerland	2.5
United Kingdom	3.2
United States	1.8
West Germany	2.4

SOURCE: Motor Vehicle Manufacturers Association, *Motor Vehicle Facts and Figures, 1986.*

ering monthly payments. The typical home in a drab neighborhood of Tokyo with only 700 square feet requires a mortgage of $310,000, with monthly payments of over $2,200, more than the typical family's monthly take-home pay.

Japan's housing crisis is centered in the Tokyo metropolitan area, where some 25 million, or 1 in 5 Japanese, live within a 25-mile radius of downtown. The problem is only marginally less in the other urban centers surrounding the Inland Sea.

The Watanabes spend little time with their son, as he attends a *juku* (a private school where supplementary study is provided to help schoolchildren get good grades as well as pass examinations required for admittance to higher schools). Mr. Watanabe is intense in his desire to see his son move up the educational ladder without having to suffer the so-called examination hell, as the extremely competitive examinations for admission to the university, or even high school, are called.

Recreation for the Watanabes is limited. Once or twice a week Masao stops at a bar on his way home and has a few drinks, and he tries to golf with friends once a month, but green fees and the trend to high-priced private golf clubs makes even that a luxury. (The average country club membership fees are nearly $40,000 per year in the Tokyo region.) Without a car the family does not have the opportunity to participate in the ritual of a Sunday drive in the country, but the increasing cost of home ownership may make them decide to obtain a car and resign themselves to a lifetime of renting.

The middle class of Japan has long accepted that they were working for the good of the nation, so they have not complained that they have less to show for their work than residents of other industrialized countries. To industrialize rapidly and to remain competitive the government has invested heavily in industry at the expense of the social sector, but the plight of the Watanabes and other middle-class Japanese may make the continuation of that policy unworkable. While Mr. Watanabe and other salaried workers see their real earnings decline in purchasing power, a growing number of businessmen are seeing theirs increase since they are able to shelter it from taxes by buying cars, golf club memberships, dinners, and so forth on expense accounts. Some observers suggest that there is emerging a two-class society in the former middle-class Japan, that of the *nyuu poor* (new poor) and the *nyuu ritchi* (the new rich). The perception that the salaried people like Mr. Watanabe, who have built the Japanese economic miracle, but are no longer sharing in the economic growth of the nation, is contributing to an undercurrent of frustration. Eventually this frustration could take the form of apathy among students and workers, causing Japan to lose their competitive edge based on harmony and homogeneity among its people. A possible solution to the problem would be for the government to invest in development of less crowded regional centers, providing jobs in towns and cities with lower land and housing costs.

Source: Based on articles in International Business, Japan Pictorial, *and interviews by the authors.*

---

problems will be solved partially through the declining birthrate in Japan, which is at a fully mature stage four in the demographic transition model. The Japanese population at present growth rates will probably stabilize at between 120 and 130 million people, allowing preservation of the existing agricultural land.

As a model Japan offers promise, but the unique characteristics of its development may prevent its adoption by other nations.

FURTHER READINGS

A Complete Atlas of Japan. Tokyo: Teikoku-Shoin, 1985.

ALLEN, G. C. *A Short Economic History of Modern Japan.* New York: St. Martin's Press, 1981.

ASSOCIATION OF JAPANESE GEOGRAPHERS. *Geography of Japan.* (Special Publication No. 4). Tokyo: Association of Japanese Geographers, 1980.

BURKS, A. W. *Japan: Profile of a Postindustrial Power.* Boulder, Colo.: Westview Press, 1981.

CHRISTOPHER, R. *The Japanese Mind: The Goliath Explained.* New York: Linden Press/Simon and Schuster, 1983.

EYRE, J. *Nagoya: The Changing Geography of a Japanese Regional Metropolis* (Studies in Geography No. 17) Chapel Hill, N.C.: University of North Carolina, 1982.

GIBNEY, F. *Japan: The Fragile Superpower.* New York: Norton, 1975.

GLICKMAN, N. J. *The Growth and Management of the Japanese Urban System.* New York: Academic Press, 1979.

GREGORY, GENE, THARP, MIKE, AND BARTHOLOMEW, JAMES. "Japan's 'Third Revolution'." *Far Eastern Economic Review* as reprinted in *World Press Review*, March 1982.

HALL, R. B., JR. *Japan: Industrial Power of Asia.* New York: Van Nostrand Reinhold, 1976.

HANE, MIKISO. *Modern Japan: A Historical Survey.* Boulder, Colo.: Westview Press, 1986.

HARRIS, C. D. "The Urban and Industrial Transformation of Japan." *Geographical Review* 72, 1982, pp. 50–89.

HARRIS, C. D. and EDMONDS, R. L. "Urban Geography in Japan: A Survey of Recent Literature." *Urban Geography* 3, 1982, pp. 1–21.

HAYAMI, Y. *Agricultural Growth in Japan, Taiwan, Korea, and Philippines.* Honolulu: University Press of Hawaii, 1979.

HOFHEINZ, R., JR., and CALDER, K. E. *The Eastasia Edge.* New York: Basic Books, 1982.

ITO, T. and NAGASHIMA, C. "Tokaido—Megalopolis of Japan." *GeoJournal* 4, 1980, pp. 231–246.

JOHNSON, C. *MITI and the Japanese Miracle: The Growth of Industrial Policy, 1925–1975.* Palo Alto, Calif.: Stanford University Press, 1982.

JOO, K. S. "Urbanization and the Urban System of Korea." *Geographical Review of Japan* 55, 1982, pp. 1–20.

JUN, U. "Japan as an Advanced Polluting Nation—Its Responsibility." *Japan Quarterly* 27, 1980, pp. 321–332.

KIRBY, S. *Japan's Role in the 1980s.* London: Economist Intelligence Unit, 1980.

KOMHAUSER, D. *Urban Japan: Its Foundations and Growth.* 2d ed. London: Longman, 1983.

MACDONALD, DONALD. *A Geography of Modern Japan.* Ashford, U.K.: Paul Norbury, 1985.

MURATA, K., and OTA, I., eds. *An Industrial Geography of Japan.* New York: St. Martin's Press, 1980.

PARK, S. "Rural Development in Korea: The Role of Periodic Markets." *Economic Geography* 57, 1981, pp. 113–126.

PITTS, F. R. *Japan.* Grand Rapids, Mich.: Fideler, 1974.

REISCHAUER, E. O. *Japan: The Story of a Nation.* 3d ed. New York: Knopf, 1981.

SAKAMOTO, H. "The Regional Distribution of Comparatively Large Farms in Japan." *Geographical Review of Japan* 55, (1982), pp. 37–50.

SCHEINER, E. *Modern Japan: An Interpretive Anthology.* New York: Macmillan, 1974.

SHISHIDO, H. "Economic Growth and Urbanization: A Study of Japan." *International Regional Science Review* 7, 1982: pp. 175–191.

SIMMONS, I. G. "The Balance of Environmental Protection and Development in Hokkaido, Japan." *Environmental Conservation* 8, 1981, pp. 191–198.

SMITH, R. J. *The Price of Progress in a Japanese Village: 1951–1975.* Palo Alto, Calif.: Stanford University Press, 1978.

STERNHEIMER, S. "From Dependency to Interdependency: Japan's Experience with Technology Trade with West and the Soviet Union." *Annals of the American Academy of Political and Social Science* 458, 1981, pp. 175–186.

TAEUBER, IRENE B. *Population of Japan.* Princeton, N.J.: Princeton University Press, 1958.

TATSUNO, SHERIDAN. *The Technopolis Strategy: Japan, High Technology, and the Control of the Twenty-First Century.* New York: Prentice-Hall Press, 1986.

TIEDMANN, A. E. *Introduction to Japanese Civilization.* New York: Columbia University Press, 1974.

TOTMAN, C. "Forestry in Early Modern Japan, 1650–1850: A Preliminary Survey." *Agricultural History* 56, 1982, pp. 415–425.

Part Three

THE WORLD OF POVERTY characteristics and challenges

One Statistic Serves To Encapsulate The Bleakness Of This [Third World] Human Condition: 600 Million Men, Women And Children, More Than The Combined Populations Of North America And Western Europe, Still Struggle To Survive On Incomes Equivalent To 15 U.S. Cents A Day.

JOHN L. VON ARNOLD, 1985

A Society That Is Not Socially Just, And Does Not Intend To Be, Puts Its Own Future In Danger.

POPE JOHN PAUL II, 1980

Inner Mongolia.

ASIA PROFILE 1990

Population (in millions)	3,100	Energy Consumption Per	
Growth Rate (%)	1.8	Capita—pounds (kilograms)	816
Years to Double Population	38		(370)
Life Expectancy (years)	61	Calories (daily)	2,260
Infant Mortality Rate		Literacy Rate (%)	56
(per thousand)	82	Eligible Children in	
Per Capita GNP	$1,020	Primary School (%)	87
Percentage of Labor Force		Percent Urban	36
Employed in:		Percent Annual Urban	
Agriculture	64	Growth, 1970–1981	4.1
Industry	17	1980–1985	3.9
Service	19		

11

Asia and the Oriental World population, poverty, and change

MAJOR CONCEPTS

The Orient / The Far East / Asia
Underemployment / Polyculture
Polyethnicism / Bustee / Tropical cyclone
Plantation / The Green Revolution
Dependency ratio / Ecumene / Riverine
population concentrations / Storm surges
Epicanthic / Monsoon / Lingua Franca
Plural society / Shifting agriculture

IMPORTANT ISSUES

Hunger, poverty, and population: the trio of issues facing the Orient

The dilemma of mushrooming urban population in underdeveloped countries

Cultural conflicts in newly independent countries

Boundary and territorial conflicts in former colonial territories

Religion and life: the challenge of transforming traditional agrarian economies

Geographical Characteristics of Asia

1. The nations of Asia are home to nearly 60 percent of the global population, and Asia (including the Soviet Union in Asia) is the largest of all the continents.
2. Less than one-tenth of the total land area of Asia is suitable for occupation because of mountains, deserts, or other harsh environmental conditions.
3. The cultures of Asia are highly diverse, and the region was the hearth area of several major religions including Buddhism and Hinduism.
4. The majority of the people of Asia still rely on agriculture for their livelihood.
5. Several of the countries of Asia have industrialized sufficiently to be categorized as newly industrialized countries.

The Orient, Asia, the Far East all evoke in the minds of Western readers visions of crowded cities, impoverished peasants, inscrutable Oriental faces, and a way of life completely different from that found in the industrialized world. But the terms are not synonymous, nor are the stereotyped views accurate. *Asia* refers to the great landmass stretching from Turkey eastward through the Soviet Union, Saudi Arabia, India, China, Japan, and Indonesia. Thus defined, Asia includes much of the Soviet Union, and contains 17.09 million square miles (44.26 million square kilometers), making it the largest of the continents. By comparison Africa is 11.69 million square miles (30.28 square kilometers); North America, 9.42 million (24.4 square kilometers); South America, 6.87 million (17.79 square kilometers); Europe (including European Russia), 3.83 million (9.92 square kilometers); and Australia, 2.97 million (7.69 square kilometers). Asia accounts for approximately 30 percent of the world's total land area and is larger than North and South America combined.

Asia: The largest of the continents, it occupies the eastern portion of the Eurasian landmass and adjacent islands.

Far East: An area commonly including the Koreas, Japan, China, and the islands belonging to them. Sometimes used to refer to all of Asia east of Afghanistan.

Orient: The countries of the Asian continent excluding the Soviet Union.

The *Far East* is a somewhat nebulous term sometimes used to refer to Japan, China, Vietnam or even India, depending on the background, perceptions, and purposes of the observer. The *Orient* is more synonymous with the part of Asia with which this chapter deals, which excludes that part of the Asian continent within the Soviet Union. (Figure 11.1).

Within the 17 million square miles (44 million square kilometers) of the Asian continent, less than 10 percent is suitable for human habitation. In the following pages we will focus on the populated areas of the Orient from Pakistan through India, Southeast Asia, Japan, and China, which lie in a band of lower, wetter lands surrounding the arid, mountainous core of the Asian continental mass (Figure 11.2). These countries contain only 7.2 million square miles (18.6 million square kilometers) and much of it is too rugged and too high to allow much agriculture. The Himalayan mountain complex, for example, rises to 29,028 feet (8,848 meters) at Mount Everest.

Although the majority of the Orient is mountainous or arid, the continent is home for the majority of the world's population. In the more humid and level regions of the south and east reside more than 3 billion of the earth's peoples. The issues presented to the world of today and tomorrow by the concentration of population in the Orient represent some of the most profound and complex we face. At the simplest but most important level is the need to provide adequate food, shelter, and jobs for a burgeoning population. This task is made more complex by the poverty that characterizes so much of the Oriental region.

THE WORLD OF POVERTY

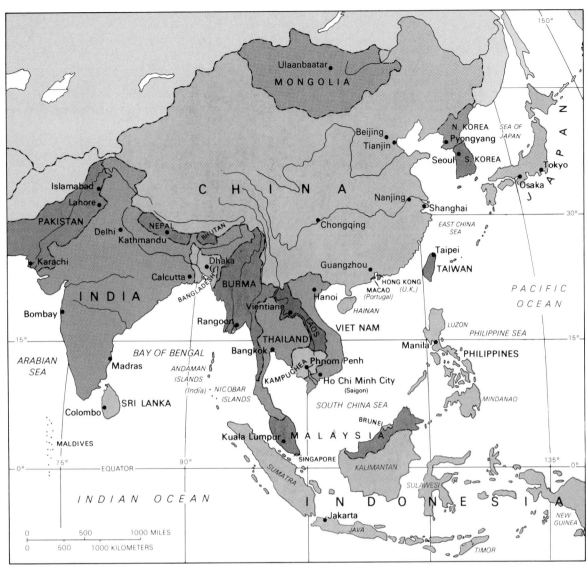

Figure 11.1 Political divisions and major cities of Asia.

The Orient typifies the characteristics and problems of the less industrialized world. Japan, and to a lesser extent South Korea, Taiwan, Singapore, and Hong Kong, share in the greater wealth and higher standard of living of industrialized countries, and China shares with them low birthrates and high literacy, but they are exceptions in the Orient. Most of the people, and most of the countries, are characterized by low literacy rates, high birth and infant mortality rates, low per capita incomes, and a high proportion of the labor force engaged in agriculture. Culturally, the extended family is the norm, providing social security, support for the handicapped, aid to dependent children, and other services that government agencies provide in the industrialized world.

Economies of individual countries vary, but all are dual economies. In each country the majority of the population is engaged in agriculture, traditional handicrafts, or reciprocal exchange markets. At the same time a small portion of the population, usually in cities, is engaged in industry, nonreciprocal exchange, and other activities characteristic of more industrialized countries. Governments tend to be impermanent in the region, and the democratic revolution as known in Western Europe is the exception in the Orient. With the exception of Japan, Thailand, and Mongolia, the countries of the region have been dominated by European colonial powers for more than a century in the past. The resulting problems are those of less industrialized countries in general, exacerbated by the immense size of the Asian population.

Other issues related to the tremendous population in the Orient result from transformations that are affecting a basically agrarian society. With a few ex-

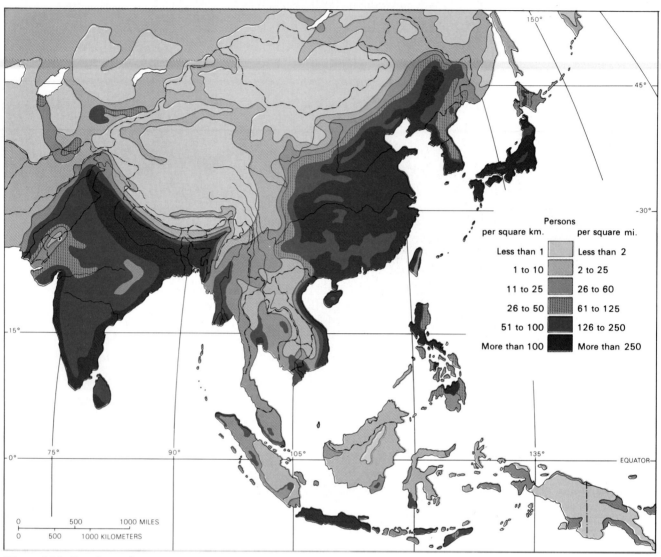

Figure 11.2 Population density in Asia.

ceptions such as Japan, oriental countries rely on agriculture, with as much as three-quarters of the population actively engaged in farming. Increasing population in the rural sectors results in fragmentation of farms, and resulting poverty and underemployment drive surplus farm population into the cities. This rural-urban migration results in rapid growth in oriental cities and presents a third set of issues that the world must resolve in the near future. More than one-third of the world's cities with populations of more than 100,000 are in the Orient; yet, cities contain only about one-quarter of the population in the Orient. In 1950 there were only 6 cities with more than 5 million people, and all but one were in the industrial world. By the year 2000 there will be at least 30 cities this large, and 18 will be in the less industrialized world. The growth of these cities as they sprawl outward in teeming slums presents one of the great challenges of the future (Figure 11.3).

A final set of issues revolves around the cultural characteristics of the Orient's population. Conflicts between different ethnic groups, religions, and political ideologies have resulted in forced migrations, death and injury, and suffering of a magnitude comparable to that of World War II and its aftermath in Europe. These migrations are in part a legacy of colonial dominance of the region, and a pressing issue today is the evolving political patterns of the oriental realm and their interrelationships.

THE POPULATION PROBLEM IN THE ORIENT

Central to an understanding of the Orient and the issues facing it is population. It is sometimes suggested that the problems of the Orient can be sum-

Mountains of the Himalayan chain have the highest elevations in the world. Mount Everest is the highest peak in the world.

marized in three words: population, population, population. Although perhaps overly dramatic, this statement exemplifies the role that the human element plays in Asia's problems. The significance of population in the Orient can be partially grasped by a simple examination of the numbers. The People's Republic of China has a population of over 1 billion. India has an estimated population in 1990 of between 850 and 875 million. Indonesia, the third largest country in the Orient, has a population of approximately 183 million.

But a simple list of numbers is insufficient to indicate the significance of the population problem facing the Orient. Consider that of every five children born in the world in a given year, one will be Indian, one will be Chinese, and one will be born into another Asian country. Approximately 60 percent of the births in the world in the 1980s occurred in the countries of Asia (Figure 11.4). But the importance of the population problem goes even further, for the high birthrates reverberate through the entire fabric of life of the residents of the Orient.

POPULATION AND HUNGER

The fundamental problem is simply feeding all the people. In a country like India, where more than 16 million new individuals are added each year, simply providing for the yearly population growth requires ever-increasing food production (Figure 11.5). Per capita food production in the world has increased in the past two decades, but only slightly. At the same

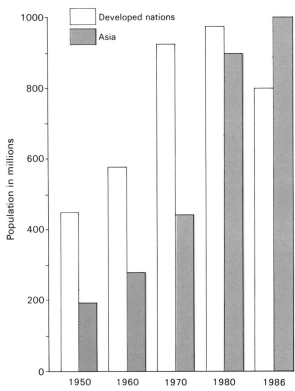

Figure 11.3 Rapid urbanization in Asia in the past two decades has resulted in an urban population whose numbers exceed that of the industrialized nations even though the majority of Asia's population is still rural.

ASIA AND THE ORIENTAL WORLD

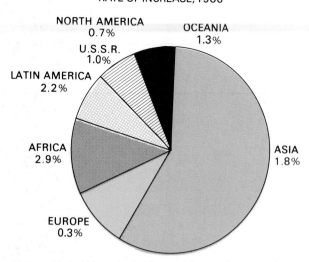

Figure 11.4 Although Asia's annual growth *rate* is lower than that of Africa or Latin America, its *proportion* of total annual growth is the highest in the world.

time, the absolute production has increased by more than a third. The ever-growing population prevents substantial gains in the per capita food availability. The tragic result is inadequate food supplies for a portion of the world's populace. The issue of hunger as related to population is obviously more concentrated in the less industrialized countries, particularly in those of southern Asia and Africa. The number of people who suffer from inadequate nutrition is impossible to verify statistically. At best, estimates can be made based on number of calories available for individual countries. Based on such extrapolations major world organizations dealing with food and population indicate a total malnourished population of between 500 million and 1 billion. Regardless of whether one accepts the high or the low estimate, the total number of malnourished individuals is an international tragedy.

This tragedy is compounded because it is the very young and the very old who suffer most from the lack of food. It is estimated that nearly three-fourths of all protein and calorie malnutrition is concentrated in Asia. This malnutrition profoundly affects the children born into the less industrialized countries. In 1988 an estimated 12 million children died from readily preventable diseases associated with malnutrition. Meanwhile, world cereal production in 1987 was 1,886,050,000 tons (1,711,024,500 metric tons) and world population was 4,762,000,000. If the cereal grains were uniformly distributed, each person on the earth could have 2.07 pounds (.94 kilogram) of grain or about 3,000 calories per day just in cereal grains without including other food like beans, vegetables, nuts, and fruits. Since only 2300 calories are believed

The role of rice in the diets and lives of a large portion of Asia's population is reflected in this scene of transplanting rice in India.

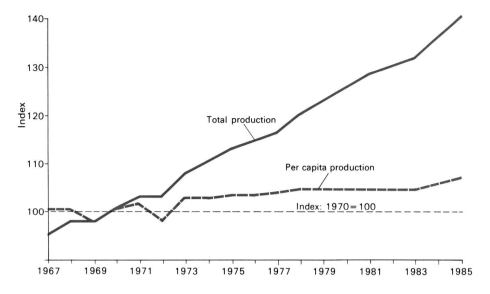

Figure 11.5 Growth of world population means that per capita increases in food production are lower than total production increases.

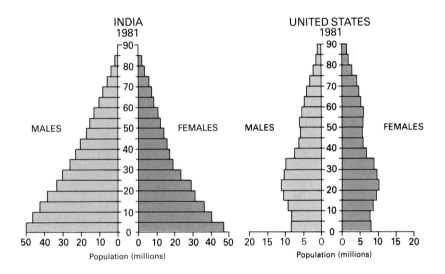

Figure 11.6 The population growth of India causes the typical pyramid shape associated with rapid population growth. The United States represents a decline in growth rate.

to be needed for an adult engaged in moderately strenuous activity, on a world scale there is enough grain and other foods produced to provide the world's peoples an abundance of food. The point is that the food and population dilemma is not one of absolute scarcity of food, but one of distribution. The issue for the world is to ensure that the distribution of the food is adequate for the well-being of the world's peoples, particularly that of the children.

POPULATION, EDUCATION, AND EMPLOYMENT: TOMORROW'S ISSUES TODAY

Assuming that the world is able to produce and distribute enough food to ensure that its population does not go hungry, the population of Asia and other rapidly growing areas presents other issues. Analysis of the population pyramid of India (Figure 11.6) reveals that, as is typical in countries in stage 2 of the demographic transition, there is a disproportionately high number of individuals age 15 and under. Comparison of the number of individuals in a society who produce more than they consume (generally ages 15 to 64) with those who are basically consumers and are dependent upon the production of others (those under age 15 and the elderly) yields the **dependency ratio.** In India and

---

Dependency ratio: The proportion of a population that is either too young or too old to be economically self-sustaining or produce a surplus. Generally the young or incapacitated old are therefore dependent on others for support.

ASIA AND THE ORIENTAL WORLD

Students in an elementary school near Shanghai in China reflect the increase in literacy that has occurred in Asia in the last few decades.

other less industrialized Asian countries as much as 50 percent of the population may be under age 15. The needs of these individuals must be met from the surplus generated by the other half of the population. By comparison, in industrial countries the pre-fifteen age group usually constitutes less than 30 percent of the total population. The dilemma of a large proportion of children for the developing countries is obvious. This massive segment of the population requires food, clothing, shelter, medical care, and education. Because they require these things today, the country's productive capability is unable to accumulate a surplus to increase the standard of livelihood for the productive sector of the economy. In practice this means that most less industrialized countries are literally running as fast as they can in order to stand still. More tragically, their best efforts may not be able to maintain the status quo, and the standard of living may actually decrease through time as a result of the burgeoning population. Since most Asian countries are unable to meet all of the needs of their growing population, the result is a lack of adequate nutrition for the younger age groups, resulting in higher infant mortality rates (Table 11.1). In addition, countries facing the tremendous total population growth of India, Indonesia, or China find it almost impossible to overcome the literacy gap through conventional methods.

A second problem of the ever-growing population of countries like India is employment. As the countries faced with population growth resolve the problems of high infant mortality, ever-growing numbers survive and enter the work force. In 1990, the developed countries had an estimated total work force of 625 million workers. This is projected to increase to 700

TABLE 11.1 Infant Mortality Rates of Representative Asian countries (1988)

Country	Rate (per thousand)
Bangladesh	135
Burma	103
India	104
Japan	5.2
Korea (North)	33
Korea (South)	30
Laos	134
Singapore	9.4
Taiwan	6.9

SOURCE: *1988 World Population Data Sheet* (Washington, D.C.: Population Reference Bureau, Inc., 1988).

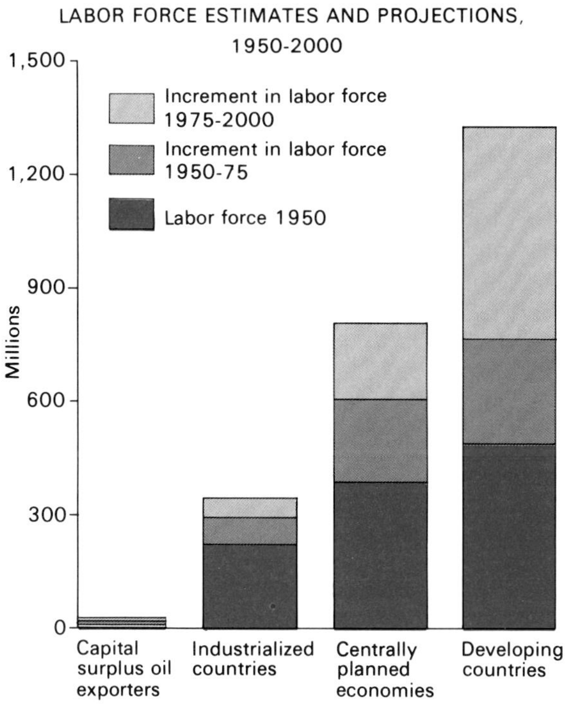

Figure 11.7 The challenge to the developing countries of providing employment for a rapidly growing population is one of the most important issues facing the world. (Source: *Finance and Development*, September 1979, p. 37. [World Bank, New York].)

million by 1995. In 1990 in the less industrial countries there were nearly 1.7 billion people in the work force, with a projected increase to 2 billion by 1995. Thus, India, China, Indonesia, Bangladesh, and Pakistan need to provide jobs for nearly one-fifth as many workers as they presently have—in only half of a decade. Not surprisingly, employment in the less industrialized countries is projected to increase at a much slower rate than the labor force (Figure 11.7). Nearly 18 percent of the labor force in less industrialized countries was unemployed in 1990, and an additional one-third was **underemployed.**

Although the figures are estimates, they point up the harsh realities of the issue of population and jobs. The dilemma of providing jobs is compounded by the need to catch up in the areas of literacy, housing, food production, agricultural modernization, and all the other services. Faced with the dual demands for basic human needs, the countries of the Orient must make a difficult decision as to whether they would be better served by investment in industrialization or in improving education and agriculture. Different countries of the Orient have approached this dilemma in different ways, but in every case the problems are exacerbated by the high illiteracy rate and the multitude of problems associated with their teeming millions.

POPULATION MIGRATION: TODAY'S PROBLEM

Multitudes of Asians are being forced from the land, which has traditionally been their livelihood. The birth- and death rates for Sri Lanka (Figure 11.8), are typical of a country in stage 2 of the demographic transition, and illustrates the magnitude of the problem. With an annual growth rate of 1.8 percent in a population of 17.2 million (1990), the agricultural land cannot possibly support all the new laborers. Cities become these people's hope for the future, and the major migrations of population in the Orient are from the villages to the smaller towns, and from the towns to the cities. Of the 25 largest cities in the world, 15 are found in Asia. Calcutta increased from approximately 4.5 million in 1950 to nearly 12 million by 1990. Bombay increased from 2.8 million in 1950 to over 12 million in 1990. The intermediate-sized cities have had an even greater rate of increase. The city of Madras increased from 1.4 million in 1950 to nearly 9 million in 1990. In China, Beijing (Peking)[1] had 4

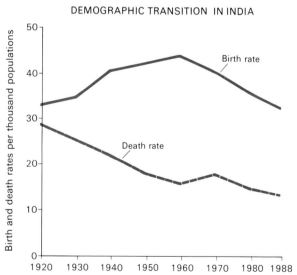

Figure 11.8 The rapid decline in death rates in Asian countries in the twentieth century coupled with maintenance of high birthrates caused rapid population growth, but birthrates have begun to fall in most Asian countries.

Underemployment: Individuals who have only part-time or seasonal work (as in agriculture) or who are engaged in labor that is only marginally productive.

---

[1]Chinese place names reflect the simplified Pinyin system of spelling, with traditional western spellings in parentheses. Chapter 13 discusses this system in detail.

THE HUMAN DIMENSION

Notes of an Agency for International Development Adviser in the Philippines

Olongapo City (population 193,539— 40 percent poor)

In Purok 6 of Pagasa Barangay, 129 families live on the garbage refuse site. (Not only on the site—but literally on the garbage.) The people live here because there is no rent and they can find enough scrap materials to put up some kind of temporary protection. Several families indicated that they have been here 3 to 5 years.

A number of pigs and rats and huge swarms of flies were helping themselves to the garbage. The stench was breathtaking but the residents reported that it was not so bad this day as the weather was good. On rainy days it is worse, they say.

It is necessary to walk over the fresh garbage that is brought in daily to gain access to many of the houses. The families carry water from a well nearby and pay 5 pesos (10 cents) each month for the privilege.

As soon as the garbage truck drives in, the families swarm out of the hovels and search through the refuse to find usable items such as bottles, plastic containers, clean papers, etc., which they carry to the buyer and receive 1 or 2 pesos per day. With this money they support their families. In some cases, it takes almost a week's income to pay a month's water bill.

One family of ten living in one tiny room said they were earning 2 pesos per day. The 17-year-old girl goes to high school at night. Her sister is pregnant with her third child. The first two babies died. Another sister has one living child. She had recently lost a 3-year-old baby boy. Her little year-old girl looked healthy but her fat appearance was only the bloat that masks the fact that she is critically malnourished. A 6-year-old nephew had scabies that completely covered his entire back and legs, leaving them open, raw, running blood and yellow matter from the untreated infection.

When asked if the public health officials ever came here, the answer was no. Not many of the children here go to school. They have not had immunizations. Those with active tuberculosis are living uncared for with families with young children.

Roxas City (population 74,337— 40 percent poor)

A number of workers were harvesting rice near the daycare center. That land was owned by an absentee farmer who permitted people to harvest on shares—6 parts for the owner, 1 part for the laborer. Five young men standing on the roadway near the field were "waiting for the wind to blow" so they could clean their rice. One lady with a 2-year-old boy was preparing to go into the field. She said she had four children and "hoped" she would have no more. The people interviewed registered no complaints in their area, each year some children were crippled by polio, "a lot" of babies died, and they had problems with drinking water. They said this was an agricultural area but no one raised gardens because of the shortage of water. The people said they did not own their land and could not get loans to improve or increase their family farming because they had no collateral.

Source: Adapted from *War on Hunger* (Washington, D.C.: Agency for International Development, 1977), pp. 26–27.

---

million inhabitants in 1960 and had swollen to nearly 15 million by 1990. Shanghai, one of the world's largest cities, had a population of less than 7 million in 1960 and approximately 16 million in 1990. In Vietnam, Ho Chi Minh City (formerly Saigon) had only 1.4 million in 1960 but over 3.5 million in 1990.

The movement of population to the cities of the Orient compounds the misery of their inhabitants. For many of the migrants the dream of a job, education, and a home is not realized. Illiterate, unskilled, and poor, they are forced to adopt the most menial tasks simply to stay alive. The results are squalid slums within and around all Oriental cities. In India these may take the form of **bustees,** hovels consisting of only a tiny cubicle with a roof made from tin cans or cardboard and partitions of burlap separating it from the next shelter. In Calcutta alone an estimated half million people live in bustees, which have no running water, no sewage facilities, no electricity, no paved streets, no educational opportunities, and little or no police protection. Other people come to the cities and are unable to find sufficient employment to allow them to construct even such a squalid hovel. Known as street people in India, there are more than 100,000 of them in Calcutta alone. They have no formal residence and spend their nights in shop doorways, under bridges, or in public parks. The problems posed by such poverty in Asian cities seem overwhelming to Western observers.

Many cities and countries of the Orient have attempted to solve the problems created by rural-to-urban migration by either formal or informal government policies. In the Communist countries of Vietnam, the People's Republic of China, and Kampuchea

---

Bustee: Any type of unsubstantial housing made from low-cost materials.

the leaders formerly attempted to force people from the cities into the country. The government of Kampuchea reduced the population of the capital city of Phnom Penh from 650,000 to only 50,000 between 1970 and 1979 but in the 1980's its population soared.

THE QUESTION OF OVERPOPULATION

The problems that the Orient's billions of inhabitants present raise the issue of overpopulation. Are there too many people in the Orient? Western observers with their tradition of high individual space requirements, demands for privacy, uncrowded conditions, and emphasis on material possessions are apt to say yes. But the value systems of Oriental cultures are not those of Western cultures, with their space and privacy demands, and individual rights are not as strongly emphasized in the Orient. For most Oriental cultures the question is whether or not there is adequate food to provide a basic livelihood for the population. Because of social and cultural values, the typical resident of the Orient demands less. To have the same impact on the environment in terms of demands for food energy and other resources as one American would take 40 Indians (Figure 11.9).

Nevertheless, even the most optimistic observers recognize that the population of the Orient is so large and is increasing so rapidly that the countries cannot make progress in raising standards of living for their people without curbing population growth. In the last half of the 1970s there began to emerge some indication that some of these countries were beginning to overcome the population explosion (Table 11.2). Singapore, Hong Kong, Indonesia, and Taiwan have all lowered their birthrates between 40 and 50 percent, and even India has cut its birthrate by more than 20 percent. (The problem is not solved, however, since death rates have declined as well.)

The decline in birthrates is a promise of the transition that these countries must make to stage 3 of the demographic transition. The lower birthrate is the result of a variety of programs. Korea has mounted a public education program with mass media techniques designed to convince the people that a family of two is sufficient and has set up free health care clinics for women, providing inexpensive birth control measures. India has tried an array of approaches to limit its population, including mass vasectomy, inexpensive contraceptive distribution, and government grants to

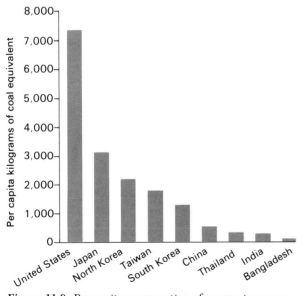

Figure 11.9 Per capita consumption of energy increases with economic development, with the United States having the highest per capita consumption.

TABLE 11.2 Growth Rates in Selected Asian Countries

Country	Average Annual Population Growth (Percent)	
	1965–1980	1980–1985
Bangladesh	2.7	2.6
Burma	2.2	2.0
Bhutan	1.5	2.2
China	2.2	1.2
Hong Kong	2.2	1.4
India	2.3	2.0
Indonesia	2.3	2.1
Japan	1.2	0.7
Korea, North	2.7	2.5
Korea, South	1.9	1.5
Malaysia	2.5	2.5
Nepal	2.4	2.4
Pakistan	3.1	3.1
Philippines	2.8	2.5
Singapore	1.6	1.2
Sri Lanka	1.8	1.4
Thailand	2.7	2.1
Vietnam	3.1	2.6

SOURCE: *World Development Report* (New York: The World Bank, 1987).

people willing to be sterilized. At one time the Indian government threatened to deny civil servants their pension and medical care if they had more than three children. China has adopted the "one-child family" program that provides incentives for couples who agree to have only one child.

As the illustrations show, small countries have had the greatest success in limiting their population growth. The sheer size of a country like India makes population control programs difficult. Nevertheless, the beginnings of the transition to stage 3 of the demographic cycle are showing up. Of greatest hope for the future is the decline in birthrates in Indonesia, Singapore, and India *before* the industrial revolution is complete. Demographers are hopeful that this trend means that countries will be able to move to stage 3 of the demographic transition before industrialization. Countries like India *must* do so because without this change each year's population increase will effectively consume all of the country's productive capability, preventing the additional capital investment needed to increase the standard of living and finance industrialization.

South Korea has cut its growth rate from 3 percent a year to 1.3 percent per year with resulting declines in population migration to the city. Although only a tiny beginning, this slowing of population growth and migration to urban areas is of worldwide significance, because it may signal the change to stage 3 of the demographic transition with its promise for higher educational standards, jobs, incomes, and life expectancy in the less industrialized countries of the Orient.

In discussing the problems presented by the population growth in Asia, we cannot overlook the causes of that population growth, which represent a major human achievement. The rapid rate of population growth in the Orient and the rest of the less industrialized world has come about not through extremely high fertility levels, but through lower infant mortality rates and longer life expectancies. For the world as a whole the average life expectancy is approximately 63 years, and even in the less industrialized countries it is about 60 years. Lest we view this achievement as a disaster, remember that it has allowed most people to live out a fairly full span of years. It has allowed the majority of children born in the world to grow to maturity. The expanding world population poses problems and challenges, but the reduced death rate represents the first step toward a better quality of life and is one of the greatest achievements of the twentieth century.

Similarly, in discussing the population problems of the Orient measurement of poverty by per capita income alone is an inadequate measure of the quality of life. Attempts to provide a more uniform measure of standard of life take several forms, but one of the most famous is the physical quality of life index (PQLI). The PQLI attempts to compare countries on the basis of literacy, infant mortality, and life expectancy. The justification for considering these variables is that they are simple, not ethnocentric, and not reflective of changing purchasing value as the gross national product (GNP) figures are. Ranked in terms of the quality of life index, India and other Asian developing countries do not emerge as poor as simple GNP would indicate. By this measure the less industrialized countries of Asia have made remarkable progress (Table 11.3).

The weakness of this measure of quality of life is that it does not consider the economic aspects of poverty. Although life expectancy has increased, literacy

TABLE 11.3 Growth in Physical Quality of Life (PQLI) and GNP in Selected Asian Countries, 1950–1981

	PQLI			Average Annual Per Capita GNP	
	Circa 1950	Circa 1960	Circa 1970	1985	Growth Rate (%), 1960–1985
India	14	30	40	49	1.7
Malaysia	47	47	67	73	4.4
Philippines	55	60	72	77	2.3
Thailand	55	58	70	79	4.0
Sri Lanka	65	75	80	82	2.9
United States	89	91	93	97	1.7

SOURCE: After Morris David Morris, *Measuring the Condition of the World's Poor* (New York, Pergamon Press, 1979), Table 13, p. 75, and calculations by the authors.

rates have been raised, and infant mortality lowered, the nature of life for the increasingly educated population of the world leaves much to be desired. The slow decline in absolute population growth in the world promises some hope for providing a higher income for the world's millions, but it is important to note that it will take at least 100 years to curb the world's population growth effectively. Since in stage 2 of the demographic transition each successive age group is larger than the preceding one, each successive generation of child-bearing females is also larger. Therefore, in high growth-rate populations the number of girls already born but under marriageable age may be as much as twice as great as the number of women presently in the child-rearing ages. Even if the fertility rate is cut in half, the same number of births will result, since the childbearing population is twice as large. This characteristic of the populations of the Orient ensures that populations in the region will not stabilize at anything less than 4 to 6 billion people.

THE GEOGRAPHICAL BASE OF LIFE IN THE ORIENT

The bulk of the Asian continent is only sparsely populated. The central core of the continent stretches from the deserts of west China through the Siberian Far East of the Soviet Union. This arid inner core ranges in elevation from 4000 feet (1200 meters) above sea level in the Soviet Union to only 300 feet (100 meters) above sea level in the central portion of China. To the south are the Himalayan mountain ranges and a series of complex hills and mountains east of the Himalayas. The Qing Zang (Tibetan) Plateau region is one of the highest populated areas in the world, with valley floors 13,000 feet (4,000 meters) above sea level. Population densities are extremely low in these highland areas (Figure 11.10).

Eastward from the core of the Asian continent lie the lowlands of Huebi (North China), the Dongbei (Manchurian) lowlands, and the North China Plains. The Dongbei (Manchurian) lowland has an elevation of approximately 500 feet (150 meters) above sea level, but suffers from inadequate precipitation. The Hubei (North China) Plain, centered on the Huang He (Hwang Ho) river system and its deltas, is an area of quite level land created by the sediments deposited by the river. The mouth of the Huang He (Hwang Ho) has migrated from north to south and then back, creating a large, extremely flat, delta region. Elevations in the plain range from 500 feet (150 meters) to the west, where it is bordered by mountain ranges, to sea level on the east. The Chang Jiang (Yangtze River) forms a part of the southern margins of the Hubei (North China) Plain, but south of the Chang Jiang (Yangtze) the land becomes more rolling and is broken by a number of river systems.

In southern China and Southeast Asia the lowlands consist of rolling hills separated by important level regions along the major rivers. The largest is the Mekong River Delta of Vietnam, Laos, and Cambodia. South of the Himalayan mountain chain lies the Indus-Ganges lowland of India and Pakistan, created by the Indus and Ganges rivers. The largest lowland area south of the Himalayas, it covers approximately 300,000 square miles (777,000 square kilometers). It stretches 1800 miles (2900 kilometers) from the mountains on the west to the Brahmaputra River on the east, and from the Himalayas to the margin of the Indus Delta some 800 miles (1290 kilometers) to the south. The Indus-Ganges lowland ranges from 500 feet (150 meters) above sea level to sea level, with the drainage divide between the Indus and Ganges rivers west of Delhi reaching over 700 feet (200 meters). Generally this lowland is between 100 and 200 miles (161 to 322 kilometers) wide.

Beyond the continental margins of Asia is a series of island arcs composed of the Japanese islands, the Philippine Islands, the Indonesian islands, and the large continental islands of Borneo and New Guinea. The physical composition of these islands is extremely complex. They are all part of the Pacific ring of fire and are subject to periodic earthquakes and vulcanism.

ASIA'S RIVERS: HOME FOR ORIENTAL PEOPLES

The great river systems, rising in the mountainous core of the continent, are an important element in the physical geography of continental Asia. Each forms the site of the densely settled core area of an Asian country. Examination of the river systems illustrates both the importance of physical features and population distribution and the role of human decisions in affecting the physical world. The major river systems of Asia include the Indus (Pakistan), the Ganges (India), the Brahmaputra (Bangladesh), the Irrawaddy and Salween (Burma), the Menam Chao Phraya (Thailand), the Mekong (Cambodia and Vietnam), the Red River (Vietnam), the Chang Jiang (Yangtze) and Huang He (Hwang Ho) of the People's Republic of China. The role of these rivers in providing transportation, irrigation water, and alluvial material for fertile agricultural land is of primary importance.

Population densities in the riverine sections of China, India, North Vietnam, Bangladesh, and Pakistan are among the highest in the world. Unfortunately the same factors that make these regions suitable for dense rural population (level land, large rivers

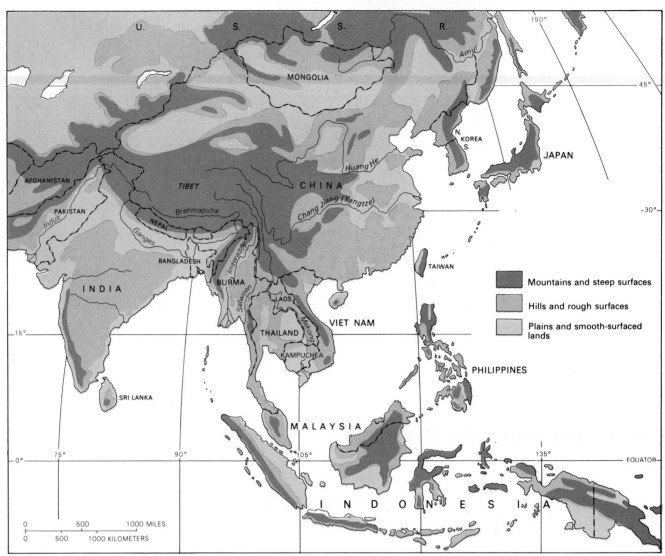

Figure 11.10 Landforms of Asia.

to supply water, and lack of physical barriers) also combine to create periodic tragedies. The most catastrophic was the *tropical cyclone* (hurricane) that struck the Bay of Bengal in 1972. Winds moving north through the Bay of Bengal produced a **storm surge,** which surged over the low-lying delta of the Brahmaputra and covered the region with as much as 10 feet (3 meters) of water. Since the land is only slightly above sea level and there are no areas of higher ground, approximately one-half million people drowned. A similar cyclone in 1985 in the Bay of Bengal resulted in the deaths of scores of thousands more.

In China's Hwang He (Huang Ho) delta, the river has deposited tremendous quantities of silt creating the fertile soils upon which the Chinese people rely.

Storm surge: Flooding caused by a storm that increases the height of the tide.

In the past as the river deposited silt in its bed, it would rise above the surrounding countryside. When spring flooding caused it to break from its channel, it would create a new channel at a lower level. Thus the river has built the relatively flat North China Plain as it has moved its channel north and south over thousands of years. In recent decades the increasing population in the plains has made such flooding prohibitively costly in terms of life, agricultural production, and livelihood. Consequently the river has been restrained within its channels and now occupies a ridge of higher ground with increasing potential for disastrous flooding.

OTHER AREAS OF CONCENTRATED POPULATION

The coastal lowlands are a second area of dense population in the Orient and the island arcs. In these areas of greater population density sedimentary de-

THE WORLD OF POVERTY

posits and/or higher precipitation are conducive to agriculture. In India the Coromandel and Malabar regions occupy this lowland. The Malabar coast on the west varies in width from 5 to 70 miles (8 to 113 kilometers). It is quite level and suitable for production of tropical crops. The eastern Coromandel Coast is very different. The width is as much as 125 miles (201 kilometers), and a number of rivers cross the lowland, creating interlocking deltas. The coastal area of Vietnam and China is densely settled because of the lower elevations and access to transportation and fertile lands. The concentrations of population in Japan are also in coastal regions and plains adjacent to the coast.

Greater population densities are also found on volcanic soils. The island of Java in Indonesia is a prime example. Soils in a tropical area are commonly highly leached, infertile oxisols. In the case of Java, however, the heavy rainfall results in erosion, but the erosion reveals younger materials, which are highly fertile because of their volcanic origin. In this case erosion is beneficial, as it removes the older, leached soil and provides more nutrients for crops.

A final area with higher population densities is found in the region of savanna climate in Southeast Asia. Away from the rivers proper, alternating wet and dry seasons in these areas have resulted in less severe leaching of the soils, and in the rolling hill country there is moderate population density.

Although the population distribution of the Orient seems to reflect the landforms, the landforms have not dictated where people should live. In many areas of the Orient, high population densities are a result of large-scale modification of the environment. In the hills of the Philippines, in the hills and mountains of China, and in the Indus plain of Pakistan, human labor has overcome harsh environments. In Pakistan and northern China irrigation provides for the high population densities. In China, the Philippines, and Indonesia, steep lands that are unsuited for cultivation have been transformed into rice paddies through terracing.

THE CLIMATES OF ASIA

As the world's largest continent, Asia stretches from equatorial to polar latitudes, and climates range from tropical rainforest to subarctic, from desert to tropical savanna (Figure 11.11). The subarctic portion is in the Asiatic portion of the Soviet Union and will not be discussed in this chapter. The bulk of the populated area of the Orient has tropical and subtropical climates. Only in northern China, Korea, and northern Japan are humid continental climates important. Indonesia and the Malay Peninsula have a tropical rain-forest climate with temperatures consistently averaging about 70°F (21°C) throughout the year. Precipitation is rarely less than 20 inches (50 millimeters) per month. The arc stretching from the Indian subcontinent through Southeast Asia has a savanna climate with alternating wet and dry seasons. Temperatures there may exceed 95°F (35°C) during the dry season. The northern part of Vietnam, much of the Ganges plain, and the bulk of southern and southeastern China have a humid subtropical climate with summer temperature maximums of 85°F (26°C), and mild winter temperatures. Precipitation totals range from 30 to 60 inches (750 to 1000 millimeters). The three southernmost islands of Japan are primarily subtropical, with only the northern half of the large island of Honshu and the northern island of Hokkaido having a humid continental climate.

Humid continental climates are also found in the Hubei (North China) Plain, Dongbei (Manchuria), and North Korea, but in total, the area of this climate is much smaller than that of the other climatic types

The ifugao rice terraces of the Philippines were created 2000 years ago and are still used today. The large populations of Asia necessitate using even lands which are seemingly nonarable.

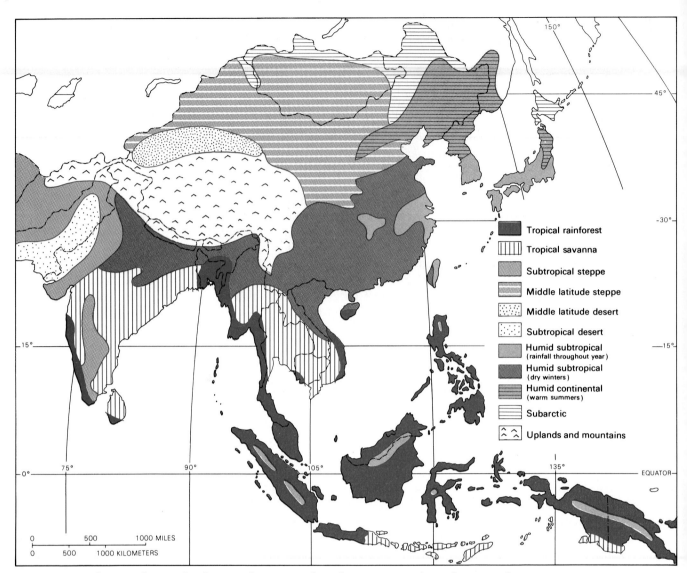

Figure 11.11 Climates of Asia. The climates of Asia range from desert to tropical rain forest to subarctic.

found in Asia. These climates, with their cold, snowy winters and hot, humid summers, contrast greatly with the prevailing tropical and subtropical conditions of Southern Asia.

Much of Dongbei (Manchuria) has a steppe climate with 10 to 16 inches (250 to 400 millimeters) of precipitation per year, a relatively cool summer season, and a cold, dry winter season. The interior of China is desert and steppe; the vast Tarim Basin is the center of the largest desert region of Asia. Another desert area is located in western India and Pakistan where the Indian desert stretches westward to the Indus River.

Vegetation in Asia reflects the climate. In the plains of the steppes are grasslands, which in the past were important for nomadic peoples. The subtropical and tropical climate have forests of such species as teak, mahogany, rubber, palm, and others of warm, humid climates. The savanna climate of Southeast Asia has a mixture of trees and shrub grassland.

THE MONSOON: THE ORIENT'S CLIMATIC DISTINCTION

The climatic pattern of the Orient is greatly modified by the monsoon. The term **monsoon** comes from an Arabic name for a season of the year and the wind that prevails in that period. In Asia the monsoon results in one season of the year in which the prevailing wind blows from the land to the sea and is dry, and

---

Monsoon: It is described as a seasonal reversal of winds, but it also implies differences in precipitation.

THE WORLD OF POVERTY

another season or seasons in which the wind blows from the sea to the land and brings moisture-laden air to the continent. Where this moist air is forced over orographic barriers, heavy precipitation occurs. Cherrapunji in the Assam Hills of India receives an average of 425 inches (10,795 millimeters) of precipitation per year. The result is that the areas with strong monsoonal influences in Asia have marked wet and dry periods. The effects of the monsoon are strongest and most obvious in India. The period of December to March, when the wind is from the land to the sea, is the dry, warm season. In May, as the monsoon begins, the precipitation comes in deluges, and temperatures are slightly lower in the summer season as a result of cloud cover.

The cause of the monsoon is complex, involving land and water relationships in Asia as well as global winds and currents, but the cause of the monsoon is less important than its impact. In India and portions of Southeast Asia and South China where its effect is pronounced, the monsoon is said to be the basis of life. It is sometimes stated that India *is* the monsoon. Although overdrawn, this statement illustrates how greatly the agrarian economy of India relies upon the moisture of the summer monsoon for crop production.

The population of the tropical regions of the Orient that relies upon the monsoon is large, but it does not necessarily constitute overpopulation. Probably even more people could be supported in the subtropical and savanna regions of Southeast Asia and in the tropical regions of Indonesia. In Indonesia only the island of Java has high population densities, although other islands (particularly Sumatra) have the capacity to support increased populations. The area of the Mekong Delta is underpopulated by the standards of the Ganges or Hubei (North China) Plain. Even the Irrawady and Menam valleys could support significantly larger populations by adoption of double cropping, in which a main crop of rice or other cereal grain is followed by a secondary crop the same year. Double cropping is little used in much of subtropical and tropical Asia, suggesting that these areas could support additional population.

THE MINERAL RESOURCES OF ASIA

Mineral resources in Asia are scattered. They include most of the major minerals, but with the exception of China, Asia is relatively deficient in coal. The most important mineral resources of the area are iron ore, tin, tungsten, and petroleum. Iron ore is found in India in several places, particularly west of Calcutta near Jamshedpur, where the world's largest deposit of high-grade ore is located. This iron ore supports India's iron and steel industry, and large quantities are exported, particularly to Japan. Smaller deposits of iron ore are found on the Malay Peninsula and in northern Vietnam. Important deposits of iron ore are found in China, particularly in Dongbei (Manchuria) and in the Sichuan (Szechwan) Basin. The iron ore deposits of Dongbei (Manchuria) are the most highly developed in China and have played a major role in the country's industrialization. Japan also has small deposits of iron ore, important in the past, but today only lower-grade ores remain (Figure 11.12).

Coal is found in significant deposits only in China and Japan. Major deposits occur in Dongbei Manchuria in the great bend of the Hwang He (Huang Ho) River, and in the Sichuan (Szechwan) Basin. The first deposits of coal to be exploited in Japan were on the island of Kyushu, but there are also limited deposits in northern Hokkaido and northern Honshu.

On a world scale one of the most important resources found in Asia is tin. Tin is concentrated in the Malay Peninsula with scattered deposits located as far north as southern China. This tin is an important strategic element, since it is found in only a few sections of the world. The petroleum reserves in Asia are of increasing world importance. Brunei, on the island of Borneo, Indonesia, China, India, and Malaysia have the bulk of the region's petroleum resource. Much of the region remains to be completely explored, and geologic indications are that the North and South China seas may both have extensive petroleum deposits. The region from Vietnam through South China has not been adequately explored, and this region may also be of potential importance for petroleum.

Other important minerals found in Asia include chrome ore in south India and the Philippines, nickel in the Celebes Islands, and copper in India near Jamshedpur. In this less industrialized region, national governments have not yet made concerted efforts to explore their territory fully for mineral wealth. Meanwhile, the limited coal resources have made it difficult for most of the Asian countries to engage in primary metals manufacturing.

THE DIVERSE CULTURES OF ASIA

The diversity of the Orient is nowhere more evident than its cultures. The designation of the entire inhabited region (**ecumene**) as the Orient may give the impression that these are all "Oriental" peoples with a homogeneous culture. Such a conclusion would be

---

Ecumene: Refers to that portion of a region that is inhabited.

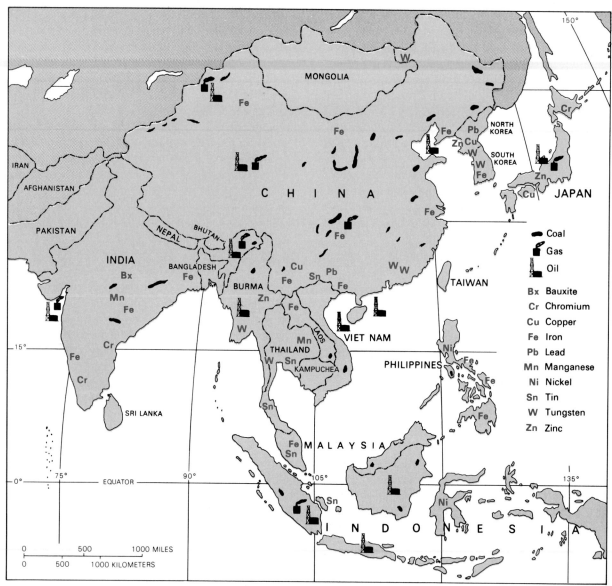

Figure 11.12 Major resources found in Asia.

completely erroneous. As would be expected in such a large landmass, the peoples in the various parts contrast greatly in language, religion, ethnic background, political institutions, legal systems, and economic systems.

ETHNIC ORIGINS

In broad terms we can divide the population of the Orient into the Caucasoid groups of Pakistan, India, and Bangladesh; the Malay groups of Southeast Asia; and the Mongoloid groups of China and East Asia. Each of these groups is separated by transitional groups, and there is little mutual agreement as to which specific ones should be included in each region.

The Indian realm is peopled by Caucasoid peoples from the same so-called racial groups as the Europeans. They have light brown skin, with prominent noses and straight hair similar to that found in the Mediterranean region of Europe. The Japanese, Chinese, and Koreans are Mongoloid. They characteristically have straight black hair, little body hair, a slight body build as compared to Caucasoids, and an **epicanthic** eyefold. The Malay people are found in

---

Epicanthic: The epicanthic eyefold is an extra fold of skin in the inner corner of the eye, resulting in the distinctive eye shape of the Mongoloid people.

THE WORLD OF POVERTY

southeastern Asia and consist of mixing of Caucasoid, Mongoloid, Negroid, and Negrito features.

The actual distribution of peoples in Asia is as complex as the physical geography of the region. Mixing and contact have taken place in this region for hundreds of thousands of years. Almost the only certainty is that northern China and Mongolia is the hearth for the Mongoloid race, which spread outward to people Korea, Japan, and most of Southeast Asia excluding the Malay Peninsula. The Indian population dominates the Ganges and Indus plains and has had a major influence on the southern part of the Indian peninsula. In the southern Indian peninsula are peoples whose ethnic background predates the Caucasoid influx into India. These peoples have more Negroid features, including dark skin and curly hair. The Malay-mixed ethnic groups predominate in the Malay peninsula and the islands of Southeast Asia excluding Japan and New Guinea (Figure 11.13).

The significance of the major population groups lies in problems resulting from ethnocentricity and feelings of ethnic superiority. Unfortunately there has been, and continues to be, conflict engendered by expansion of Mongoloid peoples from China southward, and eastward movement of Indians to Southeast Asia, as well as conflicts between various tribes and states in the region. In the twentieth century nationalism had led to adoption of myths of racial superiority among the Japanese, Koreans, Chinese, Indians, Pakistanis, Khmer (Kampucheans), and Vietnamese. The division into countries has not created states with one distinctive ethnic group, except in Japan and China,

Figure 11.13 Cultural variation in Asia. Asia's population reflects a complex mosaic of races, religions, and languages.

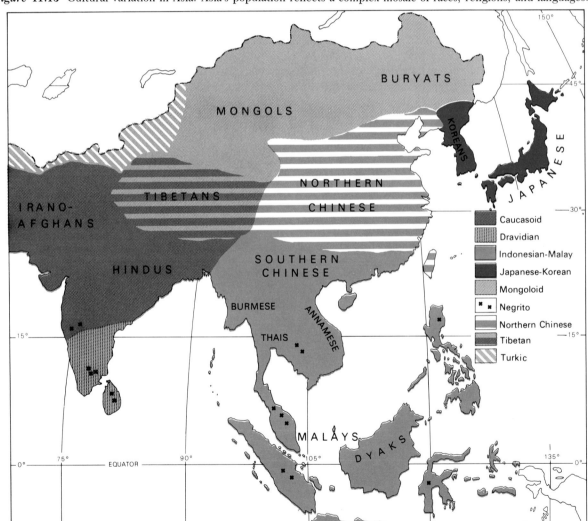

which have nearly homogeneous populations. Unfortunately this homogeneity heightened the Japanese myth of their superiority to other peoples of Asia and contributed to their justification for occupying other Asian countries in World War II.

LANGUAGE DIVERSITY

The language pattern of Asia is as complex as the ethnic pattern. Most of the countries have enclaves of peoples with distinctive languages, and few have a single language that is spoken uniformly. In the British Indian empire alone before World War II, the British government recognized 179 official languages and 544 dialects. Most were spoken by relatively small groups, particularly those isolated in the mountains in the north of India. Even today there are over two dozen languages in India that are spoken by sufficiently large groups to be politically important. Chinese has numerous dialects, and they are not necessarily mutually intelligible. Nonetheless, there has been greater language unity in China because the written language is standard, but there are two major recognizable languages, Mandarin of northern China and Cantonese of the south. In Southeast Asia the language diversity is even greater as would be expected in the fragmented island geography. Indonesia created a new Indonesian language from one of the numerous Malay tongues spoken in the area in an attempt to overcome this diversity.

Broad language divisions are indicated in Table 11.4. Northern China has always been more linguistically homogeneous than southern China where a host of languages are spoken, including Wu, Min, and Hakka. Japanese and Korean are recognized as distinct languages spoken in their respective nations. The Sino-Tibetan languages include Chinese (Mandarin), Cantonese, Vietnamese, and Thai. Other major languages include Indonesian, Dravidian (spoken in southern India), and Hindi and Urdu (spoken in India and Pakistan). In addition, English is widely used as a *lingua franca* in India, the Phillippines, and Indonesia, and

TABLE 11.4 Major Language Groups of Asia and Numbers of Speakers

Language Family and Language	Region Used	Speakers (millions)
Indo-European		
Hindi	India	325
Bengali	Bangladesh, East India	178
Marathi	India	62
Punjabi	India, Pakistan	77
Gujarti	India	37
Urdu	West Pakistan	88
Sino-Tibetan		
Chinese (Mandarin)	China, Taiwan	825
Cantonese	South China, Taiwan	71
Wu	South China	61
Min	South China, Taiwan	45
Hakka	South China	29
Vietnamese	Vietnam	54
Thai	Thailand	46
Dravidian		
Telegu	Indian Peninsula	65
Tamil	Indian Peninsula, Sri Lanka	63
Kannada	Indian Peninsula	40
Malayalam	Indian Peninsula	33
Japanese-Korean		
Japanese	Japan	124
Korean	North and South Korea	68
Austronesian		
Indonesian	Indonesia	63
Javanese	Indonesia	55

SOURCE: *World Almanac*, 1988.

French is a common **lingua franca** in Vietnam and parts of Southeast Asia.

The relative importance of these languages varies from country to country. In Japan essentially the entire population is literate in Japanese, and the same is true of Korean in the two Koreas. In India, Hindi comes closest to being a common tongue, but nearly half of India's total population neither speaks nor writes Hindi. China has standardized its language in the past two decades, and the majority of its population now is fluent in Mandarin Chinese. The use of English and French throughout Asia is common among the educated and elite classes.

RELIGION IN THE ORIENT

Just as the ethnic groups and language groups are complex, oriental religions are also an intermixture of Eastern religions and values modified in some areas by Christian missionary efforts. The distribution of religion is less complex than either the distribution of ethnic groups or languages because of the broader adoption of the major religions. Religions can be classified as tribal, compound-ethnic, universalizing, or those of complex civilizations. Tribal religions are highly correlated with specific tribal groups, like certain religions in New Guinea tribes. Compound-ethnic religions include an entire nation as in the case of Israel and the Jews. In compound-ethnic religions, religion reinforces the sense of identity as a unique people, and membership in the religion comes through birth or an extended educational process. Universalizing religions are based upon the major tenet that their religious view is applicable to all human beings. In Asia one can find all these categories, but the compound, complex religions associated with a major civilization predominate. The major religious groups include Islam, Hinduism, Buddhism, and Confucianism.

There are major differences between these religions and the Christian religions of the West. Of fundamental importance is that they are a way of life in the areas where they are found. The Hindu does not choose selectively among beliefs of the Hindu religion. Its value and belief system permeates every action of the Hindus of India. In similar fashion, a person in Thailand does not consider what aspects of Buddhism to accept; Buddhism is a life-style. It dictates roles in society, the legal system, value systems, and accepted modes of behavior. Certain aspects of Eastern religions are of central importance in national life. These religions do not place the same value on human life as Western religions and do not accept the concept that all people are created equal. The result is social stratification to an extent unknown in American society and an attitude toward human life that Western observers mistakenly label casual. Equally important are the fixed and immutable boundaries encompassing the life of an individual that these religions have formed. It is difficult for a member of a lower class to move to a higher class, in India for example, and the relative levels of the social structure were traditionally based on occupation and associated ritual purity. Religious leaders, the educated elite, the military class, and merchants have traditionally been at the top of the social structure, and farmers and those who perform the most menial tasks, at the bottom.

The distribution of the religions of the Orient includes Islam in Pakistan, Bangladesh, Malaysia, and Indonesia; Hinduism in India and scattered exclaves in Southeast Asia (including Indonesia) where Indians have migrated; Buddhism throughout Southeast Asia, Japan, and Korea; and what is known as the Chinese religion composed of Buddhism, Confucianism, and Taoism in China. In addition to these Eastern religions, Christianity is found, particularly in the Philippines where Spanish colonial dominance resulted in the implantation of the Catholic faith. There are also Christian missions and Christian minorities in essentially all countries of Asia.

HINDUISM

Hinduism is a complex of beliefs and ritual behavior; it is not a religion in the sense of having a formal dogma that is taught or a specific founder. Practices in observance of holy days, ritual behavior, visits to local temples or holy men, and the importance of offerings to the pantheon of Hindu gods vary from place to place. There are, however, some common threads that unite the Hindus, including a central god known as Shiva. Of even greater importance in its impact on the culture of India is the social stratification of the *caste system*. The caste system divides the Hindu population into a multitude of separate status groups defined by a variety of criteria. Central to the definition of caste is ritual purity. Thus, those who have anything to do with pollutions, including death or decaying matter, rank at the bottom of the caste system. The Brahman caste occupies the highest level, and at the bottom are the untouchables. One is born into a caste, and moving out of it is almost impossible. As an example, there are presently some 120 million untouchables (*harijans*) in India, and although the designation "untouchable" was outlawed when India

---

Lingua franca: A language used over a wide area as a means of communication between peoples of different languages.

The Ganges River of India is an integral part of the Hindu religion. Pilgrims from all over India journey to Varnasi (formerly Benares) to bathe in the river, to cremate their dead and scatter the ashes in the river, or to obtain water.

became independent in 1947, they still live in separate communities, drink from separate water systems, attend separate schools (if they attend school), and in general are completely segregated, particularly in the rural areas where change proceeds slowly. Only in the cities are harijans accepted as something other than an object of avoidance.

Central to the Hindu belief system is that all creatures on earth are arranged in a hierarchical order, with humans at the top of the order, and the Brahmans at the top of the humans. As part of their belief in reincarnation, Hindus believe that it is possible to break out of the cycle of birth, death and rebirth, and move into *nirvana,* or life with the gods. In order to advance in the next life, one must be a good person and serve one's caste to the best of one's ability. In practice, it is possible to advance in the next life either by following the path of duty to the caste to which you belong; or by withdrawal, mediation, and devotion to the gods; or by pursuing knowledge.

This belief in reincarnation helps explain the relative lack of public concern for the lot of the harijans. An individual's present status is a function of past deeds; therefore, the harijans have received their just reward, and to interfere would be to interfere with the eternal order of values. This also helps to explain attitudes toward cattle and other forms of life, since all are viewed as part of the continuum upon which people are found. This explanation and description of Hinduism are oversimplified, and as noted earlier there are vast differences in the extent to which the traditional value systems guide the lives of the Indian people. But it is safe to say that the vast majority of India's population, agrarian in nature, is bound to the caste system through tradition, social pressure, and belief.

BUDDHISM

Buddhism owes its origin to Gautama, who was born in the foothills of the Himalayas in Nepal in the sixth century B.C. Residing in the middle Ganges region, he became Buddha (the enlightened one) when he recognized that the path of salvation is based on the four truths. These truths are, first, recognition that life is full of suffering; second, awareness that desire is the cause of suffering; third, that happiness and satisfaction (the end of suffering) come from overcoming desires; and fourth, that proper conduct, including honesty, forgiveness, compassion, and consideration, is the means of overcoming cravings and desire. Buddhism was an attempt to reform the Hindu belief system with its emphasis on numerous gods, social stratification, and human suffering. In India, it was ultimately absorbed into the Hindu belief system, and Buddha became one of the gods. It spread into China, Japan, Korea, and Southeast Asia where it remains a dominant value system. Its importance is in its emphasis on compassion toward other individuals and honesty in interpersonal relationships. It was the state religion in Burma and Thailand; and in Laos and Kampuchea prior to Communist control of those governments.

In China, the teachings of Buddha were merged with those of Confucius who lived at approximately the same time. Confucius taught the importance of proper relationships between father and son, employer and employee, and ruler and subject and emphasized the need to treat others as they would like to be treated themselves. Confucianism became the state religion in China several hundred years before Christ and led to the development of a meritocracy, based on the concept that only the most qualified should have positions of authority. While Confucius was developing his highly formalized code of ethical conduct, Taoism was developed under the teachings of Lao-tse. Taoism teaches the need for harmony with the universe and for simplicity and tranquility in life. Buddhism mingled with both Taoism and Confucianism to create the unique Chinese belief system that dominated China until the Communist revolution of 1949.

Japan has an ethnic religion known as Shintoism, which is a mixture of respect for ancestors and nature worship. Buddhism dominated Japan until the nineteenth century when Shintoism emerged as a state religion and become a centripetal force in the emerging Japanese nation. At present, approximately one-sixth of Japan's 123 million people are Buddhists only, and the balance combine both Shintoism and Buddhism.

The significance of the religions of Asia lies in their impact on the social order, laws, and attitudes toward environment and life. In China, the influence of Confucianism affected the selection of individuals for government service. At the same time, Taoism emphasized the need to live in harmony with the environment in contrast to the Western Christian world view that exploiting the environment is the natural goal of the human population. In India, the caste system and the Hindu belief in reincarnation have combined to create a sense of fatalism in the Indian population, which has been inimical to change and modernization. In China, Korea, and Japan the influence of Confucianism on proper interpersonal relationships has led to the adoption of reverence for ancestors, which Westerners usually erroneously call ancestor worship. This attitude has been particularly important in China and Korea in fostering the extended family in which the grandparents maintain authority and importance long after their economically productive life is over. The addition of communism to the Asian religions in the post–World War II period has not ended these value systems but has grafted a new set of values onto them.

THE CULTURALLY DIVERSE ASIAN WORLD

The great variation in language and religion found in Asia, coupled with several thousand years of occupance, has resulted in a complex mixing of peoples, religions, and nations. The heterogeneous society of Asia presents several problems. In India there are at least a dozen major language groups, all normally corresponding to cultural uniqueness and awareness of regional difference. Even in Japan, Okinawans do not necessarily view themselves as Japanese, and the Japanese of the main islands certainly do not view Okinawans as their equals. Some indication of the heterogeneity of Asia is illustrated in Figure 11.14. As can be seen, the Chinese peoples have migrated

Buddhist monks in Thailand. There are full-time monks, but in addition, all men and boys may enter Buddhist monasteries for a few weeks or longer, once or several times during their lives.

Figure 11.14 Concentrations of Chinese in Southeast Asia.

throughout Southeast Asia, where they represent significant minority groups.

The problems created by the Chinese who live outside China are twofold. First, the distribution of Chinese results in a **plural society.** In the plural societies of Southeast Asia there is often a strong sense of ethnic distinction among individuals occupying the same city. In Malaysia for example, there are three major ethnic groups, the Chinese (36 percent), the Tamils from India (10 percent), and Malays (46 percent). The Chinese have a long tradition of individual industriousness, competition, and private entrepreneurship. Consequently, they often become businessmen, shop owners, bankers, or other types of individual entrepreneurs. The Malays are the largest ethnic group, are Muslims, and as a group do not have the same tradition of competitiveness and individualism. The Malays are typically found in agrarian villages engaged in small-scale agriculture or in government and the professions of law, education, medicine, or in areas other than business. The Tamils from southern India were brought in by colonial powers to serve as laborers in construction work and on the plantations. Typically, they work in mines, on plantations, and in construction. The conflicts between the Chinese culture, Indian culture, and Islamic culture of the majority occasionally result in race riots. In addition, the political powers of the various groups are not equal, and many of the Chinese cannot vote by Malay law. The Malay majority insist that the Malay language is the official language although the Chinese view it as inferior and would prefer Chinese or English. The potential for conflict in a plural society like this is ever present.

As a consequence of the widespread distribution of Chinese in Southeast Asia and their concentration in the economic area of business, they have sometimes been referred to as the "Jews of Asia." They also frequently become the focus of hostility in the developing countries as the indigenous population attempts to improve its standard of living. The problem is compounded when these countries try to form a state from previously nonorganized colonial territories. The result in some cases has been tragic for the Chinese or other ethnic groups who make up the culturally pluralistic society. In Vietnam, for example, the Vietnamese government discriminated against the Chinese after 1979, causing hundreds of thousands of Chinese to flee. The wealthiest nation of Asia, Japan, has been reluctant to accept any of the Chinese boat people. Thus continues a 500-year history of repeated pogroms against various ethnic minorities in the plural societies of Asia. In 1989 the Vietnamese lifted some of the restrictions against its Chinese minority, recognizing the country's need for their financial and entrepreneurial skills.)

Polyethnicism poses a slightly different but related problem for the countries of Asia. Polyethnic states are those with several ethnic groups, each occupying a unique territory, within their borders. A distinct ethnic group may or may not support the political state; if it does not, it may cause problems. Regional consciousness may exert more influence than national consciousness. India, and to a lesser extent China, nations with a long history and a well-developed civilization, are polyethnic countries. If an ethnic group occupies a border area adjacent to the homeland of the group, it may become an irredenta. Irredenta have been the justification for conflicts in all areas of the world, and in Asia widespread clusters of Chinese throughout Southeast Asia concern the emerging gov-

---

Plural society: In a plural society various ethnic groups live intermingled throughout a country, although often occupying distinct residential sections of cities. Each ethnic group recognizes its own distinctive cultural traits, icons, values, and life-styles.

Polyethnicism: Several ethnic groups within a country's borders, each group occupying a distinct territory.

ernments lest they become the justification for Chinese attempts to occupy territory. Examples of irredenta that have resulted in conflict include North Vietnam's claim to South Vietnam, mainland China's claim to Taiwan, Vietnam's claim to part of Kampuchea, India's and Pakistan's conflict over Kashmir, and a number of others.

The problems of irredenta, polyethnic makeup, and cultural pluralism in Asia are compounded by the historical development of the region in the past 200 years as part of the colonial empires of Europe. European colonial powers found in Asia a land of valuable raw materials and foodstuffs for their populations. European powers dominated this area for 200 years and compounded the problems of statehood and the emergence of strong political states.

Colonial activity by European powers in Asia took several forms. In China, the colonial impact was concentrated in major cities such as Shanghai and Guangzhou (Canton) where foreign powers levied demands upon the local population for trade goods. In India, it took the form of an uneasy amalgam of Hindu, Muslim, and Occidental. British governors were appointed for all regions of India, but they served with the local government leaders. Colonization by European settlers was never important in either India or China. In Southeast Asia, colonization took the form of actively developing the region for rubber, rice, and tea plantations, as well as mining franchises to obtain tin. In Southeast Asia, the colonial occupation contributed to both a plural society and polyethnicity as labor was imported for construction and operation of mines and plantations.

Throughout Asia, European colonial rule resulted in significant cultural changes and delayed the emergence of these nations as industrial and political powers in their own right. In India, the European dominance by the British resulted in grafting onto the Hindu system a civil servant class, modeled after the British government. When India became independent after World War II, the peaceful transition from British to Indian rule did not eliminate this civil service ethic, with the result that India has a distinctive distribution of the labor force between primary, secondary, and tertiary activities. Nearly 70 percent of India's population is engaged in agriculture, and an additional 20 percent is engaged in government service. Indian families traditionally have at least some of their children obtain an education to take jobs in the government service. Development in the business and economic realm has suffered, as jobs in these sectors have been viewed as having lower status.

In Southeast Asia the fragmentation that resulted from colonial occupance has resulted in continual conflict since 1939. The Japanese attempted to expand their colonial empire and succeeded in occupying parts of China and Southeast Asia from 1939 to 1944. The European colonial powers encouraged their colonies to undertake guerrilla action against the Japanese occupying forces and utilized these guerrillas in forcing the Japanese from the mainland and the islands back to Japan.

The former colonies, however, did not welcome back their European rulers. First India, then Indonesia, then China, and then Vietnam attempted to gain their own complete independence from Western powers, and the result took a variety of forms. In India, the British government, recognizing the inevitability of Indian independence after the British had been unable to devote adequate attention to them during the war years, granted independence to the Indian empire in 1947. Conflict resulted between India and Pakistan over the division of the empire, and relics of this conflict have persisted to the present. Examples include the conflict between Pakistan and India over the Kashmir, and the civil war between West Pakistan and East Pakistan in the early 1970s, which resulted in the creation of Bangladesh as an independent state.

Indonesia achieved independence after a conflict with the former Dutch colonial power, but resulting conflicts persist today. People from the island of West Timor were used by the Netherlands government as surrogate troops, and those who fought on the Dutch side were promised independence. These peoples continue to nourish the idea of a free West Timor, and skirmishes with the Indonesian government persist.

In Vietnam, the unwillingness of the French to grant independence to their former colonies resulted in a 10-year guerrilla war culminating in the withdrawal of the French. At that time Vietnam was divided into North and South Vietnam, a division unacceptable to the emerging government in North Vietnam. Conflict between the North Vietnamese and South Vietnamese resulted in the involvement of the United States in the tragic ten-year Vietnam War and finally the emergence of one Vietnamese state. At the present time, Vietnam is attempting to expand its influence in Southeast Asia, creating a potentially explosive problem for this region.

The significance of the colonial regime is that it laid the groundwork for many of the conflicts of the present-day Asia. The unwillingness of Western colonial powers to give up their former colonies or their shortsighted support of repressive and unrepresentative governments such as that of Chiang Kai-shek in China have resulted in alienation of Asian countries from the West and caused and prolonged unnecessary conflicts. The present status of Asia should be viewed

THE HUMAN DIMENSION

The Tragedy of Cambodia

The official name of Cambodia is now Kampuchea, but whatever its name, it epitomizes the cultural conflicts in Asia. Rooted in the past, these conflicts are a continuation of conflicts that predate the European colonial powers. Tragic human suffering is often the result as new countries attempt to rectify perceived old injustices or to expand their influence. In the case of Cambodia, from A.D. 800 to 1400 the Khmer civilization ruled most of present-day Thailand, Laos, Cambodia, and Vietnam. Since the fourteenth century the Thais and Vietnamese have repeatedly claimed portions of this empire, including the Mekong Delta and Saigon occupied by the Vietnamese. Present-day Cambodia was alternately ruled by Vietnamese or Thai puppet governments until the Cambodians accepted French political domination in 1863 to prevent continued interference by Vietnam and Thailand (Figure 11.15).

Following independence in 1953, Cambodia began a process of modernization, but revolutions in 1970 and 1975 and the Vietnam invasion in 1979 have defeated these efforts. The Pol Pot Communist government came to power in 1975 and began a process that has been described as returning the country to the fifteenth century. Cities were depopulated, and people were forced to return to subsistence agriculture. An estimated 3 million Cambodians were either massacred or died of the rigors of implementing this goal. The Vietnamese Communist government viewed the Pol Pot government and the Cambodians as inferior, and in December 1978 Vietnam invaded Cambodia. The surviving 4.7 million Cambodians experienced a year and a half of conflict in which few crops were planted or harvested. Tens of thousands of Cambodians fled to Thailand, tens of thousands more died of starvation, and only an international effort to deliver food from late 1979 to 1980 saved the remainder. What the future holds for the remnants of Cambodia's people is unclear, but their suffering is not over yet, since their country was occupied by the Vietnamese for more than 10 years. As late as 1989 the Vietnamese were negotiating their withdrawal, but the supporters of the Pol Pot regime were poised to attempt to regain control and commence their agenda for the country.

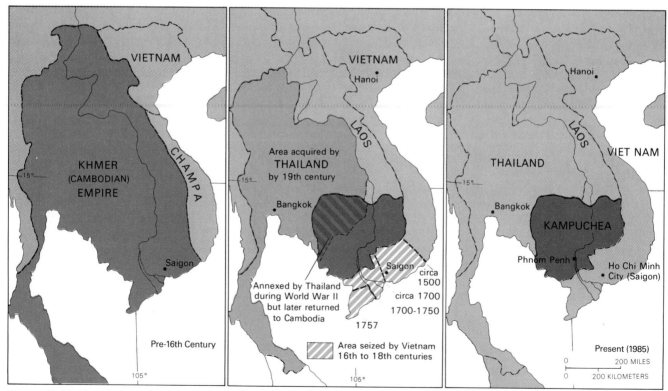

Figure 11.15 The changing boundaries in Southeast Asia reflect the historic conflict between today's Thailand, Khmer Republic (Laos) and Vietnam.

as one of flux. The emergence of Asian nations as independent countries has occurred only in the past thirty years, a short time period in which to expect stable political units. Consequently, boundary changes, conflicts, migrations of peoples, and continued tragedy and suffering for the population of Asia will probably continue. Hindered by colonial occupiers in developing centralized governments, Asian countries now face the problem of uniting diverse cultures into viable political units. Unfortunately, in areas without a strong tradition of democratic government, totalitarian centralized governments are apparently the only ones capable of leading such a transition.

THE PEASANT MASSES OF THE ORIENT

The economy and culture of the vast majority of the Asian population compound the problems of emerging countries and political units. Like other less industrialized countries, those in the Orient have a disproportionate share of their populace engaged in agriculture. It is estimated that in Bangladesh, for example, 75 percent of the population is engaged in agriculture, in China, 65 percent, in India, 70 percent, and in Indonesia, 63 percent. This presents problems as countries attempt to unify their national territory into a strong state. The villagers are characterized by low levels of literacy, strong social mores determining actions and roles in the society, and the general poverty of lack of economic development. The individual villager is oriented to the village first and possibly to an ethnic group or specific region but rarely to the central government in the capital city. Developing a strong sense of national identity in the emerging countries of the Orient is consequently difficult.

Village life in the Orient also makes progress in literacy and economic development difficult. In the absence of external force, villagers are normally content to continue in their traditional ways of life, reluctant to see female children educated, and resistant to change. Resistance to change does not represent a rural ethic based on ignorance. Rather, it is the intuitive recognition that change will destroy a system that has served to maintain life for the past 3000 years. Western observers erroneously conclude that the reluctance of Asian villagers to adopt new grain varieties, new cultivation techniques, or other Western innovations is the result of ignorance and suspicion of Western ways ingrained in an agricultural people. The real cause for this reluctance is a very pragmatic assessment of the element of risk involved. Villagers of Asia operate at the margins of existence. Agricultural production from one year is normally only sufficient to maintain the family through the succeeding non-harvest season and provide sufficient seed to plant the next year's crop. Savings cannot be expended on risky ventures that may or may not return an adequate yield. Faced with an existence in which the present system at least has in the past allowed sufficient production to maintain life, the peasant farmers of India, Thailand, and Indonesia are reluctant to adopt unproven new ways promising higher yields. Changes that do not absolutely guarantee at least an equal yield cannot be undertaken in a society in which failure to equal at least present production may mean famine and starvation. Consequently, the reluctance of some Asian farmers to undertake changes is a valid response to the high risk involved. When it is obvious that proposed changes will result in higher yields or a better standard of living, the Asian peasants are eager to adopt them.

THE GREEN REVOLUTION

An example of this adoption is the so-called Green Revolution, involving hybrid varieties of rice and wheat that under ideal conditions will yield as much as three times the original varieties. The Green Revolution had its origins in dual research undertaken in China in the 1960s and in the Philippines and Mexico shortly after. The most famous results came from the International Rice Research Institute in the Philippines, which developed the famous IR-8 variety of rice. This new variety was unlike that produced in India and Southeast Asia. The native Asian varieties of rice were selected through years to maximize the minimum yield. That is, they will produce something even in years of drought, on infertile soils, and in competition with weeds and insects. Unfortunately, these varieties do not respond well to additional fertilizer or water. Yields go up under ideal situations, but only marginally. The IR-8 rice, when given adequate water and fertilization, will yield up to three times as much as the native rices. The only drawback to the hybrid varieties is that without adequate water and fertilization, they yield less than the native varieties. The wheat grown in Asia has the same characteristics. Research in northern Mexico resulted in a hybrid wheat that is shorter in height and in which the bulk of the plant's energies is utilized to produce grain rather than straw. Both hybrid rice and hybrid wheat have been adopted wholeheartedly in Asia. It is estimated that at least three-fourths of the area sown to rice and wheat in India is now sown to the hybrid varieties. India had been unable to produce sufficient grain to feed its population in the 1960s, but it now produces a small surplus of grain in spite of a dramatically increasing population (Table 11.5).

The doubling of yields has not been accomplished without cost. Consumption of artificial fertilizer in India increased from 657,000 tons (595,900 metric tons) in 1952 to 10,021,000 tons (9,110,000 metric tons) in 1986. The requirements of the hybrid varieties for fertilizer to produce the high yields have been the basis for criticism of the Green Revolution. Some observers maintain that the profits possible with high-yielding grains encourage landlords in India and other less developed countries to replace tenant farmers with mechanized, large-scale farming. Critics conclude that although there is more food as a result, the human suffering involved in destroying the age-old tenant system is high.

AGRICULTURAL SYSTEMS IN ASIA

There are three major agricultural systems in Asia: intensive cultivation of crops, plantation agriculture, and shifting agriculture. Intensive cultivation of crops supports the majority of the population in India, where, as in the rest of Asia, tiny "mini-farms" produce rice mostly by hand labor. Typically rice is planted in a seed bed where it germinates, grows to a height of 8 to 12 inches (20 to 30 centimeters), and then is transplanted by hand to the larger growing field (paddy). This system is complex but highly efficient in terms of energy expenditure versus caloric yield (Figure 11.16).

Under ideal conditions the irrigated rice in paddies can produce more calories per acre than any other domesticated crop. At least 60 percent of Asia's population is directly involved in either growing, processing, transporting, or distributing rice. It is truly the staff of life throughout Asia. Farms are small, with individual rice paddies often less than a half acre (.2 hectare) in size. The labor-intensive cultivation of rice and allied crops in Asia is sometimes referred to as a garden type of agriculture.

The second system is **plantation** agriculture. Largely concentrated in the lowlands of India and Southeast Asia, the plantations represent the imprint of the colonial powers. Normally they concentrate on a cash crop such as rubber, tea, or coconuts. The Dutch also established a few rice plantations to provide food for the workers in their rubber or coconut plantations, and the Green Revolution is leading to large mechanized farms elsewhere, as in India. The plantation system is completely different from most garden-type agriculture found in Asia. Plantations are normally large holdings (often more than 1000 acres, or 405 hectares), they may include mechanization of certain facets of the operation, and there is often an ethnic distinction between the Asians who do the labor and European managers. With political independence of Southeast Asia, the managerial function has often remained in the hands of Europeans hired by the country, which has taken over the plantations. Increasingly, however, residents of the newly independent countries are moving into the managerial level. Plantations normally produce for export to the industrialized world.

A final broad category of agricultural systems found in Asia is that of shifting agriculture or slash and burn. *Shifting agriculture* is found throughout the less industrialized world and goes by a number of names including swidden, milpa, and ladang. Regardless of

---

Plantation: A large estate owned by an individual family or corporation concentrating on large-scale production of a cash crop for export from the tropics.

TABLE 11.5 Changes in Grain Yields and Total Production

Grain (country)	Area Harvested in Millions of Acres (hectares)			Yield in Tons per Acre (metric tons per hectare)			Production in Millions of Tons (millions of metric tons)		
	1960	1970	1985	1960	1970	1985	1960	1970	1985
Rice	18.2	20.4	23.5	.90	1.05	1.64	16.4	21.2	42.5
(Indonesia)	(7.28)	(8.1)	(9.5)	(2.05)	(2.38)	(4.05)	(14.9)	(19.3)	(38.6)
Wheat	12.3	15.6	18.3	.36	.53	.65	4.4	8.3	12.8
(Pakistan)	(4.9)	(6.2)	(7.3)	(.81)	(1.20)	(1.59)	4.0	7.4	(11.6)

SOURCE: *Production Yearbook* (Rome: Food and Agriculture Organization of the United Nations, 1962, 1972, 1985) and *Foreign Agricultural Circular* (Washington, D.C.: U.S. Department of Agriculture, February 1987).

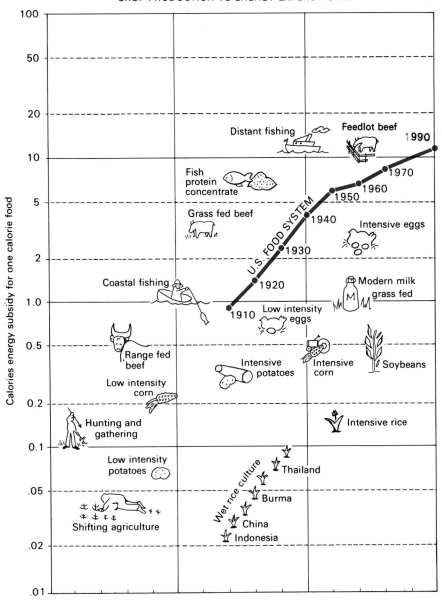

Figure 11.16 (Source: Carol E. Steinhart and John S. Steinhart, *Energy Sources, Use, and Role in Human Affairs*, Wadsworth Publishing Company, Inc., Belmount, Calif., 1974, p. 84.)

the local name, shifting agriculture involves rotation of the land rather than the rotation of crops. Concentrated in the savanna and tropical rain-forest lands where the soils are leached of fertility and nutrients are stored in the plants themselves, shifting agriculture consists of the farmer's cutting of an area of 1/2 to 2 acres (.2 to .8 hectare) of undergrowth or trees. This land is allowed to dry and is then burned just before the rainy season commences. Larger logs that do not burn are moved to the edges of the field or into piles, which are later burned again after they have further dried. Holes are then dug and root crops such as cassava or yams planted, interspersed with corn, beans, and other vegetables. This pattern of a variety of crops growing together is known as *intercropping*, or **polyculture**. Such polyculture is the opposite of the *monoculture* of the great wheat fields of the midwestern United States. Although such a system may appear inefficient to the casual observer, shifting agriculture is actually a highly sophisticated and practical adjustment to the problems of utilizing tropical environments. Because the nutrients are stored in the

Polyculture: A farming system in which numerous types of crops are grown together. Also known as intercropping.

Rice: The Staff of Life

With nearly 2 billion people depending on it as a staple in their diet, rice may be the single most important agricultural crop in the world. Despite its international significance, rice remains something of a mystery crop to the people in the United States who yearly consume only 10 pounds per capita, about half as much as the yearly consumption of ice cream. Two out of every three tons of U.S. rice are sold overseas, making the United States the second leading exporter with about 20 percent of world trade. While the United States is a major rice exporter, its export of rice accounts for less than 2 percent of the world's production.

Harvested on about 346 million acres (140 million hectares) throughout the world, world rice production in recent years exceeded 400 million tons of rough rice or 275 million tons of milled rice. The vast majority of this rice is consumed by farm families. In less industrial countries, where over half of the world's rice is produced, probably only one-fourth enters commercial channels. About 12 million tons or less than 5 percent of the rice produced is traded internationally. Approximately 90 percent of the world's rice crop is produced in Asia. Almost half of this crop is unirrigated, so the delicate balance between world rice supplies and demand depends crucially on the Asian monsoon.

There are distinct markets based on different rice types, quantities and methods of processing. Preferences for different rices are so strong in some countries that consumers refuse to eat all but their familiar quality.

Glutinous rice (also known as waxy or sweet rice) when cooked, forms a gelatinlike mass without distinct grain separation. Most rice-consuming areas in Asia produce small amounts of glutinous rice for use in desserts, ceremonial foods, and sweet dishes. In northeast Thailand and Laos, however, it is the staple food.

Aromatic or scented rice is grown mostly in the Punjab area of central Pakistan and northern India and is called basmati rice. Small quantities of it are also grown in Thailand and sold principally to Hong Kong and Singapore. The volume of aromatic rice traded is limited to about 300,000–400,000 tons annually, and it is sold at prices roughly double that of high-quality, long-grain rice. When cooked, basmati rice grains elongate to about twice their original size and remain completely separate.

Japonica rice is semisticky and moist when cooked. This round-shaped grain is grown in Japan, Korea, Taiwan, part of China, Australia, the Mediterranean area, Brazil, and California.

Indica rice is a long-grained rice grown principally in China, south and southeast Asia, and the southern United States. Indica rice when cooked becomes fluffy, and shows high volume expansion and grain separation. Indica and Japonica rice make up the vast majority of rice produced in the world.

Source: Foreign Agriculture, December 1983, pp. 14–17.

---

biomass (plants) itself, burning liberates them as ash and minerals, which are assimilated by the crops planted in the burned-over area. The multiple cropping allows each plant to absorb that portion of the nutrient spectrum required for its growth and ensures maximum use of the minerals in the ashes. The heavy precipitation leaches away the minerals that have been liberated from the biomass within one or two growing seasons, and the people will have to move to a new spot. The result is that a shifting cultivator needs at least 20 acres (8 hectares) to maintain a cycle of planting, followed by reforestation, followed by clearing and planting. So long as population densities were low in tropical Asia, this system was rational. With better medical care the population of shifting cultivators is growing, and farmers are unable to allow the land sufficient time to regrow an adequate vegetative cover to provide the necessary nutrients. This problem is accentuated by pressure from the growing population of sedentary cultivators who are clearing the forest for permanent farming and further restricting the shifting cultivation system. Successfully developing a shifting cultivation cycle that will minimize land requirements and maximize crop production while protecting the remaining rain forest is one of the major challenges facing Asian countries if this sector of their diverse culture and economy is to survive.

In all the farming systems upon which Asians rely for food, the emphasis is on hand labor. Exceptions are found in Japan and to a lesser extent in the newly industrializing countries of South Korea, Singapore, Hong Kong, and Taiwan where industrialization has drawn labor from the rural sector and led to mechanization. Mechanization of agriculture in Japan is not identical to that in the United States, which has large farms and large equipment. Because Asian farms are fragmented and small, mechanization in Japan has taken the form of small garden tillers, which allow an individual to cultivate more land more rapidly. The system remains a basically garden-type, intensive agriculture. Productivity per acre varies widely across Asia, with highest yields in Japan and the newly industrializing countries, followed by China, with India last. This is partially the result of cropping techniques, partially the result of crop varieties, and partially the result of the environment, particularly in rice production, since India has a pronounced monsoon with rainy, cloudy weather during the peak of the growing season. Rice is highly sensitive to light during growth, and the cloud cover minimizes the potential for growth. Japan and China, which have less precipitation and cloud cover during the growing season, will always be able to produce higher yields per acre of rice than India.

ISSUES FOR THE FUTURE

The problem of India's lower rice yields points up the issues and challenges facing Asia. Relying upon the monsoon for precipitation to produce its crops, India finds that the monsoon also handicaps maximization of yields. The Green Revolution, with its abundant yields during good years, has seemingly solved India's population and food problem. But reliance upon the monsoon means that eventually there will be a year in which rice production falls dramatically, and costs of fertilizer and energy are leading to consolidation of land in the hands of wealthy farmers. Unless and until India and other countries of Asia can develop an adequate food storage system, they will remain perilously close to the margins of survival. So long as the physical environment plays such a critical role, these peoples will never be fully free of the threat of famine. The continuing increase in interconnectiveness around the world has minimized the specter of famine in the past decade, but should India have a bad crop year at the same time as the Soviet Union, China, and the United States, the outcome could be disastrous.

The population dominance of Asia in the world presents it with the opportunity to become a major supplier of goods to the Western world. Labor availability in such quantities and at such low prices provides the opportunity for modernizing and raising the standard of living of Asian masses. Ethnocentricity in both Asia and the industrial West threaten the emergence of a true international economic order that would allow the Asian countries to maximize their quality of life. The future issues in Asia will continue to center on the basic questions of population, food, modernization, and cultural conflict. Changes in any of these areas seem to have the potential to occur at a dramatic rate as evidenced by the Green Revolution and China's liberalization of its attitude toward the West. With such potential for rapid change, predicting what the future holds for the population of the Orient is impossible. Nonetheless, the events of the past decade would indicate that population growth rates will continue to decline; political boundaries will change as Vietnam and China consolidate their perceived national areas; and economic development, which has had a very small and fragile beginning, will expand dramatically in the less industrialized regions of Asia. This may be accompanied by a decline in the economic/commercial dominance of Asia by Japan and to a lesser extent Korea, Hong Kong, and Singapore. Asia is in the process of correlating resources, population, and development with the realities of the oriental ecumene. Such harmonization will of necessity involve changes, changes that will often bring suffering to the individual resident of Asia. It is unfortunate that the changes that seem to be taking place do not occur in a more rational manner on a regional scale rather than through individual countries attempting to bring themselves fully into the twentieth century. The contrasting course of development being pursued by India, China, and Japan indicates both the broad range of options being considered by Asian countries and the problems they present.

FURTHER READINGS

EAST, G. et al. *The Changing Map of Asia.* New York: Harper & Row, 1972.

GINSBURG, N. *The Pattern of Asia.* Englewood Cliffs, N.J.: Prentice-Hall, 1958.

GOUROU, P. *Man and Land in the Far East.* New York: Longmans, 1975.

HANKS, L. M. *Rice and Man.* Chicago: Aldine, 1972.

MYRDAL, G. *Asian Drama: An Inquiry into the Poverty of Nations.* 3 vols., New York: Pantheon, 1968.

PANNELL, CLIFTON W. *East Asia: Geographical and Historical Approaches to Foreign Area Studies.* Dubuque, Iowa: Kendall/Hunt, 1983.

RENNER, JOHN, and PEGLER, BRIAN. *Source Book of South East Asian Geography.* London: Longman Paul, 1982.

ROBINSON, H. *Monsoon Asia.* New York: Praeger, 1967.

SAHA, K. R. et al. "The Indian Monsoon and Its Economic Impact." *GeoJournal* 3(2), 1979, pp. 171–178.

SPENCER, J., and THOMAS, W. *Asia, East by South: A Cultural Geography.* 2d ed. New York: Wiley, 1971.

YEH, S.K.H. and LAGUIAN, A. A. *Housing Asia's Millions.* Ottawa, Canada: International Development Research Center, 1979.

Agra, India.

INDIA PROFILE (1990)

Population (in millions)	850	Service	17
Growth Rate (%)	2.0	Energy Consumption Per Capita—	
Years to Double Population	35	pounds (kilograms)	462
Life Expectancy (years)	54		(210)
Infant Mortality Rate		Calories (daily)	2,189
(per thousand)	104	Literacy Rate (%)	38
Per Capita GNP	$275	Eligible Children in Primary	
Percentage of Labor Force		School (%)	90
Employed in:		Percent Urban	25
Agriculture	70	Percent Annual Urban Growth,	
Industry	13	1970–1981	3.7
		1981–1985	3.9

The Indian Subcontinent the democratic revolution and economic development

MAJOR CONCEPTS

Sufferage / Hinduism / Caste system
Salinization / Sikhism / Jainism
Reincarnation / Land tenure / Cottage
industries / Street people / Symbiotic
relationships / Asymmetrical relationships
Ritual purity / Brahmans / Nirvana
Untouchables / Dharma / Karma
Koran / Bustees

IMPORTANT ISSUES

Retardation of development due to colonial domination
Population growth
Land Tenure—ownership and fragmentation
Food resources and distribution
Rural-urban migration

Geographical Characteristics of South Asia

1. The countries of South Asia occupy the Indian subcontinent, and include more than 1 billion people.
2. The physical environment of South Asia includes the towering Himalayan Mountains, the river plains of the Ganges, Brahmaputra and Indus Rivers, and the uncertainty of the Monsoon.
3. India is the world's largest democracy, while Pakistan has experienced both democratic and autocratic governments in its short history.
4. India has some of the most diverse cultural geography in the world, with a multiplicity of languages and religions.
5. In combination, the countries of South Asia rival China in population, and higher rates of increase will make this the most populous region of the world by the end of this century.

The Indian subcontinent (India, Pakistan, Bangladesh, and Sri Lanka) is a part of the less industrialized world as measured by the standard indices (Figure 12.1). It is the single region that residents of the industrial World most commonly associate with the characteristics of the less industrialized realm. Images of overpopulation, poverty, agricultural life-styles, hunger, and deprivation are common stereotypes associated with the Indian subcontinent. Cultural values in India in particular have been widely misunderstood in the West and result in erroneous conclusions like that concerning the so-called sacred cow and its relationship to Indian poverty. For most casual observers from the Western world, India is a collage of stereotypes of cattle roaming the streets, impoverished village peasants, overcrowded cities, and unique dress styles composed of a turban and a wrapped, draped garment. The stereotypes emerge from the popular reporting on the country's problems. The mere size of India's population ensures that it will attract the attention of outside observers, but this size guarantees that simplistic statements and generalizations will inadequately convey the nature and characteristics of the region.

India is the world's second most populous country. At the beginning of the 1990s, it had a population of between 850 and 875 million (depending upon estimates). Bangladesh (115 million) and Pakistan (113 million) push the population of the region to over 1 billion. Problems of the Indian environment have resulted in periodic famines, which catapult India to the attention of the world, but rarely is the country critically assessed except in times of emergency. The image of India in the minds of Westerners is inaccurate in at least one respect and often in several. The area of India is the sixth largest in the world, and regional differences are complex and difficult to convey. As a result, simple conclusions and characterizations by Western observers of India are as likely to be erroneous as to be accurate. At the very least, the typical stereotypes of India as the home of poverty, sacred cattle, and destitute peasants residing in a rigid social structure prevent recognition of the important changes that are occurring within the Indian subcontinent.

India, Pakistan, and Bangladesh are classic examples of the less industrialized community. Colonial possessions of the United Kingdom until 1947, their economies and cultures have been shaped largely by the parasitic relationship they had with the industrial world for two centuries. The growth rate in all three countries is typical of other developing countries, with a relatively low death rate and a high birthrate (Table 12.1). Literacy rates and per capita incomes are also similar to those of other less industrialized areas. Infant mortality rates are high, industrialization is low, and the general pattern of reliance on agriculture represents the less industrialized world in its entirety. The recent independence of India, Pakistan, and Bangladesh and the limited industrial sector inherited from their colonial past combine with typical problems of agriculture in less industrialized countries to handicap development and efforts to improve standards of living.

India is by far the largest country of the subcontinent. In spite of the similarities to other developing countries, India is important because of the unique aspects of its economy and history. As a part of the British colonial empire, India received great atten-

Figure 12.1 The Indian subcontinent is part of the less industrialized world as measured by the standard indices. It is the single region that residents of the Industrial World most commonly associate with the characteristics of the less developed realm. Images of overpopulation, poverty, agricultural life-styles, hunger, and deprivation are common stereotypes associated with the Indian subcontinent. Cultural values in India have been widely misunderstood in the West and result in erroneous conclusions.

TABLE 12.1 Selected Statistics of Development for South Asia

Country	Life Expectancy	Birthrate[a]	Death Rate[a]	Literacy Rate (%)	Per Capita Income	Infant Mortality	Percentage of Labor in Agriculture	Industry
Bangladesh	50	43	17	29	$160	134	75	6
India	54	33	13	38	275	104	70	13
Pakistan	54	43	15	26	350	121	55	16

[a]per thousand
SOURCE: *World Development Report* (New York: The World Bank, 1987), *1988 World Population Data Sheet* (Washington: Population Reference Bureau, 1988); *The World Factbook, 1988* (Washington: CIA).

tion, and fear of losing control of its largest colony prompted the British to tax India's population heavily in order to construct the largest railroad net in Asia. Strategic concerns of the British rulers in a country with a population more than double the size of the United Kingdom at that time also led to the establishment of the most intensive road network in Asia. At the time of independence, slightly more than 60,000 miles (96,500 kilometers) of surfaced roads and 44,000 miles (70,800 kilometers) of railroad had been constructed in India. The combination of railroad and road networks provided emergent India with an opportunity for development not found in other newly independent countries. Because of its size, even this extent of roads and railroads is inadequate, and an estimated one-fifth of Indian villages are still connected to the outside world only by paths or roads impassable to automobiles.

India is also unique among newly independent countries in that it has a resource base that could facilitate development. Both iron ore and coal are available, and the agricultural base of the subcontinent provides important raw materials for textiles and other industries. India developed an extensive civil service under British rule, which if effectively utilized can provide the management necessary for development. The governing class speaks English, which is an accepted language within the subcontinent, allowing India to interact relatively freely with the industrial world.

The unique attributes of India as a less industrialized country are transcended by one single factor, its democratic political system. The process by which India gained its independence and the political system it subsequently developed are unlike those of any other developing region of Asia. Independence was obtained through a policy of peaceful resistance rather than armed conflict as in many newly emerging countries. The development of a democratic government based on universal and free right to vote (*suffrage*) marks India as the world's largest democracy and places it in a unique position among less industrialized countries.

THE PHYSICAL BASIS OF THE INDIAN SUBCONTINENT

The physical geography of the Indian subcontinent is complex, with a wide variety of landforms and climatic types. It ranges from the highest mountains in the world in the Himalayas of the north to tropical coastal plains in the southwest and southeast. Portions of the subcontinent, like the Thar Desert in the northwest, have truly arid conditions, while the northeast has the world's highest annual precipitation (425 inches, or more than 35 feet—nearly 11 meters). Three great river systems have played a major role in the development of the subcontinent: the Indus in the west, the Ganges in the north central portion, and the Brahmaputra in the east and northeast. Four major environmental regions are easily recognizable: the mountainous zone centered on the Himalayas in the north; the plains region of the river systems of northern and western India; the plateau region of south peninsular India; and the low tropical coastal plain surrounding the peninsula.

THE LANDFORMS OF INDIA

The most famous landform of India is the Himalayan Mountains. The Himalayas exceed 20,000 feet (6,096 meters) elevation, and the range presents a major barrier to movements of air masses north and south. Through passes in the mountains, especially the Khyber Pass in the west, invading cultural groups and ideas have diffused through the years. Movement to the north has been much more difficult, but the Karakoram Pass provides access from north central India

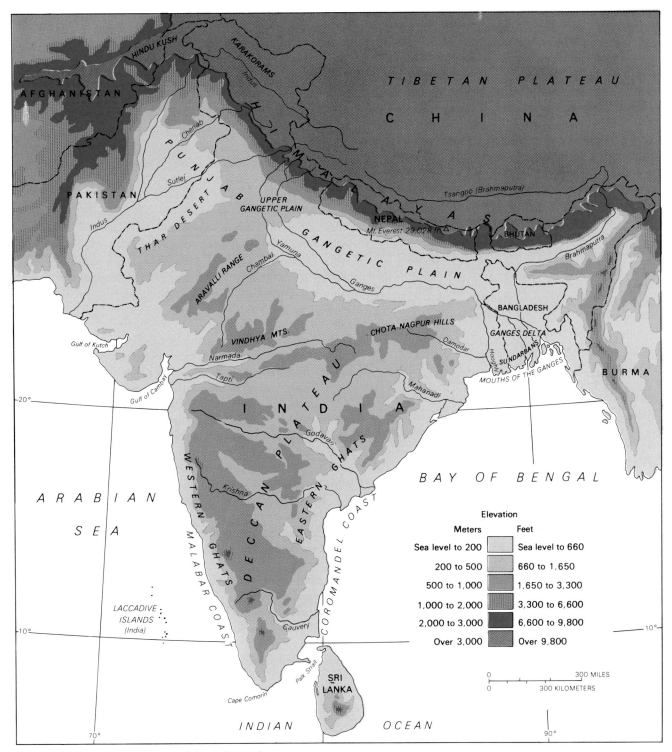

Figure 12.2 Landforms of the Indian subcontinent.

through the Himalayan and Hindu Kush complex of mountains.

Except in the more favorable valleys such as in the Vale of Kashmir and in Nepal, population in the Himalayas is limited (Figure 12.2). Nepal has a population of 19 million, separated from India by ranges of mountains. Two other mountainous states occupy the Himalayas north of India, Bhutan with 1.5 million people and Sikkim with less than a million. Sikkim has been incorporated into India and is in essence one

THE INDIAN SUBCONTINENT

of its provinces. Climates in the Himalayas vary greatly with altitude, and range from tropical lowlands to Arctic conditions in the high altitudes of Mount Everest and other tall peaks. The population of Nepal, Bhutan, and Sikkim is concentrated in the intermediate zones where the climate is relatively moderate, ranging from subtropical to humid continental. Agricultural production in these areas generally reflects the climate, with rice or wheat being the predominant grain crop, combined with a variety of vegetable and small grain-livestock combinations.

South of the Himalayas is the core area of the Indian subcontinent. The plains of the Indus, the Ganges, and the Brahmaputra Rivers have the greatest population concentrations of the subcontinent, are the economic heart of the subcontinent, and were the center of the historic civilizations that dominated the region. This plains region is from 200 to 300 miles (320 to 500 kilometers) wide through the present countries of Pakistan, India, and Bangladesh. The environment varies from arid in the west Punjab region to tropical around the Bay of Bengal. Soils are derived from alluvium, are relatively fertile, and are generally level. Centuries of irrigation in the Indus plains have created severe environmental problems through accumulation of salts in the soil (**salinization**).

South of the plains of the large rivers of the Indian subcontinent is the peninsula of India with its principal landform, the Deccan Plateau. The plateau is an uplifted area, dissected by river valleys into a complex system of level plateaus, hills, and deep ravines. The central portion of the Deccan Plateau has fertile soils developed from rich volcanic materials, and it is an important area for commercial crops, particularly cotton.

The Deccan Plateau has elevations of approximately 1000 to 1500 feet (305 to 450 meters) above sea level. It is separated from the coastal plains by the Ghats mountain ranges parallel to the east and west coasts. The western Ghats reach elevations of 5000 feet (1515 meters) in a few peaks, but the eastern Ghats are lower with few peaks over 2000 feet (610 meters). At the very southern margin of the Deccan Plateau is a cluster of hills called the Blue Mountains, which reach heights in excess of 8800 feet (2600 meters).

The eastern and western Ghats are bordered respectively by the coastal lowlands of the Coromandel and Malabar coasts. The coastal areas have a humid tropical climate with abundant rainfall from the orographic effect of the Ghats, and they are highly fertile.

THE CLIMATES OF THE INDIAN SUBCONTINENT

The climatic diversity of the Indian subcontinent rivals the landform diversity. The altitudinal variation of the Himalayas results in wide variation in temperature and precipitation in small areas. South of the Himalayas the climate is equally varied, but the specific climatic type in an area is affected by the monsoon. All over India the monsoon is the dominant climatic force. It results from seasonal wind reversals: air movement from land to sea with dry conditions in the winter, and a sea-to-land movement in the summer with humid conditions. The cause of the monsoons is complex but is a result of the shifting location of the jet stream north and south of the Himalayas with the changing seasons of the year. In winter the jet stream is divided with one part south of the Himalayas. The air movement effectively prevents moisture from the oceans from moving into the core area of India along the Ganges, and dry conditions predominate. During the summer the jet stream moves entirely north of the Himalayas in most years allowing moist air to penetrate the continent from the oceans. During the summer season, when the monsoon winds blow from the ocean, the air mass rises as it moves over the continent, causing orographic precipitation (Figure 12.3).

The rains of the summer monsoon are the basis for agriculture throughout the Indian subcontinent. Only in the dry western margins of the Punjab and the Indus does irrigation become the primary basis for agriculture. Because of the complex relationship of the jet stream to earth-sun relationships, in some years the monsoon arrives late or never arrives at all, with catastrophic effects for agriculture. In the past, the monsoon has been inadequate, on average, two out of every five years. Resulting famines have plagued the Indian subcontinent throughout recorded time.

The reliance of the majority of Indian farmers on the monsoon for crop production imbues the Indian peasant with a particular sense of fatalism. When the monsoon does not come or brings too little rain, the people suffer. When it strikes with unusual intensity, the rains may destroy the crops. Coupled with the Hindu belief system, the monsoons have played a special role in handicapping the pace of economic development in India.

The climates in India south of the Himalayas are a function of both latitude and the monsoon. The Coromandel and Malabar coastal regions have a tropical

---

Salinization: The accumulation of salts in the upper levels of soil as salts carried by irrigation water are left at or near the surface as water evaporates.

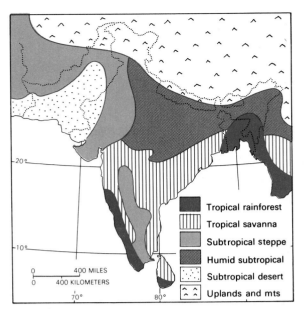

Figure 12.3 Climates of the Indian subcontinent. Much of the subcontinent is affected by a summer monsoon that particularly affects precipitation.

masses into the Delta region of India and Bangladesh, and this region also receives water from the two great river systems, making it well watered and suitable for production of rice. The Deccan Plateau has a humid subtropical climate, which is plagued by inadequate summer moisture. The western Ghats cause a rain-shadow on the Deccan Plateau, where agriculture often suffers from insufficient moisture at critical times.

The Ganges basin proper has a tropical savanna to subtropical climate that relies on the monsoon. Temperatures are uniformly high throughout the year, with winters typically having less precipitation (Table 12.2). To the west in the Indus valley the climate is drier because of the rain shadow effect of Saudi Arabia, Iran, and Afghanistan. Even the coastal regions of Pakistan fail to receive the full impact of the monsoons and have a tropical dry climate.

An important climatic element of the Indian subcontinent is its lack of freezing temperatures. Only in the Himalayas do temperatures drop low enough to prevent double cropping. Water is the critical element in determining the amount and duration of the crop season. In areas where irrigation is available, two or three crops can be harvested in a single year. Where the monsoon provides the only water, production of more than a single crop is difficult. The combination of high temperatures and erratic and unreliable precipitation provides a unique challenge to India as the country develops. If the surplus water that is lost during the monsoon season can be stored, it will be possible to increase agricultural productivity and

rain-forest climate as a result of orographic precipitation and latitudinal location. Crop yields are more dependable here than in other areas of India, since the monsoon nearly always occurs. The eastern portion of the Ganges, known as the Delta region, also has a tropical climate, but it is a tropical savanna modified by the monsoon. The Bay of Bengal funnels air

TABLE 12.2 Climatic Data for Selected Stations

Station			J	F	M	A	M	J	J	A	S	O	N	D	Year
Madras	Temp.	°F.	76	78	81	86	91	90	87	86	85	82	78	76	85
		°C.	(24)	(26)	(27)	(30)	(33)	(32)	(31)	(30)	(29)	(28)	(26)	(24)	(29)
	Precip.	in.	1.4	0.4	0.3	0.6	1.0	1.9	3.6	4.6	4.7	12.0	14.0	5.5	50.0
		mm	(35.6)	(10.1)	(7.6)	(15.2)	(25.4)	(48.3)	(91.4)	(116.8)	(119.4)	(304.8)	(355.6)	(139.7)	(1,270.0)
Cherrapunji	Temp.	°F.	53	57	62	66	67	68	69	69	69	67	61	55	63
		°C.	(12)	(14)	(17)	(19)	(19)	(20)	(21)	(21)	(21)	(19)	(16)	(13)	(17)
	Precip.	in.	0.8	1.5	7.0	23.8	67.2	115.0	96.7	71.9	46.0	17.6	1.8	0.2	449.6
		mm	(20.3)	(38.1)	(177.8)	(604.5)	(1,706.9)	(2,921.2)	(2,456.2)	(1,826.1)	(1,168.4)	(447.0)	(45.7)	(5.1)	(11,419.8)
Bombay	Temp.	°F.	75	75	79	82	85	84	81	80	80	82	81	78	80
		°C.	(24)	(24)	(26)	(28)	(29)	(29)	(27)	(27)	(27)	(28)	(27)	(26)	(27)
	Precip.	in.	0.1	0.1	0.1	T	0.7	19.1	42.3	13.4	10.4	2.5	0.5	0.1	71.2
		mm	(2.5)	(2.5)	(2.5)	(T)	(17.8)	(485.1)	(1,074.4)	(340.4)	(264.2)	(63.5)	(12.7)	(2.5)	(1,808.5)
New Delhi	Temp.	°F.	58	62	70	83	90	92	90	88	86	78	68	60	77
		°C.	(14)	(17)	(21)	(28)	(32)	(33)	(32)	(31)	(30)	(26)	(20)	(16)	(25)
	Precip.	in.	1.0	0.9	0.7	0.6	0.5	0.7	9.0	7.0	6.0	1.5	0.2	0.5	28.6
		mm	(25.4)	(22.9)	(17.8)	(15.2)	(12.7)	(17.8)	(228.6)	(177.8)	(152.4)	(38.1)	(5.1)	(12.7)	(726.4)
Begampet	Temp.	°F.	70	74	80	88	90	84	80	78	78	77	71	64	77
		°C.	(21)	(23)	(27)	(31)	(32)	(29)	(27)	(26)	(26)	(25)	(22)	(18)	(25)
	Precip.	in.	0.5	0.5	0.5	0.4	1.2	5.0	6.5	5.9	6.4	2.9	1.0	0.4	31.4
		mm	(12.7)	(12.7)	(12.7)	(10.2)	(30.5)	(127.0)	(165.1)	(149.9)	(162.6)	(73.7)	(25.4)	(10.2)	(797.6)

hopefully increase the standard of living of the Indian people.

HISTORIC BACKGROUND OF INDIA

The role of India in the world today as one of the largest developing countries is the result of its development and interaction with Europeans in the past 300 years. The present landscape, culture, economy, and political organization of a country reflect the events that created them, and it is impossible to explain the geography of the world without understanding both its temporal and spatial relationships. Less industrialized countries in particular represent but one point on a changing time-space continuum. The process of development in countries like India illustrates the futility of simply describing the geography of a place at one time. Understanding why and how the geographic patterns emerged allows prediction of future changes.

The experience of India before its independence in 1947 is significant because it mirrors the experience of most other less industrialized countries—a general pattern of exploitation, destruction of traditional cultures, and economic dependence. Examination of the colonial relationship to the British exemplifies the problem facing not only India, but other less industrialized countries as colonial powers are expelled.

HINDUISM: THE CULTURAL BASIS OF INDIA

At least a thousand years before the emergence of Europe as a dominant world force in exploration and colonization, India was the site of a flourishing civilization. This civilization developed social and economic orders that persisted until European colonialism destroyed important elements of their fabric. Central to the Indian civilization was Hinduism. *Hinduism* is called a religion, but it transcends the traditional concept of religion and embodies a way of life that still largely determines the actions of each of its followers.

Hinduism developed as seminomadic cattle herders and cultivators invaded present-day India from the northwest over several centuries. As these tribal groups interacted with the residents of the Indus River valley, they absorbed much of the culture of the advanced civilization of the Indus, including avoidance of eating beef and adoption of a hereditary priestly class of Brahmans. Subsequent additions to the value system embodied the concepts of a multiplicity of gods, the ideal of ritual purity, the development of a rigid caste system, and belief in reincarnation. These values permeated Indian civilization prior to European colonization, and elements persist to the present. The Hindu belief system combined with a social organization that made possible a **symbiotic** relationship among all members of a community, guaranteeing the perpetuation of the village and the belief system. Because Hinduism is a way of life, its major tenets must be understood if India before and after European colonialization and subsequent independence is to be understood.

Dharma in Hinduism is the individual duty of each person. Dharma is related to the rigid social order of India (the caste system), since the caste to which one is born determines the duty that must be followed. In theory there are four broad castes in Hindu society. The *Brahmans* are the teachers, religious leaders, and scholars; the *Kshatriya*, are political leaders and warriors; the *Vaisyas* are engaged in trade or farming; and the *Sudres*, who are the lowest class, provide services to support the society. Each of these four broad groups is broken down into subgroups whose relative status is dependent upon their extent of *ritual purity* (avoidance of contact with unclean objects). Those dealing with death or decaying materials were in the lowest classes of the sudres. At the very bottom of the order are the *untouchables* (or harijans), so-called because in the past (and among many Indians today) it was believed that they would contaminate others' ritual purity if there was any personal contact between them. Untouchables lived in separate communities, had separate wells for water, and in the nineteenth century were prohibited from using roads used by the other castes. The Brahmans were at the top of the caste order and were expected to devote themselves to study and knowledge.

The second aspect of Hinduism is related to the cycle of life, death, and rebirth, or *reincarnation*. For the Hindu, life is not simply a progression from birth until death but a progression in a circular fashion until freedom is obtained from the cycle. Freedom from continued reincarnation can be obtained through *nirvana*, which consists of obtaining spiritual unification with the cosmic forces, and being liberated from the human processes of death and birth. Associated with nirvana and reincarnation is the concept of *karma*, or law of the deed. Karma specifies that for each good act there will be a reward and for each evil act there will be a punishment. An individual's status in the caste system reflects actions in previous incarnations. Each individual within the caste system occupies a specific niche in relationship to his or her previous

---

Symbiotic: Mutually beneficial relationships in which both parties in a relationship are better off than operating alone.

actions; thus, every individual has an explanation for the conditions of his or her life. It is difficult to move upward in the caste system through education or acquisition of wealth or social change, since the specific caste is a result of previous actions. Doing good works or fulfilling the duty of one's caste or engaging in the pursuit of knowledge provide opportunities for advancing in the next incarnation, but upward mobility from one caste to another rarely occurred in the past and is the exception rather than the rule today.

In precolonial India the Hindu religion provided the social system that maintained a distinctive Indian life-style relatively unchanged for centuries. Of necessity the majority of the population lived in villages, there were great extremes between the untouchables and the higher castes, and distribution of wealth was uneven. The caste system that rigidly segregated individuals, also provided a system of meeting everyone's needs. Although untouchables occupied the lowest niche in society, the services that they provided in handling dead animals or disposing of waste materials were essential. The washer woman occupied a niche much lower than the Brahman priest but provided a service needed by all. The distribution of agricultural products was uneven but sufficient for each member of the village to continue his or her function. The Hindu system made each village a nearly self-sufficient unit in which all members of the society had a specific role and the basic needs of all were met. The caste group in each village was relatively small and provided a unit of socialization and social security. The rigid organization with its emphasis on duty to one's caste helped to maintain an orderly function of the society.

THE POLITICAL DEVELOPMENT OF INDIA

The development of India prior to its colonial dominance by Western Europe involved expansion of political dynasties to a zenith after which incompetency developed and the dynasty declined. Individual Indian dynasties faced invaders from the West, particularly after the development of Islam in the Middle East and its militant expansion eastward. Shortly after A.D. 700 Arabic peoples conquered the northwestern portion of the Indian subcontinent and expanded eastward to the headwaters of the Ganges River. Various kings ruled portions or all of India in the subsequent centuries, but typically southern India remained Hindu while northern and western India were dominated by Muslim leaders, changing the cultural patterns of the entire subcontinent. The most famous of the Muslim dynasties was the Mogul dynasty founded in 1526. The Mogul dynasty was financed by a system of taxation of land based on granting rights to tax to a local authority. Under the Mogul system the individual peasant retained ownership of the land. Ultimately the Mogul dynasty collapsed in 1707, leaving a kingdom without a ruler and presenting the opportunity for European domination.

EUROPEAN CONTACTS AND EMPIRE IN SOUTH ASIA

European contact with India grew out of the voyages of exploration. Portuguese, Dutch, French, and British had posts in India where they traded with Indian rulers for spices and fabrics desired in Western Europe. Before 1700 the European contacts were limited, for the standard of living in Europe was essentially identical to that of India, and in some areas, particularly in textiles, India may have been ahead. The wealth of the Indian Mogul leaders was many times the wealth of the individual countries of Europe of the time. By 1600 British representatives were maintaining that Europe was being stripped of its wealth to obtain silk, spices, and indigo (a root used to make blue dye).

BRITISH DOMINANCE OF THE INDIAN SUBCONTINENT

The British epitomized the European nations that traded with India. In 1600 Queen Elizabeth I granted to an association of businessmen and merchants known as the British East India Company exclusive rights to the spice trade of Asia. As the Mogul dynasty began to decline in the last half of the seventeenth century, British merchants were able to expand their economic dominance from the coastal cities to direct rule of large segments of the subcontinent. By 1858 nearly two-thirds of the subcontinent was ruled directly by the company, with only the less fertile or less accessible areas controlled by more than 500 princely states. Rebellion of Indian troops against the British in 1857 precipitated a year-long series of conflicts, which culminated in the decision by the crown to strip the British East India Company of its monopoly of India. In 1858 the Indian subcontinent became a crown colony of Britain.

THE IMPACT OF COLONIALISM

The impact of European influence and political and economic control on the Indian subcontinent profoundly altered the geographical patterns and the culture, civilization, and economy of what are now India, Pakistan, and Bangladesh (Figure 12.4). British impact on the Indian subcontinent can be divided into two eras, the first 100 years from 1757 to 1857 when

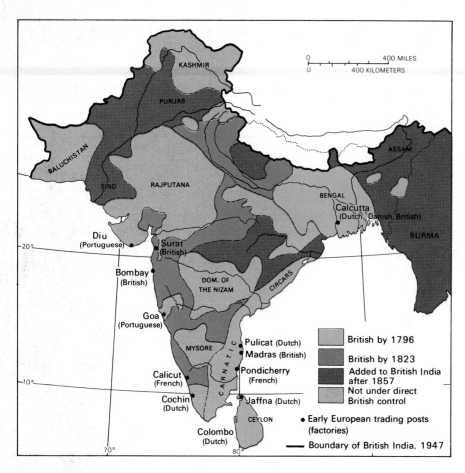

Figure 12.4 Expansion of British influence in the Indian subcontinent. Note that large areas remained under the nominal political control of the local Indian prices and that other European countries controlled small areas.

the British East India Company dominated India, and the period from 1858 to 1947 when it was a crown colony. Under the jurisdiction of the British East India Company major changes occurred in Indian villages. The first involved changing the old taxation structure of the Mogul empires and their local village representatives. Unfamiliar with the nuances of the taxation scheme, the British granted the village representatives of northern India deeds to the lands from which they had previously collected taxes. The village representatives became landlords, charging cash rent, which peasants were often unable to pay. Property ultimately ended in the hands of urban moneylenders as peasants borrowed at exorbitant interest rates to meet these taxes, and a large rural landless class emerged.

Southern India retained a true taxation system, but cash payment of these taxes was now required. The Indian peasant system broke down as loans for payment of taxes could not be repaid when drought or disease prevented full production. At the same time that land was moving from peasant ownership to the ownership of landlords, the British East India Company encouraged production of commercial products rather than food as in the precolonial society. The requirement of cash for taxes left peasants little alternative but to produce the crops that the British East India Company and its representatives would purchase. These crops, including coffee, tea, sugar, spices, cotton, indigo, and jute had a ready market in Europe. As British influence in Asia spread to China, opium also became an important crop, as it could be resold or traded in China for additional high-value spices. The introduction of these crops created much of today's agricultural pattern.

All crops encouraged by the British East India Company were for export, and an **asymmetrical relationship** developed in trade between the Indian subcontinent and the British East India Company. The beginnings of industrialization in Europe provided a ready market for India's agricultural products, and India became an agricultural hinterland for the industrial heartland of the British Empire.

A third impact on India of the British dominance after 1757 was in the potential for industrial devel-

---

Asymmetrical relationship: One in which one participant benefits more than, or at the expense of, the others involved.

THE WORLD OF POVERTY

opment in the Indian subcontinent itself. Prior to the emergence of the Industrial Revolution in Western Europe, India had a large native textile handicraft industry. As the British mechanized their own textile industry, they were able to produce a surplus of textile goods, and the export of inexpensive English textiles to India effectively limited the potential for India's own textile manufacturers to make the transition to the industrial age. By 1850, the once self-sufficient Indian society had been changed to a subservient culture with all of the attributes of today's underdeveloped economies. What industry existed was controlled by the British East India Company: a wealthy ruling class controlled land and capital; agricultural production was oriented toward items for export to the European colonial powers while the majority of the population lived in poverty and faced the ever-present threat of death from famine when crop years were poor.

ECONOMIC DEVELOPMENT OF THE BRITISH COLONY

The change of the Indian subcontinent from economic domination by the East India Company to political status as a British colony reflected British impact upon the Hindu culture. Directors of the East India Company showed a general lack of understanding of the importance and profundity of Hindu value systems and made major changes that alienated the Hindu majority. The changes individually were viewed by the British as essential to "civilize" the natives, but their effect was to shock the Hindu people whose view of life was so different from that of the British occupying power.

Changes were too numerous to catalog, but one based on European attitudes toward the preeminence of life and the lengths to which the Europeans would go to preserve life illustrates the problem. The Hindu culture with its belief in reincarnation did not place as great a weight on preservation of life, and encouraged widows of Hindu men to commit suicide on the funeral pyres of their husbands. The British made this process illegal and further exacerbated relationships with the Hindus by making it legal for Hindu widows to remarry. These two examples illustrate the general lack of awareness among the colonial powers of the pervasiveness of a religious system. In the British view both suggestions concerning the role of women were designed to provide a better quality of life for Indian women in general, but in the eyes of Hindu men and women they were viewed as intolerable trespasses into a traditional value system. These and other cultural conflicts culminated in the rebellion of 1857–1858 and transfer of the subcontinent to crown colony status.

The change from rule by the British East India Company to status of British crown colony involved important changes that have had a lasting impact on the Indian subcontinent. To administer the colony, the British government developed a civil service system largely staffed by descendants of Anglo-Indian relationships. These people did not accept the caste system and were able to obtain education and training. They did not find employment by the British abhorrent. Some members of the merchant class also encouraged their children to enter into the British civil service, and the political system that emerged ultimately became the model for the emergent countries of the Indian subcontinent.

Cultural conflicts with the Hindus in the 1800s convinced the British of the need for a transportation system to allow rapid movement of troops to potential trouble spots. Taxes were increased to provide funding for the extensive (by less industrialized countries' standards) railroad and road network built under British direction. Besides providing the transportation network and the civil service system, the British government encouraged sanitation and simple hygienic practices, which ultimately led to the beginnings of rapid population increase in India at the turn of the twentieth century, creating a problem that is still unresolved.

Funeral pyres burn constantly on the Ganges River at Varnasi. The ashes will be scattered in the river.

Except for these changes the rule of the British government was little different from that of the East India Company. Agricultural crops still focused on export to Western markets, industrialization was handicapped by the parasitic nature of Anglo-Indian trade agreements, and the process of concentration of wealth in the hands of the moneylenders and landlords continued, as the taxation scheme was not changed. The British decision to build roads and railroads required levying greater taxes, hastening the alienation of the Indian peasant from the land and reinforcing the concentration of wealth. By 1940 nearly 40 percent of the population of colonial India consisted of landless cultivators, three times the figure of forty years previously. The landless peasant remains one of the most critical issues facing the Indian subcontinent today. The British expanded land under cultivation to provide greater quantities of cotton, tea, and other desired crops but provided little incentive or technical expertise to increase food production. The persistence of British colonialism and the dominance of agriculture by a wealthy elite ensured that agricultural advances would be restricted to those areas and crops with a ready market. In consequence India's productivity performance remains the lowest in Asia today.

Industrialization in India was limited to port cities such as Calcutta or Bombay where a specific local product that would not compete with the home industries of the British could be easily produced. The major industrial activity in the Indian subcontinent was manufacture of textiles in factories owned by the British. Their construction in India represented another exploitive action as the Industrial Revolution in Western Europe moved to more sophisticated manufacturing processes.

The final effect of British colonial occupance was to maintain the fragmentation of the subcontinent. The multitude of princely states, the varieties of regional languages and dialects, and the religious distinction between Muslim and Hindu were never adequately addressed by the British. As a colonial power, British attempts at providing education were minimal, and with no universal education to overcome regional differences in dialect and culture, the fragmentation of the subcontinent persisted. Overcoming fragmentation was not in Britain's best interest, since unity would make possible a united resistance to the colonial rulers.

INDEPENDENCE OF THE SUBCONTINENT

After two centuries of British control, independence came on August 15, 1947. The newly independent Indian subcontinent was divided into two countries, and each one faced significant centrifugal forces that

Figure 12.5 Division of British India at time of Independence in 1947.

threatened its existence (Figure 12.5). Centrifugal forces included the problem of the landless laborers who had emerged under British rule, the limited literacy and educational standards of the majority of the population, the lack of an adequate industrial base, an economy that was still effectively tied to exports to Europe, and the cultural and political fragmentation inherited from the British colonial rule. Understanding the strides that India and Pakistan have made since independence is possible only if these centrifugal forces are fully understood. The cultural and economic forces are reinforced by problems of the physical environment, and together they combine to present India with a major challenge in the development process.

THE CULTURAL GEOGRAPHY OF THE INDIAN SUBCONTINENT

India has one of the most complex cultural patterns in the world. At the time of independence in 1947, the Indian subcontinent had been splintered into more than 550 princely states, nearly 900 separate dialects, and 15 major languages. The contrast between the Islamic and the Hindu population further complicates the cultural pattern. Ethnically the people of the northwest are distinct from those of the Ganges Delta or the Deccan Plateau. The tremendous diversity of culture represents one of the major centrifugal forces with which India, Pakistan, and Bangladesh have had to deal since independence. Unlike China, where only a tiny minority population is not Chinese, the new countries of the Indian subcontinent were faced with the need to develop a national identity from highly divergent cultures.

The one element that has provided an opportunity for overcoming the centrifugal forces of culture in India is the recognition of the long history of civilization in today's India. The Indian civilization predates the great age of Greece, the foundation of Rome, the birth of Christ, or the flight of the Jews from Egypt. But the very length of the Indian history has witnessed repeated invasions with subsequent changes in the cultural pattern. The first recognizable culture group in the Indian subcontinent were the *Dravidians*. The Dravidians had a civilization as highly developed as that of Egypt. They were subject to invasions by Aryans (Indo-Europeans) from central Asia from 1500 B.C. and were pushed southward along the coast of the Indian peninsula and into the Deccan Plateau. Today they constitute nearly 25 percent of the Indian population, and four distinct Dravidian languages are included among the fifteen official languages recognized by the Indian constitution.

The repeated movements of Indo-European peoples into present-day India led to development of a number of languages, primarily Hindi and related tongues of Punjabi, Bihari, and Rajasthani. These Indo-European languages are related to the Indo-European languages of Western Europe but separated by several thousand years of distinct evolvement (Figure 12.6).

The adoption of Islam by the Caucasoid peoples of the Middle East and their determination to spread Islam to their "unenlightened" neighbors culminated in the invasion of India by Islamic peoples between A.D. 800 and 1200. The Islamic peoples conquered the Indus valley and the Ganges region including the Ganges Delta and introduced both Islam and the Urdu language. Urdu is a combination of Hindi and Arabic, Persian, and Turkish languages that developed as a trade language between the Islamic conquerors and the Hindu population. The pattern of Islam and Urdu speech is complex because the advancing Islamic invaders did not maintain occupance of all the land they conquered. Today's Pakistan, the western portions of the Ganges River plains, and the eastern portion of the Ganges and Brahmaputra River Delta became predominantly Islamic, while the balance of the Ganges plain remained predominantly Hindu. South of the Ganges plain Islam was always a minority belief system.

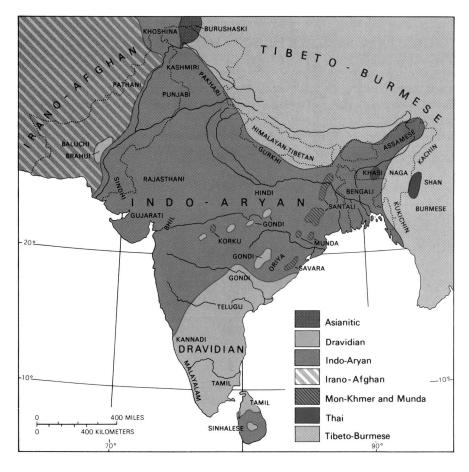

Figure 12.6 Languages of the Indian subcontinent.

THE INDIAN SUBCONTINENT

In addition to the two major religious groups there are a variety of minor groups including Sikhism, Jainism, and Christianity. Both *Sikhism* and *Jainism* were reform movements within Hinduism. The major characteristic of Jainism is its concern for all life forms and the reluctance to cause the death of any creature. Sikhism developed on the interface between Muslim and Hindu as a protest against perceived corruptions of Hinduism. Its distinctive characteristic is the style of dress of the militant segment of Sikhism in which every male adherent has the common surname of Singh and wears long hair, a full beard, a turban, and a dagger. The Sikhs shared great hostility toward the Muslims after India gained its independence. Extremists among India's 17 million Sikhs demand that the Punjab state where they are dominant be given independence as the new Sikh state of *Khalistan*. The city of Amritsar in the Punjab is the center of Sikh extremism, and the Golden Temple of Amritsar is the most sacred site in India to Sikhs. In 1984 the Indian army occupied the temple at Amritsar where extremists were defying the government and several hundred were killed. In retaliation Sikh extremists assassinated India's prime minister, Indira Gandhi, later that same year. Conflict continues between the Sikh extremists and the government to the present.

Christianity arrived as an export from western Europe and is a minor sect in the Indian subcontinent. The combination of language, religion, and the caste system creates a complex mosaic of culture in India that is difficult to display graphically, but that determines daily life in the subcontinent.

The division of the Indian subcontinent into states reflects the cultural diversity. At the time of independence the entire region was known as India, and Mohandas Gandhi, father of independence, hoped that it could become one independent political unit. Religious conflicts between Muslims and Hindus caused the partition of the subcontinent into India and Pakistan, with Pakistan divided between the two areas of concentrated Islamic population—East and West Pakistan. The division left many Hindus in Pakistan, and millions of Muslims in India. After the partition nearly 17 million people migrated between India and Pakistan. Nearly a million suffered death in riots, armed attacks by Sikhs, or starvation that resulted from disruption of the farming economy.

The two halves of Pakistan, separated by more than 1500 miles (2400 kilometers) were never effectively united, and cultural differences compounded the centrifugal effect of distance. The Punjabi language is the most common language in West Pakistan, while residents of East Pakistan speak Bengali, a language shared with the providence of Bengal in India. Contrasts between the two parts culminated in the separation in 1971 of West Pakistan from East Pakistan, which became Bangladesh.

India has been more successful in overcoming the centrifugal forces of culture since independence. It has been able to maintain its territorial integrity and has acquired control of the majority of the disputed territory of Kashmir in the northwest. In spite of the continued existence of India as a country, cultural variety remains a major centrifugal force that has absorbed the energies of the new country since its independence. After independence India was faced with the problem of uniting a diverse people whose only common heritage was Hinduism. This problem was exacerbated by the millions of Muslims who remained in India, making it the world's second largest Islamic

The Temple at Armitsar in India is the Sikhs' most sacred site and symbolizes their continued demands for an autonomous Sikh state in the Punjab of Northwest India.

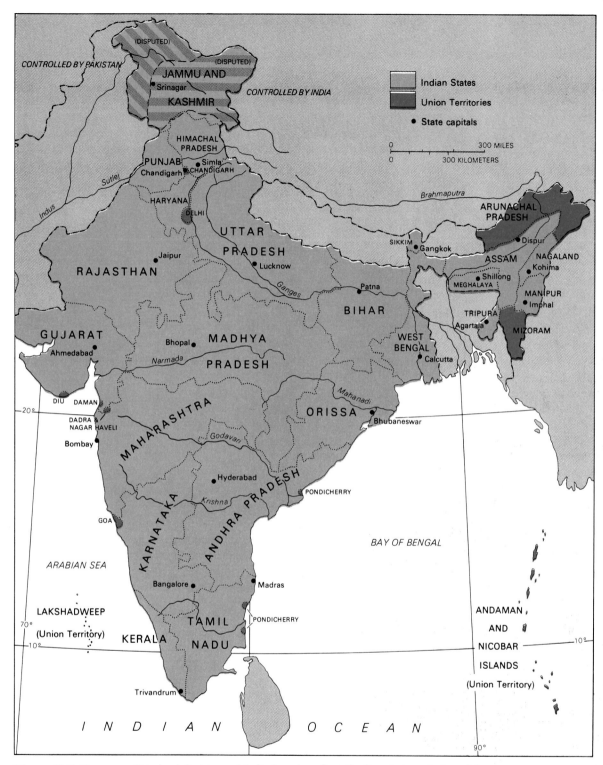

Figure 12.7 Present political subdivisions of India largely reflect the linguistic variation of the country.

state (after Indonesia). Approximately 11 percent of India's population (93.5 million) are Muslims. Leaders regarded provision of a common language for India as essential. Hindi, spoken by 30 percent of the population, was adopted as the "official" language, but English was also maintained as a second language to be used for official purposes, and 13 others were designated "languages of India."

Subsequent demands for regional autonomy based on language distinctiveness have resulted in the creation of 22 different states within India (Figure 12.7). The centrifugal force of language remains a major

THE INDIAN SUBCONTINENT

problem in India today. India's language-ethnic diversity is a good example of both polyethnicism and cultural pluralism, and there is little evidence that there will emerge a common Indian culture within the next few generations. Schools in individual regions not only teach in their native language, but in many major cities such as Calcutta there are distinct ethnic neighborhoods, and the children of each neighborhood attend schools where they are taught in their own tongue, ranging from Bengali to Punjabi to Rajasthani to Gujariti. The problem of linguistic multiplicity is one for which no easy solution is apparent.

THE INDIAN VILLAGE: A WAY OF LIFE

A cultural element that transcends the varieties of languages and religions is the role of the village in Indian life. India has always been a village society, and today there are an estimated 580,000 villages. The village is the focus of life for 75 percent of the population. These villages range in population from a few hundred to tens of thousands, but they are villages in form and function. They have seen little change in millennia except for the concentration of land in the hands of a landlord class. Labor is still performed almost exclusively by people and animals; for those peasants who own land, farm size averages less than 5 acres; per capita incomes rarely exceed $100 per year; and the life of the village reflects the agricultural demands of the particular season. Large families are the rule rather than the exception, and male children are highly desired because of their perceived value in the agricultural system. For poorer families, daughters can result in economic disaster as cash must be borrowed to pay a dowry to obtain a husband.

The role of male and female in the Indian village is specified by long tradition. Men do the plowing, threshing, and heavier labor; women grind grain, do household chores, gather firewood and water, and plant, harvest, and weed crops. The Hindu belief system permeates village life, with offerings of food to animals viewed as sacred even during times of famine. The pervasiveness of village life and the isolation of many of the villages has made unification of India more difficult and contributes to the centrifugal forces that perennially threaten the country.

INDIAN DEVELOPMENT: RURAL AND URBAN

At the time of independence in 1947, the seemingly insurmountable problems of cultural diversity, economic backwardness, national poverty, and limited education were of a magnitude experienced by only one other developing nation—China. China and India have both attempted development since the late 1940s, but they have done so by different methods. China opted for a Communist system initially modeled upon the Soviet Union. India attempted to transform its society and economy through a democratic process. At the time of independence islands of development in Indian cities were surrounded and nearly drowned by the underdevelopment and poverty of nearly 80 percent of the population. India has begun the process of transforming its society into a major world economic power, and by 1990 India was the tenth greatest industrial country as measured by value of industrial output. This accomplishment must rank as one of the major achievements of the twentieth century, but it still leaves India far behind China in development. In part this slower rate reflects the absence of an authoritarian government, which can make and enforce unpopular decisions even if they result in individual deprivation for a period of time. In part it is the result of the fatalism of Hindu culture, which will tolerate survival if there is sufficient food and shelter to sustain life. Potential resources have also affected India's development and industrialization.

RESOURCES FOR DEVELOPMENT

The major resource that affects India's development is agricultural land. It remains the basis for important crops such as cotton, jute, tea, and to a lesser extent tobacco. These crops support important industries and exports, but the majority of manufactured items based on agricultural products (textiles, burlap from jute, and so on) are consumed within the country. India is the second largest producer of grains in the world, but until the 1960s was a net importer of grain. Periodic droughts still cause grain shortages in parts of India in some years, and while minor quantities of Indian rice and wheat have been exported to neighboring countries in the past few years, grains offer little possibility for economic development.

Minerals for development offer some major opportunities in India. High-quality iron is abundant, including the largest deposit of high-grade iron ore in the world (Figure 12.8). In Bihar state alone a single range is estimated to hold nearly 3 billion tons of iron ore. Alloys for steel are also available, and India produces more than a million tons of manganese, 80 percent of which is exported. Known reserves of petroleum are limited, but the greater part of the subcontinent has been inadequately explored. Discoveries in the Bay of Bengal in 1980 have been de-

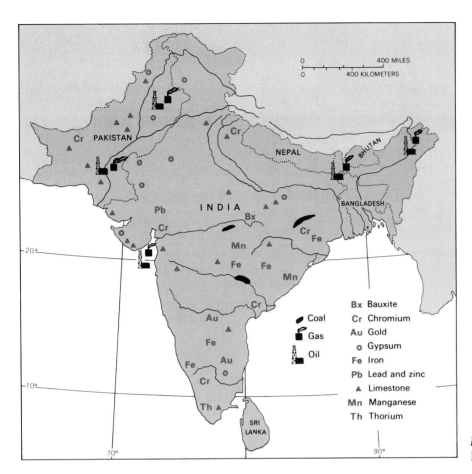

Figure 12.8 Resources of India.

veloped so that today India produces approximately 75 percent of its petroleum needs. Coal is widespread through the eastern and northeastern portion of India, but the Damodar fields of northeastern India account for over 50 percent of coal production. India's coal is primarily subbituminous or lower-grade bituminous, which handicaps expansion of the steel industry. Coal does provide fuel for railroad locomotives, heating, and thermal electrical generation.

India also has a great potential for hydroelectric power if dams can be built on the large rivers. It has important deposits of uranium and other ores that could be used for nuclear power. Minor deposits of lead, zinc, copper, and asbestos are insufficient to meet India's present needs. A major deposit of phosphate, essential for mineral fertilizer, is found in the Thar Desert in the northwest. The central portion of the Deccan Plateau and the eastern margins of the Coromandel coast produce abundant manganese, much of which is exported to Japan. Chrome ore, gold, industrial diamonds, bauxite, and mica are also produced and serve as important exports to Japan and other countries. In general India has the necessary mineral resources to support a much higher level of industrial activity than it presently has.

INDUSTRIALIZATION IN THE INDIAN ECONOMY

Industrialization in India proceeds along an uneasy path. Gandhi championed development of the **cottage industries** that existed prior to the intervention of Britain. He argued that the large-scale, Western-style, automated industries would be self-defeating, since they would eliminate rather than create jobs. Nevertheless, attempts at economic development focused on industrialization for the first 30 years after independence. Successive five-year economic development plans emphasized expansion of India's steel industry and development of an industrial base adequate to provide for the needs of a country approaching a billion people.

INDUSTRIAL REGIONS IN INDIA

Major industrial centers exist in old cities with industries dating from the colonial era and in cities near

---

Cottage industries: Cottage industry involves small-scale production using high labor inputs.

THE INDIAN SUBCONTINENT

THE HUMAN DIMENSION

An Indian Woman

Kalyani, a 35-year-old agricultural laborer belonging to one of the scheduled castes [as opposed to the untouchables], lives with her husband and five children in a small thatched hut in a squatter settlement on the outskirts of the city of Bombay.

Kalyani collects her wages in cash. She generally earns Rs. 7 [approximately 90¢ U.S.] a day, whether she is working in the fields or on construction. Her children usually know where she has gone to work and can guess when she will return home. If Kalyani has gone for transplanting or weeding, she is back around 5 P.M.; if she has gone to a construction site, she returns home around 6 P.M. Vani, her second daughter, aged nine, often waits for her mother at the road junction near the squatter settlement where there is a small market. She carries a small basket for rice and groceries and two bottles, one for kerosene and another for coconut oil.

Since Kalyani has mortgaged her ration card [which allows purchase of set quantities of rice at a subsidized price], she must buy all her rice in the open market, where the price (Rs. 2.50 per kg.) is currently about 50 percent higher than at the fair-price shops. Since she buys 1 3/4 kg. of rice every day, this purchase alone costs Rs. 4.40.

Her daily shopping needs also include fish for Rs. 1.00–1.50; coconut oil for 25 paise (Rs. 1 = 100 paise); raw coconut for 40 paise; onions and spices, including tamarind, coriander, and chiles, for 50 paise; and kerosene for 25 paise. She spends a total of Rs. 7–8 per day, depending on whether soap has been included. She buys a bar of soap every third day. The days both Mosha (her husband) and she are out of work, she does not buy any fish. Although the grocer and coffee shop give her credit, the fish vendor does not. Most of what she buys is used in preparing the evening meal for the entire family; there is always a little cooked rice and rice gruel left for the children's breakfast and lunch.

From her daily wage of Rs. 7, Kalyani has to pay 60 paise for her breakfast and 50 paise for her midday snack and tea. On working days when she is not doing transplanting and has a full rice lunch, she has to pay Rs. 1. She must also purchase the quantity of betel leaves and nuts she consumes daily. While on other days she spends 50 paise on betel leaves, she spends almost twice this amount on transplanting days. Thus, she is usually left with only Rs. 5 from her own wages for daily shopping. On the days she buys soap or talcum—the latter is a must even in the humblest cottages, for men, women, and children dust themselves liberally with it after bathing—she must cut down on her daily food purchases.

On the days Mosha has work, however, he gives Kalyani Rs. 5 or more. Indeed, on days both of them have work, the household can be run smoothly. Problems arise because work is not available every day for both Kalyani and Mosha. For every day Kalyani works, she spends at least one day without a job, however hard she may try. When Mosha was in good health, he fared better: for every day he went without work he had two days of employment; so on average, at least, one of the two was always working. In practice, there are days when both are with or without work, creating problems of management that Kalyani seems to find very hard to solve.

To the delicate question of why Mosha spends more than twice the amount Kalyani spends (Rs. 5 compared with her Rs. 2 every working day), Kalyani's answer is emphatic. Mosha is a man and should have some freedom to spend his money as he likes. He does a far more strenuous job than she does and does not get home until 8 P.M. so he has to eat a lot more, and oftener, outside. Moreover, all men in the neighborhood eat out on working days. Mosha does not drink and only smokes bidies. The whole neighborhood considers him extremely well behaved. Still, the fact remains that Mosha spends almost half of his wage eating out.

This year the monsoon came early and in force. The first showers of the season were so heavy that Kalyani's thatched roof, which had not been replaced for two years, and part of the walled structure, gave way. Kalyani's immediate concern is to repair the house and move back as quickly as possible. If she had not already mortgaged her ration card, she could pawn it now to borrow Rs. 100 to buy new dried palm leaves and some bamboo and areca poles.

Source: Reprinted with permission of the Population Council from Leela Gulati, "Profile of a Female Agricultural Laborer," Studies in Family Planning 10, no. 11/12 (November/December 1979): 416–417.

---

important mineral deposits (Figure 12.9.) The largest industrial center is Calcutta, which under British control manufactured jute. After the separation of India and Pakistan in 1947 the jute-producing region in today's Bangladesh was cut off from the manufacturing plants in Calcutta. Today the city is a center of diversified industrial activities focusing on such items as bicycle manufacture, textiles, food processing, electrical motors, and other light industry.

The most important heavy industry complex is in the Jamshedpur region, 150 miles (240 kilometers) west of Calcutta. Coal and iron ore resources in the region were the basis for Tata Steel Works, which is the single largest steel-making complex in India. Major heavy industry has grown up around the steel production in the Jamshedpur region, and the iron and steel producing area is sometimes referred to as the Indian Ruhr. Although the steel production of Jam-

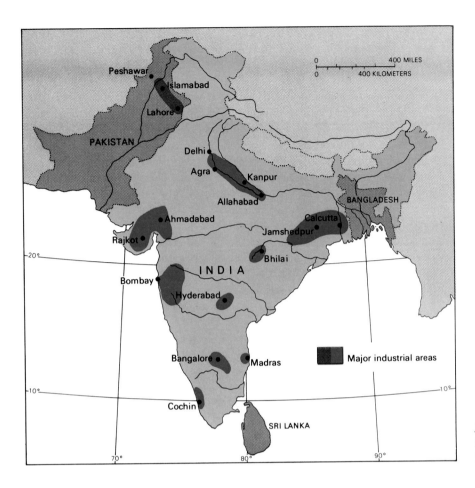

Figure 12.9 Major industrial regions of the Indian subcontinent.

shedpur is only a fraction of that of Germany, the nickname is accurate in describing the importance of the Jamshedpur iron and steel production to the Indian economy.

Bangalore is another major industrial center that was originally a cotton textile manufacturing center but today supports diversified electrical manufacturing, light industry, machine tools, the construction industry, and food processing. On the west coast Bombay is similar to Bangalore; it was originally a cotton textile center but has diversified into manufactures of all types, including automobiles and petrochemicals. Ahmadabad has a similar industrial makeup, and was also a textile manufacturing area initially. Together the five manufacturing regions centered on Calcutta, Jamshedpur, Bombay, Bangalore, and Ahmadabad produce two-thirds of all manufactured products in India.

Below the level of these five regional concentrations of industrialization are a large number of regional centers where cottage industries and processing of agricultural products dominate. These cities are large in their own right, reflecting their location as junctions on the road and railroad networks established by the British.

Manufacturing remains a minor element in the total work force, employing only 13 percent of the people. Nevertheless, since 1947 India has become nearly independent of imported manufactured items. The ubiquitous American-made items common in other less industrialized countries are rarely found in India. The manufacturing sector can provide the country's basic necessities without the need to spend precious capital on unnecessary imports, but the society is very different from that of Western industrial nations. To keep India's self-sufficiency in perspective, remember that a family owning a motor scooter is middle-class, and the proliferation of automobiles and other consumer goods common to industrialized countries is not found in India.

The latest five-year plan (1984–1989) recognizes that Indian industrial development has accomplished its goal of providing basic goods but has failed to provide jobs for the population. The unemployment rate in the 1980s was estimated at between 15 and 20 percent, and additional industrialization will make only a marginal impact on this figure so long as it emphasizes Western-style, automated industry. The Indian government has now placed restrictions on establishment of new industrial plants in the fields of textiles, leather goods, shoes, oil milling, and safety matches. (Safety matches are a much more important element in the Indian economy than in an industrial society as tremendous numbers of safety matches are used for lighting fires, candles, and so forth. Oil milling is the process of grinding oil seed to make edible oils, a process that can be accomplished at a small scale.) The reason for these restrictions is that these sectors of the economy are suitable for cottage industry. Indian planners estimate that cottage industries employ 40 individuals for every one employed in a large automated factory producing the same products. Production of shoes, weaving of textiles, milling of oil seeds, and manufacture of safety matches will employ a massive number of people in fully developed cottage industries.

At the same time that cottage industries are being encouraged for a limited number of economic activities, an estimated 750 products have been designated as amenable to production in small industries, defined as those using not more than the equivalent of $100,000 of capital. Areas in which small industries are focusing include radio receivers, hand tools, plumbing fittings, padlocks, and a variety of other items that can be produced in a nonautomated fashion. The ultimate aim of the Indian drive to develop cottage and small-scale industries is to utilize India's most abundant resource, its rapidly increasing population. The Indian government is not engaging in a retreat from the Industrial Revolution but simply attempting to achieve an increased standard of living and provide jobs and necessities for the population with an alternative development form: the Indian development program. If fully implemented, the proposal to decentralize the manufacturing process will change the present pattern of industry in India.

AGRICULTURE: THE ULTIMATE ARENA OF DEVELOPMENT

India's achievements in industrialization since independence are noteworthy, but the numbers of individuals who will ever be employed in industry in India will of necessity be a small proportion of the total population. India remains an agricultural society with 70 percent of its population either directly or indirectly engaged in agriculture in 1988. In the agricultural sector the changes since independence are perhaps even more significant than the industrial development. India's great success is that it has moved from a food-deficit region in which famine and food rationing were common, to general self-sufficiency and food surplus. With a population that increases by about one and one-half million individuals per month, India has been able to increase agricultural productivity enough to feed all its people and export rice and wheat to Bangladesh and surrounding countries. These exports are but a tiny fraction of the total production, but they are indicative of the agricultural gains that India has made. In spite of these gains, India's traditional agricultural systems have yields per acre that are among the lowest in the world.

The simple fact of providing adequate food masks important problems in Indian agriculture. India is still one of the twenty poorest countries in the world. In 1990 the gross national product per capita was only an estimated $275, and 45 percent of the country's total population lived below the low poverty line designated by the Indian government. Three hundred million people are barely surviving, but they are surviving.

One of the major problems contributing to the poverty of this large segment of the Indian population is the failure of adequate land reform. *Land tenure* (ownership) is still heavily weighted in favor of larger landholders. Although the government has adopted a rule limiting ownership of "good" land to 18 acres (7 hectares) per family, wealthy landowners have circumvented the law by distributing their holdings to relatives. In India in 1990 nearly two-thirds of rural farm families have less than 5 acres (2 hectares) of land or no land at all. There are an estimated 140 million landless tenants in India. Because India is a democratic country, it seemingly cannot strip the wealthy landowners of their land and apportion it among the landless peasants. The lack of significant change in land tenure since independence will continue to be one of the major limiting factors in increasing the standard of living of all people in India.

The Green Revolution. The increases in crop production in India have come as a result of the Green Revolution. Adoption of "miracle" grains of wheat and rice in the 1960s resulted in doubling and tripling the yields of grain under ideal circumstances. The miracle grains respond well to fertilizer and irrigation water, but without these inputs they yield less than the native varieties. The Green Revolution provided the means for increased productivity, and the government recognized that industrialization had failed to resolve the country's dilemma. After 1966, it increased its emphasis on agricultural development. Capital was made available for drilling wells, acquiring fertilizer, purchasing tractors and other farm implements, and improving road systems to facilitate transport of agricultural products to the urban markets. In spite of the tremendous increases of grain production under the Green Revolution, India must feed a much larger population on a smaller food base than the United States (Figure 12.10).

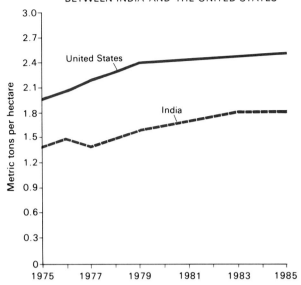

Figure 12.10 Comparison of wheat yields between India and the United States.

Some observers see the Green Revolution, with its high-yielding grains, as merely a temporary solution to India's problems. The country's population growth necessitates increasing the grain yields if per capita availability of food is not to decline. Overall food production must increase by more than 2.5 million tons (2.3 million metric tons) per year in order to maintain a minimal consumption level. Although India exports small quantities of rice, the perennial threat of drought makes continued surpluses questionable. The increased availability of grain has not eliminated hunger in India, since individuals with low incomes still cannot purchase as much as they need. The typical Indian peasant, whether a landless farmer or a laborer in village or city, operates at the threshold of survival.

The problems underlying the agricultural problem in India have not been resolved, only postponed by the Green Revolution. Small farm sizes, fragmented parcels of property, lack of capital for fertilizer or expansion of irrigation systems, and concentration of landownership will effectively negate the impact of the Green Revolution in the long run. Recognition of the need to increase agricultural production continually has fostered increased development efforts to expand irrigation, making India the location of the world's largest irrigation networks, particularly in the drier parts of the northwestern Ganges plains. Government projects in the last decade have focused on small-scale projects designed to benefit the individual village or farmer including provision of irrigation water through tube wells and pumps. Government programs to teach farmers better methods and to introduce better varieties of crops and better methods of fertilization have been effective in increasing production in much of India. But unless population is ultimately controlled, all these efforts may prove to be inadequate.

Agricultural regions of India. The agricultural pattern of India is complex, but, based on the predominant crop, there are several distinct agricultural regions (Figure 12.11). The primary food grains in India are rice, wheat, and millet. Wheat is concentrated in the drier areas of western India, while rice is concentrated in the eastern delta region of the Ganges and Brahmaputra and around the tropical coastal lowland. Millet is grown widely in the Deccan Plateau where it will produce under moisture-deficient conditions. Peanuts generally coincide with millet production in the southern Deccan Plateau, and jute is concentrated in the Delta region of the Ganges and Brahmaputra. Coconut and tea production are concentrated with rubber in the coastal regions of the southern peninsula of India. All of these crops are produced primarily for internal consumption, and exports, except for tea and jute, are minor. Coconut is dried to make copra, which is also exported to Western countries where coconut oil is extracted for use in soaps and other items. Cotton is concentrated in the Deccan Plateau but is also grown in northwestern India. The majority of this cotton is used for cloth for the large internal market. India is the world's largest producer of sugar cane, primarily in the Ganges plains, but almost all of it is consumed within the country.

Livestock production in India varies according to the climate. India has more livestock than any other country, with nearly 200 million cows, 60 million water buffalo, 60 million goats and sheep, and 5 million horses, donkeys, camels, and elephants. Sheep are of major importance in the drier west where the greatest Islamic population is concentrated. Water buffalo are dominant in the Ganges Delta and coastal regions of India, and cattle (particularly the Brahman or Zebu breeds) are found throughout India. The great numbers of cattle in India in the past gave rise to the belief that the Hindu taboo on killing animals and resultant view of cattle as sacred handicapped food production. Such a view is at best misguided and at worst completely in error. Cattle are an integral element of the Indian agricultural economy. They are the primary source of draft power for the majority of India's agrarian population—plowing, pulling carts, grinding grain, and a host of other tasks. During the dry season, in nonirrigated areas, cattle and sheep are permitted to graze the limited forage, which would otherwise be

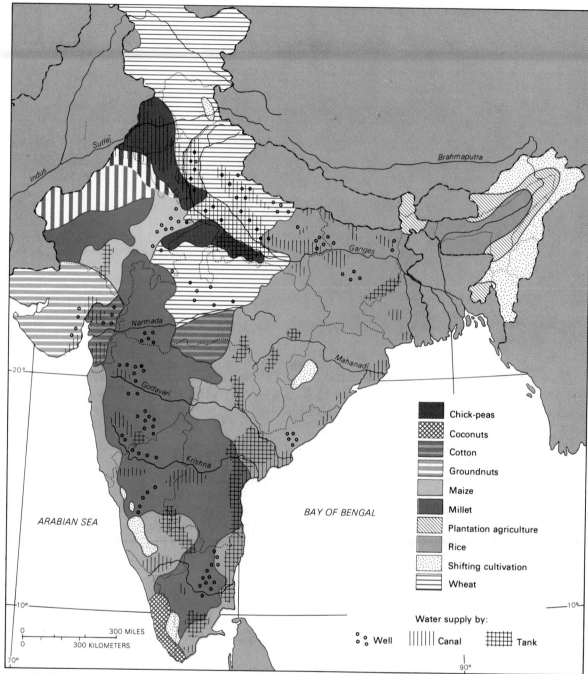

Figure 12.11 Generalized agricultural regions of India. Within each region a variety of crops are produced, but one crop generally dominates.

wasted. Cattle also consume secondary agricultural by-products such as straw, rice husks, and corn stalks. Of even greater importance cattle produce an estimated 850 million tons (771 million metric tons) of cow dung, the principal source of domestic fuel. In addition dung is mixed with mud and used for plaster, and it is a major source of fertilizer. Cattle also produce most of India's milk (the bulk of which comes from water buffalo), providing a major source of protein in the vegetarian Indian diet. When a cow dies, it is consumed by the untouchables (who have no prohibitions about consuming beef when it is available) or the large Muslim population. Cow hides are a major source of leather and an integral part of the economy. Some anthropologists who have studied the Indian veneration of the cow conclude that the maintenance of the large numbers of cows and buffalo is a completely rational activity in the Indian agricultural economy.

Indian agriculture suffers from a host of other, more basic problems including farm fragmentation. Farm sizes in 1961 averaged 7.47 acres (3 hectares), but farms as small as one one-sixtieth of an acre were common. Land reform laws have distributed approximately 11.5 million acres (4.6 million hectares) among landless laborers, but by 1990 the average farm size in India was less than 6 acres (2.4 hectares), smaller than in the past. Other problems include an illiteracy rate estimated at 70 percent in rural areas. Most Indians in rural areas are undernourished, and their diet is deficient in proteins, fat, and minerals. Drinking water, obtained from open wells or contaminated ponds, is often polluted. According to recent studies 80 percent of the agricultural workers in the northeastern and southern parts of India suffer from hookworm and anemia.

POPULATION AND INDIA'S FUTURE

The problem of transforming the Indian economy is compounded by the country's tremendous population increase. As of 1990 the absolute increase in numbers each year in India was greater than any other country in the world, totaling a net gain of 17 million persons per year—more than live in all but a few of the most populous states of the United States. The additional capital investment in industry and agriculture that is required simply to feed and clothe the additional population effectively prevents India from making significant economic gains. India's population problem is

Throughout India, the streets and sidewalks are a focus of activities not seen in western cities. Crowded conditions in the tenements of India's cities make the sidewalks an effective part of the living space.

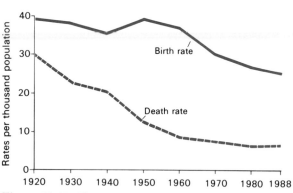

Figure 12.12 Changes in birth- and death rates in the Indian subcontinent still result in high growth rates.

typical of that of less industrialized countries. The birthrate has been declining for several decades, but the death rate has declined at an even faster rate (Figure 12.12). The population of India increased from 250 million in 1920 to over 850 million in 1990. The present increase of 17 million per year is greater than the total population of Australia, which has two and one-half times the area of India. At present growth rates, there will be more than 1 billion Indians by the turn of the century. At present India's population is increasing at a rate of 2 percent per year, or 1.4 million per month, or 46,000 every day. One Indian planner describes the effect of this continual increase in numbers as making all development efforts like writings in the sand. Beautiful pictures may be drawn in the sand through great effort, but the incoming tide destroys them. Expanding the industrial base, increasing

While a small portion of India's population have no homes except the streets, the majority live in tiny, cramped quarters, as in this street in Calcutta.

THE WORLD OF POVERTY

THE HUMAN DIMENSION

Water for Life

The shop run by Sri Isamuddin in Kohir village is a simple, rather ramshackle stall, whose whole inventory of crumbling cigarettes, fleshy green betel leaves and silvery spices would fit inside a large paper bag. It is a stall like countless others throughout India, and its location next to a shady teashop on the main street is in no way remarkable. But Sri Isamuddin has set it up there for another reason. It keeps him near his other job: caretaker and overseer of Kohir village handpumps and water supply system. There are four handpumps to serve the 15,000 people in the village and the one near his shop is the most heavily used.

Sri Isamuddin is only 21 years old, has received a mere two days training for his part-time job, and is unrewarded except by a tool kit and a certificate that confers on him a modest rise in status. But his services as a do-it-yourself maintenance man keep the village handpumps in working order and, when one breaks down, Sri Isamuddin acts as the link with the next tier in the maintenance system.

As well as his maintenance duties, Sri Isamuddin also functions as an informal public-health official, trying to ensure that the water used by mothers and children in Kohir is kept as germ-free as possible. The area around the handpump near Sri Isamuddin's shop is immaculate. All dirt and refuse has been swept away and a small channel in the concrete area at the handpump's base is draining off the excess water.

Sri Isamuddin is a bright young man and is conscientious about his duties. He seems to have fully absorbed the various elements of his training that took place a few months ago at a two-day camp. Together with 95 others from the district, he was taught rudimentary preventive maintenance—greasing bolts, cleaning the pump head—and was made aware of the connection between contaminated or stagnant water and the spread of disease.

One of the main problems in the past was the pump itelf. Before 1974, the type of pump invariably installed was an old-fashioned cast-iron pump that, although it might have the apparent advantage of costing very little, broke down with monotonous regularity. This kind of pump, patterned on types used years ago in the rural Western world, was intended for use by a single family. Under the pressure of use by the two or three hundred families living in an Indian village, the pump's strength soon gave out. With aid from the United Nations, the Indian government developed the India Mark II Handpump, an all-steel pump that is produced under strict quality control and that, in the past seven years, has gained a reputation for technical reliability. Properly installed, the Mark II handpump can function for considerable lengths of time without the need for major repairs.

The combination of the Mark II handpump and a village caretaker is having some definite effect. One survey in a nearby district carried out within the last 12 months discovered that over 95 percent of the newly installed Mark II handpumps were working. The equivalent figure of a few years ago at any one moment would have been around 30 percent. Gradually, the health impact of clean water supplies in Indian villages is bound to begin to make itself felt.

Source: Development Forum, Sept.-Oct., 1982, p. 14.

---

yields per acre of grains, bringing new lands under cultivation, expanding medical care facilities, or providing more schools is similarly absorbed by the tide of growing population. The perpetually increasing numbers of Indians help to keep India one of the poorest countries of the world. The per capita income in India in 1990 is estimated at $275, but as in other less developed countries the actual distribution means that a significant proportion of the population receives much less. It is estimated that more than 300 million people in India's urban slums and rural areas live below the level necessary to maintain health.

The impact of India's population on attempts to transform the economy is pervasive and insidious. The literacy rate of India is estimated at 38 percent, and the logistics of overcoming the mass illiteracy even without population growth would tax the resources of the entire subcontinent. Simply providing one elementary school for each of the 580,000 villages is beyond India's ability at present.

The increasing population has destined many people to a life of deprivation unlike anything found in the Western countries. In the slums of Calcutta an estimated 300,000 of the city's inhabitants are known as *street people*, who have no homes of any kind but sleep under bridges, railway overpasses, in doorways, or wherever they can find a spot. Slightly better is life in the slums of India's worst areas where homes consist of hovels made of burlap, cardboard, or other scrap material known as *bustees*. An estimated 2 million people exist in bustees with average dimensions of 6 by 10 feet by 5 feet (2 by 3 by 1 1/2 meters) in height. The bustees have no running water, no sewage system, no garbage collection, and only marginal police protection by the government. Even for the residents of India's major cities who are fortunate enough to have a job and a permanent, masonry home, more than half of the families have only a single room in which to live.

Urbanization in India. Ironically the worst poverty in India is found not in the villages, but in the cities such as Calcutta, Bombay, or Old Delhi. The cities of India are increasing rapidly in size but only 25 percent of the Indian population resides in urban places of 20,000 or more. With its large population, 25 percent translates to more than 212 million people, and at present growth rates, within two decades there will be more urban residents in India than the total population of the United States. The movement of rural people to urban areas is marked in India, but only one-fourth of urban population growth results from rural to urban migration, less than in other Asian countries. The balance of the growth in India's cities is a result of natural increase among the urban population. Life expectancy in the cities is lower than in the rural areas because of less access to food, more congestion and unsanitary conditions, and the breakdown of the village social system.

An indication of the nature of these problems can be obtained from the population densities in Calcutta. Calcutta averages 36,000 persons per square mile (13,900 per square kilometer) for its entire 400-square-mile (1036-square-kilometer) area, compared to New York's average density of 24,000 per square mile (9264 per square kilometer). Other major cities of India (Bombay, Delhi, Madras, Hyderabad, Ahmadabad, Bangalore, Kānpur, and Poona, all exceeding 1 million) are lower than Calcutta's, but rival those of New York and other major metropolitan areas. The poverty and deprivation among the poorest half of India's population equal that in the poorest countries in the world.

While the majority of India's population exists in poverty, there is a significant wealthy elite. These are the business leaders, industrialists, political leaders, and large landholders who have managed to increase their wealth since independence and engage in conspicuous consumption similar to that found in elite classes in other less industrialized countries. This group stands in stark contrast to the life of the majority. The existence of the dual society epitomizes the impact of the Indian fatalism and ability to accept seemingly intolerable contrasts in wealth and poverty because of the underlying influence of Hinduism's emphasis on duty and to one's own caste. The contrasts that in other countries would foster violent revolution and class warfare are limited in India because of the religious tenets of the Hindu majority population.

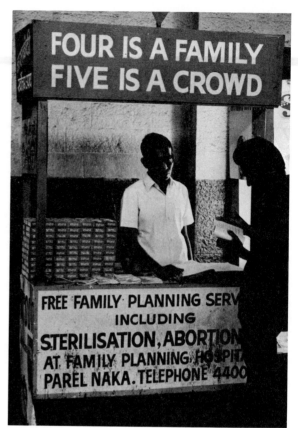

The graphic message of this family planning booth is representative of Indian efforts to curb population growth.

RESOLVING THE POPULATION ISSUE

Because population growth is central to all of India's development efforts, the country has adopted programs designed to curtail the ever-increasing growth rate. Initially these programs were modest, since Gandhi believed that India could develop a society in which people's needs were met uniformly without the great variation in wealth that deprives a sector of the society of its fair share. The escalating population in the 1950s caused the adoption of a family planning program in the first five-year plan. In 1952 the government of India adopted family planning as a national policy. In the first years birth control clinics were established and information on family planning was provided to those who sought it. By 1961 there were 4165 such clinics. Although large, this number was inadequate for the large Indian population, and ultimately the country moved to a mass education program encouraging people to have smaller families. Billboards show couples with two children and the statement that the family of two is a happy family while larger families create problems for the entire country. As part of the program to limit births in the 1960s and 1970s voluntary sterilization programs were established in many states of India. As many as 5 million male Indians were sterilized under this program, but their numbers were inadequate to prevent continued population increase.

Social changes among the Hindu majority attempted to limit population growth further, including raising the minimum age for marriage for females to 15. The continued increase in population ultimately led the country to adopt a national population policy in 1976, which set goals for cutting population and adopted relatively harsh measures to ensure its success. The major elements of the 1976 policy included increasing the age of marriage for females to 18 years and for males to 21 years, tying financial grants from the national government to the state governments to their performance in limiting births, provision of sex education in schools, expansion of compensation for voluntary sterilization, and use of incentives by governments to encourage people to limit their family size. The latter took harsh forms in some states, including cutting retirement incomes for civil servants who had more than two children.

The 1976 population policy was viewed as intolerably harsh and led to the defeat of the government leader, Indira Gandhi (no relation to Mohandas Gandhi). The government leaders from 1977–1980 retreated from the 1976 policy and made participation completely voluntary in 1977. The program of 1977 still maintained that sterilization for men and women would be offered free of charge, that marriage age should be increased to 18 and 21 years for females and males respectively, and that 8 percent of the central government assistance to the individual states would be linked to performance and success in the family planning program.

India has made some headway in attempting to curb the birthrate, and plans are being made for a community health worker for each of the 580,000 villages. More than 30,000 such workers have been trained in the provision of basic medical care and family planning. The 1979–1984 plan was designed to lower the growth rate to 25 per thousand by 1984, and although Indira Gandhi was returned to power in 1980 the birthrate was only 34 in 1985. Since the assassination of Mrs Gandhi in 1984 and the election of her son as president, India has not adopted more dramatic measures to limit the growth rate. Consequently the population will continue to be the major factor affecting the country's attempts to transform its economy.

INDIA IN RETROSPECT

India's accomplishments since its independence are indeed great. Simply maintaining the territorial integrity of the state in the face of the numerous centrifugal forces threatening to destroy it is of worldwide significance. The continued existence of a democratic form of government, in which a woman was able to dominate the political system for more than a decade, is remarkable by world standards. The increase of food production, which has allowed India to be transformed from the world's largest food importer to a surplus producer, must rank as one of the twentieth century's greatest achievements. India's ability to steer a middle course in political affairs in the modern world without being subservient to either the industrial powers or the centrally planned economies such as the Soviet Union is also noteworthy. The development of industry allows India to provide bicycles, radios, safety matches, textiles, and other limited consumer items for its huge population. This achievement equals the advances made by industrializing Western Europe during the same time span. In the 40-plus years since independence, India's progress in industrial development has rivaled that of Britain during its formative industrializing years. The Indian economy is one of the ten largest in the world, and its trained and skilled work force is third in size only to that of the United States and the Soviet Union.

Compared to its achievements, India's failures stand out dramatically. India remains among the twenty poorest countries in the world in terms of per capita gross national product. Nearly 50 percent of the country's total population lives below the poverty line. This poverty is apparent even to casual visitors. When taxis stop at a traffic light in any Indian city, begging hands appear almost from nowhere through open windows. In rural areas when a car of a foreigner stops at a railroad crossing, begging hands are again thrust in. The poverty of India is so intense that it must be seen to be appreciated. Life expectancy of India's population is only 54, one of the lowest in the world. Literacy rates remain below 40 percent, and the pressing problems of landownership and employment remain critical. The question concerning India's experience is whether the maintenance of a democratic society with its attendant slow rate of development justifies the continued poverty of India's masses.

India remains one of the most complex countries in the world. The unifying theme of Hinduism is insufficient to overcome the centrifugal forces that affect the country economically, culturally, and intellectually. It will remain a country of contrasts, and it is important to remember that the generalizations of this chapter refer only to the majority. There are exceptions to each of the country's problems. India's greatest hope for the future is its people themselves. Although the large number of people presents a problem, it also provides an alternative because of its size and cultural heritage. India has absorbed many invasions of people and ideas and combined them into the unique Indian culture. Its goal is to industrialize but not to westernize, and to integrate socialism in a democratic society as Gandhi advocated. The Indian willingness to suffer patiently rather than to riot or revolt may provide India the opportunity to make the transition to a developed country peacefully. If India can maintain its national identity and make the apparently slow transition to a higher standard of living in a peaceful and democratic manner, it will be indeed a unique example of development, one unrivaled in the developing world.

THE INDIAN PERIMETER

Surrounding India are five countries that are relatively small in area but share the population and economic problems of India. In area they range from Pakistan's 319,867 square miles (828,456 square kilometers) to Bhutan's 18,147 square miles (47,000 square kilometers). With a total area only one-seventh as large as the United States, these five small countries have a population that is its equal. Bangladesh, for example, has an area roughly equivalent to that of the state of New York, but with a population of more than 116 million it has rural population densities among the highest in the world. With an average per capita income that averages only $260 per year (1990) and with a population that will be double within 26 years at present rates of growth, these countries typify the less industrialized world. The individual countries share a common heritage of development with India, but each has its own unique aspects.

PAKISTAN AND BANGLADESH: DIVISION FROM UNITY

Pakistan. Pakistan and Bangladesh share a common acceptance of Islam, but otherwise are quite different. Pakistan shares the Indian problem of multiple languages, with the official language of Urdu spoken by less than one-tenth of the population. Punjabi, the language of over 50 percent of Pakistanis, dominates in the north, and Sindhi in the south. Since over 98 percent of the people of Pakistan are Muslims, religion provides the unity within the country just as it once served to unify Pakistan and Bangladesh into one country. In Pakistan today, Islamic law derived from the **Koran** is the basis for an attempt to create an Islamic Republic.

The geographic characteristic that dominates Pakistan is drought. With the exception of the higher elevations in the north, agriculture is reliant upon irrigation. The rivers, particularly the fertile alluvial plains of the Indus River, provide the basis for the economy of the country. Major crops include wheat and cotton in the north, and rice in the south; minor crops include a host of less widely grown grains, fruits, and vegetables. Pakistan's mushrooming population exceeds the country's food production, making it a net food importer. Industrially, Pakistan has embarked on a modernization program that includes a large integrated iron and steel mill at Karachi based on imported iron ore and coal. Economic development has been hindered by limited resources as Pakistan has only minor petroleum resources and no major iron ore or coal reserves. Political concerns further handicap development as uncertainty over government nationalization intentions have affected the textile and other existing industries. Pakistan's growing population has been one source of economic growth as tens of thousands have gone to work in the capital surplus oil exporting countries of the Middle East and sent money back to Pakistan. A nation of farmers, with a limited resource base and a growing population, struggling to find its political destiny, Pakistan seems destined to continue as an archetypical member of the less industrialized world.

Bangladesh. Bangladesh shares with Pakistan the dominance of the Islamic faith, but its physical geography is different. Where Pakistan's geography is shaped by drought, Bangladesh is faced with problems of flooding associated with the monsoon. Bangladesh is a nation of farmers, farmers occupying the fertile floodplain and deltas of the Brahmaputra-Ganges River system. The year-round growing season and abundance of water and fertile alluvial soils allow production of three crops of rice annually in Bangladesh, supporting one of the highest rural population densities in the world. Less than 20 percent of the people are urban, and agriculture is the focus of life. Even industry has traditionally relied on agriculture, as the jute industry has been the economic mainstay. Jute plants yield a fiber used to make rope, sacking, and

---

Koran: Writings of the prophet Muhammad that Muslims accept as divine revelations from Allah.

carpet backing, but this industry has never fully recovered from separation from India and the subsequent war of secession with Pakistan.

Bangladesh has a per capita income of $160 (1990), which places it among the five poorest countries in the world. Food production is inadequate for the population, a population that at the present rate of growth will reach 160 million by the year 2000, 200 million by 2010. A farming country, it is a country in which 50 percent of the rural population is now landless. Repeated divisions of small holdings have made the farms of those who do own land miniscule, and further population growth will result in even greater fragmentation. The villages where more than 80 percent of the Bangladesh people reside are isolated from one another and from the two urban centers of Dhaka and Chitagong. Transportation typifies the less industrialized world as boats, bicycles, and foot are the most common forms. Wages are low, averaging 60 cents to $1 per day in construction, 60 cents for agricultural work. Rice, the staff of life in Bangladesh, averages 15 to 20 cents per pound (30 to 40 cents per kilogram), making adequate nutrition an impossibility for most residents. Unless and until Bangladesh can slow its population growth and develop a stable and conscientious government, it seems destined to remain one of the least industrialized countries of the world.

THE SMALL STATES OF THE SUBCONTINENT'S PERIPHERY

In terms of population, Sri Lanka, Nepal, and Bhutahn are small compared to the rest of the subcontinent since they have less than 38 million total people. Nepal and Bhutan occupy the central portion of the Himalayan Mountains. As with other less developed countries these countries are rural, poor, and lacking industry. Bhutan has only 1.5 million people and is a semicolony of India. Nepal has 19 million people in an area as large as Bangladesh. Nepal never came under the domination of Islamic expansion or the influence of Christians and with Bhutan has historically been a place of refuge, and a buffer between India and China. The two countries are landlocked and, even with the advent of aircraft, are among the most isolated areas of the world. Population, towns, and agriculture of the two mountain kingdoms are concentrated in the foothills and valleys of the Himalayas adjoining India. The Himalayas themselves are an attraction for hikers and climbers, and tourism to such isolated and exotic places as Katmandu is one of the few forms of interaction they have with the rest of the world.

Sri Lanka (formerly Ceylon) is another former colony of the United Kingdom. Lying just off the southeast coast of India, Sri Lanka has an area about equal to that of West Virginia, but a population of more than 17 million. In many ways Sri Lanka is a microcosm of the less industrialized world. Its economy is based on production of tea, rubber, and coconuts in the southwestern portion of the island. The southern part of the island has elevations over 8000 feet (2500 meters), and population and commercial agriculture are concentrated here. Plantations were established in the southwestern portion of the southern highlands, with coconut and rubber production in the hot moist zone up to about 2000 feet (600 meters), and tea at higher elevations. Export of these three products accounts for the majority of Sri Lanka's export income.

Culturally Sri Lanka is unlike India, Pakistan, or Bangladesh as it is neither Hindu nor Islam, but Buddhist (75 percent). The population is polyethnic, with Tamils and Sinhalese being the dominant group. The Sinhalese majority (70 percent) of Sri Lanka are descendants of migrants from northern India who occupied the island over 2000 years ago. The Sinhalese were historically concentrated in the northern half of the island where they had an extensive civilization based on irrigated rice cultivation. Subsequent migrants of Tamils from south India became the dominant ethnic group in the northern lowlands, forcing many Sinhalese to the southern lowlands. More recently, British-sponsored Tamils migrated to the southern mountain core to work on the plantations. Friction between Tamils and Sinhalese fostered an agreement with India in 1978 to relocate the "plantation" Tamils back to India. In 1983 Tamil extremists supporting an independent Tamil state in North Sri Lanka rioted, triggering Sinhalese retribution with destruction of life and property among the Tamils that continues sporadically to the present. The riots and destruction further compounded the problems common to less industrialized countries such as Sri Lanka.

The government has attempted to capitalize on its large labor supply, and some Japanese firms have established plants to manufacture china, glassware, and ornaments. Most industrial activity other than that associated with agricultural goods is oriented to providing building materials, clothing, housewares, or food products for the local market. Faced with rapid population growth, a stagnant economy, and ethnic conflict, Sri Lanka, like most less industrialized countries, is making only slow progress toward economic development, but the quality of life as reflected in the social indicators of health, education, and life expectancy is better than most less industrialized countries.

FURTHER READINGS

Bhardwaj, S., ed. *Hindu Places of Pilgrimmage in India: A Study in Cultural Geography.* Berkeley: University of California Press, 1983.

Chakravati, A. "Green Revolution in India." *Annals of the Association of American Geographers* 63, 1973, pp. 319–330.

De Silva, K. M. *Managing Ethnic Tensions in Multi-Ethnic Societies: Sri Lanka 1880–1985.* Lanham, Md.: University Press of America, 1986.

Dikshit, R. *The Political Geography of Federalism: An Inquiry into Origins and Stability.* New York: Wiley, 1975.

Dutt, A. "Cities of South Asia." In *Cities of the World: World Regional Urban Development,* pp. 324–368. Edited by S. Brunn and J. Williams. New York: Harper & Row, 1983.

Dutt, Ashok K. and Geib, Margaret. *An Atlas of South Asia.* Boulder, Colo.: Westview Press, 1987.

Farmer, B. *An Introduction to South Asia.* London: Methuen, 1984.

Frankel, F. R. *India's Political Economy, 1947–1977.* Princeton, N.J.: Princeton University Press, 1978.

Ganguly, Sumit. *The Origins of War in South Asia: Indo-Pakistan Conflicts Since 1947.* Boulder, Colo.: Westview Press, 1986.

Geddes, A., ed. *Man and Land in South Asia.* New Delhi: Concept Publishing, 1982.

Hall, A. *The Emergence of Modern India.* New York: Columbia University Press, 1981.

Hardgrave, R. L., Jr. *India: Government and Politics in a Developing Nation.* 3d ed. New York: Harcourt Brace Jovanovich, 1980.

Huke, Robert E. "The Green Revolution," *Journal of Geography* 84, November–December 1985, pp. 248–254.

Johnson, B.L.C. *India: Resources and Development.* London: Heinemann, 1983.

——— *South Asia.* London: Heinemann, 1969.

——— *Bangladesh.* London: Heinemann, 1975.

——— *Pakistan.* London: Heinemann, 1979.

Karunatilake, H.N.S. *Economic Development in Ceylon.* New York: Praeger, 1971.

Kaufman, M. "Across India, A Rural Tide Is Engulfing the Cities." *The New York Times,* February 19, 1982, p. 2.

Kayastha, S. L. "An Appraisal of Water Resources of India and Need for National Water Policy." *GeoJournal* 5, 1981, pp. 563–572.

Khan, A. "Rural-Urban Migration and Urbanization in Bangladesh." *Geographical Review* 72, 1982, pp. 379–394.

Lodrick, D. O. *Sacred Cows, Sacred Places.* Berkeley: University of California Press, 1981.

McLane, J. R. *India: A Culture Area in Perspective.* Boston: Allyn and Bacon, 1970.

Michel, A. A. *The Indus Rivers.* New Haven, Conn.: Yale University Press, 1967.

"Monsoon and Its Economic Impact." *GeoJournal* 3, 1979, pp. 171–178.

Moorhouse, Geoffrey. *India Britannica.* London: Harvill Press, 1983.

Murton, B. "South Asia." In *World Systems Traditional Resource Management,* pp. 67–99. Edited by G. Klee. New York: Halsted Press/V. H. Winston, 1980.

Myrop, R. *Area Handbook for Ceylon.* Washington: U.S. Government Printing Office, 1971.

National Atlas of India. 8 vols. Calcutta: National Atlas and Thematic Mapping Organization, 1982.

Newman, J. and Matzke, G. *Population: Patterns, Dynamics, and Prospects.* Englewood Cliffs, N.J.: Prentice-Hall, 1984.

Noble, A. G. and Dutt, A. K., eds. *India: Cultural Patterns and Processes.* Boulder, Colo.: Westview Press, 1982.

Norton, James H. K. *The Third World: South Asia.* Guilford, Conn.: Dushkin, 1984.

Raju, S. "Regional Patterns in the Labor Force of Urban India." *Professional Geographer* 34, 1982, pp. 42–49.

Schenk, Hans. "Residential Immobility in Urban India." *The Geographical Review* 76, April 1986, pp. 184–94.

Schwartzberg, J. E., ed. *A Historical Atlas of South Asia.* Chicago: University of Chicago Press, 1978.

Sharma, U. *Women, Work, and Property in Northwest India.* New York: Tavistock, 1980.

Singh, G. *A Geography of India.* Delhi: Atan Ran, 1976.

Singhal, D. P. *Pakistan.* Englewood Cliffs, N.J.: Prentice-Hall, 1972.

Sopher, D. E., ed. *An Exploration of India: Geographical Perspectives on Society and Culture.* Ithaca, N.Y.: Cornell University Press, 1980.

Spate, O.H.K. and Learmouth, A.T.A. *India and Pakistan: A General and Regional Geography.* London: Methuen, 1971.

Stevens, W. "Bangledesh Threatens to Burst at the Seams." *The New York Times,* April 6, 1983, p. 4.

——— "English Gains as India's Chief Tongue." *The New York Times,* May 19, 1983, p. 4.

——— "Rioting Reduces Sri Lanka's Hope to Ruins." *The New York Times,* August 4, 1983, pp. 1, 6.

Tambiah, Stanley J. *Sri Lanka: Ethnic Fratricide and the Dismantling of Democracy.* Chicago: University of Chicago Press, 1986.

Veit, L. A. *India's Second Revolution: The Dimensions of Development.* New York: McGraw-Hill, 1976.

Vesiland, P. J. "Monsoons: Life and Breath of Half the World," *National Geographic.* December 1984, pp. 712–747.

VISARIA, P. "Poverty and Unemployment in India: An Analysis of Recent Evidence." *World Development* 9, 1981, pp. 277–300.

WADE, ROBERT. *Village Republics: Economic Conditions for Collective Action in South India.* Cambridge: Cambridge University Press, 1988.

WHYTE, R. O. *Land, Livestock, and Human Nutrition in India.* New York: Praeger, 1968.

WOLPERT, S. *India.* Englewood Cliffs, N.J.: Prentice-Hall, 1965.

Tiananmen Square, Beijing.

CHINA PROFILE (1990)

Population (in Millions)	1,123	Energy Consumption Per Capita—	
Growth Rate (%)	1.4	pounds (kilograms)	1,771 (805)
Years to Double Population	49	Calories (daily)	2602
Life Expectancy (years)	66	Literacy Rate (%)	75
Infant Mortality Rate (per thousand)	44	Eligible Children in Primary School (%)	96
Per Capita GNP	$325	Percent Urban	29
Percentage of Labor Force Employed in:		Percent Annual Urban Growth 1970–1980	3.1
Agriculture	65	1980–1985	3.3
Industry	22		
Service	13		

13

China: modernization and communism

MAJOR CONCEPTS

Four Modernizations / Zero Population Growth (ZPG) / Spheres of influence / Long March / Cooperatives / Producer's cooperatives / Dynasty / Peoples Communes Barefoot doctors / Humid China / China Proper / Terra incognita / Confucian meritocracy / Mandarins / Cooperatives Treaty ports / Free Enterprise zones / Arid China / Cultural revolution / Contract System

IMPORTANT ISSUES

Chinese-Western interaction
Transforming a society
Development of a strong agricultural base
The speed and character of mechanization
Absolute population growth

Geographical Characteristics of China and East Asia

1. China is the world's third largest country in area and the largest in population.
2. China can be divided into humid eastern China where the population is concentrated, and arid western China with only sparse settlement.
3. China has always been a rural society and in spite of the revolutionary changes of the post–World War II era, remains so today.
4. The Chinese people are highly ethnocentric, reflecting the evolution of the Chinese society in isolation from the balance of the world.
5. China's economic development has been handicapped by isolation, foreign intervention, and internal conflict, and even its present rapid rates of industrialization are questioned by many who prefer the guarantees of the Communist system during the lifetime of Mao.

China has more than one-fifth of the world's population, and its area is exceeded only by the Soviet Union and Canada (Figure 13.1). Slightly larger than the United States in area, China's population of more than 1.1 billion is nearly five times as large. Although the single largest representative of the less industrialized world, China, like most of Asia, is *terra incognita* (unknown land) to most Western observers. Western ethnocentricity and belief in ethnic superiority because of the technical achievements of the Industrial Revolution have been compounded by a misunderstanding of the Chinese Communist revolution, which culminated in 1949, and subsequent events including the government-ordered massacre of an estimated 1 to 2 thousand students in Beijing in 1989.

To Western observers in the pre-World War II era, China was a land of peasant farmers with little ambition or inclination to change their lives. The confrontation between the Soviet Union and the Western powers at the end of World War II in Western Europe led to a public hostility toward communism, and the Communist revolution that brought Mao Zedong (Mao Tse-tung) to power in China after World War II resulted in transferring the hostility for Soviet communism to the emerging Chinese Communist government. Ironically, this hostility toward the Chinese Communists was not felt among contemporary observers from the United States and other Western nations. A study by the U.S. War Department (today's Department of Defense) in 1945 pointed out that:

practically all impartial observers emphasize that the Chinese communists comprise the most efficient, politically well-organized, disciplined, and constructive group in China today. This opinion is well supported by facts. It is largely because of their political and military skill, superior organization, and progressive attitude, which has won for them a popular support no other party or group in China can equal, that they have been expanding their influence throughout the past seven years.[1]

Exactly how Mao and the Chinese Communists viewed their relationship with the Soviet Union and the United States and Western powers at the time of their accession to power in 1949 is unclear. What is clear is that they did attempt to form friendly relations with the United States before the actual success of their revolution as well as immediately afterward. Mao himself offered to go to Washington, but the Western fear of communism and Western financial interests, who resented the loss of their investment in the revolutionary China, prevailed. The result was hostility toward the new Communist government, culminating in the British government's dispatch of a warship up the Chang Jiang (Yangtze River). Finally, 30 years later in the 1970s, President Richard Nixon tentatively accepted the friendship that Mao had extended toward

---

[1] Lyman P. van Slyke, ed., *The Chinese Communist Movement: A Report of the U.S. War Department, July 1945* (Stanford, Calif.: Stanford University Press, 1968), pp. 7–8.

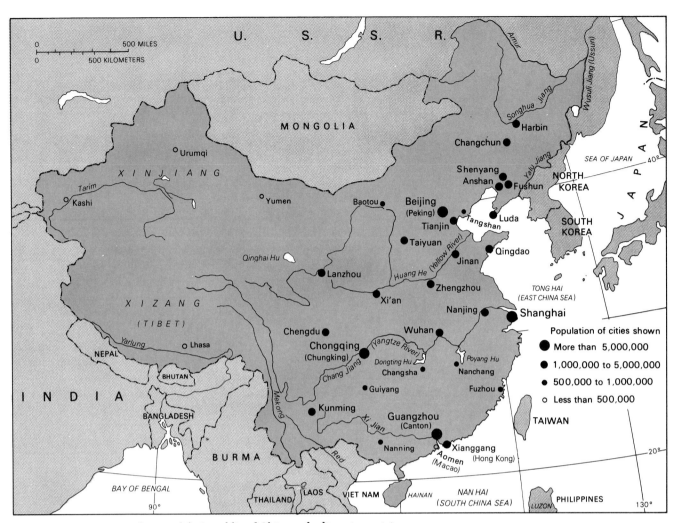

Figure 13.1 The People's Republic of China and adjacent countries.

the United States in 1947 and visited China. Since then relations between China and the United States have become somewhat more friendly, and western industrial nations are attempting to capitalize on China's immense population as both a market and a source of cheap labor.

CHINA BEFORE THE REVOLUTION

It is not the purpose of this chapter to detail the history of China, but it is important to place the Chinese Communist revolution in perspective. This perspective can be obtained only through at least a general understanding of the Chinese experience for the past 2000 years and the issues and forces for change that faced China at the time of the Communist revolution. What is now the country of China had its origin in the hearth area of the Huang He (Hwang Ho) or Yellow River area of northern China. This area developed an agricultural system based on irrigating the fertile loess soils found along the Yellow River. A series of governments ruled this hearth area and periodically expanded influence to other areas of present-day China. The term *dynasty* refers to these governments, which lasted for long periods of time. The Qin (Chin) dynasty, from which the name China is derived, united China in the period from 221 to 206 B.C. (Figure 13.2). The Han dynasty followed and during its 400-year rule adopted *Confucian meritocracy*, establishing fundamental governing characteristics that persisted until the Chinese Communist revolution. The fundamental governing principles of Confucianism were a series of proper relationships between leader and follower, master and servant, father and son.

Important elements of this political system were based on the teachings of Confucius (554–447 B.C.). Confucius taught that the emperor was responsible for the well-being of the Chinese people. He was

CHINA

357

THE CHINESE DYNASTIES	
Xia (HSIA)	2205–1766 B.C.
SHANG	1766–1122 B.C.
Zhou (CHOW)	1122–770 B.C.
Chunqui (SPRING & AUTUMN ANNALS)	770–476 B.C.
Zhanguo (WARRING STATES)	476–221 B.C.
Qin (CHIN)	221–206 B.C.
HAN	206 B.C.–A.D. 220
Sanguo (THREE KINGDOMS)	A.D. 220–265
Tsin (TSIN)	A.D. 265–420
SOUTHERN & NORTHERN	A.D. 420–589
SUI	A.D. 589–618
TANG	A.D. 618–907
Wutai and Shiguo (FIVE DYNASTIES & TEN KINGDOMS)	A.D. 907–960
Song (SUNG)	A.D. 960–1280
YUAN	A.D. 1280–1368
MING	A.D. 1368–1644
Qing (CHING)	A.D. 1644–1911
The Republic of China	A.D. 1912–1949

Figure 13.2 China has existed as a distinct country for over 5,000 years.

viewed as superior to all other monarchs, both within and outside of China, and the success of the emperor was measured by lack of war, drought, famine, or general disorder. Peace and prosperity were an indication that a particular emperor had been approved by heaven. Ruling officials who served under the emperor were drawn from a scholarly elite who exercised all power and controlled disposition of political favors, jobs, and potential wealth. Known as *mandarins*, the scholarly class was made up of men who studied the writings of Confucius for as long as 20 years and then took a series of examinations from which the top 2 or 3 percent were allowed to move into the scholarly-administrative order. The power of the mandarins lay primarily in their right to collect taxes and fees and to dispense justice. The aspiration of every peasant family was that one of their sons arrive at such a prestigious position. Even a community could benefit greatly if one of its members was selected to the scholarly Confucian elite.

From 200 B.C. until A.D. 1911, this general system of government in China persisted. No single dynasty ruled during this extended period, and there were famines, civil disorders, and human suffering, but the Confucian ideal dominated during those 2000 years. The mandarins lived in idyllic wealth, a small group of landlords owned the bulk of the land, and the masses of the peasants endured alternate periods of adequate subsistence and famine as a result of crop failure, political unrest, or changes in landlord-tenant relationships. The scholarly leaders were responsible not only for collecting taxes to operate the government, but for seeing that enough was left for those over whom they had jurisdiction.

The landlord was responsible for advancing credit to the tenant farmers, so that they could obtain seeds and grain for each year's crop. In turn, the peasant farmers were responsible for fulfilling their obligations to the landlord. In practice this formal relationship often led to excesses by the landlords or mandarins and resulted in slavery or essential slavery because the peasants were always in debt.

Periodically, as a result of natural disasters that indicated heaven's displeasure with the emperor, the governments changed. In the 2000-year period before 1949, an estimated 25 distinct governments ruled China. Changes in government led to new dynasties, based on either changing Chinese rulers or invasions by groups such as the Mongols, who established the Yuan dynasty from A.D. 1279 to 1368 after China had been invaded by Genghis Khan. The Ming dynasty, which grew out of Chinese resistance to the Mongol control of their country, followed the Yuan dynasty. The Ming dynasty lasted from 1368 to 1644, and during its rule Chinese government leaders turned inward. Anything from outside was regarded as barbaric and culturally inferior. The Qing dynasty, from 1644 to 1911, was the last of the great Chinese dynasties, and again it was imposed by outside forces. The Qing dynasty, which grew out of invaders from today's Dongbei (Manchuria) was responsible for the problems that ultimately led to the Communist Chinese revolution.

CHINA UNDER THE QINGS

The Qings, who came to power in China in the mid-1600s, were familiar with Chinese administrative techniques, and they came to dominate the most prosperous and complex society in the world at that time. The Qings inherited a China that had been slowly expanding its population and cultivated area, its arts and education, and its social and governmental organizations for more than 1500 years. Each passing century had resulted in expanding the cultivated area, the length of canals, and the road and transportation network. At the base of the Qing dynasty was the Confucian ethic and the mandarins who administered it. Central to the view of the mandarins, and the Chinese peasants as well, was that China was the epitome of civilization and cultural advancement. The Chinese referred to themselves as the *Middle Kingdom*, the central and most important kingdom on earth. To a certain extent, such a view was accurate. European countries had only recently emerged from the Dark Ages as independent nations during this period. As measured by individual standards of living, material goods, roads and postal services, or other measures of quality of life, China had a more universally high standard of living than any other civilization then on the earth. Tragically, between the mid-1600s and the beginning of the twentieth century, the Chinese standard of living fell from one of the world's highest to one of the lowest. The causes of this decline are complex, but they are largely responsible for the changes in the Chinese civilization in the twentieth century under the new Communist leadership.

CHINESE-WESTERN INTERACTION

The European voyages of discovery that predated the Industrial Revolution were partially a response to rumors of great wealth in China. Marco Polo and overland trade had brought to Europe spices and tales of the Orient, and the explorers who inaugurated the European age of discovery and exploration hoped to find and exploit these riches. Initial interaction with European nations focused on exporting tea and silk to Europe. The Europeans really had very little to offer China in the way of trade goods, since their technology was viewed as inferior. The Europeans ultimately concentrated on exports of textiles and opium and other medicinal or luxury goods to China. The opium trade became the key leverage in China, and the British government's profits were sufficient to justify (in their view) the Opium War of the 1840s, in which the British prevented Chinese leaders from limiting imports of opium.

European colonial intrusions into China never took the form of colonies for agriculture or mining. From enclaves in major cities such as Guangzhou (Canton), Shanghai, and others, Europeans controlled the economic activities that benefited them, particularly the opium trade. Portugal sent ships to Guangzhou between 1510 and 1520 and established a permanent settlement at Aomen (Macao) in 1557. The bulk of the European involvement came after 1800, beginning with the establishment of *spheres of influence* by the British, the French, the Germans, the Russians, and the Japanese. These spheres of influence involved special concessions from the Chinese to individual foreign countries that allowed them to set up a compound in a Chinese port city or along the Yangtze River. The first of these **treaty ports** was established in 1842 and the last in 1898 (Figure 13.3).

Treaty ports: Selected ports where western powers established commercial enclaves from where they traded with China. Initially five ports were opened in 1842 but eventually over 100 were designated.

The Europeans had an impact on China in two important ways. First, foreign control of key elements of the economy affected the economic development of China. In addition, British insistence that the Chinese accept the opium in spite of the desire of the emperor to end the opium imports into his country fostered widespread addiction that affected the fabric of society. The second major impact of the Europeans was to reinforce the Chinese view of their own superiority and to intensify the inward orientation of the Chinese leaders. The refusal of China's political and economic leaders to consider the technological advances of the West played a major role in the ultimate success of the Chinese Communist revolution. In the 2000 years prior to European establishment of spheres of influence, dynasties had been formed, had flourished, and had been replaced as a result of changing economic and environmental conditions within the country. At the time the Europeans were making their greatest impact on China, the Qing dynasty was in a period of decline. The dynamic Qing leadership, which had founded the dynasty in 1644, was replaced in the 1800s by inept emperors. In 1851 the Taiping rebellion began in Kwangtung and lasted until 1865. Initially supported by the Europeans because its leader was a Christian messiah, it resulted in the death of 20 million people and was ultimately ended only when the Europeans gave financial and material aid to quell the rebellion. Had the Europeans not done so, the rebellion might have culminated in replacement of the Qing dynasty by a new emperor.

CHINESE SOCIETY IN THE PRE-COMMUNIST PERIOD

As a result of European political and economic pressures, the Qing dynasty rapidly deteriorated, and the last Chinese emperor abdicated his throne in 1911. Then followed four decades of chaos in which local warlords governed their respective areas, the Communists attempted to develop a new government, and in the East, Chiang Kaishek fought the Communists and ultimately defeated them. In the 1930s the Chinese were invaded by the Japanese, who placed a puppet government in power under the former emperor. During the period of turmoil at the end of the nineteenth century and the first decades of the twentieth, the Chinese people who had enjoyed the highest standard of living in the world in the seventeenth century had suffered an absolute decline to the lowest standard of life in Asia, rivaling the poorest nations of the world.

Land and life before the revolution.   China now is, and has always been, a primarily agrarian economy. In the past, between 80 and 90 percent of the pop-

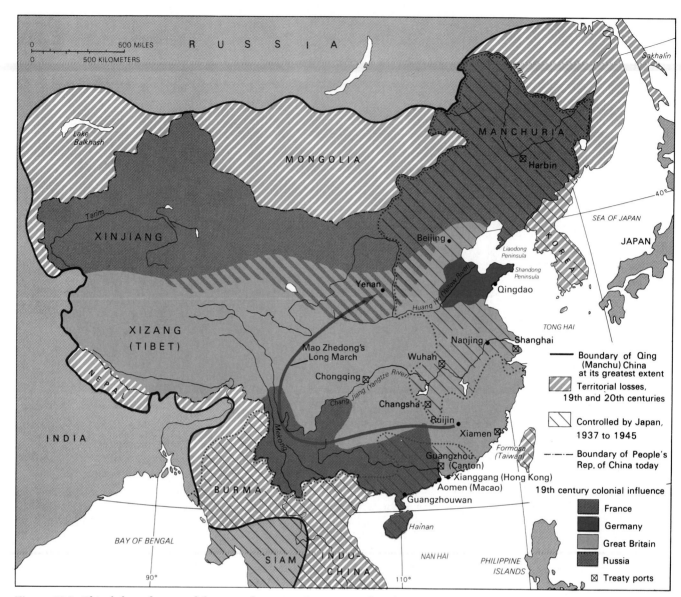

Figure 13.3 China's boundaries and foreign influence in the nineteenth and twentieth centuries.

ulation of China had been actively engaged in agriculture. Under the Confucian Mandarins, the peasants could share the productivity of their labors, but by the end of the Qing dynasty there emerged a landlord class that mocked the Confucian ideals. In the first decades of the twentieth century landlords who owned land that they rented out and did not cultivate themselves represented 8 percent of the rural population, but they owned 80 percent of the land. The other 92 percent of the population were either poor peasants who owned a fraction of an acre of land but had to rent other land to grow sufficient crops for their needs, or hired laborers who owned no land at all (Figure 13.4).

The landlords rented their property at exorbitant rates to the landless laborers and to peasants who had insufficient land. The landlord advanced seed and supplies, which the farmer was expected to pay back from the production of the farm. Interest on such loans ranged from 100 to 200 percent for the season, and it was rarely possible for the tenant farmer to escape from debt to the landlord. This resulted in such excesses as the tenant farmer selling his children to the landlord as slaves to work in his household or bonding his sons as insurance to repay the debt, resulting in a lifetime of servitude to the lord.

Even more tragically, when periodic famines struck northern China, the peasant was first required to pay the landlord from the yield, even if it meant inadequate food for his family to survive the coming year. Since the landlord required that the peasant pay him immediately at harvest time, the peasant was paying

THE WORLD OF POVERTY

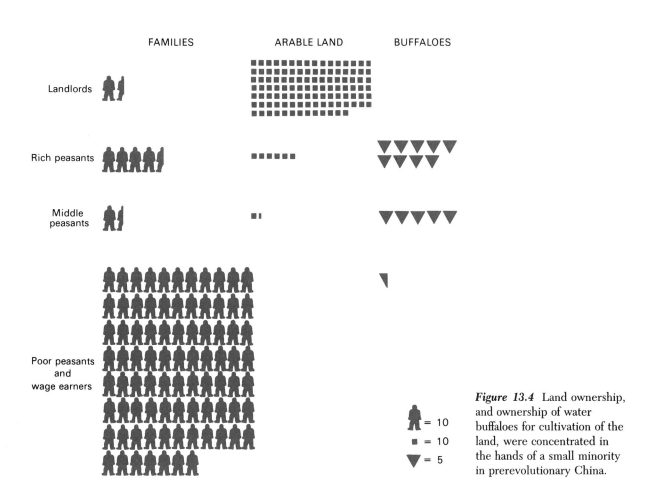

Figure 13.4 Land ownership, and ownership of water buffaloes for cultivation of the land, were concentrated in the hands of a small minority in prerevolutionary China.

when there was an abundance of grain or other crops and the prices were low, while the landlord with his surplus was able to sell during the winter months when there was a shortage of foodstuffs. The greatest tragedy occured during the times of famine when the landlords who had grain were unwilling to make it available to the peasants who had no money and as a result either had to move away or starve. In 1877 an estimated 9 million people died in the great famine that affected the Huabai (North China) Plain. In times of drought landlords seized the peasants' tiny holdings as partial repayment of their debts when the crop failed. Millions of Chinese were driven from the land to the cities of southern China, where they attemptd to sell their daughters into slavery or to obtain some type of menial employment to survive.

By contrast, the landlords and the scholar-bureaucrat class lived lives of affluence and luxury. The scholar-bureaucrats taxed the people to maintain the government, but they also had the right to charge other taxes and fees that benefited them personally. The landlords practiced conspicuous consumption and were a law unto themselves insofar as their peasant workers and tenant farmers were concerned. The Qing government failed to make reforms to correct the excesses of the warlords, the scholar-bureaucrats, and the landlord class, setting the stage for Mao Zedong and the promises of the Communist revolution.

THE COMMUNIST REVOLUTION

The China of today owes its origins to the tumult of the first 50 years of this century. With the abdication of the emperor in 1911, the Republic of China emerged under the direction of Sun Yat-sen and his disciple Chiang Kaishek. Sun Yat-sen was trained in the West, and Chiang Kaishek was the son of a middle-class merchant. Chiang Kaishek controlled the military and in the early 1920s was able to wrest control of the Chang Jiang (Yangtze River) and the northern area, including Beijing (Peking), from the local warlords with the help of the Communists. After the death of Sun Yat-sen, Chiang Kaishek turned on the Communists and killed the majority in unexpected purges. Mao Zedong survived this initial bloodbath and rallied the survivors in central Hunan province (Figure 13.10).

THE HUMAN DIMENSION

Life in Feudal China

My ancestors had passed down two *mou* of land. [A mou is about one-sixth of an acre, or .07 hectare.] After the birth of my father and his two brothers, Grandfather could not support his family with the meager income from the crop in this small plot. Although he should never have done it, he leased several *mou* of land from the big landlord Yang Chen-kang. He worked hard, the suffering increased, and the back taxes accumulated into a large debt. Yang Chen-kang had the nickname of "Black Snake," and this venomous viper coiled around our family. He wanted to recover the debt by seizing our two *mou* of land. My grandfather could never give up that land, so he had to work for Yang Chen-kang and pay back his debt with his wages. He toiled for more than 20 years, but he was unable to clear the debt. When my grandfather died, we still had that two *mou* of land, which was still insufficient to feed us. Like my grandfather, my father leased a plot of land from Yang Chen-kang. The same things happened again. My father incurred a debt and had to work for Yang Chen-kang and repay his debt with his wages. And so we slaved for the Yang house from one generation to the next.

After the birth of my second younger brother, Ch'uan-pao, which meant another mouth to feed, my mother worried all the more. One night my father returned from work and stood by the k'ang and stared at me and my two brothers. . . . My mother thought he looked strange and asked about it. But he would not say anything. At that time, my father often got together with some of his poor friends. Often he did not come back until midnight. We did not know what he was up to. As soon as someone called him from the yard, he hurried off. This added much to my mother's worries. She feared something would happen. . . .

Early on the morning of December 20, someone knocked at the door and cried: "Oh no, Chin-pao's father has been murdered!" Immediately, my mother pulled me and carried my younger brothers to the riverbank. Oh god! My father's body was there. . . .

Yang Chen-kang had said that he wanted to arrest the culprit, but in fact the culprit was Yang Chen-kang himself. My father and his friends had aroused his suspicion. He was fearful lest the poor people in the village unite against him. He decided to "show his strength by making the first move." In order to kill one to warn a hundred, he ordered one of his lackeys to murder my father. He also planned to seize our two *mou* of land.

When Grandmother heard this story, she grew both angry and vengeful. Her heart was overflowing with the death of her husband and her son, but she could not even tell the world. She lay on the k'ang mortally sick. She died a few days later.

My brothers and I helped Mother till our land. We worked from before dawn till after dusk. We were cultivating two plots of land: one was our own two *mou* of land; the other was the four *mou* of land that my father had leased from Yang Chen-kang. . . .

Who would have imagined that God himself would have his eyes closed to us? It did not rain for more than three months, and the wheat sprouts were becoming yellow and wasted. My mother's hopes for paying off our debt was crushed, and she fell sick from worry.

After the autumn harvest, Yang Chen-kang seized our two *mou* of land and took two large earthen vessels and a large wash basin away from us, which in no way diminished our debt. Yang Chen-kang also sent someone to fetch me back so that I could work for him as a shepherd and thereby work off our debt with my wages. My mother was infuriated and she swore at the man: "Black-hearted black snake! You taxed us for using a few *mou* of your land. You tortured my father-in-law to death. You murdered my husband. You seized our land. Now, you're going to lay your hands on my son!" She recalled my grandfather's words: Never make a living by filling the rice bowl of the Yang house. But if she did not send me, they would have no mercy. So, mother finally sent me off to him with her tears. . . .

Source: Peter J. Seybolt, THROUGH CHINESE EYES, Volume I: Revolution: A Nation Stands Up (New York: Praeger Publishers, 1974) pp. 24–26.

In the Hunan area more than a million Chinese peasants lived in communal agricultural cooperatives, and Mao believed these communities to be the basis for a revolution. The Chinese Communists gained local support in south central China when Mao began to redistribute land. After five extermination campaigns by the government to destroy Mao's movement, the Communists embarked on the famous *Long March*, a 6000-mile (9650-kilometer) trek to northwest China (Figure 13.3). Approximately 90,000 Communist Chinese began this march in December 1934. The army averaged 24 miles (39 kilometers) per day and fought an average of one minor battle per day, plus 15 major battles with the government troops of Chiang Kaishek. They arrived in northwest China 368 days later with only 30,000 soldiers. The Long March was important for the Communists because en route it publicized the revolution and the right of the peasants to overthrow the rich landlords and to control the land themselves. The promise of reform was significant, but the Communists also redistributed the land and other property of the landlords to the peasants, leaving in their wake thousands of peasants to train other guerrilla forces. The Long March is one of the most important epics in Chinese Communist history and is celebrated in song, drama, and literature.

The Communist philosophy of Mao Zedong was based on the premise that the peasants of China should be the owners of the land they cultivated and that their labor should benefit their families rather than the landlords. The Japanese invasion of China in 1937 prevented Chiang Kaishek from further attacks on Mao's remaining armies as the Japanese pushed Chiang's Nationalists back to the mountains of Sichuan (Szechwan). After the end of World War II and the Japanese withdrawal, the Chinese Communists defeated Chiang Kaishek and his followers. The continued poverty and misery in both urban and rural areas, which had not been alleviated under the Japanese occupation, aided the Communist cause. Increased taxes by both Chiang Kaishek's forces and Mao's forces to support their various armies only exacerbated the poverty.

The Communists were also aided in their victory by recurrent famines that plagued China. It is estimated that there had been at least one famine a year on the average in some part of the country throughout Chinese history. In northwestern China between 1928 and 1930 between 3 and 6 million people died as a result of a famine. In 1943, 3 million people died from famine in the province of Hunan alone. At the same time that peasants were dying or selling their children for the price of a meal, landlords were hoarding grain and taking advantage of shortages to raise prices and further enrich themselves. Mao's promises that the revolution would end the exploitation by the government and landlords was important in gaining the support of the peasants. Part of the significance of the Chinese revolution is that it is the first revolution by the peasant masses rather than by disenchanted middle-class or elitist groups.

The Chinese Communists spread their influence through the countryside, distributing land as they went and attracting more supporters. In major battles in 1948 a Nationalist army was surrounded and destroyed in Dongbei (Manchuria), and in central China in 1949 an additional half-million Nationalist troops were killed or captured. By the end of 1949 Chiang Kaishek and his remaining followers had fled to the island of Taiwan, the historic refuge for rulers who had been deposed on the mainland of China.

AFTER THE REVOLUTION: CHINA IN CHANGE

Upon assuming power, the new Communist government of China was faced with a number of pressing problems. The old bureaucracy had been largely destroyed, the nation was exhausted from nearly 30 years of intermittent warfare, and the economy was near total collapse. The new Communist leaders consequently implemented a number of changes that have had far-reaching effects for the Chinese nation. The first focused on agriculture, the traditional basis of Chinese life, and the focus of Communist promises for land reform. The first step was to fulfill the promise of land distribution.

LAND REFORM

From 1949 to 1952 the Communists based their land reform program on the economic standing of individual Chinese farmers. They recognized four categories based on landownership and rental versus private cultivation: large landlords with 8 or more acres (3 or more hectares) who rented out their land; rich peasants with 5 to 8 acres (2 to 3 hectares); small farmers with less than 5 acres (2 hectares); and the landless laborers.

The rich peasants and the landlords were the groups who had exploited the other Chinese peasants, and many of them suffered persecution and even death from local peasant mobs of Communist groups. Many rich peasants and landlords who had supported the Nationalist party and had given aid to the Japanese occupying forces were tried as criminals. An estimated 1 million lost their lives during the process of land redistribution in which approximately 120 million acres (49 million hectares) of land were distributed to about 350 million peasants. The resulting fragmentation of land did not eliminate the problems of hunger in China, as the surplus that had traditionally gone to the cities to support the urban populations was no longer available.

The farms that were given to the Chinese peasants averaged less than one-half acre (.2 hectare), and few peasants had the capital for draft animals, seed, fertilizer, or the other inputs necessary for successful farming. This fragmentation was compounded by the distribution of the landlords' wealth, which left some peasants with tools, others with land, but few with both. The individual peasants still faced the same age-old problems. Most labor was done by hand, equipment was nonmechanized, and rainfall rather than irrigation was the main source of water. The Chinese Communist government, recognizing this problem, determined that the peasants could never accumulate the capital necessary for improvement of agriculture from their micro-farms. The government response was the establishment of **cooperatives.** Initially these co-

---

Cooperatives: Members of a village owning their own land cooperatively cultivated it to benefit from larger-scale operation, sharing tools, labor, and draft animals.

Massive numbers of people were mobilized for land-reclamation projects by Chinese communes in the 1960s.

operatives consisted simply of pooling labor and sharing equipment and draft animals during planting and harvest seasons. The success of these measures quickly led to adoption of formal encouragement for community-wide cooperation. During the cooperative period, land still remained in the hands of the peasants, but planting, tilling, and harvesting were shared. Members were compensated on the basis of the amount of work they contributed to the cooperative. The process of organization into cooperatives was completed by 1956. During the process ever-greater cooperation led to formation of agricultural *producers' cooperatives,* in which residents of a village of 30 to 50 households combined their resources, and their land was controlled by the cooperative.

During the process of creating the cooperatives, the Chinese government embarked on *collectivization* of the cooperatives in a fashion similar to that which created the Soviet kolkhoz. The cooperatives were grouped together to form collectives, and by the end of 1957 there were 700,000 collective farms in China, comprising approximately 90 percent of the peasant households. Private ownership was eliminated except for garden plots, and collective farms were made up of one large village or a few small villages combined. The large labor force in these cooperatives could expand the amount of land under cultivation, repair irrigation systems, create more rice paddies in hill country, and during the nonplanting and harvest seasons, improve roads and found schools for the largely illiterate Chinese peasants.

After the first eight years of the Chinese Communist experience, the new nation was still barely able to feed its population. As in the Soviet Union in the early years, the peasants were not creating sufficient surplus to free large numbers of people to engage in industrialization in cities. From 1950 to 1957 the Communist government had exported food through Hong Kong to pay for needed imports of equipment, fertilizer, and technical expertise. By 1957 the Chinese population of 540 million, which the Chinese Communists had had to feed in 1949, had became 600 million.

INDUSTRIAL REVOLUTION UNDER MAO

In the industrial area, the Chinese Communists faced problems similar to those in agriculture. As in the Soviet Union, an emergent industrial base had grown rapidly after 1911. In essentially every major Chinese city from Beijing to Guangzhou, between 1912 and 1937, almost every type of industrial enterprise was developing. The Japanese invasion in 1937 critically handicapped this emerging industrial base but led to

THE WORLD OF POVERTY

Pinyin and Chinese Places

One of the challenges facing geographers is the number of names applied to an individual place as a result of language differences. This challenge is compounded when there are regional differences in spelling and pronunciation as there has been in Chinese. Westerners dealing with languages that do not use the Roman alphabet must decide what Roman letter sounds most like the sound in the language they are Romanizing. In the case of China, the most commonly used system for transforming Chinese place names into English in the past was the Wade-Giles system. In 1958 the Chinese adopted a standard system of transcribing Chinese characters into western languages called *Pinyin*. Letters in Pinyin are generally pronounced the same as they are in English. Major exceptions include Q which in Pinyin is pronounced like ch (as in cheer), X like sh (as in sheep), Z like ds (as in beds), and ZH pronounced like J (as in Japan). In this text the Pinyin spelling will be used, followed by the Wade-Giles in parentheses at the first usage. Some familiar names with both spellings are listed as examples below:

Pinyin	*Wade Giles*
Beijing	Peking
Shanghai	Shanghai
Chang Jiang	Yangtze River
Huang He	Hwang Ho River
Nanjing	Nanking
Guangzhou	Canton

Chinese Leaders' names:
| Mao Zedong | Mao Tse-tung |
| Deng Xiaoping | Teng Hsiao-ping |

expansion of important heavy industry to supply the Japanese. The primary center of the iron and steel industry in China was northeast of Beijing in Dongbei to take advantage of the rich coal resources there. In addition, textile mills, bicycle factories, armament factories, and all the others common to an industrializing country still existed in the coastal cities, and to a lesser extent in the cities along the Chang Jiang and Huang He, and in the interior cities in eastern China.

The Chinese Communists, from 1949 to 1957, expanded industrial production dramatically. Under the first five-year plan, industrial production, utilizing the existing industrial base, increased nearly 250 percent. Such a figure is misleading in a way, since the conflicts that drove the Japanese from China and subsequent conflict between the Communist and Nationalist forces had handicapped industrial production in the pre-1949 period. The achievement of such an increase was partly the result of the existing industrial base and assistance from technicians provided by the Soviet Union. The Communist achievement in spreading industrial production away from the traditional concentrations in Dongbei and the coastal cities was just as important as the increase in total industrial production. Nearly two-thirds of the industrial projects that the Communists began during the 1953–1957 period were located in cities away from the coast. By 1957 the new China had a very different industrial map from that which existed prior to the revolution (Figure 13.5). In addition to Shanghai and Dongbei, industrial centers emerged on the coal fields located between Beijing and Zhengzhou (Chengchou) to the south. The Sichuan Basin had emerged as a major coal-producing area to provide coke for steel indus-

China is relying on joint ventures with foreign firms to help modernize its economy, as in the Shanghai Volkswagen Automotive Plant which produces the Santana car.

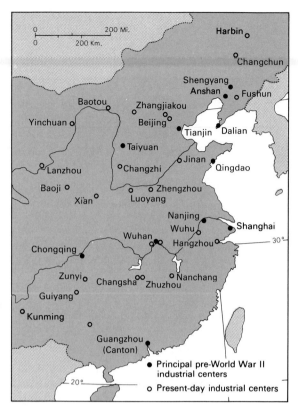

Figure 13.5 Prior to World War II, China's industrial centers were concentrated in the coastal area. Since the revolution industrial centers have emerged over much of humid China.

tries in the area. The coal in areas outside of Dongbei was the basis for steelworks at Chongqing (Chungking), Wuhan, Lanzhou (Lanchou), and Baotou (Paotou), north of the great bend of the Yellow River in Mongolia.

THE GREAT LEAP FORWARD

In order for China to continue its rate of industrial growth, by 1958 it was essential to have ever-greater production from agriculture and to increase the production of iron and steel and other industrial products. As a result, the Chinese government under Mao suggested a new revolution. Entitled the *Great Leap Forward*, it was designed to provide an agricultural surplus, to produce significantly greater amounts of iron and steel, and to rededicate the peasants to revolution. In increasing production the Great Leap Forward took two forms. In agriculture it involved transforming the cooperatives into *people's communes* during 1958 and 1959. The commune consisted of grouping cooperatives together to form a massive farm with great labor potential. The largest communes had as many as 100,000 people, although the average consisted of only 5,000 households. The Communist leadership viewed the commune as the basis for the social, economic, and political organization of China. It corresponded very closely with the Chinese administrative units of pre-Communist China and provided not only agricultural production, but schools, stores, factories, nurseries, health care, and any other service the residents required.

The second thrust of the Great Leap Forward was to utilize the labor of the communes to accomplish land reclamation, road building, and industrial production. To accomplish Mao's goals, peasants were expected to work on the commune during the day and then engage in cottage industrial production at night. This cottage industrial production consisted of small blast furnaces like those of a rural blacksmith. Over one-half million of these small, backyard furnaces were developed, and steel production increased dramatically. Agricultural production also increased as a result of the intense efforts of the communes to expand the acreage under cultivation. The increases were somewhat misleading, since much of the steel was obtained by melting down hinges, kitchen implements, farm tools, or other existing iron and steel.

On paper the accomplishments of the Great Leap Forward appeared great. Steel production increased from slightly more than 5 million tons (4 1/2 million metric tons) to 8 million tons (7 million metric tons), but government planners admitted that 3 million tons of this steel was of inferior quality. Agricultural production increased only marginally, from the 185 million tons (168 million metric tons) of 1957 to approximately 200 million tons (181 million metric tons) in 1958. Use of the labor of the communes did result in some important changes, particularly in expanding irrigation systems, and building reservoirs, wells, and roads. Unfortunately, other activities of the communes were ill-conceived, poorly thought out, and detrimental in the long run. Typical was the program to exterminate sparrows and other small birds that were perceived to be destroying grain crops. An estimated 1 billion sparrows were killed, but the result was increased insect populations, which probably caused greater crop loss than the small birds had. More important, a series of poor crop years, which greatly reduced the production of food grains, followed after 1958. Drought affected most of the Huabei Plain from 1959 to 1962, and total production fell below that of 1957.

Peasant resistance to the communal life of the communes intensified, and major changes were made in 1962. The fundamental one was to reduce the actual administration of the commune to a more manageable level based on *production teams*. These production teams consist of approximately 30 to 50 families and correspond well with the old Chinese villages.

In retrospect, the Great Leap Forward and the massive efforts of the communal labor from 1958 to the early 1960s can be viewed as a Communist experiment that had mixed results. Faced with the dilemma of overcoming poverty, providing food, modernizing the nation, providing basic human rights of jobs, literacy, and medical care, Mao and the Chinese Communists felt they had to try something. A number of activities were less than successful. In light of the problems facing them, however, what they did can be explained as part of the attempt to transform a society of 600 million people completely in only a decade. When viewed in this light, the relative success of the Chinese Communists is great.

TRANSFORMING CHINESE SOCIETY

The most fundamental and sweeping change that accompanied the Chinese Communist revolution has been in Chinese society. Pre-Communist China had all the characteristics of less industrialized countries: high birthrates, high death rates, high illiteracy rates, unequal distribution of wealth, beggars, high crime rates, and a host of other social problems. The Chinese Communists have eliminated many of these problems and have made a beginning in solving most of the others. Inherent in their efforts was the concept of ensuring minimum standards for all people. The basic tool for transforming Chinese society was education. In order to implement a program to overcome the national illiteracy, the Communists developed a simplified Chinese alphabet. Workers in factories were taught before and after work, and adult extension schools were established for farm workers. Through a national program of having the equivalent of high school graduates become primary school teachers, the Chinese Communist government expanded the enrollment to essentially all school-age children by the end of the 1950s.

To prevent loss of productivity from farmers attending school and industrial workers taking classes, the Chinese used two methods. The first was simply an extension of the work day so that factory workers worked 10 to 11 hours plus several hours in school. Students who were being trained in high school, technical school, and university were expected to devote part of their time to constructive labor, usually manual labor on farms or in factories. This requirement of Chinese education would also prevent the emergence of a scholar-elite class, like that of pre-Communist China, which was unfamiliar with manual labor.

Because of the existence of the high death and illness rates among the peasant population of China, the Communists also worked on improving health standards. The difficulty of providing sufficient doctors for such a massive population prevented the Chinese from utilizing traditional physicians who needed years of intensive training. As in education, the Chinese opted to have less highly trained students provide basic services. Female students who excelled in secondary school were then sent to one year of intensive medical training to create a group referred to as **barefoot doctors**. In essence they are the equivalent of highly trained practical nurses. They can recognize common illnesses, teach hygiene, give innoculations, and provide medication for common illnesses such as measles, mumps, flu, or other diseases. They are also trained to recognize the symptoms of more

---

Barefoot doctors: Roughly equivalent to practical nurses in the United States. Provide basic medical services of innoculations, simple medicines, and birth-control techniques.

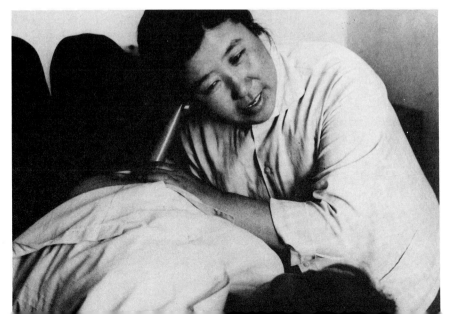

A "barefoot" doctor examining an expectant mother.

severe medical problems, such as appendicitis. When they find someone with a disease or an injury beyond their capacity, they send the patients to regional centers where medical doctors provide the more specialized service. Barefoot doctors have been tremendously effective in increasing the health standards of the Chinese people.

A final area of transformation was in granting equal rights to men and women. In pre-Communist China, peasant women were the chattels of the family whose son they married. The son of a Chinese family purchased a daughter of another family through an intermediary, and this daughter was expected to move into her husband's family where she was at the whim of her mother-in-law. Although equality of sexes is not complete in China today, as a result of the Chinese revolution they have achieved more equality than in many other less industrialized countries of the world. As in other areas of the world, change has been slowest in rural areas. The custom of foot binding has been abandoned, as have the rights of the mother-in-law and the custom of sale of daughters into marriage.

The accomplishments of the Chinese Communists in providing educational opportunities, health care, and emancipation are notable. Their achievement in raising the standard of living through increasing agricultural and industrial productivity is likewise significant. Although China still has a per capita income of only $325 per year, it has made tremendous strides. In the last decade China has undertaken another major experiment, the modernization of their economy, and society. Whether other countries can emulate the Chinese experience in development, with or without a Communist government, remains to be seen. Examination of what the Chinese Communists had to work with as they established their revolution and the process by which this transformation of a society has been accomplished provide insights into answering this question.

THE PHYSICAL GEOGRAPHY OF CHINA

The physical environment and resource base of China is complex, as would be expected in the third largest country in the world. China consists of 3.7 million square miles (9.6 million square kilometers). The nation has 15,000 miles (24,000 kilometers) of land boundaries and 9,062 miles (14,500 kilometers) of coastline. It extends from 18° to 57° north latitude, climatically from the warm, humid subtropics to the same latitudes as southern Alaska. Since it is located on the eastern margins of the world's greatest landmass, the climate in most of China is continental. In combination, the factors that affect China's climate and landforms severely restrict the amount of usable territory. Only an estimated 11 percent of China's total land area is suitable for cultivation, while 78 percent is desert, waste, or urban (Figure 13.6).

Of the 89 percent of the land that is not cultivated 32 percent is wasteland, mountains, or other areas unsuited for expansion of agriculture. Only 8 percent of China's total land area is forested at the present time. Even with the geographic limitations facing the Chinese government, the large land area provides some distinct advantages. The potential for resources in large areas of China, which have been only cursorily explored, the potential for reclamation and expansion of settlement, and the vast areas that serve to buffer the populated regions of China from its neighbors are all important. Conversely, the large land area is a problem for defense against foreign encroachment and hinders transport and communication.

THE TOPOGRAPHY OF CHINA

China's topography (Figure 13.6) is highly complex but several broad divisions can be recognized. In the west and southwest is a series of plateaus. The largest and highest is the Qing Zang (Tibetan) Plateau, consisting of plateaus over 12,000 feet (3,500 meters) above sea level. Rising above the plateaus are mountains reaching elevations of over 20,000 feet (6,000 meters). North of the Qing Zang Plateaus are a series of desert basins, the largest of which is the Tarim Basin with elevations as low as 505 feet (130 meters) below sea level. East of the Qing Zang Plateau are two plateau regions averaging 4,000 to 7,000 feet (1,200 to 2,200 meters) above sea level, separated by the Quinling (Chin Ling) mountains. South of this mountain range are the Yungui (Yunnan) and Guizhou (Kweichow) plateaus. Between these two plateaus and the Quinling Mountains is the Sichuan Basin, surrounded by mountains.

North of the Quinling range is an area of great plains. The Dongbei Plain in the northeast is a large basin surrounded by mountains on the north, west, and southeast. South of the Dongbei Plain is the Huabei (North China) Plain with its leveler lands along the Huang He River. In southeastern China, south of the Huabei Plain, is a hilly country with limited areas of level plains. In southeastern China the only large areas of level land are in the central lakes, the plains area, and along the lower Chang Jiang (Yangtze River). The coastline of China reflects the topography. North of Shanghai ports are limited because the coast is an extension of the Huabei Plain with a relatively smooth coastline. South of Shanghai the coastline is indented and there are numerous sites for harbors, but their hinterlands are very small.

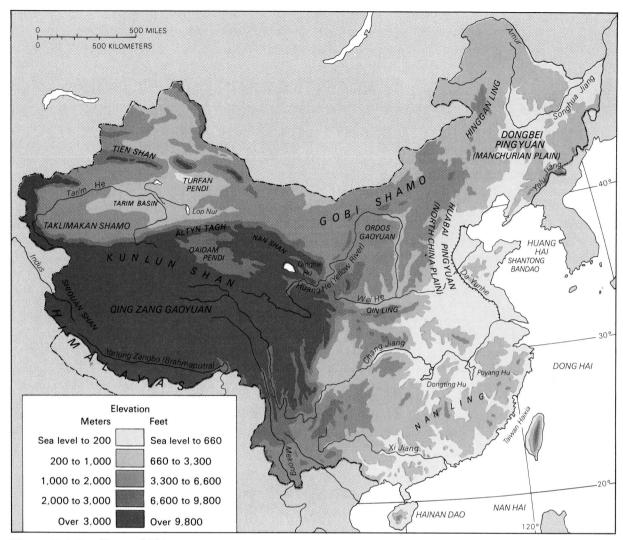

Figure 13.6 Landforms of China.

In summary, then, China has two contrasting regions: western China, which consists of either high plateaus or low desert basins; and eastern China, with gently rolling lowlands as in southeastern China or a relatively level plain as in northern China.

THE CLIMATES OF CHINA

The climates of the various landform regions of China affect their utility. The climatic controls that determine China's climates are

1. the continental location
2. the location on the eastern margins of the continent
3. the wide latitudinal range of the country
4. the resulting wind systems

The overriding factor that affects China's climate is the distance of most of the country from water. Except for the eastern coastal areas the country is isolated from potential sources of moisture by either high mountains or the great Eurasian landmass. As a result of the continental influence, the great interior basins of China are deserts. In addition, because the large landmass cools greatly in winter, northern China and Mongolia have cold, dry winters.

The coastal areas of China have a more humid climate as a result of the effect of the summer monsoon. Just as the winter's cold temperatures result from the cooling of the landmass, summers in all of China are hot because of continental heating. As the landmass heats in the summer, a monsoon effect is set up, and eastern China is humid during the summer months. Unlike the Indian monsoon, the Chinese monsoon does not result in a sudden onset of a persistently rainy season. The monsoon in China simply brings precipitation to the mainland and results in a summer maximum (Figure 13.7). The precipitation of the summer monsoon is insufficient for reliable production of crops in northern China, but is much heavier and more reliable in southern China. On the basis of the greater precipitation of the summer monsoon, the eastern portion of China is commonly called *humid China*, while the area west of the Yungui Plateau along with northern China and Dongbei are referred to as *arid China*. Even within humid China there are great seasonal variations in precipitation, causing periodic droughts. Inland from the coast, precipitation declines.

On the basis of the temperature and precipitation, it is possible to recognize five distinctive climatic types in China (Figure 13.7). The largest is the desert of western China. In Mongolia and north central China there is a transitional zone of steppe climates with their typical grasslands. In the northern part of humid China, north of the Chang Jiang (Yangtze River), is found humid continental climate. Dongbei has extremely cold winters with continuous snow cover, but farther south is found humid continental, warm-summer climate. South of the Chang Jiang humid China has a humid subtropical climate with its hot, humid summers and mild, humid winters. Xizang (Tibet) is typical of mountain regions with a highly undifferentiated climate varying as a function of elevation and exposure.

CHINA'S RIVER SYSTEMS

The river systems of China reflect the climatic regimes in which they originate, as well as the landforms from which they flow. The Huang He (Yellow River) of northern China originates in high plateaus of west central China. At an elevation of 13,000 to 15,000 feet (4,000 to 4,500 meters) above sea level, its source region is a high grassland. From its source to its mouth in the Huang Hai (Yellow Sea) the Yellow River is 3,388 miles (5,464 kilometers) long and drains an area of 290,519 square miles (752,443 square kilometers) that contain more than 49 million acres (20 million hectares) of agricultural land and a population of over 200 million people. The Huang He has always been important to the development of China. The ancestral home of today's China was in the Huang He basin, but then and now the river has also posed unique problems for successful occupance of its land.

From its origin in Xizang (Tibet) the Huang He flows north to form the Ordos bend in the Ordos deserts and steppes of Nei Monggol (Inner Mongolia). The loess soils of this region are highly susceptible to

Figure 13.7 Major climatic regions of China.

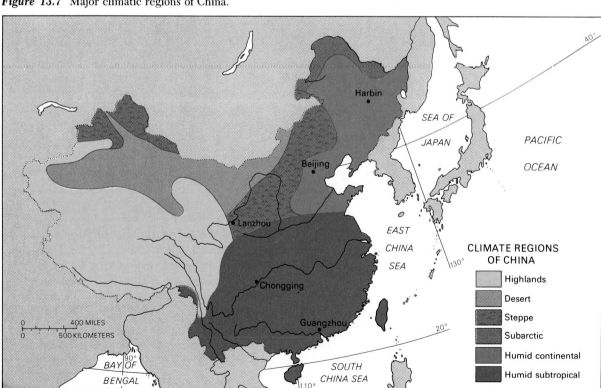

Controlling floods, providing water for irrigation, and generation of electric power are all benefits from hydroelectric projects in China. This project in Southwest China was completed in 1982 and generates 3.3 million kilowatt hours of electricity per year.

erosion. The river carries one of the heaviest silt loads in the world and floods frequently. As material carried by the river is deposited as alluvium in the lower reaches of the river, the bed builds up above the surrounding land. In most places the surface of the river is 6 to 10 feet (2 to 3 meters) above the surrounding lowlands, but in some places it is said to be as high as 32 feet (10 meters) above the surrounding land. As a result the Huang He has always flooded in its lower reaches. Twenty-six major changes of the course of the river and 1500 breaks in the dikes protecting the farmland have occurred in the past 2500 years. It is estimated that in two out of every three years there has been a flood of some type along the Huang He with associated loss of property and life. One-third of the total drainage area of the Huang He is subject to these periodic floods. The river is commonly known as "China's Sorrow" because of its repeated flooding.

The periodic flooding of the Huang He has handicapped use of its fertile, alluvial floodplain, but the river has been the focus of major, massive reclamation efforts since the revolution in 1949. Among these projects have been a 1120-mile (1800-kilometer) dike to protect agricultural land from floods, diversion dams for irrigation, drainage of low-lying floodplain, and other projects to utilize the Yellow River's potential more fully. In spite of the dike and the giant Liujiaxia Dam completed in the upper portion of the river in 1975, flooding still occurs. The most recent flood occurred in 1983, causing both immediate loss of property and life and reduced crop yields. The floodplains of the Huang He are the home of about the same number of people as live in the United States and they represent nearly one-fourth of China's total population.

Even larger than the Huang He in both size and importance, is the Chang Jiang (Yangtze River), which is 3430 miles (5488 kilometers) long. Slightly more than one-quarter of China's population live in its drainage basin. The river is highly navigable, with ocean liners able to go as far as 575 miles (920 kilometers) upstream, and small river steamers like those used on the Mississippi can travel 150 miles (240 kilometers) beyond Chongqing in Sichuan Province. The region along the Chang Jiang is much different from that along the Huang He. The river flows through a more humid area, which does not have the easily eroded loess to create high sediment loads and can be divided into three sections: the Sichuan Basin, middle, and lower river areas. Agricultural development is very intensive in the middle and lower Chang Jiang valleys with rural densities exceeding 1000 persons per square mile (386 per square kilometer). In the Sichuan Basin 1000 miles (1600 kilometers) from the mouth of the Chang Jiang, live an additional 100 million people. The three parts of the Chang Jiang are very different, with the Sichuan Basin having a mild climate but hilly terrain. The lower and middle Chang Jiang Plain is an area of gentle relief as it reaches the ocean. The head-link city for the Chang Jiang is Shanghai, located along the *estuary* of the Chang Jiang. Shanghai is the largest city in China, and its 15 million inhabitants make it second only to Tokyo in Asia. The Chang Jiang gives Shanghai a vast hinterland, and the city accounts for one-fourth of all of China's exports and one-sixth of all industrial production.

The third major river and focus of settlement in China, is Xi Jiang (Hsi River), with a length of 1590 miles (2560 kilometers). This river flows through a relatively small plains area with a high population. The Xi is used for navigation in those areas that are level enough and is the basis for irrigation and hydroelectric power. The head-link city for the Xi basin is Guangzhou, a city of more than 3 million people. An estimated 12 million people live in the lower basin of the Xi River, making it the third most populous area of China.

SOILS, VEGETATION, AND RESOURCES IN CHINA

The soils and vegetation patterns of China initially reflected climate, altitude, and latitude. In the mountains of Dongbei in the north and east the limited evaporation leaves sufficient moisture for forest growth. Although precipitation totals are less than 20 inches (80 millimeters) in the Mongolian Plains and in the Tarim Basin, the vegetation was initially grassland. In southeastern China the humid subtropical climate originally supported a mixed deciduous forest with some coniferous trees in sandy areas. In Xizang, climates vary with altitude. At elevations up to 6500 feet (2000 meters), the vegetation is a steppe grassland. Above this level the mountains were originally forested. Above the timberline, where it is too cold for forest to grow, is a zone of moss and lichens similar to the tundra.

The soils reflect the climate and vegetation. The grasslands of northern China have a mollisol soil type. Although fertile, soils of the grasslands are also plagued by inadequate precipitation and require irrigation, which sometimes results in alkaline soils. In humid eastern China the soils are generally alfisols and inceptisols. In the arid Tarim Basin are found aridisols, soils that require both fertilization and irrigation to produce crops. In general the most fertile soils are those of the plains of Dongbei and northern China, the areas of loess, the alluvium of the rivers, and those of the Sichuan Basin. Less fertile are the soils in areas of poor drainage near the coast or in the subtropical or desert areas.

Forest, soils, and grasslands have been modified by generations of occupance. Little of the original forest cover remains except in inaccessible highland regions. The soils were modified by draining waterlogged soils, fertilizing the soils of Dongbei, and even changing the alluvium of the river floodplains through centuries of cultivation. The grasslands of the steppes were extensively grazed by the flocks of the Mongols and other nomadic groups, causing erosion of much of the loess in highland regions. The vegetation has been removed in southern China to meet demands for building materials, fuel, and expanding agricultural land. As a result, the Chinese landscape that the Communists inherited after the 1949 revolution was only vaguely similar to the pristine environment when the Chinese civilization began more than two millennia earlier.

MINERAL RESOURCES

Coal is the dominant industrial resource, and with 140 billion tons (127 billion metric tons), China ranks third in the world in proven reserves. More than 60 percent of the coal is located in the northern and northeastern portions of China, in the great bend of the Huang He and Dongbei (Figure 13.8). Major production areas include Tangshan (Taishang Shan), Shenyang (Mukden), and Baotou (Paotou). Outside the old northeast, coal is the basis for important iron and steel production at Wuchan, Yichang (Ichang), and Chengdu (Chengtu). Adequate coal for several decades of use is also scattered throughout all the other provinces. The significance of these new resources is that coal provides an estimated 80 percent of the energy used in China. Coal is the basis for charcoal, for cooking and home heating, for China's steel industry, and other industrial uses. Petroleum and natural gas have not been as important in China because of the limited use of internal combustion engines in industry and transport, but is growing. China produces an estimated 150 million tons (136.75 million metric tons) of petroleum per year, which is sufficient for its needs and allows for some minor exports to Japan. Petroleum has been discovered in several areas of China, the most important in Dongbei at Daqing (Taching), and on the lower Huang He. These new discoveries are located close to populated centers where they can be readily used. Drilling is presently taking place in the North China Sea under a joint venture agreement between China and western firms. Geologic evidence indicates major petroleum deposits will be found there. Natural gas is of limited importance at present.

Iron for heavy industry is scattered in several areas of China although there are no massive reserves similar to those of India. These widely scattered deposits were the basis for early industrialization in Dongbei and subsequently elsewhere in the country. China has a wide array of other minerals, including large reserves of manganese, tungsten, molybdenum, and to a lesser extent tin, sulfur, and mercury.

CHINESE PEOPLES: THE ULTIMATE RESOURCE

The population of China is much more homogeneous than that of the Soviet Union, the first nation to undertake development through communism. The overwhelming majority—approximately 94 percent—of China's population is Han Chinese. They are concentrated in humid eastern China in what was known as **China Proper** in the past. China Proper is the humid core area of China south of the Great Wall and en-

---

China Proper: The name applied to humid eastern China focusing on the three rivers of the east.

compassing the Huang He and Chang Jiang with their productive agriculture. China Proper represents the oldest continuously governed area of the world. The people and culture of humid China Proper are the modern descendants of more than 2000 years of continuous cultural evolution as one dynasty has followed another since the beginning of recorded history in China. Although there are regional variations in the appearance of the Han Chinese, they are recognized as one group. In the past they constituted the Chinese nation.

There are a number of national minorities, with an estimated total population of between 55 and 65 million in 1990, primarily in the arid and mountainous parts of China. The Chinese government recognizes a total of 55 national minority groups. The largest are the Tai groups in the rugged interior hills west of Guangzhou, and the largest Tai group is the 13 million Zhuangs, the largest minority group in China. Other significant groups include the Hui, Kazakh, Tibetan, and Korean (Figure 13.9).

The official Chinese government policy is to recognize all nationalities through administrative subdivisions. At present there are five autonomous regions (equivalent to provinces), 31 autonomous administrative regions, and 69 autonomous areas (equivalent to counties). The autonomous regions are located near national borders where they provide autonomy to minority groups in the same way such groups make up the individual republics of the Soviet Union (Figure 13.10). The official government declaration of 1950 states that all national minorities are to have equal rights with the Han majority, are to be part of the provincial and county governments in those areas where there is a mixture of Han and minority groups, the culture, religion, and customs of the minorities will be respected (including not being compelled to learn to speak or write the Han language unless they

Figure 13.8 Resources of China. Note the coal deposits in the great bend of the Huang He River.

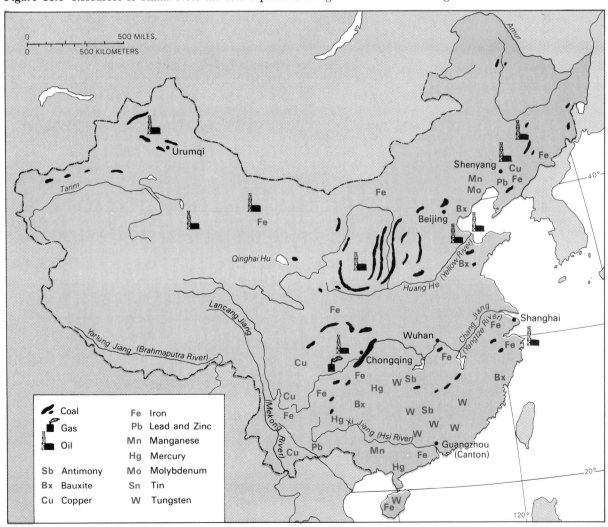

CHINA

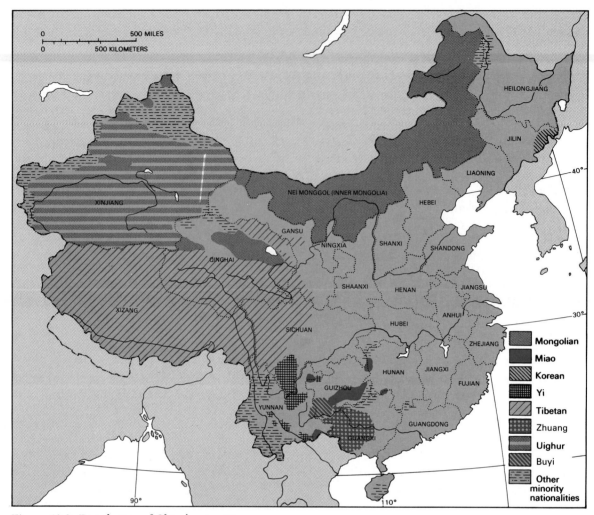

Figure 13.9 Distribution of China's minorities.

so desire), and the traditional sense of superiority of the Han would be eradicated to teach all residents of China the equality of all peoples.

One of the immediate results of the Chinese revolution for the minority groups was provision of written languages for the 30 national groups, numbering more than 13 million people in 1949, who had no written language. In practice these written languages for the minority groups are based on Han Chinese. The justification is that the vast majority of the people already understood Han Chinese and that the minority groups should at least share a language based on the same characters. This has been facilitated by the adoption of a written form of Chinese based on the Latin alphabet, which was approved in 1957 but was primarily used in conjunction with the older Chinese pictographs to indicate pronunciation.

The importance of the standardization of pronunciation and the adoption of a simplified written language cannot be overstated. By eliminating the need to master a minimum of 3000 Chinese characters, the Chinese language can be taught much more rapidly using the latinized alphabet. Adoption of a common alphabet will be a major factor in the continued integration of the national minorities and various regions into a unified Chinese state.

CHINA SINCE MAO: REVOLUTION AND CHANGE

TRANSFORMING CHINESE AGRICULTURE

The death of Mao Zedong in 1976 began a new period in China's history. Mao's role as head of the Chinese government covered nearly 30 years and witnessed revolutionary changes in Chinese society, economy, and political life. Changes under Mao reflected a continued commitment to Marxist theories of economic

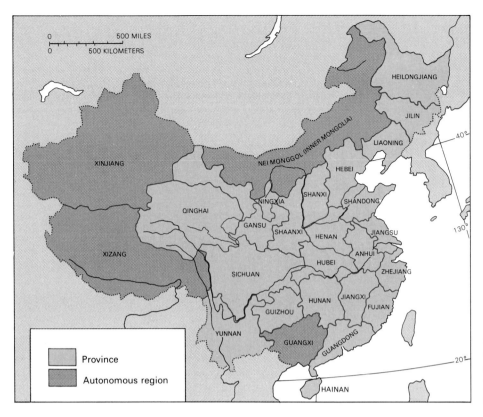

Figure 13.10 Political divisions of China. Major minority groups are given political representation as an autonomous region.

development with state ownership and control. Since Mao's death China's new leadership under the direction of the pragmatic Deng Xiaoping has made major changes in economic policies as they experiment with alternative methods to hasten economic development. The pace and extent of change from the centralized Stalinist-based Marxist economy of Mao has been so great that it is sometimes said that China is "reinventing itself." The official voice of the Chinese government, *The People's Daily* newspaper stated on December 7, 1984: "We cannot use Marxist and Leninist works to solve our present-day problems." Although subsequently modified to read "all present-day problems," it is symbolic of the abrupt change in China after Mao.

The underlying policy of change is called the *Four Modernizations* and refers to modernization and mechanization of agriculture, modernization and expansion of industry, modernization and development of science, technology and medicine, and modernization of the military. The modernizations in agriculture are a variation of the experimentation begun under Mao, experimentations necessitated by the need to feed China's nearly one-fourth of the world's population.

Because the vast majority of the Chinese population relies on agriculture for its livelihood, it was the focus of much of the Chinese Communist experimentation. The process by which land was taken from landlords, given to peasant farmers, changed to cooperatives and then to communes has already been discussed. The communes occupy 96 percent of the total agricultural area of China, and represent some of the Communist Chinese government's major attempts at change. Most of these changes were organizational. Instead of land being held by landlords and a few poor peasants with large numbers of landless peasants as before the revolution, today on the communes land is held collectively in the name of the state. Instead of a landlord—small farmer—landless peasant trichotomy, today all farmers are nominally equal. Each commune allows individual farm families to have a private plot to cultivate as they desire. The products belong to the individual farmer. The land in such private plots, however, still belongs to the commune and cannot be sold, rented, or transferred to other people. As in the Soviet Union the collectives pay a percentage of their yield as taxes to the government and pay wages based on the amount of time each individual devotes and yields that the farm attains.

The Chinese also introduced the equivalent of the Soviet sovkhoz, the state farm, in marginal areas that did not have a large existing rural population. They are primarily designed to reclaim additional land for agriculture in sparsely settled areas. These state farms

Traditional methods of planting rice continue to be the rule in China, with manual labor in planting, harvesting, threshing, and transportation the norm.

comprise only 4 percent of the cultivated area and produce less than 2 percent of the total grain output. Nonetheless, they represent one of the major changes attempted by the Communists, as they are the focus of mechanization efforts, additional fertilization, and technological changes designed to increase their productivity. In spite of this additional capital investment, the yield per acre is lower than that on the communes, where the high yield is the result of intensive labor applied to relatively small areas. Such garden-type agriculture nearly always produces higher yields per unit of land than does mechanized agriculture.

Beginning in the 1980s the policy of modernization prompted an important organizational change in agriculture. The large agricultural communes adopted what the Chinese call the *responsibility* or *contract system*, a system that a few locations had experimented with since 1978. Under the responsibility system the production brigades that carried out the communes' agricultural practices were given more independence. Brigades contracted with smaller work groups, who were entitled to keep all of the production over that required by the contract quota. The concept of the work group contracting to operate a specific parcel and share the excess over the quota profit was expanded to include the *household contract*, in which one family was given a specific parcel of the collectively owned land to cultivate. These parcels are not private plots that supplement the major agricultural production of the commune, but are collective lands operated by individual farm families who receive all of the production above their quota-contribution to the work group. (See *The Human Dimension: A Contract with Chinese Peasants.*) By 1990 the Chinese government estimated that more than 90 percent of farm families were farming under a contract system. While the land still belongs to the collective, the contract system has effectively introduced free enterprise incentives into Chinese agriculture. Many farmers produce not only their agreed-upon quota, but specialize in sideline production of chickens, ducks, vegetables, or some other product in high demand in nearby markets. Farm families in humid China have incomes that may exceed industrial workers, and the majority have television sets, bicycles, and other consumer goods. Many farm families or villages also have begun small manufacturing plants of their own, in some cases creating a new group of wealthy entrepreneurs.

In spite of the change in organization and land-ownership, the rural agrarian majority of China lives a life only moderately different from that of their ancestors. The main difference is that the strong central government with its commitment to food storage and elimination of famine ensures that there have been no more of the great famines of the past. When crop yields have been lower, the government has been willing to import additional grain, and the lack of a landlord group to exploit grain shortages prevents food hoarding, which would jeopardize hundreds of thousands of lives.

Mechanization of Chinese agriculture is proceeding slowly for several reasons. First, the government must develop armaments, transportation systems, and urban infrastructure before a massive effort can be devoted to providing mechanical equipment for the agricultural sector. Second, the sheer size of the Chinese population poses the problem of what to do with the millions of individuals who would become surplus labor if farming were mechanized. With an estimated 600 to 700 million people directly engaged in agriculture replacing even 15 percent of the labor needs with machines would free nearly the equivalent of the total employed labor force in the United States. A final reason for proceeding slowly with mechanization is the size of many of the fields. Only 11 percent of Chinese land is suitable for agriculture, and much of this has been made so by terracing hillsides with tiny fields. Traditional agricultural equipment is too large for such farms, and only the garden tiller/type tractor with small attachments is useful in these areas. In steeper areas it would be difficult to use even such small implements because of the difficulty of carrying them up the hills from one terrace to another. In spite of these difficulties, small garden tractors are in use on an estimated 20 percent of the total cultivated area.

The Chinese have increased agricultural production since the Communist revolution. Grain yields have gone from a reported 143 million tons (130 million metric tons) in 1950, to an estimated 400 million tons (362 million metric tons) in 1988 (Table 13.1). These increases have been accomplished primarily through raising productivity per acre, especially by use of chemical fertilizer. Prior to the Communist revolution very little chemical fertilizer was used, but since the revolution the use has increased from 2.5 million tons (2.3 million metric tons), in 1960 to 36 million tons (32–66 million metric tons), in 1988. More than 8.5 million tons were imported in 1988, making China one of the world's largest importers of chemical fertilizer. China has started building chemical fertilizer plants to ensure that the farmers will be able to increase inputs even further. China's use of fertilizer is much less than that of industrial nations but much higher than that of other less developed countries.

In terms of total production, rice is the dominant crop, representing over half of total grains. Wheat and corn are approximately equal and combine for a total of 43 percent of grain production. The balance is made up of millet, grain sorghum, and barley.

TABLE 13.1 Grain Production in China

	Production	
Year	Million Tons	Million Metric Tons
1950	143	(130)
1955	198	(180)
1960	172	(156)
1965	213	(194)
1970	267	(243)
1975	312	(284)
1979	318	(289)
1980	328	(290)
1983	348	(316)
1985	374	(340)
1988	400	(362)

SOURCE: Harold C. Hinton, *The People's Republic of China: A Handbook* (Boulder, Colo.: Westview Press, 1979), and *Foreign Agriculture Circular* (Washington, D.C.: U.S. Department of Agriculture, February 1984). *Beijing Review*, Vol. 31, No. 52 Dec. 26, 1988–Jan. 1, 1989.

On the basis of the primary agricultural products and major agricultural techniques it is possible to recognize specific regions within China. Xizang (Tibet) emphasizes production of sheep, yaks, and other hardy grazing animals (Figure 13.11). Agricultural crops are restricted to oats, rye, barley, and other crops that mature rapidly in the short summer season in the high altitude of the Xizang mountains and plateaus. North of Xizang, in western arid China and Nei Monggol, agriculture concentrates on grazing livestock, with or without grain production under irrigation.

There are major differences between northern humid China and southern humid China based on the predominance of wheat and coarse grain north of the Huai and upper Hanshui River valleys. Northern humid China focuses on the Huabei (North China) Plain, the largest unbroken area of arable land in China. The Huabei Plain comprises approximately 125,000 square miles (323,750 square kilometers), equal to the state of Kansas plus two-thirds of Oklahoma, but contains 75 million residents, fifteen times as many as in Kansas and Oklahoma. The climate, too, is similar to that of Oklahoma and Kansas and has traditionally been the area of famine, famine whose impact is intensified by population densities that are among the highest rural densities in the world.

Prior to collectivization, individual farms were the equivalent of about .9 acre (.4 hectare) per capita of farmland. The major crop is winter wheat, but corn is increasingly important. In addition, as the climate becomes warmer and moister toward the east and

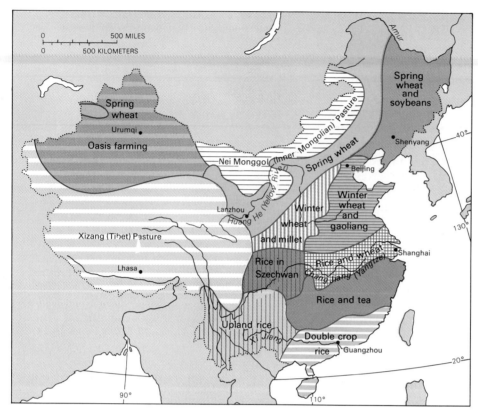

Figure 13.11 Agricultural regions of China.

south, soybeans become an important crop. Sorghum is also grown in this area, as are a variety of vegetables and either Irish potatoes or sweet potatoes. Sugar beets are a major industrial crop, and cotton is grown in the southern and eastern margins. Although not the dominant crop, rice is grown wherever conditions permit.

Southern humid China includes the humid subtropical, hilly region from the Chang Jiang Basin southward. The predominant crop is irrigated rice rather than wheat. In the northern portion only one crop of rice is grown, with wheat sown in the fall as a second crop. Cotton is also important in the transitional zone. Farther south, major crops include double croppings of rice, with as many as three crops in southeastern China. Tea, vegetables, sugar cane, sweet potatoes, and other crops are grown in conjunction with the rice. The sweet potato is a common foodstuff in this area.

This simple dichotomy is not intended as anything other than a device to illustrate the great variety within China. There are wide variations of crops within each of the regions, and there are a great variety of vegetables and small grains produced throughout China that have not been discussed.

The most important accomplishment of the Chinese Communists in agriculture has been in elimination of the starvation that in the past resulted from floods and droughts. No longer do families need to fear that they will be unable to feed all their children through the year. No longer are the peasant farmers at the mercy of rich landlords who exploit their labor and condemn them to a life of near slavery. The difficulty that the Chinese Communists have faced in attempting to overcome the perennial problems of poverty, famine, and maldistribution of agricultural goods in the rural areas has focused upon the ever-increasing population. Since the revolution, China's population gain has averaged some 2 percent per year, adding an additional 16 million mouths each year. As a consequence, although food production has increased nearly two and one-half times, the per capita production of food has increased only marginally. During the 1950–1990 period the population increased from one-half billion to more than 1.1 billion. The need to control the burgeoning population remains one of China's major challenges.

TRANSFORMING CHINA'S INDUSTRY

The Chinese Communists who came to power in 1949 faced a task of industrialization even greater than that facing the Soviet Union in 1917. Although incipient

THE WORLD OF POVERTY

THE HUMAN DIMENSION

A Contract with Chinese Peasants

Yan Jinchang, a peasant in Xiaogang village, had a contract with his production team for 1988 outlining the acreage under his cultivation, the planned output for 1988, his grain, cotton, oil-bearing crops and pig quotas, farm tax, collective (which refers to production team) accumulation funds, public welfare funds, management fees and subsidies for team cadres, and the quantities of chemical fertilizers he will be supplied at state prices, according to the contracted acreage.

The records attached to the contract showed that Yan paid his farm taxes and funds for the collective. These were his economic obligations assigned by the state and the collective. The records also showed that Yan fulfilled all his quotas. That is his contribution to the state. Income from the sale of products belongs to Yan, and above-quota products are placed at his own disposal.

"We feel quite at ease now that the contract system is in force," Yan said. "After fulfilling our task for the state and the collective, all surplus products belong to individuals. Who wouldn't work hard? Anyone who doesn't is certain to lose out. The responsibility system really meets the needs of farm production in the Chinese countryside."

The responsibility system ensures more pay for more work. In this way, the peasant's concern about direct benefits has turned into concern for the final economic results. This concern is what motivates the peasants to produce more.

In the years before the system was introduced, Yan got so little grain and money for his labor that he had to borrow money to buy commodity grain. Over the past 20 years, the Xiaogang production team bought only one water buffalo, three calves, and several wood or iron farm tools. In the five years since the responsibility system was implemented, 7 out of the 21 households (including Yan's) have bought small tractors. Seven others have saved enough money to buy their own.

In Taicang County, Jiangsu Province, 99.5 percent of the production teams have adopted the contract system. The local economy has since developed rapidly. Cao Bingsheng, a 60-year-old peasant in Taicang, said, "The thing about the household contract system which satisfies me most is that it grants us decision-making powers.

"In the past, if we wanted to ask for leave or borrow some money from the production team, we had to have the permission of the team leader. We had no time, not even a bit, to do any sideline production. It seemed that we worked only for our team leader. But now things have really changed. We can arrange and do everything according to our own specialities.

"My family's net income was 1020 yuan in 1982, when the group contract system was introduced. In 1983, when the household contract system was introduced, my family earned 2065 yuan. Every household in our county is earning more now. This year, we've got another three new products: cultured pearls, mushrooms, and garlic. In this way, we're sure we can get another 1000 yuan."

In Kouzhen village, Laiwu City, Shandong Province, many households specializing in commodity production have become well-off. The per-capita income for one family can be close to the national average.

Wang Yu, 47, has a 60-square-metre, five-room tile-roofed house and a courtyard. Strings of dried maize ears hang on the walls. On the east and west sides are cooking stoves, chicken coops, pigsties and rabbit hutches. In the courtyard, planted with more than 20 trees, are a stone mill and small water pump. The rooms are well furnished. The only thing reflecting the influence of modern living in this typical self-sufficient household is the 14-inch black-and-white TV set.

The family of five comprises husband and wife, an 18-year-old son working in a township-run paper mill, a 14-year-old daughter at school and a 20-year-old daughter working on the farm with her parents on a 0.24 hectare of contracted land producing wheat, maize, peanuts and vegetables.

In 1987 Wang Yu's family had a total income of 2,758 yuan, including a cash income of 1,555 yuan and an income of 1,203 yuan earned from farm and sideline products. After deducting production costs of 449 yuan (including expenses for tools, seeds, fertilizer, insecticide and fodder), 25 yuan in agricultural tax and 60 yuan in public accumulation funds for the collective, the family's net income came to 2,224 yuan; the per-capita income was 445 yuan, which shows a balance between income and expenditure, with a slight surplus.

Wang Yu said his family was eating better food now than in previous years, but foodgrain still formed a large part of their diet and the non-staple foods they ate were mostly medium- and low-grade. A fried dish could be guaranteed for lunch and supper every day; at the same time, homemade pickles were indispensable as a side dish.

Of the food consumed by the Wang family, 542 yuan worth or 53.3 percent of the foods consumed was self-produced. The family spends 4 yuan a year on a newspaper, *The Rural Public* and 3 yuan on haircuts and baths.

One to two film shows are projected each month in the village; the money for this is drawn from the collective's accumulated funds. Admission is free. Electricity costs 24 yuan per year, and the family spends 15 yuan on firecrackers, paper, and joss sticks (incense) for festivals and prayers.

Farmers do not pay for the well or tap water which they draw from the public utility. Wheat and maize stalks and wild plants are important sources of fuel for farmers, reducing the need to spend money on fuel. Last year, the Wang family only needed to pay the state 25 yuan in agricultural tax.

Source: Beijing Review, *44, October 29, 1984, p. 21, July 4–10, 1988, pp. 21–23 and Oct. 17–23, 1988.*

iron and steel industries had been established by the Japanese in the northeast and textile industries had been developed in coastal cities by outside powers, China was basically a nonindustrialized nation. As one of the goals of the Chinese Communists, industry became a focus of major investment and development activity. (The Soviet Union assisted the new Communist state until the mid-1960s when differences between the two led the Soviet Union to withdraw its technical expertise). At the time of the revolution, aside from steel, armaments, and textiles, most industry was small-scale and oriented to local subsistence needs.

The Communists' major accomplishments in industry have been in the fields of expanding iron and steel production, expanding production of transportation equipment, expanding production of agricultural equipment, and development of aircraft, automobile, bicycle, and other industries necessary to transform the society into a developed nation. The major center of iron and steel and related heavy industry is still in the northeast (Dongbei) in the area developed by the Japanese prior to World War II (Figure 13.12). Geographical relationships that benefit this area center on the iron ore and coal of the region. Anshan is the largest steel-producing center

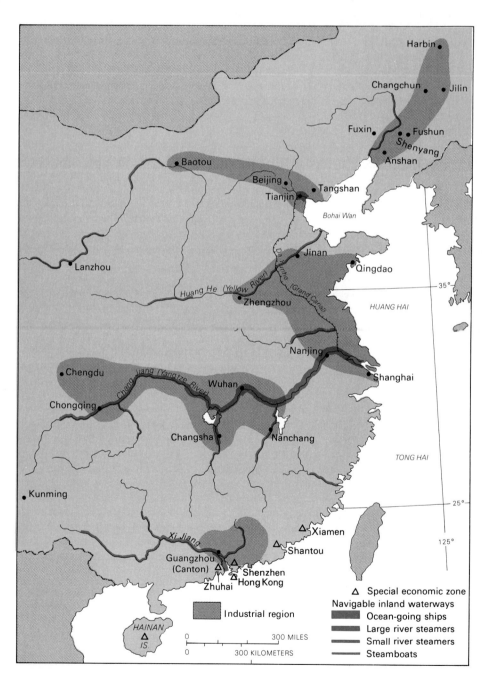

Figure 13.12 Major industrial regions and special economic zones of China.

THE WORLD OF POVERTY

380

Handicraft industries such as this rug factory in eastern China represent the beginnings of industrialization of the country.

in China, with production approaching 2 million tons per year. Shenyang is the largest city of the northeast (with an estimated 6.5 million) and produces steel and a variety of industrial goods for China including machine tools.

Originally developed by the Japanese to capitalize on the resources of coal and iron ore, the northeast has expanded greatly since the 1949 revolution. New industrial developments at Harbin, Jilin (Kirin), and Fushun (Fushan) have made each of these cities important producers in their own right. This region is China'a leading producer of steel, machinery, petroleum, and electric power. The entire northeast is analogous to the Ruhr in the role it plays for China, and its geographical relationships indicate it will remain so in the forseeable future. In terms of total value of production the northeast produces slightly less than one-fifth of China's goods.

A second concentration of industry in China is the coastal area from the mouth of the Chang Jiang (including Shanghai) to Qingdao (Tsingtao), that is referred to by the Chinese as the east. This coastal zone is the most densely settled area of China, with more than 300 million people residing in the region (Figure 13.12). It has the greatest total value of industrial production, presently producing just over one-third of China's industrial goods. Shanghai is the dominant city of this region, and of China. Its location on the Chang Jiang and its hinterland of the most intensive agricultural region in China make it the major port for both exports and imports in the country. Shanghai's locational advantages made it the focus of foreign activities prior to the revolution, and recent development as China relaxes its self-imposed isolation are bringing foreign firms back into the city.

A third important industrial region lies between the coastal region and the northeast. The Chinese call this region North China, and it centers on the region from Tianjin (Tientsin) to Beijing in the north. Small in area, it is a major coal-producing region because of the reserves in the great bend of the Yellow River and has iron and steel mills at several cities, including the new steel center at Baotou. The cities in this region also produce transportation equipment and bicycles, sewing machines, and other consumer-oriented items, totaling one-fifth of China's industrial output.

The Chinese call the fourth largest industrial area the central-south region. It consists of two nuclei of industry, one centering on Guangzhou, and the other on the Sichuan and middle basins of the Chang Jiang. Major industries are located in Guangzhou, where light industry focusing on sugar processing, silk textiles, and paper are concentrated. Iron and steel and heavy industry have been established at Wuhan, Changsa (Changsha), and Luoyang (Loyang).

The other two regions of China, the northwest and southwest, are just beginning industrial development but are much less important than these four. Both regions are sparsely inhabited, and the costs of developing industry are higher. The goal of the Chinese government is to expand industrialization in these areas to make them regionally self-sufficient in most manufactured items.

MODERNIZING INDUSTRY

Achievements from the revolution of 1949 to the death of Mao in 1976 transformed the Chinese society through the industrialization process. Although changes in agriculture were largely organizational, changes in industry have transformed much of the urban scene in China. From the regional concentration in Dongbei, industrialization has spread to other areas. Although not the world leader in any major industrial item, China ranks in the top ten in many, including iron and steel, coal, iron ore, and related items. China is exceeded by many smaller countries in the production of consumer items such as automobiles, televisions, bicycles, and other items that the Chinese have not regarded as essential to their development (Table 13.2). There remains a significant gap between the truly industrialized nations of Western Europe, Japan, the United States, and the Soviet Union and China as shown by the fact that China's steel production totaled only 40 million tons in 1986, but is estimated at 55

TABLE 13.2 China: Production of Selected Consumer Goods (000)

Year	Bicycles	Cameras	Radio Sets	Television Sets	Watches
1950	21	.....	.....	.....	.....
1955	335	.....	151	.....	.....
1960	1,840	.....	1,500	.....	650
1965	1,792	.....	1,500	5	1,200
1970	3,640	.....	4,600	15	.....
1975	5,460	.....	18,000	205	.....
1980	13,020	.....	30,040	2,948	17,633
1982	24,200	.....	17,240	5,920	33,010
1984	28,613	1,262	22,203	10,038	38,071
1986	35,700	2,150	.....	14,470	64,450

SOURCE: Harold C. Hinton, *The People's Republic of China*, (Boulder, Colo.: Westview Press, 1979); and John L. Scherer, ed., *China: Facts and Figures*. Vol. 10, 1987. (New York: Academic International Press, 1988), *Europa Year Book*, 1988.

million tons in 1990. However, the increase is significant when compared to the pre-1949 peak of 3 million tons of iron and steel.

The process of industrializing during Mao's lifetime was erratic because of changes in the government's emphasis and social experimentation such as the Cultural Revolution from 1966 to 1976 that closed universities and fostered social unrest. After Mao's death the Four Modernizations program greatly affected industry, bringing change in four particular areas: acceptance and inflow of technology, training, and capital from the industrial world; joint ventures with Western firms; establishment of **Special Economic Zones** (SEZ), and change to a market regulated economy for much of industry. As of 1990 China had entered into joint ventures with Japanese, U.S., and European firms to build complete industries, and thousands of Chinese are being trained in industrial countries. Firms from industrial countries are also engaging in manufacturing activities in the SEZ established by the Chinese. Enterprise zones allow non-Chinese firms to establish factories in China, using Chinese labor, with either foreign or Chinese management. In 1984 the entire eastern coastal region of China was designated as part of the SEZ but of the four major SEZ, Shenzhen across the border from Hong Kong is prospering most. The foreign firms control operations, prices, wages, sales, and so forth. As in agriculture, decentralization is occurring in industry as local industrial managers are given power to respond to market conditions. As of 1990 the decentralization and move to a market economy has been restricted mainly to consumer goods. Heavy industry such as iron and steel, fuels, and machinery remain under central control but success in decentralization of consumer goods may lead to a market orientation for heavy industry as well.

As in agriculture, Chinese workers in industry are no longer guaranteed a set wage regardless of effort, but receive incentives as part of a contract system similar to that used in agriculture. Small businesses can be established by individuals with government control restricted to a tax on profits.

The impact on China's industry and economy of the modernization program is profound. Where the programs of decentralization have been practiced for the past several years annual economic growth rates approach 10 percent. If the program is extended to all of China this rate may fall, but it is apparent that China is entering a period of rapid economic change as part of their program of creating a unique Chinese form of economy and society. The nature and extent of that change remains unclear after the events of 1989. Conservative leaders ordered the massacre of students who were occupying Tiananmin Square in Beijing in support of greater democracy to accompany economic modernization. Consequently, China may slow its economic development because of fear of Western influence and the withdrawal of Western investment.

Special Economic Zones: specific cities where foreign firms are allowed to establish factories to capitalize on Chinese labor.

TRANSPORTATION: A CHINESE DILEMMA

One of the factors that has handicapped industrial expansion in China has been the lack of adequate transportation. Prior to the revolution, the use of human porters to move goods of all types accounted for fully 50 percent of all transportation. Reflecting the technology of the time and the terrain, the internal transportation that the Chinese emperors developed was a network of roads with steps that could be used by humans or animals but were unsuitable for wheeled vehicles. Only in central and southern China and in the North China Plain did canals and rivers become an alternate transportation form. These waterways were the major means of moving bulk items in inland locations. The most important of these canals is the Da Yunhe (Grand Canal), which extends from Hangzhou south of Shanghai to Tianjin south of Beijing. A total of 40,000 miles (64,360 kilometers) of navigable waterways on rivers and canals were developed prior to the revolution. Railroads built by European colonial powers connected major cities, but by 1937 there were only about 10,000 miles (16,000 kilometers) of railroad. Today there are 39,000 miles (53,000 kilometers), representing a major building project by the Chinese Communists. (The United States, which is approximately the same size, has 171,875 miles, or 275,000 kilometers, of railroads.)

Road transportation is equally poor, with only an estimated 121,794 miles (196,000 kilometers) of roads that have been improved with asphalt. (The United States has more than 3.75 million miles, or 6,000,000 kilometers.) China has an additional 475,000 miles (760,000 kilometers) of improved or unimproved earth roads and tracks, or graded dirt roads 6 to 16 feet wide. Although the Chinese have developed major routes between the important cities of humid China, much of the rest of the country remains connected by roads passable only by all-terrain vehicles or human or animal traffic. The government states that 70 percent of all villages are served by roads at the present. In airfields the contrast between China and an industrial nation is also apparent. China has an estimated 325 airfields, compared to more than 15,000 for the United States. The great difference between the United States and China in transportation is somewhat misleading, since China does not have the tremendous number of automobiles, which require the intense highway network of the United States. Nonetheless, transportation remains a critical factor handicapping industrial development.

TRANSFORMING THE CHINESE SOCIETY

The transformations in agriculture and industry have had a profound impact on the Chinese society. This impact has been intensified by changes that the government has attempted in order to overcome some of the traditional problems of Chinese society. Major changes include equality for the sexes, education, minimum standards of livelihood, eradication of most crime and vice, and a decrease in the size of families. The gains in education alone are extremely noteworthy. Educational programs begun after the revolution continue to the present, with an estimated 146,164,200, or 96 percent, of the young people aged 7 to 11 enrolled in grade schools. (Although an emerging problem is associated with rural families who keep children from school to work on the family's contracted land, or send them to work in rural or urban factories.) The near-universal education presently available in China stands in stark contrast to the pre-revolutionary era in which only the elite were given the opportunity to gain an education.

Achievements in medicine have been equally impressive, exceeding that of any less industrialized country in the world. A total of 1.5 million "barefoot doctors" serve the rural population, a ratio of one barefoot doctor to every 550 rural residents. There are a total of 2.5 million medical and health workers in China according to government figures, of whom 250,000 are college-trained physicians and another 250,000 are traditional herbalist doctors. An additional 450,000 are mid-level medical practitioners who have had some college training but not the equivalent of an American M.D. The large numbers of medical doctors and partially trained medical practitioners provide Chinese citizens with an acceptable health care program when compared to that of other less industrialized countries. Free basic health care is provided to all workers, and cooperative medical care is available to all others through the commune in which they live or in their city neighborhood.

Another major accomplishment of the Chinese revolution has been in the area of population control. The rate of population growth is markedly lower than at any time in the recent past, although still higher than the government desires. This change has come about through changing traditional Chinese society. In traditional Chinese society numbers of male sons were the primary measure of an individual's success. Large families resulted, families in which female children were of little value except as laborers. To overcome this attitude toward large families, the Chinese government has instituted a number of programs. China's population grew from 540 million in 1949 to 1.1 billion in 1985. The increase of 560 million represented an annual average growth rate of 2 percent for the 35-year period (20 per thousand), and during the 1960s the growth rate reached 40 per thousand.

After the failure of the Great Leap Forward, China embarked on a massive program to curtail this population growth, and since 1971 has cut its population growth more dramatically than any other country has

This billboard in Guangzhou advocating only one child per family, illustrates China's efforts to limit family size to slow population growth.

been able to do in such a short time. By 1988 the crude birthrate had declined to 21 per thousand, and the annual growth rate had declined from 23.4 to 14 per thousand. This dramatic decrease has been one of the major changes in Chinese society. The Chinese government utilized a variety of methods to curtail population growth after 1971. Central to all was the sequence of (1) learning, (2) criticism, (3) remembering, and (4) lecture to encourage people to have smaller families. Learning emphasized the relationship between the theories of Marx, Lenin and Mao, and family planning. The learning phase involves greater awareness that revolution can come only as society is changed. Criticism is based on motivating people to criticize feudalism and anarchy. Anarchy is defined as unplanned and disorderly occurrences, including the birth of an unwanted or unplanned child. Chinese people were also encouraged to criticize the idea that more children brought more happiness and that male children were somehow more desirable than female. Remembering reminded the people of the suffering of the majority of the people before the revolution, suffering on the part of both children and parents and the miserable conditions that both endured. The lecture portion emphasized five reasons for having a smaller family:

1. to create a socialist country
2. to plan births in order to have healthy children and mothers
3. to allow more time for the mother to work and study
4. to lower the number of births so that education can be provided for the next generation
5. to limit family size to benefit the domestic economy

The methods of limiting family size were encouraging late marriage, sterilization, and use of contraceptives. In the rural and urban areas, both sterilizations and abortions are free, and individuals who are sterilized are given either bonus work points or days off from work. The legal age for marriage is 20 for women and 22 for men, but the desirable marriage age is at least 23 for women and 25 for men, a 10- to 12-year difference between China and many less industrialized countries. China began implementing its program of population control before 1980, but the decline in birthrates was insufficient to solve China's problems. A natural increase of only 13 per thousand results in an addition of 14.3 million people per year because of the large population base. In other words, it adds a larger population than that of more than half of the countries in the world. This large number of births even with a low rate of increase will be compounded in the next generations because of the increased number of females entering the childbearing years as a result of the high birthrates during the 1960s. It is estimated that there will be 22 million people (11 million couples) entering the marriage age each year in the next decade.

As a result of the pressures of such absolute population increase, China began an even greater program of population control in selected provinces in 1979 and adopted the program for the entire nation in 1980. Basically the program consists of changing the three reproductive norms of later marriage, three years' spacing between the first and second births, and no more than two children, which had been postulated under the program used in the 1970s, to one encouraging families to have only one child. The new goal for China was to reduce the natural increase rate to 5 per thousand by 1990. The end goal is **zero population growth** (ZPG) by the end of the century.

To achieve this short-term demographic transition, China embarked on a program of economic incentives. In cities, couples who have only one child and proved that they took measures to prevent additional children were given a one-child certificate. This certification entitled them to a cash grant every month until the child reaches age 14. The single-child family is entitled to the same amount of living space as a two-child family, and receives preferential treatment in obtaining housing. The children of single-child families are to be given priority in admission to schools and jobs, and when the parents of a single-child family retire, they will be entitled to larger pensions than what they would otherwise receive. A similar program was designed for rural areas where one-child couples can receive either additional monthly work points, the basis for payment on the communes, until their child reaches age 14, or larger plots for contract farming. The couples will get the same grain ration as two-child couples, and all couples regardless of family size will get the same size private plot for cultivation and same size housing lot. Should the child of a single-child family die or become disabled, the family will be permitted to have another child and continue to enjoy the same benefits.

To ensure that people do not utilize the economic incentives unfairly, there are penalties for those who break their agreements. A third child born after January 1, 1980, is not eligible to participate in the worker's family medical scheme. The grain ration for a third child in urban and rural areas will be calculated according to a higher price than that obtained under the rationing system until the child is 14 years old. In addition, if a family receives the bonuses and incentives of a single-family certificate and then has a second child, all the bonuses and bonus work points awarded them must be returned.

The answer to families' concern over having only one child is that the socialist system is superior to the feudalistic system, since the state will take care of families and individuals, and there is no need to have numerous children for labor or social security. For the childless widow or widower, there is the *five-guarantees* system: guarantees of food, clothing, shelter, medical care, and a decent funeral. Everyone is entitled to the minimum grain ration of about 500 to 550 pounds (225 to 250 kilograms) of grain per year, plus a ration of other basic foods, plus subsidies for those who need help. Each commune or factory has "respecting the old houses" where the elderly are cared for and supported by the public welfare fund. If an elderly person adopts the "respecting the old house" option, there is no need of the five-guarantees program, as all of his or her needs are provided. A childless elderly person using the five-guarantees program stays in his or her own house and is cared for by relatives or neighbors with the five basic needs provided by the commune or factory.

Ironically, the four modernizations program seems to be counteracting the one-child family goal. The advantages of a larger family for farming is once again a benefit under the contract-system of agriculture. With nearly four-fifths of China's population rural, the population growth has increased slightly since 1980 as farm families evade the one-child restrictions. China will not meet its goal of stabilizing its population at 1.2 billion at the turn of the century, but the population will be much lower than it would have been if growth rates of the early 1970s had continued. Unfortunately, the indications now suggest a population of 1.32 billion by the end of the century, 120 million more than China had planned for. (See *Human Dimension.*)

In another major break with tradition, in the 1970s the Chinese government changed the pattern of residence of newly married rural couples. Traditionally in China and most other countries the bride went to the household of the groom. In China the government is encouraging a program in which the daughter of a family that has no sons brings her husband to her household upon marriage. The law requires that the relatives not interfere with the bride and groom; that the village to which the bride belongs does not discriminate against the new groom in job assignments,

---

Zero population growth: ZPG represents a stable population that basically maintains itself without increasing or decreasing.

wages, housing, or admission of their children to school or work. In addition, the groom cares for the bride's parents when they become old and inherits their property upon their death.

The significance of the social changes in China is overwhelming. In the space of 30 years the illiterate, nonindustrialized, impoverished, feudalistic society of China has been transformed into one in which most basic needs are met, industrialization has begun, literacy is becoming universal for the young, and the country is moving into stage 3 of the demographic transition.

THE CHINESE EXPERIENCE IN RETROSPECT

Compared to prerevolutionary China, the average member of Chinese society has a better standard of living. The threat of periodic famine seems to have been eliminated. The feudalistic system of landlord and bonded sharecropper has ended, the role of women transformed to allow at least a modicum of equality with males, and the perennial population problem is seemingly beginning to be resolved. It should be pointed out, however, that this new standard of living in China is far below that of the Western industrialized nations as measured in material items. Per capita incomes in China are low, with the average gross national product under $400 per capita per year. Employment of all the youth remains a problem, as an estimated 10 million persons enter the labor market yearly according to Chinese sources. At present creating sufficient jobs for the young people entering the labor market is beyond China's financial ability. In part to counteract this, the government encourages urban workers, especially women, to retire at age 50, to free their positions for their child. In education, the Chinese are unable to do all that they would desire. Five percent of the primary school-age children are not enrolled in schools, 12 percent of the graduates of primary schools cannot go on to junior high school, and half of the graduates of the junior high schools cannot go on to senior high schools. Only 5 percent of the graduates of senior high schools are admitted into institutions of higher learning.

China has attempted to limit rural-to-urban migration to minimize the social and environmental problems found in large urban areas in other less industrialized regions. Nevertheless, China has the world's largest urban population as a result of its large population base (Table 13.3). The urbanization of China has not been completely spontaneous, but has resulted in part from direct government planning ef-

TABLE 13.3 Estimated Population of China's Major Cities

City	Population	Date
Shanghai	15,000,000	1990
Beijing	13,500,000	1990
Chongqing	8,000,000	1990
Tianjin (Tietsin)	9,750,000	1990
Wuhan	6,000,000	1990
Shenyang	6,500,000	1990
Jinan	5,500,000	1990
Nanjing	5,150,000	1990
Guangzhou	6,400,000	1990
Harbin	5,000,000	1990
Hangzhou	5,850,000	1990
Chengdu	5,000,000	1989
Changchun	6,750,000	1990
Qingdao	5,000,000	1990

SOURCE: Compiled by authors from a variety of official and unofficial Chinese population estimates.

forts. Although only 29 percent of China's population is urban, this proportion equals more than 300 million people, more than the total population of all but two countries (China and India). Since the Communist revolution, urbanization has focused on development of inland cities, which have grown much more rapidly than coastal cities. Urban growth in China is tied to the development of industrial cities located close to raw materials. Since coal is the basic fuel, the important coal-mining areas have become one focus of urbanization. Cities based on production of steel and other industrial centers have been developed following the model of the Soviet Union. In order to minimize population growth in large urban areas, the Chinese government has established a complex set of administrative procedures to hinder rural to urban movement. The four modernizations and SEZ have been accompanied by rapid growth in most cities.

In comparison to India, the other giant less industrialized country of the world, the apparent success of China in the areas of population control, education, medical care, and provision of basic needs stand out. Whether it will be a model for other developing countries remains to be seen. China is faced with the need to control its population, expand industrialization, and develop the scientific and technical expertise necessary to expand the well-being of its people. The problems of population and poverty that the Chinese face are even more severe than those that the Soviet Union faced. China is also in the unusual position of facing a hostile Communist state—the Soviet Union. In spite of these problems, China seems to be in a position to expand its development rapidly as a result of the long tradition of meritocracy inherited from the previous Chinese dynasties. Adoption of the four modernizations and the decision to allow foreign investment and some capitalistic activities in both agriculture and industry indicate China's continual commitment to economic development. The long tradition of work, excellence, and diligent industry seems to indicate that if China is successful in solving its population problem, in the twenty-first century the Chinese revolution will result in major changes in the standard of life for the peasants who compose the majority of China's population. This will occur only if government fears of loss of control by the communist party leadership do not cause them to once more isolate China from the rest of the world.

TAIWAN: THE OTHER CHINA

There is another China lying 85 miles (140 kilometers) off the southeast coast of the mainland. Officially entitled the Republic of China and informally called Taiwan, it occupies the island called Formosa by Europeans. This island has been claimed in the past by the Chinese mainland governments as a part of Chinese territory. It has repeatedly served as a place of refuge for deposed rulers or those who were temporarily out of favor with the ruling emperors. It became the home of the Nationalist Chinese when the Communists came to power in 1949. Today the People's Republic of China still claims it as an integral part of its territory. The Nationalist leaders on Taiwan and many non-Asian countries regard it as an independent, free China. This China is quite different from mainland China in standard of living, level of industrialization, and population makeup (Figure 13.13).

CLIMATES AND RESOURCES OF TAIWAN

Taiwan has a subtropical climate. Major resources include the forests, which cover an estimated 55 percent of the island. In addition there are small coal deposits in the north end of the island around the main city of Taipei.

POPULATION OF TAIWAN

Taiwan has a population of approximately 20 million people, with a growth rate of 1.1 percent per year. The population consists of primarily ethnic Taiwanese, who comprise 84 percent; 14 percent mainland Chinese;

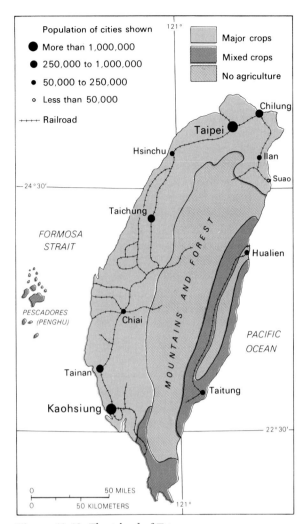

Figure 13.13 The island of Taiwan.

and 2 percent aboriginal peoples. The official language is Mandarin Chinese as a result of the imposition of the mainland Chinese Nationalists, although Taiwanese is also widely spoken and the Hakka Chinese dialect is also used. Ninety percent of the population of Taiwan is literate. Mainland Chinese dominate the country politically and economically, controlling the government and the important positions in industry.

ECONOMY

The economy of Taiwan is very different from that of mainland China. The labor force is distributed very differently, with only 20 percent in primary activities of fishing, farming, and lumbering. Forty-one percent of the labor force is in secondary activities, including manufacturing and construction. An additional 39 percent work in tertiary industries in commerce and services. The per capita gross national product is more than 10 times that of mainland China, averaging $3500.

Agriculture focuses on production of rice, sweet potatoes, sugar cane, and tropical and subtropical fruits and vegetables. Major industries include textiles, clothing, chemicals, wood products, electronics, and shipbuilding.

Taiwan is rapidly changing from a developing to a developed country, and represents one of the newly industrialized countries (NIC). Although the per capita income is far below that of the mature industrial economies of the West, it is far above most of those in the less industrialized countries of Asia. It is important to note that the great strides in industrialization and standard of living in part reflect the tremendous amount of foreign aid that the United States has given Taiwan. Taiwan has also enacted favorable legislation to attract foreign companies that have capitalized on its low wage rates. The result is the emergence of Taiwan as a major exporter of items requiring a high labor input.

TAIWAN: CHALLENGES FOR THE FUTURE

Changing events in the 1980s raised numerous questions concerning the future of Taiwan. At the end of the 1970s, the People's Republic of China was formally recognized as the Chinese government, replacing Taiwan in the United Nations. The United States, the major supporter of Taiwan, changed its attitude toward the mainland Chinese government, recognizing it as the official representative of the Chinese people. Most countries with whom Taiwan trades have also recognized mainland China as the official Chinese government. This shift presents a major challenge to the Chinese minority who rule Taiwan, as they maintain that Taiwan is China. One of the major potential conflicts of the next few decades may center around mainland China's attempt to regain control of Taiwan. The industry of the island would be a great benefit to the to the mainland Chinese.

A second challenge facing the Taiwanese government is the trade upon which the country relies. Japan, the United States, and Western Europe are all attempting to increase their trade with the mainland Chinese, the world's greatest potential market. As trade with mainland China increases, Taiwanese trade may well decline if mainland China insists on discriminatory measures against Taiwan as a basis for granting trade agreements. Events since the formal recognition of the People's Republic of China, however, tend to indicate that this perceived threat may not be as severe as the Taiwan government had feared. Industrial and trade development in Taiwan has increased at an even more rapid rate since the recognition of mainland China. Although this could change with time, it is doubtful that Taiwan will suffer economically as a result of the recognition of mainland China.

Taiwan is often used as an example of how a less industrialized country can be transformed into an industrial country without communism. Although true, it should be remembered that this transformation required massive inputs of capital from industrialized nations. If the Western world is concerned about form of government and individual freedoms, they will have to embark on income transfers to less industrialized countries to promote economic development without a totalitarian or centrally planned government.

HONG KONG: CHINA'S WINDOW TO THE WORLD

Hong Kong represents a relic of colonialism in Asia. It is a crown colony of the United Kingdom occupying the northeast side of the broad estuary of the Xi (Hsün) River. Hong Kong is one of the most crowded areas in the world, with 5.8 million people in its 400 square miles (1036 square kilometers) of territory. The majority of the population is concentrated on the island of Hong Kong itself which has only 32 square miles (82 square kilometers). Geographically it is part of China, and even the population is 98 percent Chinese. Economically it lacks natural resources and relies on China for food and water. Its reason for existence is its excellent harbor.

Hong Kong island was ceded to the British in 1841 as one of their treaty ports, but portions of Hong Kong on the mainland were acquired on a 99-year lease, which expires in 1997. In the fall of 1984 China and the United Kingdom agreed that Hong Kong would become a part of China under the concept of "one country, two systems." China maintains that Hong Kong will be able to maintain its present economic rights, freedoms, and life-styles for 50 years after rejoining China in 1997. The colony is an anomaly. It is a newly industrialized country, with the majority of the labor force engaged in manufacturing, particularly textiles, electronics, plastics, and light metal products. Hong Kong has been the destination of more than a million refugees from China and a focus of Western economic activity. Hong Kong provided China access to the West during the life of Mao when the government nominally refused relations with the industrial world. Contacts with Western firms were handled through the firms of Hong Kong Chinese, enabling China to gain needed technology without loss of face. Since the change in view of the Chinese government after Mao's death, China does not rely on Hong Kong as much, but the colony still provides an

The harbor at Hong Kong is one of the busiest in Asia.

important area for trading with the West. Hong Kong's own industrial productivity provides important technology for China and makes it one of the rapidly industrializing countries of the world. The continued importance of Hong Kong to mainland China as a supplier of manufactured goods and technology, as well as the potential for China to show Taiwan that former territories can rejoin China and still have local autonomy, suggest that China will fulfill its promises concerning the status of Hong Kong within China after reunification in 1977.

THE TWO KOREAS: A NATION DIVIDED

South of Dongbei (Manchuria) in China is the Korean Peninsula, homeland for a nearly homogeneous population of nearly 67 million Koreans (Figure 13.14). These people speak the same language, in the past had the same religious heritage based on Confucianism and Buddhism, and were a united nation until after the Japanese were driven from the area at the end of World War II. At that time the peninsula was divided into North Korea (the Democratic People's Republic of Korea) and South Korea (the Republic of Korea). North Korea has the larger landmass, but has only half the population of the smaller South Korea (Table 13.4). Geographically, the landforms of the two areas are quite similar—a hilly landscape with relatively little level land. There is a coastal plain along

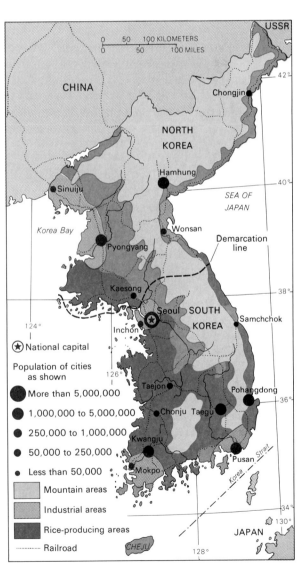

Figure 13.14 The Korean Peninsula.

CHINA

389

TABLE 13.4 Comparison of the Two Koreas

	North Korea	South Korea
Population (millions)	23.1	43.8
Per Capita GNP	$910	$2400
Average annual population growth rate	2.5	1.3
Percentage of labor force employed in:		
Agriculture	43	36
Industry	30	27
Service	27	37
Percent urban	64	65
Life expectancy at birth	68	68
Per capita calories (daily)	3151	2841
Energy consumption, per capita pounds (kilograms)	6301 (2864)	3439 (1563)

SOURCE: *World Development Report* (New York: The World Bank, 1987), and *World Population Data Sheet* (Washington, D.C.: 1988).

the west coast of Korea, but it also has hills within it. Few of these hills are extremely high, and most are under 7000 feet (2000 meters). All of North Korea has a humid continental climate, as does nearly half of South Korea. Only in the southern south-facing valleys of south Korea is there a humid subtropical climate.

KOREA'S UNFAVORABLE LOCATION

Korea has suffered throughout its history from its location between the Chinese mainland and the Japanese islands offshore. As a result, it has been the routeway for attempts by the Japanese to gain control of part of the Asian landmass. The eventual result of the continued interaction between Japan and China was to turn the Korean people inward. After sixteenth-century invasions by Japan, Korea accepted a status as a tributary to the Chinese Qing dynasties but proceeded on a quasi-independent path. Its desire to be free from the problems of its central position caused it to seek isolation for the next 300 years. Isolated from changes imported from the West, Korea became an even more feudalistic society than mainland China. Social changes were virtually nil, and the government became increasingly dictatorial and demanding of taxes from the people. The intensity of the isolation of the Korean peoples led to their designation as "the Hermit Kingdom."

Korea's isolation lasted until the end of the nineteenth century, when Japan occupied Korea and annexed the peninsula. From 1895 on, Korea was exploited as a basis for industrial development, food, and raw materials for the glorification of the Japanese, who viewed themselves as a superior people. Upon gaining independence in 1946, Korea was partially occupied by Soviet-supported troops and partially by American-supported troops. The resulting governments reflect the input of the two groups. North Korea opted for a Communist form of government based on the Soviet model and implemented by Koreans who had lived in the Soviet Union. South Korea initially opted for a democratic form of government loosely modeled after the American system, but with numerous political parties. The division of the two regions was cemented after a conflict in the early 1950s, which involved American and other Western forces in support of South Korea against Chinese and Soviet-supported North Korea. This divison is unfortunate because the two halves of Korea complement one another very well.

Mineral resources in the two Koreas are varied and widespread. Major minerals include coal and iron ore, tungsten, and small deposits of manganese, molybdenum, nickel, lead and zinc, graphite, and gold. If Korea were one unified nation, its resources would be sufficient for industrialization, with the exception of petroleum. The division found the bulk of the best coal and iron resources, more than half of the hydroelectric potential, and many of the other mineral resources in North Korea. Just as the two Koreas complemented one another in minerals, so too the agriculture of the two countries is complementary. The 67 million people of Korea rely on only 13 million cultivated acres (5.3 million hectares) for their support, and the bulk of this land, and the best land, is located in South Korea.

The complementary relationships of Korea were recognized by the Japanese conquerors, and North Korea was developed as an integral part of the industrial base of Japan, while South Korea became a source of rice and other food for Japan's growing urban population. Since the division of the two nations their contrasting character has lessened; yet they are still markedly different in agriculture, climate, and political organization.

NORTH KOREA

North Korea adopted the centrally planned government and economy based on the Soviet model. Agriculture was reformed in a fashion similar to that of the Chinese, moving through land distribution, to a cooperative phase, and then into rural communes. Major crops in North Korea include corn, rice, soybeans, Irish potatoes, millet, and barley. In addition, North Korea produces a wide variety of vegetables and fruits. Major industrial cities include Pyongyang (1.5 million), Chongjin (300,000), and Kinchaek (265,000).

Although North Korea had the advantage of the Japanese impetus in industrial development, the Korean War destroyed or damaged much of the industrial base. Since the end of the Korean War, North Korea has rebuilt much of its industry, but the standard of living has not increased as rapidly as that of South Korea or even as rapidly as in the Soviet Union. North Korea remains a Communist, centrally planned society, with the basic necessities of life, food, shelter, education, medical care, and a job provided for all. It lacks opportunities for the bulk of the people to increase their standard of living by very much in the short term.

SOUTH KOREA

South Korea has approximately the same amount of cultivated land (7 million acres, or 3 million hectares) as North Korea, although it is a smaller country. More important, this land is suitable for intensive wet rice cultivation and double cropping (Figure 13.14). Major crops include rice, barley, soybeans, and cotton. In addition, there is widespread production of beans, sweet potatoes, and corn. Animals are relatively unimportant in both Koreas. Industry in South Korea has expanded rapidly as a result of both foreign aid from the United States and government legislation favoring foreign companies. The industrial productivity of South Korea is now much greater than that of North Korea. In agriculture South Korea has followed a completely different role. Since there had developed in what is now South Korea the typical landlord and tenant laborer division, agrarian reform measures were introduced after 1945 by American occupation forces. The Americans expropriated nearly 700,000 acres (283,300 hectares) of land that had formerly belonged to Japanese and provided farms for approximately

South Korea is an example of the newly industrializing nations of East Asia, capitalizing on their highly skilled population to produce a variety of industrial products for world markets.

600,000 peasant families. There still remains in South Korea a landlord class and a landless laboring class, as well as tiny peasant farms in which the farmer tills his own land and also labors for a large landholder. Typical farm sizes in South Korea range from 2 to 4 acres (.8 to 1.6 hectares).

South Korea's rapid growth in industry and standard of living reflects the foreign aid that has come into the country, as well as its role as a producer of industrial items that require a great deal of labor. Major industries include textiles and clothing, food processing, chemical fertilizers, chemicals, plywood, and electronics.

At present, North and South Korea are tentatively discussing the potential for unification, but such a possibility seems remote. South Korea continues to expand its industrial base, and increase its standard of living, and in spite of the widespread variation in income and standard of living, there seems little justification or reason for it to join with the North Koreans except on the basis of shared culture. The development of the South Korean economy since the end of the Korean War makes reunion for economic reasons unjustifiable. Only the homogeneous population would justify reunification, and the division between the essentially atheistic North Korea and the religious South Korea would be difficult. At present the contrast between North and South Korea is an interesting example of the potential for development between a centrally planned and a noncentrally planned economy.

FURTHER READING

APPEL, B. *Why the Chinese Are the Way They Are*. 2d ed. Boston: Little, Brown, 1973.

BARNET, A. DOAK and CLOUGH, RALPH N., eds. *Modernizing China: Post-Mao Reform and Development*. Boulder, Colo.: Westview Press, 1985.

BAUM, R., ed. *China's Four Modernizations, the New Technological Revolution*. Boulder, Colo.: Westview Press, 1980.

BRIDGES, BRIAN. East Asia in Transition: South Korea in the Limelight. *International Affairs* [Guildford] Vol. 64, no. 3, Summer 1988. pp. 381–392.

BUCHANAN, K. *The Chinese People and the Chinese Earth*. London: G. Bell and Sons, 1966.

———. *The Transformation of the Chinese Earth*. London: Bell and Sons, 1970.

"Capitalism in the Making." *Time*, April 30, 1984, pp. 25–35.

CHANG, S. "Modernization and China's Urban Development." *Annals of the Association of American Geographers* 71, 1981, pp. 202–219.

CHUANJUN, WU. "Geography in China." *The Professional Geographer* 36, November 1984, pp. 179–180.

CHISHOLM, M. *Modern World Development*. Totowa, N.J.: Barnes and Noble, 1982.

CHIU, T. and So, C., eds. *A Geography of Hong Kong*. London: Oxford University Press, 1983.

CHU, D.K.Y. "Some Analyses of Recent Chinese Provincial Data." *Professional Geographer* 34, 1982, pp. 431–437.

CLAYRE, ALASDAIR. *The Heart of the Dragon*. London: Harvill Press, 1982.

CRESPIGNY, R.R.C. *China: The Land and Its People*. New York: St. Martin's Press, 1971.

DING, C. "The Economic Development of China." *Scientific American* 243, 1980, pp. 152–165.

GINSBURG, N. "China's Development Strategies." *Economic Development and Cultural Change* 25, 1977, pp. 344–352.

GUOHUA, YANG. "Curbing Births Key to Future." *Beijing Review* 31, March 7, 1988, p. 58.

HARDILL, IRENE. "The Shenzhen Experiment." *Geography* 71, April 1986, pp. 146–148.

HINTOM, H. C., ed. *The People's Republic of China: A Handbook*. Boulder, Colo.: Westview Press, 1979.

HO, S.P.S. *Economic Development of Taiwan*. New Haven, Conn.: Yale University Press, 1978.

HOFHEINZ, R. and CALDER, K. *The East Asia Edge*. New York: Basic Books, 1982.

HSIEH, CHIAO-MIN. *China: Ageless Land and Countless People*. New York: Van Nostrand, 1967.

HUANG, P.C.C. *The Development of Underdevelopment in China*. White Plains, N.Y.: M. E. Sharpe, 1980.

HSU, MEI-LING. "Growth and Control of Population in China: The Urban-Rural Contrast." *Annals of the Association of American Geographers* 75, June 1985, pp. 241–257.

JAO, Y. C. and LEUNG, CHI-KEUNG, eds. *China's Special Economic Zones: Policies, Problems and Prospects*. New York: Oxford University Press, 1986.

KIRBY, RICHARD J. R. *Urbanization in China: Town and Country in a Developing Economy, 1949–2000 A.D.* New York: Columbia University Press, 1985.

KNAPP, R., ed. *China's Island Frontier*. Honolulu: University Press of Hawaii, 1980.

KOLB, A. *East Asia*. London: Methuen, 1971.

LEUNG, C. and GINSBURG, N., eds. *China: Urbanization and National Development*. Chicago: University of Chicago, Department of Geography, Research Paper No. 196, 1980.

LEUNG, C. et al., eds. *Hong Kong: Dilemmas of Growth.* Canberra: Australian National University Press, 1980.

LIANG, E. *China: Railways and Agricultural Development, 1875–1935.* Chicago: University of Chicago, Department of Geography, Research Paper No. 203, 1982.

LOHR, S. "Four New Japans' Mounting Industrial Challenge." *The New York Times,* August 24, 1982, pp. 1, 35.

MA, L. "Preliminary Results of the 1982 Census in China." *Geographical Review* 73, 1983, pp. 198–210.

MA, L.J.C. and HANTEEN, E., eds. *Urban Development in Modern China.* Boulder, Colo.: Westview Press, 1981.

MA, L.J.C. and NOBLE, A., eds. *The Environment: Chinese and American Views.* New York: Methuen, 1981.

MURPHEY, R. *The Fading of the Maoist Vision: City and Country in China's Development.* London: Methuen, 1980.

MURPHEY, RHOADS et al., eds. *The Chinese: Adapting the Past. Building the Future.* Ann Arbor: University of Michigan, Center for Chinese Studies, 1986.

MYERS, R. *The Chinese Economy: Past and Present.* Belmont, Calif.: Wadsworth, 1980.

NGOK, L. and CHI-KEUNG L., eds. *China: Development and Challenge,* vol. 2. Hong Kong: University of Hong Kong, 1979.

PANNELL, C., ed. *East Asia: Geographical and Historical Approaches to Foreign Area Studies.* Dubuque, Iowa: Kendall/Hunt, 1983.

PANNELL, C. W. and MA, L.J.C. *China: The Geography of Development and Modernization.* New York: Wiley, 1983.

PANNELL, C. W. and WELCH, R. "Recent Growth and Structural Change in Chinese Cities." *Urban Geography* 1, 1980, pp. 68–80.

PARISH, W. and WHYTE, M. *Village and Family in Contemporary China.* Chicago: University of Chicago Press, 1978.

"Peasants' Revolt." *The Economist* 306, January 30, 1988, p. 27.

ROBOTTOM, J. *Twentieth-Century China.* New York: Putnam, 1971.

SALISBURY, C. *China Diary After Mao.* New York: Walker and Co., 1979.

SONG, JIAN et al. *Population Control in China: Theory and Applications.* Westport, Conn.: Praeger, 1985.

SONGQIAO, ZHAO. *Physical Geography of China.* New York: Beijing: Wiley, 1986.

TAN, K. C. "Small Towns in Chinese Urbanization." *The Geographical Review* 76, July 1986, pp. 265–275.

TAWNEY, R. H. *Land and Labor in China.* New York: M. E. Sharpe, 1977.

TREGEAR, T. *China: A Geographical Survey.* New York: Halsted Press/John Wiley, 1980.

———. *A Geography of China.* Chicago: Aldine-Atherton, 1965.

TUAN, YI-FU. *China.* Chicago: Aldine-Atherton, 1969.

WALKER, TONY. "Dismantling China's Communes." *The Age* as reprinted in *World Press Review,* June 1982.

WANG, JICI BRADBURY, JOHN H. "The Changing Industrial Geography of the Chinese Special Economic Zones." *Economic Geography* 52, October 1986, pp. 307–320.

WHITE, T. H. "China: Burnout of a Revolution." *Time,* September 26, 1983, pp. 30–49.

YOUNG, GRAHAM, ed. *China: Dilemmas of Modernization.* Beckenham, U.K.: Croom Helm, 1985.

Borobudur, Java.

PROFILE (1990)

Population (in millions)	460	Energy Consumption Per Capita—	
Growth Rate (%)	2.0	Pounds (kilograms)	805 (375)
Years to Double Population	35	Calories (daily)	2385
Life Expectancy	63	Literacy Rate (%)	67
Infant Mortality Rate (per thousand)	68	Eligible Children in Primary School (%)	93
Per Capita GNP	$730	Percent Urban	31
Percent of Labor Force Employed in:		Percent Annual Urban Growth, 1970–1981	3.9
Agriculture	58	1980–1985	3.3
Industry	13		
Service	29		

14

Southeast Asia and the Pacific the colonial heritage

MAJOR CONCEPTS

Atolls / Elongated state / Paddy / Tropical ecosystem / Sawah / Indochina / Insurgent state / Territorial shape / Superimposed boundaries / Plantation economy / Confucian ethic / Plebiscite / Domino theory / Economic takeoff / Fragmented state / Nationalization / Trust territories / Transfer payments / Prorupt state / Compact state

IMPORTANT ISSUES

Destruction of tropical ecosystems

Boundary conflicts and former colonial borders

Population migration and cultural conflicts

The domino theory and conflict in Southeast Asia

The impact of European colonialism in today's world

Adoption of the Western model of industrial development versus traditional values

Geographical Characteristics of Southeast Asia and the Pacific

1. This broad region occupies the margin of the Asian landmass and the vastness of the Pacific Ocean.
2. The entire region is characterized by tropical climates and tropical ecosystems.
3. This region was dominated by European colonial powers until the Japanese invasions during World War II, and the individual countries gained independence only after the war.
4. Southeast Asia has important reserves of tin and petroleum, but the Pacific Islands have few resources other than their physical setting.
5. Some countries of Southeast Asia are newly industrialized, whereas others remain among the least industrialized. The Pacific Islands are not industrialized, but some have developed as tourist destinations for wealthy residents of the industrial world.

Sandwiched between the Asian giants of India and China is Southeast Asia and its geographic extension, the Pacific islands (Figure 14.1). Contrasts between Southeast Asia and the Pacific islands are apparent to even a casual observer, but the general level of development and history as colonial possessions provide a thread of unity. The region as a whole illustrates the difficulty of explaining the world's present pattern of wealth and poverty and the impossibility of stating the causes of underdevelopment definitively. European interest in Southeast Asia and the Pacific realm antedates the Industrial Revolution and subsequent changes in Western Europe. European desire to gain access to the spices of Asia spurred advances in navigation by the Portuguese and voyages of exploration by Europeans (including Columbus).

At the time of first contact, the mercantile interests of Europe wanted to capitalize on Southeast Asia's spices, but the exploitation that accompanied the Industrial Revolution was absent. The relative levels of development in Southeast Asia and the Pacific and in Europe in terms of quality of life and economy in the fifteenth and sixteenth centuries were roughly the same. In both regions the predominant economic activity was agriculture, cities were small, and manufacturing consisted of handicraft and cottage activities. In subsequent centuries, the European nations emerged as industrialized powers, their societies and economies were revolutionized, and their impact on regions like Southeast Asia and the Pacific became more exploitative. The causes of the European development and industrialization and the Southeast Asian and Pacific lack of development are complex and include elements of the physical geography, relative isolation and accessibility, and the availability of resources in Southeast Asia and the Pacific.

The Industrial Revolution in Europe was preceded by the earlier agricultural and economic revolutions, which allowed concentrations of population in large industrial centers. Southeast Asia did not experience the same revolutions, in part because the tropical climate that prevails throughout the region handicapped overland transportation and development of trade connections between the individual countries. Europeans also did not face the problems that the tropical environment presented to agriculture, particularly in the production of livestock. Diseases that thrive in the high temperatures of the tropics were another element hindering modernization efforts. Lack of an adequate overland transportation system limited contacts between nations to the cities at coastal or riverine locations. The complexity of the landforms of the region further contributed to physical isolation, as the coastal lowlands and river valleys are separated by rugged mountains covered with dense forests.

Coal and iron ore, the resources upon which the European Industrial Revolution was based, are largely absent from the region, and the mineral resources that are available do not provide the same basis for modernization. While it is incorrect to conclude that lack of development in Southeast Asia and the Pacific is environmentally determined, characteristics of the environment in Southeast Asia and the Pacific made

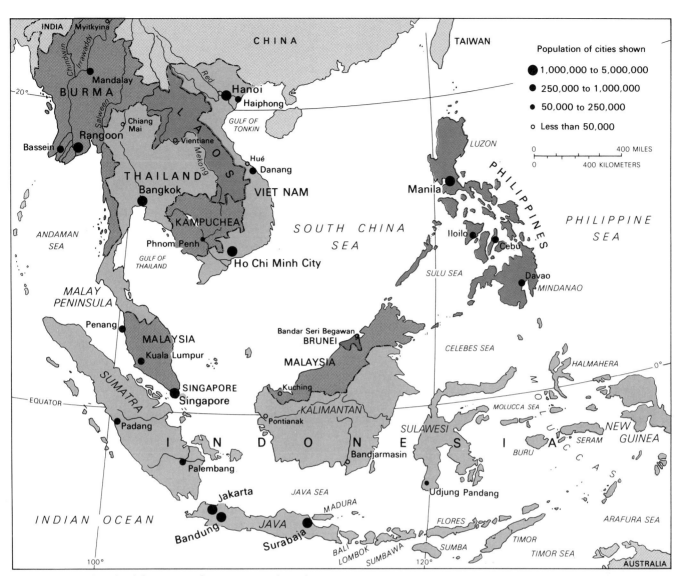

Figure 14.1 Political subdivisions and major cities of Southeast Asia.

development more difficult, and combined with the isolation of the region from the rest of the world, effectively prevented the region from achieving the same transformation as northern and western Europe.

REGIONAL DEFINITIONS

The boundaries of Southeast Asia and the Pacific are defined to the north and west by Asia's two great civilizations: China and India. The southern and eastern margins are more indeterminant as the islands and *archipelagos* (island groups) of Southeast Asia become more scattered to the south and east in the Pacific. Geographically the region is made up of the continental margin of Asia and the islands and archipelagos of Southeast Asia and the Pacific.

CLIMATIC UNIFORMITY

The unifying characteristics within this area are climate and vegetation. The climate throughout the entire region is tropical, with temperatures exceeding 60°F (15°C) throughout the year. Although portions of the region have a dry period from a monsoon or a more savanna-like climate, there is no truly arid region in Southeast Asia and the Pacific (Figure 14.2). Prevailing winds in the islands and along the coasts of the continent cause rain shadows in some locales, but the region generally receives high precipitation. Vegetation reflects the precipitation, with tropical rain

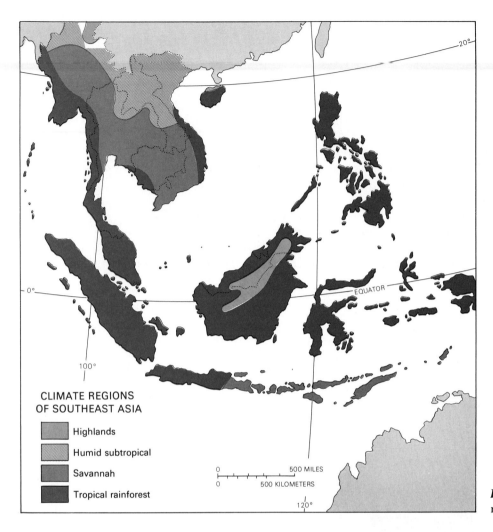

Figure 14.2 Climate regions of Southeast Asia.

The Tropical savannah climate region in Southeast Asia has soils fertile enough for sedentary agriculture. In Thailand irrigation systems provide water during the dry season.

THE WORLD OF POVERTY

forest predominating, but there are also areas of savanna.

CULTURAL VARIATION IN SOUTHEAST ASIA AND THE PACIFIC

The unity of the tropical ecosystem is in marked contrast to the cultural variation and political fragmentation in this region. Successive migrations of peoples have moved through Southeast Asia and into the Pacific islands, and the resulting cultural diversity has become extremely complex. Elements of Indian, Chinese, Middle Eastern Islamic, and Western culture after the age of exploration can be found in the region. The language diversity is illustrative, with four major language groups: the Austronesian family (including Malay, Javanese, Samoan); the Sino-Tibetan family (Thai, Burmese, Mandarin Chinese, Cantonese); the Austro-Asiatic family (Khmer, Vietnamese), and the Papuan family (New Guinea, New Hebrides). In these families there are hundreds of distinct languages and dialects, and only rarely do all the people of any one country speak a common language. The culture diversity of the area represents one of its major centrifugal forces (Figure 14.3).

There are also broad variations in economic development. Some of the countries are among the poorest of the world, while others have reached the stage of newly industrialized countries and at least one (Brunei) is a member of the Surplus Capital Oil Exporting nations. Overall, the peoples of the region share in the general pattern of underdevelopment with its pov-

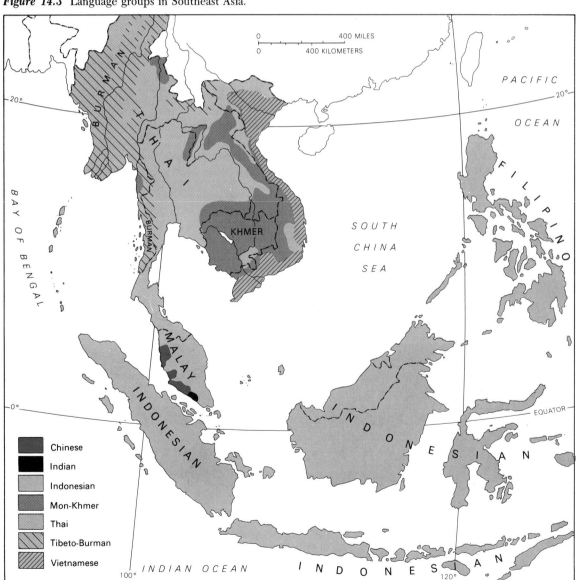

Figure 14.3 Language groups in Southeast Asia.

erty, illiteracy, high infant mortality rates, inequitable distribution of wealth, inadequate services, and rapid population growth. Politically the countries of the region are also diverse, ranging from quasidictatorships to parliamentary governments, to Marxist-oriented Communist governments. Analysis of the geography and developmental experience of this region illustrates the difficulty of prescribing a common developmental path for the less industrialized countries of the world.

GEOGRAPHICAL CHARACTERISTICS

The geographical characteristics of Southeast Asia and the Pacific center on the maritime location of the entire region. Only Laos is landlocked (Figure 14.1). The region covers a vast area of the earth's surface, stretching 8,000 miles (12,800 kilometers) from western Burma to the Tahitian islands. Most of the vast area is water, but there are also nearly 2 million square miles (5,180,000 square kilometers) of land. Mainland or continental Southeast Asia contains 807,000 square miles (2,090,130 square kilometers) in Burma, Thailand, Kampuchea (Cambodia), Vietnam, Malaysia, Singapore, and Laos. There are 915,000 square miles (2,369,850 square kilometers) in the islands and archipelagos of Southeast Asia proper. By comparison, the scattered islands of the Pacific total less than 10,000 square miles (25,900 square kilometers) of land area.

The geographical characteristics of the region can probably be most easily understood by considering, first, those of the continental margins and then those of the islands or insular portions of the region. The continental margins of Asia are dominated by rugged and complex topography. Mountains rarely exceed 10,000 feet (3,000 meters), but they are steep and covered with dense forest and are sparsely populated. The important geographical features of the mainland are the river valleys. In these valleys lives the population core of each individual country: the Red River of northern Vietnam, the Mekong River and its valleys, the Irrawaddy River basin, and the central lowland of Thailand.

The islands and archipelagos of Southeast Asia are even more complex. They form part of the Pacific Ring of Fire, and unlike the continental margins of Southeast Asia, suffer periodic volcanic activity. Indonesia, for example, has more than 145 volcanoes, and the island of Java alone has more than 50 that are active. Vulcanism and subsequent erosion and deposition have created a landscape of great complexity, and the human use of the insular portions of Southeast Asia reflects this great variation.

The landforms of the Pacific islands are of two general types, volcanic and coral. Volcanic islands tend to have a mountainous core surrounded by lowlands of eroded material, while the coral islands form low atolls.

The landform diversity of Southeast Asia and the Pacific is in marked contrast to the homogeneity of the climate of the region. No other area of the world of comparable size has such a uniform climate. The temperature is typical of tropical realms, and there are no marked seasonal extremes as found in the middle latitudes. The only distinction is between tropical savanna and tropical rain forest. Within approximately

The volcanic islands of Tahiti are important for tourism from the industrial countries.

10 degrees of the equator the climate is a true tropical rain forest with no dry season. Abundant rain falls every month, but it comes in the form of convectional precipitation with brief, heavy showers each day. North of the tenth parallel are the tropical savanna regions with alternating wet and dry seasons. Unlike India where there is a prolonged drought during the dry season, the savanna lands of Southeast Asia generally receive scattered precipitation, with only certain areas of the mainland in a rain shadow receiving less than 60 inches (1,500 millimeters) per year. The wet season in the savanna lands may consist of a single wetter period in the high-sun months from May to September, or there may be two wet seasons with a dry period between. In either case the wet season is characterized by heavy convectional precipitation modified by orographic barriers.

The rainfall tends to be torrential and, with the constant high temperatures, causes extreme erosion and leaching of the soils to a degree unknown outside the tropics. These problems limit the agricultural practices of the region. The luxuriant tropical rain forest can grow on these infertile soils, and its presence has sometimes led observers from the industrialized Northern Hemisphere to conclude that tropical soils are more fertile than they actually are. The **tropical ecosystem** is a delicately balanced phenomenon that stores the nutrients necessary for maintenance of the forest canopy in the plants until the death or destruction of the trees. Tree roots absorb nutrients freed by leaf fall or decaying trees and recirculate them into the system.

Clearing these forests destroys the delicate tropical ecosystem and intensifies soil erosion. The use of the tropical rain forest is restricted to three activities: harvesting of forest products such as teak or mahogany lumber and rubber from the *hevea* rubber tree; shifting agriculture in which land rather than crops is rotated; or sedentary agriculture in locations with higher soil fertility. An estimated two-thirds of the mainland and the islands of Southeast Asia are still covered with forest of one type or another.

MINERAL RESOURCES

The difficulties that the tropical environment poses for agriculture in Southeast Asia are nearly matched by deficiencies in mineral resources. Although much of the region has been only superficially surveyed for economically useful minerals, there probably are not substantial mineral deposits awaiting discovery. Of the significant minerals in the region, only tin is of worldwide importance. Tin-producing earths extend from southwestern China in a discontinuous zone southward through Burma, Thailand, Malaya, and some of the Indonesian islands (Figure 14.4). Tin is by far the most important mineral produced in Southeast Asia, and since World War II the region has accounted for more than half of the world's total new production. Peninsular Malaysia is the largest single producer of tin in Southeast Asia, typically accounting for more than one-third of the world's total production.

Relatively large quantities of iron ore are located in the Philippines and the Malay Peninsula, and the tonnage (but not the value) of iron ore production exceeds that of tin. A variety of other important minerals is found in Southeast Asia including manganese, tungsten, bauxite, copper, and to a lesser extent gold, silver, zinc, and chromium. In the islands of the Pacific are important deposits of bauxite, nickel, and manganese. The importance of the Pacific islands for phosphates used in fertilizer is well known, particularly the tiny island of Nauru, which consistently ranks as a major producer.

The fossil fuels of oil and coal have become increasingly important in Southeast Asia. Coal is found in the Philippines, Indonesia, Thailand, and Vietnam. The Tonkin coalfields of northern Vietnam are the most important and are the basis for nearby iron and steel industry. The coal resources of the other countries are small and scattered and are primarily of sub-bituminous quality. Petroleum is important in many Southeast Asian countries, and Brunei on the north coast of Borneo produces enough to be classified a capital-surplus, oil-exporting country. Indonesia is another major producer of petroleum, and Vietnam and Malaysia are actively exploring for petroleum. The shallow waters of the region from southern Vietnam to the island of Java offer great potential. At present Southeast Asia produces as much petroleum as China, and the total production of the region far exceeds that of India. Natural gas fields recently discovered in the Gulf of Thailand indicate that major resources may yet be found in the region.

THE CULTURAL GEOGRAPHY OF SOUTHEAST ASIA AND THE PACIFIC

The continental rim of Asia, the islands of Southeast Asia's archipelagos, and the Pacific islands have been the home of humans for countless millennia. Fossil remnants indicate that half a million years ago early

---

Tropical ecosystem: The collection of all living organisms in the tropical environment and their inter-relationships.

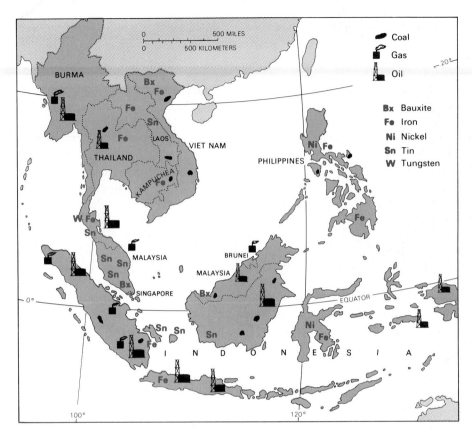

Figure 14.4 Resources in Southeast Asia.

ancestors of human beings lived in the islands of Indonesia. The occupance of the region by modern humans is postulated by some scholars to have led to the earliest plant and animal domestications. Carl Sauer hypothesized in 1952 that the first sedentary agriculture was practiced in Southeast Asia rather than in the Middle East as the archaeological records suggest.[1] Subsequent archaeological finds indicate that the domestication of plants and animals in Southeast Asia may have occurred 5000 years earlier than in the Middle East. The present economy of Southeast Asia is derived from a combination of the earliest plant and animal domesticates and cultivation practices and the subsequent diffusion of crops and techniques from China, India, and Europe.

The people involved in the development of the Southeast Asia and Pacific region are of four major groups: Australoid peoples, Negrito peoples, Melanesoid peoples, and Mongoloid peoples. The Australoid and Negrito peoples seem to be short in stature (approximately 5 feet, or 1 1/2 meters) and dark skinned. The Australoid people have a long head and wavy hair while the Negrito people have a rounder head and wooly hair. The Melanesoid are somewhat taller (5 feet 5 inches) but are also dark skinned. The process of migration and interaction leading to these three groups is so complex and poorly understood that we will not attempt to document it here. The fourth group, the Mongoloid people, are migrants from east-

A fisherman from Thailand.

[1] Carl Sauer, *Agricultural Origins and Dispersals* (Cambridge, Mass.: M.I.T. Press), 1969.

THE WORLD OF POVERTY

Schoolboy in a mission school in the New Hebrides Islands.

Father, daughter, grandaughter in Indonesia.

their ethnic background is uncertain. Their ancestry clearly includes Australoid and Melanesoid peoples, but some observers postulate a Caucasoid influence as well.

PRECOLONIAL SOUTHEAST ASIA

The peoples of Southeast Asia and the Pacific developed an economic system based on the early domestication of plants and animals in the region. Prior to European intervention the economy was primarily agricultural with some small handicraft activities and metalworking with local tin and copper ores. The vast majority of the population relied on agriculture, and the agriculture was generally uniform, reflecting the generally similar climate of the entire region. Agriculture consisted of either sedentary villages cultivating rice as a staple crop supplemented by a variety of vegetables and fruits; or *shifting* (slash and burn) agriculture in the less favorable areas. These two forms of agriculture remain dominant throughout Southeast Asia and the Pacific today, although landholding patterns and the crops cultivated have changed since contact with Western powers.

The most important agricultural system in terms of acreage and population supported has always been the continuous cultivation in sedentary villages known as *sawah* agriculture. The term *sawah* is an Indonesian word for a rice *paddy*. Technically the term *paddy* refers to unhusked rice, but it is also the name for the flooded fields in which the rice is grown. Permanent cultivation in Southeast Asia relies heavily on pro-

Samoan woman.

ern and southern Asia and are today the distinctive Vietnamese, Malays, Khmers, and other peoples of the mainland. The intermixing of these groups over at least the past 15,000 years has created the very ethnically complex region with several hundred recognizable groups of people. On the islands of the Pacific farther from the Asian mainland, the particular combination of peoples is referred to as Polynesians;

Rice is the dominant food crop throughout Southeast Asia. Its cultivation, harvesting, processing, and distribution occupy the majority of the region's population. In this scene, women separate the grain from the chaff using a centuries-old process.

duction of irrigated rice, using either artificial irrigation systems or natural rainfall. The water serves to protect the flooded soil from erosion and excessive leaching and combines with the nutrients released by rice production to maintain soil fertility over millennia. Rice is the predominant crop of the *sawah* system, with more than 50 percent of the permanently cultivated land of Southeast Asia devoted to it. By comparison China devotes only one-third of its farmland to rice, and India only one-quarter. Rice is so important that it is estimated that as many as 3000 varieties, with growing seasons ranging from 60 days to one year, are produced in Southeast Asia.

Along with the production of rice in Southeast Asia there have long been garden plots producing a wide variety of fruits and vegetables. Production from these permanently cultivated lands has always been primarily for local farm family consumption, but a surplus has also been produced to provide for the relatively small towns that existed prior to the European contact with the region. The agricultural base of Southeast Asia before European contact involved relatively few animals, as farms were small and the steeply terraced hillsides were often inaccessible to the primary form of draft power, the water buffalo. Poultry and pigs are more common, as they can forage for themselves and live on waste products, but their care was incidental to the production of rice from the *sawah*.

The other major form of agriculture prior to European involvement in Southeast Asia was slash and burn (or shifting) agriculture, which was practiced widely in both Southeast Asia and the Pacific islands. To Europeans familiar with crop rotation rather than land rotation, shifting agriculture seemed wasteful and destructive of the forest, but in reality it is a highly effective means of using the tropical environment. Before Europeans became involved in Southeast Asia, the low population densities allowed the shifting agriculturalist to farm land for one or two years and then leave the land to rejuvenate for 20 or 30 years and regrow the forest necessary to support another cycle of cultivation. Population increases in the past decades have placed greater pressure on the shifting agricultural system, limiting the amount of time the land can lie fallow and threatening the ecological balance of the newly industrialized countries of Southeast Asia.

In the original form of shifting cultivation, farmers cut the forest in a half-acre to 1-acre (.2 to .4 hectare) parcel and allowed the trees to dry during the dry season. The lumber and underbrush were burned just prior to the onset of the rainy season, and a variety of crops was planted in the ashes. Traditionally the tools were simple, most often involving simply a stick or a hoe to make a hole for a seed or tuber. Planting just at the beginning of the wet season was essential for two reasons: to allow the crop to benefit fully from the nutrients of the ash left from burning; and to provide early germination to give some protection from soil erosion.

The great variety of crops planted under shifting cultivation seems to lack pattern, but in reality they reflect the accumulated wisdom of the past 10,000 years. Crops are sown in those areas best suited to each type, whether the root crop cassava (manioc), perennials such as bananas or yams, or even rice. The great multiplicity of crops (polyculture) ensures that something will produce even during times of drought or insect infestation. The plots have sometimes been compared to the natural multiplicity of plants of the uncleared forest. The complex of plants reflects the high degree of rationality of a system that to Europeans appeared irrational. The necessity to move to newly cleared land every one or two years required a large land area, and Europeans rarely recognized that what seemed to be virgin forest was actually part of a cycle of cultivation.

In nearly all countries of Southeast Asia, fishing supplemented farming. The shallow waters of both Southeast Asia and the Pacific islands provide an important source of protein, and coastal fishing has always been important in the region, probably antedating the development of agriculture. Tree crops are also important in shifting cultivation systems, particularly in the Pacific islands where the coconut palm provides a mainstay of the diet. The shifting cultivator today produces some crops for sale, but the system was initially one of subsistence.

POLITICAL DEVELOPMENT BEFORE EUROPEAN CONTACT

The cultural diversity of the Southeast Asian and Pacific region combined with the physical barriers separating the populated core areas from each other hindered the development of great empires like those of India or China before European intervention. King-

A slash-and-burn farmer uses a hoe to plant tubers. The felled trees in the background are what remain after the first burning. They will be piled and burned again.

doms existed in the area of Thailand, Burma, Kampuchea, Vietnam, and Indonesia, but there was no effective organization of the separated populated core regions into one unit. In the Pacific islands the small land area separated by large expanses of ocean effectively limited the political organization to individual islands or groups of islands. Even the strongest kingdoms in Thailand and Kampuchea were handicapped by the difficulty of transportation and communication within their territories. The primary transportation was by water, and in many instances the existence of population in isolated locations effectively prevented their complete political control. The environmental difficulties of political organization were compounded by the cultural fragmentation and diversity, reflecting the location of the region between India and China. The relatively small Southeast Asian political units with their weak central governments and poorly defined boundaries were transformed by the colonial powers after the fifteenth century.

THE EUROPEAN IMPACT: COLONIALISM AND ITS RELICS

The European association with Southeast Asia and the Pacific from the age of exploration until the present exemplifies the problems and potentials presented to the less industrialized world by long association with the West. Initially the European explorers came seeking spices and developed a *symbiotic* (mutually beneficial) trade relationship with the Southeast Asian peoples. As the Industrial Revolution progressed in Western Europe, the individual nations increased their interest in Southeast Asia, ultimately dividing the whole area except Thailand into colonies. The European colonial system has affected the fabric of life and the landscape throughout Southeast Asia and the Pacific. It is not limited to the development of exploitive activities before the twentieth century but persists today in conflicts between non-Asian powers to obtain influence in the region. European influence was felt in agriculture, transportation, city development, population and cultural patterns, and modern political patterns. Conflicts between the European powers as they expanded their trading and territorial designs into Southeast Asia and the Pacific are reflected in today's conflicts in the region.

Although the Portuguese were the first Europeans to reach the area in the early sixteenth century, by the mid-seventeenth century the dominant power in the region was the Netherlands. Britain and France did not become as actively involved until after the beginning of the nineteenth century, being content to establish parasitic trading cities prior to that time. British expansion eastward from India and French expansion westward from present-day Vietnam led to agreement in principle to maintain the country of Thailand (then called Siam) as a buffer between the two imperialist powers. Ultimately the British annexed portions of western Thailand and the French acquired parts of the former Thai empire east of the Mekong. Partition of territories among the Dutch, the Portuguese, the French, and the British further complicated the political map of Southeast Asia, particularly as the Polynesian islands of the Pacific did not have items that the Europeans perceived as valuable. The European powers divided their territories strictly for ease of administration. The boundaries were all **superimposed** (Figure 14.5).

The French government organized its possessions in Southeast Asia into the Indochina region. Indochina was divided into Kampuchea, Laos, Tonkin (North Vietnam), Annam (South Vietnam), and Cochin China. This division split the Annamese people of Southeast Asia among three different colonial administrative units. In the present Indonesian archipelago, the Dutch united the diverse peoples into the Netherlands East Indies. Although these boundaries did not coincide with the cultures, their effective enforcement by European colonial powers provided the basis for subsequent demands for independence along the colonial boundaries.

The European desire to obtain important tropical products changed the urban organization of Southeast Asia dramatically. In Burma, for example, the populated core of the country was in the savanna zone nearly 500 miles (800 kilometers) inland from the sea around today's Mandalay. The British developed Rangoon as the political, administrative, and trade center for the Burma colony and ultimately reoriented the country from the interior to the coast. In other regions European powers created entirely new cities, such as Singapore established by the British in 1819 because of its strategic location to control trade through the Straits of Malacca. Ho Chi Minh City (Saigon) is basically a French creation, Manila reflects its development as an administrative center for the Spanish in the Philippines, and Djakarta was developed by the Dutch. All the largest cities of Southeast Asia today except Bangkok and Hanoi were coastal ports developed by Europeans to exploit the region's resources.

To obtain the tropical products that had initially lured the Portuguese explorers to Southeast Asia, the European powers changed the agricultural and mining systems of Southeast Asia for their benefit. Initially the Europeans simply used the existing institutions to obtain locally produced spices and other tropical crops that had attracted them to the region. As they attempted to introduce rubber and other products that were not native to the area, they began the process of change in the agricultural system that still presents problems. Introduction of the rubber tree and coffee took three major forms: adoption as a part of the shifting cultivation cycle where they were planted after the first year of vegetable production; enforced production through specifying that a percentage of the permanently cultivated land (sawah) be devoted to the designated exports; and expropriation of communal property for use in large plantations owned and operated by European-based companies. The first use of rubber and coffee trees by individual shifting cultivators was a positive diffusion to the region. The trees were compatible with the tropical environment and provided a cash crop that did not require constant attention. But the inability of the Europeans to control the quantity and quality of the product when there were so many producers restricted this method of production. Even so, the individual producer was the primary source for rubber and other goods until the development of corporations in Europe provided the necessary capital for a new enterprise to survive until the trees (which provided most tropical exports) began to produce.

Once European corporations became actively involved in production of tropical crops in Southeast Asia, dramatic changes came to the rural society and landscape. Throughout Southeast Asia and in the Pacific islands, communal ownership of land was the rule rather than the exception before European dominance of the region. Each community owned lands that were permanently cultivated in the case of alluvial plains or other suitable lands, and other land that was cultivated intermittently as a part of the shifting cultivation system. The European emphasis on private ownership led the colonial powers to grant the lands not permanently cultivated to corporations on generous terms and to recognize village elders or others responsible for allocating the communal sawah lands as the absolute owners of the property. Consequently an elite class of landowners was created from a formerly self-reliant, subsistence-oriented, egalitarian community. Where the pattern of the agrarian society prior to European colonization emphasized production for survival, the Europeans introduced the concept of individual profit maximization, and through granting absolute rights to one class of the local population, created another, larger group of landless tenants. Emphasis on production of cash crops for export led to food deficits in the traditionally self-sufficient

---

Superimposed boundaries: Superimposed boundaries are those placed over the existing political and cultural patterns, fragmenting cultural groups and joining peoples of diverse background.

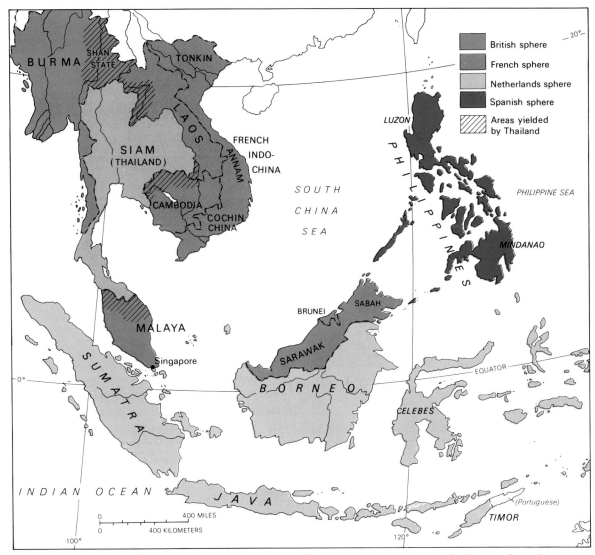

Figure 14.5 Colonial holdings of European powers in Southeast Asia prior to 1945. Only the Kingdom of Siam (Thailand) retained its independence from complete foreign control.

agricultural centers of Southeast Asia. Since not all areas of Southeast Asia are suited to production of the major crops of rubber, coffee, tea, and coconut palm, the colonial powers developed quasi-specialization of production. Areas in the true humid tropics became the focus of rubber production, while the Mekong Delta and the Irrawaddy alluvial plains developed as food-producing areas for the nonfood-producing plantations.

The plantation as an agricultural system differed dramatically from the traditional sedentary and shifting agricultural systems of Southeast Asia. The plantation economy emphasized production of cash crops and was based on large-scale operations, with management controlled by nationals of the various foreign powers. Above all, the plantation represented the implanting of a Western cultural element into the Southeast Asian cultural realm, a factor that has had profound significance in the postcolonial period.

The necessity to provide labor for the plantations led to another major impact of long-term significance. Millions of Chinese and Indian laborers were recruited to work on the plantations of Southeast Asia, adding a major component to the ethnic diversity of the area. Although there had been small groups of Chinese in Southeast Asia prior to European movement into the area, the large numbers of Chinese in today's Southeast Asian countries date largely from the nineteenth-century population movements instigated by the Europeans. By the end of the colonial period, some 32 million individuals had participated in migrations to provide labor. Many of them were temporary residents, such as Indians who worked in Malay or Burma. The Chinese came primarily as permanent settlers to the southern part of today's Vietnam around Ho Chi Minh City, in the Bangkok area of Thailand, in Singapore, and on the island of Java.

Most Chinese lived in the colonial cities and ultimately came to have a symbiotic relationship with the Europeans. The movement of the Chinese from mainland China to the colonial areas of Southeast Asia broke their traditional ties to their land and allowed them to move from their role as peasant agriculturalists to a new position in the economy. Although they first came as laborers on the plantations, their **Confucian ethic** of hard work and emphasis on education quickly led to their movement into middle management positions, between the Europeans and the indigenous peoples. The Chinese families with their strong organization facilitated amassing capital to become shop owners in the cities and to move to a dominant role as the merchant class of Southeast Asia. In part because of their close ties with the European colonial powers, the Chinese have suffered discrimination since the period of independence since the end of World War II. The Chinese tendency to occupy a specific sector of cities, to marry other Chinese, and become merchants further exacerbated their relationships with other cultural groups.

The European impact on Southeast Asia also transformed the traditional mining of tin, copper, and other metallic ores. The pre-nineteenth-century production of tin was concentrated in the Malay Peninsula where the indigenous population used open-pit methods to obtain ore from the shallow, easily accessible ore bodies. Growth of industrial activities in Western Europe quickly raised demand to levels above the supply of tin, and the European powers brought Chinese and Indians to work the mines. This European influence led to the development of large communities such as Kuala Lumpur as mining centers. Ultimately Kuala Lumpur developed as an administrative center as well. The European demand for tin led to two methods of mining, replacing the small-scale traditional methods of the pre-European period. One relied on the Chinese migrants who quickly amassed their capital and, utilizing their Confucian work ethic, were able to compete successfully to mine tin on a fairly large scale. The Chinese relied on cheap labor imported from mainland China and introduced techniques from China, which allowed them to mine deeper ores in large quantities. Not until the end of the nineteenth century did hydraulic mining and use of dredges replace the labor-intensive Chinese system. By the beginning of World War II the mining was completely dominated by the Europeans with the use of equipment that minimized the need for Chinese laborers.

SOUTHEAST ASIA AT THE BEGINNING OF WORLD WAR II

The cumulative effect of European interaction with Southeast Asia from 1500 to 1940 had created major changes in the organization and use of the Southeast Asian landscape. The European cities established to exploit the local tropical crops or mineral resources replaced the original capital cities, which had served primarily as repositories for the great cultural monuments of the individual nations of Southeast Asia. The population makeup had changed dramatically as an estimated 32 million people had moved into the region under the stimulus of European activity. While many subsequently returned to their homelands, many died within the region, and others remained on a semipermanent basis. The traditional self-sufficient agricultural society had been greatly changed by European regulations designed to promote greater production of selected tropical crops for export. National boundaries had been established based on European whim, further fragmenting the peoples of the region. The movement of Chinese into the area had produced a new merchant class in which the Chinese were a dominant group in much of the Southeast Asian region.

SOUTHEAST ASIA IN THE POSTCOLONIAL ERA

The colonial forces that transformed the economy, political organization, social structures, and ethnic makeup of Southeast Asia remained intact until after World War I. By then, after several hundred years of European dominance, there were enough educated and trained individuals in each of the European colonies to provide a nucleus for an independence movement. Both the British and Dutch unwittingly encouraged this movement by providing a degree of local autonomy and political involvement. World War II provided the major impetus for growing independence movements in Southeast Asia, as the Japanese were able to defeat the colonial powers seemingly without effort. During the four years of Japanese control of Southeast Asia, major changes were made in the European administrative boundaries. The old French possessions of Indochina were divided into present-day Vietnam, Laos, and Kampuchea (Cam-

---

Confucian ethic: Chinese and other societies that adopted the teachings of Confucius are felt by some observers to be more work and goal oriented.

bodia) as independent political units. The Thai government was encouraged to expand its borders to regain territory it had lost to European colonial powers.

As Western forces began pushing the Japanese from Southeast Asia near the end of World War II, the Japanese supported local independence movements, which became the nucleus for the subsequent independence of the region. In Indonesia nationalist leaders declared an independent republic in 1945. Although the Netherlands tried to prevent actual independence, after four years of political negotiation and armed revolutionary conflict, Indonesia became independent in January 1950. Vietnam developed a strong, Communist-oriented, nationalist movement for independence in the Red River area of northern Vietnam, but the French were adamant in their attempts to reincorporate Indochina into the French empire. The French were never able to defeat the nationalist movement, and as the Vietnamese nationalist leader Ho Chi Minh became openly more Communist in his outlook, the Soviet Union and China began supplying the revolutionaries with arms. In 1954 the Geneva Peace Treaty divided the country into North and South Vietnam with the commitment to a **plebiscite** to determine a national government for a united Vietnam by July 1956.

Western fears that the creation of a Communist nation in North Vietnam was the beginning of the spread of communism from the Soviet Union and China into Southeast Asia were embodied in the *domino theory*. The domino theory postulated that if South Vietnam were united in a Communist country with North Vietnam, all of Southeast Asia would fall like dominoes under control of Communist-oriented governments. In response the United States became involved in supporting the dictatorial government of Ngo Dinh Diem in South Vietnam. Ultimately the United States was drawn into armed conflict with North Vietnam and by extension its supporters, China and the Soviet Union. This conflict continued until 1975 when the United States capitulated and Vietnam emerged as a unified country.

The independence movements in other areas of Asia are variations on the experience of Indonesia and Vietnam. In Burma, the British, rather than facing a prolonged guerrilla war, granted independence in 1948. Laos and Kampuchea gained independence in the 1950s, but internally the centrifugal force of strong minority Communist groups has made development of stable nations difficult. The conflicts and turmoil that followed independence in the present states of Southeast Asia have been of varying intensity and impact. In former Indochina Vietnam attempted to capture territory from its former French colonial neighbors, and now tries to disengage from its costly occupation of Kampuchea and other areas. Indonesia suffers from unstable governments and stagnant economic development.

All of Southeast Asia has been handicapped by cultural conflicts and the legacy of the colonial era. None of the new states has a homogeneous population, and in many instances the allegiance of their citizens is not to the state at all. In some states, such as Burma and Malaysia, certain ethnic groups (in these cases, the Chinese) are not allowed to participate fully as citizens, or may be forced to leave the country, as the Chinese of Vietnam. The continued revolution and political upheaval that have plagued the region since the end of World War II reflect both the impact of the previous three centuries of European domination and the cultural development that predated it.

Recurrent conflicts in the post–World II era affecting the countries of Southeast Asia in general and specific countries in particular have handicapped the development process of the region. Based on the degree to which they have adopted Western ideas and technologies, several of the countries have become newly industrialized and outward-looking, while the bulk of Southeast Asia remains typical of the less industrialized, inward-looking countries of the less developed world.

THE NEWLY INDUSTRIALIZED STATES OF SOUTHEAST ASIA

Several countries of Southeast Asia have made significant economic gains since independence. Per capita income, literacy rates, life expectancy, infant mortality rate, death rates, and the percentage of population employed in industry have increased to a level analogous to those of portions of southern and eastern Europe. Although some still have high birthrates, they have begun the move to stage 3 of the demographic transition, and have reached the stage in economic growth known as *economic takeoff*. One theory of economic growth postulates that at a certain point of industrialization and economic development an economy literally takes off and continues to grow and develop, based on its own production. The newly industrialized countries of Southeast Asia include the Philippines, Singapore, Malaysia, and at a lower level, Thailand. They have relatively little in common to explain their development, except for a somewhat democratic form of government, relatively stable political atmosphere, and an outward-looking society

---

Plebiscite: A plebiscite is simply an election by the voters of a country or region to determine what course of action they will follow.

Territorial Shape

Each of the new states of Southeast Asia is plagued with geographical centrifugal forces matching the cultural centrifugal effect of their diverse peoples. Indonesia, composed of the islands of the Indonesia archipelago and portions of Borneo and New Guinea, is a *fragmented* political state. Ninety percent of Indonesia's population is concentrated on the island of Java, which has only 10 percent of the country's territory. The isolation created by the geographical fragmentation makes unification of the diverse peoples of Indonesia even more difficult.

Vietnam is an example of an *elongated* state. In elongated states transport and communication are costly, political control is difficult, and it is difficult to overcome the ties the peoples may have to neighboring territories, particularly in Southeast Asia where the national boundaries were superimposed by European powers and do not coincide with cultural boundaries. Burma and Thailand are roughly circular in their core area but have an extension, or *proruption*, that brings all the problems of the elongated state. Providing adequate transportation and other linkages to integrate these extensions into the core area is costly and in the case of Burma has never happened. These extensions provide fertile ground for secessionist movements or for political control by outside forces.

Kampuchea (Cambodia) is the only country in Southeast Asia that has a *compact* shape, which minimizes the centrifugal forces associated with spatial organization. Unfortunately, Kampuchea lies between areas of the historically powerful Vietnamese and Thai people and has been unable to emerge as a recognizable nation in its own right.

---

willing to adopt what they perceive as the best of Western ways while maintaining their own culture.

Singapore's development reflects the combination of its strategic location and its emergence as a city-state with a majority Chinese population. Some observers have argued that the Confucian work ethic is similar to the Protestant work ethic, which was significant in the Industrial Revolution in Europe. These observers point to Singapore, Hong Kong, Japan, Taiwan, and South Korea as examples of the Confucian work ethic leading to development. The Chinese who make up the majority population of Singapore believe not only in work, but their separation from their homeland provided the revolution in attitudes, expectations, and social organizations that allowed them to transform their entire society and view of life. The uprooting of the Chinese and their movement to Southeast Asia is analogous to the Meiji revolution of Japan with its emphasis on Western development.

Thailand has not developed as rapidly as Singapore, but because it was not a colonial possession, it has been more willing to adopt selected values from the West. As with the Japanese, these adoptions have been transformed to become part of Thailand's own approach to development. The Philippines has developed in part because it was closely associated with the United States during the early twentieth century and adopted many American ideas of government and economics. Malaysia was essentially unoccupied until after European intervention and expansion of tin mining, and the movement of Chinese, Indian, and Malay immigrants into the area provided a break with the migrants' past, allowing them to revolutionize their society and set the stage for development. In all newly industrializing countries of Southeast Asia, apparently the important element has been the transformations that have allowed adoption of selected elements of Western culture as opposed to the complete resistance of Western influence in the less industrialized, inward-looking countries.

The newly industrializing countries of Southeast Asia are engaged in a variety of economic activities that tie them closely to the industrialized, developed world. Many Western corporations are moving manufacturing and assembly functions to these countries. The impact of the Western companies on countries like Singapore, which is a major assembler of manufactured components, is a mixed blessing, bringing economic development but further revolutionizing the traditional societies. Thailand and the Philippines are important for textiles and apparel manufacture, representing the early stages of industrial revolution. Malaysia has used the capital accumulated from its tin mines and rubber exports to become an incipient industrial power. These countries have adopted a range of light manufacturing activities and have attempted to utilize their labor and location to develop their economies.

The emergence of the newly industrialized countries of Southeast Asia has created a dual economy. While manufacturing of electronics or textiles or assembly of cameras or television components is com-

THE HUMAN DIMENSION

Cheaper Than Machines

Kim Hai Kwee, 19, sits tensely hunched over her microscope. Jaunty music plays in the background and her head aches dizzily from hours of soldering tiny gold wires to almost invisible chips of silicon. Kim is a worker in the export-oriented electronics industry. Some 90 percent of the workers are young women aged 18 to 23. They work 8 to 10 hours a day, 6 days a week. Many developed severe eye problems during their first year of employment: chronic conjunctivitis, nearsightedness, and astigmatism. Virtually anyone who stays on the job more than 3 years must wear glasses, and is called "grandma" by her friends.

In one investment brochure the Malaysian government touts the "manual dexterity of the oriental female" as an incentive for Western investors. In one factory, manual dexterity consists of dipping assembled units in large open vats of sulphuric and nitric acid. Heavy fumes are everywhere and the floors are wet and slippery. The women wear boots and gloves that sometimes leak, causing burns. The workers in these plants are exposed to some of the most dangerous acids and solvents, such as trichlorethylene, xylene and benzene, which cause nausea and dizziness. They have also been linked to cancer, and to liver, kidney and lung disease.

"I am sick because of acid concentrated," one woman, Maznah, revealed in broken English. She had been dipping components in acid rinse for 3 years. The company refused to allow her to transfer to other work. When she visits the camp doctor, he tells her she has "flu."

Over 40 percent of the women come from rural areas to work for the $2-a-day wages. When asked why they had left their families, they say there are no jobs back in the village. Most had never worked before. The electronics industry continually recruits young, rural women—which does nothing to help relieve the steady buildup of jobless in the cities. Women workers have an average stay in the Malaysian factories of 1–2 years; the turnover rate is as high as 80 percent.

The electronics firms glamorize Western female stereotypes, stressing female passivity and emphasizing a paternalistic discipline. One company arranges cosmetics lectures and organizes sports teams. Another offers free lipsticks for reaching production goals. Beauty contests ("Miss Motorola," "Miss National Semiconductor") are common and reinforce the Western ideals of consumerism and modernity over traditional Muslim values.

To meet production goals, many factories operate round the clock. Production quotas goad the women to ever higher targets. "If they say one hundred and we can do it, then next week they give us a lot more to do," Azizah, 23, told one investigator.

These pressures pay off in profits. According to one plant manager in Malaysia, "One worker working for 1 hour produces enough to pay the wages of 10 workers working one shift, plus all the costs of materials and transportation."

Factory managers praise the women workers highly. According to one personnel manager at an instrument assembly plant: "This job was done by boys 2 or 3 years ago. But we found that girls do the job as well and don't make trouble like the boys. They're obedient and pay attention to orders. So our policy is to hire all girls."

It's no wonder the workers are praised for their docility. There is little alternative when strikes are forbidden and the death penalty invoked for inciting labor unrest. Performance bonuses, often as much as one-fifth of the monthly wage, are often revoked for one day of tardiness, sickness or even personal leave or vacation.

Some companies state they will be automating their assembly plants in the next 10 years. But as long as wages are low and workers available and docile, the chances of being replaced by automated bonding machines are slim. As Mae-fun, a Hong Kong assembly worker put it: "We girls are cheaper than machines. A machine costs over $2,000 and would replace only two of us. And then they would have to hire a machine tender, for $120 a month."

Diana Roose, *Cheaper than Machines* (Copenhagen: World Conference of the United Nations Decade for Women, 1980).

---

mon in the **primate cities,** the manufacturing activity in regional centers focuses on processing of the traditional tropical crops for export to industrialized countries. The dual nature of the economies of the newly industrialized countries of Southeast Asia is evident in any of the *primate cities* from Singapore to Manila to Bangkok. The cities seem relatively affluent except for the slums, which are being replaced by planned housing in the larger cities. The towns are teeming with traffic, consumer goods, and a prevailing sense of economic development. Rural areas in Thailand and the Philippines particularly continue to share most of the characteristics of less industrialized regions in general. Lack of mechanization, inadequate

---

Primate city: A city that dominates the urban network of a country, often being five times or more larger than the next largest city. Normally they are the focus of government, industry, and economic development.

Apartment buildings in Singapore stand in stark contrast to the traditional housing found in most of Asia.

transportation, and inequitable land distribution associated with the less industrialized world prevail in these areas. Nevertheless, Western ideas have diffused even to these rural areas. In spite of this dichotomy, the NICs of Southeast Asia are evidence that newly emergent states can transform their economies and join the industrialized, developed world. The great strides in Malaysia, Thailand, Singapore, and the Philippines are evidence that the less industrialized world does not need to remain permanently underdeveloped.

THE LESS INDUSTRIALIZED COUNTRIES

Indonesia, Vietnam, Kampuchea, Laos, and Burma share the characteristics common to less industrialized countries. Birthrates are high and death rates are relatively low, resulting in a high population growth rate; infant mortality rates are high; per capita income is low; and the literacy rate is significantly below that of the NICs. The majority of the population is engaged in agricultural pursuits, and manufacturing employs but a small percentage (Table 14.1). Two characteristics of these countries seem to be important in explaining their relatively low level of development. First, colonialism fostered a strong nationalistic antipathy toward Western culture. Second, political stability has been repeatedly shattered by revolutions and conflicts. These factors explain the relatively low level of Western investment in these countries and concomitant low level of industrialization. Economic growth is stagnant, and in some cases the standard of living is lower than it was under the colonial powers. Political aberrations like the Pol Pot regime of the 1970s in Kampuchea were primarily concerned with a warped view of the national heritage of the individual states, and development efforts were not addressed. In Kampuchea (Cambodia), Pot attempted to turn back the hands of time to a previous life-style that was perceived as idyllic and bucolic. In the process, hundreds of thousands of people were forced from the cities to undertake agricultural activities for which they were unprepared, hundreds of thousands were killed outright, and more than a million others, particularly the

TABLE 14.1 Development Data for Selected Less Industrialized Countries

Country	Population (Millions)	Population Birthrate	Infant Mortality	Per Capita Income	Literacy Rate	Percent of Labor Force in Agriculture	Percent in Manufacturing
Burma	43	34	103	200	66	53	19
Indonesia	183	27	88	500	62	57	13
Laos	4.2	41	122	180	85	76	7
Kampuchea	7.3	40	134	—	48	80	4

SOURCE: *World Population Data* (Washington, D.C.: Population Reference Bureau, 1989); and *World Development Report* (New York: World Bank, 1988).

elderly and children, died as a result of the upheaval in the food-producing system. Subsequently occupied by Vietnam, Kampuchea still has neither a stable government or economy upon which to base development.

Vietnam ended its 30-year conflict with the Western powers of France and the United States only 15 years ago and is only now beginning to focus on economic development rather than territorial aggrandizement. Laos is like Kampuchea in that its independence drive led to a socialist orientation and a Communist government. Nationalistic in outlook, Laos has been unable to transform its economy from its traditional agricultural emphasis.

Indonesia has the potential for modernizing its society, but a rapid birthrate, which has pushed the population to more than 183 million, makes development efforts ineffective. Almost two-thirds of Indonesia's population is concentrated on the island of Java, where population growth has pushed the population of the capital Djakata from one-half million in 1945 to 10 million. With an annual growth rate of 1.7 percent the economic advances Indonesia has made are absorbed by its swelling population. Indonesia does have a good resource base, including petroleum, timber, major tin deposits, and tropical crops of rubber, tea, palm oil, and copra. Industrialization efforts are accompanied by efforts to establish new agricultural settlements on the less densely settled islands of the nation and to increase agricultural productivity on Java. (See Human Dimension.) The population problem has been compounded by the nationalism of the early governments of postindependence Indonesia.

TOMORROW'S ISSUES: THE FUTURE OF SOUTHEAST ASIA

The issues that face Southeast Asia as it enters the last decade of the twentieth century are a combination of relics of colonial rule, geographical challenge, and the general problems facing the less industrialized world. The issues that present the greatest challenge are those directly related to the former colonial domination of the region. The arbitrary division of the region into administrative units is the primary factor in today's continued conflict. All former French territories have suffered war or insurrection since the end of World War II. The prevalence of such hostilities has led to the development of the concept of the **insurgent state**. In Vietnam, for example, Communist forces under Ho Chi Minh had a political organization that the French after World War II declared illegal, but it continued to exist as a guerrilla movement and ultimately gained control of all of Vietnam. The demands of the Communist Vietnamese have been repeated by insurgents in Indonesia, portions of the Philippines, and Malaysia.

A second problem is that the centrifugal forces created by inadequate transportation linkages within the various countries aids guerrilla movements behind insurgent states. A major issue in most Southeast Asian

Insurgent state: Political and territorial control over part of a country by insurgent (guerrilla) movements that oppose the existing government.

Rural villages in Southeast Asia are usually built along rivers and canals, which have historically been transport arteries, water supply, and sewage system.

THE HUMAN DIMENSION

An Indonesian Farmer

Suprapto has been a farmer in Indonesia all his adult life, and still is. The difference is that now, for the first time, he is able to make enough income from it to give his family something better than a hand-to-mouth existence.

The reason lies in the mound of earth on which he is standing: the high bank of an irrigation trench coursing across a green, flat, valley floor that stretches like a lush carpet from one volcanic-ridged horizon to the other. This is the Brantas River basin in East Java—an important granary of the world's fifth most populous country (and largest rice importer), yet overburdened by population and, until recently, abused by the vagaries of the elements.

Suprapto is one of some 57,000 farmers in this part of the valley. Their average cropland is a miniscule 0.26 hectare, or less than two-thirds of an acre. And the two low-yielding crops that they used to grow each year (rice in the wet season, and in the dry season *palawija*, a term encompassing maize, soybeans, peanuts, cassava, and assorted vegetables) often generated earnings insufficient to meet basic family needs.

Suprapto proudly tells a visitor of the changes brought by the arrival of irrigation in the late 1970s. No longer dependent on rainfall, he now raises 3 crops per year (2 crops of rice and 1 of palawija) instead of 2. Paddy yields have jumped from 2–3 tons per hectare to 7–9 tons, and palawija yields have doubled from 1 ton per hectare to 2 tons. His income has risen accordingly.

"Before," he says, motioning toward the thick paddy being weeded behind him, "the land was dry and there would be no crops at this time of year."

---

countries is the need to unify their people, to tie rural areas to the primate cities, and to spread the urban benefits of education and medical care to the countryside.

Another issue facing the Southeast Asian countries is nationalism. Nationalism led to the creation of the independent states of Southeast Asia, but in some cases this nationalism acts as a barrier to their development. The years of colonial rule and oppression have led some Southeast Asian states to detest any Western influence. The countries of the former Indochina, and Indonesia and Burma to a lesser extent have emphasized this view, and it has handicapped their economic growth. Refusal to adopt the better aspects of Western life has effectively prevented their movement from the ranks of the less industrialized to that of newly industrializing countries. Particular problems have been created by the intense nationalistic spirit and poverty that led to *nationalization* (government ownership and operation) of transport systems, industry, and other important elements of the economic system. Although nationalization does not necessarily imply less efficiency and progress, it has had this effect in these Southeast Asian countries. Development has proceeded more slowly because of lack of capital, skilled workers and management, and recurrent conflicts.

Another major issue that threatens the viability of many Southeast Asian countries is *neocolonialism*. Although these states are independent, they have continued to be affected by foreign powers. The location of Southeast Asia near China, India, and Japan, the two most populous countries of the world and one of the three greatest economic powers, places important strains on the future of Southeast Asia. The former colonial powers also affect the economies of these states through their trade and political relationships, and the United States, the Soviet Union, and their respective allies affect their economy through foreign aid and military grants. Capital from the industrial world was largely cut off after the independence of Southeast Asian states, but the domino theory of the spread of communism has prompted Western nations to pour massive amounts of money into some of the countries for development and military purposes. The issue for the Southeast Asian countries is how to capitalize on this Western concern, maximize it in the form of capital investment to modernize their economies, and at the same time prevent themselves from being caught up in conflicts among the industrial powers and the democracies and Communist powers of the world.

The problem of tension and conflict in the complex cultural mix of these countries is related to both continued conflict in the region and to European colonization. The roots of this tension go back to the ethnic complexity the Europeans found in the region, but it is compounded by the mass movement of Indian and Chinese people into the area under the impetus of European interests. This tension takes many forms, from periodic armed conflict in Indonesia, to the Vietnamese effort to expel the Chinese from the Saigon area when they gained control of Vietnam. More than a million Chinese boat people were forced to leave Vietnam from 1978 to 1981 (Figure 14.6). Many migrated to Singapore, Malaysia, or Indonesia, but Indonesia in particular has been reluctant to accept more

Chinese because of its already low standard of living. Tens of thousands of Chinese from Vietnam have moved to the United States, but ironically the wealthiest country of Asia, Japan, has been unwilling to accept more than a token number of these refugees.

Another issue related to the former colonial powers is the contrast between the primate cities and the rural areas. The primate cities in the newly industrialized countries tend to have most of the characteristics of affluence associated with large cities in the Western world—crowded streets, high buildings, and an abundance of consumer goods. They also have large, sprawling slums, but even here life is more desirable than in the rural areas where increasing population makes survival difficult. The necessity of raising the standard of living in rural areas presents one of the greatest challenges facing Southeast Asia. Financial assistance from wealthy industrial countries can be effective in increasing the standard of living of rural residents if such development aid is sufficient and used wisely. (See Human Dimension, page 414.)

A final issue is population growth, poverty, and external dominance typical of the less developed world in general. Some countries of southeast Asia have population growth rates rivaling the highest in the world. These increases create the traditional problems of rural-to-urban migration with the teeming slums of Manila, Djakarta, or Bangkok as prime examples. They also present potential for revolution as the quality of life in rural areas actually declines with increasing population. The challenge for the less industrialized Southeast Asian countries is to gain control of their rapid population growth so that they can effectively provide a reasonable standard of living for their people.

The problems facing Southeast Asia are not insurmountable, as Singapore and Malaysia have shown. Both have managed to overcome the centrifugal forces of ethnicity and fragmentation and to begin the process of industrialization and increasing standards of living. Singapore represents the only example of a country entirely in the tropics that has raised the standard of living of its population above the level of the less industrial nations. With a total land area of only 225.6 square miles (584.3 square kilometers), it is less than 20 percent the size of the state of Rhode Island. The

Figure 14.6 Non-Palestinian refugees of the world were estimated at 7 million in 1988. There were also an estimated 5 million Palestinian refugees.

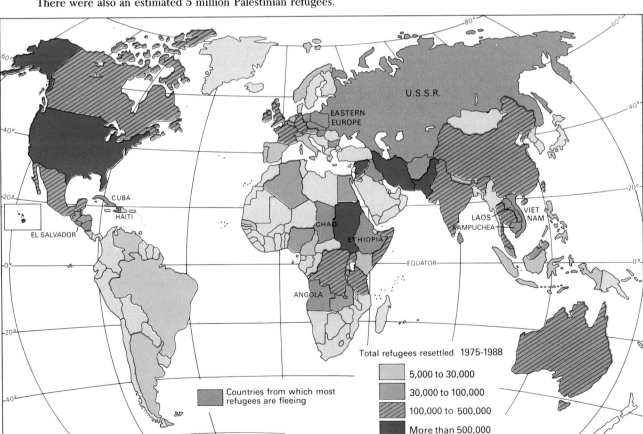

THE HUMAN DIMENSION

Slowing Population Growth in Indonesia

In Indonesia, family planning is not merely a medical matter discussed in clinics; it is part of the fabric of daily life. Contraceptives are available everywhere and widely used. Motivational techniques include the playing of a family planning song for pedestrians at traffic lights and for motorists at railroad crossings.

As a result of a coordinated and sophisticated family planning campaign, the birthrate has sharply decreased and the annual rate of population growth has slowed significantly. The government estimates that two-thirds of its reproductive age adults now use some form of birth control. In the process, the family planning program has become a model for other less developed nations.

Stemming the population tide is a matter of utmost significance to Indonesia's future. The nation is the world's largest archipelago, made up of more than 13,000 islands straddling the equator between India and Australia. With a population of 183 million in 1990, Indonesia is the fifth most populous nation in the world. Sixty percent of its inhabitants live on the island of Java, which contains only 7 percent of the nation's land.

With the advent of the current New-Order Government in 1966, ushered in by President Soeharto, family planning and population control became top priorities on the public agenda. A national family planning agency (BKKBN) was established, and Indonesia launched its first full-scale family planning program in 1970. The results, after two decades, according to a report issued by the Department of Demography at the Australian National University, have been "dramatic."

Not only is fertility declining in Indonesia, but the pace of the change is increasing, even in those areas that had experienced relatively large declines in the early 1970s. The apparent annual rates of fertility decline for the early 1980s are nothing short of extraordinary. Australian research shows that between 1967 and 1985 the fertility rate dropped by 41 percent, effectively slowing the country's annual population growth rate to 1.7 percent.

These impressive results have been achieved by the two-pronged approach of widespread availability of free contraceptives, buttressed by a comprehensive educational and informational campaign aimed at co-option, rather than coercion. Contraceptives are distributed through government-funded medical clinics, called Puskesmas, located in every town, village, and hamlet. A separate tier of armed forces hospitals—much like the Veterans' Administration hospitals in the United States—also dispenses contraceptives, as does a network of privately owned and operated clinics providing obstetrical and gynecological care.

More than 35,000 family-planning field workers have been trained and dispatched by BKKBN. Supported by an estimated 100,000 volunteers, they not only instruct women in the mechanics of the various contraceptives available, they try to provide the psychological framework necessary for acceptance of the concept. Tapping into Indonesia's highly organized society, with cultural values that put the needs of the group—the family, the community and, by extension, the nation—ahead of those of the individual, field workers explain that Indonesia's limited resources cannot support an ever-growing population.

Religious leaders were made part of the process. Indonesia is 95 percent

---

population density is over 10,000 per square mile (4,000 per square kilometer), in the city of Singapore, the center of the republic. The population is overwhelmingly Chinese, and the Confucian work ethic has combined with a developmentally oriented government leadership to transform the tiny state into one of the most important manufacturing centers in Asia. Singapore's ability to make this transition suggests that other countries could accomplish the same transformation.

The promise of Singapore may not reach fruition in the short term in other countries of Southeast Asia because of the particular leadership, location, and geographic characteristics of the other parts of the region. The continued conflict in former Indochina and the continued ethnic tension in most Southeast Asian countries seem to indicate that Singapore and Malaysia may well be the exceptions in the region. The Philippines and Thailand may well reach the takeoff stage in their economic development and enter a period of rapid transformation of their economies if they are not caught up in ethnic problems or armed conflict that divert their capital to unproductive sectors. The prospect for Southeast Asia outside the NICs is not optimistic, as their location and real or imagined strategic and political significance may cause them to be the continued center of world political conflict.

THE PACIFIC WORLD: VARIATION IN DEVELOPMENT

East of Southeast Asia the Pacific Ocean has a number of island groups. This region is sometimes known as the South Pacific or Oceania and can be divided into three island groups: Melanesia, Micronesia, and Polynesia (Figure 14.7). Melanesia is the part closest to

Muslim, a religion that has no specific decree against preventing contraception. Muslims do, however, have a taboo against cutting the body. Indonesia compromised, agreeing it would not actively promote surgical sterilization if religious leaders would assist in providing legitimacy to the program.

Explaining that the continuing support of religious leaders has been a cornerstone of the program's effectiveness, the director of family planning in Indonesia suggests the importance of properly framing the request. "There is not a religion on earth that will actively preach against the health of the family," he says with a smile.

And he points out that among the incentives the government offers to long-term, family-planning acceptors is an all-expense-paid trip to Mecca—the pilgrimage all Muslims should make once in a lifetime. Other rewards include scholarships, silver and gold-plated pins, discount cards for shopping, and even a visit with president Soeharto.

In the small Indonesian village of Dampit, East Java, Ibu Ismaryono gets a 10 percent discount at the local grocery store. She has three children and has been taking birth control pills. Discount cards are provided by the National Family Planning Coordinating Board (BKKBN) on the basis of consistent contraceptive use. Cinemas, insurance firms, and most merchants honor the card. It is one innovative approach in the Indonesian government's varied repertoire aimed at encouraging people to have fewer children.

Ibu Ismaryono pays for her purchases with, among other currency, a 5-rupiah coin. On one side of the coin is a picture of a mother, father, and two children with the inscription "family planning, a small family is a happy family." Local businesses, service establishments, retail stores, physicians, and professionals give discounts on their goods and services to the card-bearers. A pretesting of the discount card demonstrated that a 10 percent discount was more than compensated for by the increase in gross sales to the participating merchants and professionals.

The duration of the discount services is contracted for one year at the end of which the establishments may choose whether or not to renew their discounts.

To qualify for the discount, members of small families must meet the following requirements:

Those with two children must use permanent methods: tubectomy, vasectomy, or implants.

Those with one child must have used the IUD, injections, birth control pills, or condoms for at least five years and do not plan to have any more children.

Meanwhile, the government continues to concentrate its energies on education. Just outside Bangdung, Indonesia's third largest city, there is a phrase neatly set with white stones into the lush, green hillside. The words can be easily read by everyone using the well-traveled road. They read: DUA ANAK CUKUP—Two children (are) enough.

"Popline," December 1988, pp. 2–4.

---

the Southeast Asian archipelagos, and the islands are large with a tropical climate similar to that of the mainland rim and the Indonesian archipelago. Melanesia extends from the Southeast Asian mainland to Australia and consists of a number of large islands, the largest being New Guinea. (New Guinea consists of Indonesian Irian Jaya and the Pacific island nation of Papua New Guinea.) Micronesia consists of a few volcanic and complex high islands and many tiny **coral atolls**; Guam, a volcanic island, is the largest with only 210 square miles (550 square kilometers). Many of the atolls rise only a few feet above the high-tide level, but a few higher islands of volcanic origin reach elevations of over 2600 feet (800 meters).

Coral atolls: A coral island or islands with a reef surrounding a lagoon.

Polynesia covers the largest area of the South Pacific, but the total landmass is extremely small. Physically this group includes both low coral atolls and volcanic islands. The most important resource of these islands at present is their claim to 200 miles (322 kilometers) of surrounding ocean. The issue of territorial extent would seem to be decided, but the problem of borders of political states where they meet the ocean has proven to be nearly intractable. The United Nations Law of the Sea convention has been meeting for the past decade to attempt to devise ways to resolve conflicting claims to a country's bordering seas. Although initially most countries of the world claimed only a 3-mile (5-kilometer) territorial limit, today the United Nations recognizes a 200-mile (322-kilometer) claim by bordering states for exclusive economic development (Figure 14.8). This 200-mile (322-kilometer) zone gives the oceanic islands exclusive economic control over 5 million square miles (12,700,000 square

THE HUMAN DIMENSION

Aiding the Poor

Slipping off his shoes at the bottom step with as much respect as if he were entering a sultan's palace, 29-year-old Harun Mamat pads barefoot up a set of open wooden steps and plops down on a bench on the tiny porch of his house. The place is a single-story, unpainted, timber-plank house with a steeply pitched roof, perched chest-high above the ground on posts: Not much, perhaps, but to this young father of three, it's a dream house. Having owned little more than the clothes on his back for most of his life, he now owns his own home in Rasau Kerteh, Malaysia.

Relaxing in jeans and T-shirt after a day's work in a nearby oil palm plantation—10 acres of which he also owns, and tends communally with 23 other families who each own adjoining 10-acre plots—the little porch affords a breezy refuge from the sun.

As Harun squints out across his front yard, flowering bushes clinging to the sand, he recalls the harsher life he led just a few years ago as a fisherman in a coastal village 25 miles from here on the South China Sea. Fishermen in villages such as his earn the equivalent of about U.S. $75 per month—well below the poverty line. New-town settlers working in oil palm plantations make $500–$650 a month.

The struggling fisherman got his chance for a more promising life through that currently much-maligned economic instrument—multilateral development aid. He heard about a series of five new towns being carved out of the jungle in Trengganu (a state on the eastern coast of peninsular Malaysia that ranks as one of the country's least developed states) with the aid of a $16-million loan from the Asian Development Bank (ADB), a Manila-based multilateral lending institution financed by industrial nations.

Forsaking his all-too-often empty fishing nets, he moved to Rasau Kerteh and was introduced to the oil palm whose clusters of reddish fruit are processed into cooking oil, soap, and other household products.

After three years as a renter, he exercised the option to buy his house (over 17 years, with his 3 years of rent considered as down payment). The cost: about $900.

The sea, too, can be coaxed to yield more than a poverty-level livelihood to those who spend long, hard, and risky hours fishing its depths. Just ask R. M. Chandradasa, a Sri Lankan now 38 who has been fishing the Indian Ocean commercially since he was 16.

For most of that time, it was a financial struggle. The 900 fishermen in the ancient southwestern port of Galle—putting out to sea in wooden outrigger boats like those used for centuries, except for the addition of small outboard motors—brought home the average equivalent of only about $20 a month. For Mr. Chandradasa's household of six (his own family, plus his mother), it often simply wasn't enough.

Help came from an ADB project, and those picturesque but outmoded boats were replaced by 200 23-foot, fiberglass-hulled, diesel-powered vessels equipped with new nets and other modern gear. Credit was extended on affordable terms to help skippers buy their boats.

One of the new skipper-owners is Mr. Chandradasa. With 30 percent larger catches and a 30–40 percent larger income, he paid off the loan on his 4-man boat in just 19 months.

The sarong-clad skipper cites other evidence of his new-found prosperity. He has built a new home, and bought a second, smaller fishing boat.

Like the nearly 100 Galle skippers who also have become owners of their modern boats—catapulting them from struggling fishermen to middle-income property owners—Mr. Chandradasa is enjoying a privileged taste of a regrettably rare phenomenon: upward mobility among Asia's down-trodden.

Peter C. Stuart, "Beyond the projects people profit," Development Forum, *May 1982, p. 16.*

---

kilometers). At present little has been done in this area, but it may provide potential for future development of the region.

ECONOMIC DEVELOPMENT IN THE PACIFIC

The level of economic development in the Pacific islands varies greatly. In Melanesia the levels are among the lowest in the world. Subsistence agriculture is the prevailing economic activity, and commercial agriculture is similar to that of the Southeast Asian mainland, with coconuts or sugar cane raised on plantations the predominant cash crop.

Micronesia and Polynesia have higher standards of living, but except for Nauru none are economically independent. Many of the other areas of Micronesia and Polynesia were **trust territories.** The United Nations devised this arrangement because it felt that these islands with their tiny areas and limited resources could not develop without external assistance, and under United Nations supervision the industrial northern countries could have a benevolent, positive

---

Trust territories: A trust territory is a former colonial holding assigned by the United Nations to one of the industrialized nations for development assistance.

impact. In practice the arrangement has not always worked out this way, as the small populations have tended to encourage welfare dependency status for the islands. Trust territories are almost a political anachronism as most have received some type of independence today.

LOCATION AND FRICTION OF DISTANCE IN THE PACIFIC

The islands of Oceania suffer a number of climatic and locational handicaps. They are scattered across the broad Pacific Basin and isolated from the rest of the world by long distances. Their relative isolation is further increased by their small size, which hinders development of major manufacturing, resources, or agricultural commodities to justify more transportation linkages with the rest of the world. As a result, these areas have had relatively little opportunity to share in the development found in the newly industrialized countries of Southeast Asia. Increasing costs of ships, fuel, and labor have made the islands relatively more inaccessible in the modern age. Fewer ships call at most of the islands of the Pacific today than did 50 years ago. Within the island chains themselves there is less interisland transportation for the same reasons. Ironically, as the rest of the world industrializes and sees its standard of living rise, increasing the standard of living of many Pacific islanders is impossible, as exporting the coconuts, coffee, cacao, or sugar cane that they can grow is becoming increasingly unprofitable. Transport costs to assemble the agricultural products at the port city for each island group often exceed the value of the product itself, negating efforts at development.

PROBLEMS IN PARADISE

The islands of Micronesia and Polynesia and the smaller islands of Melanesia have tropical climates moderated by their maritime setting. This has given rise to the view of the tropical islands as a veritable paradise. In fact the majority of these islands suffer from the general characteristics of less industrialized countries, but the problems they face are different from those of the Southeast Asian countries. The islands of Micronesia and Polynesia usually do not have the population problem of Southeast Asian countries. In a few of the islands of Polynesia and Micronesia there has been an absolute decline in population in the last decade as the result of migration to the United States or other trust parent nations. At the other extreme several nations such as Kiribati are beginning to face population pressure, and high growth rates in other countries will cause similar problems in a generation or two.

Economic development of the islands is generally based on improving traditional agriculture, developing island mineral or biological resources with funding from the industrial nations, through tourism and cash remittances from island residents working in the industrial world or from industrial countries involved with the former trust territorial arrangements.

Agricultural systems in the islands can be divided into plantation and subsistence types, and both have changed since World War II. The large plantations of former colonial powers are now generally owned by local individuals or the government. Subsistence agriculture is increasingly involved in a market exchange system based on small-scale production of either fruits and vegetables or crops traditionally produced on plantations. Development of the smaller subsistence farming sector is handicapped by common traditional societal restraints of the less industrial world. In the subsistence sector the ownership of land is vested in the community, and individual rights may not have the long-term status necessary to justify large investments of capital. The requirements of commercial agriculture are difficult to meet in the traditional society of the highly organized island states where it is more important to maintain one's position within the village than to increase income. Thus, in the past the welfare system has consisted of trading labor with other small farmers to ensure that your own products could be planted and harvested should you be unable to work. It is difficult to replace this system with a fully com-

Slum residences along a railroad line in Indonesia's capital.

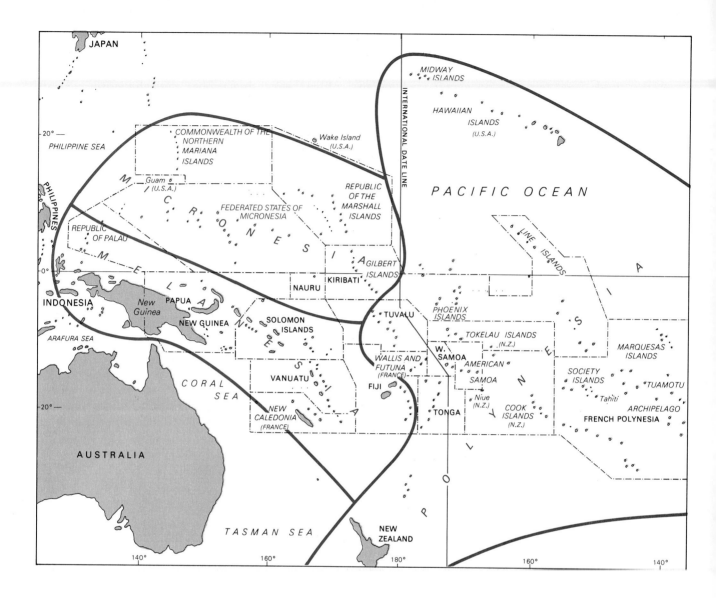

mercial system so long as the village remains the focus of activity.

The second form of development is based on **transfer payments** from the industrial world. Migrants who have gone to the United States or Western Europe may return part of their salaries to their parents and families. These remittances form an important part of the economy of many of the Pacific countries. The second basis for wealth transfer is tourism. Tahiti, Fiji, and to a lesser extent Samoa have been able to capitalize on their favorable natural setting to attract the wealthy residents of the industrial world. Tahiti has attracted developers who have built hotels and other tourist attractions that attract international tourists, especially French. Tourism seems to be an ideal way to raise the standard of living within a country because the hotels and other tourist services provide jobs and income without the need of exports.

Unfortunately this form of development has a darker side in its impact on the societies of the island peoples. The native cultures of all the islands have been contaminated by Western influence. This influence is subtle and pervasive, and embodies changing values, demands for consumer items, and attitudes toward individual and society. At one level the impact of these changing values has prompted migration to Europe or the United States, or to the larger towns of the islands. Recognition that goals such as higher

---

Transfer payments: The process of transferring wealth from one area to another. It takes many forms, from welfare within countries to economic aid or tourism between countries.

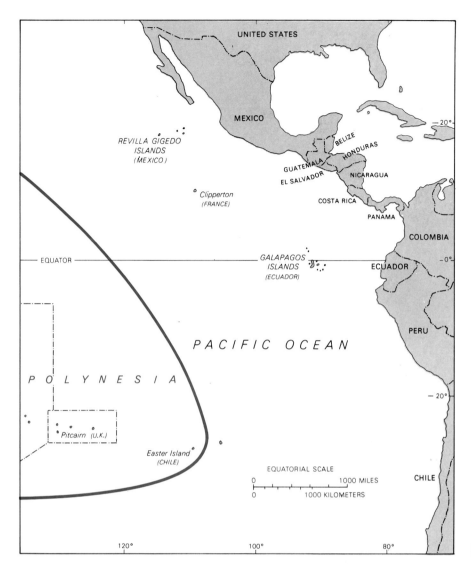

Figure 14.7 The Pacific Region.

incomes, escape from the traditional demands of society on time, property, and behavior, and better employment or better education can be found in the United States or in the larger towns and cities is causing a genuine problem of labor supply in the rural areas. Coupled with the low rates of population growth in most of this region, it assures that the area is becoming more reliant upon the industrialized world for food.

Major changes have occurred in the food habits of the Pacific islanders since the end of World War II, and today's residents demand imported wheat flour and polished rice for their staple diet. Refined white sugar and canned meat, fish, and vegetables are replacing the traditional staples that have supported these peoples for millennia. In order to obtain the funds to pay for increased imports of such items, more of the land is converted to commercial agriculture and the production of coconut for copra, sugar cane, coffee, or other products in demand in the industrial world. The location of the islands raises the cost of importing Western items. Although most imports come from Australia and New Zealand, which are seemingly nearby, the higher cost of production in Australia and New Zealand and the cost of bringing the items there from the industrialized world raise the transportation cost for these imports.

Particularly alarming to the older generation in the Polynesian islands is the changing value structure of the youth of the cities and returned migrants. The traditional communal society with its emphasis on reciprocity is replaced by a cash-exchange society in which tradition and community position are not valued. The problems that these changes in attitudes and behavior cause are obvious and are viewed with grave concern by both the elderly and those who have chosen to maintain their traditional life-styles. The future of the islands is uncertain, since goods that they pro-

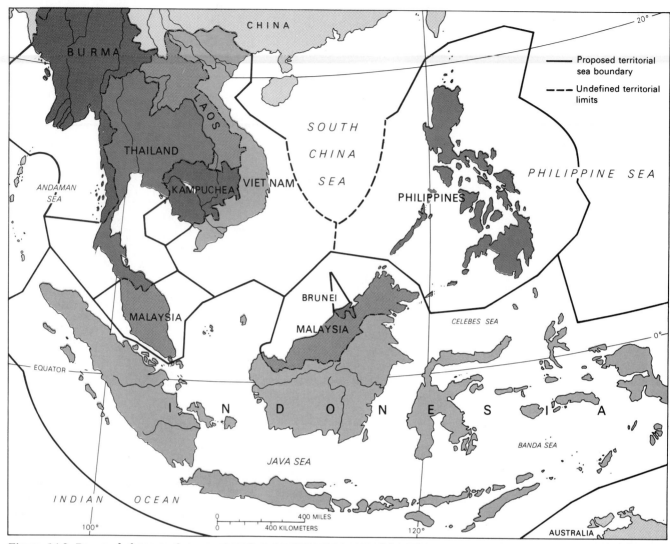

Figure 14.8 Proposed claims to the ocean based on the 200-mile territorial boundary.

duce for the world economy are insignificant by any measurement. Even the tremendous phosphate deposits of Nauru, if unavailable, could be replaced readily by larger deposits in other areas. The islands of the Pacific world are in the unenviable position of being caught in the dilemma of development. Faced with exposure to the consumer-oriented, wealthy lifestyle of the West through their status as dependent trust territories and tourist destinations, they have little in the way of apparent resources to avail themselves of economic development. The way in which they meet this challenge may well determine the relative affluence and importance of the Pacific in years to come.

FURTHER READINGS

ABDUL RAHIM MOKHAZANI, D. B. *Rural Development in Southeast Asia.* New Delhi: Vikas, 1979.

ABLIN, DAVID A. and HOOD, MARLOWE, eds. *The Cambodian Agony.* Armonk, N.Y.: M. E. Sharpe, 1987.

AIKEN, R. S. "Squatters and Squatter Settlements in Kuala Lumpur." *Geographical Review* 71, 1981 pp. 158–175.

AIKEN, S. ROBERT, LEIGH, COLIN H., LEINBACH, THOMAS R., and MOSS, MICHAEL R. *Development and Environment in Peninsular Malaysia.* Singapore: McGraw-Hill International Book Company, 1982.

BEE, OOI JIN. *The Petroleum Resources of Indonesia.* New York: Oxford University Press, 1982.

BLAKE, GERALD H., ed. *Maritime Boundaries and Ocean Resources.* Totowa, N.J.: Rowman and Littlefield, 1987.

BROOKFIELD, H. C., ed. *The Pacific in Transition: Geographical Perspectives on Adaptation and Change.* New York: St. Martin's Press, 1973.

CHAY, JOHN and ROSS, THOMAS E., eds. *Buffer States in World Politics.* Boulder, Colo.: Westview Press, 1986.

COUPER, A. "Who Owns the Oceans?" *Geographical Magazine,* September 1983, pp. 450–457.

———., ed. *The Time Atlas of the Oceans.* New York: Van Nostrand Reinhold, 1983.

DOBBY, E.H.G. *Southeast Asia.* London: London University Press, 1969.

DUTT, A. K., ed. *Southeast Asia: Realm of Contrasts.* Dubuque, Iowa: Kendall-Hunt, 1974.

FISHER, C. A. *Southeast Asia: A Social, Economic and Political Geography.* New York: Dutton, 1966.

FRYER, D. W. *Emerging Southeast Asia: A Study in Growth and Stagnation.* New York: McGraw-Hill, 1970.

———. *Southeast Asia: Problems of Development.* New York: Halsted, 1978.

GLASSNER, M. and DE BLIJ, H. J. *Systematic Political Geography.* 3d ed. New York: Wiley, 1980.

HILL, R. D. "Integration in Agricultural Geography with Some Southeast Asian Examples." *Philippine Geographical Journal* 25, 1981, pp. 98–107.

HUNTER, G. *Southeast Asia: Race, Culture, and Nation.* New York: Praeger, 1967.

JONES, G. W. "Population Trends and Policies in Vietnam." *Population and Development Review* 8, 1982, pp. 783–810.

JOHNSTON, DOUGLAS M. and SAUNDERS, PHILLIP M., eds. *Ocean Boundary Making: Regional Issues and Developments.* Beckenham, U.K.: Croom Helm, 1987.

KOLB, A. *East Asia.* London: Methuen, 1971.

KUMAR, RAJ. *The Forest Resources of Malaysia: Their Economics and Development.* Singapore: Oxford University Press, 1986.

LEE, D. *The Sinking Ark: Environmental Problems in Malaysia and Southeast Asia.* Kuala Lumpur, Malaysia: Heinemann, 1980.

LEE, Y. L. "Race, Language, and National Cohesion in Southeast Asia." *Journal of Southeast Asian Studies* 11, 1980, pp. 122–138.

LEINBACH, T., and ULACK, R. "Cities of Southeast Asia." In *Cities of the World: World Regional Urban Development,* pp. 370–407. Edited by S. Brunn, and J. Williams. New York: Harper & Row, 1983.

LEVINE, STEPHEN. *Pacific Power Maps: An Analysis of the Constitutions of Pacific Island Politics.* Honolulu: Centre for Asian and Pacific Studies, and Social Science Research Institute, 1983.

LEVISON, M., WARD, G., and WEBB, J. W. *The Settlement of Polynesia.* Minneapolis: University of Minnesota Press, 1977.

McCLOUD, DONALD G. *System and Process in Southeast Asia: The Evolution of a Region.* Boulder, Colo.: Westview Press, 1986.

MILNE, ROBERT S. and MAUZEY, DIANE K. *Malaysia: Tradition, Modernity, and Islam.* Boulder, Colo.: Westview Press, 1986.

Natural Resources Planning for Economic Development: The AID/NPS Project for Managing Natural Resources on a Substantial Basis in LDCs. Washington, D.C.: U.S. Agency for International Development: U.S. National Park Service, 1985.

PRESCOTT, J. *The Political Geography of the Oceans.* New York: Halsted Press, 1975.

PRYOR, R., ed. *Migration and Development in Southeast Asia.* Kuala Lumpur: Oxford University Press, 1978.

PURCELL, V. *The Chinese in Southeast Asia.* New York: Oxford University Press, 1965.

ROBISON, RICHARD, HEWISON, KEVIN, HIGGOTT, RICHARD, eds. *Southeast Asia in the 1980s: The Politics of Economic Crisis.* Sydney: Allen and Unwin, 1987.

SALITA, D. C. and ROSELL, D. Z. *Economic Geography of the Philippines.* Bicutan: National Research Council of the Philippines, 1980.

"Southeast Asia's Boom Belt." *Wurtschafts Woche* as reprinted in *World Press Review,* February 1983.

SPENCER, J. E. *Shifting Cultivation in Southeastern Asia.* Berkeley and Los Angeles: University of California Press, 1976.

STEWART, IAN CHARLES and SHAW, JUDITH. *Indonesians: Portraits from an Archipelago.* Jakarta Selatan: Paramount Cipta, 1983.

TAP, D. V. "On the Transportation and New Distribution of Population Centers in the Socialist Republic of Vietnam." *International Journal of Urban and Regional Research* 4, 1980, pp. 503–515.

TAYLOR, A., ed. *Focus on Southeast Asia.* New York: Praeger, 1972.

THRIFT, NIGEL. "Vietnam: Geography of a Socialist Siege Economy." *Geography* 72, October 1987, pp. 340–344.

ULACK, RICHARD and LEINBACH, THOMAS R. "Migration and Employment in Urban Southeast Asia: Examples from Indonesia and the Philippines," *National Geographic Research* 1, 1985, pp. 310–331.

WHEATLEY, P. "India Beyond the Ganges—Desultory Reflections on the Origins of Civilization in Southeast Asia." *Journal of Asian Studies* 42, 1982, pp. 13–28.

"When Will the Peace Begin? An Exclusive Look Inside Troubled, Still Divided [Vietnam]." *Time,* April 25, 1983, pp. 82–85.

WILHELM, D. *Emerging Indonesia.* Totowa, N.J.: Barnes and Noble, 1980.

YEUNG, Y. and LO, C., eds. *Changing South-East Asian Cities: Readings on Urbanization.* Singapore: Oxford University Press, 1976.

ZEGARELLI, P. "Antarctica." *Focus,* September-October 1978.

Isfahan, Iran.

MIDDLE EAST AND NORTH AFRICA PROFILE: 1990

Population (in millions)	279	Energy Consumption Per Capita—	
Growth Rate (%)	2.8	pounds [kilograms]	2,226
Years to Double Population	25		[1,012]
Life Expectancy (years)	60		(11,453)*
Infant Mortality Rate			([5,206])*
(per thousand)	91	Calories (daily)	2,685
Per Capita GNP	1,900		(3,265)
	(11,250)*	Literacy Rate (%)	52
Percentage of Labor Force		Eligible Children in Primary	
Employed in:		School (%)	87
Agriculture	42		(94)
Industry	22	Percent Urban	49
Service	36	Percent Annual Urban Growth,	
		1970–1981	4
			(8.2)
*( ) oil-surplus countries		1980–1985	4
			(5.8)

15

The Middle East and North Africa cradle of civilization, center of conflict

MAJOR CONCEPTS

Kibbutzim / Desert pavement / Judaism
Hejira / Migratory genocide / Basin irrigation
Hajj / Monotheistic religion
Tribal religion / Moshav / Levant
Hydraulic civilizations
Universalizing religions / Qanats
Transhumance / Horizontal nomadism
Ethnic religions / Dry farming / Diurnal
Sunnite / Fertile Crescent / Shiite
Zionism

IMPORTANT ISSUES

Zionism and the Israeli state
Necessity and location of a Palestinian homeland
World economic and political power of oil-producing countries
Land reclamation and environmental degradation
Maintenance of traditional values in a changing world

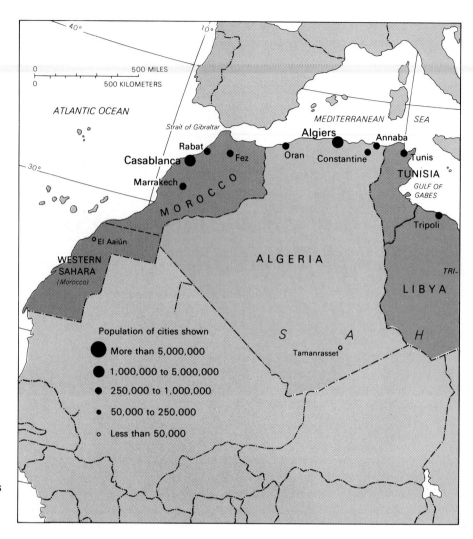

Figure 15.1 Countries and major cities of North Africa and the Middle East.

The Middle East and North Africa is a widely misperceived region (Figure 15.1). In the Western industrial world the term *the Middle East* evokes images of oil-rich sheikhs, conflicts between Arab and Israeli, and nomadic Bedouins crossing the desert in camel caravans. Popular stereotypes of the region are misleading, and even definitions of the boundaries of the region rarely agree entirely. The terms applied to the area from Afghanistan to the northwestern corner of the African continent were the result of European perspective. Various parts of the region have been called by different names. The Arabs called the northwestern area of Africa the *Maghreb*. The Europeans lumped the Maghreb and today's Libya together as the North African region. To the east, lands from Egypt to Afghanistan were called the Near East to separate them from European colonial possessions in India and other areas referred to as the Far East. At the beginning of the twentieth century the term *Middle East* was coined and applied to the area from Egypt through Afghanistan. This term has gained popular acceptance for the entire region.

There are certain characteristics that justify separate consideration of this region. Most of it has a dry climate with precipitation totals under 20 inches (508 millimeters) and in many places less than 10 inches (254 millimeters). Culturally it is a single region because of the influence of the Islamic religion. The Middle East is the hearth of the Islamic religion, and at least 80 percent of the people are Muslim. The population density in all countries of the Middle East is low. The most suitable areas for human occupance are widely scattered, and most of the lands of the Middle East and North Africa have high population densities only in more favorable sites (Figure 15.2).

These unifying factors are important, but they are largely subjective and are applied by the outside world. The apparent homogeneity of the area is more perceived than real. The inclusion of such a diverse area of peoples, economies, and environments under the

THE WORLD OF POVERTY

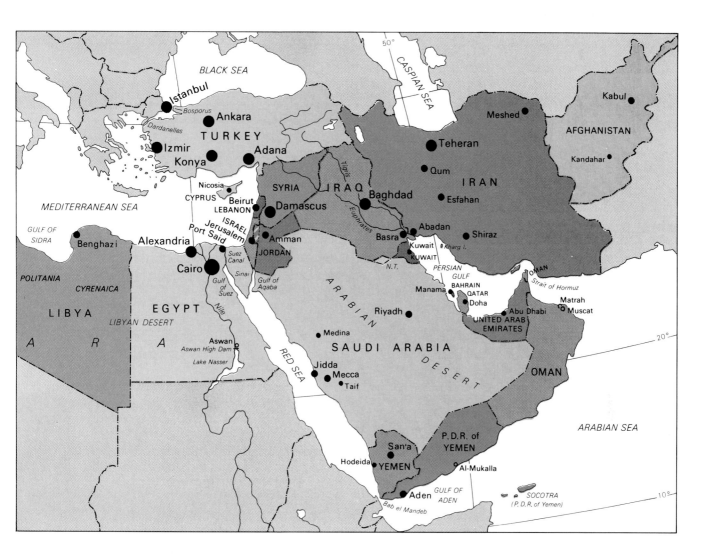

broad heading of Middle East is the result of European attempts to regionalize the world. For the people of what we know as the Middle East, such a definition is purely arbitrary.

Even the image of the Middle East and North Africa as a desert region inhabited by nomadic herdsmen mounted on camels is false. Population is concentrated in riverine or oasis locations, and the vast majority of the peoples have little to do with nomadism, camels, or the Sahara. The view of the region as the home of wealthy Arab sheikhs engaging in conspicuous consumption from their oil wealth is equally erroneous. Not only is the oil wealth concentrated in only a few countries, but within them a significant portion of that wealth is being channeled into development efforts to benefit a broader segment of the population. Even the aridity of the area is modified by elevation and increased precipitation in the highlands. The idea that the Sahara is entirely sand is an error, since most of it is covered by a rocky gravel surface known as **desert pavement** rather than sand. The importance of the Middle East and North Africa involves much more than the common misperceptions of the region, the oil wealth, the potential for conflict over contrasting cultures, or the strategic location of the region. The importance of the region can only be understood in relation to its broad historical development.

THE WORLD IMPORTANCE OF THE MIDDLE EAST

The importance of the Middle East and North Africa to the world centers on its vast reserves of oil and oil-

---

Desert pavement: land surface composed of pebbles and rock fragments, often cemented together by calcium carbonate or other salts left at the surface through evaporation.

THE MIDDLE EAST AND NORTH AFRICA

Geographical Characteristics of North Africa and the Middle East

1. The region of North Africa and the Middle East shares the twin characteristics of dominance by the Islam religion and arid climates.
2. North Africa and the Middle East contain the majority of the world's known petroleum reserves, concentrated in the Persian Gulf countries.
3. The Middle East was the hearth area for three of the world's great religions, many plant and animal domesticates, and important early civilizations and inventions, including the alphabet.
4. The region is a focal point of conflict in the modern world, with conflicts between and among numerous culture groups and countries.
5. The environment of the Middle East and North Africa limits the utility of much of the region for agriculture, and population is concentrated where there is water.

related wealth, its role as a cultural hearth for the world's major religions of Judaism, Christianity, and Islam; the importance of the area historically as a cradle of civilization; and its present strategic location and repeated conflicts that threaten to draw other regions of the world into global war. Of these factors, oil is popularly perceived to be the most important to the Western industrial world. The oil wealth of the Middle East and North Africa is indeed great. Reserves in the entire region are estimated at nearly 60 percent of the total proven reserves of the world. In the area around the Persian Gulf (including Saudi Arabia) are more than 50 percent of known world petroleum resources. Saudi Arabia alone has a quarter of the world's reserves. The region also has more than one-fourth of the world's natural gas reserves (Figure 15.3). The importance of these reserves to the world cannot be overstated, especially to Japan and Western

Figure 15.2 Distribution of population in the Middle East and North Africa.

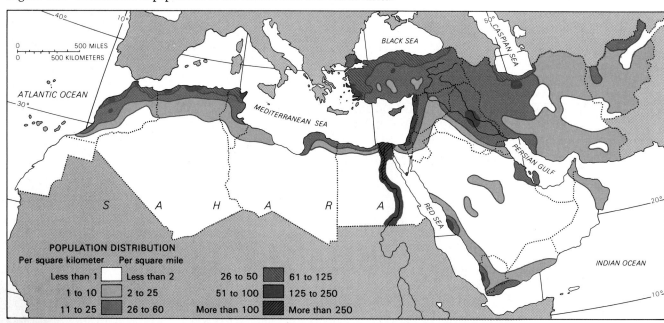

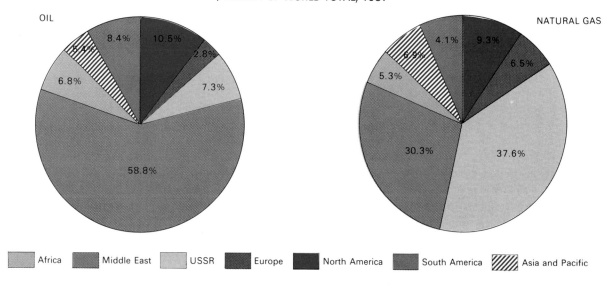

Europe, which rely on the Middle East and North Africa for nearly 75 percent of their petroleum needs. The United States obtains approximately 20 percent of its oil needs from this region. Only recently have Middle Eastern oil countries capitalized on the importance of oil to the Western world and the power that this importance generates. The great increases in petroleum prices of the 1970s reflected the efforts of these oil-rich Middle Eastern countries to receive a better price for their resources. While oil prices are lower now, they are still far above pre-1970 levels. In the process there has been a major transfer of funds from the industrial countries to oil-surplus countries.

Equally important to the rest of the world is the role of the Middle East and North Africa in geostrategic and conflict problems. Strategically the region has been important not only because of its oil but in the past because of the Suez Canal as a transportation link. The Suez Canal allows ships traveling between Asia and Europe to avoid the long journey around Africa. The conflicts in the region in the past hundred years have involved recognition of the strategic importance of the region. Conflicts have included those between French and British, British-American-French forces versus Germans and Italians, and more recently between Soviet and American forces, which have military outposts in the region. In the post–World War II era the most important conflicts have involved Israel and neighboring Arab states. There have been wars in 1948, 1956–1957, 1967–1970, 1973, and 1982, and tension between Arab and Israeli continues. Conflicts in the region that have not involved Israel in-

The changes brought to the oil-rich countries of the Middle East are typified by this scene in Saudi Arabia.

clude Syria and Egypt in 1961, Syria and Jordan in 1970, North and South Yemen and Egypt in 1961–1967, Iran and Iraq in 1980 to 1988, Lebanon and Syria beginning in 1982, and Libya and Chad in 1983–1984, and involvement of the Soviet Union in Afghanistan 1979 to 1989. Because of the relationship of the protagonists to outside powers, any of these disputes could conceivably involve the major world powers if there is not successful resolution.

One of the critical factors affecting the conflict in the Middle East and North Africa is based on its long-enduring role as a crossroads, cultural exchange, and hearth of civilization. Most archeological evidence indicates that 8,000 to 10,000 years ago the region was the place of original domestication of many plants and animals used throughout much of the world, including wheat, barley, sheep, goats, cattle, horses, the olive, and a variety of smaller grains and vegetables. The domestication of plants and animals and the dramatic change in the occupancy of the earth that it permitted we call the *Neolithic Revolution*. The Neolithic Revolution produced a food surplus, which allowed some individuals to be freed from the constant task of food gathering. These individuals ultimately developed what we describe as civilization with written laws, political organization, and formalized religions. The ancient civilization of the Egyptians, the Sumerians, Babylonians, and Assyrians made notable technological advances upon which subsequent development was based. Invention of such seemingly mundane items as the alphabet, methods of smelting iron, and the complex code of laws of Hammurabi in Babylon are indicative of the cultural legacy of this area, which ultimately affected most of the world.

JUDAISM AND CHRISTIANITY

Judaism was the first of the modern religions to evolve in this area as tribal groups came into contact with the civilizations of the Middle East. Some 3000 years ago Judaism developed as a **monotheistic religion** (worshiping a single god) with a complex code of behavior partially based on the code of Hammurabi of Babylon.

Judaism became distinctive because of its commitment to a single god and belief that the Jews were a chosen people. Judaism was a **tribal religion** that evolved into a complex religious system associated with a nation-state. The belief in the uniqueness of Judaism gave the Jewish population a strong unity, allowing them to defeat other tribes and develop a civilization which lasted until 586 B.C. when they were conquered by the Babylonians and sent into exile. After a brief rebellion in A.D. 70 they were dispersed through Europe and the Middle East where they were often the focus of persecution and discrimination. In 1948 a state based on Judaism once again emerged in the form of Israel.

The location of the core area of the Middle East on the trade routes between Africa and Europe, Europe and Asia, and Africa and Asia ensured a constant flow of ideas into and out of the region. The constant flow of ideas ultimately led to the development of two offshoots of Judaism, Christianity and Islam, each with many more adherents than Judaism. Christianity developed around the man his followers believed to be the promised Messiah to the Judaic peoples, but certain teachings of Jesus Christ led to its change from the exclusive belief system of the Jews to a **universalizing religion**. Its adoption in western Europe and subsequent expansion with European explorers and colonists spread Christianity far beyond the boundaries of its hearth.

THE WORLD OF ISLAM

The second great religion derived from the general principles of Judaism is Islam, which dominates the culture of the Middle East today. Islam originated in Saudi Arabia about A.D. 600. Like Christianity it was a universalizing religion, viewed as a means of obtaining salvation for all peoples, and trade and warfare spread it far beyond the boundaries of its origins. Islam has had a profound and pervasive impact on the Middle East and North Africa. Approximately 80 percent of their people today are Muslims. (Islam is the religion; a member is a Muslim.) Today more than 950 million people are Muslims, and Islam is the world's fastest growing religion. Muslims make up the majority of the population in 40 countries, and all have rapidly growing populations, because the number of children each woman has ranges from a low of 3.3 in Albania to more than 8 in North Yemen. The emergence of Islam as a dominant force in the world between A.D. 620 and 750 is one of the most important factors affecting the Middle East and North African regions (Figure 15.4).

Monotheisistic religion: a religion which believes that there is only one God.

Tribal religion: A religion associated with a specific tribe or group and a specific place.

Universalizing religion: One which has broken its ties to a specific space and maintains its beliefs are appropriate for all people.

Religion and Culture

The beliefs, values, sentiments, and attitudes shared by groups of people are the primary elements of culture. Variations in culture are one of the important aspects that contribute to the unique sense of place of each locale in the world. Among the aspects of culture, religion is perhaps most important since it is generally the underlying basis for the beliefs and value system of a group. Religion affects laws and institutions, dress styles, foods, language and many other aspects of a place in ways that are often subtle and unrecognized. Its pervasive effect makes it one of the major variables of the world's geography.

An important and useful distinction among religions is that between *universal* and *ethnic* types. Ethnic religions refer to the shared beliefs and practices of a group, which are acquired by being a member of that group. Ethnic religions do not seek worldwide acceptance, as membership comes through birth, marriage, or adoption into the group. They include *tribal* religions associated with a specific tribe such as the Navajo, or those associated with a nation (as the Jews), or a civilization (as the Hindus). Universal, or universalizing religions, aim at worldwide acceptance and actively proselytize for new members. Membership in the major universalizing religions—Buddhism, Christianity, or Islam—is open to all who make some type of symbolic commitment.

The Middle East is the hearth for Judaism, Christianity, and Islam. All have diffused widely from their original hearths, with each religion having many more adherents outside of the region than are found within it. The greatest numbers of Christians are today in Europe and the Americas, while comparatively few remain in the Middle East and North Africa. Although Israel was established as a homeland for Jews, there are more than twice as many in the United States as in Israel. While most of the states of the Middle East and North African region are more than 80 percent Islam, the countries with the greatest Islamic populations are Indonesia, Bangladesh, and Pakistan. The specific religion or religions that dominates each individual place within the Middle East and North Africa largely determines the character of society, economy, and landscape of the place.

Muhammad, the founder of Islam, lived in Mecca in what is now Saudi Arabia and was involved in the trade between the Orient and Europe. A member of the middle class, Muhammad began preaching in Mecca in A.D. 610. His teachings emphasized the monotheistic nature of God, the universal brotherhood of all people, and the necessity for individuals to submit themselves to the word of *Allah* (the Islamic name for God). In A.D. 622 Muhammad and his followers fled to the town of Medina some 200 miles (322 kilometers) away to escape persecution. This flight is known as the *Hejira* and marks the beginning of the Islamic era and calendar. At the time of Muhammad's death in A.D. 632 his followers had spread through most of the Arabian peninsula. One of his most important teachings is the concept that the brotherhood of all Arabs is more important than blood relationships. This belief helped overcome the tribalism that resulted from allegiance to blood relatives and provided the first effort for Arab unity.

Following the death of Muhammad his son-in-law assumed the leadership of the Islamic world. Succession in Islamic leadership was controversial because Muhammad did not designate a new leader (caliph). Ultimately the dispute over leadership split Islam into two major groups, the *Sunnites* and the *Shiites*. About 90 percent of Muslims today are Sunnites, but the Shiites form a majority in Iran (where they constitute 93 percent of the population), and Iraq (where 55 percent are Shiites).

The Islamic faith spread rapidly from its homeland in Arabia, and by A.D. 650 the new religion had spread throughout the Middle East except North Africa and Turkey. By A.D. 730 Islam had spread to all of North Africa, across the Mediterranean into Spain, and east into what is now India. This rapid spread was based on two characteristics of Islam. The first was acceptance of both Christians and Jews as believers who were allowed to maintain their beliefs if they paid a tax to the conquering Islamic armies. Other peoples were given the option of converting to Islam or suffering death. In reality few people were forced to convert, as the conquering armies were normally willing to allow non-Islamic peoples to continue their own culture so long as they paid their taxes. Islam, with its emphasis on the brotherhood of humanity and the equality of all before Allah, attracted numerous converts, and its influence expanded as Arab traders traveled farther east to today's Southeast Asia. In most Muslim areas, Islam is the basis for a code of life and laws governing society and the daily actions of Muslims, and countries such as Iran and Pakistan are trying to create a new political order based on the Islamic religion. In most countries of the region the conflict between Islam and secular influences is a major issue (Figure 15.5).

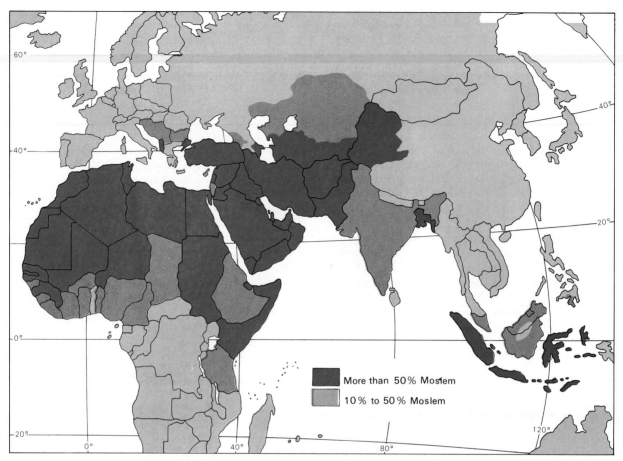

Figure 15.4 Maximum extent of areas dominated by Islamic religion.

THE ISLAMIC WAY OF LIFE

The teachings of Muhammad are based on revelations he received, recorded in the *Koran*. The content of the Koran is very similar to the Old and New Testaments. Many prophets of the Old Testament are discussed in the Koran, and there are several passages in the Koran that are identical to those found in the Old Testament. The Koran is accepted as the word of God and the basis for answering all questions of individuals. Since translating the Koran was officially discouraged, Arabic spread as the common language for those who were literate. Not until the 1920s did the government of Turkey, which was the first government to allow the translation of the Koran into another language, allow its translation into Turkish. The content of the Koran is used as the sole text in Muslim theological colleges, as it is believed to be the basis for all important knowledge. It is the basic text for students in schools that belong to Islamic mosques (churches) throughout the Islamic world. Interpretations by jurists are the basis for the major schools of Islamic law (known as *Sharia*). In the Sunni world there are four major schools, each named for a particular jurist. They range from the liberal interpretation found in Egypt to the most rigid interpretation found among the Wahabis of Saudi Arabia. Many of the contrasts between Saudi Arabia, where it is illegal to sell or consume intoxicating beverages or for women to appear without a veil, and the relaxed attitudes toward Islamic details found in Egypt can be traced to these variations.

The belief and values of Islam. Although interpretation of some actions varies between groups within Islam, there is little difference of opinion over the fundamental duties required of members. The five pillars of Islam specify the action that the devout must follow. The first is the public profession that "There is no God but Allah and Muhammad is the messenger of God." Closely allied to this profession is ritual prayer and worship, to be carried out five times daily between daybreak and evening. At the prescribed hours Islamic faithful bow toward Mecca to show their belief. The importance of prayer, particularly public prayer in a mosque, is reflected in the landscape of Islamic cities where the mosque dominates the urban

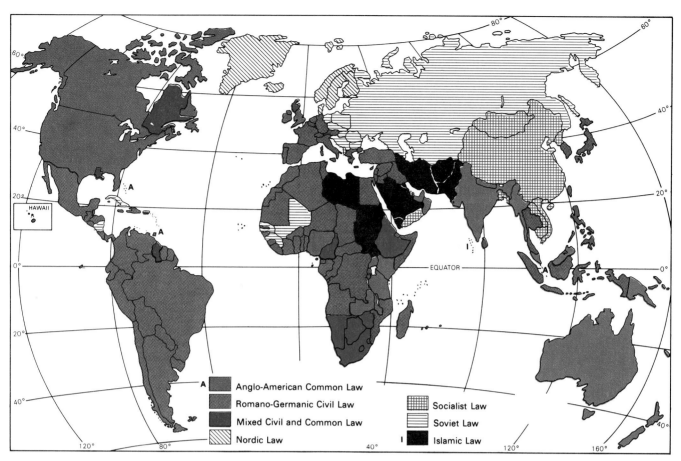

Figure 15.5 Distribution of world legal systems.

Traditional views of social roles and dress of men and women in Islamic countries are illustrated by this man and his two wives in Kuwait.

scene. It would be virtually impossible to provide sufficient mosques to accommodate the total Islamic population for prayers if everyone appeared at the mosques but many Islamic people either pray privately at home or ignore prayers because of their duties or lack of belief. It is very important to participate in prayers at noon on Friday in the central mosque of the city. Therefore, in every city there is one large mosque (Friday Mosque) and many small mosques. The traditional business district of Islamic cities (the bazaar) was located around the Friday Mosque.

Another important teaching of the prophet Muhammad was that of giving alms for the poor. Muhammad specified the rate of almsgiving at 2 1/2 percent to be used to aid the poor and construct public buildings. Ultimately the breakdown of Islamic political power removed the taxation element of almsgiving, but it remains an important source of funding for schools, hospitals, orphanages, and other charitable institutions and individuals in the Islamic world.

Fasting during the month of Ramadan is a fourth duty that is widely observed throughout the Islamic world. Muhammad specified that during the period from dawn to sunset no food, drink, medicine, or

THE MIDDLE EAST AND NORTH AFRICA

tobacco should be consumed, and sexual relationships were prohibited during this period. The extent of commitment to the fasting varies from country to country; in some countries restaurants and other food sources are closed from dawn to dusk during the month of Ramadan. The Islamic calendar has 12 months, but the month of Ramadan contains 29 days as the Islamic calendar is based on a lunar month, so that the days of each month in comparison to the Western calendar are different each year.

The final requirement of Islam is for the believer to make a pilgrimage to Mecca at least once during his or her life. Those who are too poor are excused from making this trip, but the pilgrimage, known as the **hajj,** focuses on the *kaaba,* a sacred shrine in Mecca which predates the establishment of Islam. The pilgrimage is of major geographical importance, because it brings between 1.5 and 2 million believers to Mecca during the month of Ramadan. It also provides a mechanism for reinforcing the unity of the Islamic world and an important economic resource for Mecca, Medina, and the port city of Jidda in Saudi Arabia. Because it draws Muslims from throughout the Islamic world, the *hajj* has served as a source of diffusion of ideas and beliefs throughout the Islamic world and in the past also spread disastrous diseases (Figure 15.6).

THE CULTURAL DIVERSITY OF THE MIDDLE EAST AND NORTH AFRICA

The variety of human occupation in the Middle East and North Africa, the continual interaction with other regions of Africa, Europe, and Asia; and the development of a number of major religions and a host of splinter groups of each make the Middle East and North Africa one of the most culturally complex regions of the world. The isolated nature of much of the settlement of the region necessiated by widely scattered water sources has further complicated the pattern of culture by allowing development of isolated groups. Each isolated society consisted of self-sustaining settlements only tenuously connected with the outside world by trading groups. Attempts to generalize about this complex pattern obscure many important exceptions, but such general understanding

Hajj: The pilgrimage to Mecca. The Islamic belief system includes the belief that each believer must make the pilgrimage at least once in his/her lifetime.

In this rare photo, pilgrims are shown reciting verses from the Koran inside the great mosque of Mecca.

THE WORLD OF POVERTY

THE HUMAN DIMENSION

The Hajj: An Introduction

The Hajj—the Pilgrimage to Mecca—is essentially a series of rites performed in and near Mecca, the holiest of the three holy cities of Islam—Mecca, Medina, and Jerusalem. As the Pilgrimage is one of the five pillars of Islam—that is, one of five basic requirements to be a Muslim—all believers, if they can afford it and are healthy enough, must make it at least once in their lives.

The Hajj must be made between the eighth and the thirteenth days of the twelfth month (called *Dhu al-Hijjah*) of the Muslim lunar year.

Donning the Ihram

In a general sense, the Pilgrimage begins with the donning of the *Ihram*, a white seamless garment, which is a symbol of the pilgrims' search for purity and their renunciation of mundane pleasures.

At the moment of donning the Ihram the pilgrims enter a state of grace and purity in which they may not wear jewelry or other personal adornment, engage in any disputes, commit any violent acts, or indulge in sexual relations.

Uttering the Talbiyah

In donning the Ihram the pilgrims also make a formal Declaration of Pilgrimage and pronounce a devotional utterance called the *Talbiyah*—"Doubly at Thy service, O God"—a phrase that they will repeat frequently as an indication that they have responded to God's call to make the Pilgrimage.

Entering the Haram

After donning the Ihram—and only after—the pilgrims may enter the *Haram*, or Sanctuary. In a sense, the Haram is merely a geographical area that surrounds Mecca. But because its frontiers were established by Abraham and confirmed by Muhammad, the Haram is considered a sacred precinct within which man, undomesticated plants, birds, and beasts need fear no molestation, as all violence, even the plucking of a wild flower, is forbidden.

For the duration of the Hajj, Mecca and the Sanctuary that surrounds it have a special status. To cross the frontiers of the Haram—which lie outside Mecca beween 3 and 18 miles from the Ka'bah—pilgrims from outside Saudi Arabia must now have a special Hajj visa in their passport; this entitles pilgrims to travel only within the Haram and to certain other places that pilgrims must, or customarily do, visit. Non-Muslims are strictly forbidden to enter the Haram under any circumstances.

Going to Mina

On the eighth day of Dhu al-Hijjah the assembled pilgrims begin the Hajj by going—some by foot, most by bus, truck, and car—to Mina, a small uninhabited village 5 miles east of Mecca; there they spend the night—as the Prophet himself did on his Farewell Pilgrimage—meditating and praying in preparation for "the Standing" (*Wuquf*), which will occur the next day and which is the central rite of the Hajj.

Standing at 'Arafat

On the morning of the ninth, the pilgrims move en masse from Mina to the Plain of 'Arafat for "the Standing," the culmination (but not the end) of the Pilgrimage. In what is a basically simple ceremony the pilgrims gather on the plain and, facing Mecca, meditate and pray. Some pilgrims literally stand the entire time—from shortly before noon to just before sunset—but, despite the name of the ceremony, are not required to do so.

Going to Muzdalifah

Just after sunset, which is signaled by cannon fire, the pilgrims gathered at 'Arafat immediately proceed to a place called Muzdalifah, a few miles back toward Mina. There, traditionally, the pilgrims worship and sleep under the stars after gathering a number of pebbles for use during the rites on the following days.

Stoning the pillars

Before daybreak on the tenth, again roused by cannon, the pilgrims continue their return to Mina. There they throw seven of the stones which they collected at Muzdalifah at one of the three white-washed, rectangular masonry pillars. The particular pillar which they stone on this occasion is generally thought to represent "The Great Devil"—that is, Satan, who three times tried to persuade Abraham to disobey God's command to sacrifice his son. The throwing of the pebbles symbolizes the pilgrims' repudiation of evil.

Performing the sacrifice

Now begins the greatest feast of Islam: the *'Id aladha*—the Feast of Sacrifice.

After the throwing of the seven stones the pilgrims who can afford it buy all or a share of a sheep or some other sacrificial animal, sacrifice it, and give away a portion of the meat to the poor. The Sacrifice has several meanings: it commemorates Abraham's willingness to sacrifice his son; it symbolizes the believer's preparedness to give up what is dearest to him; it marks the Muslim renunciation of idolatrous sacrifice; it offers thanksgiving to God; and it reminds the pilgrim to share his blessings with those less fortunate.

Doffing the Ihram

Now having completed a major part of the Hajj, men shave their heads or clip their hair and women cut off a symbolic lock to mark partial deconsecration. At this point the pilgrims may remove the Ihram, bathe, and put on clean clothes.

Making the Tawaf

The pilgrims now proceed directly to Mecca and the Sacred Mosque, which encloses the Ka'bah (Kaaba), and, on a huge marble-floored oval, perform "the Circing," or *Tawaf*. The Tawaf consists essentially of circling the Ka'bah on foot seven times, reciting a prayer during each circuit.

Kissing the Hajar al-Aswad

While circling the Ka'bah the pilgrims should, if they can, kiss or touch the Black Stone (the *Hajar al-Asad*), which is embedded in the southeastern corner of the Ka'bah and which is the precise starting point of the seven circuits. Failing this, they salute it. Kissing the Stone is a ritual that is performed only because the Prophet did it and not because any powers or symbolism are attached to the Stone per se.

Source: Condensed from ARAMCO WORLD Magazine, Vol. 25, No. 6, Nov.–Dec. 1974, pp. 2–6.

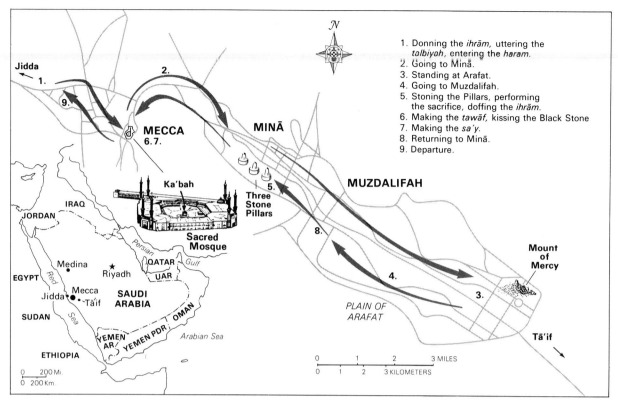

Figure 15.6 Pilgrimage space in Mecca.

is necessary to grasp the full importance of the developmental changes now occurring in the region.

In general terms there are three broad cultural groups, based on language, occupying the Middle East and North Africa: the Semitic, Turkish, and Iranian (Persian). Within each are numerous subdivisions, and it is impossible to associate each with a specific cultural or ethnic group, since their languages and their variations transcend cultures.

THE SEMITES

The Semites are the largest cultural group. Their principal languages are Arabic and Hebrew (Figure 15.7). The origin of the modern Arabic and Hebrew script was the alphabet of the Phoenicians, who occupied the eastern end of the Mediterranean (the *Levant* Coast) historically. Arabic and Hebrew are consequently quite similar in grammar, vocabulary, and sentence structure. Modern Hebrew as a primary language is concentrated in Israel. Although normally associated with the Jewish religion, not all Jews speak Hebrew, and it is the native language of only half of the population of Israel. The migration of the Jews to Israel after 1948 and adoption of Hebrew as the country's official language caused the majority of the migrants to learn it, but the total number of users of Hebrew (including Palestinians in Israel) in the Middle East is only about 4.5 million.

Of much greater importance as a unifying cultural trait is Arabic, spoken by approximately 150 million people. The people who use Arabic often refer to themselves as Arabs; yet, there is tremendous diversity in their language and way of life. The term *Arab* does not denote a nationality, for there are many Arab countries, nor is it synonymous with the Islamic religion, although it is nearly so. There are Christian Arabs in Lebanon, Egypt, and Syria, and the Arabs of the region constitute only one-fifth of the total Islamic population of the world. The term *Arab* refers to the language, but is increasingly accepted as an indication of a culture that emphasizes the mission and teachings of Muhammad, the founder of Islam, and is concerned with the issues related to the Arab-Israeli conflict. Within the Arabic language there are four major dialects, which are quite different in their pronunciation: Moroccan, which is spoken in the northwest area of the Maghreb; Egyptian, which is the basic language family in Libya, Tunisia, and Egypt; Syrian, which is spoken in the Arab sections of the Levant in Jordan, Israel, Syria, Lebanon, and Iraq; and the most pure Arabic, spoken in Saudi Arabia, Kuwait, and neighboring Arab countries. Written Arabic is common to all Arabs and is derived from the

The mosque at Qom in Iran illustrates the size and architectural style of Islamic mosques.

Koran, the holy book of Islam. Within the four major language dialects spoken Arabic is distinct to individual countries, with the language of Egypt as different from that of Tunisia as French is from Italian.

THE TURKIC

The second largest group in the Middle East is the Turkic language family. The Turks are descended from tribal groups who moved south out of central Asia nearly a thousand years ago. The majority of the Turkic people were converted to Islam, and the present Turkish-speaking population of the Middle East is essentially 100 percent Islamic. The Turkic language stems from the same sources as Hungarian and Finnish, and the Turkic people are concentrated in Turkey, with subgroups in Afghanistan and the U.S.S.R. Turkic influence in the past has affected other languages such

Figure 15.7 Language groups of North Africa and the Middle East.

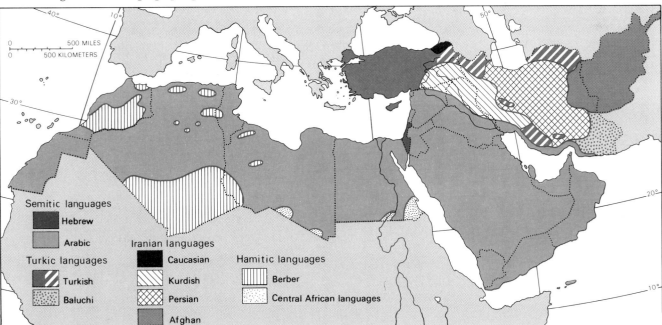

THE MIDDLE EAST AND NORTH AFRICA

A market scene from Tunisia represents the traditional bazaar of both Africa and the Middle East.

THE WORLD OF POVERTY

The Islamic city is characterized by narrow streets impassable for automobiles, as in this view of an Algerian town.

as that of Iran. An estimated 60 million residents of the Middle East have languages strongly affected by the Turkic tongue.

THE IRANIANS

The Iranians speak *Persian*, which is the third largest linguistic group in the Middle East and North Africa. The Persian of Iran is an Indo-European language, which has evolved over the past millennia in today's Iran. Persian-speaking peoples are found in many countries in the Middle East, but they are the dominant group in Iran. Linguistically related to Iranian is the language of the Kurds in northeastern Iran, northern Syria, Iraq, and Turkey. There are an estimated 9 to 11 million Kurds, with the greatest concentration (4 million) in Turkey. More than 3 million are found in Iran and more than 2 million in Iraq. Although related linguistically to the Persians, the Kurds have developed their identity separately through time and view themselves as distinct from Persian-speaking peoples. The primary importance of the Kurds today is their separatist movements. The distinctive Kurdish language and culture have prompted repeated attempts to create a Kurdish state, but there is little prospect of such an eventuality.

The language diversity indicated by the three major linguistic groups in the Middle East and North Africa fails to show fully the great cultural diversity of the region. The languages related to the major tongues are found in a variety of places, including such Indo-European languages as Pathan, Kafiri, or Armenian. Regional differences within Arabic are widely recognized and present significant problems for attempts to create Arab unity. The linguistic and ethnic mosaic is further complicated by the centuries of European colonial dominance with the diffusion of European languages, which have become important in elite groups.

LIFE AND LANDSCAPE IN THE ISLAMIC WORLD

The world of Islam that developed after the death of the prophet Muhammad based on the Koran and its principles embodied three major life-styles: the rural farm population, the townspeople, and the nomadic peoples and their associated landscapes. Only 20 to 25 percent of the population of the Islamic world lived in cities before the twentieth century. The Islamic city was the focus of education, religious activity, and commercial functions. Islam specifies that the complete worship service must include the Friday assembly in the Friday mosque. Building a Friday mosque was reserved for settlements that had a bazaar (market) where the prayer rugs, incense, and other materials necessary for worship could be obtained.

City life. The form of the Islamic city was different from that found in the Western world because of the Koran and interpretations of it. The focus of the Islamic city was the mosque. Next to the mosque was the bazaar with merchants selling those articles necessary for worship closest to the mosque, the skilled artisans working with silver or gold adjoining them, and farthest from the mosque the activities that were least compatible with the worship service such as textile manufacture or food preparation (Figure 15.8). The cities themselves had narrow streets, since the Koran did not provide for public space greater than that needed for a laden camel to pass. Streets were not oriented at right angles to one another except in towns predating Islam. The construction of buildings in the Islamic city reflected the strong acceptance of individual need for security and privacy. Islamic architecture was interior- rather than exterior-oriented, with a central court bounded by the rooms of the home and no windows in the exterior walls. Individual sections of the city were set apart for different groups such as Arabs or Jews. Each quarter was separated from the others by walls and strong gates for protection of the minority groups. The rulers lived on the edges of the city to facilitate retreat when necessary.

The cities became the home of major institutions of higher learning throughout the Middle East and

THE MIDDLE EAST AND NORTH AFRICA

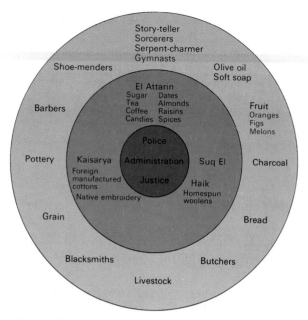

Figure 15.8 Plan of a large *suq*. (market).

North Africa, and in the core area of Baghdad, Istanbul, Cairo, and other great Islamic cities there emerged a highly sophisticated group of scholars who were responsible for many scientific advances from A.D. 750 to 1400. The cities themselves housed a large variety of individual cultures, but the Islamic peoples normally dominated.

The contrast between the very wealthy and the very poor was enormous. Like most less industrialized countries today, there existed a wealthy elite in each city, but the majority was poor and lived in conditions roughly equivalent to those found in poor countries today. Problems of sanitation, water supply, unsafe housing, and access to adequate food contributed to a low standard of living among the urban poor. In spite of the problems of city life, many rural people migrated to the city in the past as population or natural or social catastrophes made life in the countryside untenable. The breakdown of the traditional village organization, which had provided security for these migrants, increased the difficulty of life in the cities. Cities in the Islamic world were generally small, with only three—Cairo, Baghdad, and Istanbul—exceeding 250,000 people by the mid-1800s.

Village life: home of the majority. The focus of Middle Eastern life in the past was the individual village. Village life has always focused on production of crops of wheat, dates, barley, and other small grains and vegetables for local consumption and for trade with other areas. The villages had all the characteristics of less industrialized societies. The extended family, variations in poverty and wealth, and control of wealth and lands in the hands of a few were common. The location of the villages themselves depended on the critical resource of the Middle East—water. Water was the basis not only for life but for social organization in the villages. The technology of the time that Middle Eastern villages were established limited their size to relatively small, compact settlements. The landscape of the typical village varied from the adobe construction of the Nile to the reed houses

In the marshes of the tributaries of the Tigris and Euphrates Rivers, swamp dwellers live on artificial islands in houses made from reeds.

of the marshes of southern Iraq. Houses tended to be small, without any running water or other conveniences. In contrast to the homes of the peasants were the larger, more sumptuous homes of merchants and the most imposing structure of the local *sheikh* (ruler). Other elements of the village landscape were the mosque, the market (bazaar), and the fields surrounding the village. In highlands, where more abundant precipitation allowed nonirrigated cultivation (**dry farming**) of grain, the community was surrounded by the grain fields. In riverine locations, above and below the community were the fields, while in the drier regions was grazing land for livestock. In the oases of the true desert areas of the Middle East the homes might be surrounded by date palms, which provided an important source of food.

Wherever the village was located, water was the paramount factor in village life. Methods for controlling and distributing water to families and flocks and fields dominated the activity of the community and was reflected in the social order. Where the water source was a spring or oasis, the social stratification extended from the headwaters downward. The sheikh occupied the highest position where the water was most pure. Downstream were the merchants or landowners, while farthest downstream were the sharecroppers and landless laborers. Location on the water source was important for two reasons: water quality and water availability. Downstream locations tended to have contaminated water with resulting sanitary problems that periodically decimated infant populations. The quantity of water was also affected by location, and those farthest downstream tended to have the least reliable water supply.

Development of social organizations to ensure adequate water for the villages of the Middle East and North Africa fostered the rise of political organizations and civilizations such as those associated with the Tigris and Euphrates rivers and the Nile. The relative abundance of water in these great riverine valleys necessitated a system of water diversion that would be at least partially equitable. These civilizations were known as *hydraulic civilizations* because of the emphasis upon water and the canals and laterals to facilitate distribution. Management of water distribution fostered development and improvement of mathematics, engineering, and soil science. In areas where water was less abundant, the necessity for ob-

Traditional methods of raising water to higher levels for irrigation purposes require high labor inputs to move relatively small quantities of water.

Dry farming: *A method of cultivation in which crops are grown intermittently and cultivation techniques developed to minimize water loss in fallow years are used.*

taining water supplies required tremendous labor outlays.

The most unique means of acquiring water in the Middle East and North Africa is the *qanat*. The qanat is an underground canal dug laterally from a village into the base of adjacent mountains. The underground canal is maintained by a series of vertical shafts and in some cases is as long as 30 miles (48 kilometers). Qanats provide water for as much as one-third of the total irrigated land in Iran. They are used throughout North Africa, Saudi Arabia, in Iraq and Iran, and Afghanistan. Other efforts to obtain water are the Persian waterwheel, the shaduf (a balanced beam for dipping water), and the waterwheel. Water rights were zealously guarded, and the entire social fabric of the village was oriented toward maintaining the orderly provision of water.

Villages were basically self-sufficient, but surplus grains were shipped to the urban centers. Much of the land in the Middle Eastern village was owned by landlords who lived in the cities. Overseers in the village acted as supervisors for the landowner. Land tenure agreements throughout the Middle East and North Africa were based on the *fifth system*. In the fifth system one-fifth of the harvest was given to the one providing the labor, one-fifth for providing the seed, one-fifth for provision of draft animals and tools, and one-fifth to the one providing water. Peasants were rarely able to provide anything other than their own labor, so that landlords took 80 percent of the harvest. Such a generalization obscures important examples of villages in which landownership was private, but it illustrates the general nature of peasant life in the Middle East and North Africa. Because of this system, which ensured that the bulk of the agricultural product left the village to enrich the urban landlord, villages of the Middle East and North Africa have traditionally experienced a low standard of life. The one element that helped to maintain the fabric of this life in the face of pressures for change was Islam.

The nomadic sector. The third major group is the nomads. For many Western observers the nomad is the most familiar stereotype of the Middle East, but nomads have always comprised only 5 to 10 percent of the total population. Two major types of nomadism have been practiced in the Middle East in the past, *transhumance* (seasonal movement of animals to the mountains) and *horizontal nomadism* (movement of herds over large areas following periodic moisture). Nomads utilized a part of the environment that would not otherwise be productive and provided a source of meat, cheese, leather products, and animals for the village dwellers.

Over time a symbiotic relationship emerged between village and nomad in which the villager sold the nomads dates, grains, salt, and other items. Nomads primarily herded sheep and goats, but camel nomadism is also found in portions of the Middle East. The largest area of nomadism is in central Saudi Arabia, the Sinai peninsula, and South Jordan where the Bedouins breed camels and horses in addition to keeping flocks of sheep and goats. In present-day Iran and Iraq the nomads primarily herded sheep and goats, although some cattle and horses provided transportation and additional milk.

Camel nomadism was centered on the drier portions of the Middle East bordering the Sahara and the deserts of Saudi Arabia. The true desert areas are rarely capable of supplying the grazing needs even of camels, and the nomads tended to occupy the margins of the deserts in their migrations. Nomadic Bedouins often extracted a tax from the villagers in the form of a percentage of crops or subjugated the villagers and maintained control of the land outright. In some areas of the Middle East the villagers were slaves of the nomads and farmed the land on their behalf.

Transhumance has been practiced in the past primarily in the mountain and plateau regions of Turkey, Iran, Iraq, Afghanistan, and the Maghreb. Unlike the camel nomads, these people of the mountains relied on sheep and goats. They moved the flocks into the mountains in the summertime when snow had melted and the grass was available and back into the valleys during the winter when cooler temperatures and increased precipitation provided grass and water. All nomads had some kind of contact with the villagers or city dwellers who provided them important items in trade or as tribute.

THE DEVELOPMENT OF NORTH AFRICA AND THE MIDDLE EAST

The process of economic development of the Middle East and North Africa from the time of the Neolithic Revolution to the end of the twentieth century reflects the experience of other countries of the less industrial world. Development in pre- and postcolonial times was the result of the location and physical and cultural geography of the region. The central location of the core area of the Middle East is at least partially responsible for the emergence of the Neolithic Revolution and subsequent historical events. As a focus of trade routes from Asia, Africa, and Europe, the Middle East was the recipient of ideas and concepts that aided the development of the early civilizations. It became the cultural hearth for the expansion of knowledge that ultimately culminated in the modern technological and industrial revolutions. The Middle East was a major center for diffusion of the concept of plant and animal domestication, and from the Middle East the cultural values and traditions of Islam and Christianity were diffused to much of the world (Figure 15.9).

The centrality of the Middle East, which facilitated acquisition of innovations and ideas, ultimately became a threat to the region. The crossroads location of today's Israel, Lebanon, Syria, Jordan, Egypt, and Saudi Arabia made this region important to expanding political systems. The Islamic religion, which developed after A.D. 600, is a prime example. The last Islamic empire was the Ottoman Empire, which expanded eastward from Turkey after A.D. 1300. The Ottomans ultimately controlled portions of three continents, Asia, Europe, and Africa, from Vienna to Iran, and encompassed North Africa westward to and including Morocco (Figure 15.10).

The Ottoman Empire remained the dominant political force in the Middle East and North Africa until the nineteenth century. Istanbul was the center of the Ottoman kingdom, and local leadership carried out many political functions in the areas under Ottoman control. The primary concern of the central government was to maintain the flow of taxes and tribute to Istanbul to support the lavish life-style of the ruling elite and to preserve the territorial integrity of the

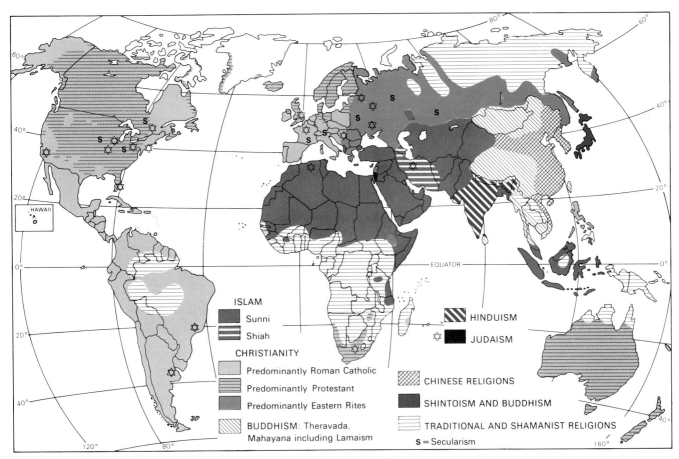

Figure 15.9 Distribution of major religious systems of the world.

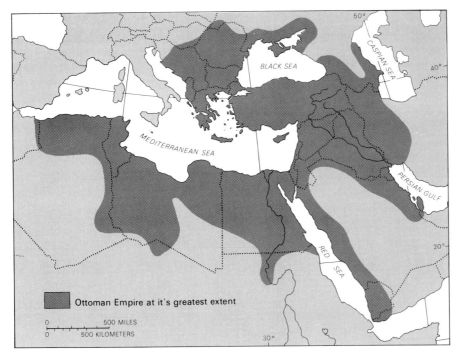

Figure 15.10 Territorial extent of the Ottoman Empire.

THE MIDDLE EAST AND NORTH AFRICA

443

empire. The Ottoman Empire began to decline after loss of today's Hungary and Yugoslavia to Austria between 1686 and 1718.

EUROPEAN INFLUENCE

Expansion of European influence after the beginning of the Renaissance brought contact between the Islamic world ruled by the Ottomans and the emerging mercantile states of western Europe. The movement of ideas was the reverse of the diffusion from the Middle East during the previous 8 millennia. The ideals and ideas of the Christian nations of western Europe began to infiltrate the Islamic world beginning with the granting of agreements for merchants in 1536. These agreements, granted first to the French, ensured that the Ottoman Empire would treat French merchants in the same manner that the French government treated Ottoman merchants. Such agreements were granted to other governments of Westen Europe as they spread their trading interests into the Ottoman Empire.

The increasing power of western European nations signaled the destruction of the Ottoman Empire as Poland, Austria, and Russia acquired portions of it. By 1774 Russia defeated the Ottoman armies and acquired significant lands around the Black Sea. Ottoman leaders tried to reverse the tide of territorial loss by using Western methods, which had been successful against the Ottomans themselves. By 1808 the Ottoman government introduced Western military techniques and weapons and established military and medical colleges where Western ideas were taught. Islamic leaders in outlying provinces, who had gained semiautonomy over the previous centuries, were more resistant to Western ideas, but by the middle of the nineteenth century the process of European domination of the Middle East was well advanced.

The major method that Western powers used to control of portions of the Ottoman Empire was granting loans to the *sultan* (king). The breakdown of political control had hindered the flow of taxes to the capital, and the sultan and his government required increasingly large loans to maintain his lavish life-style and military forces. Lands were also sold to Western interests by both the Sultan and local leaders, including the sale of portions of modern Israel to Western Jews who wanted to create a Jewish homeland. These individuals, referred to as *Zionists* because of their commitment to creating a modern-day Zion, reintroduced the issue of a Jewish homeland to the Middle East with all its consequent political and social repercussions. The continued decline of the Ottoman Empire culminated in 1881 when the failure of the sultan to pay his debts to the West placed the empire in bankruptcy. Although the Ottoman Empire persisted for a while, for all intents and purposes the Middle East had become an economic colony of Western Europe.

EUROPEAN IMPACT ON THE MIDDLE EAST AND NORTH AFRICA

The European impact on the Middle East and North Africa can be classified into five general categories: European ideas, economic change, population growth, European migration, and petroleum exportation. The educated elite of Middle Eastern cities adopted European liberal ideas. These Western ideas emphasized democracy and individual rights and generated the idealism that ultimately precipitated the emergence of independent Middle Eastern states. The second European impact affected the economies of parts of the region. The traditional self-sufficient village system was replaced by an economic structure designed to produce goods for European markets. For example, as industrialization began in Europe, the need for cotton for the textile mills prompted replacement of the traditional crops of the Middle East with industrial monoculture of cotton, making Egypt essentially a cotton plantation for England.

The production of cotton in Egypt for English mills had its counterpart in the production of tobacco in Turkey and coffee, oranges, or dates in other areas. The production of industrial crops did not benefit the industrial development of the Middle East and North Africa. The European colonial powers ensured that manufacturing stayed in the ruling countries. The resulting pattern of development was an asymmetrical, parasitic relationship in which the colonies were maintained as suppliers of cheap raw materials and as markets for manufactured goods from the industrialized lands. The development of roads, railroads, and other infrastructure was undertaken only when it benefited the colonial power. Railroads were built to transport industrial crops from agricultural areas, but a rational program of road building to unify individual peoples or countries received little attention. The resulting infrastructure with its characteristic parasitic cities ensured that the standard of living among individual Middle Eastern countries remained among the lowest in the world. Under European control, literacy rates in countries like Algeria or Egypt were as low as 5 percent of the adult population.

Another major effect of European influence was to improve sanitation and medical facilities and precipitate the same tremendous population increase that occurred in other underdeveloped areas of the world. The increasing population was not matched by investment in schools or manufacturing, further encouraging the expansion of poverty. In the colonial possessions administration remained in the hands of

Europeans and a small local civil servant class. The great majority of the population lived in poverty similar to that found in the worst areas of today's less industrialized countries. Growing populations in Europe added to the cultural complexities of the Middle East and North Africa, as more than 1.5 million French migrants colonized portions of the Maghreb. These French settlers developed an intensive agricultural economy, especially in production of wine for export to France. Increasing industrialization in Western Europe and the development of the internal combustion engine led to European development of Middle Eastern petroleum after the late nineteenth century.

The general impact of the Europeans can be summarized as transforming a subsistence economy into a dependent economy, ensuring that industrial development did not occur within the individual countries, and exploiting the resources of the Middle East. Not until the end of World War I did the countries of the Middle East and North Africa begin the movement that culminated in independence in the area. European development did provide some medical care, sanitation, and transport systems to exploit local resources, but did not materially increase the well-being of most of the people.

NATIONALISM IN THE ISLAMIC REALM

The educated elite of the Middle East and North Africa with their awareness of the liberal ideals of democracy in western Europe were always opposed to European dominance. Until World War I these educated people rarely had more than a minor impact on the European colonial powers. Then, involvement of the industrial nations in an internal conflict distracted their attention and allowed nationalistic movements in Middle Eastern countries to flourish. The greatest success was in Turkey, with Turkish forces under the leadership of Ataturk (Mustafa Kemal). The crippled Ottoman Empire supported Germany in World War I, and the Western allies initially planned to divide up the remains of the Ottoman Empire, but the Turkish nationalists under Ataturk's direction drove the English, Italian, and French forces from Turkish land and declared an independent state. Under Ataturk's direction, Turkey began a process of industrialization based on the mineral resources within the country's boundaries and began to westernize the culture.

Other areas of the Middle East and North Africa were less fortunate in their attempts to develop. Iran, the core of the old Persian empire, obtained independence in 1921 and a westernization program began. Utilizing the French as a model, the new Iranian rulers began to adopt industrialization, expand the road system, and attempt transformation of the economy. The political system did not include the democratic ideals of the West, as Reza Khan, the army officer who had led the successful military revolt in 1921, was crowned Shah and the traditional monarchy was retained until Islamic revolution overthrew it in 1979. From 1979 to 1988 Iran was involved in a destructive war with Iraq, at the same time developing an Islamic Republic based on a fundamentalist interpretation of the Koran. Economically Iran is less developed than it was prior to 1979.

In the balance of the Middle East and North Africa political control remained in the hands of European countries under mandate (a grant of political power) of the League of Nations after World War I. During the interwar years, the British granted increasing political independence to areas over which they held a mandate, including Egypt and Iraq. The French were more reluctant to give up their former colonies because of the number of French settlers in Algeria and Tunisia. Ultimately the French were forced to leave this area after prolonged and bitter civil war finally drove them from their last possession in Algeria in 1962. Spanish possessions in North Africa did not gain their complete independence until the end of the 1970s.

The process of independence that began in Turkey and spread throughout the Middle East and North Africa did not resolve the problems of isolation and subsequent colonial rule of this region. In many countries dictatorial leadership emerged, and individual rights as known in the Western world are largely unknown. The lack of a democratic revolution in many Middle Eastern and North African states continues to handicap development efforts. The new independence has prompted another major problem, as the Islamic majority in some countries resists attempts to transform their economy because of centuries of fear and distrust of ideas borrowed from *infidels* (nonbelievers). The present pattern of the Middle East and North Africa is a complex mosaic combining all elements of typical less industrialized lands with tiny islands of development. The complex cultural pattern even within the Islamic realm further complicates this pattern, as some Islamic schools of thought accept industrialization and modernization as being within the teachings of the prophet Muhammad, while others insist that the Western influences will only pervert and destroy the Islamic tradition.

MIDDLE EASTERN LANDS: CHALLENGE OR OPPORTUNITY

The physical geography with which the Middle East has been endowed is a curious combination of rich and harsh lands. The common stereotype of the Middle East and North Africa is that of a desert dominated by the Sahara, which bounds this area to the south. The Middle East and North Africa are not as arid as commonly assumed, but in general the region does have a dry climate. The predominant climatic type is desert or steppe with precipitation totals of 10 inches (254 millimeters) or less per year (Figure 15.11). The driest portions are the Sahara and the great deserts of the Arabian Peninsula. Steppe climates are found where increasing elevation increases orographic precipitation.

In much of the desert region years may pass without noticeable, measurable precipitation, followed by periods of intense rainfall from cloudbursts. The Sahara itself is used only marginally by the nomadic peoples but has isolated oasis settlements. Steppe climates in the Middle East may receive as much as 20 inches (508 millimeters) of precipitation per year and are the focus of nomadic activities as well as dry farming of grain crops in higher elevations, which have greater precipitation.

Mediterranean climates border the Mediterranean Sea in northwestern Africa and the Levant. Important areas of Morocco, Algeria, Tunisia, Turkey, Syria, Israel, and Jordan have Mediterranean climates. Precipitation totals of 15 to 35 inches (380 to 889 millimeters) make farming more profitable in these areas, particularly fall-sown crops such as wheat and barley, which utilize the winter maximum of rainfall (Figure 15.12).

The temperatures of the Middle East are almost uniformly high during the summer months. The highest average daily temperatures during the summer nearly always exceed 85°F. (30°C) and in Libya and Egypt and the lowlands of the Persian gulf may exceed 97°F. (35°C). Summer maximums exceed 110°F (44°C). Winter temperatures are much more variable, with the highland regions of Turkey, and the mountains of Lebanon, Syria, Iran, and Iraq receiving snow. The desert lands have a high daily (*diurnal*) range of temperature. The lack of large water bodies and the dry atmosphere allow the earth's heat to be radiated rapidly into the atmosphere at night. Nighttime temperatures may fall to 50°F (10°C) even where daytime temperatures reach 111°F (45°C).

Because of the dominance of dry climates in the region, the critical environmental factor is the presence or absence of water. Population is concentrated along the rivers, in the highlands, or near highlands where moisture is available. The mountains of the Middle East are the principal geographic factor affecting the distribution of water and hence population. In the Maghreb, the Atlas Mountains provide an orographic barrier, and the water from the mountains is the lifeblood of the agricultural region of northwestern Africa. In Turkey, Iran, Afghanistan, and Pakistan the Taurus, Zagros, and other mountain ranges reach elevations of over 18,000 feet (5,400 meters) (Figure 15.13). These mountains are the source of streams and rivers that provide water for the Tigris-

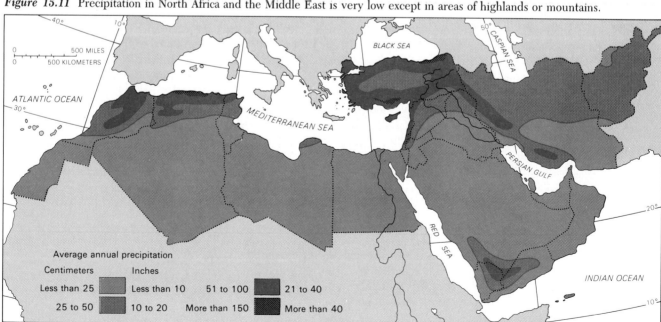

Figure 15.11 Precipitation in North Africa and the Middle East is very low except in areas of highlands or mountains.

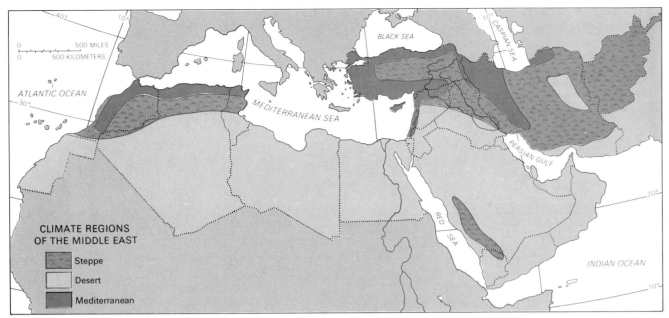

Figure 15.12 Climate regions of North Africa and the Middle East.

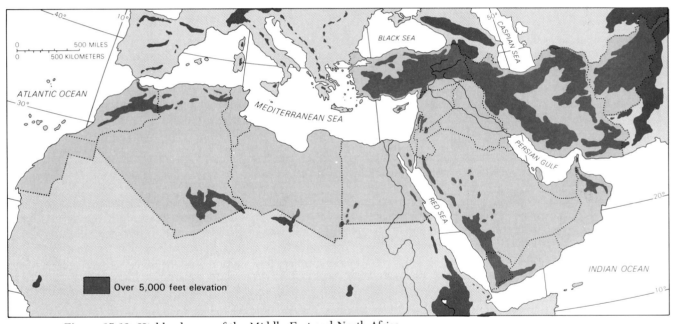

Figure 15.13 Highland areas of the Middle East and North Africa.

Euphrates rivers as well as a host of smaller ones. South of the region defined here as North Africa are the mountains of Ethiopia and the savanna climates of Africa south of the Sahara, which provide the water for the Nile River.

Aside from the areas made suitable for settlement by rivers or oases, there is a small area of subtropical climate on the Iranian coast of the Caspian Sea. This climate is warm and moist and is a complete anomaly in the entire region.

The lack of water is mirrored by the absence of other major resources in this region. Fuel resources are commonly perceived to be abundant, but petroleum is concentrated in a relatively few places (Figure

THE MIDDLE EAST AND NORTH AFRICA

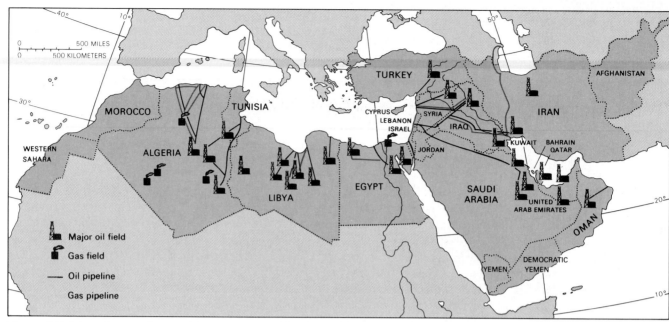

Figure 15.14 Petroleum and natural gas in North Africa and the Middle East.

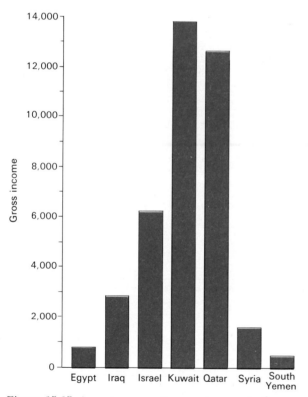

Figure 15.15 Average per capita gross incomes in the Middle East. If revenues from oil exports were evenly divided, the residents of the surplus capital oil-exporting nations would have the highest standard of living in the world, in stark contrast to the other countries of the Middle East.

15.14). Saudi Arabia, Kuwait, Bahrain, the United Arab Emirates, and other states around the Persian Gulf have most of the oil wealth of the Middle East. Other countries, such as Algeria and Libya, have petroleum, but their larger populations and smaller oil reserves prevent accumulation of the great wealth of the Persian Gulf states (Figure 15.15). The distribution of petroleum shows minor deposits beyond the Gulf in Iran, Iraq, Egypt, and other areas, but only in Iran do they rival the reserves of the Persian Gulf countries. Coal is found in only a few areas, with Turkey having the best resources. Even wood, a characteristic source of fuel in less industrialized countries, is available only in the highlands. The demands of the growing population for fuel seriously jeopardize the few forests that remain.

Other resources are limited. Iron ore is primarily found in the mountains of Algeria, Iran, and Turkey, and other metallic minerals are scattered widely across the region. Chrome, concentrated in Iran and Turkey, is the region's second most important mineral in terms of value. Phosphate and potash are produced around the Dead Sea in Israel and Jordan, and manganese and lead are produced in a number of areas, with Morocco the leading producer. Of greater importance than any resource other than petroleum is the agricultural land of the Middle East, which is highly fertile if adequate water is available, and many of the incipient industrial projects in the region rely on agricultural products as their major input.

THE WORLD OF POVERTY

THE MODERN MIDDLE EAST AND NORTH AFRICA: DEVELOPMENT AND CONFLICT

The development of the individual Middle Eastern and North African countries reflects both the physical geography and the past and present cultural events that affect them. Those countries with an abundance of petroleum, such as Saudi Arabia or Kuwait, have developed differently from those like Egypt with only limited petroleum. In each case the available resources and their development have been critically affected by the past relationship between the country and the former colonial power that dominated it and the relationships within and between the Middle Eastern and North African countries themselves. On the basis of oil wealth or lack of it, there are three general types of countries in the region: least industrialized countries, capital-surplus, oil-exporting countries, and industrialized countries.

ISRAEL

The only industrialized nation is Israel. Israel illustrates the potential of the region, but also exemplifies the problems of transforming the economy of the Middle East and North Africa. The modern-day state of Israel had its origin in land purchases by Jewish Zionists in the nineteenth century and related Jewish immigration. Most of the Jewish immigrants lived in cities, but land was purchased from willing Arab landowners including swamps, marshes, and dune areas.

The British government tacitly supported a Jewish state in the Balfour Declaration of 1917, which said that it favored a national home for the Jewish people in Palestine provided that establishment of such a Jewish homeland would not "prejudice" the rights of non-Jewish communities there. After World War I, Jewish immigration to Palestine increased, and the European migrants purchased from local sheikhs land that the Arabs had not used intensively. The European Jewish migrants were often subsidized by the National Zionist Movement allowing them to purchase lands at prices far above what the local farmers could afford. Since nearly one-third of the Palestinian farmers were landless laborers or sharecroppers on property owned by absentee landlords, the potential for conflict between the newly arrived Jews and the Arab majority was great. Riots among the Arab population caused periodic British opposition to additional Jewish immigration to the country, but the numbers of migrants continued to increase.

At the end of World War II, many European Jews who had survived Hitler's holocaust clamored to be admitted to Palestine. Faced with conflict between Arab and Jew, the British attempted to prevent additional Jewish immigration, but ultimately abandoned the Palestinian question to the newly formed United Nations. In November 1947 the United Nations proposed the division of Palestine into an Arab and a Jewish state with Jerusalem and its suburbs as an international zone. Only 1.5 percent of the total population of the proposed Arab state was Jewish, but 45 percent of the population of the proposed new state of Israel was Arab.

As fighting mounted between Arab and Jew, the new state of Israel was declared independent by Jews on May 14, 1948. The Arab states of Egypt, Syria, Lebanon, Jordan, Iraq, and Sudan declared war on Israel, and the fighting continued until 1949. The smaller Jewish force was able to defeat the numerically superior Arab forces because of a strong unity of purpose and organization. The Jewish settlers were committed to obtaining a homeland and were willing to sacrifice individually to that end. The Arabs had no strong central leadership and lacked the combat experience that some of the Jewish soldiers brought with them from World War II.

In 1949 an armistice was reached, but the 1948 Arab-Israeli War had profound impact. Seven hundred and fifty thousand Palestinian Arabs from the new state of Israel fled to neighboring Arab countries as a result of the conflict. The original borders suggested by the United Nations for the new nation of Israel were expanded by approximately 20 percent (Figure 15.16).

The refugee question and the issue of control of the territory of Israel became fundamental sources of conflict, which persist to the present. The attitude of many Middle Eastern Arabs is that Israel represents the ultimate example of European imperialism. A non-Arab, non-Muslim population has been imposed upon a region that had been Arab and Muslim for hundreds of years. The development of a commitment to eliminate the Israeli state caused repeated confrontation between Israel and its Arab neighbors. Both sides armed extensively and mobilized their forces.

In the Six Day War in June 1967, Israel destroyed the air forces of Egypt, Jordan, Syria, and Iraq. The Israelis expanded their control of lands in the Middle East to the Suez Canal and seized the west bank of the Jordan River from the kingdom of Jordan. Hundreds of thousands of additional Palestinian Arab refugees fled into Jordan, Lebanon, and Syria, adding to those who had been living in refugee camps since 1949.

Repeated conflict after the peace agreement ending the 1967 war culminated in the 1973 war in which Arab forces from Egypt and Syria were able to defeat the Israeli forces initially. After a few days of retreat

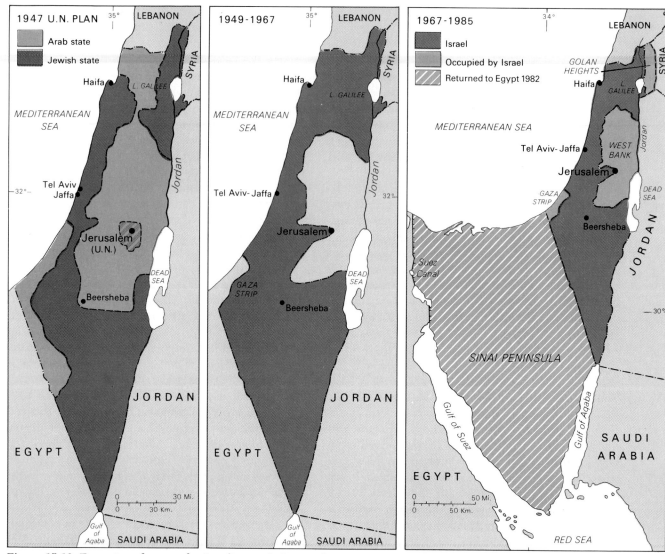

Figure 15.16 Expansion of area under Israeli control. During the 1967–1980 period Israel also occupied the Sinai peninsula of Egypt.

however, the Israelis counterattacked and regained the lost territories, and the Israeli armies advanced into Egypt and Syria. A peace was negotiated by Western interests, but the Arabs claimed a victory from their initial successes.

The conflict between Arab and Jew in the Middle East remains one of the critical factors affecting the development of the Israeli nation and the adjoining Arab states. Since 1978 Egypt has seemingly become receptive to the idea of a permanent Jewish state and has accepted peaceful coexistence, culminating in Israel withdrawal from the Sinai peninsula of Egypt by 1982. Other Arab states have been less willing to accept the existence of Israel, probably because they have not borne the cost of the repeated conflicts. The potential for resolving the Arab-Israeli conflict is made more difficult by the Palestinian Liberation Organization (PLO), which was founded in 1962. The PLO maintained an official policy of elimination of the state of Israel until 1988 when its leader, Yasser Arafat, stated that the PLO was willing to accept the idea of both a Palestinian and Jewish homeland. The PLO receives financial and moral support from several Arab countries and has carried on a guerrilla campaign, which has caused death to civilians and damage to nonmilitary targets. Ironically many of the tactics of the PLO are based on the Jewish guerrilla opposition to the British before Israel's independence. The rapidly growing Palestinian population in Israel increases the urgency of resolving the question of a Palestinian homeland since riots, civil disobedience, and armed resistance by Palestinians continue to increase in Israel.

In spite of the problems that have surrounded the development of Israel, it is the most industrialized country in the Middle East. Development has been based on skilled migrants from the industrial world, transfer payments from Western European and American governments and foreign Jews, and the unity of the people in the face of constant opposition of the Arab world. Agriculture is modern and mechanized. Some of the farms are **kibbutzim,** a style of ownership found only in Israel. Kibbutzim are true collective farms that were founded by migrants to Israel from eastern Europe and the Soviet Union. These migrants were mostly poor idealists who wanted to create a true communitarian movement. Lands for their farms were purchased by the International Zionist Movement, and land is owned by the movement rather than by the individual farmers. The kibbutzim are operated by cooperative action in farming, education, and the growing industrial activity that some of them carry on. Another form of landownership in Israel is the **moshav,** a type of cooperative with individual ownership but a high degree of cooperation in marketing, purchasing, and even occasionally cultivation.

The agriculture of Israel is the most productive in the Middle East, and one of the most productive in the world. Adoption of Western technology has allowed reclamation of both swamps and arid land. Israeli use of irrigation water is widely recognized as among the most efficient systems in the world. Unlike the traditional irrigation found in much of the Middle East, the Israeli system emphasizes such advances as drip irrigation, which provides water directly to the individual tree or plant root zone, and a nationwide reclamation scheme. Major products include oranges and vegetables for export, and a variety of grains, vegetables, and meat products for local consumption.

The industrial development of Israel is also the most advanced in the Middle East. Israel lacks almost every resource for industrialization but has capitalized on the skills of the immigrants. It has become a world center for cutting and polishing of diamonds, as well as being a leader in technical innovation in areas such as computers. All manufacturing activities common to industrialized countries are found in Israel, and the standard of living for Israelis is typical of industrialized countries. Manufacturing and population are concentrated in the largest centers (Tel Aviv–Jaffa and Haifa) and the coastal lowland between them.

The expansion of Israel's borders has brought 1.8 million Arabs in occupied lands under Israeli rule, threatening the continued development of the state. Israeli attempts to maintain control over the West Bank and other occupied areas have taken many forms, most important the establishment of Jewish communities in the occupied lands to insure Israeli control (Figure 15.17). The Arab inhabitants of the occupied lands are given local autonomy under overall control by the Israeli government and a large military presence. The Arabs have been integrated into the Israeli economy, where they provide cheap labor. Their standard of living has been greatly increased since the end of the 1967 war, but they have not been effectively integrated socially into Israel, and they remain a constant threat to the peace of the Israeli nation. The continued existence of the PLO and the demands of the nearly 2 million refugees and their descendants from previous wars for repatriation to Israel, combined with Arab claims to the holy city of Jerusalem and to the waters of the Jordan River, make the future of Israel, Jordan, Lebanon, Syria, and the Palestinians uncertain.

The increasing importance of Middle Eastern oil to Western countries has caused many of Israel's most vocal supporters in the West to reduce their support in the face of Arab threats to limit oil exports. As the most industrialized nation in the Middle East, Israel's economy is threatened by a high inflation rate and high taxes to maintain a strong military force. An increasingly worrisome problem to the Israelis is the outward migration of skilled population in the face of the high cost of living and of political uncertainty in the nation. The beginnings of a movement toward peace between Arab and Jew indicated by the actions of Egypt in the past few years may eliminate these uncertainties, but in all probability Israel will remain a focus of conflict in the Middle East.

THE LESS INDUSTRIALIZED COUNTRIES OF THE MIDDLE EAST AND NORTH AFRICA

Because there are more than two dozen individual countries in the Middle East and North Africa, it is impossible to discuss each in detail. The general experience can be illustrated by examining the geography of several of the less industrialized countries. In the northern tier of states from Turkey to Afghanistan, the mountains provide water and a variety of microclimates in the mountain valleys. All these states are agrarian, with Turkey being the most industrialized and Afghanistan the least. The contrasts within

---

Kibbutzim: A collective farm or settlement in Israel.
Moshav: A cooperative farm or settlement in Israel.

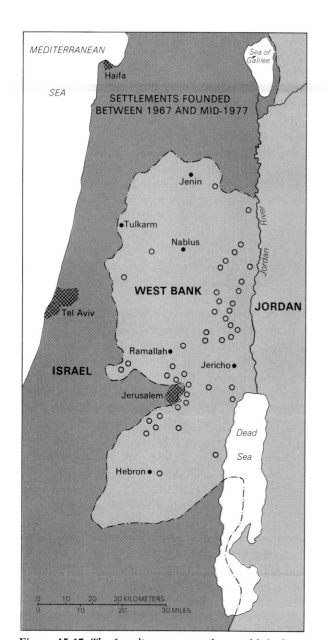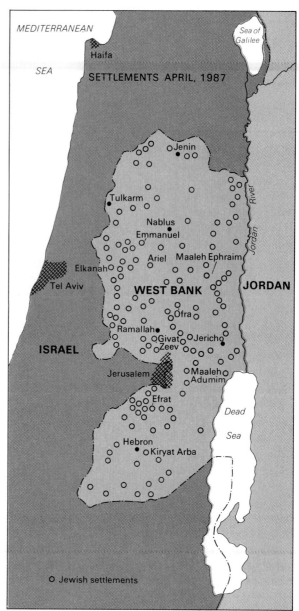

Figure 15.17 The Israeli government has established communities and settlement areas in the Israeli occupied West Bank region of the Jordan River.

just these states typify the problem of characterizing the development or geography of such a complex region.

Turkey. Under the direction of Ataturk, who changed many of the customs of traditional Islam in the interests of modernization, Turkey entered the process of development. Under Ataturk's direction religious schools were replaced by state-supported secular schools, and the Arab script was supplemented by Latin characters. Slavery and polygamy were outlawed, and women were given legal rights as full citizens, but the society under Ataturk and subsequent Turkish leaders still has many characteristics of the less industrialized countries. Industrial production is greater in value than agricultural products, but the majority of the population is still engaged in agriculture. Major manufacturing activities include textiles, food processing, and simpler industries that rely on a high labor input. There is one center of heavy industry associated with the iron and steel mill at Karabuk, using Turkish coal and iron ore.

Per capita incomes of $1200 in Turkey are among the highest in the non–oil-producing countries of the Middle East. The literacy rate is less than 60 percent, and the birthrate is 2.2 percent per year, typical of less industrialized societies. The population is concentrated along the plains and valleys of the Black Sea

THE WORLD OF POVERTY

A Palestinian refugee family in the Jabalia Camp, Gaza Strip, Israel. Many refugees have never known life outside the refugee camps.

and Mediterranean sea coast. The current population is 54 million and the rapid growth rate represents a major problem for the country. The major agricultural products in Turkey are cotton, tobacco, wheat, barley, corn, sugar beets, and livestock products. Unlike many Middle Eastern and North African countries Turkey is self-sufficient in foodstuffs. Turkey's problems as a less industrialized country are alleviated somewhat by its location near Europe. An estimated 1.5 million Turkish people work in Europe, relieving some of the population pressure.

Turkey is an associate member of the Common Market, and this special arrangement provides potential for even greater development. The Turkish government is a democratic, popularly elected republic consisting of a parliament, a president elected by the parliament, and a prime minister appointed by the president from the members of parliament. Suffrage is universal for males and females 21 and older. Like most less industrialized countries, Turkey's major problems for the future are the increasing cost of energy imports in a predominantly agrarian society and the difficulty of achieving sufficient import payments from low-value agricultural exports.

Iran. Under the leadership of the Shahs, Iran followed a parallel development path to that of Turkey. The Shahs instituted important reforms in their attempts to modernize the Iranian economy and utilized Western experts as advisers. Unlike Turkey, Iran has an excellent source of energy in the oil fields around the Persian Gulf, allowing the country to expend millions on Western technology. The income from oil and the incipient industrialization pushed the per capita income in Iran to nearly $2000 by 1979, but it is less today, and the population growth of 3.2 percent handicaps efforts to eliminate poverty. Iran is primarily an agricultural country, but industry in the major cities of Tehran, Tabriz, and Meshed could provide the basis for transforming the economy.

Industrialization and modernization under the Shah were based on importation of Western technology, and the conflict between traditional Iranian Islamic views and the increasingly Western view of the Shah's government resulted in the overthrow of the Shah and the monarchy he represented in 1979. The overthrow was followed by formation of an Islamic republic, with a freely elected parliament that has power to dismuss the president and his cabinet. The one factor that provides unity for Iran is the predominance of the Shiite Islamic group. The country has important subgroups of Kurds in the north who are unwilling to accept the leadership of the Shiite leader (the Ayatollah), and there is increasing resentment among the urban middle class of the fundamentalist and simplistic views of some Islamic religious leaders who also have important political positions. Since the advent of the revolutionary Islamic government, Iran's economy has been declining. Per capita income is lower today than it was in 1980. Agricultural, industrial and oil production are down. The war with Iraq compounded the economic problems of the country. Production of oil for export continues to be the dominant economic activity, followed by production of wheat, barley, rice, and other agricultural products. The burgeoning population of Iran has changed the country from a grain surplus to a grain importer in the past 20 years.

Conflict in Iran seems certain in the future. Events there from the 1970s to the present reflect the problems of transforming traditional economies when significant social changes must be made. The Shah's attempt to modernize and westernize his country foundered in the face of a largely illiterate peasantry (the nation has only 50 percent literacy for adults seven years of age or over) who were unwilling to accept ideas from the infidel world. The war with Iran added an even greater cost to Iran's new leaders. An entire generation of men have been decimated or crippled and maimed, and the country's economy prostrated by the war.

Whether or not Iran will be able to transcend the problems of the Islamic revolution and war with Iran and develop in a uniquely Islamic way remains to be seen. Clearly in Iran, as in many less industrialized countries, the benefits of development were not widespread, and the majority of the Iranian population continued to live lives of poverty in the villages. It appears from Iran's experience that the Western model of economic development is not welcome in the region.

Afghanistan. Afghanistan is in even worse condition than other countries in the mountainous northern tier of the Middle East. The per capita income of Afghanistan is estimated at less than $100 per year, the lowest in any country in the Middle East or North Africa. Afghanistan has been described by some observers as a country rushing headlong from the twelfth century into the thirteenth. It is one of the least industrialized countries of the world, with a population of 15 million and a growth rate of 2.4 percent. The infant mortality rate is 182 per thousand; otherwise, the growth rate would be even higher. Literacy rates are estimated to be 20 percent for the adult population. The physical environment is mountainous and arid to semiarid with few resources. The location of Afghanistan near the Indian Ocean and near the oil-producing countries has attracted the attention of non-Arab countries. Most recently it was occupied by military forces from the Soviet Union in late 1979, and its future remains unclear. The Soviet Union faced fierce opposition from the illiterate Muslim tribesmen who were even more hostile to communism than to Christians or Jews. The Soviet Union's policy of subjugation forced nearly 4 million of Afghanistan's population to flee to Pakistan. Deaths and reduced birthrates among these refugees has led to the concept of **migratory genocide,** the elimination of a people through the hardship of being homeless. The Soviet Union was unable to subjugate the tribespeople completely, and withdrew in 1989. It is too early to predict the outcome after the departure of Soviet troops, but in all probability Afghanistan will continue to be a focus of conflict between rival forces wanting to lead the country.

Iraq. Iraq has adopted a form of development based on a socialist government. After obtaining its independence, Iraq was ruled by a monarchy until a *military coup* (revolution) in 1958 brought a socialist group to power. Iraq's major source of export income is its oil, which provides a gross national product equal to about $1,600 per person per year. Otherwise, Iraq is similar to other less industrialized countries, with a birthrate of 45 per thousand and a growth rate of 3.5

Migratory genocide: Term used to refer to destruction of a group through forced migration, under harsh conditions.

The Nile River is sometimes referred to as the "Lifeblood of Egypt" because of its importance in irrigation and transportation.

percent per year. The literacy rate is over 50 percent, and only 38 percent of the population is engaged in agriculture. Sixty-eight percent of the population is urban, and the primate city of Baghdad has more than 4 million people.

The irony of Iraq is that all evidence indicates that the villagers enjoyed a higher standard of living in the past than they do now. The great Tigris and Euphrates river systems provide water for expanding the presently irrigated land of Iraq by a factor of five. The irrigation network prior to the Islamic expansion into this area was greater than it is today. The present government of Iraq is attempting to expand irrigation by building major dams for water control on the erratic rivers and draining the marsh areas near the mouth of the Tigris and Euphrates. Aricultural products include wheat, rice, barley, and subsistence crops. Iraq also produces the majority of the world's dates. Industrial activities focus on production of crude petroleum, with only handicrafts and minor industrial development. The potential for the development of Iraq lies with its oil wealth, but the continued population growth tends to negate development efforts. The destructive war with Iran handicapped economic growth, but financial assistance from other Arab countries allowed Iraq to maintain the war without the economic destruction Iran suffered.

Egypt: the gift of the Nile. The problems of the less industrialized countries of the Middle East are epitomized in Egypt. Egypt had a population of approximately 6 million at the time the British began trans-

Population Crisis in Egypt

Hosni Mubarak, the first Egyptian president to take a strong public stand in favor of family planning, is continuing to speak out about his country's population problem.

In February 1982, President Mubarak startled his advisers when he held his first meeting to discuss Egypt's economic future. He devoted most of the session to emphasizing the futility of reform unless the nation's population growth rate was brought under control.

Although Mubarak's position represents a dramatic change from the viewpoint of previous Egyptian leaders, it remains to be seen whether he can mobilize the country toward effectively reducing its rapid population growth because of the continuing negative attitudes of many Egyptians.

In a 1959 interview the late President Nasser said: "I am not a believer in calling on people to exercise birth control by decree or persuasion. Instead of teaching people how to exercise birth control, we would do better to teach them how to increase their land production and raise their standard (of living).

". . . We live in and make use of only 4 percent of the area of our country. The rest is neglected and desert. If we direct our efforts to expanding the area in which we live instead of concentrating on how to reduce the population, we will soon find the solution."

Nasser's comments are significant because today, 25 years later, most Egyptians, including religious and government leaders continue to echo those sentiments.

Meanwhile, Egypt has one of the world's most severe population problems and family planning programs have been relatively ineffective.

Half of the Egyptian population is under age 15. Illiteracy in terms of absolute numbers is increasing. Job opportunities are dwindling. Low-cost housing cannot be built fast enough and by the year 2000, Egypt's population, which is 48 million today, is expected to reach 75 million.

The country's Ministry of Health gives family planning a low priority, and the Ministry of Planning gives it no priority at all.

Egypt's National Committee for Population Problems had established eight family-planning clinics in Cairo and Alexandria by 1955. But the government prohibited publicity about the facilities and few Egyptians used them.

Two decades later official national policy under President Anwar el-Sadat, who succeeded Nasser after his death in 1970, set forth several population goals to be achieved by 1982. These goals included: reducing the birthrate from 33 per 1000 to 23; reducing the population growth rate from 2.1% to 1.1%; insuring that the population does not exceed 41 million.

But trends all moved in the opposite directions. By 1985 the birthrate had increased to 37 per 1000; population was growing at the rate of 2.7% a year, and had exceeded 48 million. Had the goals been achieved, the population in the year 2000 would be 50 million, about 25 million fewer than current projections indicate.

If Egypt achieves its current goal of reducing its birthrate from the current 37 per 1000 to 20 per 1000 by 2000, and if fertility continues to decline thereafter—reaching levels just below replacement—the population will be 63 million in 2000, instead of the currently projected 73 million, and about 84 million in 2030, instead of the 183 million now projected. Past experience in Egypt does not inspire confidence that they will reach their goals.

Source: Popline, *Population Institute*, Washington, D.C., November 1984, p. 3.

---

forming it into a cotton-producing farm for British industry. The population in Egypt had remained relatively stable for the previous thousand years, and the agricultural system was little changed from that which had existed at the time of the Pharoahs. The vast majority of the population lived in villages, often in conditions of poverty unknown to Europeans. Ninety-six percent of the population was (and remains), concentrated along the Nile. The living standards were particularly bad, since the primary source of water was the Nile, and the river also served as the sewer for villages upstream. Common diseases such as bilharzia and malaria as well as hookworm and trachoma (resulting in partial or total blindness) were widespread. Cholera epidemics spread through Egypt several times, and the diet of as many as 75 percent of the population was inadequate. The development of Egyptian cotton production for European needs was accompanied by a declining death rate and rapid expansion of the Egyptian population. By 1900 there were 10 million in Egypt, by 1940 nearly 17 million. By 1990 the population of Egypt had grown to over 56 million, and the birthrate of 38 per thousand combined with a declining death rate caused the population to increase at a rate of 2.8 percent per year. (See Box: Population Crisis in Egypt.) The concentration of available agricultural land in the Nile Valley coupled with the rapid population increase causes farm sizes in Egypt to be extremely small. More than 90 percent of farm owners in Egypt own one acre of land or less. These tiny plots are cultivated intensively on a double-cropping system based on changes in irrigation that accompanied the British occupation of Egypt.

The traditional pattern of irrigation along the Nile was known as **basin irrigation.** Runoff from the summer rains of the savanna climate in the headwaters of the Nile in the Sudan and Ethiopia fills the Nile to flood stage. These floods on the Nile used to be periodic, predictable, and rather benign in impact. Lands were cultivated in Egypt prior to the onset of the high water, and each field was surrounded by a mud dike. The annual flooding brought silt-laden waters to the fields along the river, and as the floodwaters receded, the dikes trapped the water and the silt, allowing the land to be saturated and the fertile silt to be deposited uniformly on it. Supplementary irrigation was provided during the growing season by waterwheels and shadufs. The British transformed the irrigation system by introducing the use of low dams along the river. These allowed water to be diverted from the Nile to irrigate during the nonflood season, increasing crop production.

Basin irrigation: A system of irrigation in which floodwaters of a river are trapped in basins created by building small dikes around fields to retain the silt-rich floodwater.

A street scene in Cairo, Egypt, illustrating the narrow, crowded streets of much of the city.

After Egypt's independence, a monarchy was established and perpetuated the poverty of the peasants while the monarch lived in luxury. The peasant population of Egypt in 1948 was estimated to have a per capita income of only about $50. The Egyptian peasants lived in villages whose lands were controlled by absentee landlords, whose share of crop production was small, and who saw any increase in productivity go largely to the landlord. Such conditions did little to encourage transformation of either the stagnant Egyptian peasant economy or to develop a middle class or industrialization. In 1952 the Egyptian monarchy was overthrown in a military coup and a revolutionary government installed. Since that time the government has had strong authoritarian overtones, but there have been elections, and there has not been the repression of individual freedoms found in other countries.

Since the revolution Egypt has made important efforts at development, but the increasing population growth has effectively prevented per capita rises in the standard of living. The per capita income, only $800 per year, is among the lowest found in the Arab region. Changes to try to modernize the economy include land reforms and government support of industrialization. At the time of the revolution only one-half of 1 percent of the population owned over one-third of the arable farmland in the country. After the revolution, size of holdings was legally limited to slightly more than 200 acres (81 hectares) per individual with a maximum of 300 acres (121 hectares) per family.

This land reform involved only one-tenth of the cultivated area of Egypt but signaled a major social change in the country. The second reform law of 1961 reduced the maximum size of holdings to slightly more than 100 acres (40 hectares) per person, and in 1969 the maximum size per individual was reduced to slightly more than 50 acres (20 hectares). Although many large landholders were not forced to comply with the new size limitations, the transfer of land marked an important development in attitude of the peasants, who for the first time became individual entrepreneurs with complete equality before the law. The revolutionary government also attempted to introduce cooperatives to provide agricultural assistance and credit to the rural peasants. The government also instigated consolidation of cropping patterns for redistributed lands in which the lands of a village are divided into three large blocks, each assigned to one of three main groups of crops yearly. This triennial crop rotation system provides important benefits for the Egyptian farmer. Now all crops grown within the block are similar and require water and fertilization and spraying at similar times. Irrigation water can be used more efficiently, and yields have increased as much as 20 percent. Land is still cultivated individually, but the designation of the annual type of crop has important implications for the country. Participation in this crop consolidation scheme is mandatory for recipients of the redistributed land and is encouraged for other villagers.

Another major change instigated since the revolution has been the expansion of irrigated acreage, particularly through construction of the high Aswân Dam. The high Aswân is situated near the southern border of Egypt on the Nile, 4 miles (6 kilometers) upstream from the older low Aswân Dam built in 1902 and later heightened. The high Aswân Dam forms Lake Nasser, which can hold all of the yearly floodwaters and prevent flooding of lands downstream. It provides irrigation water, which can be released throughout the year, adding approximately 20 percent to the irrigated acreage of Egypt. The dam also provides hydroelectricity, has stabilized the level of the Nile below the dam, and has improved navigation on the Nile. The high Aswân has been a center of controversy although it has provided Egypt essential agricultural products directly and industrialization indirectly through the power. But by eliminating seasonal flooding, the dam has increased the need for artificial fertilizer to maintain soil fertility. Lower silt content in the water below the dam has affected fishing on the Nile by limiting the nutrients available in the water, and the increased speed of the water from the lowered silt load has resulted in serious erosion problems downstream. Evaporation loss from Lake Nasser is great, and the high silt load of the Nile is causing rapid siltation of the lake itself. It is impossible to conclude whether Lake Nasser and the high Aswân will be beneficial in the long term, but in spite of the criticisms leveled against its impact on the environment of the Nile River, its contributions to maintaining an expanding agricultural production have been essential to Egypt's existence. Even though Egypt has a low standard of living, without the high Aswân it would be even lower. Hydroelectric power from the dam has assisted Egypt's economic development through providing energy for industrialization.

Industrialization in Egypt has proceeded rapidly by Middle Eastern standards, and Egypt today is the most industrialized country of the Middle East or North Africa after Israel. Textiles are the dominant industrial activity, but food processing, automobile assembly, and chemical production are major industries. Industrialization is concentrated in the major cities of Cairo (11 million) and Alexandria (3 million), which continue to attract rural migrants. Egypt's continued high population growth threatens the economic advances that have been made and seems to dictate its continued role as a less developed country.

THE OIL-RICH COUNTRIES OF THE MIDDLE EAST

A unique group of countries occupies the area around the Persian Gulf and to a lesser extent includes Libya in North Africa. These countries have all the characteristics of less industrialized societies but extraordinarily high per capita gross national products. Libya, for example, has a birthrate of 39 per thousand and a death rate of only 8 per thousand. The nearly 3.1 percent annual growth rate gives a doubling time of only 22 years for Libya's population. The per capita income is nearly $8600 per year in Libya, and in the Persian Gulf countries it is even higher. In Kuwait, a population of 2 million with a growth rate nearly identical to that of Libya has a per capita income of over $14,000 per year. The United Arab Emirates with only 1.5 million people have a per capita income of $15,000 per year. These figures are misleading, since much of the wealth is not distributed uniformly but is maintained by the local sheikhs.

Aside from the tremendous quantities of petroleum and resulting wealth, these countries have many of the characteristics of less industrialized societies. Literacy rates range from 25 percent in Saudi Arabia to over 60 percent in Kuwait. A few individuals reside in luxury while the majority live lives that are better than those in most less industrialized countries, but often only marginally so. The unique aspect of the Persian Gulf countries and other capital-surplus oil producers is their small population in comparison to their tremendous mineral wealth.

The Middle East has the world's largest petroleum reserves. The increasing demand that led to exploitation of the region's oil wealth by Europeans in the early twentieth century now provides a massive infusion of capital to these countries. Prior to 1973 the oil wealth was largely controlled by Western interests, but since that time the Organization of Petroleum Exporting Countries (OPEC) has encouraged individual countries to take control of their oil industry, and prices have increased 1500 percent. This oil wealth is the basis for a modernization attempt in the Persian Gulf countries never before equaled. The capital flow is beyond comprehension in terms of absolute dollar values. Oil revenues exceeded $200 billion in the region in 1980, and although lower today, the use of much of this capital to modernize the economies of Saudi Arabia, Kuwait, the United Arab Emirates, and Bahrain has resulted in important industrial gains. Capital is used to purchase entire industries, to build

THE HUMAN DIMENSION

Korean Workers in the Middle East

The migration of Korean workers to the Middle East can be described as voluntary, organized temporary migration—due to newly emerging economic opportunities and occurring within the context of the labor contract between the individual workers and the construction companies employing them. The term of contract is 1 year, although it can be extended for a few months or renewed on a yearly basis. The main form of recruitment is that of direct hiring by one of the private companies in Korea that have their own business concerns in the Middle East. Only a small portion of migrant workers is hired by foreign companies through a Korean Government agency, the Korean Overseas Development Corporation.

According to a sample of Korean construction workers, the average age of the migrants at the time of departure for the Middle East is about 32. When we compare the age composition of the migrant workers with that of domestic males employed in production and related works, it is clear that the labor migration to the Middle East is selective in favor of the young adult. Average educational levels of the migrants to the Middle East appear to be higher than those of the domestic employees.

About 60 percent of the migrants are married, and about 86 percent are either household heads or eldest sons who are usually responsible for the management of household affairs. Yet their economic assets are poor, as indicated by the fact that about 54 percent of the migrants did not have their own houses and about 68 percent had economic assets valued at less than U.S. $17,000. Data on the educational level of the fathers of the migrants, indicate that most migrants were from lower social strata and were rural-to-urban migrants. The majority of the migrants were engaged in the same occupation they had before migration, because the most important criterion of the recruitment to the Middle East is the kind of skills in demand.

The physical and social environment under which Korean migrant workers must live for 1 or more years is entirely different from that of their homeland. The Middle East region is much drier and hotter than Korea, and the temperature difference between day and night is much greater in the former. As a result, the Korean migrants often suffer from fatigue, loss of appetite, sunstroke, and cold.

Islamic culture also differs greatly from Korean culture, prohibiting alcoholic beverages and eating of pork, which most Koreans enjoy. It also imposes a strict code for social behavior and behavior between the sexes. Any violations against these codes are subject to punishment, even within the work camp.

Most Korean camps are located near the work site but are distant from the residential areas of the local population. They are usually enclosed by some sort of physical structures, fences or natural barriers. The accommodation in such work camps is similar to that provided in military barracks.

The contract states that the employee may be subject to reprimand, reduction of salary, or dismissal for committing such offenses as misconduct against national dignity or policy; gambling or other harmful behavior against public morals leading to deterioration of discipline and work ethic; forgery of documents; and work absenteeism of more than 12 days without legitimate permission.

The workers' everyday life is tightly scheduled. They must get up early in the morning while they go to bed only late at night. They start to work at around 6 A.M., take a 2- to 3-hour siesta break after lunch and finish their regular working day at about 6 or 7 P.M. However, nearly all of the migrants work overtime for 2

---

homes for the citizens of the countries, and hopefully to transform their societies so that when their oil resources are exhausted, they will have provided the groundwork for a stable economy.

The development efforts being undertaken by the oil-rich countries are handicapped by the geography of their areas. The environment of Saudi Arabia and the countries with surplus oil capital bordering the Persian Gulf is among the driest in the world. It is primarily a desert receiving less than 3 inches (76 millimeters) of precipitation per year. Only the southern portions have mountains high enough to produce orographic precipitation. The southern desert is the Rub al-Khali or Empty Quarter covered with sand dunes and inhospitable to people. Only 1 percent of the total area of the Arabian peninsula is under cultivation, and this land is concentrated near springs, wells, or qanats where water is available. One of the projects for which the oil-rich countries have used their capital is to drill hundreds of wells to supplement water sources and to develop desalinization plants to provide water for extremely dry areas such as Kuwait.

The development projects of the capital-surplus countries affect other less industrialized countries. (See The Human Dimension.) Nearly one-third of the labor involved in the development projects of Saudi Arabia and Kuwait is drawn from the two Yemens, and transfer payments by Yemenese in Saudi Arabia and Kuwait to their families is the primary source of income in these two countries. Other major contributions to the labor force include Pakistanis, Koreans, and workers from Bangladesh (Table 15.1). The long-term importance of the developments taking place in the oil-surplus countries relates to both their impact upon

or more hours after dinner, since the overtime allowance is 50 percent higher than the regular hourly wage. On Sundays they do not work but spend most of the day in the work camp. The employers provide various recreation facilities and occasional entertainment such as movies and outdoor games.

Before departure for the Middle East, the migrant workers follow a series of complicated procedures: application for overseas employment, physical examination, education from their employers as well as from the Korean Overseas Development Corporation, and application for passport. Upon departure they are required to carry a set of standardized personal belongings. However, they are not allowed to take such items as the Bible, pornographic literature, alcoholic beverages, and playing cards. Interviews with a number of returnees suggest that upon departure most of the migrants feel a strong sense of anxiety.

In contrast to the staff of the work camp, who form a totally different group, the workers are less educated, low-paid skilled and unskilled workers who are employed only temporarily. They are a relatively homogenous group in terms of socioeconomic status, who rationalize their migrant worker status only in terms of economic benefit.

According to the data, the rate of premature return is about 15 percent. Such a high rate of premature return suggests that the migrant's adjustment to the working and living conditions is unusually difficult. The most prevalent personal and familial reasons for the premature returns among the migrants appear to be the unfaithful behavior of their wives or lovers and misuse of the remittance money by members of their families.

Return movement from the Middle East to the homeland means a change from the total institution to an open labor market situation. Our interview data show that about half of the returnees were jobless for more than three months after returning and that more than one-third had to change their jobs. About half of them reported that their earnings after return decreased more than 40 percent. For this and other reasons, one-third of them want to re-migrate to the Middle East.

The monthly average earnings for the temporary employees (base wage plus overtime allowance) have increased from about U.S. $520 in 1973 to about U.S. $756 in 1982. The overtime allowance comprises 46 to 55 percent of the total earnings. This average earning is about 2.5 times greater than that of male workers in the construction industry in Korea with similar experience.

According to the regulations, all the migrant workers who are employed by Korean overseas companies are required to remit over 80 percent of their total earnings to Korea via the Korean banking system. It is generally known that in fact nearly all the migrants remit about 90 percent of their earnings to Korea. They are able to remit such a large portion of their earnings not only because of government regulations to discourage consumption at destination but also because their traveling costs and living expenses in the Middle East are paid by their employers.

On the average, the returnees can make only about half of what they made in the Middle East. But their earnings after return are greater than the earnings before migration, perhaps because some of them improved their skills while others invested their remittance in a profitable business.

Adapted from United Nations University Project on the Global Impact of Human Migration, 1988.

---

the Persian Gulf region and to their increasing political and economic power. The lack of arable land means that this region may well have to rely upon some type of infusion of capital for the forseeable future. When the oil is exhausted, the government leaders hope that their industrial base will provide the exports to maintain their population.

CHANGE AND CONFLICT IN THE MIDDLE EAST AND NORTH AFRICA

The great contrasts in individual Middle Eastern countries between the oil-surplus, capital-rich countries of the Persian Gulf and the poorer, less industrialized, traditional societies based on agriculture exemplify the problems of the less industrialized world. Saudi Arabia and Kuwait are an anomaly because of their tremendous wealth, which allows them to modernize and proceed at a pace never before equaled in the world. Countries like Egypt are also modernizing and industrializing, but the traditional enemies of development in less industrialized countries, social organization and population growth, have to date prevented major changes in standard of life. The fact that per capita incomes, birthrates, and literacy rates are typical of those of other less industrialized regions of the world should not obscure the important changes begun in individual countries of this region.

Land reform has been started in nearly every country, beginning with Egypt in 1952 and continuing through the 1970s as individual governments in Syria, Iraq, Algeria, and Iran undertook to change the com-

TABLE 15.1 Percent of Foreign Workers in the Work Force of Selected Arab Oil-Exporting Countries (in %)

	1960	1965	1970	1975	1980	1985
Iraq	NA[a]	NA	NA	2	14	25
Kuwait	73	78	74	71	78	70
Libya	2	6	16	33	34	28
Saudi Arabia	10	17	27	40	53	50
United Arab Emirates	NA	15	67	84	89	80
Percent foreign workers in region	5	9	14	23	37	36[b]
Foreign work force (in millions)	0.167	0.328	0.655	1.413	2.997	3.1[b]

SOURCE: *Finance & Development*, December 1984, p. 35, and *The World Factbook 1988*
[a] Data not available.
[b] estimates

mon pattern of absentee landlord ownership of land. In many countries of the Middle East these reforms have followed a revolutionary change in government, which has overthrown the monarchies that replaced European colonial power. (The monarchy existed in most cases before independence.) Although these land reforms provided landless peasants with their own property, in many cases the farms are too small and the farmers too undercapitalized to allow full benefit from individual entrepreneurship in the rural areas.

Slums in Cairo, Egypt, share the problems of sanitation, disease, and dilapidation of slums everywhere.

THE WORLD OF POVERTY

THE HUMAN DIMENSION

From Suqs to Supermarkets

It's a familiar scene. Along most of Saudi Arabia's narrow, crowded streets, the shoppers stroll, pause, walk some more and then, from worn sidewalks, turn into shaded suqs (shops) to inspect, sniff, squeeze and eventually choose from overflowing wooden crates jammed haphazardly together their fruits and vegetables for the day.

The next step is equally familiar. They hand their choices to the shopkeeper who pops their bags onto one side of an old brass balance scale, nimbly picks out a combination of weights to balance the trays, and makes change from an old wooden drawer while simultaneously greeting friends, bartering with customers and loudly advertising today's loss leader.

Further on, in the meat suq and the fish suq, or in a small shop stocking canned goods, you'll see similar scenes. They're part of Saudi Arabia's economy and heritage: a familiar way of shopping—and of life. But today, that scene is changing swiftly. Nothing, in fact, shows the swift modernization of Saudi Arabia more vividly than the emergence of the giant supermarket.

Overnight, it seems, the kingdom has gone from suq to supersuq, from the homey corner store to the computerized supermarket with its steel-frame racks of smartly packaged international foods, and from the shopkeepers' old brass scales to electronic registers that total bills, calculate refunds, and give printouts of sales.

With products like fresh camel meat selling at $1.36 a pound and lamb at $160 per whole lamb—both cut, prepackaged, weighed, and displayed alongside the meat from other parts of the world, supermarkets have drawn large numbers of Saudi Arab customers.

In these stores, shopping begins when electric eyes sweep doors open for customers and display what in some cases is a Disneyland array of brightly colored packages: fresh vegetables and fruits, meats in shining trays, and rows of decorative and functional household goods beneath high ceilings overlooking thousands of different products bathed in fluorescent light, fresh Caribbean coconuts, Greek parsley, Australian beef and lamb, Swiss chocolates, English biscuits, Japanese oysters, Danish caviar, and Saudi dairy products and breads.

Most of the food in the supermarket is imported, and imported items are regulated. One regulation affects all products containing pork or alcohol—both forbidden to Muslims in the Koran. Under a 1973 law on labeling, all oils and shortenings, creams and milk products, juices, tomato paste and tea and coffee extracts must show the name of the food, ingredients, volume, name and address of the producer, date of packing, and country of origin—in Arabic. And under a 1979 amendment, jams, tuna, salt, soups and canned meats were added to the list. In addition, new foods are put through laboratory tests to check for pork and alcohol content.

As authorities enforce the labeling law by checking goods when they arrive at Saudi ports, some products have been banned because there is not enough information on them. Manufacturers, for instance, couldn't adequately show the origin of the animal bones—from which the gelatin in Jell-o is made—so Jell-o, for a time, couldn't be imported since it was unclear if bones came from cattle or hogs.

In addition, regulations require that shipments of meat be certified that the animals were slaughtered according to Islamic requirements—including proclaiming "Bismillah" and "Allah akbar" ("In the name of God" and "God is great"). Saudi consulates or their representatives issue the certificates in the exporting countries.

Source: *Aramco World Magazine*, Jan–Feb 1981, p. 8–16.

The reforms rarely stripped landlords of all their possessions, and the landlord class still enjoys a much higher standard of living than the peasant.

Another important change that has begun to benefit the rural majority in the Middle East and North Africa is irrigation and land reclamation schemes. The expansion of land in Egypt as a result of the high Aswân Dam has parallels in other countries where a host of reclamation projects has allowed expansion of the cultivated land. This expansion has provided farms for individual rural peasants in Iraq, Algeria, Egypt, Saudi Arabia, and most other Middle Eastern countries. The cost of such expansion is very high and because of the arid nature of most reclaimed land, it has created important environmental problems in some areas.

In portions of the Tigris and Euphrates River valley where irrigation has been expanded, the soils have become saline after only a few years, greatly reducing yields. Soil salinity can be overcome by installation of drains in the soil and periodic flushing of the salts by massive applications of water, but the cost is high. In other areas like Turkey and Syria the yields per acre on reclaimed land are low, reflecting their marginal nature. But more critical than the nature of the environment or the soils in the reclaimed areas is the population problem, which makes each advance in cultivated land a temporary gain that is quickly sur-

Resettlement projects for nomads, such as this project in Egypt, are rapidly destroying the nomadic culture.

passed by the expanding population. The expansion of land under cultivation by nearly one-fourth throughout the Middle East and North Africa has not equaled population growth, and the region has changed from a net exporter to a net importer of agricultural goods in the past quarter century.

The expanding population has contributed to the growth of urbanization in the region, with all the problems of urbanization in emerging lands. Each country is dominated by one or two primate cities, which are magnets to village residents. Cairo and Alexandria in Egypt, Istanbul (3 million) in Turkey, Tehran (6.5 million) in Iran, and Baghdad (3.5 million) in Iraq exemplify the problems of the primate city. The perception that life in the city is better is only part of the reason for the flow to cities. The crowded rural areas and low standard of village life have forced or pushed many of the rural residents from the small, overpopulated villages to the cities.

Any migration results from both push and pull (expelling and attracting) forces. The expelling forces driving migrants from the rural villages are a lower standard of living and large families of the small farmers and general lack of economic opportunity. The attracting forces are the promise of jobs, medical care, freedom from the uncertainties of agriculture, and the general perception that only in the cities can an individual make significant economic advances. For some of those who enter this migration this image is correct. An incipient middle class of merchants, skilled service occupations, and industrial workers is beginning to emerge in nearly every Middle Eastern and North African country.

For the majority who migrate to the cities, however, the reality is quite unlike the dream. Large portions of Middle Eastern cities are slums in which the rural migrant actually lives in worse conditions than in the rural areas. The city destroys the extended family of the villages, with its mutual support function, and eliminates even the possibility of the family producing part of its food needs. In spite of the problems of urban life, the cities remain magnets to the rural poor because of their perceived opportunities.

Another change that is occurring in the Middle East is the settlement of the nomadic tribes-people. The nomads who in the past roamed freely across the northern Sahara and the Arabian deserts and practiced transhumance in the northern mountain zone are faced with eventual settlement. As individual countries of the region gained independence, they increasingly placed controls on the migration of nomadic peoples across international borders. Many of the national governments have attempted to resolve the problem of the nomads by drilling wells or expanding irrigation to provide land for their permanent settlements. The technology of the nomads, which in the past allowed them to dominate the villagers of the region, is no match for the modern armaments of Iraq, Iran, or Saudi Arabia. They have been stripped of their role as taxers of the village and reduced to village life themselves. Less than 5 percent of the population of the region still practices nomadism, and the numbers of nomads are constantly decreasing.

Settling the nomads has often involved major conflicts as between the Kurds of northern Iraq and the Iraqi government, Bakhtiari peoples of Iraq and Iran versus their respective governments, and related conflicts. The conflict between tribal leaders and individual countries is but one focus of conflict in this region. The conflict between Israel and its neighbors is well known, but others exist between Iran and Iraq, North and South Yemen, Egypt and Libya, and so on. These periodic confrontations reflect the culture and history of individual nations in a region of seemingly homogeneous culture. The physical geography of the Middle East and North Africa prompted development of the core populated areas of each state in isolation. The unifying factor of Islam was never able to overcome this reality of isolated development. The repeated attempts by Islamic millennialists to create an Arab nation have foundered on the political realities of the Middle East and North Africa. The contrast between the Sunnites and the Shiites illustrates the broad division within the area, but even within the Sunnites there are differences in interpretation of Islamic law and cultural practices. The future for the Middle East is seemingly one of continued conflict as the individual countries attempt to transform their economies and create a unified nation within their state boundaries.

The attempt to create an Arab nation that transcends the state boundaries is probably a mirage. Past attempts that have created such anomolous names as the United Arab Republic failed to overcome the state boundaries that had been largely superimposed by the European powers. The United Arab Republic was born in 1958 when Egypt and Syria tried to begin the process of creating an Arab nation by uniting their governments, but it failed when Syria withdrew in 1961. In 1970 Egypt agreed to unification with Libya and the Sudan, but the attempts by Egypt in 1979–1980 to reach peace with Israel have apparently destroyed any potential of this agreement to develop. An agreement between Libya and Morocco in 1984 to join their countries was but the latest of these attempts. Just as Christianity failed to be a unifying force transcending individual cultural and traditional values in Western Europe, so Islam seems incapable of uniting the disparate peoples of the Middle East and Northern Africa peacefully.

The Middle Eastern and North African countries are now at a point of transition in their economic development, and their futures may well depend on actions beyond the borders of their region. The reliance of the capital-surplus oil countries on the economies of the industrialized West makes their developmental efforts subject to technological changes beyond their control. If the Western economies are able to free themselves from dependence on the oil resources of the region, development in these countries would be dealt a severe blow. In addition, countries such as Saudi Arabia have invested billions of dollars in Western economies, which further ties their future to the West. The non–oil-rich countries of the Middle East are perhaps even more subject to outside influence in the form of foreign aid, power conflicts between major industrial countries, and tariffs affecting the imports and exports from less industrialized countries in general. The less industrialized countries must overcome the problem of population in addition to dealing with these externalities if they are going to be successful in transforming their economies. The hope of the leaders of the Middle Eastern and North African countries is that their efforts to overcome their problems will prevent their region from being drawn into major conflicts and prevent them from adopting the negative aspects of industrial countries, while allowing the development of a unique Islamic society.

FURTHER READINGS

ABU-LUGHOD, J. *Cairo: 1001 Years of "The City Victorious"*. Princeton, N.J.: Princeton University Press, 1971.

———. *Rabat: Urban Apartheid in Morocco*. Princeton, N.J.: Princeton University Press, 1980.

AHMED, AKBAR S. *Discovering Islam: Making Sense of Muslim History and Society.* Boston: Routledge and Kegan Paul, 1987.

AL FARUQI, ISMAIL R. and AL FARUQI, LOSI L. *The Cultural Atlas Of Islam.* New York: Macmillan, 1986.

AL FARUQI, I. and SOPHER, D., eds. *Historical Atlas of the Religions of the World.* New York: Macmillan, 1974.

ALLAN, J. A. *Libya: The Experience of Oil.* London: Croom-Helm, 1981.

BARBOUR, K. M. *The Growth, Location and Structure of Industry in Egypt.* New York: Praeger, 1972.

BEAUMONT, P., BLAKE, G. H., and WAGSTAFF, J. M. *The Middle East: A Geographical Study.* London: Wiley, 1976.

BLAKE, G. H. and LAWLESS, R. I., eds. *The Changing Middle Eastern City.* London: Croom-Helm, 1980.

BONINE, M. E. "The Morphogenesis of Iranian Cities." *Annals of the Association of American Geographers* 69, 1979, pp. 208–224.

BUTZER, K. W. *Early Hydraulic Civilization in Egypt: A Study in Cultural Ecology.* Chicago: University of Chicago Press, 1976.

CLARKE, J. I. and BOWEN-JONES, H., eds. *Change and Development in the Middle East: Essays in Honour of W. B. Fisher.* London: Methuen, 1981.

CLARKE, J. I. and FISHER, W. B., eds. *Populations of the Middle East and North Africa.* Cambridge, England: Heffer, 1972.

COHEN, B. *Jerusalem: Bridging the Four Walls.* New York: Hertzl Press, 1977.

COHEN, SAUL B. *The Geopolitics of Israel's Border Question.* Boulder, Colo.: Westview Press, 1987.

COLE, D. *Nomads of the Nomads: The Al Murrah Bedouin of the Empty Quarter.* Chicago: Aldine, 1975.

COTTRELL, A., ed. *The Persian Gulf States: A General Survey.* Baltimore, Md.: Johns Hopkins University Press, 1980.

CRESSEY, G. *Crossroads: Land and Life in Southwest Asia.* Philadelphia: Lippincott, 1960.

FAHIM, H. *Dams, People and Development: The Aswan High Dam Case.* Elmsford, N.Y.: Pergamon, 1981.

"A Fever Bordering on Hysteria: After Five Years, Khomeini Still Seems in Full Control of Iran's Revolution." *Time,* March 12, 1984, pp. 36–39.

FISHER, W. B. *The Middle East: A Physical, Social and Regional Geography.* 7th ed. New York: Methuen, 1981.

FOROUK-SLUGLETT, MARION and SLUGLETT, PETER. *Iraq Since 1958: From Revolution to Dictatorship.* London: KPI Ltd. 1987.

GILBERT, M. *Atlas of the Arab-Israeli Conflict.* New York: Macmillan, 1974.

GISCHLER, C. *Water Resources in the Arab Middle East and north Africa.* Cambridge: Menas, 1979.

GORDON, D. *The Republic of Lebanon: Nation in Jeopardy.* Boulder, Colo.: Westview Press, 1983.

GRADUS, Y. "The Role of Politics in Regional Inequality: The Israeli Case." *Annals of the Association of American Geographers* 73, 1983, pp. 388–403.

HARRIS, GEORGE S. *Turkey: Coping with Crisis.* Boulder, Colo.: Westview Press, 1985.

HARRIS, LILIAN C. *Libya: Qadhafi's Revolution and the Modern State.* Boulder, Colo.: Westview Press, 1986.

HARRIS, WILLIAM WILSON. *Taking Root: Israeli Settlement In the West Bank, the Golan and Gaza-Sinai, 1967–1980.* Chichester: Wiley, 1980.

HAYNES, K. E. and WHILLINGTON, D. "International Management of the Nile—Stage Three?" *Geographical Review* 71, 1981, pp. 17–32.

HILDORE, J. J. and ALBOKHAIR, Y. "Sand Encroachment in Al-Hasa Oasis, Saudi Arabia." *Geographical Review* 72, 1982, pp. 350–356.

HOLZ, R., ed. *The Surveillant Science: Remote Sensing of the Environment.* 2d ed. New York: Wiley, 1984.

HULDT, BO and JANSSON, ERLAND, eds. *The Tragedy of Afghanistan: The Social, Cultural and Political Impact of the Soviet Invasion.* London: Croom-Helm, 1988.

KARAN, P., and BLADEN, W. "Arabic Cities." *Focus,* January–February 1983.

KARMON, Y. *Israel: A Regional Geography.* New York: Wiley, 1971.

KOLARS, J. "Earthquake-Vulnerable Populations in Modern Turkey." *Geographical Review* 72, 1982, pp. 20–35.

LONGRIGG, S. *The Middle East: A Social Geography.* 2d ed. Chicago: Aldine, 1970.

MALLAKH, RAGAEI EL and MALLAKH, DORTHEA H. EL., eds. *Saudi Arabia, Energy, Developmental Planning, and Industrialization.* Lexington, Mass.: D. C. Heath, 1982.

MANNERS, I. R. "The Middle East." In *World Systems of Traditional Resource Management.* Edited by G. A. Klee. London: Edward Arnold, 1980.

MELAMID, A. "Urban Planning in Eastern Arabia." *Geographical Review* 70, 1980, pp. 473–477.

MIKESELL, M. "The Deforestation of Mount Lebanon." *Geographical Review* 59, 1969, pp. 1–28.

———. "Tradition and Innovation in Cultural Geography." *Annals of the Association of American Geographers* 68, 1978, pp. 1–16.

MORRIS, BENNY. *The Birth of the Palestinian Refugee Problem, 1947–1949.* Cambridge: Cambridge University Press, 1987.

MUNSON, HENRY, Jr. *Islam and Revolution in the Middle East.* New Haven, Conn.: Yale University Press, 1988.

NIV, D. *The Semi-Arid World.* London: Longman, 1974.

PERETZ, D. *The Middle East Today.* New York: Holt, Rinehart and Winston, 1978.

PERLMUTIER, AMOS. *Israel: The Partitioned State.* New York: Scribner's, 1985.

ROBERTS, H. *An Urban Profile of the Middle East.* New York: St. Martin's Press, 1979.

RUEDISILI, L. and FIREBAUGH, M., eds. *Perspectives on Energy.* 3d ed. New York: Oxford University Press, 1982.

SHAW, R. PAUL. *Mobilizing Human Resources in the Arab World.* Boston: Routledge and Kegan Paul, 1983.

SHWADRAN, B. *Middle East Oil: Issues and Problems.* Cambridge, Mass.: Schenkman, 1977.

SMITH, C. G. "Water Resources and Irrigation Development in the Middle East." *Geography* 55, 1979, pp. 407–425.

TAYLOR, A., ed. *Focus on the Middle East.* New York: Praeger, 1971.

WALTON, S. "Egypt After the Aswan Dam." *Environment* 23, 1981, pp. 31–36.

WEINBAUM, M. G. *Food, Development, and Politics in the Middle East.* Boulder, Colo.: Westview Press, 1982.

WHEATLEY, P. "Levels of Space Awareness in the Traditional Islamic City." *Ekistics* 253, 1976, pp. 354–366.

ZARTMAN, I. W., ed. *Man, State and Society in the Contemporary Maghreb.* London: Pall Mall, 1973.

Nigeria.

AFRICA PROFILE: 1990*

Population (in millions)	519	Energy Consumption Per Capita—	
Growth Rate (%)	2.9	pounds (kilograms)	631
Years to Double Population	24		(287)
Life Expectancy (years)	50	Calories (daily)	2125
Infant Mortality Rate		Literacy Rate (%)	35
(per thousand)	115	Eligible Children in Primary	
Per Capita GNP	500	School (%)	70
Percentage of Labor Force		Percent Urban	30
Employed in:		Percent Annual Urban Growth,	
Agriculture	73	1970–1980	5.6
Industry	11	1980–1985	5.3
Service	16		

*South of the Sahara

16

Subsaharan Africa
The least industrialized region

MAJOR CONCEPTS

Natural boundaries / Physical boundaries
Cultural boundaries / Head-link cities
Tribalism / Continental drift / Lateritic soils
Kwashiorkor / Carrying capacity / Apartheid
Cataracts / The Sahel / Desertification
Desert pavement / Schistosomiasis
The Sudd / Federal Government

IMPORTANT ISSUES

Tribalism and the emergence of independent African countries

Desertification and environmental degradation

Preservation of African wildlife and environments

Racial conflicts in South Africa

Centrifugal and centripetal forces in African countries

Tropical diseases and economic development

Geographical Characteristics of Subsaharan Africa

1. Subsaharan Africa is the poorest and least industrialized continent in the world.
2. Subsaharan Africa was divided into colonies by European powers in the nineteenth century, and present boundaries reflect those former colonial boundaries.
3. Africa is a series of level plateaus separated by escarpments. Climates are tropical and often have too much or too little moisture, while soils are generally leached and infertile.
4. Africa has an abundance of resources, but the fragmentation of the continent among more than 50 independent countries prevents any one country from having an adequate assemblage of complementary resources.
5. Tribalism, unstable governments, and colonial relics such as South Africa make economic development difficult in Subsaharan Africa.

Africa is one of the largest continents on the globe but for many reasons is the most sparsely inhabited of any except Australia. In spite of its sparse population it is the scene of repeated famines, especially in countries immediately south of the Sahara Desert. The common division between Africa north and south of the Sahara basically reflects cultural contrasts. (Figure 16.1). Africa south of the Sahara is dominated by negroid peoples while caucasoid characteristics are dominant in North Africa. North Africa has an almost uniformly Islamic culture, while southern Africa does not. Islam is dominant in the countries that border the Sahara, but farther south the influence of Islam fades rapidly. The contrasts in culture are in part the result of the inaccessability of much of the African continent in the past. Although Arab and Persian traders have always crossed the Sahara to trade with kingdoms bordering the Sahara or sailed south along Africa's east coast, for most of recorded time the interior of the continent has been effectively isolated from the rest of the world by distance, the dense vegetation and diseases of tropical environments, and falls and rapids on the rivers. Contacts with traders from north of the Sahara and the Persian Gulf brought Islam to Africa south of the Sahara, but the technological and social changes that took place in western Europe after the eighteenth century did not follow. The people are ethnically distinct, for the region is the homeland of the Negroid, or black peoples. Even today, it is primarily populated by black peoples, with only scattered Caucasian or Asian populations forming minorities in most areas. In the past Africa south of the Sahara was dominated by traditional tribal religions associated with only one tribe.

Africa's isolation and harsh environment have been reinforced by colonial domination and exploitation to make Africa the least developed region in the world. Africa is a land of poverty, with the absolute lowest per capita incomes in the world, in some countries $100 a year or less. Low literacy rates, high birth- and death rates, high infant mortality rates, and the predominance of rural village residence indicate the level of development of the continent.

EUROPEAN INFLUENCE

European contacts with Africa south of the Sahara began with Portuguese attempts to reach Asia in the fifteenth century. For the next 400 years contact was restricted to coastal locations as the slave trade developed in West Africa, and Cape Town was established at the southern tip of Africa as a way station for ships making the months-long voyage to Asia. Profits from the slave trade combined with the power of the African middlemen who sold the slaves to prevent European movement into the interior of the continent. The European demand for slaves reoriented trade from North Africa to the coastal slave ports of empires south of the Sahara. It also effectively weakened the

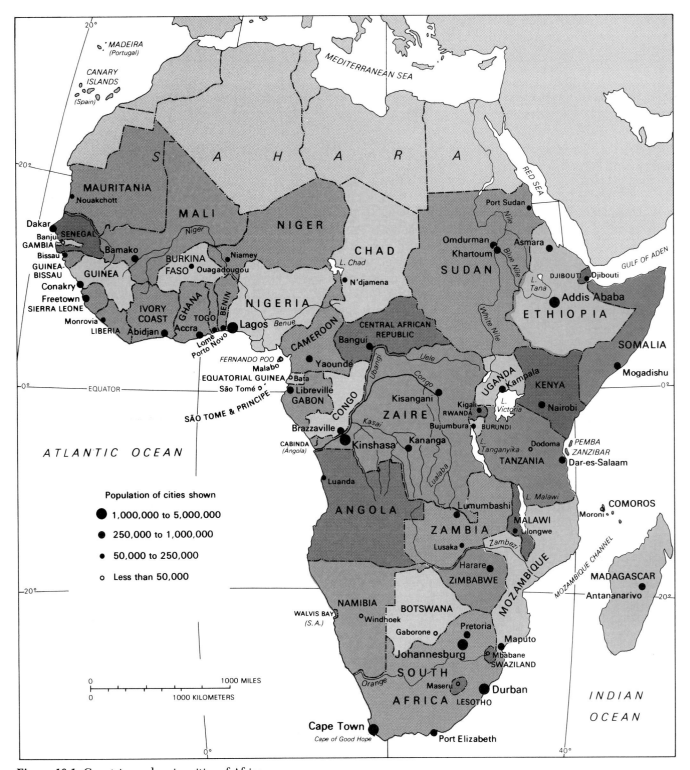

Figure 16.1 Countries and major cities of Africa.

African civilizations by repeated slave raids on their populations.

European exploration of the interior of the African continent did not begin until the mid-nineteenth century. Motivation for expeditions to the interior was fourfold: financial gain, Christian missionary work, expansion of geographical knowledge of the world, and world status for countries controlling colonies. The emergence of European societies dedicated to expansion of geographical knowledge in the late eighteenth and early nineteenth centuries led to the funding of several expeditions to explore Africa. The growing idea in western European cultures that Christianity was the universal religion necessary for salvation and that it was the duty of those who knew of Christianity to share it with non-Christians, motivated other explorers, like the famous David Livingstone. The decline of the slave trade after the mid-nineteenth century and the increasing industrialization of Europe also fostered interest in expanding contacts with Africa for potential profit. The emergent European industrial states of Great Britain, Germany, Belgium, the Netherlands, and France were all determined to have a share in any potential resources or markets that the unknown African continent might produce. In addition, possession of colonies was perceived as a sign of the relative world importance of European countries. Conflict between European nations, particularly Britain and France, over control of Africa led to the Berlin Conference of 1884. All major European nations, the United States, and the Ottoman Empire were represented at the conference, but Africans were not. The main purpose of the conference was to ensure that countries other than Britain, France, and Belgium gained colonies in Africa, but it also adopted resolutions to provide for the welfare of the African people. One provision of the agreement stated that effective occupation of an area was necessary before a European nation could claim a territory.

Although the agreements of the conference were never actually implemented, it signaled a mass rush to divide up Africa among the colonial powers. Each individual European nation attempted to expand its influence over Africa through military, missionary, or economic control. The result was the partitioning of Africa into a series of European colonies (Figure 16.2).

The immediate effect of the creation of these colonies was to mark out rigid boundaries around each one, thus superimposing the European concept of territoriality upon the African continent. These boundaries had little relationship to cultural or physiographic features but rather were arcs of circles, lines

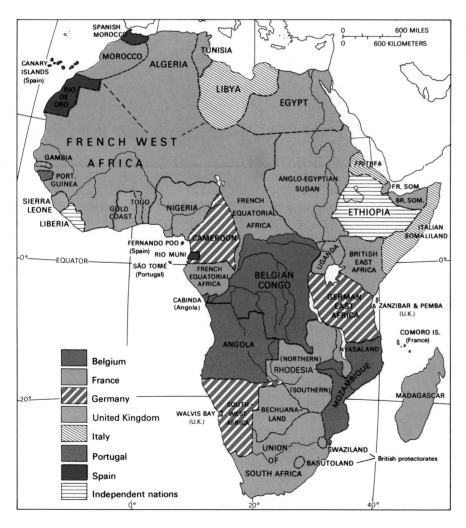

Figure 16.2 Representative examples of European Colonial boundaries in Africa.

of latitude, or meridians. Such boundaries were simple for the European diplomats to draw, but they cut through the tribal and cultural groups of the African continent. Nearly 200 tribal groups were bisected by national political boundaries based on the colonial boundaries, and an estimated 75 percent of all boundaries did not respect cultural differences.

COLONIAL BOUNDARIES AND MODERN AFRICAN COUNTRIES

There are several recognizable classes of boundaries on the land surface. *Physical boundaries* are those that follow mountains, deserts, the ocean, or other physical features that are difficult to traverse. Such physical boundaries are particularly effective if they are inhospitable to settlement and thus have few if any inhabitants. A second type of boundary can be defined as a *cultural boundary*. This type of boundary is based on cultural variation and reflects the change from one language group to another, one religion to another, or other easily recognizable cultural feature. Such boundaries are effective when groups have cultural homogeneity and a strong regional consciousness. A third type can be described as a superimposed boundary. These boundaries reflect neither the culture nor the physical geography, but are superimposed over existing physical and cultural differences.

The potential for conflict that can result from superimposed boundaries is obvious. The resulting political states are faced with strong centrifugal forces and few centripetal forces. In Africa the result has been twofold. Ethnic or tribal divisions in individual states may resort to conflict to ensure the stability of the state or their position of power (Figure 16.3). As an example, in Burundi the majority Hutu people are dominated by the Tutsi, who constitute only about 15 percent of the population. Introduction of colonial boundaries separated the territory of the Hutu and Tutsi between Rwanda and Burundi and introduced the democratic concept of one person, one vote. In order to maintain their political dominance, the Tutsi

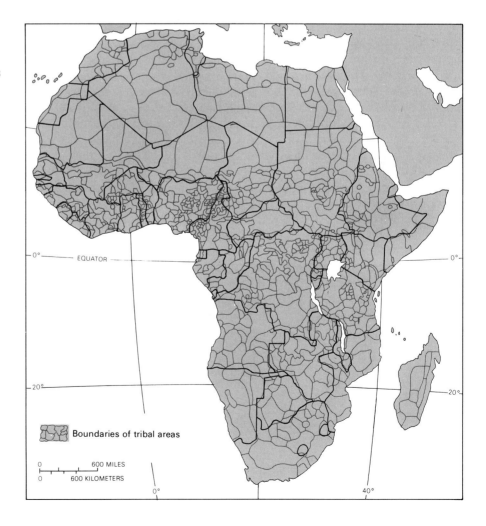

Figure 16.3 Tribal boundaries in Africa. The numerous tribal groups of Africa do not have boundaries that correspond to the boundaries of African countries.

massacred the educated, literate Hutu, whom they viewed as a threat to their continued dominance. In 1988 the Tutsi, who still dominate the country, again attacked Hutu villages, killing more than 5,000 and leaving more than 100,000 homeless. The Tutsi-dominated government maintains that the massacre occurred because exiled Hutu tribal members promoted violence.

At the time of independence in Uganda all Asian residents were given the option of accepting British citizenship, which most chose. The Asians controlled most business activity, and the issue of foreign ownership became a source of conflict. In Uganda in the mid-1970s a former boxer and army corporal by the name of Idi Amin came to power. In order to maintain a semblance of unity within the country and to ensure the dominance of his tribe, Amin first stripped the large Asian minority population of their lands and businesses and then entered a five-year period of persecution and murder of other tribal groups. Amin's reign of terror was ended only when exiles from his country, with the support of Tanzanian troops, forced him to flee in 1979. The subsequent government has been accused of similar atrocities.

In Nigeria there are more than 250 tribal groups. After several internal conflicts as groups other than the majority Hausa, Yoruba, and Ibo tribes were granted representation in the government, the Ibo declared the independent state of Biafra, and civil war began. This Biafran-Nigerian war continued until 1970 and resulted in the deaths of tens of thousands of Nigerians. In African countries that are attempting to develop a strong central government, a common alternative has been for one ethnic or tribal group to dominate the country while the other tribal groups are relegated to a second-class role. In Sudan a civil war has raged since 1983 between the Muslim majority in the north and the largely non-Muslim south. This conflict has occurred intermittently since 1960 as the Muslims insist on an Islamic state and the non-Muslim southerners resist.

In spite of the problems created by the superimposed colonial boundaries, the independent African states continue to attempt to establish themselves within them. Even though the boundaries may not be rational in terms of physical or cultural features, the European colonial development corresponded to them and the resulting circulation pattern reinforces them today.

In most of Africa Europeans founded a single city that served as a **head link** to the rest of the colony. From there roads and railroads gave access to the minerals of Africa or to areas producing tropical crops of palm oil, cacao (cocoa), cotton, or peanuts. As a result the European-based city became the focus of each colony and rapidly emerged as a primate, parasitic city. The major cities and capitals are still those that Europeans established to exploit the African continent, from Dakar in Senegal to Lagos in Nigeria, to Luanda in Angola, to Dar es Salaam in Tanzania. Today they are the largest cities in the newly emergent African states, normally having more than 50 percent of the small urban populations in these countries. All are the centers of commercial, cultural, and political activity in individual countries and are growing rapidly.

The long-term effect of the external focus of the African countries is to tie individual African states more closely to their former colonial rulers than to their African neighbors. Even today foreign trade from African states is primarily with European or other industrial nations rather than with other African countries (Figure 16.4). In 1988, 58 percent went to countries of Europe, while only 9 percent went to other African countries. The European-imposed boundaries, noncongruent with the ethnic and tribal diversity of Africa, have resulted in political states with an economic orientation that is a centrifugal force for Africa as a whole.

Faced with the tremendous cultural diversity in Africa, the emergent states have a series of options as they attempt to create unity from diversity. The first alternative is to establish a strong, totalitarian central government, either Marxist-oriented as in Angola or Ethiopia, or an apolitical dictatorship as in Uganda under Idi Amin. As an alternative to such dictatorship, a political state with numerous ethnic groups may opt for a **federal government** based on popular sovereignty. In a federal government, groups elect representatives to govern the country. In Nigeria the resulting state required a civil war to enforce the rule of the central government and prevent the various tribal groups from forming independent countries of their own. Such a federal confederation can be successful only through force or through true cooperation and use of the resources of the country to benefit all groups.

Another alternative for overcoming the fragmentation caused by the numerous ethnic groups is to create economic organizations. Such economic organizations, possibly modeled after the Common Mar-

---

Head link: A city that serves as a link to the rest of the world for a country as imports and exports flow through it.

Federal government: A political framework where the various groups in a country are represented by a central government, but retain their own identity through local customs, language, law, and such.

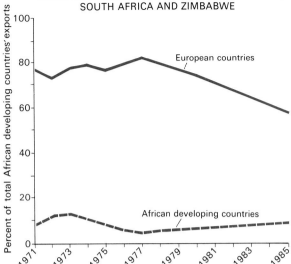

Figure 16.4 The exports of African countries go primarily to European countries rather than to other African countries, often reflecting former colonial relationships.

ket, allow African resources to be utilized for the development of the continent rather than industrial nations in Europe, America, and Asia. A final alternative is the creation of a multitude of microstates. The dilemma that microstates pose is exemplified by Rwanda and Burundi. Although each contains only a few tribal groups, their small size and lack of resources make development difficult. Consequently microstates based on recognition of individual ethnic groups will probably not be a viable alternative for Africa.

In all probability the African states will face continued conflict in the next decades as they attempt to create a sense of nationalism among their diverse inhabitants. The development of national icons, a national heritage, and loyalty to the state instead of the tribal or ethnic group will come only after the provision of mass education, distribution of wealth, and provision for the basics of life for all peoples in each state.

To accomplish this task, the African countries must face the challenge of transforming their traditional economies. The majority of the population in these countries is involved in agricultural pursuits, traditionally in subsistence or semisubsistence crop or nomadic livestock systems. There are also significant numbers of Africans who are engaged in hunting and gathering to provide their livelihood, including the pygmies of the tropical rainforest, the bushmen of the Kalahari of Southwest Africa, and groups in Ethiopia.

Regional contrasts between the various areas of Africa south of the Sahara create wide variation in the general pattern of subsistence agriculture. The poorest countries include those on the southern margins of the Sahara. In Ethiopia, Chad, and similar areas the per capita income is less than $150 per year. In these countries only a tiny minority of the population is part of the interconnected modern world, while the balance is engaged in traditional agricultural pursuits. The wealthiest countries are in the southern portion of the continent where resources and foreign colonial development have changed the economy to a greater

The chief in a village of Burkina Faso supervises his wives as they bring in the harvest. The family, village, and tribe are far more important than the national government for most African people.

reliance on industrial and commercial activities. South Africa has a per capita income of approximately $1800 if the total wealth of the country were evenly distributed. Other areas of Africa, including portions of Nigeria, which is a major source of petroleum, have a higher standard of living because of resource development. Within each country of Africa, there are islands of wealth around the urban areas or sites especially suited for agriculture or resource development. In spite of these islands of wealth (or because of them), Africa is generally a continent of poverty and underdevelopment, as the changes of the Industrial Revolution arrived late in this area.

THE AFRICAN ENVIRONMENT

One of the primary factors affecting development in Africa is the physical geography of the continent, and one significant element of Africa's geography is its tremendous size. The entire continent totals 11,685,000 square miles (30,264,000 square kilometers), second only to Asia. The part of Africa south of the Sahara is itself a large area, totaling approximately 9 million square miles (23,310,000 square kilometers) or roughly three times the size of the United States. At its widest extent, this region is more than 4500 miles (7200 kilometers) wide and extends more than 3000 miles (4800 kilometers) from the Sahara to the tip of South Africa. The sheer size of this landmass has played a major role in its relationship with the rest of the world. In the days before railroads such size handicapped transport and communication, which remained concentrated along coastal areas or navigable waterways.

THE CHALLENGE OF AFRICAN ENVIRONMENTS

Most of Africa lies at elevations of 650 feet (200 meters) to 6500 feet (2000 meters). True lowlands, from sea level to 300 feet (100 meters) comprise only a small area. Of all the continents, Africa has the smallest amount of coastal plain readily accessible from the sea even though there are no major mountain ranges like those in Europe, Asia, and the Americas. The landforms are a relatively level series of plateaus imposed one upon the other with sharp escarpments where they descend to the narrow coastal plain. Consequently the rivers have rapids and falls (*cataracts*) a short distance inland that effectively prevent navigation to the interior. Access from the sea is further handicapped by the absence of natural harbors along the coast. The wave action of the ocean causes heavy surf and creates sandbars across the mouths of the rivers, further limiting access even to the coastal lowland areas. The reefs along the tropical coasts of East Africa are an additional hazard for approaching ships. Only a few areas of Africa have been accessible to the outside world.

LOWLAND AFRICA

It is possible to divide the landforms of tropical Africa into low-level plateaus and high-level plateau regions (Figure 16.5) with the dividing line extending from Angola north to the Red Sea. Lowland Africa rarely exceeds 2000 feet (600 meters) in elevation. It includes the river systems of the Congo and the Niger as well as the bulk of the drainage of the Nile. Highland Africa occupies the southern and eastern portions of the continent and is 3000 feet to 4000 feet (800 to 1200 meters) above sea level. Volcanic mountains like Mount Kilimanjaro in Tanzania (19,340 feet or 5,895 meters) rise from the high-level plateaus but are only of regional importance.

The great plateaus of lowland Africa consist of the core of both the ancient *Gondwanaland* and *Pangaea*, the united landmass that existed before continental drift separated the continents. (Figure 2.4) Africa is made up of ancient crystalline rock and subsequent sedimentary deposits that form relatively flat basins superimposed one upon another like a stack of dishes, each separated from the one below it by an escarpment. Lowland Africa has undergone considerably less geological folding and faulting than other continents, and important mineral deposits are readily accessible. They contribute to the economy and importance of this region.

HIGHLAND AFRICA

A major feature of the landforms of highland Africa is the Rift Valley system of East Africa, which stretches from southeastern Africa to the Jordan River valley of the Middle East. In this area of faulting and folding, mountains have been thrust up while between them large continental blocks have sunk. These great rift features stretch one-seventh of the way around the world, creating the high mountains of East Africa as well as its great lakes—Lake Tanganyika between Zaire and Tanzania and Lake Nyasa between Malawi, Tanzania, and Mozambique, as well as Lake Edward, Lake Albert, and Lake Turkana farther north. The Red Sea and Gulf of Aden as well as the Dead Sea of Israel are also part of this great rift valley system.

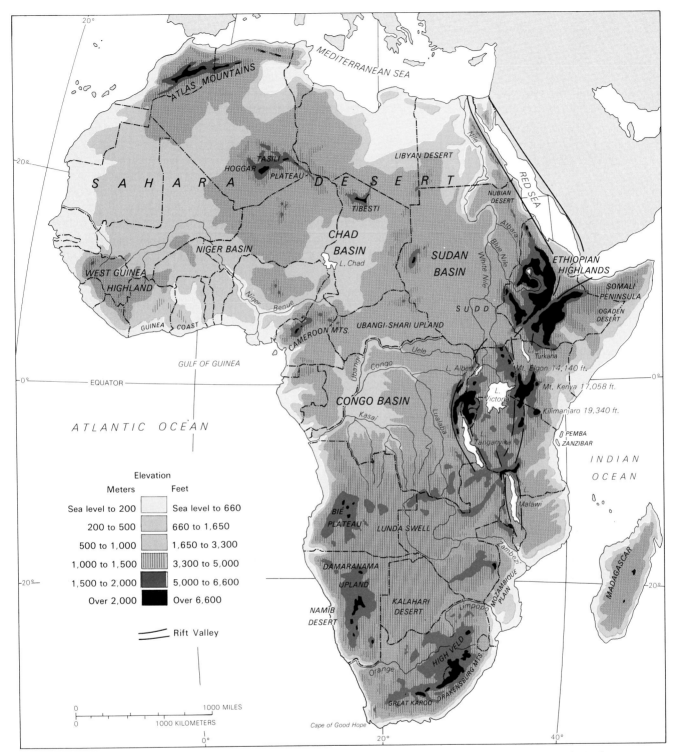

Figure 16.5 Landforms of Africa. Africa is dominated by plateaus and upland plains above 500 meters (1650 feet) elevation.

THE CLIMATES OF AFRICA

The climates of Africa generally have too much, too little, or poorly timed precipitation. The location of the landmass within 30 degrees of the equator means that the entire area is warm or hot. The only exceptions are the mountains and highlands of the rift zone of East Africa. The Congo Basin, centered on the equator, is the center of the tropical rain forest (Figure 16.6). Extending approximately 5 to 8 degrees north and south of the equator, this area has low annual *ranges* in temperature and precipitation. Daytime temperatures average between 70° and 80°F (21° and 27°C) throughout the year, and diurnal ranges rarely exceed 15°F (8°C). Precipitation generally exceeds 45 inches (1143 millimeters) per year, and much of the region exceeds 60 inches (1524 millimeters) annually.

The vegetation of this tropical rain forest represents one of the largest, most imposing plant complex association on the earth. It is possible to recognize three general levels of vegetation: one under 50 feet (15 meters), an intermediate level from 50 to 100 feet (15 to 30 meters), and one above 100 feet extending to 150 feet (46 meters). In a true rain forest, the vegetation provides a continuous canopy that shields the floor of the forest from the direct rays of the sun. Thus, the floor of the true rain forest is fairly open, with limited underbrush. Only when the canopy is broken, as by a river or burning to allow the sun's rays to penetrate, do vines, creepers, and shrubs grow up with young trees to create the Western image of a jungle of impenetrable vegetation.

In spite of luxuriance of the vegetation, soils in the tropical rain forest are relatively infertile. The vegetation itself, although luxuriant, is easily destroyed, and once destroyed, restoring it is difficult. The assemblage of plants in the tropical rain forest is a result of hundreds of years of growth and change. When groups that want to settle there burn the forest, the shallow soils are easily destroyed. When exposed to the sunlight, these soils become **lateritic** (bricklike). The term *laterite* is often applied to tropical soils. Nutrients are in short supply in tropical soils and are stored in the vegetation. When the vegetation is cleared

---

Lateritic soils: Highly leached soils of the tropics that have high concentrations of iron and aluminum.

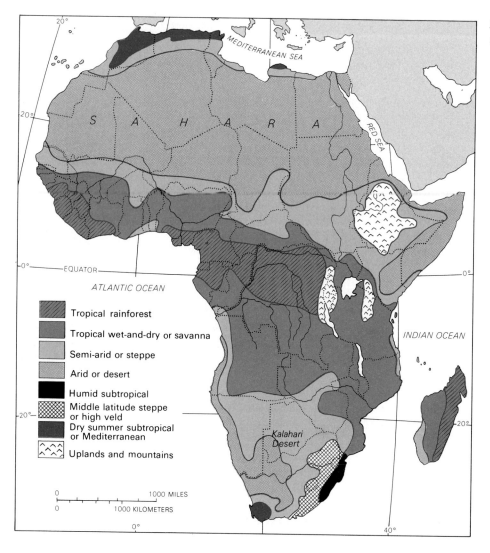

Figure 16.6 Climate regions of Africa.

Savanna lands in Kenya. The tall grass of the world's savanna climates has been the home of deer, antelope, zebra, and other grazing animals and their predators.

and burned, the nutrients are quickly leached from the soil, leaving it infertile and unproductive. Consequently, the continued population pressure on the margins of the tropical rain forest constitutes one of the great ecological threats for the world. An area as large as the state of Massachusetts is cleared of forest each year as plantations, peasant farmers, and woodcutters ravage the rain forest.

The savanna climates of Africa occupy a zone north and south of the tropical rain-forest climates, extending to 15 to 20 degrees north and south of the equator. The eastern highlands of Africa along the equator also have a savanna climate, as a result of temperature and precipitation changes of higher elevations. The savanna climate produces vegetation ranging from the tall grasslands of Nigeria, Sudan, Ivory Coast, and Kenya, to a forest different from the tropical rain forest only in density and number of species of trees. Portions of the savanna areas of West Africa seem to be the result of repeated and persistent burning by both the present and former occupants of the area. These savanna regions have tall trees typical of the tropical rain forest interspersed in the grasslands.

Soils in the tropical savanna are somewhat more fertile than in the tropical rain forest because they are less leached. They are important for the production of the major agricultural crops of Africa south of the Sahara, but are also subject to many of the problems of the soils in the tropical rain forest. Peanuts, cotton, and a variety of staple crops such as millet, sorghum, manioc, yams, and corn are produced in the savannas.

The savanna lands are also the home of the last great herds of wild animals and their predators. A great ecological challenge for the world is the preservation of these herds in the face of increasing population pressure and competition for the lands they occupy. For Africa's rural farmers the wildlife is simply another challenge to be overcome. Besides occupying land that could be farmed, the animals transmit disease and destroy crops. Many Africans feel that Western pleas to save the wildlife represent another form of colonialism. Western nations have destroyed much of their wildlife as they have built farms, factories, and cities. To some Africans, Western attempts to preserve the great animal herds of the continent simply maintain the existing dichotomy of development and wealth in the world. Residents of the industrialized world want to preserve the wildlife, but they do not offer sufficient financial aid for the African countries to provide for their populations without encroaching on the range of the animals.

THE STEPPE AND DESERT CLIMATES

North and south of the tropical savanna lands is a transition zone of steppe climate. The transitional region immediately south of the Sahara is known as the *Sahel*. Precipitation totals between 7.5 to 20 inches (190 to 508 millimeters) yearly, and temperature maximums are constantly in the range of 80° to 100°F (27° to 37°C). The diurnal temperature range is great, with nighttime lows falling to 50° to 60°F (10° to 15°C). The typical vegetation is short grass and brush. This steppe region extends nearly the full width of the African continent but is relatively narrow in north south extent. In the past it has been plagued by recurrent drought. There is evidence that the Sahara has been expanding southward into the steppe land as a result of the general drying trend of global climatic changes and as a result of the destruction of the vegetation of the steppe margins by overgrazing. Drought has persisted since the late 1960s, with tremendous impact on the life-styles and economy of the region. The expansion of desert into more humid regions is known as **desertification** and is an important environmental issue in the world.

South of the southern zone of savanna in Africa is another belt of steppe land. It is concentrated in Botswana, Zimbabwe, and Zambia, but portions extend into South Africa and Angola. This region is similar to the Sahel and has suffered from similar drought and environmental problems. Soils in these steppe lands are marginal, but they are quite fertile if irrigated. They can also be used for dry farming of grains, but the yield is low.

North and south of the steppe lands are the true deserts of Africa. In the north is the Sahara, the world's largest, and in the south is the Kalahari. These desert areas are characterized by temperature extremes, limited precipitation, and isolated settlements. The desert areas of Africa have recorded the world's highest official temperatures, 136°F (58°C). Precipitation ranges from 2 to 6 inches (51 to 152 millimeters) per year in most places, making extensive settlement difficult. Population is restricted to oases or valleys of rivers such as the Nile, which bring water from the tropical savanna and tropical rain-forest areas. Other rivers penetrate the margins of the Sahara and are the basis for such settlements as Tombouctou (Timbuktu) on the Niger, or Kaedi on the Senegal River on the border between Mauritania and Senegal.

Most of the Sahara is covered with scattered brush, grass, and other xerophytic vegetation, rather than being one vast expanse of sand as is popularly perceived. The diurnal range of temperatures is extremely high, with daytime highs exceeding 125°F (52°C), while nighttime temperatures fall below 60°F (15°C). Soils are essentially nonexistent, as the typical surface is stones and hard-packed gravel eroded by the wind, called desert pavement. Incessant heat and wind combine with salts left by evaporation to make this layer of gravel and hard dirt impermeable.

In southeastern and southwestern Africa, increased elevation or influence from prevailing winds results in important areas of subtropical climate. The southwest tip of Africa around Cape Town has a Mediterranean, or dry-summer, subtropical climate. The southeastern coast has a true humid subtropical climate caused by modification of the steppe lands by higher elevation. It is divided into the High Veld and the Low Veld on the basis of elevation. The High Veld is an open brush and grassland where the higher elevation gives the humid subtropical climate a distinctive dry season similar to a savanna during the period of the low sun. Mean annual temperatures rarely exceed 70°F (21°C). Soils in the High Veld are fertile if adequate water is available. The Low Veld lies at 1000 to 2000 feet (300 to 600 meters) with a hotter climate and a distinctive woodland vegetation.

THE PATTERN OF DISEASE IN AFRICA

Less obvious than the problems of drought, floodings, or accessibility, is the challenge of disease and insect pests. Africa has a range of disease-carrying pests, but major ones are mosquito vectors carrying malaria, the tsetse fly carrying *trypanosomiasis* (sleeping sickness), and intestinal worms. Malaria is endemic in the tropical parts of Africa, and since mosquitoes breed in standing water and swamps, people living near riversides face the greatest danger. Malaria is rarely fatal, since genetically many African people have the abnormal hemoglobin of *sickle-cell anemia*. The sickle-cell hemoglobin is found in as much as one-quarter of the African population in malarial areas, as it offers protection against death from malaria although not from the chills and fever of malarial attacks. Africans in areas where malaria is common suffer periodic attacks of debilitating fever, particularly at the beginning of the rainy season at the very time when they need their strength for planting their crops. In addition to malaria, mosquitoes transmit yellow fever, which is fatal, but vaccination offers protection. Mosquitoes also transmit small parasitic worms (*filariae*), and *dengue,* the so-called break-bone fever because of the intense pain it produces.

The tsetse fly carries another major disease, sleeping sickness (trypanosomiasis), to humans and a destructive disease known as *nagana* to cattle and horses. The distribution of the tsetse fly is from approximately 15 degrees north latitude to 15 degrees south latitude (Figure 16.7). It is common in the savanna lands with more than 30 inches (760 millimeters) of precipitation and the wooded savanna tropical rainforest margin. Because of the destructive effect of the tsetse fly on both animals and humans, in the past areas where it was endemic were only thinly populated. Today inoculations make living in tsetse fly areas possible. Since the tsetse fly normally feeds on animals, a common method of eliminating the threat of sleeping sickness is to destroy the wild animal population upon which the flies feed by clearing bush land, thus changing the entire environment. Although still a problem, sleeping sickness is far less common than it was in the past.

In attempting to make the environment more hospitable, the inhabitants of the tsetse fly areas of Africa are destroying one of the world's major ecology types. Destruction of the typical hosts of the tsetse fly (zebra,

---

Desertification: The expansion of the desert to moister areas along its margins. Plant cover and soils are destroyed through overgrazing and erosion and possibly through long-term climatic changes.

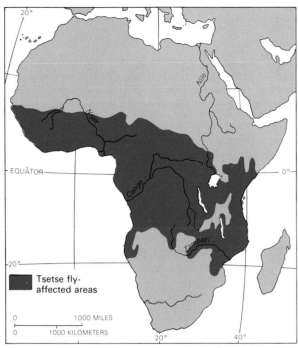

Figure 16.7 Much of tropical Africa is affected by Tsetse flies, which can transmit disease to both humans and livestock.

reinfection. Most of the common parasitic worms of Africa are transmitted by unsafe water supplies, a problem that is now receiving the attention of the World Health Organization. The lack of pure water remains one of the most important problems in Africa at the present time. Other diseases caused by parasitic worms include *onchocerciasis* (river blindness), which afflicts millions in Africa, and **schistosomiasis,** which seems to be increasing as irrigation projects or other developments that result in more standing water spread the disease. In addition to these diseases are leprosy, yaws, pneumonia, and a host of others associated with the tropics. In addition, Africa's population is affected by diseases associated with inadequate diets.

Africa also suffers from periodic invasions of locusts. Large swarms of locusts that destroy all vegetation in their path have been a traditional plague of Africa. Today such locust swarms are sprayed, but they still cause extensive damage to crops. The locust invasion of 1988 was the worst in Africa in a generation.

Many of the diseases that plague Africa can be successfully eradicated. Infestation by intestinal and other parasitic worms can be prevented through better sanitation and control of water quality. Malaria and sleeping sickness can be controlled through spraying the breeding areas, and diseases like smallpox can be prevented through inoculations. Because of the hot, tropical climate, however, the success of preventative programs relies on repeated treatments. Since the bulk of the African population is rural, with low levels of income, education, and literacy common to less industrialized areas, it is difficult for an individual government to deal successfully with disease and pests. Only as the wealthy, industrialized countries make available an adequate amount of aid for research and medical care in Africa can the people in the tropical

buffalo, wart hog, bush pig) may result in the extinction of these animals unless game preserves are maintained for them. People in the tsetse fly belts in Africa, faced with the prospect of sleeping sickness in themselves and nagana in their animals, are reluctant to spare the animals that contribute to the disease. The range of the tsetse fly in tropical Africa is particularly important in explaining the absence of draft animals. Even today cattle can survive in tsetse fly infested areas only by repeated inoculations.

Parasitic worms, particularly the hookworm, are nearly universal among the people of tropical Africa. Individuals infected by hookworms suffer anemia and debilitation. Although worms are easily treated, the lack of adequate sanitation and hygiene almost ensure

Schistosomiasis: Disease caused by infestation with parasitic worms commonly found in the blood stream and liver. Also called bilharzia or liver flukes.

Locust invasions periodically wreak havoc in many African countries with arid and semi-arid climates.

climates of this continent be liberated from the persistent disease and parasitic infections that have traditionally plagued them.

AFRICAN RESOURCES

The abundance and array of minerals in Africa constitute an important base for potential development (Figure 16.8). The petroleum of North Africa and the Persian Gulf areas has already been discussed. Major petroleum deposits are also found in Nigeria, Gabon, and Angola, but Nigeria is the most important producer. Coal is limited, and major coal resources are concentrated in Zimbabwe and South Africa. It is estimated that less than 3 percent of the recoverable coal reserves of the world are found in Africa. As a result of the limited energy resources and the agricultural nature of most African economies, per capita consumption of fossil energy resources is low (Table 16.1). Today, wood supplies 80 percent of the continent's energy needs, and its role as the principal energy source increases the rate of desertification.

Africa is the world's leading producer of diamonds, cobalt, and gold (Table 16.2). Diamonds are concentrated in South Africa, Zaire, and Sierra Leone, gold in South Africa, and platinum and chromium in South Africa. Africa is also a major producer of copper, iron ore, tin, and uranium. Copper is concentrated in the so-called copper belt of Zambia and Zaire. Iron ore is found in Liberia and to a lesser extent in South Africa. The iron ore in West Africa consists of high-grade ores, which are of great importance to the iron and steel industry of Western Europe. Bauxite is of central importance to the countries of Guinea, Ghana, and the Ivory Coast.

If you examine the map of mineral resources, you can see several major regional concentrations of mineral production. The first and most important is in South Africa, which is a leading producer of a variety of minerals. It has the best complement of minerals and is one of the few areas in Africa that is industrialized. South Africa is the world's leading producer of gem diamonds and gold and is Africa's leading producer of platinum, manganese, chromium, asbestos,

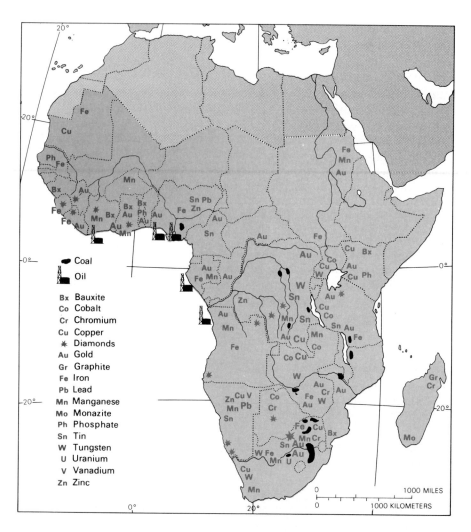

Figure 16.8 Resources in Africa.

THE WORLD OF POVERTY

480

TABLE 16.1 Annual Consumption of Energy in Selected African and Industrialized Nations

Nation	Per Capita Energy Consumption (in kilograms of coal equivalent)
Burkina Faso	20
Zaire	73
Mozambique	86
Kenya	103
Liberia	345
Nigeria	165
Zimbabwe	427
Italy	2,606
Japan	3,116
United Kingdom	3,603
Netherlands	5,138
Sweden	6,482
United States	7,278

SOURCE: *World Development Report* (New York, The World Bank, 1987).

coal, uranium, and antimony. The major center for the production of minerals in South Africa is Johannesburg, at the center of the *Witwatersrand* or *Rand* area.

A second major mineral area is north of South Africa in the countries of Zaire, Zambia, and Zimbabwe. The major mineral in terms of value in this region is copper, produced in northern Zambia and southeastern Zaire in what is known as the Shaba. Neither Zambia nor Zaire is the world leader in copper production, but both are normally among the top ten. In addition to copper, this region produces diamonds and cobalt, and lead, zinc, and gold recovered in conjunction with the copper mining activities.

Zaire is the world's leading producer of industrial diamonds. The major producing area is west of the Shaba region in the southeast. Zaire and Zambia are also major producers of cobalt, with Zaire being the source of nearly two-thirds of the world's supply in typical years. An important alloy in making steel, this strategic element is in demand in industrial nations. In addition, Zimbabwe has the only important coal deposits in Africa outside South Africa, with the major producing area in the west around the city of Wankie.

Another major mineral region stretches from Angola through Mauritania. This area has a variety of minerals, including petroleum, iron ore, lead and zinc, tin, and gold. One of the major mineral products is the high-grade iron ore that is exported from Liberia, Mauritania, and Sierra Leone. In addition to the petroleum from Nigeria and Angola, Nigeria is a major exporter of tin, and Ghana exports manganese.

In total, Africa's minerals are sufficient for an industrial economy, but the geographical relationships between and among the individual countries handicap this development. With the exception of South Africa, few countries have complementary mineral resources. Because of the political organization of Africa, the minerals have been distributed among a large number of individual countries and economic development lags.

ECONOMIES AND POTENTIAL FOR DEVELOPMENT

Africa has a dual economy, resulting from European colonial actions. One aspect of this dual economy is exploitive activities that Europeans established, focused on the parasitic cities they founded. These activities include mineral extraction, production of tropical crops, or energy development. Whatever their form, the exploitive activities and primate cities, ex-

TABLE 16.2 Mineral Production in Africa as a Percentage of the World

Mineral	Africa's Percentage of World Production
Antimony	13
Asbestos	8
Bauxite	16
Cadmium	2
Chromite	40
Copper	11
Cobalt	25
Gems	65
Gypsum	10
Gold	41
Lead	5
Nickel	4
Phosphate	21
Silver	5
Tin	1
Tungsten	1
Uranium	25
Vanadium	29
Zinc	4

SOURCE: U.S. Bureau of Mines, *Mineral Yearbook 1987*.

cept those in southern Africa, are externally oriented. Economic benefits from these parasitic activities and cities have only a marginal impact on the majority population of Africa.

The majority of Africans are engaged in agriculture, the other segment of the African economy. The large size of the continent and the variety of climates produces great diversity of agricultural activities, ranging from small hunting and gathering groups, to nomadic herdsmen, to sedentary cultivators who produce crops for export.

AFRICAN PASTORAL NOMADISM

The agricultural activity of the drier parts of Africa is normally herding cattle, sheep, goats, or camels. The use of grazing animals is normally restricted to areas with less than 35 inches (900 millimeters) of precipitation as in the steppes and savannas south of the Sahara and south of the equator. Sheep and goats are the most numerous livestock among the pastoralists, as they can live in dry areas and reproduce rapidly. Cattle or camels are the preferred livestock and are prestige animals that are utilized wherever environmental or income restrictions are not present. The economy of the nomadic African pastoralists in the past was based on movement of animals, families, and possessions within a definite range. The rotation of the animals reflects a seasonal availability of forage, with highlands or wet savanna lands grazed during the summers and low areas or margins of more humid regions grazed during the dry season.

The pastoral societies include the Tibu of the south central Sahara, the Fulani in the savanna zone of West Africa, the Nuer of the southern Sudan, the Masai of Kenya, and the Baggara, who herd cattle between the White Nile and Lake Chad. Traditionally African pastoralists relied on trade with sedentary groups for grain or other things they could not produce because of constant movement with their flocks and herds. The nomadic peoples are increasingly settling permanently in villages or towns, with livestock herding carried on by only a portion of each family, usually the males. This transformation of their society is a result of modern medicine and resulting population growth in Africa. As population increases among the sedentary subsistence farmers, cultivation expands into the more arid zones. The pastoralist is cut off from former grazing lands, and in order to ensure forage for the animals, participates increasingly in farming. Farmlands are cultivated during part of the season, and the stalks and fodder of the plants are grazed after harvest.

The pastoral economy of Africa remains primitive by Western standards. Animals are still primarily regarded as wealth rather than as crops, and the pastoralist is more interested in numbers of cattle than in their quality. Animals are sold only to provide essential needs rather than to accumulate capital for development purposes. African cattle have been increasing in quality as ideas of selective breeding diffuse to the herders. Improvement of breeds combines with inoculation of cattle to cause more rapid growth of herds and resulting overgrazing of pasture lands. Drilling of wells allows use of cattle where previously they could not survive. Pressure from larger numbers of cattle, along with the growing population of African subsistence farmers, is forcing pastoralists either to move to the harshest environments or change to a sedentary agricultural life-style. Whichever option they choose, African pastoralists are often among the poorest people of the continent. Per capita incomes among them rarely exceed $50 per year. The environmental impact of the pastoral nomads is much greater than their small number would indicate. Growing numbers of livestock are causing both overgrazing and destruction of wildlife. Desertification from overgrazing in the semiarid and savanna zones causes irreversible environmental changes. The destruction of wildlife, particularly in East Africa, has reached crisis proportions and raises fears of extinction of some species. Wildlife are generally viewed as a competitor or a hazard by African farmers.

Kraal of pastoralists in Niger. The distinctive circular huts and fences of African kraals are an important element of the region's landscape.

THE HUMAN DIMENSION

Drought in Africa

The recurring droughts of the past two decades have seriously disrupted the life ways of nomadic peoples in Mali, West Africa. Many have been forced to move from the country to the city in hopes of better survival. One such person is Moussa.

Moussa is a Tamachek herder who left his nomadic life on the edge of the desert and moved to the bedraggled city of Gao, on the Niger River. "Prior to the drought my life was with many animals," Moussa said. "Then our animals died, so we made gardens. But now there is no water for gardens."

The ongoing drought conditions have foiled the traditional herding and farming lives of countless people throughout Africa's Sahel region, turning many of them into refugees and beggars. Standing in the shade of a tree in 100-degree heat in the desert south of N'Djamena, Chad, Al-Hajj Umar, a Muslim husband and father, looked thin, drawn, but indestructible. In white skullcap and black beard, he began to tell how he had trudged for 11 days from his devastated northern village with his three wives and six small children.

It's increasingly a tragically common story in Chad and across the eight countries of the Sahel: an already precarious balance between life and death upset by drought—too many people and too many animals turning marginal land into sand. The result—a new breed of famine refugee, walking to live: Al-Hajj Umar. He came from Bokoro, 155 miles east of N'Djamena. Once he owned 17 cows and six goats, but the only animals he set out with were two donkeys.

One wife sat on one. The others walked. "In our village, it hadn't rained properly for seven years," he said stoically. "The cows died of hunger. I sold the goats to buy food. I tried to plow. I tried to grow sorghum and groundnuts [peanuts]. I left late last year with 15 other families."

The trek was terrible. They walked day and night. Two children in friends' families died. Now his family has been brought here by truck from a holding camp near the shrunken Chari River, ready to work in new fields carved out of dry soil behind us, hoping someone would provide a small pump to lift Chari water to them. It all looked bleak, dry, and wretched, but Mr. Umar thanks Allah he is not still back in Boroko. "If I am here it means that Allah has saved me," he said. He thought for a moment. "Even if the food still isn't enough for us. . . ." Al-Hajj Umar cannot read or write. But he can still work. And hope.

Adapted from "The Art of Survival", in The Christian Science Monitor, *September 2, 1988, p. 23.*

SUBSISTENCE FARMING SYSTEMS

The bulk of the African population is engaged in subsistence agriculture on small farms with low yields per acre. Subsistence farmers produce primarily for family consumption. Basic crops include millet, sorghum, cassava, and corn. Other crops such as yams, oil palm, cocoa, rice, or cotton are grown in areas with favorable soils and climate. There are two main types of subsistence agriculture in Africa, shifting and sedentary.

Shifting agriculture. Shifting agriculture in Africa is both similar to and different from that of Asia. It also varies within Africa, depending on the particular problems in specific areas. It may involve simply rotating the fields being cultivated, while maintaining a fixed location for the village; or it may include periodically moving the entire community. A concentrated area of shifting cultivation is the tropical forest of central Africa and the Ivory Coast region, and the savanna lands of West Africa and highland East Africa (Figure 16.9). The typical shifting agricultural system in the tropical rain forest involves an extended family in a small village, usually numbering fewer than 50 people. Land is cleared in a specific site each year and planted with the staple crops of manioc, corn, yams, and other vegetables. Crops are grown on the cleared space for 1 to 3 years, depending on the nature of the soil and the type of crops.

As the limited soil fertility provided to the land by burning of the vegetation is depleted, a new site is cleared and planted. This type of shifting agriculture should not be viewed as unsophisticated, as the actual cultivation practices are highly advanced. The African peasant farmer plants specific crops to ensure that there will be a food supply maturing at different times throughout the year and to ensure that each distinctive microclimatic and soil regime will be utilized by the crop that can best use its resources. Thus bananas are planted on the most fertile soils, yams on the poorly drained areas, and grains on the drier areas of this minifarm.

A second type of shifting cultivation is practiced in the savanna lands or drier margins of the humid tropics. Here fairly large villages may have populations of as many as 2000 or 3000. Each individual peasant farmer in these villages has an area near home that is planted yearly. Soil fertility on these plots, which are used for kitchen gardens to produce frequently used vegetables, is maintained by application

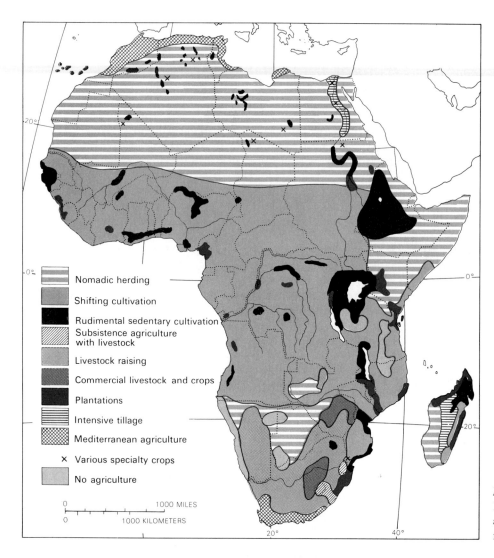

Figure 16.9
Distribution of agricultural economies in Africa.

of animal manure and plant wastes. Larger areas are cleared and planted using the techniques of shifting cultivation. Major crops include sorghum, millet, or corn where there is enough moisture. Planting of two or more crops (polyculture) in the same field is common to ensure at least partial success in the event of drought. In portions of Africa the fertility of the shifting plots is maintained for 3 to 5 years by letting the animals graze them part of the year and manure them at the same time.

Incomes are low among all shifting cultivators, even though most produce something for sale. The commercial production of these subsistence systems is minor, possibly consisting of certain tree crops, such as palm oil or cacao, which are protected when the other vegetation is burned. People engaged in such shifting cultivation systems have a high rate of malnutrition and infant mortality, and low rates of literacy. Children are especially affected by the heavy reliance on crops that produce primarily carbohydrates and starch. High incidence of **kwashiorkor,** a disease resulting in anemia and learning disabilities in children, is associated with a diet that provides only minimal calories and little protein.

Sedentary grain cultivators. A larger percentage of the total agricultural population of Africa is engaged

---

Kwashiorkor: Severe malnutrition resulting in anemia, loss of skin pigmentation, hair loss or color change, and protruding stomachs. Kwashiorkor has traditionally been viewed as the classic example of a protein deficiency causing a specific disease. Recent research suggests it is not caused by lack of protein but is a specific disease caused by a fungus known as aflatoxin. Aflatoxin forms in badly stored crops in hot, damp conditions for reasons still poorly understood because of the lack of research.

THE WORLD OF POVERTY

The nomadic peoples of Africa are still common, although pressures from increasing populations, drought, and government efforts to control their borders are decreasing their numbers. These nomads are from Morocco.

in sedentary agriculture with grain the dominant crop. Sedentary agriculture is concentrated in a belt between the humid and arid regions north and south of the equator. Although the total area occupied by sedentary farmers is smaller than that devoted to shifting cultivation or pastoralism, the higher yields per acre enable the area of sedentary farming to support a higher population total (Figure 16.10). Sedentary cultivation seems to have resulted from increasing population pressure. When population densities exceed approximately 15 per square mile (6 per square kilometer), each plot must be used more frequently and there is insufficient time for native vegetation to reestablish itself between periods of cultivation. Burning cannot provide the necessary soil fertility for effective shifting agriculture. Thus the increased population pressure dictates the use of animal wastes or other artificial means to maintain soil fertility.

The dominant grain crop among the subsistence and semisubsistence economies of the sedentary farmers reflects climate conditions. North of the equator in Senegal, Sierra Leone, Liberia, the Ivory Coast, and parts of Nigeria it is rice. Farther east precipitation decreases, and corn, sorghum, and millet, which will produce in drier locations, become the dominant grains. The actual array of crops in each specific area depends on the combination of precipitation and soil type.

The society based on the sedentary agricultural system of Africa is markedly different from that in other areas of the world. The roles of men and women are clearly defined, and the woman is often responsible for most of the field work (Tables 16.3 and 16.4). Estimates by the United Nations indicate that in Africa 60 to 80 percent of all agricultural work, plus 100 percent of food processing, is performed by women. The highly specific gender roles dictate that women do all the chores related to the household as well. Much of the women's work must be accomplished while pregnant or nursing a child. Tragically, the choicest food is often reserved for the men, further exacerbating the debilitating effects of labor, pregnancy, and malnutrition for women in Africa and other less industrialized regions. Improvements in agricul-

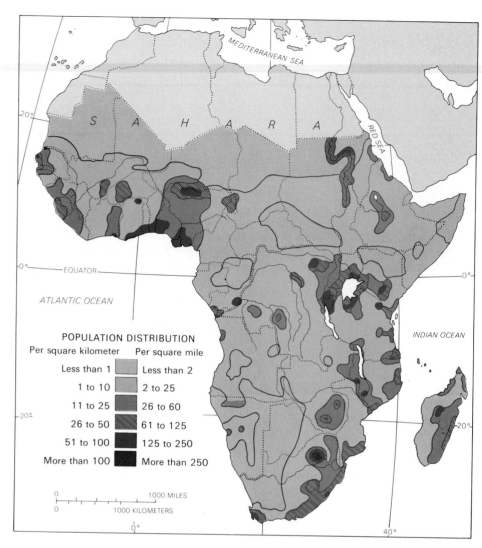

Figure 16.10 Population distribution in Africa.

ture often cause even greater handicaps for women. In many African countries the beginnings of mechanization of agriculture take the form of tractors for plowing and planting. Unfortunately, tractors allow more land to be cultivated, making more work for the woman, who must still weed by hand. Even the Green Revolution with its higher yields may cause more work for the woman in harvesting, threshing, and marketing the grain. At the same time changes in the traditional system create social problems as the traditional role of men as warriors and herders is decreasing. Many men migrate to cities seeking acceptable employment, placing a greater burden on the women left behind, and handicapping efforts to increase food production.

Landownership in the subsistence, sedentary farming sector is not normally based on individual property rights; land is held by the extended family or community. Traditional African agriculture is nonmechanized and has in the past relied heavily on the hoe as an implement rather than the plow of the middle latitudes. Changes are occurring in the traditional system as draft animals and tractors replace the traditional hoe. Such changes are inhibited by the division between the pastoralist with the animals and the village dweller who rarely had any animals other than sheep, goats, or small fowl. A true mixed farming system is emerging in which the peasant cultivator has his own animals to produce meat, milk, and draft power, but it is restricted to a relatively few areas of the continent.

Another change is that more and more farmers are selling part of their crops to nonagricultural areas. The African villager is aware of the superiority of certain implements such as metal tools, radios, factory cloth, and other manufactured items that can be obtained only with cash. Directly or indirectly the peasant farmers of Africa are being drawn into the interconnected world system. At present, this interconnected system operates within individual countries, but certain items are entering international

TABLE 16.3 A Day in the Life of a Rural Woman of Africa

Daily Activities	Time Spent (hrs.)
Waking up (5 A.M.)	—
Walking up to the field 1/2 to 1 mile—often with baby on back	0:50
Plowing, planting, weeding (until about 3 P.M.)	9:50
Collecting firewood on way home	1:00
Pounding or grinding grain	1:50
Fetching water—often at long distances	1:25
Lighting fire and cooking meal for family	1:00
Dishing out food and eating	1:00
Washing children and clothes	1:00
Going to bed (9 P.M.)	—
	17:55

Summary:

Hours of work	17
Hours of eating	1
Hours of rest	0
Hours of sleep	6
Total	24

SOURCE: UNECA: Women of Africa, Today and Tomorrow (Addis Ababa: UNECA 1975), p. 6.

trade, and some of those the peasant farmer buys come from foreign lands. The forces for change affecting the peasant farmer are many and include the establishment of individual property ownership to replace communal property rights. (See The Human Dimension.)

Commercial agriculture. There is another agricultural sector in Africa that involves fewer people but is the source of major exports from the continent. There are three main areas of commercial export agriculture: West Africa, the highlands of Kenya and Uganda, and the Nigerian savanna. West Africa developed in response to European demand for palm oil, cacao (cocoa), bananas, rubber, and peanuts. The actual commercial system takes a variety of forms, but the primary arrangement is large, foreign-owned plantations or small farms created from old plantations. In Liberia, for example, the Firestone Corporation has large rubber plantations, which are cultivated by migrant labor. In the Ivory Coast and Ghana, by comparison, coffee and cacao are produced on small farms averaging less than 10 acres each. The combination of commercial plantations with European administration and small African-owned farms means that many Africans living beyond the more fertile areas suitable for commercial agriculture migrate to the commercial lands or to the cities where the export of tropical products provides jobs. These migrants typically stay only for a season and then return to their families.

In East Africa there are two areas of commercial agriculture, but Europeans dominated its development to the virtual exclusion of Africans during the colonial period. One is in the highlands of Kenya around Nairobi and was developed by European farmers attracted by the temperate climate. This area produces coffee, tea, sisal, and cattle. Much of the agricultural land has been transferred to African ownership since Kenya's independence, but the plantations and large ranches are still owned by Europeans who employ black laborers. A second commercial agricultural region in East Africa is the northern shores of Lake Victoria, around Kampala. This area has been developed by individual African farmers producing cotton and coffee as commercial crops.

A third region of intensive commercial agriculture is in the savanna lands of Nigeria. Sedentary farmers

TABLE 16.4 Division of Labor — Africa

	Percentages of Total Labor in Hours	
Activities	Men	Women
Cuts down the forest: stakes out the fields	95	5
Turns the soil	70	30
Plants the seeds and cuttings	50	50
Hoes and weeds	30	70
Harvests	40	60
Transports crops home from the fields	20	80
Stores the crops	20	80
Processes the food crops	10	90
Markets the excess (including transport to market)	40	60
Trims the tree crops	90	10
Carries the water and fuel	10	90
Cares for the domestic animals and cleans the stables	50	50
Hunts	90	10
Feeds and cares for the young, the men, and the aged	5	95

SOURCE: UNECA: Women of Africa, Today and Tomorrow (Addis Ababa: UNECA 1975), p. 6.

THE HUMAN DIMENSION

Women of the Less Industrialized World

Women are the majority of the world's food producers. They make up 60 to 80 percent of agricultural workers in Africa and Asia and more than 40 percent in Latin America. Women all over the world have always worked in agriculture and in food preserving, preparing, and cooking. They plant, weed, supply water for irrigation, harvest, thresh, winnow, tend poultry and animals, store foods, grind flour and meal, preserve foods as sauces, syrups, juices and in many other ways.

The work they do depends not only on where they live but on their place within the rural economy: whether they are landless or landowning, tenant farmers or sharecroppers, members of a cooperative or communal farm; the size of their landholdings; whether they have their own plots, their own income from the cooperative, or whether these are reserved for a male "head of the family." These are some of the factors which determine women's work.

A characteristic common to most of these women is a long, hard day. The African Training and Research Centre for Women (ATRCW) describes a farmer's day like this:

She rises before dawn and walks to the fields. In the busy seasons, she spends some 9 to 10 hours hoeing, planting, weeding or harvesting. She brings food and fuel home from the farm, walks long distances for water carrying a pot which may weigh 20 kilograms (50 pounds) or more, grinds and pounds grains, cleans the house, cooks while nursing her infant, washes the dishes and the clothes, minds the children, and generally cares for the household. She processes and stores food and markets excess produce, often walking long distances with heavy loads in difficult terrain. She must also attend to the family's social obligations such as weddings and funerals. She may have to provide fully for herself and her children. During much of the year she may labor for 15 to 16 hours each day and she works this way until the day she delivers her baby, frequently resuming work within a day or two of delivery.

In spite of these long hours, women have very little control, very little say in decisions about food production. They produce the world's food, cook it and serve it, yet they are malnourished. Food is distributed unequally, not only among countries and social classes, but within the family. Men eat first; women and children get the leftovers, in many places. Women's nutritional needs are greatest because of their work, childbearing and breastfeeding, but they get less food, fewer calories, less of the best available than men.

From Marilee Karl, "Women, Land and Food Production," in Women in Third World Development, Sue Ellen M. Charlton, editor (Boulder, Colo.: Westview Press, 1984): 73–74.

---

produce cotton and peanuts, cacao, copra, and palm oil for sale, and traditional crops of millet, beans, and other grains for subsistence. A final area of commercial agriculture is in southern Africa. Developed by European migrants, it produces sugar cane, cotton, dairy products, and vegetables and grains for the more wealthy South African population rather than for export.

POPULATION DIVERSITY OF AFRICA

Of a total African population of approximately 660 million, 519 million live south of the Sahara. Because of the limited carrying capacity of most African agricultural systems, the entire region is only sparsely populated. Areas of denser settlement are the southeastern portion of the continent, the east central highlands, and the Gulf of Guinea or Ivory Coast area. African countries have demographic characteristics typical of less industrialized countries, with birthrates as high as 53 per thousand and death rates as low as 6 per thousand. The annual growth rate for most of Africa is approximately 2.9 percent and would be higher

Area of heavy overgrazing in Kenya. The continuous grass cover that once protected the soil has been destroyed, and serious overgrazing has been accompanied by the invasion of less useful plants.

A woman of Mali pounds millet grain to make a crude flour. The woman's home, tools, and possessions are only marginally different from those of her ancestors.

Mechanization of agriculture in Africa is only beginning, as in this scene of planting millet in Tanzania with a Chinese tractor.

if infant mortality rates were not the highest in the world. The average for the continent is 110 and some individual countries like Sierra Leone and Mali have infant mortality rates of 175 or more per thousand.

Africa's growth rate is still the highest of all the major world regions (Figure 16.11). At present rates there will be over one billion people in Africa before the year 2005. Literacy rates can only be estimated but

Figure 16.11 Number of Births in a Woman's Lifetime.

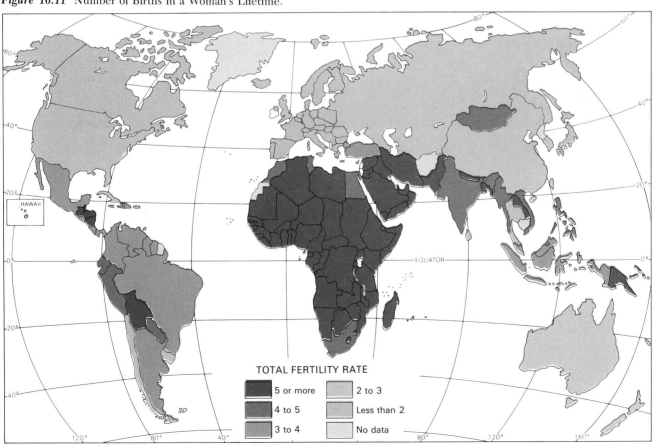

SUBSAHARAN AFRICA

489

are believed to be between 25 and 40 percent for most African countries. The small urban population reflects the low level of development; with Burundi having only 5 percent of the total population living in urban areas.

POLITICAL FRAGMENTATION IN AFRICA

There are 53 independent or quasi-independent political states in Africa ranging in size from the 118 million people inhabiting the 356,669 square miles (923,772 square kilometers) of Nigeria to the 100,000 inhabitants occupying the 145 square miles (375 square kilometers) of the Seychelles. Sudan (967,500 square miles or 2,505,800 square kilometers) and Zaire (905,567 square miles or 2,345,410 square kilometers) are the largest in total area, but Nigeria has by far the largest population. Zaire has only 35 million people, and the Sudan only 25 million. Whatever the size of the countries, political fragmentation presents a major problem for economic development of the tropical African region. It constitutes a centrifugal force, but an even greater centrifugal force is tribal loyalties.

TRIBALISM: CHALLENGE TO AFRICAN UNITY

The majority of the population of tropical Africa can be divided into several major, recognizable racial groups, a variety of language groups, and more than 500 tribal groups that consider themselves distinctive. The complex mosaic of language and tribal groups in Africa presents a major challenge for the development of these regions. It has been estimated that there are as many as 800 African languages, some spoken by millions of people, others by only a few hundred. Although it can be argued that many of these languages are simply dialects, the tremendous diversity indicates the problems facing the emergent African states.

The African peoples generally fall into the broad groups of black (Negroid) peoples, pygmies, and bushmen. The Negroid peoples are the numerically dominant group. Pygmies are found mainly in the rain forest of the Congo Basin. They are unique in the world in their physical appearance, with an average adult height of under 5 feet. The bushmen differ from the Negroid peoples in having a light brown skin with thin lips instead of everted lips of the Bantu majority. In addition to these major groups, there have been subsequent invasions by Europeans and Asian Indians. Both represent the impact of colonial powers on Africa, and both are small in total number. Europeans number less than 2 percent of the total population of tropical Africa, while Indians and other Asians constitute less than 1 percent. Europeans are concentrated in Kenya, Zambia, Zimbabwe, and most of all, South Africa. Asians are concentrated along the east coast of Africa in Kenya and South Africa.

Social and economic stratification exists to a marked degree in Africa, with the Europeans typically having the dominant position in terms of wealth, professional activities, and social status. The Asian and Negroid peoples have varying social positions depending upon the individual country, but in the majority of African states, the Negroid population has a dominant social and political position, with the Asians serving as merchants. In South Africa the white government has decreed that Asians have greater social standing and access to job opportunities than the Negroid majority. The bushmen (!Kung San) and pygmies (Mbuti) are generally at the bottom of the social and economic order in the areas where they are found, but generally live in isolated areas and are thus outside of the social structure.

The language distribution of Africa is even more complex than its racial composition (Figure 16.12). Because languages are distributed by independent tribes, it is impossible to show all of them adequately. For example, in Nigeria major languages include Hausa, Yoruba, and Ibo; in Zaire, Lingala, Swahili, Kikongo, and Chiliuba; and in each there is a host of lesser languages. Because there are so many languages in Africa, it is almost impossible to regionalize the language distribution, but in extremely broad and simplistic terms, in an area immediately south of the Sahara, Arabic dialects are common, along the Ivory Coast Hausa is common, and in the balance of the continent either tribal languages predominate, a foreign colonial tongue is used as an official language, or Swahili is common. Swahili, a trade language based on elements of Arabic and local dialects, is used throughout much of East Africa. The languages of the Congo-Kordofanian and Afroasiatic groups cover most of Africa, but Nilo-Saharan and the Khoisan of the bushmen (!Kung San) are also important.

The problems posed by the multiplicity of languages in the emergent African states is obvious. The individual African pays more allegiance to his or her tribe than to the political state. The tribal languages are a major centripetal force that ties the Africans to their tribal groups. They also serve as centrifugal forces preventing unification and emergence of strong central governments. Consequently, many African countries have adopted a colonial language as an official language, as in Nigeria where English is the official tongue. The justification for adopting such Western languages is the need to have a uniform education system, the need for a commonly understood language that can be used for trade and communication, and

THE HUMAN DIMENSION

Women of Botswana

In Gaborone, the men in the government are sighing with relief as they watch the rain soaking into their perfect green lawns. It marks the end of a drought in Botswana: the cattle's grazing will be saved now. And cattle mean wealth and security. By selling just one good ox a man can feed his family for a year.

But for the 48 percent of Botswana families who don't own any cattle the effects of last year's drought will linger on until the next harvest in June.

Six months seems a long time to someone like Nteapa. She is one of the 33 percent of women in Botswana with no men and no cattle to help support her family. Her husband died of TB five years ago, leaving her to provide for four children and their grandparents. If the rain falls at the wrong time, she will face another year of crippling poverty. Neither of her two sons is old enough to plow her land, even if she had the cattle, and strong traditions prevent her from plowing herself, or from performing any of the agricultural tasks for which a relatively decent wage can be earned.*

She is forced to wait until she can persuade a neighbor or relative to plow for her. But they are busy on their own land, and by the time they get around to helping her, the best of the rains may be over. She manages by getting what she can from "majako" (an informal arrangement where agricultural work is done for payment in kind), beer brewing, and by raising a few chickens.

She is not the only one in this plight.

In a skimpy miniskirt and bare feet, Nnyoro passes the village bottle store on her way home. She looks up in response to shouts and jeers from the group of young men lounging outside. But she shakes her head when they call her over.

Her small figure, bulky with the traditional tartan blanket tied around her waist, is a familiar sight. She is never still; collecting water from the nearby standpipe; sweeping her immaculate, beautifully decorated compound; repairing the walls of one of her three round mud huts (rondavels); returning from a six-mile walk into the bush for firewood. Or disappearing for a week with her family to work on a neighbor's land. For a day's bird-scaring or weeding she and her older children can earn 72 thebe (65 cents) between them. It is not much, but it buys a small bag of mealimeal—just enough to keep body and soul in tenuous contact with each other for another few days.

Nnyoro is angry. If only one of those men would marry her instead of just fooling around. . . .

With nearly 50 percent of men between the ages of 20 and 29 away in South Africa's mines, a man in Botswana can afford to pick and choose. He likes to know that his woman can bear children before he commits himself. But, for a woman, an illegitimate child is no guarantee of a husband.

This acute shortage of men hits Botswana's rural womenfolk doubly hard. Firstly, they are left without labor during crucial times of the year. Secondly, their self-esteem is measurably damaged by having to compete with one another for the fleeting favors of a man who may decide to leave at any moment.

Susannah has proved her fertility twice—to two different men. But they are under no obligation to marry her or support her children. But Susannah is philosophical: "This my child is like a purse of gold. Am I to say 'to whom does gold belong'? No, she is *my* purse of gold." The thin baby is called "Mosetsanagape"—literally "another girl"—a cruel illustration of how vital men are to women in rural areas. From the moment they are born little "basimane" (boys) have an easier time then "basetsana." They are fed the choicest food, and have little to do except play football until they are 8 years old, while their sisters are already performing all the chores of womanhood. Then they are sent out to herd cattle and goats to and from the watering holes. In this job, the boys have milk and meat from the animals to supplement their diet of sorghum porridge. At home in the village, little girls fare much worse. In the queues outside the malnutrition clinics, twice as many girls as boys are found to be pathetically thin.

Their early work with cattle far from home means that many boys miss out on their education. Up until the ages of 14 and 15 there are more girls than boys at school. A few are lucky enough to get a government job in Gaborone. But for others, less fortunate, there is no alternative but to continue to bend over the iron-hard land. Hoping for a son. Waiting for rain. Praying for a man.

---

*The women of Botswana face an unusual problem caused by the migration of men to the mines and factories of South Africa, but the agricultural problems are similar to other African countries. In other African countries migration of men to the primate cities creates a problem similar to that in Botswana.

---

the necessity to overcome the centrifugal effects of the multitude of languages and tribal groups. Other countries rely on unofficial use of a European language as a lingua franca.

AFRICA UNDER EUROPEAN DOMINANCE

As a result of the numerous tribal groups with their loyalty to tribes, the political map of Africa is a recent creation and is still in a state of flux. The majority of the present African states have come into existence since the end of World War II. Of the 53 states that are either independent or claim independence (three of the quasi-independent states are in South Africa and have a status similar to the Indian reservations in the United States), only six gained independence prior to 1956. The oldest is Ethiopia, which was independent or quasi-independent before European contact. Liberia was created in 1847 as a home for freed slaves from the United States, and Sierra Leone was created

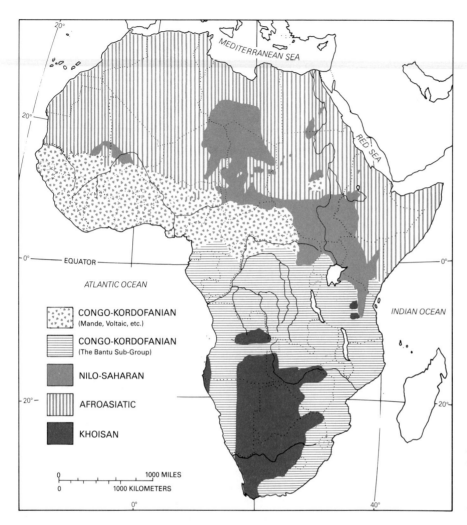

Figure 16.12 Major language families in Africa. Each broad language family has numerous individual languages and dialects within it.

by the British in the nineteenth century for similar purposes. The other three countries came in the twentieth century, with the Union of South Africa gaining independence from the United Kingdom in 1910, Egypt becoming independent from the United Kingdom in 1922, and Libya in 1951. Between 1956 and 1960 an additional five states gained their independence, and between 1960 and 1964 26 states became independent. Between 1964 and 1970 an additional eight states gained their independence, and in the 1970s and 1980s several states have either become newly independent or changed their name or government. The relative newness of African countries reflects their dominance by European colonial powers in the nineteenth and twentieth centuries. The impact of the Europeans on Africa not only retarded the emergence of the African states, but permeated all aspects of the social and economic development process.

The major influence of the Europeans during the first four centuries of their contact with Africa involved diffusion of corn, metals, and firearms to the African peoples; and export of slaves from the continent. The slave trade from Africa was not new, as Arab peoples from North Africa had captured or purchased slaves and marched them overland to North Africa or sent them by ships as far east as India for at least 3500 years before the Europeans became involved. The Europeans provided a much larger market for slaves, and over a 400-year period brought population growth in West Africa to a virtual halt. Slaves were captured by warring tribal groups and brought to the European coastal ports where they were sold to the slavers. The primary source area for slaves was the so-called Guinea Coast of Africa extending from the present-day country of Ivory Coast to Cameroon. In the organized African civilization of this area the capture and sale of slaves was an important source of wealth for African kings in the interior of the continent.

The total number of Africans moved in the slave trade can only be estimated, but it is believed on the basis of existing records that 8 to 10 million black Africans were brought to the American colonies between 1600 and 1850.[1] The total numbers involved may have been from two to three times this number. From one-half to two-thirds of those captured in Africa were killed during the conflicts leading to their capture, died during transport to the coast, or died in the slave ships en route to the Americas. Most of these slaves were brought to the Caribbean and tropical Latin American coastal regions, where they worked in sugar cane plantations, or later to the southeastern United States where they worked in tobacco, cotton, and rice production. The relatively high mortality rate among the transplanted slaves and the overwhelming preponderance of males in the slave population exported from Africa ensured a continual demand for additional slaves. The existence of slave trade was the first major exploitation of the African people and continent by Europeans and in terms of numbers probably represents the greatest mass migration of people and example of inhumanity in recorded history.

With the virtual abolition of the slave market after the United States Civil War ended in 1865, European exploitation of Africa took other forms. Until the Berlin Conference in 1884 precipitated colonial claims to the whole continent, the Spanish and British were most active in Africa, primarily in exploring and spreading Christianity. After 1884 the European nations carved up the continent and began the exploitation and colonial rule that lasted until the mid-1970s when the French freed the last of their colonial territories in West Africa. The actual methods varied among the European colonial powers, but all involved "pacification" of the Africans and profit to the colonial powers. The impact of the European imperialism on African boundaries has been discussed, but colonialism also dramatically changed African society.

Europeans regarded African colonies as storehouses of needed raw materials. To obtain the needed volume of exports of agricultural and mineral products from Africa, the Europeans forced many Africans into commercial agriculture. In nearly every colonial possession Africans were taxed to pay for roads and railroads and to assemble exports, and since failure to pay the tax in cash resulted in punishment, the Africans had to grow goods for sale. The major commodities exported included cacao, peanuts, coffee, tea, spices, sugar, cotton, and rubber. Many Africans were also forced to work on plantations for European owners, further disrupting the centuries-old social and economic patterns of the continent. Colonial efforts of European powers followed similar patterns, including appropriation of village lands, forced migration of population to exploit minerals, and establishment of a dual society with the European minority dominating and the African majority confined to menial tasks.

CHANGES IN THE POST-COLONIAL ERA

Political independence of African countries began after World War I, but the majority of Africa remained in colonial status until after World War II. The level of development of African states at independence was lower than that in any other former colonial region in the world. European colonialism had destroyed existing African societies, increased tribal antagonisms, and exploited a continent without creating viable alternatives. The African population was generally kept in a state of fear and despair while the European claim of African inferiority precluded hope for the future. The Europeans had established few institutions of higher learning for Africans, health care and transportation were extremely limited, and there was little other internal infrastructure. At the time of independence disease was widespread and illiteracy pervasive. The Europeans had not adequately trained Africans for technical and management positions, and their departure handicapped factories and businesses in many countries.

Since independence, development has been slow in most of Africa as the individual states attempt to resolve the problems of their colonial legacy. Changes that have resulted from development efforts in individual states include individual ownership of land to replace the traditional communitarian tenure in a few states such as Kenya; the restriction of movement of pastoralists to grazing lands as a result of political boundaries; and government attempts to place former nomads in permanent communities where medical and educational facilities can be provided. The degree of change is directly correlated to the distance from the major urban areas. Those near the large cities have a ready market for agricultural products and a source of things that people perceive they need but cannot produce themselves. Increasing distance from the headlink cities minimizes the forces for change because of the limited transportation system in most African states.

Even in the villages, however, forces for change have come with itinerant traders, periodic markets, and government programs. In the most isolated villages the peasant pastoralist or shifting agriculturalist or sedentary farmer is aware of the supposed advantages of life in the city. Young men traditionally have migrated to the cities where European exploitive economic activity offered jobs in mine, factory, or plantation. This migration persists in Africa today, result-

---

[1] Robert W. Fogel, *Time on the Cross* (Boston: Little Brown and Co., 1974).

ing in as many as half the adult male population being absent for extended periods of time. Such migratory labor has been a fact of life in Africa since the end of the nineteenth century, when European development of African resources really got underway. It is entering a new phase as men who have migrated to the cities for work encourage their wives and families to join them. As population grows in Africa's rural areas, this migratory trend will increase. (See the Human Dimension, "Women of Botswana.")

DEVELOPMENTAL AND REGIONAL VARIATION IN AFRICA

The challenges that the countries of Africa face as they attempt to raise the standard of living of their people center around a number of environmental problems, problems associated with transforming agrarian economies, the challenge of industrialization, and racial conflicts. Each is found to a greater or lesser degree in most African states, but some epitomize both the problems and their proposed solutions.

THE CHALLENGE OF THE AFRICAN ENVIRONMENT

Challenges to development of Africa's resources reflect the varied geographical endowment of the continent. Problems of development in Africa's tropical regions differ from those of desert lands. Periodic drought has decimated animals and people across the Sahara borderlands and in the desert and savanna regions south of the equator. An example of these challenges is the Sahel region.

The word *Sahel* is Arabic for *shore*, referring to the lands that form the "shore" of the great "sand-ocean" of the Sahara. The Sahel extends 3000 miles across Africa and is the home of 70 million people in nine countries: Ethiopia, Senegal, Gambia, Mauritania, Borkina Faso (Upper Volta), Mali, Niger, Chad, and the Cape Verde Islands (Figure 4.7). It is an area of poverty, with an average life expectancy of about 43 years prior to the drought of the early 1980s. Less than 10 percent of the population is literate, and over 90 percent of the people are engaged in farming and grazing. In spite of its poverty, the Sahel has potential. There are large areas of cultivated land that could be utilized more intensively in the river basins of the Senegal, the Niger, the Volta, and the Gambia rivers, as well as in the basin of Lake Chad.

Precipitation in this area is erratic and rarely exceeds 20 inches (508 millimeters). A period of decreased rainfall began in the early 1960s, but the population continued to increase. Periodic drought occurred in the 1970s and 1980s, and led to mass starvation of the cattle and camels on which the people depend, destruction of grazing and woodland in attempts to provide feed for animals, migration and resettlement of as many as 20 million people, and death to an estimated 300,000 people.

Even before the recent droughts the Sahel faced monumental challenges because of growing populations of humans and livestock. Since wood provides 90 percent of the energy needs of the residents of the Sahel, the region was already denuded of brush and timber. Destruction of forest and brush combined with overgrazing caused soil erosion, which destroyed the land for future use. The Sahel was overpopulated in relation to the **carrying capacity** of the land. In the Sahel, rural densities are only a fraction of those in Bangladesh, but the harsh environment and the land use practices exceeded the carrying capacity of the land. Periodic droughts increased the suffering of its residents.

From 1968 to 1973, the Sahel suffered its longest, most severe drought in half a century. The major rivers, including the Senegal, the Niger, and the Volta, dwindled to mere trickles. Lake Chad declined to one-third of its former size, leaving former fishing villages more than 10 miles (16 kilometers) from the new shoreline. Animal losses equaled 75 percent in some areas. Over the entire region, an estimated 40 percent of livestock died, and 300,000 people died as a direct result of famine-related disease, malnutrition, or starvation.

Massive relief efforts were mounted by the industrialized world to replace the lost food production in the Sahel, enabling the majority of the population to survive. But as many as 20 million people were forced to migrate southward. The challenge for the Sahel lies in developing an economy that will allow the region to support its people without fear of recurrent drought and famine. Overcoming the problems of desertification caused by expansion of the Sahara southward, soil erosion, deforestation, disease, and malnutrition will require a major development program. The proposed solutions to the problems of the harsh environment of the Sahel focus on two major approaches.

The first involves establishing most of the population in sedentary life-styles. Former nomadic herd-

---

Carrying capacity: The carrying capacity of an area is the number of animals, crops, or people it can support on a continual basis without degrading the environment. The carrying capacity varies with technology, land use techniques, and geographic characteristics.

Africa: The Hungriest Continent

More than 20 years ago, the biggest movement of food between two nations in history took place when the United States used 600 ships to provide India with grain. The monsoons had failed in India in 1965–66, and the country was desperate.

Today India is self-sufficient in grain. The reasons: more incentives to farmers, better weather, thousands of new irrigation wells in the Ganges Plain—and a Green Revolution that produced new "miracle" hybrid seeds and new chemical fertilizers by the ton to boost their yield.

The Green Revolution also lifted Latin American food production. But so far the revolution has not come to Africa. Why?

African countries are poorer and more diverse. Sub-Saharan Africa alone is three times as large as the United States. It contains some 46 countries ranging in population from a few hundred thousand to 118 million (Nigeria).

Extended drought in the Sahel and down the length of East Africa has dried up water tables, blighted crops and lifted valuable topsoil into dust clouds of erosion. Still-high petroleum prices make it difficult for African countries, deep in foreign debt, to buy petroleum-based chemical fertilizers.

India and Latin America have large numbers of trained manager-class civil servants; Africa does not. Even simple record keeping and accounting skills are often lacking.

African populations are rising so fast—"the fastest growth of any continent in history," according to Lester Brown of the Worldwatch Foundation in Washington—that food gains are being wiped out.

Too many nations have preferred "show" projects such as dams and highways rather than promoting agricultural reforms and incentives.

Farming methods are still primitive even in relatively advanced nations such as Kenya, where 84 percent of all cropland is cultivated by hand, usually by women using digging sticks and hand hoes. Twelve percent of Kenyan land is worked by oxen. Only 4 percent is tilled by tractor.

Soil is poor in many areas. No Ice Age enriched it eons ago. Overgrazing and erosion are frequent in the Sahel, in Ethiopia, and elsewhere.

Twenty-seven of the UN's 37 "least-developed" countries are in Africa. Twenty-six African countries earn more than half their foreign exchange on a single mineral or export crop, for which prices are now low.

Thirty-seven countries depend on one or two items for as much as 80 percent of overseas earnings. By the 1980s 75 percent of Senegal's export earnings came from peanuts, and four-fifths of Chad's were generated by raw cotton.

Migration by men to the cities leaves women and children to farm, further reducing production.

Urgently needed to generate an African Green Revolution:

1. More basic research into high-yield, drought-resistant sorghum and millet seeds. Promising work is being done in Ibadan, Nigeria, at the International Institute for Tropical Agriculture (IITA) into new farming systems and on improvements in rice, maize, cowpeas, yams, sweetpotatoes, and cassava.
2. Better training of farmers, as the U.S. and the World Bank are trying to do.
3. Fewer rigid government monopolies on buying and selling food, and more private initiative.
4. More incentives to farmers to grow subsistence food as well as cash crops for export.
5. Better farm management on a continent where agriculture still employs more than 70 percent of the population.
6. More irrigation canals, but only if Africans can be trained to manage and maintain them. Irrigation is expensive. The failure and under-use rate of African projects is high, donors ruefully report.

Source: Christian Science Monitor, Nov. 28, 1984, p. 22–23. Reprinted by permission.

---

ers are being settled in villages created in sparsely settled areas of the river basins, which had been avoided because of disease. Concomitant international efforts to eradicate disease in these areas of the Senegal, Niger, and other rivers are designed to allow the present and future population of the Sahel to create an agricultural economy less subject to environmental whims. As part of the goal of expanding sedentary agriculture, studies show that several areas could support increased irrigation, but costs will be high. At the present time only about 561,400 acres (227,300 hectares) are under full or partial irrigation. It is estimated that over the next 75 years almost 6.18 million acres (2.5 million hectares) could be reclaimed for irrigation. To achieve the total acreage, however, would require a cost of as much as $10,117 per acre ($25,000 per hectare). Smaller projects in the best areas could be developed much more cheaply, however, and hold promise for the short term. The problem with all these projects is the change in life-style that would be required of the former nomadic peoples of the desert. In addition, increased irrigation provides greater health problems from both bilharzia and onchocerciasis.

Another plan to restore the Sahel proposed massive reforestation projects and the development of small wells to support smaller herds of livestock. Such a solution is handicapped by the attitude of African nomads who treat animals as a source of wealth in and of themselves and by the long time period necessary for reforestation to succeed. All development efforts proposed will be inadequate to provide for the projected population increase by the end of the century. The dilemma of the governments of countries in the Sahel is compounded by the fact that the rains were more reliable after 1973, and there is a tendency to forget the harsh realities of the drought. Severe drought struck again from 1983 to 1988 in the Sahel and included the arid and semiarid lands south of the equator as well. All of the development projects of the previous decade failed to provide adequate food, and once again famine, mass migration, and human misery dominate life in Africa's harsh lands. The continued challenge of life in the Sahel typifies the harshness of much of the African environment, a harshness intensified by ever-growing populations that exceed the carrying capacity of the land.

TRANSFORMING AGRARIAN ECONOMIES

Methods of utilizing the African environment for the benefit of the human inhabitants take a variety of forms. The tropical rain forest and wooded savannas of the Ivory Coast (Côte d'Ivoire), for example, have supported a profitable economy based on production of coffee, cacao, and peanuts. The Ivory Coast has experienced two decades of success in developing harsher lands by emphasizing small, independent commercial farms.

There are about 450,000 farmers in the Ivory Coast growing coffee and tobacco on farms of less than 10 acres in size. The success of the Ivory Coast in producing a profitable item from the tropical rain forest has depended on a stable government during the three decades since independence, extension services provided for the farmers by the government to increase yields and thus incomes, and major expansion of transport infrastructure linking rural areas with the port cities. The Ivory Coast's success is all the more remarkable because many migrants have come into the country, cleared the tropical rain forest, and produced coffee and cacao. It is estimated that more than a third of the country's population is made up of foreigners. Although there is some ill will between native citizens and the migrants, to this point there has been sufficient land for all, with the result that ethnic conflict has been avoided.

As a model for development of other African countries, the Ivory Coast may be less than ideal. The Ivory Coast was fortunate to have an educated and progressive leader who took power and remained president for more than two decades after independence from France in 1960. As a result, the 65,000 Frenchmen and 100,000 Lebanese who had immigrated to the Ivory Coast when it was a colonial possession of France have remained. Unlike many newly emergent African states, the Ivory Coast never lost its technical and managerial sector, and the export trade with France was never broken. Consequently it was able to begin the transition from colony to independent state without the turmoil and suffering that others have experienced. Its reliance on one crop makes its economy highly vulnerable to outside economic change, and it was hard hit by the world recession of the early 1980s, when per capita incomes decreased 25 percent (Figure 16.13).

Another African country that is attempting to transform its agricultural system through a diametrically opposed system is Ethiopia. Ethiopia is one of the poorest countries in Africa, with a per capita income of less than $125. Its 51 million people have existed as an independent nation for hundreds of years. Because of geographic characteristics, Ethio-

International assistance for refugees from natural and human-caused disasters in Africa has been common in the past few decades. This refugee camp in Ethiopia was established by the International Red Cross in 1985.

THE HUMAN DIMENSION

African Farmers

Listen to the voices of the African farmer.

"I am proud to be a farmer," says Mwalabu Ndonye, a Kenyan. "Let anyone come, even experts on agriculture, and ask me how I do things on my farm, and I will tell them."

"We now believe in ourselves," says Ibrahima Seck, a Senegalese peasant leader. "We believe in our possibilities, in our ability to take our own development into our own hands."

"Farming," says Nigerian Igwe Fred Uziogwe, "is the only road to success."

Seven out of every 10 people living in Subsaharan Africa are small-scale farmers. In the hands of these peasant farmers, says the ambassador of the Organization of African Unity to the United Nations, Oumarou Youssoufou, "rests the key to rehabilitation and to any successful agricultural program." Youssoufou goes on to call agriculture "the cornerstone of the [economic] recovery strategy" initiated by African governments in the mid-1980s.

Who are these men and women—Africa's farmers—who are increasingly being recognized as the key to the successful development of Africa's economies; indeed, to the very future of Africa?

First, they are resilient, overcoming extraordinary circumstances and confronting formidable challenges. In "The Role of Farmers in the Creation and Continuing Development of Agricultural Technology and Systems," Robert E. Rhoades says, "Rather than being conservative, bound by tradition, and simple, [small-scale farmers are] experimenters, risk takers [and] innovators; intensifiers and diversifiers; colonizers or pioneers; addicts for new information; practitioners of great common sense; social and economically rational beings."

They are smallholders of land—two-thirds of all land holdings on the continent are under 2 hectares (5 acres), and nearly 96 percent are fewer than 10 hectares (25 acres). They are often among the poorest of the poor. Continent-wide, urban incomes are 4 to 8 times higher than incomes in agriculture.

And, significantly, they are overwhelmingly women. Women, working on small farms, now produce 90 percent of the food consumed locally. In most countries, more than 80 percent of those who earn their living as farmers are women. On average, 22 percent of all African farm households are headed by women.

Until the 1980s the full potential of African farmers went largely untapped. Neglected in favor of other priorities, the farmer was hampered by low prices for crops, lack of access to necessary inputs, and failed government policies. Per capita food production declined annually at an average rate of about 1 percent. Africa's capacity to feed itself dropped from 98 percent in the 1960s to around 84 percent by 1980. The potential and challenge facing African farmers is indicated by the experiences of two farmers.

Mavis Katumba, a 50-year-old farmer and mother of ten, has 5 acres of land in Zimbabwe, but she grows crops on less than 2. She belongs to a farmers' group of 100 people. The nearest place they can buy supplies is 10 kilometers away, and transport is so bad that they used to waste whole days getting there and back. Now they pool their orders and send just one or two people.

In 1985 she used the full package of fertilizers and hybrid seeds, and ensured herself by planting three maize varieties—one short-maturing, in case of poor rains, one medium and one long, which would do well if the rains were good. She got 33 bags of maize—about 2.5 tons a hectare. "The price is so good now," she explains, "you can bank the money and draw on it when you need it. I don't need loans anymore."

After eight years of schooling, which included an agricultural course, Yotham Motoka decided to leave his family's farm and try his hand at growing tea in the West Ankole district of Uganda. "In late 1965," says Motoka, "I was making enough money to put some aside, so that I could marry. Then there was another pair of hands to help with the farm and the house."

By 1977 Motoka had an established home, family, and farm and was playing an increasing role in local affairs. In that year, however, political events resulted in a disruption of the tea processing plant on which he depended. In the following year the situation became so bad that he and his neighbors stopped harvesting altogether, and Motoka's acreage began to turn into a small wood of overgrown tea bushes.

He was able to feed his family, but his income dried up. In 1982 there was an improvement in the political situation. Motoka tackled the formidable task of re-establishing his 16,000 tea bushes. "By 1984," Motoka says, "the political situation deteriorated once again, and once more I sadly abandoned my tea plants. All my hard work was undone." Today, the processing plant is back in operation, but twice burned, three times shy. Motoka is not hurrying to jump into the enormous task of making his tea acres productive again.

Motoka is philosophical, but disappointed. He has a strong and happy marriage, his younger children are all in school, his oldest daughter was married last year and he is now a grandfather. His family has enough to eat, but that was not the extent of his ambition when he took the bold step of becoming a pioneer tea farmer 24 years ago. An avid reader, he knows that small-scale tea farmers in neighboring Kenya have prospered over the years, as they have enjoyed unbroken collection services and fertilizer deliveries. Yotham Motoka's efforts have been frustrated through no fault of his own, but by political events outside his control.

Adapted from African Farmer *(1) 1988, pp. 6–7, 11.*

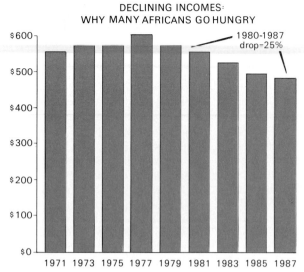

Figure 16.13 Change in per capita income in Subsaharan Africa (in U.S. dollars).

pians have been isolated from the rest of the world and have not been a major part of the interconnected world of the twentieth century. The birthrate in Ethiopia is approximately 46 per thousand, while the death rate is 15. The literacy rate for Ethiopia as of 1988 was estimated at 20 percent. Ninety percent of the population is engaged in peasant agriculture or pastoral systems. The remaining 10 percent work primarily in government, military, and trade organizations.

For 45 years under the rule of Emperor Haile Selassie there was little advance in standard of living. In 1974 the emperor was overthrown by a military coup, and since then the government has been a quasi-democracy controlled by the military. The development approach has followed the Marxist-Leninist model used by the Soviet Union. The present leaders seem determined to transform Ethiopia's impoverished rural populace. Since the primary export has been coffee (75 percent of exports) and the bulk of the population has been simply subsistence agriculturalists, accomplishing the transformation may take a long time.

The government is attempting collectivization of agriculture to try to maximize capital and access to government extension programs. To date Ethiopia's efforts have not been very successful. The lack of mineral resources and the harsh environment with its recurrent drought seem to indicate that achieving the standard of living of the Ivory Coast may not be possible without extensive foreign aid. The drought that began in 1983 was particularly devastating to Ethiopia as collectivization has limited the traditional methods of surviving droughts, and governmental and social changes have destroyed traditional mutual support systems and economies.

A third country that is only now undertaking an agricultural transformation is the Sudan. The Sudan extends from the middle of the Sahara to the tropical savannas of the headwaters of the Blue Nile and the White Nile. Faced with recurrent drought of the Sahel, the Sudan has a major resource in the waters of the two Niles. One of the most important development projects is now taking place in an area of the White Nile known as the *Sudd*, an immense swamp the size of Florida. *Sudd* means barrier in Arabic, and the swamp separates the northern, arid Sudan with its Moslem majority from the southern, humid Sudan with its Christian and native religions. As the Nile River flows into this great marsh and swamp area, it spreads out, losing an estimated 1483 billion cubic feet (42 billion cubic meters) of water a year through evaporation and seepage. If it could be saved, this tremendous water resource would enable irrigation of enough land to make the Sudan the breadbasket of North Africa and the Middle East. If fully developed, a million acres (404,700 hectares) of new farmland will be irrigated by waters saved from the Sudd.

The heart of the development program consists of a 173-mile (278-kilometer) navigable drainage canal through the Sudd to prevent the water from spreading into the swamps. An estimated one-third of the wasted water will be saved and made available for irrigation. One hundred individual projects are involved in this scheme, most of them financed by oil money from Arab states. One of these, the Rahad project, completed in 1979, irrigates 300,000 acres (121,400 hectares) of farmland in an area extending from Mataza to the junction of the Rahad River and the Blue Nile. It created 12,800 farms of 22 acres (9 hectares) each to produce cotton and peanuts, and 4000 farms of 5 acres (2 hectares) to grow fruits and vegetables. Against these benefits are environmental costs, including the concern that a unique ecosystem is being destroyed in the process of reclamation.

The development project includes financial assistance to farmers to establish themselves in the project and a government-supervised scheme for farming, processing and marketing facilities, and extension agents. It provides homes for 100,000 people, with drinking water, electricity, schools, medical clinics, and roads. If the Sudd project is completed as planned, it will be an example of what can be done with sufficient capital and resources in the harsh African environment. Developed by Arabs and Muslim and Christian Africans, it will be a monument to cultural cooperation, but to date it has not fulfilled the expectations that prompted its development. Ongoing conflict between the Muslims of the North who dominate the government, and the Christians and **animists** of the South have seriously handicapped completing the goals of the project.

RACIAL CONFLICT: THE LEGACY OF COLONIALISM IN SOUTH AFRICA

South Africa is an anomaly in Africa. Climatically it has more areas of middle-latitude climatic types than any other portion of the continent. Culturally it has the largest concentration of people of European ancestry. It is the major site for actual European settlement in the continent. South Africa has one of the highest per capita incomes in Africa, although the wealth is not uniformly shared between black and white population. South Africa is the most industrialized country of Africa, producing more industrial goods than the rest of the continent combined. It has the richest mineral deposits in the continent, ranked first in the world in gold, platinum, chrome, and vanadium, and is second in diamond, uranium, antimony, asbestos, and phosphate reserves. It controls 75 percent of the rail network of the 20 countries of southern Africa. South Africa has the highest literacy rates, with an estimated 100 percent of the white population and 50 percent of the Africans literate. South Africa contains approximately two-thirds of the Europeans in Africa today, and they have created a landscape patterned after Europe's. The middle-latitude Mediterranean and humid subtropical climates were similar enough to those the Europeans had left to present none of the difficulties found in the humid

---

Animists: People who worship naturally occurring phenomena such as mountains, trees, or animals.

tropics and the humid savannas. South Africa today has large cities such as Johannesburg, Pretoria, Cape Town and Port Elizabeth, East London, and Durban, which are almost identical to large European cities. Tall office and apartment buildings in the downtowns are surrounded by industries and suburban housing complexes. The outward similarity to Europe of the structural landscape is in complete contrast to the cultural makeup of the country.

South Africa consists of 37 million people of four major ethnic groups. The largest is the native Africans, who comprise 70 percent or 6 million people. The white population is second largest at 17.5 percent, or 6.5 million people. The third largest is called *coloured*, and comprises 9.4 percent, or some 3.5 million people. The coloured population is descended from a mixture of the African population with early European visitors to South Africa and immigrant Malays. The final group is the Asians, who comprise about 3 percent (1.1 million) of the population. The racial issue in South Africa is in part a demographic one. In the 1850s over 50 percent of the population was white, but the wealthy white population has had a low growth rate. The other groups have the demographic characteristics of less industrial countries, and their growth rate has hovered near 3 percent per year. The ethnic groups in South Africa reflect the colonial history of the country. South Africa was first established by the Dutch East India Company in 1652 as a way station for its ships engaged in trade with Asia. Later, Dutch farmers (*Boer* in Dutch means farmer) settled around Cape Town. These migrant Dutch Boers used black slave labor to expand their farms inland.

After 1800 the colony was acquired by Great Britain, and in 1833 the British abolished slavery in their empire. The Boers subsequently moved farther north into the High Veld. Here they came into contact with the Bantu African occupants of the region and in a

Segregation remains a fact of life in South Africa, although some of the most petty aspects of the past laws have been removed.

The manicured landscape of South Africa's Cape Province reflects the European origins of the settlers and is in stark contrast to the rural landscape of most of Africa.

series of battles defeated them. The Boers were able to subjugate the Bantu primarily because destructive conflicts between tribal groups in the 1820s and 1830s had scattered and weakened other tribal groups. When the Boers arrived, the High Veld was nearly uninhabited, but it had still belonged to the Bantu.

British colonists moved into the southern portion of the present-day South Africa after 1800, and the discovery of gold and diamonds in the High Veld region and conflict between the Dutch Boers and British gold miners and businessmen ultimately led to the Boer War in 1899, in which the British defeated the Boers. They established the Union of South Africa in 1910 as a British self-governing, constitutional monarchy under the direction of the British crown. Since then, South Africa has been developed by the white minority population to benefit their own interests. Blacks have been and are treated as second-class citizens, subject to an overwhelming set of rules and regulations designed to destroy the black cultural heritage and pride. The basis of the race relationships in South Africa is the concept of *apartheid* (literally, separateness). It combines practices used in the United States in the nineteenth century to segregate blacks and isolate Indians on reservations. The concept of apartheid and the government's legal enforcement of it intensified after 1960 when the white population voted to make the country fully independent of Great Britain. On May 31, 1961, the Republic of South Africa was declared an independent state by the white minority, who were the only people legally entitled to vote.

Today, the white minority is proceeding in its avowed program of apartheid, a program that the government euphemistically calls "separate development." The South African racial policy separates the white population from the other three groups. Coloureds, Asians, and Africans were required to carry an identity card and work pass with them at all times, until 1986. Legally, the cities of South Africa are the province of white only, and other ethnic groups live there only at their discretion. Servants or others who work in the cities must have a work pass to prove they are entitled to live in black-only, Asian-only, or coloured-only suburbs (townships) around the white cities. One result of this policy is that the huge township of Soweto ("Southwest Township") located near Johannesburg is the second largest city in Africa south of the Sahara, and yet legally it is not even a city. Residents could not own their own homes within it until 1980.

Another part of the South Africa racial policy is to create "homelands" for the black population. To date, this policy has taken the form of creation of four reserves, which South Africa maintains are independent countries (Figure 16.14). The Transkei was created in 1976, Bophuthatswana in 1977, Venda in 1979 and Ciskei in 1982. The South African white minority created these four homelands and proposed an additional six for the balance of blacks to guarantee white control of most of the rich minerals and lands. Once a homeland is created for a tribe (the Tswana in Bophuthatswana, for example), tribal members become citizens of that purported country and are viewed as only temporary residents in white South African dormitory communities. Legally they can own no land in areas outside of their so-called homeland. The homelands as originally proposed would contain only 13 percent of the land area of Africa but would be home for 22 million black Africans today and an estimated 37 million by the year 2000. At the same time, the minority white population would have not only the

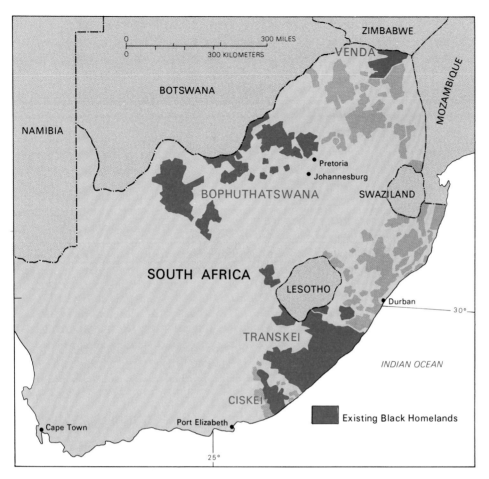

Figure 16.14 Location and areal extent of black homelands established or planned by the South African government. The homelands contain only 13 percent of the area of South Africa, whereas the black population comprises nearly three-fourths of the population.

great majority of the land but the majority of the mineral wealth of South Africa. The four homelands created to date are in reality very similar to the Indian reservations of the United States. They have inadequate resources for their population and rely upon South Africa for the bulk of their electricity, communications network, medical care, and other governmental services. The fiction that they are independent is accepted only by the ruling South African white minority party. In 1985 the South African government indicated they would establish the other homelands. Blacks not already given citizenship in a homeland simply became South African citizens, but without the right to vote.

In spite of the demeaning treatment that they receive in South Africa, hundreds of thousands of Africans migrate to white cities to meet the needs of the white residents and for labor in farm, factory, and residential and urban settings. An estimated 1.5 million blacks come from South Africa's homeland areas and from neighboring countries such as Mozambique to work. The South African government restricts the types of jobs that can be held by nonwhites, but labor shortages are forcing liberalization of these restrictions.

The dilemma of the white minority of South Africa is how to maintain its own wealth while raising the standard of living of the blacks sufficiently to prevent revolution. At the present time black South Africans have incomes, educational opportunities, and health care that are better than those of most other countries of Africa but still only a fraction of that enjoyed by the white minority.

Neighboring Zimbabwe with a population of 10,400,000 composed of over 10,000,000 Africans and only 100,000 Europeans, had a similar situation until 1980. At that time a transfer of power to the black majority of the population was formalized, ending the worst aspects of colonial rule. There is some indication that South Africa is beginning to liberalize the treatment of the African majority within its boundaries, and many South African whites recognize the necessity and equity of such a move. The white population of South Africa consists of 60 percent descendants of the Boers, who today speak *Afrikaans* (a derivative of Dutch), and 40 percent descendants from Britain, who speak English. The Afrikaans people are more resistant to the concept of greater equality for the African majority, but political realities may force them to encourage moderation of the treatment of the

THE HUMAN DIMENSION

The Struggle for South Africa's Cities

Home to Elizabeth Kmanye is shack no. 2008 in Soweto's Mshenguvillie squatter camp. A tiny, one-room structure, Mrs. Kmanye's house is a crazy quilt of corrugated tin, wood, and cardboard. Inside, it is choked by smoke from the coal-burning stove during winter, and soaked by rain that pours through the cracks in summer.

Although this is not Kmanye's idea of an ideal abode, she has few options. For 14 years, she has been on waiting lists for a real home in the teeming townships set aside for South Africa's blacks under its segregationist apartheid laws. "I'm tired of waiting for a house," says Kmanye, almost shouting. "I want something permanent, and I'm not moving from here until I get it."

She may have no choice. The government recently proposed tough new laws that critics contend will give it sweeping powers to evict the nearly 4 million people living in squatter conditions—almost one-seventh of the total black population. And relatively few will receive housing in return, critics maintain. The legislation is only one piece of wide-reaching new laws that political analysts say will allow the government to determine the number and location of blacks in cities. The other bills, also pending in Parliament, would make it easier to evict blacks and others living in residential areas set aside for whites.

Analysts say the laws are an attempt by the government to reassert control over the black urban population and, to a lesser extent, to play up to the far right. The main issue is control. The laws will be used to regain control of blacks in a way that will make previous methods look tame according to these observers. Government officials dismiss the gloom-and-doom predictions, however, insisting that the legislation will benefit blacks. By providing for "orderly urbanization," Pretoria can better begin to tackle the pressing problems of land and housing, they say.

These are dilemmas that have been decades in the making. As a way of enforcing its pass laws—which sought to control the influx of blacks to cities—Pretoria for years did not allocate new land or build additional houses in townships. Driven by employment prospects, blacks nonetheless poured in from the impoverished tribal "homelands," created for them by the government as part of apartheid.

The government finally abolished the much-hated "pass laws" in 1986, after two years of violent black upheaval. Right-wing opponents predicted a virtual flood of blacks into metropolitan areas. But Lawrence Schlemmer, director of the Center for Policy Studies, says that while squatter camps and backyard shacks mushroomed, they mostly resulted from a 2.6 percent annual population growth rate—not migration.

Take Transvaal Province, for example, where up to 2 million of the region's 5.5 million blacks live in squatter-like situations, according to a study Mr. Schlemmer recently completed. Schlemmer says the majority were born in the area—children who lived in real houses, then moved out to establish families and were unable to find conventional accommodations.

A young woman lives in the white-city section of Soweto, Johannesburg's sprawling black township, knows the problem all too well. Her mother sleeps on a cot in the postage-stamp-size living room, while she shares a bed with her grandmother in a closet-like bedroom; four brothers sleep on the floor. The place is bursting with clothes and bedding stuffed into corners and piled onto chairs. "If I want to get married, I'll have to move out," she says mournfully. "But it's impossible to find anything in Soweto."

Thus, the squatter proliferation. While anti-squatting laws exist, lawyers have been able to exploit loopholes to stave off evictions and, in some cases, halt them altogether. All that could change, however, with the new Prevention of Illegal Squatting Bill. Some contend the bill would give the state blanket powers to move squatters where it pleases.

But providing suitable alternative housing for evicted squatters will not be simple. The white government has identified 50,000 acres to be used for black housing, but critics say only 18,000 acres are usable. In addition, the government wants private builders to construct the housing, yet few commercial builders are building low income homes, the only kind that 70 percent of squatters can buy. Critics say that the bill will only allow blacks with high income jobs to move to the designated black urban areas, creating a class of blacks around the white cities that is more likely to cooperate with the white government.

Sarah Mthombeni, for one, concurs with the conclusion. Mrs. Mthombeni lives with her husband and three children in a one-room Mshenguvillie shack. She has to walk a quarter mile to the nearest water tap and toilet—and then wait in a long, winding queue. Her kids play in the deep canyons of sludge and sewage that criss-cross the camp. She wants out.

But Mthombeni says she cannot find a house under construction in Soweto for less than about $13,000—well beyond the salary her husband earns as a supermarket shelf-packer. "We're not asking for luxury," she declares, "just a place where our kids can be comfortable. Where in this country are our houses?"

Adapted from The Christian Science Monitor, *July 13, 1988, pp. 1 and 8. (Used by permission.)*

black majority. Unless South Africa with its white-supremacist government makes the necessary changes to bring the African minority into the mainstream of economic development, it will become the focus of armed revolution and bloodshed.

The well-being of the South African whites is based on mineral and industrial wealth. South Africa is the world's leading producer of gold from the Rand region, which is centered on Johannesburg. The Rand is a zone 115 miles (190 kilometers) east to west and ranging in width from 5 to 30 miles (8 to 45 kilometers). Other gold-producing areas are now equaling the Rand in output, but the Rand is still the center of white urban population. A by-product of gold mining in South Africa is uranium sufficient to place South Africa in the top three producers in the world. Diamonds have been mined at Kimberly, Pretoria, Cape Province, and in the trust territory of Namibia (Southwest Africa).

South Africa is the leading producer of iron and steel in Africa, with iron and steel plants centered in the Johannesburg region. Coal comes from the Witbank and New Castle area, and iron ore comes from Thabazimbi and Sihen. Manufacturing is concentrated in the Johannesburg region, and the entire range of items associated with an industrialized country is produced. In addition to iron and steel, auto assembly from imported and domestic components, tractors, chemicals, textiles, and electronics are all manufactured. Because of a lack of petroleum resources, South Africa makes synthetic gasoline from coal. The cost of petroleum has focused world attention on South Africa's coal conversion efforts.

The wealth of South African industry and mineral resources is matched by the productive agriculture of the white farming areas. The High Veld is the leading area of crop and livestock production with corn the major crop. South Africa is also a major producer of cotton, grapes, tobacco, citrus fruits, and other agricultural products. The bulk of the wealth from the South African economy remains with the white minority population; very little goes to the African laborers. Nevertheless, black Africans in South Africa enjoy the highest per capita income in any African country except the oil-rich areas.

Zimbabwe presents a different picture from South Africa. The black majority population of Zimbabwe was dominated by the tiny white minority until the black majority replaced the white government of Rhodesia and renamed the country in 1980. Over half the white population fled, but unlike many other newly independent African nations Zimbabwe has begun successfully to transform its economy through the efforts of Africans. Unlike many African nations that have gone from a position of food surplus to one of import dependency in the past two decades, Zimbabwe has a surplus. This surplus has been achieved largely by increased production from small farms, many of which were created from large European estates after independence. While the large commercial farming sector is still dominated by white farmers, the small farmers who once accounted for only 10 percent of grain produced in Zimbabwe now produce more than 50 percent. Zimbabwe has accomplished what some describe as a modern agricultural miracle by recognizing that agriculture is the basis from which development efforts must proceed in Africa. The government has distributed land to many landless rural residents, guaranteed prices for small farmers at the same level as for large, and provided basic training and simple tools. Women farmers are given equal land rights and training as male farmers. Zimbabwe's accomplishments are an exception among the countries of Africa, but they demonstrate that Black Africans can develop their countries if given the opportunity.

INDUSTRIALIZATION ATTEMPTS IN AFRICA

One African country that seems to be beginning to transform itself is Nigeria. With a population of 118 million and a growth rate of nearly 3 percent, Nigeria faces many of the same problems as other less industrialized countries. Literacy is estimated at only 35 percent, and the variety and complexity of ethnic divisions have resulted in English being declared the official language. Because of its petroleum resources, Nigeria is now in the process of modernizing its economy, and per capita incomes rose to above $800 per year before petroleum prices fell. Since the end of

Huts for the residents of the black township of Leauadi in Bophuthatswana typify the poverty of the black homelands.

THE HUMAN DIMENSION

Removing the "Black Spots"

Kwa Ngema, Driefontein, and Mogopa are towns designated by the white South African government as "black spots" in the 87 percent of South Africa designated as white land. The black residents of numerous small rural settlements as well as specified urban districts are being removed to make the geography of South Africa conform to the government's allocation of territory on the basis of race. For the black residents it means leaving ancestral homes and land that may have been in the family for generations. It means removal to new locations selected by the white South African government, locations that rarely have the geographical amenities that prompted establishment of the black farms and towns originally. All of the new black towns are in the black homelands, where they represent a hidden geography. Hidden because they are known only to the government workers involved in the relocation. Hidden because they are out of sight and off-limits to even South Africa's white population.

Examine any atlas. See if you can find towns such as Indermah, or Onverwacht. These are black "towns" designed for over 100,000 people, yet they have none of the services commonly associated with towns. These towns, like other relocation settlements in the black homelands, have more in common with temporary refugee camps associated with people fleeing drought, famine, and war than they do with the urban geography they replace. As with the American Indians forced to relocate in harsh settings by the advancing European settlers, the blacks of South Africa forced to relocate to the homelands face a grim prospect. Uprooted from familiar space they are expected to recreate their human geography in inferior locations. Like the American Indian, the blacks of South Africa are losers in a confrontation with a technologically more advanced group.

The distribution of black African farms, homes, and towns may someday be replaced as Indermak, Onverwacht and other black towns are developed, but the disruption of African society and life will reverberate through generations yet unborn. In their efforts to deprive present and future generations of black South Africans of citizenship and human dignity the white South African government may well have only sown the seeds of greater conflict in the future. The American Indian represented a tiny minority, unable to ever rectify the loss of their lands. The black South Africans are a majority, a majority whose growing numerical dominance is reinforced by the rapidly growing population of the whole of Africa who view white South Africa as a "white spot." It is doubtful that the white spot of Africa can continue to indefinitely discriminate against the blacks who form the fabric of the continent's geography.

For urban blacks who are essential to the continued prosperity of white South Africa some of the more petty discrimination is ending. Races mix in shops and most public places, and beginning in 1985 the beaches and swimming pools of Capetown became the first to remove racial restrictions. Such minor changes should not obscure the fact that the great majority of the people of South Africa have or are in the process of losing their citizenship and can now only visit the white land of South Africa by permission.

---

the civil war in the early 1970s, Nigeria has developed a dual economy. Based on the free-enterprise, capitalistic model, it has resulted in an emergent middle and elite class who have access to the wealth and benefits from the oil revenue that the country has realized in the last decade.

A military government ruled the Nigerian people until the end of 1979, when a freely elected president took control of the country. The Nigerian political system is consciously modeled after that of the United States, with a congress, a president, and governors for the 19 Nigerian states, but the military seized power again in 1984. Economic development has been centered in the cities, with resultant migration to share in the perceived jobs and well-being found there. The largest city, Lagos, has a total population of between 4 and 5 million, many of whom are living in shanty towns constructed of tin, mud, and scrap lumber. Escape from such urban poverty, and the magnet attracting rural people, is education. The government has completed thirteen universities, but they are inadequate for the numbers desiring formal education. An estimated 20,000 Nigerian students are enrolled in American universities as a result of shortage of educational space in Nigeria.

The country is dedicated to industrialization, and major industries process resources or agricultural products—palm oil, peanuts, cotton, and petroleum. Manufacturing industries include textiles, cement, building materials, and food products. Approximately 54 percent of the active laboring population is engaged in agriculture. The main industrial centers are Lagos, Port Harcourt, and Benin City. The government is attempting to transform the outward focus of the former colony through the coastal cities to development of the interior. With a land area of 357,000 square miles (924,630 square kilometers), one-quarter of which is suitable for cultivation, Nigeria must develop the rest of its land area. As part of this effort, the government proposes to build a new capital near Nigeria's geographic center at Abuja.

The rapid increase in oil revenues in the 1970s and the migration of rural dwellers to the cities are creating havoc with Nigeria's traditional economy. Once able to feed its population, today Nigeria must import increasing amounts of food. Inflation makes the cost of living in cities such as Lagos among the most expensive in the world. Families without access to gardens or relatives in farming areas find it difficult to earn enough to buy sufficient food for their families. Falling petroleum prices in the decade of the 1980s caused Nigeria to experience a negative economic growth rate.

AFRICA'S CHALLENGES TOMORROW

The inflation that comes with development is but one of the challenges African countries must resolve. With economies tied to a limited number of agricultural or mineral exports, nearly every African country is vulnerable to market changes that result from technological changes, crop or mineral production in other continents, or world political affairs. Such vulnerability to external markets and market forces makes the African countries neocolonial holdings. Each country still trades extensively with its former colonial ruler, and so long as Africans rely on this trade, it is difficult for them to overcome poverty and illiteracy and to begin the process of development.

A second problem in Africa results from persistent environmental and social problems. The greatest social problem is that of population. At present rates of growth there will be more people in subsaharan Africa by 2015 than in China today. The average woman of Africa will have more than five children, and in Kenya and Burundi, eight. Population growth of this magnitude makes overcoming African poverty difficult, if not impossible. Growing populations mean that no amount of increase in agriculture can keep up. Between 1960 and 1988 total food production in Africa doubled, whereas the per capita production fell by 20 percent. Continued population growth is compounded by problems associated with refugees. There are an estimated 5 million refugees in Africa at the present time. The United Nations estimates 1 million refugees in Sudan and Somalia, and 500,000 in Zaire. Many of them have fled because of open warfare, which has threatened their families and their homes. Others have fled because of conflicts with the existing political system in their country or drought and famine. Whatever the cause, they represent a mass of human suffering rivaled only by the disasters in Southeast Asia. One-half of all the world's refugees are in Africa, many the result of the growing inability of African countries to feed their populations even in good years. (See box.) Famine remains a major factor forcing people to become refugees.

The existence of these displaced persons is symptomatic of another major African problem, the need to create strong, viable, stable governments. The wealth of Africa and the instability caused by ethnic fragmentation within individual countries has made creation of stable governments a difficult challenge. This challenge is made more difficult by the relics of colonialism and the expansion of neocolonial ambitions by the Soviet Union. Faced with recurrent revolution and overthrow of governments, plagued by poverty and underdevelopment, several African states have turned to the Soviet Union, China, or Cuban Communists for assistance. Although foreign Communist forces number less than 100,000, they constitute another source of friction and instability in Africa. Unless the African countries can freely develop their own political systems to a viable standard, it will be nearly impossible for them to begin resolving the basic problem of increasing the standard of living of their people.

The potential for conflict, whether from hostility between non-African forces, internal African problems relating to demands for power and representation by varying tribal groups, the demand for equality in racist governments such as South Africa, or simple revolution to attempt to increase the standard of living, constitutes one of Africa's greatest challenges. The intense poverty of the continent is beginning to be overcome. Countries from the Congo to Botswana, from Kenya to Nigeria are facing their transportation problems; provision of medical services is increasing through use of midwives and nurses; and development of manufacturing has started in nearly all African countries. Nevertheless, Africa remains the poorest continent in terms of the human standard of living, and this poverty, combined with the other problems facing Africa, raises the potential for conflict and suffering. The major challenge facing Africans is to increase the standard of living while harmonizing their traditional values with the needs and forces of the interconnected industrial world. Only as the African countries control their own destiny in a rational and coherent form will they be able to free themselves fully from domination by outside forces and resolve the poverty and underdevelopment that plague them. Achieving this goal may take many decades.

FURTHER READINGS

ADALENRO, I. A. *Marketplaces in a Developing Country: The Case of Western Nigeria*. Ann Arbor: University of Michigan Press, 1981.

"Africa's Woes." *Time*, January 16, 1984, pp. 24–41.

AKENBODE, A. "Population Explosion in Africa and Its Implications for Economic Development." *Journal of Geography* 76, 1977, pp. 28–36.

ALTSCHUL, D. R. "Transportation in African Development." *Journal of Geography* 79, 1980, pp. 44–56.

BARBOUR, K. "Africa and the Development of Geography." *Geographical Journal* 148, 1982, pp. 317–326.

BERNARD, F. and THOM, D. "Population Pressure and Human Carrying Capacity in Selected Locations of Machakos and Kitui Districts (Kenya)." *Journal of Developing Areas* 15, 1981, pp. 381–406.

BEST, A. C. G. "Angola: Geographic Background on an Insurgent State." *Focus* 26, 1976, pp. 1–8.

BEST, A. C. G. and DE BLIJ, H. J. *Africa Survey*. New York: Wiley, 1977.

BEYER, J. "Africa." In *World Systems of Traditional Resource Management*, pp. 5–37. Edited by G. Klee. New York: Halsted Press/V. H. Winston, 1980.

BROWN, LESTER, R. and WOLF, EDWARD. "Food Crisis in Africa." *Natural History*, June 1984.

BURLEY, T. and TREGEARR, P. *Africa Development and Europe*. New York: Pergamon, 1970.

CARTER, G. M., KARIS, T., and STULTZ, N. M. *South Africa's Transkei: The Politics of Domestic Colonialism*. Evanston, Ill.: Northwestern University Press, 1967.

CHRISTOPHER, A. *Colonial Africa: An Historical Geography*. Totowa, N.J.: Rowman and Allanheld, 1984.

———. *South Africa*. London: Longman, 1982.

CHURCH, R. J. H. *West Africa*. New York: Wiley, 1969.

CLARKE, J. I. and KOSINSKI, L. A. *Redistribution of Population in Africa*. London: Heinemann, 1982.

CRUSH, J. "The Southern African Regional Formation: A Geographical Perspective." *Tijdschrift Voor Economische en Sociale Geografie* 73, 1982, pp. 200–212.

DE BLIJ, HARM and MARTIN, ESMOND, eds. *African Perspectives: The Economic Geography of Nine African States*. New York: Methuen, 1981.

DOSTERT, P. *Africa 1984*, Washington: Stryker-Post Publications, 1984.

FAGE, J. "Slaves and Society in Western Africa, c. 1445-c. 1700." *Journal of African History* 21, 1980, pp. 289–310.

FAIR, T. *South Africa: Spatial Frameworks for Development*. Cape Town: Juta, 1982.

FINCHAM, R. "Economic Dependence and the Development of Industry in Zambia." *Journal of Modern African Studies* 18, 1980, pp. 297–313.

FLOYD, B. and TANDAP, L. "Intensification of Agriculture in the Republique Unie du Cameroun." *Geography* 65, 1980, 324–327.

FRANKE, R. W. and CHASIN, B. *Seeds of Famine*. Montclair, N.J.: Alanheld, Osmun, 1980.

"Geographical Perspectives on the Crisis in Africa (Symposium)" *Geography* 73, January 1988, p. 47–73.

GOLIBER, THOMAS J. "Sub-Saharan Africa: Population Pressure on Development," *Population Bulletin* 40, 1985, pp. 1–46.

GOUROU, P. *The Tropical World: Its Social and Economic Conditions and Its Future Status*. 5th ed. Translated by S. Beaver. London: Longman, 1980.

GRAUBARD, S. et al. "Black Africa: A Generation After Independence." *Daedalus* 111, 1982, pp. 1–273.

GRIFFITHS, I., ed. *An Atlas of African Affairs*. London: Methuen, 1984.

GROVE, A. *Africa*. 3d ed. Oxford: Oxford University Press, 1978.

GROVE, A. T. *Africa South of the Sahara*. London: Oxford University Press, 1970.

HANCE, W. A. *Population, Migration and Urbanization in Africa*. New York: Columbia University Press, 1975.

———. *The Geography of Modern Africa*. New York: Columbia University Press, 1975.

HARRISON, PAUL. "A Green Revolution for Africa," *New Scientist*, May 1987, pp.35–59.

HIBBERT, C. *Africa Explored: Europeans in the Dark Continent, 1769–1889*. London: Allen Lane, 1982.

HIDORE, J. J. "Population Explosion in Africa: Some Further Implications." *Journal of Geography* 77, 1978, pp. 214–220.

HULL, GALEN SPENCER. *Pawns on a Chessboard: The Resource War in Southern Africa*. Washington: University Press of America, 1981.

KAHIMBAARA, J. A. "The Population Density Gradient and the Spatial Structure of a Third World City: Nairobi, A Case Study." *Urban Studies* 23, August 1986, pp. 307–322.

KESBY, J. D. *The Cultural Regions of East Africa*. London: Academic Press, 1977.

KILSON, M. L. and ROTBERG, R. I., eds. *The African Diaspora: Interpretive Essays*. Cambridge: Harvard University Press, 1976.

KLOOS, H. and THOMSON, K. "Schistosomiasis in Africa: An Ecological Perspective." *Journal of Tropical Geography* 48, 1979, pp. 31–46.

KNIGHT, C. G. and NEWMAN, I. L. *Contemporary Africa: Geography and Change*. Englewood Cliffs, N.J.: Prentice-Hall, 1976.

LEARMONTH, A. *Patterns of Disease and Hunger*. London: David and Charles, 1978.

LELE, U. "Rural Africa: Modernization, Equity, and Long-Term Development." *Science* 211, 1981, pp. 546–553.

LEVI, J. and HAVINDEN, M. *Economics of African Agriculture*. Harlow, England: Longman, 1982.

LEWIS, LAWRENCE and BERRY, L. *African Environments and Resources*. Boston: Allen & Unwin, 1988.

MOSLEY, P. "Agricultural Development and Government Policy in Settler Economies: The Case of Kenya and Southern Rhodesia, 1900–1960." *Economic History Review* 35 1982, pp. 390–408.

Natural Resources and the Human Environment for Food and Agriculture in Africa. Rome: Food and Agriculture Organization of the United Nations, 1986.

NWAFOR, J. "The Relocation of Nigeria's Federal Capital: A Device for Greater Territorial Integration and National Unity." *GeoJournal* 4, 1987, pp. 359–366.

O'CONNOR, A. *The Geography of Tropical African Development: A Study of Spatial Patterns of Economic Change Since Independence*. 2d ed. Elmsford, N.Y.: Pergamon, 1978.

———. *The African City*. New York: Holmes and Meier, 1983.

O'CONNOR, A. M. *The Geography of Tropical Africa Development*. New York: Pergamon, 1971.

OLIVER, R. and CROWDER, M., eds. *The Cambridge Encyclopedia of Africa*. Cambridge: Cambridge University Press, 1981.

OGUNTOYINBO, J. "Climatic Variability and Food Crop Production in West Africa." *GeoJournal* 5, 1981, pp. 139–149.

OSEI-KWAME, P. *A New Conceptual Model of Political Integration in Africa*. Lanham, Md.: University Press of America, 1980.

PAGE, MELVIN E., ed. *Africa and the First World War*. Houndmills, United Kingdom: Macmillan, 1987.

PIRIE, G. "The Decivilizing Rails: Railways and Underdevelopment in Southern Africa." *Tijdschrift Voor Economische en Sociale Geografie* 73, 1982, pp. 221–228.

PIRIE, GORDON H. "Urban Population Removals in South Africa." *Geography* 68, October 1983, pp. 347–349.

POMEROY, D. *Tropical Ecology*. New York: Wiley, 1988.

PROTHERO, R. M. *People and Land in Africa South of the Sahara*. New York: Oxford University Press, 1972.

RICHARDS, P. "The Environmental Factor in African Studies." *Progress in Human Geography* 4, 1980, pp. 589–600.

SMITH, D. M., ed. *Living Under Apartheid*. Winchester, Mass.: Allen & Unwin, 1982.

SMITH, DAVID MARSHALL. "Conflict in South African Cities." *Geography* 72, April 1987, pp. 153–158.

"Special Representative Named for Sudan, With Over 11 Million Affected by Drought; Camel Caravan Aids Isolated Tribesmen in Drought-Affected Sudan." *UN Chronicle* 22, No. 5, 1985, pp. 56–59.

STAMP, L. C. and MORGAN, W. T. M. *Africa: A Study in Tropical Development*. New York: Wiley, 1972.

STOCK, ROBERT. "Disease and Development 'or the Underdevelopment of Health': A Critical Review of Geographical Perspectives on African Health Problems." *Social Science and Medicine* 23, No. 7, 1986, pp. 689–700.

TAYLOR, A., ed. *Focus on Africa South of the Sahara*. New York: Praeger, 1973.

THOMPSON, L. and BUTLER, J., eds. *Change in Contemporary South Africa*. Berkeley: University of California Press, 1975.

UDO, R. K. *The Human Geography of Tropical Africa*. Exeter, N.H.: Heinemann, 1982.

UNGAR, SANFORD J. *Africa: The People and Politics of an Emerging Continent*. New York: Simon & Schuster, 1985.

WATTS, MICHAEL. *Silent Violence: Food, Famine, and Peasantry in Northern Nigeria*. Berkeley: University of California Press, 1983.

WELLINGTON, J. H. *South West Africa and Its Human Issues*. London: Oxford University Press, 1967.

WILLIS, DAVID K. "Africa: Blueprint for Survival," *Christian Science Monitor* Reprints, October 1985.

WINTERS, C. "Urban Morphogenesis in Francophone Black Africa." *Geographical Review* 72, 1982, pp. 139–154.

Andean village in Peru.

LATIN AMERICA PROFILE: 1990

Population (in millions)	447	Energy Consumption Per Capita—	
Growth Rate (%)	2.2	pounds (kilograms)	2866
Years to Double Population	32		(1300)
Life Expectancy (years)	66	Calories (daily)	2550
Infant Mortality rate:		Literacy Rate (%)	79
(per thousand)	57	Eligible Children in Primary	
Per Capita GNP	$1720	School (%)	95
Percentage of Labor Force		Percent Urban	68
Employed in:		Percent Annual Urban Growth,	
Agriculture	37	1970–1980	2.3
Industry	28	1980–1985	3.5
Service	35		

17

Latin America focus of change in the Western Hemisphere

MAJOR CONCEPTS

Land tenure / Ejidos / Encomienda
Encomendero / Haciendas / Junta
Population explosion / Altitudinal zones
Hollow frontiers / Effective national territory
Latifundia / Minifundia / Structural changes
Cultural changes / Revolutionary changes
Effective National Territory Bagasse

IMPORTANT ISSUES

Economic domination of resource development by foreign nations

Cultural conflicts created by diffusion of European culture

Extension of southern European institutions of aristocracies, one-party system, army, and church

Agricultural exploitation by landed gentry and developed countries

Economic susceptibility caused by Latin American nations' reliance upon a few exports

Geographical Characteristics of Latin America

1. Latin America is a part of the less industrialized world, but the general level of development is higher than in Africa or Asia.
2. Rapid population growth and related problems associated with urban growth, inflation, education, and employment pose serious challenges for the economy of the region.
3. Latin America was the focus of colonization by Spain and Portugal after the voyage of Columbus, and today southern European cultural traits of language and religion dominate the region.
4. The cultural groups found in Latin America represent a combination of Native American, European, African, and Asian ancestry, with the countries of the Andes having the highest proportion of Native Americans.
5. Brazil, which dominates Latin America in terms of area, population, and economy, still presents the dichotomy between rich and poor typical of countries that are only beginning to move to an industrial economy.

Latin America refers to the American landmass south of the United States border with Mexico (Figure 17.1). For most residents of Anglo-America, and particularly those of the United States, the term *America* is synonymous with the United States or, generously, the North American continent. Rarely do Anglo-Americans recognize that Latin America is also America, and the people of Latin America are Americans. Although both are American, the contrasts between Anglo-America and Latin America are numerous and pervasive, ranging from climate to culture; from standard of life to life expectancy. Together the contrasts between Anglo-America and Latin America are significant because nowhere in the world is there such a marked and abrupt change from the industrial, wealthy industrial world to the relatively poor less industrialized world as that between the United States and Latin America at the U.S.–Mexican border.

In general, Latin American countries have the characteristics of the less industrialized realm. The climates are primarily tropical, the majority of the landmass is located in the Southern Hemisphere, there is a high incidence of poverty in most of the independent countries, birthrates remain high, the disparity between the tiny, wealthy elite and the bulk of the population living at the margins of existence is an unbridgeable gulf, and the dominance of primate cities and the existence of dual economies is the norm rather than the exception. The standard of living varies greatly from country to country within Latin America, with some countries like Mexico, Brazil, or Argentina newly industrialized, while others such as Bolivia or Guatemala are among the poorest of the less industrialized countries. The region as a whole is more industrialized than either Africa or Asia (excluding Japan), and industrialization has proceeded rapidly in many places in the post–World War II era.

Understanding the relative level of development in Latin American countries and the changes and challenges that the future will bring necessitates recognition that Latin America is not a homogeneous unit but a composite of independent states with differing geographical sizes and resources, levels of economic development, traditions, cultures and attitudes (Table 17.1). The seeming homogeneity suggested by *Latin* reflects the role of Spain and Portugal, relics of the Roman Empire with their Latin languages, in colonization and colonial dominance of the area. Latin America is generally divided into three regions: Middle America from Mexico through Panama, the South American continent, and the islands of the Caribbean. The designation of the whole region as *Latin* America hides the important diversity of cultures and economies throughout the region.

Variation in economic development is illustrated by Mexico, Brazil, and Argentina representing industrializing, technologically advanced countries that are rapidly progressing to full-scale industrial states.

Figure 17.1 Countries and major cities of Latin America.

TABLE 17.1 Comparisons of Latin American Regions

Region	1990 Population (Millions)	Rate of Natural Increase (%)	Birthrate (per 1,000)	Death Rate (per 1,000)	Life Expectancy	Per Capita Income
Caribbean	34	1.8	26	8	68	$1620
Middle America	116	2.5	32	8	66	1640
Middle-Latitude South America	49	1.5	23	8	70	2050
Tropical South America	248	2.2	30	8	65	1680

SOURCE: World Population Data Sheet (Washington, D.C.: Population Reference Bureau, 1988).

Each country has a substantial industrial sector although agriculture remains the major source of employment. Some countries, including Venezuela, Ecuador, Trinidad-Tobago, and Mexico, have substantial petroleum reserves to fuel their economies. Some countries such as Chile, Bolivia, Brazil, and Mexico have important mineral resources for export. All the countries in Latin America have extensive agricultural sectors, with production of tropical foods and products dominating the export economies of the individual countries. Fifteen commodities are presently responsible for one-half of the total exports by value for the entire region (Table 17.2).

TABLE 17.2 Exports of Latin America (1986)

Country	Major Export	% of Country's Total Exports
Bolivia	Natural Gas	51.5
Brazil	Tobacco	21
	Coffee	9.2
Chile	Copper	46.1
Colombia	Coffee	62.3
Costa Rica	Coffee	34.45
Cuba	Sugar	76
Dominican Republic	Sugar	18.5
El Salvador	Coffee	66.7
Ecuador	Oil	41.7
Guatemala	Coffee	32.2
Haiti	Coffee	33.4
Honduras	Coffee	37.7
	Bananas	30
Peru	Oil	17.4
Trinidad/Tobago	Oil	79.1

SOURCE: Europa Yearbook 1988 (London: Europa Publications Limited, 1988).

The heavy reliance on export of staples to the United States, Western Europe, and Japan reflects the pattern of economies in less industrialized countries. It dates from the European dominance of Latin America from the time of Columbus until the eighteenth century, and the quasicolonial relationship of Latin America to the United States since the mid-nineteenth century. Latin America was the first region of the world to which the Europeans expanded their political control. The Spanish and Portuguese set the example for colonization that was followed by other European nations in subsequent centuries as they agreed to a division of the Americas for exploitation. The Treaty of Tordesillas (1494) divided the Western Hemisphere at 50 degrees west longitude. East of that line Portugal was to have possession, and lands to the west of that line were to be Spanish.

The Spanish obtained the portion of Latin America with the complex civilizations of Indians, highlighted by the Aztecs and Mayas of Middle America and the Incas of the Andes. Wealth obtained from these Indian civilizations aided the development of Europe in general and prompted further exploration in hopes of obtaining wealth for the ruling country. Development by Europeans from 1500 to the independence in the early 1800s emphasized production of goods with a high value: sugar, coffee, cacao, silver, and gold. Relics of the colonial era are found in a variety of areas, but the reliance upon a few tropical crops and minerals for export earnings is one of the most important.

While the European powers were occupied with the internal conflict of the Napoleonic wars, their colonies in Latin America were able to declare themselves independent and began the process of development. The new countries reflected the administrative boundaries of the Spanish and Portuguese colonial powers and the infrastructure of the beginning of the nineteenth century. Subsequent to their declaration of independence, the emerging United States proclaimed the Monroe Doctrine of 1823, which has prevented the colonial activities carried on by Europeans in Africa, Asia, and the Pacific from becoming dominant in Latin America. The Monroe Doctrine also had the side effect of making Latin America a quasi-colony of the United States.

The individual countries of Latin America had only slow development prior to World War II as the emergent countries simply crystallized the social organizations and structures that existed under the Spanish and Portuguese. The past four decades have brought major changes in the level of economic activity in individual Latin American countries as a result of changing political, social, and economic organizations and attitudes.

THE BEGINNINGS OF MODERNIZATION IN LATIN AMERICA

The major changes in Latin America at the present time offer promise of an increasing standard of living for the half billion people who will occupy the region at the turn of the century. Advances in the industrial sector in this period have made Brazil and Mexico important producers of iron and steel, automobiles, and a broad variety of manufactured products associated with the industrial world. Table 17.3, which shows the change in population and labor force in Latin America since 1950 and projects it to the year 2000, gives an indication of the magnitude of this change. The population engaged in agriculture has declined for the entire region, while that in industry has increased significantly. Measured in constant 1978 dollars, the gross domestic product (the total value of all goods manufactured and produced) nearly tripled between 1960 and 1979. During the same period the average per capita GNP, the value added by manufacturing, and foreign trade increased proportionally. Unfortunately growth rates in the 1980s have been lower, and accompanied by high inflation (Table 17.4). In many of the countries of Latin America the rapid industrialization has been accompanied by an actual decrease in the quality of life of individual citizens. In Mexico, for example, the income of the average worker as measured by purchasing power has declined by more than half in the last 10 years.

The increase in industrial productivity has been associated with a rapid increase in the standard of living for Latin America in general. Death rates for the entire region have fallen dramatically, while life expectancy at birth has increased. Lower infant mortality rates reflect the decline in death rates and the increase in life expectancy (Figure 17.2). Provision of safe water has done much to add to the increasing health standards, with an increase from 60 percent to 78 percent of urban residents and an increase from 6 to 34 percent of rural residents receiving safe drinking water. These gains have taken place in a period when the population has nearly doubled, making the total number of people receiving treated water even more significant.

School enrollments and literacy rates have increased in the past 20 years, with the enrollment rate among children in primary grades (one to six) increasing from 50 percent in 1960 to over 90 percent in all but two countries in 1990. In the same time span the proportion of high school age youth attending school increased from 13 to 40 percent, and those attending universities increased from 2 to 4 percent of the 18- to 25 year age group. In 1960 illiteracy rates in the seventeen countries of Latin America varied between 8 and 62 percent. In four countries, more than half the population was illiterate, and in twelve more than a quarter. By 1990 illiteracy rates had dropped in all countries, and only in Haiti was the rate more than half the population. The increasing literacy of Latin American countries and related gains are a major accomplishment. Nonetheless, poverty, unemployment, and malnutrition continue to confront significant portions of the region's population, particularly in rural areas. The agricultural sector, which still employs 37 percent of the total population of Latin America, has shown the least progress, with illiteracy and poverty concentrated among rural residents or the slum-dwelling migrants from rural areas.

Latin America continues to face the serious problems of rapid population growth typical of less indus-

TABLE 17.3 Population and Labor Force, Estimates and Projections, 1950–2000

Total (thousands of persons)	Population	Labor Force
1950	154,295	54,178
1970	269,140	84,948
1980	354,945	111,808
2000	596,858	189,041
Cumulative growth (%)		
1950–1970	74.4	56.8
1970–1980	31.9	31.6
1980–2000	68.2	78.0
Growth rates (% per year)		
1950–1970	2.82	2.27
1970–1980	2.81	2.79
1980–2000	2.63	2.93

SOURCE: *Development Forum* 9 (April 1980) (New York: United Nations), p. 12.

TABLE 17.4 Inflation in Major World Regions, 1969–1988

	Average 1969–78	1980	1982	1984	1986	1988
Industrial Countries	7.8	9.3	7.3	4.3	3.4	3.4
United States	6.7	9.1	6.5	3.9	2.6	3.4
Other industrial countries	8.6	9.4	7.7	4.5	4.1	3.4
Developing Countries	16.7	27.2	25.0	39.4	28.6	29.5
By Region						
Africa	11.6	16.4	11.4	20.3	14.8	10.5
Asia	8.7	13.1	6.3	7.2	5.9	5.6
Europe	11.7	37.9	23.7	28.0	27.4	16.6
Middle East	10.8	16.8	12.7	14.9	11.1	9.9
Latin America	31.0	54.6	68.4	129.3	86.5	98.8

SOURCE: *World Economic Outlook, 1988* (Washington International Monetary Fund, 1988), p. 125.

trialized countries (Figure 17.3). The population growth rate in Latin-America is 2.2 percent per year, presenting all the individual countries with the entire gamut of population problems. Of critical importance is the overwhelming number of young people in the population, which ensures continued high rates of population growth. For Latin America as a whole 38 percent of the population is age 14 or younger. In 1990 there were a total of 170 million children in the region. The magnitude of this problem is illustrated by the 80 million school children between the ages of 6 and 12. Providing educational opportunities for their burgeoning population is a critical problem throughout the Latin American region. By the end of the twentieth century, there will be more than 185 million children under the age of 15 if present trends and population projections are accurate. Over 12 million additional births are recorded in the region each year, and even though the birthrate per thousand continues to decline, declines in the death rate maintain the growth rate.

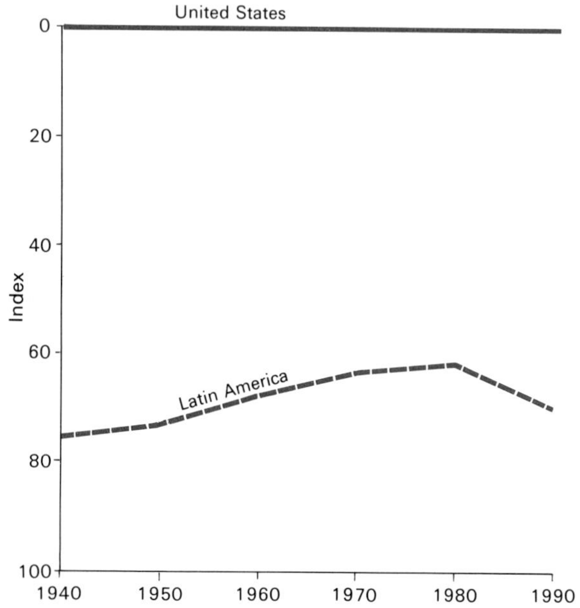

Closely related to the problems of providing adequate education for their growing populations is the countries' need to provide employment opportunities. With an estimated increase of the labor force from 54 million in 1950 to nearly 120 million at the turn of the century, providing jobs is a critical issue for the Latin American countries. The demand for jobs and a better quality of life contributes to mass rural-to-urban migration throughout the region, particularly to the primate cities. The dominance of the primate city in each country is illustrated by the fact that of the total urban population of Latin America, 75 percent live in cities of more than 100,000 inhabitants, and in most countries over half of the urban population resides in the single largest city. By the end of this century, there will be nearly 50 cities in Latin America with a population of 1 million, and four of them (Mexico City, Buenos Aires, São Paulo, and Rio de Janeiro) will rank among the ten largest cities in the world (Table 17.5).

Other problems facing the developing Latin American region are the traditional ones of food and nutrition for the poor, political instability and demand for revolutionary change, and the gulf between rich

Figure 17.2 Standard of living of Latin America compared to that of the United States. (0 = no gap in standard)

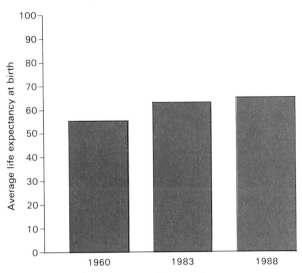

Figure 17.3 Changes in population indices in Latin America indicate the general improvement in quality of life in the region.

and poor in each country. The middle class of most Latin American countries is small but growing, but the continued dominance by a wealthy elite is a major obstacle to development inherited from the centuries of European control and exploitation.

DISCOVERY AND CONQUEST: EUROPEAN EXPANSION INTO LATIN AMERICA

The discovery and conquest of Latin America by Spanish and Portuguese explorers at the end of the fifteenth century reflected advances in navigation, sailing technology, and knowledge of the world's geography combined with European desire to find new routes to the spices of Asia. While the Portuguese were attempting to sail around Africa to reach Asia, Columbus suggested that Asia could be reached by sailing west. His attempts to gain support for such a journey of exploration in the 1480s were dismissed by the Portuguese, who recognized that his estimate of the size of the earth was grossly inaccurate. Columbus then attempted to obtain support for such a voyage from the Spanish. Spanish scholars recognized his inaccurate assessment of the distance to Asia, but Queen Isabella ultimately supported his famed journey of exploration in the fall of 1492. He believed his landfall on the island of San Salvador in the Bahamas justified his perception of the world's geography, and began a process of European expansion that continued until after World War II.

Columbus' return to Spain with some gold, spices, and American Indians provided the impetus for a larger expedition of colonization the following year. The Spanish viewed the New World as a potential source of wealth and power. To direct the exploitation of the new lands, the Council of Indies was established in Spain to prepare the guidelines under which individuals were granted rights of exploration, conquest, and colonization. One of the principles the council developed was establishment of the **encomienda** system

Encomienda: A large grant of land including its Indian occupants. It was designed to "civilize" the Indians and protect them from European exploitation by placing them under the protection of a Spaniard who was to teach, feed, and clothe them in return for their labor.

TABLE 17.5 Growth of Major Urban Centers in Latin America

	Estimated Population (in millions)					
City	1950	1960	1970	1980	1990	2000
Mexico City	3.1	5.2	9.2	15.0	21.3	26.3
São Paulo, Brazil	2.8	4.9	8.3	12.8	18.8	24.0
Rio de Janeiro, Brazil	3.5	5.1	7.2	9.2	11.4	13.3
Buenos Aires, Argentina	5.3	7.0	8.5	10.1	11.7	13.2
Lima, Peru	1.1	1.7	2.9	4.6	6.8	9.1

SOURCE: United Nations, Department of International Economic and Social Affairs, *Estimates and Projections of Urban, Rural, and City Populations, 1950–2025*. (New York: United Nations, 1985).

of land tenure to protect the Indians while providing initiative to individuals in Spain to undertake the hazardous task of exploration and colonization.

The Indians of Latin America held their lands in common, and while the encomienda system did not technically grant this land to a Spaniard, in the Spanish view, lack of private ownership signified absence of a bona fide owner and over time the Spaniard receiving the encomienda grant gained absolute control over much of the land. In turn the Indians on the lands granted to the **encomendero** were required to labor on their land for his benefit. The encomendero was required to teach the Indians European culture, the most important aspect of which was the Roman Catholic religion.

Failing to find abundant supplies of gold in the Caribbean islands, the Spanish developed an agricultural economy, but the introduction of European diseases to which the Indians had no resistance decimated the population. Smallpox killed over half the Indians as it spread through the population. The European colonists continued their search for gold and reached the mainland around Vera Cruz, Mexico, in 1518. Under the leadership of Hernán Cortes, a small group of soldiers moved inland to the basin of Mexico where today's Mexico City is located and discovered the Aztec civilization with its high degree of political organization and wealth. Cortes' success in conquering the Aztec federation provided the Spanish with the wealth they had been seeking in the New World.

Cortes and his small groups of soldiers were successful in defeating the Aztecs for a variety of reasons, including the Spanish superiority in military technology, especially their use of the horse; the political organization of the Aztec federation, in which a variety of peoples were required to pay taxes and tributes to a strong central government; and according to some observers the Aztec government's weakness and a lack of strong military commitment to the central power that the government had enjoyed in the previous century.

The Spanish replaced the Aztec rulers as the leaders of a politically organized Indian population of more than 12 million in the highlands of Mexico. The Spanish quickly discovered the mines that were the source of the gold that had been paid in tribute to the Aztec leaders and utilized the encomienda system to provide labor to expand mining activities. As in the Caribbean, the Spanish impact was disastrous, as epidemics of smallpox, measles, and typhus fever decimated the Aztec population. By the end of the sixteenth century the Indian population is estimated to have fallen below 3 million.

Cortes' success prompted a host of other Spanish explorations to find the wealth they were convinced existed in the New World. Expeditions were undertaken northward and southward from the Mexican highlands, but the only other discovery of similar magnitude was that of the Inca empire in Peru by Pizarro in 1532. The Spanish found in Peru an empire that was even more sophisticated than that of the Aztecs in Mexico. The mineral wealth of the highlands of the Andes Mountains had been accumulated by the Inca rulers at their capital at Cuzco, and the Spanish were able to replace the Inca leaders as the controlling agents of the Indians of the Andean highlands. As in other areas settled by the Spanish, the encomienda system granted control of Indians and their lands to Spanish nobles and soldiers, but the majority of the mineral wealth was exported to the home country. The impact of the Spanish on the Inca civilization was disastrous. Under the Incan emperor, a civilization had been developed that provided the basic needs of all of the people through a strong central government that enforced production of food surpluses. European diseases destroyed nearly half the population, while the Spanish emphasis on production of valuable metals siphoned labor from the agricultural sector into the mining sector and disrupted food production.

Ultimately the Spanish pushed their explorations southward into Chile, Argentina, and Paraguay and northward into the western and southern United States. To the east the Spanish found Indians who were not in organized groups and not easily subdued, and a tropical climate in the lowlands of the Amazon. In combination they prevented Spanish expansion. The Portuguese explorers who landed on the east coast of Brazil found no large Indian civilizations that could be subjugated. The settlement pattern of the Portuguese and Spanish in the New World became one of enclaves. Enclaves grew up in port locations around the perimeter of the continent, relying on items produced in the local area for export to the Old World, or linked to another series of interior population enclaves based on the mines or agricultural production of Indian populations predating the European conquest (Figure 17.4).

THE EUROPEAN IMPACT UPON LATIN AMERICA

The Europeans completely changed the society and economy of the Indians. The decline of the Indian population in the Andes, in Mexico, and in the Carib-

Encomendero: The Spaniard to whom an encomienda was granted.

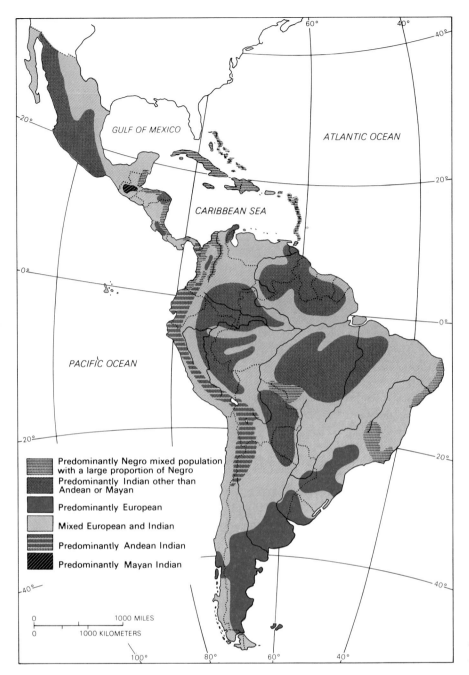

Figure 17.4 Population groups in Latin America.

bean islands had a profound impact upon the development of the entire region. The transfer of landownership from Indian villages to Spanish grantees provided the basis for land tenure problems that have yet to be resolved in most Latin American countries. Beginning with the encomienda system, and later supplemented by outright grants of large tracts of land to Spaniards, much of the best land was divided into large estates owned by Spaniards. The legacy of these grants are today referred to in South America as **latifundia** (or as **hacienda** in Mexico). The Indian population in the rural sector is concentrated on **minifundia**.

Latifundia: The latifundia are large estates owned by individuals of European descent, who typically control most of the wealth in the individual countries.

Hacienda: Another term referring to large estates in Latin America, commonly used in Mexico.

Minifundia: The minifundia are tiny farms that provide a subsistence existence upon which a significant portion of the Indian population of Latin America depends.

A religious festival in Peru illustrates the role of the church in the lives of the people of many of the countries of Latin America.

The Europeans also transformed the cultural values and patterns of Latin America. The European conquerors found a continent and its islands dominated by a wide variety of self-sufficient Indian groups. In their desire to obtain riches from their colonies they turned to production of tropical crops for the world market. World demand for sugar prompted development of sugar production in the Caribbean Islands and along the northeast coast of Brazil. The decimated Indian population was too small to provide sufficient labor for these plantations, so the Europeans turned to slave labor from Africa, adding another major cultural element to Latin America. The Europeans themselves remained a minority of the population throughout Latin America, since their overwhelming attitude was one of exploitation with the goal of someday returning to Spain or Portugal. A European-born elite persisted in all Latin American countries until after independence, and even today the European component of population in each Latin American country tends to have most of the wealth, political power, and prestige.

The European desire to exploit their colonies in the New World has also handicapped development by fostering a boom-and-bust psychology. In response to world market conditions, the individual Latin American countries exploit one commodity during times of boom and abandon it when prices fall. A wide variety of crops and minerals ranging from bananas to silver, from iron ore to petroleum has gone through this cycle of exploitation, and in each case it has tended to handicap long-term development.

In addition to cultural changes from the introduction of Europeans and Africans to Latin America, the Spanish and Portuguese introduced the Catholic religion. The close relationship of the rulers of Spain and Portugal with the Roman Catholic church prompted a commitment to bringing the Catholic version of Christianity to the indigenous population of America. The role of the Catholic missionaries in teaching the Indians represents one of the great cultural diffusions of the last millennium. The old religions of the large Indian empires were modified to an external observance of Catholicism, but all retained important elements of their old beliefs as well. Outwardly however, Catholicism is the dominant cultural feature throughout Latin America, providing an element of homogeneity. The construction of elaborate cathedrals in all major and many minor cities has provided some of the most magnificent architecture in Latin America today. In the process the wealth of the local Indians may have been severely diminished, and certainly the adoption of the Catholic values system as taught to the Indians has had a negative impact on economic development. The Catholic system as it emerged in Latin America in the past taught submission to leaders, fatalistic acceptance of social conditions, and recognition of the immutability of the social system with its extremes of wealth and poverty.

The final major cultural impact of the Spanish was in the organization of towns. Under the Council of Indies guidelines for cities were established under guidelines that called for a central plaza with the church, government, and military headquarters around it. The

THE WORLD OF POVERTY

city was laid out in a rectangular grid pattern, resulting in an urban form that facilitated the primary function of the Spanish towns, administration of the lands and Indians around them.

CONTRIBUTIONS OF LATIN AMERICA TO THE WORLD

The riches that the New World sent to Spain and Portugal in the form of gold, silver, and other valuable metals had a negative impact upon the economies of Spain and Portugal. Increasing wealth prompted expansion of military capabilities and ultimately led to disastrous wars and conflicts, which drained the treasuries of both Spain and Portugal.

The New World also provided other forms of wealth of worldwide significance in the form of domesticated plants and animals that are the basis for the livelihood of many people even today. Scores of plants and animals were domesticated in Latin America, but five have diffused widely and are of worldwide significance: corn (maize), peanuts, manioc, sweet potatoes, and several varieties of beans. Corn was apparently domesticated in the Andes but was the staple crop throughout Latin America and among the Indians of North America by the time of Columbus' voyage. Corn has a number of advantages that have made it of worldwide importance in both production and export, ranking after wheat and rice as the third most important grain produced in the world. Corn has a high protein and oil content, mature grain stores well, and with fertilization it produces abundantly even in tropical climates. It is well suited to slash-and-burn agriculture, since it can be cultivated on an individual-plant basis. The bulk of the world's corn, however, is grown in the industrial countries of the world where most of it is fed to animals.

Peanuts were first domesticated in Brazil or Paraguay but were grown in Peru and other areas of Latin America in pre-Columbian times. Peanuts are important because they are high in protein, do well in tropical and subtropical conditions, and are a major source of oil. The protein-rich peanut and peanut oil are major sources of food in parts of Asia and Africa. Manioc (cassava) came from Brazil and after Columbus' voyage was spread by the Portuguese to much of Asia and Africa. Manioc grows well in tropical conditions and has the added advantage of being harvestable over an extended time period. It is primarily grown for subsistence, but some Asian countries produce manioc flour, which is exported to make tapioca for use in the industrial countries. At the present time, it is a major component of the diet of tens of millions of people in Asia, Africa, South America, and the Pacific islands.

Both the white potato and the sweet potato originated in Latin America. The sweet potato is important because it is higher in vitamins and minerals than the staple crops of the less industrialized countries of Asia, Africa, and Latin America. It is highly palatable but requires relatively fertile soils and is used as a dietary supplement to rice, corn, and cassava throughout the less industrialized world. It is particularly useful because it grows in subtropical conditions where the white potato will not. The white potato is apparently a native of the Andes. It was domesticated in the highlands and grows well in cool, damp climates; thus it was rapidly adopted in Europe. Potatoes yield high tonnage and are useful for both food and production of industrial products such as alcohol. Today the major producers are in northern and western Europe and the Soviet Union.

Important varieties of beans domesticated in Latin America include the kidney, lima, navy, pinto, string beans, and other garden beans. Beans have a high protein content and in the less industrialized world often provide most of the protein of the diet. In the industrial world, where meat is more readily available, beans are less significant, but for the world as a whole they represent a major contribution of Latin America.

Other plants that came from Latin America and have had worldwide impact include cotton, tobacco, cacao (cocoa), and rubber. The rubber tree used in Latin America was not a true domesticate, but the Indian use of latex from the wild rubber trees was the basis for subsequent tapping by the Spanish in the 1700s and ultimate development of the rubber industry. From Latin America rubber trees were diffused to Southeast Asia where they became an important part of the European colonial economy. Cotton was known in the Old World prior to the voyage of Columbus—it was grown in ancient Egypt—but the New World varieties account for most cotton produced in the world today. Cacao was grown in Latin America and used to make a drink important in Aztec religious ceremonies. Cacao trees are native to Central and South America and were diffused to Africa where the world's major production now occurs. Tobacco was being used by the Indians of Caribbean and other areas of both North and South America at the time of Columbus, and in subsequent centuries it has spread throughout much of the world. It represents a major crop in many regions, often displacing food crops which are in short supply.

The majestic architectural relics of the Mayan, Incan, and Aztec cultures represent some of the wonders of the world. The art forms developed by these civilizations have unique components and styles not found in other regions. The relics of the pre-Columbian era represent a major cultural heritage of the present-day countries of Latin America. Economically they are important as magnets for tourists from the industrialized world.

The ruins at Machu Picchu in Peru are a major tourist attraction. Their size and apparent past function are indicative of the high level of development among Latin American peoples long before contacts with European peoples.

MODERN LATIN AMERICA: VARIATION IN DEVELOPMENT

Latin America epitomizes the complexity of the developmental process and the subtle and intricate interrelationship of the wide range of variables. Economic development seems to involve the physical resources of a country; the relative location and accessibility of a place; the attitudes; social, political, and economic organizations; and the time at which development efforts are undertaken. Some argue that the primary ingredient in successful developmental efforts involves *structural changes*. The structural view of development argues that it occurs as changes are made in formal structures such as land tenure, political organization, or economic relationships with other areas. Other observers maintain that changes in *cultural* values are the primary determinants of development: a strong work ethic and emphasis on individualism are required for success. Still others maintain that development is the result of *revolutionary changes*, arguing that no development can occur until the existing social-political order is overthrown and replaced by a completely new order and value system. Compounding this argument is the relative importance of resources and trade relationships in development processes. What is certain is that no one single element is the primary determinant of development in all cases, as the relative importance of each variable is a function of the time and place (geography) of an individual country. Examination of the Latin American experience illustrates the disparate nature of the variables in the developmental process.

THE LATIN AMERICAN ENVIRONMENT

The Latin American environment is highly complex and diverse, but the landforms can be simply described. A high mountainous backbone extends from the United States border to the tip of Cape Horn, two major river basins are separated by lower hills, and an arc of islands borders the Caribbean (Figure 17.5). The mountain chain is the dominant physical feature, with the Andes stretching 4500 miles (7200 kilometers) from the tip of southern South America to the northern part of Colombia, with other mountains extending through Central America and Mexico. The Andes are high mountains, with elevations in excess of 22,000 feet (6,600 meters) in the southern portion, and elevations of over 18,000 feet (5,400 meters) common in many locations. They provide an important barrier between the west coast of Latin America and the balance of the continent. In some places the range is narrow, but in other areas such as Bolivia and Peru it widens to a series of parallel chains separated by high basins. Most passes across the Andes are high and have not readily facilitated movement of peoples and goods.

North of South America discontinuous mountain ranges comprise Middle America. These are rarely as

Figure 17.5 Landforms in Latin America.

high as the Andes, and in several locations low passes make movement across the continent easier. At the Isthmus of Panama the distance from the Atlantic to the Pacific is less than 50 miles (80 kilometers); here the Panama Canal is a strategic routeway.

The largest lowland basin of Latin America is the basin of the Amazon and its many tributaries. The Amazon Basin as a whole is as large as the conterminous United States, covering more than 2 million square miles (5.18 million square kilometers). Elevations throughout this entire region are low. The Amazon is the largest river in the world, carrying one-fifth of all river water in the world. The southeastern portion of South America is drained by the Paraná-Paraguay-Plata river system. Separating these two systems and dominating the eastern portion of Brazil are the Brazilian Highlands where elevations reach only to 2000 to 3000 feet (600 to 900 meters). Lowlands of varying width and importance are found in the coastal areas.

CLIMATES OF LATIN AMERICA

Tropical conditions dominate the climatic map of Latin America. A belt of tropical rain-forest climate extends across the core of Latin America in the Amazon Basin. North and south of the tropical rain forest are savanna climates, and in the southern portions of Brazil and in Argentina are subtropical climates (Figure 17.6). The tropical rain-forest, savanna, and humid subtropical climates cover approximately 80 percent of Latin America. They all have high temperatures throughout the year with relatively infrequent periods of frost. None of the subtropical regions, tropical rain forests, or tropical savannas of Latin America experience the extreme winter temperatures found in North America.

The other climatic types of Latin America consist of a small area of Mediterranean climate in central Chile, and dry steppe and desert climates in Argentina, Paraguay, northern Chile, Peru, and northern Mexico. These dry climates, especially those in Argentina and Mexico, are important for livestock ranching, introduced by Europeans.

The rather simple pattern of climatic distribution centered on the equator is made complex by the impact of the landforms. Of particular note is the presence of *altitudinal zones* of climate in the mountains. Just as climates change poleward from the equatorial regions, so increasing elevation brings climatic changes paralleling increasing latitude. In the Andes there are four altitudinal climatic zones. The lowest zone, up to 3000 feet (approximately 900 meters) in the tropical regions, is the *tierra caliente*, or hot land. This zone has all the characteristics of tropical rain-forest climate, and population densities are low. Higher than the tierra caliente is the *tierra templada* (temperate land). The tierra templada has cooler climates with conditions analogous to the humid subtropical climates. Above 5000 to 6000 feet (1500 to 1800 meters) the tierra templada changes to the *tierra fria* or cold climates with year-round temperatures in the range of 50° to 60°F (10° to 16°C). Higher than the tierra fria is the *páramos*, with periodic freezing conditions and no forest growth.

The altitudinal zonation of climate in Latin America has affected European use of Latin American resources and resultant settlement patterns. The tierra caliente tends to be devoted to the commercial production of tropical crops for export, including sugar cane, rice, and cacao. European settlement has concentrated in the tierra templada, where cooler conditions have been more desirable to the Europeans and the major economic activity is production of coffee. In the tierra fria are found the Indian populations of Latin America with their traditional societies based on subsistence agriculture. European influence in the tierra fria has been restricted to the European cities such as La Paz or to mining activities such as those in Peru and Bolivia.

THE MINERAL WEALTH OF LATIN AMERICA

The gold and silver that attracted the Spanish are scattered through the highlands of Latin America from Mexico through Peru (Figure 17.7). They are important in the economies of Mexico and Peru even today, as the two countries normally produce more than 30 percent of the world's silver. Bolivia is a major producer of tin, ranking after Southeast Asia as the world's primary producer. Bauxite is produced in Jamaica and the Guianas, and in the past Jamaica was the world's leading producer and exporter, but development of Australia's bauxite has reduced Jamaica's rank to number two. Iron ore is found in a number of areas, with major deposits in Chile, Brazil, Venezuela, and Mexico. The most important areas are in the eastern portion of Brazil (which has the largest reserves in Latin America), in the Cerro Bolivar mines of eastern Venezuela, and in newly discovered deposits in the Amazon Basin. Mexico and Chile have iron ore deposits that have been important in their own industrial development but are relatively small on a world scale.

Of greater importance than the metallic minerals are the energy resources in Latin America. Coal is found only in limited quantities in Colombia and Mexico, and the lack of coal is a major challenge to development attempts of the region, but there are important deposits of petroleum, particularly in the Caribbean islands and countries surrounding the Gulf

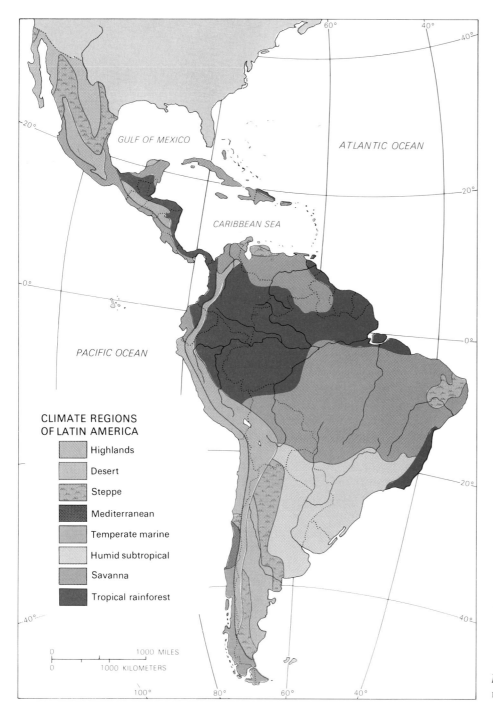

Figure 17.6 Climate regions of Latin America.

of Mexico (Figure 17.8). The gulf coast of Mexico is an extension of the oil fields in the gulf coast of the United States (or vice versa) and has been an important producer for the past 50 years. Venezuela has major petroleum resources around Lake Maracaibo, and other smaller deposits are found elsewhere. In the past, Venezuela has been the most important producer and exporter of petroleum in Latin America, but discoveries in southern Mexico are changing this picture.

New discoveries in Mexico indicate it may have the fourth largest reserves in the world and have made it the world's sixth largest petroleum producer. How large Mexico's reserves may ultimately be is still unclear, since only 12 percent of the land has been explored and drilling has been restricted, but the Mexican government claims 100 billion barrels of proven reserves of petroleum (compared to 150 billion for Saudi Arabia). Since the discoveries of Mexico's large oil fields in the late 1970s, Venezuela has also an-

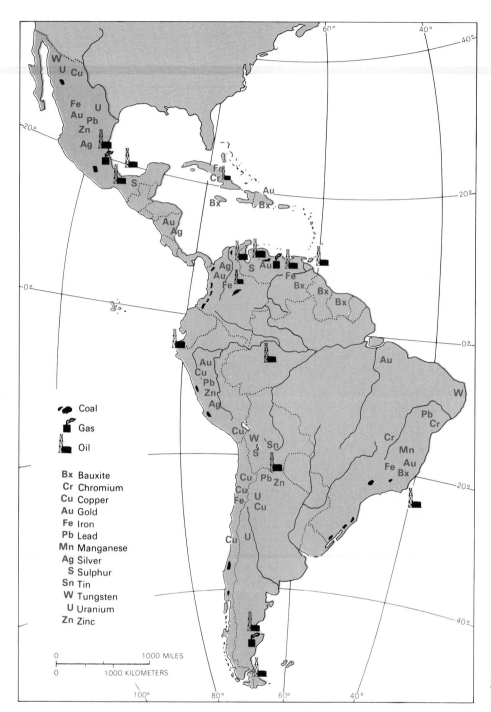

Figure 17.7 Resources of Latin America.

nounced new discoveries in the Orinoco River area. Initial enthusiasm maintained that there were 500 billion barrels in this field, nearly equal to the world's total known pertroleum reserves. Subsequently, the Venezuelan government has lowered its estimate to 50 billion barrels of proven reserves, but this rivals the Mexican finds.

The other areas of Latin America are less fortunate in terms of petroleum. Trinidad and Tobago have important petroleum deposits, which are an extension of the Venezuelan fields. Colombia, Ecuador, Peru, Bolivia, and Argentina all have sufficient petroleum to meet their own needs. Brazil and Chile of the newly industrialized countries lack petroleum but, particularly in the case of Brazil, additional exploration may yield significant reserves, since oil is found in the Amazon headwaters of Peru. Except for petroleum, the resources of Latin America are adequate to support industrialization far beyond that which presently exists within the region.

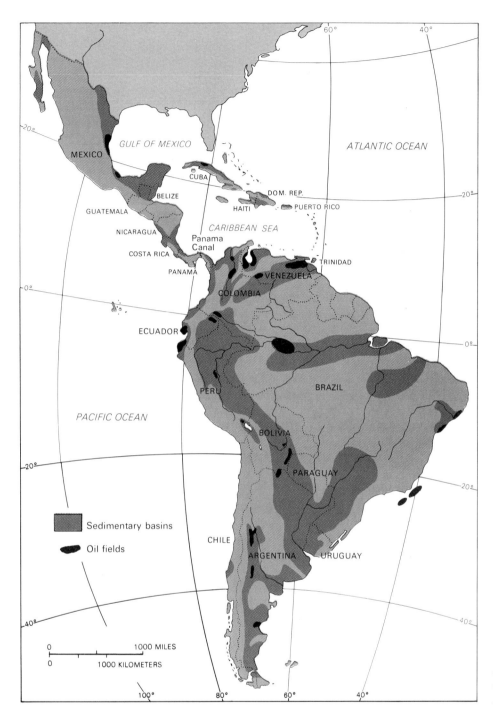

Figure 17.8 Petroleum deposits in Latin America.

CULTURAL AND INSTITUTIONAL SETTINGS

The institutional frameworks and cultural mosaic that comprise the Latin American countries are viewed by some as a handicap to development. The cultural groups in Latin America include the remnants of the original Indian inhabitants, the descendants of mixed marriages between Indian and European (mestizo), the black population descended from slaves imported to work on European sugar plantations in the Caribbean and the northeast coast of Brazil, the descendants of mixtures of European or mestizo and black (zambos), and the descendants of Europeans. There are also small numbers of Asians in Guyana and Trinidad where Asian Indians were imported for work, and in the Amazon area of Brazil where Japanese migrants have established farms producing tropical crops. The countries with large Indian populations are Guatemala (55 percent), Ecuador (55 percent), Bolivia (51 percent), Peru (40 percent), and Mexico (33 percent). The Indians are overwhelmingly rural or village residents who are out of any mainstream developmental attempts. These rural Indians constitute one element of a dual economy and society in nearly every Latin American country. The rural Indian or mestizo is often illiterate, does not participate in the economic advances taking place in the country, and is bound by a lifestyle based on the patterns of previous centuries.

Mestizos are dominant in most Latin American countries. The term *mestizo* is more than a description of ancestry; it also implies a set of cultural values and position in society. Mestizos speak Spanish, are more apt to live in cities, and are the source of the emerging middle class. The mestizo is less likely to be tied to the traditional Indian village and is usually more willing to accept revolutionary changes. The black and zambo populations occupy the same basic cultural niche as the mestizo population.

The European-born and those of European ancestry (creole) tend to dominate the economies of most Latin American countries. The Europeans and their descendants typically own the large estates and control the banking, emerging industrial activities, and other sources of wealth. Europeans and creoles are normally the best-educated, have the most contact with the balance of the world, and may constitute a negative force for development as they attempt to protect their own special interests.

Overshadowing the cultural differences among Indian, mestizo, European, and black populations is the dichotomy between rural and urban. The rural resident, whether mestizo or Indian, tends to be poor, have a lower income and fewer opportunities for education or other forms of upward mobility, and is isolated from not only the events transpiring in his or her own country but even more so from those in the world. Urban people of Latin America, even those in the slums, have a different view of life. The urban resident is more aware of events within and without the country that affect an individual's destiny, and by virtue of urban residence has greater potential for education, wage-labor positions, and opportunities for increasing standard of living. As in other developing regions, this contrast between urban and rural is the magnet that pulls people from the countryside to the cities.

INSTITUTIONAL FRAMEWORKS AND LEVEL OF DEVELOPMENT

The institutional framework of Latin America countries reflects their colonial history. Of central importance since the independence of Latin American countries in the early 1800s has been the close connection of government, church, military, and large landholders. The relative importance of these four has varied spatially and temporally, but all have affected each country. Independence movements freed the Latin American countries from European political control, but the political institutions remained oriented toward maintaining the privileged position of the wealthy elite. Political systems have varied from dictatorships to nominal democracies, but in each case they have maintained the position of the ruling elites. The military emerged as the supporter of governments or as the basis for overthrowing governments. The church and the landowners supported the political powers who maintained their own positions. The importance of these institutions lay in their cementing the gap between rural and urban. The political center became the focus of economic activity prior to the independence movements, but postindependence development in nearly every country solidified this position and dominance. In many cases the independent states of Latin America had only nominal control over areas farther away from the political center. The **effective national territory** was restricted to a small area around the political center.

LATIN AMERICA ECONOMIES

The economies of Latin America reflect the institutions and the urban/rural dichotomy that characterize the individual countries. Traditional exchange systems dominate the rural economy. The mestizo and Indian populations are engaged in subsistence or semisubsistence agriculture, while Europeans, creoles, foreign companies, and other members of the elite dominate commercial agriculture.

Peasant farmers exchange or sell their products at periodic or daily markets little changed from previous centuries. Buses provide more rapid access to the market in many areas, but the markets themselves are typical of traditional economies anywhere in the world. There is relatively little connection with the broader external world, as even the commercial crops of bananas, coffee, cacao, and other products that small farmers produce for export are purchased by a middleman who wholesales it to the exporters. Except for

---

Effective national territory: That part of a country that is controlled by the central government, contributes to the economic base of the country and feels itself part of the country.

The peoples of Latin America are diverse, and a range of ethnic groups, from Native Americans to Europeans, can be seen in many countries.

bananas, the majority of the production of commercial crops for export takes place on plantations rather than in the traditional rural economy.

The other half of the dual economy is that of the cities of Latin America. Urban commercial exchange is heavily oriented to world markets. Ironically, the primary exports from Latin America continue to be agricultural in nature, and even though the products (bananas, coffee, cacao, sugar cane, rubber, and so on) are produced in rural areas, the persistence of latifundia owned by the wealthy urban elite makes their production a part of the commercial economy. Workers on the latifundia are involved in the commercial exchange economy, since they are paid cash wages and then must buy their food from retail sources or grow it on their own minifundia plots.

One or two exports dominate the commercial sector of each country (Table 17.2). The reliance of Honduras on bananas, Colombia on coffee, Chile on copper, or Cuba on sugar makes the commercial sector of Latin American countries heavily subject to conditions beyond their boundaries. The production of staple items for export relies upon external demand and supply conditions over which the countries have little control. The resulting cycles of boom and bust have played a major role in the economic development of individual Latin American countries.

Part of the commercial economy is the emerging industrial base in Latin America. Industrial workers make up only slightly more than one-fourth of the labor force, but are responsible for 40 percent of the economic output. Closely associated with those who actually work in industry are the increasing numbers in service-related activities who produce 45 percent of the economic output. Both industrial workers and service employees are integrated into the commercial exchange economy, and their existence indicates the capacity of these countries to transform their standard of living. The concentration of industrial and service jobs in the primate cities of Latin America maintains the commercial dominance of the urban centers. In Mexico, for example, the Mexico City metropolitan area contains nearly 60 percent of the industry, service jobs, and government employees of the country. The overwhelming dominance of the primate center in each country in industry, service, and commercial activity reinforces the dual nature of the economies. The effects of this dual economy are pervasive and affect all aspects of the Latin American societies.

THE WINDS OF CHANGE IN LATIN AMERICA

The traditional characteristics of Latin America are beginning to change, and since the beginning of the twentieth century some important alterations have occurred in the institutional framework of several Latin American countries. Revolutions in Latin American countries have occurred as governments have attempted to resolve the issues of dual economies and wide variation in wealth and poverty. Foreign control of the export economy of several Latin American countries has been eliminated through nationalization, giving the individual states more opportunity to capitalize on their own resources. Mexico expropriated foreign-owned oil industry in 1938, and Venezuela followed in the 1970s. Control of banana production by the United Fruit Company of the United States is coming to an end as new arrangements for the production and marketing of the bananas will allow individual farmers to benefit from the profits. One of the last major vestiges of colonialism ended when the United States granted control of the Panama Canal Zone to the Republic of Panama in 1978. A few of the Caribbean islands are still under colonial rule, but even these tiny states are rapidly gaining independence.

Associated with, and partially as a result of, the changes in institutions in Latin America have been changes affecting the fundamental issues facing all less industrialized countries. Latin America is only now beginning to enter the final stage of the demographic transition, which portends a major change in the re-

TABLE 17.6 Distribution of Labor Force by Sector of Activities, 1950–1980 (in percent)

Sector of Activities	Labor Force		
	Males	Females	Total
1950			
Agriculture	59.0	27.4	54
Industry	18.6	20.0	19
Service	22.4	52.5	23
1970			
Agriculture	47.6	15.7	45
Industry	23.1	17.1	21
Service	29.3	67.2	34
1986			
Agriculture	—*	—	37
Industry	—	—	28
Service	—	—	35

*Information not available by sex.
SOURCE: *Development Forum* 9 (April 1980) (New York: United Nations), p. 12, and World Development Report (New York, New York, 1983).

Periodic market in Mexico.

gion. The present population growth rate remains the second highest of any continent in the world, but in an important break with the past, predominantly Catholic Latin America is beginning to recognize the need to control population. In the mid-1970s the archbishop of Mexico publicly offered support for the Mexican government's program to curb population growth. Catholic leaders in other Latin American countries are making similar statements as recognition of the need to curb population in order to increase standard of living and protect the environment from further degradation becomes a major issue.

Priests in Latin America are becoming more politically active in other ways as well. In many Latin American countries they no longer support the old status quo of rich versus poor, but are vocally and actively supporting revolutionary movements fighting to obtain a better percentage of the wealth for the poor. Some of these revolutionary movements have spread from Cuba, where in the early 1960s Fidel Castro overthrew the dictatorial Juan Batista government, which had been supported by U.S. economic interests. Castro has since developed a socialistic economy that has provided at least minimal literacy and medical care, food and shelter, and the basics of life to the people of the island. Other Central American states including El Salvador and Nicaragua have experienced similar revolutions recently, and indications are that other Latin American states will attempt a similar course. Ongoing conflicts in Peru with the Sendero Luminoso (Shining Path) and in El Salvador with the Farabundo Marti National Liberation Front (FMLN) typify the problem of revolution in Latin America. In each case the revolutionaries have the greatest influence in a specific region of the country, creating an *insurgent state* within the country. In the insurgent state the revolutionaries control the territory, administer the law, run schools, and perform the other functions of a government. Their actions reflect the conviction of many Latin American activists that it is impossible to change the standard of living of the poor of Latin America without a revolutionary change in government and social organizations.

Partially in response to the renewed threat of revolutionary overthrow, several governments in Latin America are engaged in active attempts to integrate the rural sector of their countries more fully into the national economy. In the past, this has been difficult to achieve because of the lack of adequate transportation systems. The barrier of the Andes, the location of the major rivers far from populated centers, and the lack of incentive to provide transport routes have served to isolate the rural population socially and geographically from the urban residents. Brazil exemplifies present attempts to overcome this centrifugal force as it builds new highways to the Amazon Basin and the other interior areas (Figure 17.9). At the same time, many countries of Latin America are increasing their expenditures on education, medical care, and other social services. Whether this trend will be sufficient to offset the rising demands for complete change in institutional and social organization remains to be seen. Latin America is apparently entering a period of rapid change, and whether the old institutions will survive is doubtful. The pressure to provide a higher standard of living for the poor is a powerful impetus that increases yearly as the world's poor recognize their power to achieve change.

The efforts of the Latin American countries to transform the quality of life of their people are handicapped by the entire range of factors that work to the disadvantage of less industrialized countries. Of cen-

Figure 17.9 Major transport routes in Latin America.

tral concern to many Latin American countries is the control of their revenue by foreign actions and interests. The inability of countries that rely heavily on a single export to control their own destinies acts to limit long-term, steady growth. Latin American countries products have not increased in value as rapidly as the goods they must purchase from industrial countries. World inflation is having a dramatic impact upon even newly industrialized countries. In Brazil, the inflated price of imported manufactured products is in stark contrast to the slowly rising (or even falling) value of their exports of coffee, cacao, petroleum, or other products. The increasing price differential has created rampant inflation, which threatens the progress they have achieved to date. In the least developed countries, such as Bolivia, the impact of world inflation and high prices has had an even more dramatic impact.

Another important change affecting Latin America is the changing role of the military. Many younger military leaders have had training in the United States and Western Europe and have absorbed ideas about the need to eliminate the gap between the rich and poor in their countries. In several instances the military has supported overthrow of governments that perpetuated the excesses of wealth and poverty. The changing role of the church and the military in Latin America gives promise for the development throughout the region.

Industrialization is increasing, particularly in Mexico, Brazil, Argentina, Colombia, Venezuela, and Chile. There are marked contrasts among countries in the level of industrialization, as each has its own characteristics and the individual pattern of development provides sharp contrasts. These contrasts reflect differences in population size and skills, geography, climate, natural resources, and the national priorities of governments in power. It is impossible to discuss adequately the level of development and degree of change in each country in this book, but comparison of the experiences of countries that have attempted a revolutionary change and contrasts between the major subregions provide some insight into the degree of change and development in Latin America.

VARIATIONS IN DEVELOPMENT: REGIONAL CONTRASTS

The easiest region to define in Latin America is the Caribbean. Separated from the mainland, the area still has most of the classic attributes of Latin America. With 34 million people inhabiting the islands, the Caribbean has the highest population density of any region in Latin America. The lowest level of development is in Haiti. Haiti is among the 25 poorest countries of the world, and the level of development is similar to that of Africa. Industry has been built by foreign multinational firms, which manufacture baseballs, textiles, and other goods relying upon the cheap labor of the island. The overwhelming majority rely on agriculture, and overpopulation and associated destruction of Haiti's forests for fuel and building material has resulted in country-wide erosion on a scale unknown elsewhere. Life for the average Haitian is worse today than it was in the past, despite the overthrow of the dictatorial Duvalier government. Exports are still from agriculture, with coffee and sugar dominating.

At the other end of the spectrum is Puerto Rico, which has been a possession of the United States since 1898. Puerto Rico is officially a commonwealth of the United States, and its people are American citizens. American industry has moved to Puerto Rico, providing a large industrial base.

Between the two extremes lie the bulk of the islands of the Caribbean. The major economic activity in many of the smaller islands, particularly in the Bahamas and the Virgin Islands, is tourism. The mild winter climates and warm Caribbean waters combine with beautiful beaches to attract millions of tourists from the United States and Western Europe. With some 12 million visitors in 1990, tourism is a mixed blessing to the islands, providing revenue for one sector of the society, while the influx of tourists from the wealthy industrial world emphasizes the poverty of the majority of the region's residents. Tourism is further criticized for its tendency to destroy the integrity of local culture by requiring local people to pander to the tourists. Stereotypes of the culture are staged for visitors, but the values and traditions that give vitality to the Caribbean cultures are threatened by such trivialization of society. Tourism has fostered some economic development, including handicraft industries, but the region in general is part of the less industrialized world, with average per capita GNP of under $2500 in thirteen of the seventeen countries in the region.

Of major interest in the Caribbean is Cuba, the first country of Latin America to opt for a Communist model of government. Cuba is a former Spanish colony that was under U.S. control from 1898 to 1901. From 1901 until the 1930s Cuba was a protectorate of the United States, which maintained the right to intervene in its internal affairs. After the 1930s Cuba developed a dictatorial government led by General Juan Batista, who was overthrown by Fidel Castro in July 1959. At the time of the overthrow of the Batista

THE HUMAN DIMENSION

Latin America: The Children's Voices

I went to live on the street because my mother is very poor. There are eight children. So I went off to get food for myself. I looked after cars for about 50 or 60 pesos a day. I liked to earn money. Sometimes I stayed out all night looking after cars outside the disco. It is difficult living on the street. There is just nowhere to wash. I didn't like asking for money from strange people. On the street I slept with my friends. I often got cold at night, and sometimes got sick. When I grow up I want to get a job as a driver to help my mother. *Jesus Antonio Pinella (10) Bogota, Colombia.*

Right now I don't have no shoes and we go hungry. I feel better must come one day. I just have a mind that better will come one day. I must look forward. In four years or five years I would try to make myself more happier. In 20 years time I wish I'm in nice job with my wife and children and house, you know. You have the mind to go up further ahead you must go further ahead, but if you have mind to go down, if you don't think of future time to come, you will never have no better to come. *Vinton Faulkner (14) Jamaica.*

People should have as many children as they can look after. Some people have children and just take them to a park and leave them there, and when the children look round the parents have gone. *Nadine Housen (15) Jamaica.*

I sell cigarettes in the street. I sell cigarettes from 8:00 in the morning to 6:00 at night usually without taking a rest. I buy the cigarettes on the black market at San Andresito. I usually sell about 20 packets a day. I get 5 pesos on each packet. All the money we can get goes to my mother to buy food and clothes. My father tries to get work as a handyman. He only knows how to paint. I earn more than he does. If I stayed at school I could have been a lawyer. Nobody criticizes a lawyer the way they do if you are selling cigarettes on the street. *Luis Tacheco (14) Bogata, Colombia.*

What I would really like to do is go to university and live in the town. Here in the countryside you get very lonely. There is too much silence. It's more cheerful in the town. *Aurelio Vidarte, Peru.*

Source: International Planned Parenthood Federation, Forum, August 1979, pp. 35–36.

---

government, Cuba epitomized the problems of the single-export economies of Latin America. The primary economic base of the island was sugar, which accounted for more than 80 percent of all exports, and sugar prices were subject to the vagaries of world markets and agricultural problems. Periodic hurricanes destroyed large segments of the harvest; insect pests and disease decreased yields in other years. The economy and life of Cuba's people were typical of less industrialized countries, with high rates of illiteracy, infant mortality, birthrates, and great disparities between the rich and the poor. After the revolution Castro began turning for economic support to the Soviet Union and expropriated American agricultural enterprises in Cuba. Cuba's relationship with the Soviet Union has continued to intensify, and in 1988 two-thirds of all Cuba's exports went to the Soviet Union. Cuba has joined the **CMEA** (Council for Mutual Economic Assistance) composed of the Soviet Union, Cuba and the Eastern European Soviet bloc nations.

Modernization of the Cuban economy has focused on industrialization in an attempt to free the country from reliance on imports of manufactured items. The share of agriculture in total gross national product has declined to only 16 percent, while industry provides nearly 70 percent. In 1978 Cuba began operating its own assembly plant for sugar cane harvesters. The Cuban economy has benefited from attempts to diversify, and today nickel (Cuba has the world's fourth largest reserves of nickel) accounts for 6 percent of exports. Production of citrus fruit has increased tenfold in the last 20 years, but still comprises only 1 percent of exports. Production of cattle is increasing, and attempts are being made to utilize more fully the by-product of sugar cane, **bagasse.** Bagasse is the fiber that remains after the juice has been extracted, and Cuba has developed a plywood factory using bagasse as a raw material and a plant producing cattle feed from molasses left over from sugar production. Plans are underway for a paper factory utilizing bagasse as the raw material.

The Cuban economy has been transformed in the past 30 years, but severe problems remain. The growth of industry has caused labor shortages in some sectors, particularly in sugar production, and the continued reliance upon sugar for export earning (sugar still provides 70 percent of export earnings) places the Cuban economy in potential peril. At the present time, Cuba's economy provides for all of the population only because of continued subsidies from the soviet Union in the form of higher-than-world prices for sugar. The

---

Bagasse: The fiber that remains after the sweet juice has been removed from sugar cane.

subsidy has been inadequate to replace the lost revenue from tourism that existed prior to the Castro revolution. Tourism has long been one of the major sources of income in the Caribbean islands, but American sanctions against Cuba have prevented tourists from going to the island since 1959.

Cuba remains in a state of flux, firmly commited to a Communist, centrally planned state economy and the goal of exporting this system to other countries. As a model for Latin America, it has not been widely adopted because many Latin American countries have raised their standards of living higher than Cuba's without resorting to a Communist revolution. It will be interesting to see what form Cuba's economy takes as it continues attempts to transform the one-crop agricultural economy into an industrial country based on the Soviet model.

MIDDLE AMERICA

Middle America is the area from Panama to northern Mexico. It is important to distinguish this term from Central America, which refers to the countries of this region except Mexico. Middle America was the first area where the Spanish imposed their political rule and began the process of developing an exploitative economy. There are major differences within the region, particularly between Mexico and the other seven countries. Mexico, the third largest independent state in area in Latin America, has the second largest population (nearly 87.5 million people).

Mexico is the only country of the less industrialized world that borders directly upon an industrialized nation. The contrasts between the United States and Mexico in standard of life have made the United States a major magnet for migration from Mexico, both legal and illegal. Mexico remains the single largest source for legal migrants into the United States, a dominance that is multiplied when illegal migrants are added. The development of Mexico's economy has in part reflected this location, with Anglo-American investment of particular importance in the northern states of Mexico.

Geographically Mexico is complex, ranging from the arid and semiarid north to the moderate climates of the highlands of central Mexico where the majority of the population lives, to the tropical conditions of the southern and eastern lowlands. More than half of the population is concentrated in the valleys and basins of central Mexico, particularly around Mexico City proper. The presence of a relative abundance of minerals within Mexico has facilitated industrialization, but the dual economy still exists.

Mexico's economic development reflects its cultural background under Spanish colonial rule and its location and geographic characteristics, as well as the political changes of the last century. The heritage of Spanish dominance was epitomized by the large estates (*haciendas*) and the role of the Catholic church. Having lost their land, the peasants in the villages lived in grinding poverty. After the war of independence (1810–1821) Mexico experienced political and military strife until 1877, when the dictator Porfirio Diaz came to power. He provided stability but widened the gap between rich and poor. Large estates were expanded as remaining communal Indian lands were occupied. Diaz used his power to grant public lands to friends or foreigners, further handicapping the peasants. At the same time his government encouraged foreign investment and the beginning of industrialization and commercialization of coffee and other crops for export.

In 1911 came the Revolution as the impoverished peasants supported guerrilla warfare, which culminated in the establishment of today's government of Mexico. This Revolution, which was more than just a change of dictators, began in the regions of Mexico where Indians lived in greatest numbers, and the reaction of the central government in sending troops to destroy the villages simply drove more Indians into the guerrilla ranks of Emiliano Zapata. After nearly a decade of strife, a new constitution was adopted and major structural changes attempted in the economy. Of fundamental importance was land reform, which broke up the large estates and granted the former communal landholdings to the Indian villagers. This movement created the *ejidos* of Mexico, a form of communal landownership in which plots are granted to an individual farm family for life, but cannot be sold.

Since the Revolution, Mexico has developed a government based on a one-party, semisocialist, state-planned economy. Basic industries have been nationalized, unions have been recognized, and the federal government has undertaken major land reforms designed to increase the standard of living of the people.

The poorest Mexicans are those classified as Indian, not necessarily by ethnic background, but by culture. Indians speak a native language, even though they may also speak Spanish, emphasize Indian rather than European customs, and are primarily engaged in agriculture. Approximately 20 percent of Mexicans remain illiterate, and nearly one-third live in poverty. Indians practice their traditional agriculture on even the steepest slopes in the highland valleys. Production of corn, beans, squash, and other traditional crops remains important for this sector of the agricultural economy. There is also a major commercial agricultural sector. In northern Mexico irrigated vegetables, cotton, and cattle are produced for Mexican consumption and export to the American markets.

THE HUMAN DIMENSION

Haiti—"The Problems of Life"

Mrs Aidenaire Carm lies in Port-au-Prince, and makes what must be euphemistically called her living buying bananas in the market, and reselling them, peeled. Her only customers are her neighbors in La Saline, one of the city's poorest slums, who like herself are trying to make ends meet in the most marginal of occupations. She is the only provider for her four children. Her husband, she says: "has abandoned his homeland. Who will help me now?" she laments. "Who will give me food?"

Aidenaire Carm came to Port-au-Prince with her family when she was a small girl. Her parents were forced off the land, victims of the relentless erosion which has blighted Haiti's hills and valleys and depleted the soil's fertility. The family fetched up in the city's slums in search of a livelihood.

La Saline is one of the worst. Like many of the other slums, it is situated on reclaimed land only just above sea level. When it rains heavily, the water pours down the steep hillsides of uptown Port-au-Prince, sweeping the city's detritus before it, and ends up feet deep among the shacks and hovels.

Charcoal and its residues permeate every corner of La Saline. The ankle-deep mud is half composed of ash and cinder, and the tin huts are black with the grime of countless fires of cooking braziers. More unpleasant still, and more hazardous to health, are the pools of stagnant water and piles of putrefying garbage—and worse—that litter the ground. Plaintively, a crooked notice beside an open ditch reads: "Help us to keep the neighborhood clean."

Aidenaire Carm is illiterate and unequipped for any job requiring skills and training. Her 15-year old daughter has turned to promiscuity as a way of helping out. Her brothers and sister, like most of the children that play barefoot in the narrow alleyways, all manifest some signs of the neighborhood's ill-effects: sores, a skin rash, unusual thinness, a persistent cough. Small wonder that Aidenaire Carm cites "the problems of life" as her most persistent preoccupation.

Source: UNICEF Annual Report 1983, p. 23 (Washington D.C. United Nations, 1984).

Most Mexicans have been unable to obtain lands either in the ejido communities or in the national lands being developed through irrigation in the north and drainage in the southern lowlands. These people are finding work in services and manufacturing in Mexico City, Guadalajara, and the other major cities. Major industries include textiles, steel, automobiles, electrical products, and food processing. The dominant manufacturing center is the Valley of Mexico where more than 50 percent of the country's industrial laborers and more than 60 percent of the value added by manufacturing are concentrated. This concentration of industrial activity in the Valley of Mexico has created urban problems on a scale rivaled in only one or two other cities of Latin America. Mexico City is estimated by some to be the largest city in the world, with estimates of its population as high as 20 million in 1900. Mexico's industrial revolution has propelled it into the front ranks of the newly industrialized countries, and it is one of the few countries of Latin America to have a well-developed transportation infrastructure.

In spite of Mexico's achievements in industrialization, the country faces serious problems in bringing the Indian population fully into the mainstream of development, curbing its population growth, which is typical of less industrialized countries, providing a means of increasing agricultural productivity and in-

The rapidly growing population of Mexico necessitates using even marginal land. Steep slopes such as this are often cultivated without regard to erosion potential, resulting in loss of soil and destruction of fields.

comes from minifundia of the ejidos, and providing adequate jobs and housing for migrants to urban regions. The underlying problem remains Mexico's rapidly growing population, which handicaps development efforts and prompts between 3 and 5 million Mexicans to migrate to the United States yearly. Unless population-control efforts are successful, Mexico's development efforts will be unable to overcome the poverty that characterizes the life of rural Mexicans and the slum dwellers of Mexico's cities.

The countries of Central America have many similarities. Their physical geography is dominated by east-west trending mountain ranges, and the basic climatic pattern is heavy rains during a hot summer with decreased rainfall in the winter, but no truly dry season in most of the region. Culturally Central America typifies the less industrialized region of Latin America, with reliance on a few exports, high illiteracy rates, and great disparities between rich and poor. In Guatemala, for example, 2 percent of the landowners control 70 percent of the land. The economies of these countries rely heavily on tropical agricultural exports, especially bananas and coffee. A large share of their economic development has been based on foreign capital, such as the United Fruit Company and its subsidiaries.

In all countries of Central America, foreign trade is predominantly with the United States. All are examples of colonial and foreign exploitation in which tropical food crops are exported while the people are unable to produce adequate food for their own needs. The reliance on food imports could easily be overcome if more land were devoted to production of basic foodstuffs rather than luxury exports. These countries lack good communication systems, capital for investments to develop more rapidly, and political stability.

The contrasting wealth and poverty makes the countries subject to revolutions, and El Salvador and Nicaragua have both experienced revolutionary changes in government in the past decade. In Nicaragua in 1979, the Marxist Sandinistas overthrew the government controlled by the Somoza family. In 1984 the Sandinistas gained an overwhelming victory in a general election although some observers feel that Sandinista tactics prevented opposition candidates from having a chance of success. The Sandinista government has introduced land reforms that initially expropriated only the 20 percent of the country's land held by the Somoza group. Reforms initially loosely followed the Soviet model but have recently changed to emphasis of cooperatives, but over 50 percent of all agricultural land is still in private ownership.

El Salvador's land reform in the late 1970s has been stalled by changing governments. In El Salvador and Nicaragua, active opposition to the present governments is a daily part of life. It is too early to predict the future of the area, but it is clear that the great gap between rich and poor in Middle America will continue to encourage conflict. The exception to this general pattern is Costa Rica, where over 80 percent of the population is of European ancestry, with a stable government and broad-based landownership by individual small farmers.

THE COUNTRIES OF THE CENTRAL ANDES

In the Andes the three altitudinal climatic zones of tierra calienta, tierra templada, and tierra fria are most clearly discernible. The Indian and mestizo population is clustered in the tierra fria, the European population is more highly concentrated in the tierra templada, and in the tierra caliente lives a combination of Indian, mestizo, and black population, descendants of workers imported to exploit mineral and agricultural potential. The region's landforms consist of three types: the Pacific coastal plain, the Andes with their slopes and high valleys, and the eastern foothills and plains drained by the Amazon, Orinoco, and Parana river systems.

The richest country of the Andes is Venezuela, whose primary wealth comes not from the mountains but from the oil deposits around Lake Maracaibo, but even here the average per capita GNP is only about $3000 (1990).

The majority of Venezuela's population is urban and the largest city (Caracas) and the greatest concentrations of population are located in the Venezuelan Highlands in the north and west of the country. Those engaged in agriculture are concentrated in the central highlands where cocoa, coffee, cotton, and sugar cane are produced in the tierra templada zone on plantations owned by Europeans and mestizos. The oil wealth of Venezuela, combined with its abundance of iron ore, has led to important efforts to industrialize. Maracaibo, the second largest city in Venezuela, is the center of oil-related activities. The majority of the population in Venezuela is mestizo (67 percent), but slightly more than 20 percent are European, only 2 percent are Indian, and 10 percent are black.

Colombia shares some of the characteristics of Venezuela, except that it does not have as much petroleum and iron ore. The ethnic makeup of the 32 million population is similar, with 58 percent mestizo, 20 percent European, 18 percent mulatto or black, and 4 percent Indian or zambo. Colombia is a large country, but its population is concentrated in the valleys of the highlands. The economy today relies heavily on agriculture, with more than 25 percent of the labor force engaged in agriculture compared to 21 percent in manufacturing. Major crops include coffee, rice, corn, sugar cane, bananas, cotton, and tobacco. Manufacturing is concentrated in the textiles, food processing, clothing, and footwear segments. The north coastal area of Colombia is the site of the Cerrejon coal mining development, which is one of the largest bituminous coal mines in the world. Colombia is also the source for a major part of the cocaine illegally smuggled into the United States. Illegal production of cocaine is a major social problem in Colombia, as drug "lords" have killed government officials from the justice minister and attorney general of the country to local mayors. These drug lords have corrupted police and other government officials with bribes, diverted peasant labor from food crop production to cocaine production, and established a class of super-rich who are beyond the reach of the legal system.

Ecuador, Peru, and Bolivia are all less developed and share the major problems of other similar regions. They have predominantly Indian populations, mostly concentrated in the highlands. Spanish colonial rulers transferred the lands of these countries from communal Indian ownership into large latifundia. Revolutions have been attempted in both Bolivia and Peru to return the land to the Indians, with mixed results.

In Bolivia in 1952 a revolution began with the tin miners, and by the end of the year a new government had nationalized the mines. In 1953 the government adopted a land reform law based on the principle of nationalization of land and land distribution modeled after Mexico's ejido program. The agrarian reform law of Bolivia succeeded in breaking up the large estates and redistributing the land to the agricultural workers, but it has been criticized as simply replacing the old poverty of the latifundia with a new poverty of minifundia. For example 8,000 hectares (about 20,000 acres) in one valley was divided among 8,000 families. The plots of 2 1/2 acres (1 hectare) each are insufficient to provide capital or the size necessary for mechanization and commercialization, thus ensuring the continued prevalence of subsistence agriculture in Bolivia. The government has attempted to colonize the lands of the tierra caliente in the eastern Andes, but to date the success has only been nominal.

Peru also attempted a land reform movement, as the land tenure in the country reflected the worst of the latifundia system. In 1961, 3400 holdings (0.4 percent of total holdings) occupied 74.4 percent of the arable land. Peasant opposition to such dicrepancies in holdings led to the promulgation of a land reform law in May 1964, requiring breaking up of large holdings, but properties belonging to companies engaged in industrial processing of agricultural products and sugar cane plantations were excluded.

In 1968 a military coup led to a new agrarian reform law eliminating the exemption for industrial companies. Between 1968 and 1970 the new government attempted widespread land reform, particularly in the coastal areas where European commercial agriculture was concentrated. The European latifundias were turned into production cooperatives in which the state was directly involved in ensuring continued production. In the mountains the latifundia were divided among the laborers, sharecroppers, and community members. Then a new government came to power in the early 1970s and began returning agricultural operations to private hands. Today the Indians of the Peruvian Andes are generally poor, subsistence farmers producing corn, potatoes, barley, or other small grains. East of the Andes in the Amazon is a frontier region whose oil resources are being shipped across the Andes to the urban areas of the west coast. The coastal area is economically the most important area of Peru with fish, cotton, and sugar important exports. As in Colombia, cocaine is a major crop, creating the same types of problems.

Ecuador is the smallest of the Andean countries, with an area almost identical to that of the state of Nevada, and a population of only 10.8 million. Although Ecuador has the same three climatic zones and the three types of landforms found in Peru and Bolivia, and the same combination of Indian, Mestizo, Black and European population, it differs in important aspects. The major difference is the preponderance of small, privately owned farms in the coastal region that makes Ecuador one of the largest banana exporters in the world, and the export of petroleum from the coastal plains that makes it the second largest petroleum exporter of Latin America. In spite of these differences, Ecuador is similar to the other Andean countries in that a high proportion of the population is still engaged in agriculture, and the Indian majority of the highlands still live in poverty while the European owners of the latifundia enjoy a life of wealth.

MID-LATITUDE LATIN AMERICA

The countries of Chile and Argentina share the southern Andes, but climatically they differ from Andean countries to the north. Chile and Argentina share predominantly mid-latitude climates with Uruguay and Paraguay. Mediterranean, marine west coast, humid subtropical, steppe, and desert rather than tropical climates predominate. With the exception of Paraguay which is 95 percent Indian or Mestizo, European influence is stronger here than most of Latin America, and the region has the highest per capita income of any region in Latin America.

Chile is a long (2500 miles or 4000 kilometers), narrow country that epitomizes the elongated state.

THE HUMAN DIMENSION

A Tale of Poverty, Water, and Roses

For most of his 45 years, Rosario Genaro Gonzales Ruiz has farmed the arid land of Peru's Chau Valley near the city of Trujillo. When times were good, his efforts supplied the barest essentials. But often, his efforts were not enough, and Gonzales Ruiz, his wife, and their 10 children could only hope that next year would be better. Gonzales Ruiz lived a life that was not his own. For one thing, if the rains did not fall in the mountains, the nearby river would remain a dry bed. There would be no water to irrigate his crops.

For financing, he had to rely on the local, informal, sources of capital. He had to split his profits with them after the lender deducted the costs. "If I had a good crop, I would get something," he explained. "If not, I got nothing. I can't read or write, or do mathematics. So I never knew exactly how the lender figured it out." Gonzales Ruiz was also beholden to a large landowner, on whose estate he worked. Three days during the week he tended the landowner's crops, and three days he worked a parcel of land that was lent to him.

This went on, year after year, until the mid-1970s. Then, Gonzales Ruiz began to gain control of his own destiny. Not that things improved dramatically; for the time being, they actually got worse. The first big change took place when the landowner moved out of the country and gave Gonzales Ruiz the land on which he worked. With agrarian reform, he became a legal property owner. Now he could avoid the moneylenders by applying instead for a loan from Peru's Agricultural Bank. He received a $600 credit through an Inter-American Development Bank (IDB)-financed program.

Just when good fortune seemed assured, the riverbed turned dry. A period of severe drought had set in, and Gonzales Ruiz could barely eke out one corn crop from four hectares. His other seven hectares were idle. Once again, Gonzales Ruiz was left with no crops, no money, and a hungry family. In addition, he had a big unpaid loan. The drought continued for two years, and then someone in the valley had an idea: if they could dig wells and store the water in reservoirs, they would be relieved of dependence on the fickle river. The new irrigation scheme seemed promising. But it would take a sizable investment. Gonzales Ruiz and his neighbors were in no position to put up the money.

Then, good fortune came again with a second IDB loan for another agricultural credit program. Even though Gonzales Ruiz was in arrears on his original loan, the Peruvian Agricultural Bank had faith in him and in the new scheme. It made him a new loan of $2,000 to build his reservoir and drill the wells. The scheme worked, and the Chau Valley turned green again. Gonzales Ruiz now could put all of his land in production: seven hectares in corn, two hectares in papayas, with the remainder in orange and lemon trees. In September he planted a secondary crop of melons.

Already, Gonzales Ruiz has paid off his first loan, and is current on the second one. He works hard, and his labor gives him a good return. His days of sharecropping and poverty are over, and no longer must he fear a dry season. But Gonzales Ruiz was not prepared to relax after one success. His farm stands at a fork in the road, and he decided that the trucks and busses that passed by could be potential customers. So he opened a small convenience store and a restaurant, which is operated by his wife, children and relatives.

One final problem remained. Gonzales Ruiz was now sending his children to Trujillo for their education, and they needed money for lunches, books, and bus fare. Looking around his farm, he noticed a small plot of idle land. There he planted roses. So every day, before his children board the bus for Trujillo, they pick the blossoms, which they hawk in the city. The proceeds are their pocket money.

Conde, Carlos. IDB News *(Inter-American Development Bank; Washington, D.C. 20577), August 1984. pp. 4–5.*

The Andes to the east provide a defensive barrier and the Pacific Ocean provides transportation, minimizing the typical problems of an elongated state. North Chile has a desert climate, central Chile has a Mediterranean climate, and south Chile has a marine climate. More than 85 percent of Chile's 13 million people live in Central Chile. Economically only 19 percent are engaged in agriculture, and 81 percent are urban residents. Agricultural products include grapes, wine, fruit, vegetables, wheat and other crops typical of Mediterranean lands, as well as cattle, sheep, and potatoes. Chile is a net food importer, and unstable governments continue to handicap the economic development.

Land reform was attempted in Chile from 1964 to 1973, with major progress being made between 1970 and 1973 under the leadership of the Marxist government of Salvador Allende. By the middle of 1973, the latifundia of Chile had been redistributed among the farm workers, but a military coup led to a return to the former policies and restoration of the land to the latifundia owners. Today Chile is beginning to industrialize as urban centers develop and the present government attempts to control inflation, which during the early 1970s ran as high as 1200 percent per year. Although there are numerous mineral resources in Chile, the country's future is centered in the Mediterranean-like central valley.

Argentina is the largest of Latin America's mid-latitude countries and its 1.1 million square miles (2.8 million square kilometers) is second only to Brazil in Latin America. Argentina's 33 million population is the third largest in Latin America. The geographical base for Argentina's development is the grasslands, or *pampas*, with their fertile soils. The presence of this fertile grassland region combined with political, economic, and social conditions in Europe in the late nineteenth century to attract more than 5 million migrants from Italy and Spain to this area. An additional 2 million migrants from other European areas including Germany have followed and created a landscape of farms producing wheat, corn, and soybeans, and ranches producing cattle in the drier regions of the pampas. Outside the pampas, Argentina has important agricultural regions specializing in irrigated production of deciduous fruits and grapes, sheep grazing, and mixed agriculture.

The grasslands of Argentina's Pampas have emerged as one of the premier food-surplus regions of the world, and competes with North America and Australia for world markets. As in North America, only a small proportion (13 percent) of the population is engaged in agriculture, reflecting the more widespread use of mechanization in agriculture. After the Pampas, petroleum is Argentina's most important natural resource. Argentina has a producing oil field near Comodoro Rivadavia in the south central coastal area, and potential reserves in the northwest along the border with Paraguay.

The bulk of the population of Argentina is urban (84 percent) with the Buenos Aires conurbation containing more than 11 million people. Buenos Aires is a primate city like most other capital cities of Latin America and dominates the political, economic, and cultural aspects of the country. Industrial development is concentrated in Buenos Aires and Cordoba (1.3 million) the second largest city. Industrial products range from automobiles to electrical appliances, but are primarily for domestic consumption.

The combination of resources, especially fertile agricultural land, gives Argentina the potential to become a modern industrialized state. Unfortunately, Argentina has also had problems of political leadership. A military **junta** ruled Argentina from 1976 to 1984, borrowing heavily as their mismanagement contributed to inflation rates exceeding 100 percent annually. The military government mercilessly eliminated thousands of opponents and led the country to war with the United Kingdom over the Falkland Islands (known as the Malvinas in Argentina) in 1982. The present government was freely elected, but the potential of another military coup is ever present in the country. The return of democratic rule is seen as a positive sign that Argentina is once again on the path to general prosperity, a prosperity that its geographic relationships suggest should already have been attained were it not for governmental mismanagement. The emerging manufacturing sector produces the entire range of manufactured products and makes Argentina a member of the newly industrialized countries.

Paraguay is one of the least developed countries of Latin America, and has one of the highest population growth rates in the world. Ninety-five percent of the population is mestizo or Indian, and the balance is the European elite. Although nominally a republic until overthrown by elements of the military in 1989, the president was an authoritarian leader who had done little to transform the backward economy of the country. Paraguay has one of the most moderate climates of Latin America, with temperatures rarely falling below 60°F (16°C). Summer temperatures may exceed 100°F (38°C), but precipitation exceeds 50 inches (1,270 millimeters) in the area of Asunción where the majority of the population is concentrated. The country has great potential for agriculture, but political problems to date have handicapped its development.

Originally developed by Spanish Jesuit missionaries, Paraguay was a prosperous colony self-sufficient in foodstuffs and an important exporter of cotton, tobacco, tea (made from the quebracho tree), meats and hides and wool. Spanish landowners precipitated the expulsion of the Jesuits, and Paraguay's economy was ruined. A series of wars with neighboring countries, culminating in a disastrous conflict with Brazil, Argentina, and Uruguay in the 1860s, decimated the population. At the end of the war only about 28,000 adult males were still alive. Paraguay accepted numerous German migrants at the end of World War II. Today Paraguay's political instability, poverty, and internal problems have led many people, especially the educated elite, to migrate to the neighboring countries of Brazil and Chile.

Uruguay is the opposite of Paraguay. Uruguay was first settled by Spanish ranchers who established large landholdings, but subsequently these tracts were subdivided into smaller private farms. The typical landowning class enjoying wealth while the balance of the population lives in poverty that is typical of so much of Latin America has not developed in Uruguay, and the country enjoyed a government with strong social concerns. It has the highest standard of living, as

---

Junta: A group of military officers controlling the government, generally obtained by forceful seizure of the government and often exercised dictatorially.

Large scale farms and modern methods characterize agriculture in some parts of Argentina and Brazil, illustrating the process of change associated with their movement into a more industrial society.

measured by the physical quality of life index, in Latin America. Population growth rates are low, literacy is nearly universal, and the standard of living is the highest of any South American country. The population is overwhelmingly European in origin, and the economy has developed as a producer of agricultural products for export to the European mainland.

BRAZIL: LATIN AMERICAN COLOSSUS

Brazil is important to Latin America, and its size and mineral resources make it of world importance. Culturally Brazil differs from the balance of Latin America in its Portuguese colonial heritage. It covers nearly one-half of the area of South America and possesses one-third of the population. With its 3.3 million square miles (8.5 million square kilometers), it is the fifth largest country in the world, being roughly equal to Canada, China, and the United States (that slightly exceeds it). Yet with only 150 million people (1990), it is only sparsely settled when compared to the United States or China. Nearly half of Brazil consists of the Amazon Basin, the largest area of tropical rain forest in the world. The rest of Brazil consists of plateaus and low hills with only a limited coastal plain. The earliest area of settlement and intensive European activity in Brazil was in the northeastern coastal regions, where the Portuguese developed a prosperous sugar industry utilizing imported labor from Africa. For more than three centuries this was the cultural heart of Brazil as the wealth of the large landowners was used to develop universities and cultural centers and to develop a political leadership that affected all of Brazil.

Inland the northeastern portion of Brazil becomes drier and is called the *sertão*. The drier plateau produces cotton on its eastern margins, but recurrent drought makes it subject to great fluctuations in productivity. Farther inland, as precipitation decreases, the sertão supports pastoral activities producing cattle, goats, sheep, and donkeys. Subsistence farming occupies more favorable sites, where coffee, beans, corn, sweet potatoes, and manioc are grown for local consumption. A high rate of population growth, coupled with the recurrent climatic hazards in the northeast, causes constant migration of population, particularly to the southeast where industrial and urban opportunities are greater.

Southeastern Brazil occupies less than 11 percent of the total area of the country but contains 45 percent of the total population. The large urban centers of Rio de Janeiro and São Paulo constitute the most industrialized region of Latin America. The two states in which the cities are located account for two-thirds of all of Brazil's industrial production. At Volta Redonda, Brazil has an integrated iron and steel complex utilizing Brazilian raw materials with imported coal. The entire gamut of industrial products is manufactured in the southeast, ranging from automobiles to airplanes. This urban region continues to attract migrants from other areas of Brazil, and the cities embody the problems of developing countries with their ring of slums and their wealthy elite.

Southern Brazil has rapidly developed a diversified agricultural economy in the last three decades in contrast to the traditional boom-and-bust cycle in the rest of Brazil. Migrants from Germany, Italy, and Portugal have settled this area and developed a viable agricultural economy without the great contrasts in wealth typical of the rest of Latin America.

The Brazilian interior from the sertão through the Amazon lowlands has been only lightly colonized. Repeated waves of exploitation have swept across the region capitalizing on minerals, rubber, or other extractive activities. The repeated colonization of this region in response to new explorative opportunities has created **hollow frontiers** rather than complete oc-

---

Hollow frontier: The frontier is the zone where pioneer settlement is occurring. In North America frontiers evolved into permanent towns, farms, and industry as the frontier moved further west. In Brazil the frontier has traditionally moved on as the exploitative economy collapsed, creating hollow frontiers rather than permanent colonization.

cupation of the land. The Brazilian government wants to increase the population of the interior, and the location of the new capital of Brasília reflects this goal. Brasilia was planned as an attempt to pull population from the Rio de Janeiro and São Paulo regions. It has been only partially successful in this regard, but it is indicative of the Brazilian commitment to develop the interior.

Elsewhere in the interior developmental efforts have focused on expanding the road network, providing assistance in colonizing lands to produce rice, soybeans, sugar cane, and other agricultural crops. Brazil has emerged as one of the world's leading producers of soybeans, normally ranking second after the United States. The government has emphasized improving transportation along the Amazon to allow growth of cities in the Amazon Basin. The Amazon is navigable as far inland as Iquitos in Peru, but settlement is concentrated in a few cities like Belém and Manaus. Japanese colonists have undertaken settlement in some areas in the Amazon delta where they produce pepper and other crops for export, but the economy of much of the Amazon Basin is still based on extraction of rubber, mahogany, and other tropical products. Extensive deposits of minerals have recently been discovered in the Amazon lowland, offering promise of greater development, but the leached and infertile soils of the tropics combine with the isolation and climatic handicaps to slow population growth in this area. Expansion of mining continues the boom and bust cycle of Brazil's economy. The Carajas iron ore deposit in the Tocatins River (a tributary of the Amazon) for example, is one of the largest in the world. To develop it for export to Europe, Japan, and the United States, a railroad, a $48-million bridge over the river, and a port facility at Sao Luiz had to be constructed. Jobs provided during construction ended when the railroad began operation in 1985.

Today, Brazil is a Latin American colossus, ranking first in the world in several agricultural exports, including sugar (world's top producer), coffee (world's leading producer), orange juice, soybeans, and cocoa. Sixty percent of Brazil's export earnings come from manufactured goods, including steel, aircraft, and autos. Brazil produces 80 percent of its petroleum needs, and has developed a major industry manufacturing fuel from sugarcane (gasohol). Brazil has been industrializing at a rapid rate, but at least half the population is still in poverty, and Brazil's debt threatens its economy (Table 17.7).

LATIN AMERICA IN RETROSPECT

The problems of development facing Latin America are in part the result of its colonial heritage with its emphasis on the encomienda and its successor, the latifundia. Development of this region will require major changes. The reliance of much of Latin America's population on agriculture seems to dictate that part of this change must involve the agricultural sector. Recent studies indicate that 73 percent of the families of Latin America live at no better than subsistence level, and that the latifundia with 1200 or more acres (500 or more hectares) comprise less than 3 percent of the total landholdings but occupy 67 percent of all land. The minifundia constitute 76 percent of all farm operations but comprise only 7 percent of the total area under cultivation. In 1965, 94 percent of all arable land in Latin America belonged to 7 percent of the landholders, while only 6 percent of the available land was divided up among the remaining 93 percent. The need for structural changes in land tenure continues to be one of the major problems facing Latin America.

Coupled with the need for reform in the agricultural sector is the necessity to reduce the mono-export economy of many Latin American countries. The persistence of production of tropical crops or minerals for export whose prices and demand are controlled by outside forces makes most Latin American countries vulnerable to changing trade regulations of the industrialized countries. Protectionist policies of industrialized countries, which place restrictions on foreign trade in agricultural products, handicap the development of the Latin American countries. The impacts of such restrictions are compounded by the lack of firm commitment on the part of most Latin American countries to carry out the reforms necessary to transform the agricultural sector.

TABLE 17.7 Total External Debt in Billions of U.S. Dollars

	1988
Brazil	$120.2
Mexico	107.3
Argentina	62.7
India	60.7
Indonesia	57.6
U.S.S.R.	43.6
Poland	40.7
China	40.7
Turkey	40.6
Egypt	35.2

SOURCE: Institute of International Finance.

The second problem facing Latin America, as in all less developed regions, is control of population. Latin America has one of the highest growth rates of any region in the world, and the necessity to slow this rate of increase to allow developmental efforts to increase the standard of living is a paramount issue. Fortunately the Catholic church in nearly every country has become supportive of attempts to limit population growth, and most Latin American countries have adopted a comprehensive population control program.

Another issue facing the Latin American countries is the necessity to develop a stable and balanced industrial base to allow them to enter the takeoff stage of industrialization. This issue is becoming more complex for countries without petroleum, particularly Brazil, who are facing geometrically increasing costs for their petroleum imports. The countries of the tropics such as Brazil may in the long run be in a fortuitous position with respect to energy if they are able to utilize the biomass productivity of the tropics to produce alcohol to replace petroleum. Brazil anticipates producing sufficient alcohol in the 1990s to provide one-quarter of its energy needs. As petroleum resources are exhausted, the renewable tropical resources may actually enable these countries to provide their energy needs more cheaply than those that rely upon the stock resource of petroleum.

A final issue facing Latin America is the necessity to expand transportation to incorporate all of their population into the national economy. The absence of adequate road and railroad networks has effectively prevented the rural population from participating in gains in education and medical care and handicapped utilization of the entire country. As Latin American countries are able to integrate their population more fully into the national economy and society, they will be able to benefit more fully from the economic gains they are now experiencing.

FURTHER READINGS

AGUILAR, L. *Latin America 1984*. Washington: Stryker-Post Publications, 1984.

BAILEY, H. and NASATIR, A. *Latin America: The Development of Its Civilization*. Englewood Cliffs, N.J.: Prentice-Hall, 1973.

BARNET, R. J. and MULLER, R. E. *Global Reach: The Power of the Multinational Corporations*. New York: Simon and Schuster, 1974.

BLAKEMORE, H. and SMITH, C., eds. *Latin America: Geographical Perspectives*. 2d ed. London: Methuen, 1983.

BLOUET, B. and BLOUET, O., eds. *Latin America: An Introductory Survey*. New York: Wiley, 1982.

BROMLEY, R. and BROMLEY, R. *South American Development: A Geographical Introduction*. New York: Cambridge University Press, 1982.

BROWN, P. and SHUE, H., eds. *The Border That Joins: Mexican Migrants and U.S. Responsibility*. Totowa, N.J.: Rowman and Littlefield, 1983.

BROWNING, D. *El Salvador: Landscape and Society*. Oxford, England: Clarendon Press, 1971.

BURNS, E. *A History of Brazil*. 2d ed. New York: Columbia University Press, 1980.

CAVIEDES, C. *The Southern Cone: Realities of the Authoritarian State*. Totowa, N.J.: Rowman and Allanheld, 1984.

CHOUCRI, NAZLI. *Energy and Development in Latin America*. Lexington, Mass.: D. C. Heath, 1982.

CLARK, C. G. *Kingston, Jamaica: Urban Development and Social Change, 1692–1962*. Berkeley: University of California Press, 1975.

COLE, J. *Latin America: An Economic and Social Geography*. 3d ed. Totowa, N.J.: Rowman and Littlefield, 1976.

CROWLEY, W. and GRIFFIN, E. "Political Upheaval in Central America." *Focus*, September–October 1983.

DICKENSON, J. P. *Brazil: Studies in Industrial Geography*. London: Dawson, 1978.

DOOLITTLE, W. E. "Aboriginal Agricultural Development in the Valley of Sonora, Mexico." *Geographical Review* 70, 1980, pp. 327–342.

DOOLITTLE, WILLIAM E. "AGRICULTURAL EXPANSION IN A MARGINAL AREA OF MEXICO." *The Geographical Review* 73, July 1983, pp. 301–313.

"Falklands: 150 Years British." *Geographical Magazine*, January 1983, pp. 30–39.

GAUHAR, ATLAF, ed. *Regional Integration: The Latin American Experience*. Boulder, Colo.: Westview Press, 1986.

GILBERT, A., ed. *Latin American Development: A Geographical Perspective*. Baltimore: Penguin, 1974.

GILBERT, A., with HARDY, J. E., and RAMIREZ, R., eds. *Urbanization in Contemporary Latin America*. New York: Wiley, 1982.

GRIFFIN, E. and FORD, L. "Cities of Latin America." In *Cities of the World: World Regional Urban Development*, pp. 198–240. Edited by S. Brunn and J. Williams. New York: Harper & Row, 1983.

GRIFFIN, E. and FORD, L. "A Model of Latin American City Structure." *Geographical Review* 70, 1980, pp. 387–422.

GRIGG, D. *The Agricultural Systems of the World: An Ev-*

GWYNNE, ROBERT N. *Industrialization and Urbanization in Latin America.* Baltimore: Johns Hopkins University Press, 1986.

HALLAJ, MUHAMMAD. "Revolt in Occupied Palestine." *American-Arab Affairs* [Washington, D.C.] No. 24., Spring 1988.

HELMS, M. W. *Middle America: A Culture History of Heartland and Frontiers.* Englewood Cliffs, N.J.: Prentice-Hall, 1975.

HENNESSY, A. *The Frontier in Latin American History.* London: Edward Arnold, 1978.

HOY, D. and MACFIE, S. "Central America: A Bibliography of Economic, Political, and Cultural Conditions." Athens: University of Georgia, Department of Geography, 1982.

"Jamaica—Special Issue." *Journal of Geography* 82, September–October 1983, pp. 89–152.

JAMES, P. E. *Latin America.* New York; Odyssey Press, 1969.

KATZMAN, M. T. *Cities and Frontiers in Brazil: Regional Dimension of Economic Development.* Cambridge: Harvard University Press, 1977.

KRAUSE, W. and MATHIS, F. J. *Latin America and Economic Integration: Regional Planning for Development.* Iowa City: Iowa University Press, 1970.

LA FEBER, W. *The Panama Canal: The Crisis in Historical Perspective.* New York: Oxford University Press, 1978.

LANG, JAMES. *Inside Development in Latin America: A Report From the Dominican Republic, Colombia, and Brazil.* Chapel Hill: University of North Carolina Press, 1988.

LEVINSON, J. and DE ONIS, J. *The Alliance That Lost Its Way: A Critical Report on the Alliance for Progress.* Chicago: Quadrangle Books, 1970.

LEWIS, P. "Axioms for Reading the Landscape: Some Guides to the American Scene." In *The Interpretation of Ordinary Landscapes: Geographical Essays,* pp. 11–32. Edited by D. Meinig. New York: Oxford University Press, 1979.

LOWENTHAL, D. *West Indian Societies.* New York: Oxford University Press, 1972.

MCGRATH, P. "Paraguayan Powerhouse (Itaipu Dam)." *Geographical Magazine,* April 1983, pp. 192–197.

MACPHERSON, J. *Caribbean Lands.* 4th ed. New York: Longman, 1980.

MERRICK, T. and GRAHAM, D. *Population and Economic Development in Brazil: 1800 to the Present.* Baltimore: John Hopkins University Press, 1979.

MORAN, E. *Developing the Amazon.* Bloomington: Indiana University Press, 1981.

MORAN, EMILIO F., ed. *The Dilemma of Amazonian Development.* Boulder, Colo.: Westview Press, 1983.

MORRIS, A. *Latin America: Economic Development and Regional Differentiation.* Totowa, N.J.: Barnes and Noble, 1981.

———. *South America.* 2d ed. Totowa, N.J.: Barnes and Noble, 1982.

MORRIS, A. S. *Regional Disparities and Policy in Modern Argentina.* Glasgow, Scot.: Institute of Latin American Studies, 1975.

MORRIS, ARTHUR. *South America.* 3d ed. Totowa, N.J.: Barnes and Noble, 1987.

ODELL, P. R. and PRESTON, D. A. *Economies and Societies in Latin America: A Geographical Appraisal.* 2d ed. London: Wiley, 1978.

PARRY, J. H. *The Discovery of South America.* New York: Taplinger, 1979.

PEARCY, G. E. *The West Indian Scene.* New York: Van Nostrand Reinhold, 1965.

PERLOFF, H. S. *Alliance for Progress: A Social Invention in the Making.* Baltimore: Johns Hopkins University Press, 1969.

PICO, R. *The Geography of Puerto Rico.* Chicago: Aldine, 1974.

RENNER, JOHN, ed. *Source Book of South American Geography.* 2d ed. London: Longman, 1982.

RICHARDSON, C. B. *Caribbean Migrants: Environment and Survival on St. Kitts and Nevis.* Knoxville: University of Tennessee Press, 1983.

SILVERMAN, MARILYN. "Agrarian Processes Within Plantation Economies: Cases From Guyana and Coastal Equador." *The Canadian Review of Sociology and Anthropology* 24, November 1987, pp. 550–570.

SMITH, N. *Rainforest Corridors: The Transamazon Colonization Scheme.* Berkeley: University of California Press, 1982.

STERNBERG, H. *The Amazon River of Brazil.* Wiesbaden, Germany: Geographische Zeitschrift, Monograph no. 40, 1975.

STOHR, W. *Regional Development Experiences and Prospects in Latin America.* The Hague: Mouton, 1975.

TATA, ROBERT J. *Haiti: Land of Poverty.* Washington: University Press of America, 1982.

TAYLOR, A., ed. *Focus on South America.* New York: Praeger, 1973.

THEROUX, P. *The Old Patagonian Express: By Train Through the Americas.* Boston: Houghton Mifflin, 1979.

WAGLEY, C., ed. *Man in the Amazon.* Gainesville: University of Florida Press, 1974.

WEBB, K. *Geography of Latin America: A Regional Analysis.* Englewood Cliffs, N.J.: Prentice-Hall, 1972.

WEBB, K. E. *Changing Face of Northeast Brazil.* New York: Columbia University Press, 1974.

WEST, R. and AUGELLI, J. P. *Middle America; Its Lands and Peoples.* 2d ed. Englewood Cliffs, N.J.: Prentice-Hall, 1976.

Part Four

THE WORLD OF TOMORROW
issues for today

We Must Proclaim A Bold Objective—That Within A Decade No Child Will Go To Bed Hungry.
U.S. SECRETARY OF STATE HENRY A. KISSINGER, (1974)

Today's Rate and Scale Of Change Is Perceived By Many To Be Transcending The Adaptive Capacity of Individuals, Governments, Industry, And Nature. Caught In The Currents Of Turbulence, Man Searches For A Generic Cause To Give Reason To The Unreasonable And Hence A Basis Of Action. Like Fourasti's Bewildered Bee In A Moving Train, Man Appears To Have Forgotten Where He Comes From And Has Lost The Sense Of Where He Is Going.
CRAWFORD S. HOLLING, UNESCO COURIER

Lisbon, Portugal.

18

Tomorrow's Issues
world society
or world conflict

MAJOR CONCEPTS

Social inertia / Conspicuous consumption
Resources / Capital transfer / Alternative
foods / Acid rain / Fertility rate / Growth
rate / Direct consumption / Ozone depletion
Rain-forest destruction

IMPORTANT ISSUES

Relative economic development
Poverty
Resource allocation
Population growth
Deprivation
Distribution of economic systems
Development of alternatives
Job opportunities
Energy consumption
Adaptation of Western technology
World conflict or world prosperity
Alternative food sources
Energy demand and supply
Distribution of health care

> ### Geographical Characteristics of Tomorrow's Issues
>
> 1. Population increase will remain a major issue for the foreseeable future because of the momentum of growth associated with the large numbers of youth in less industrialized regions.
> 2. The need to increase food production in countries with rapidly increasing populations presents the need to both increase output from traditional crops and adopt new species or varieties.
> 3. Resources for economic development remain an issue because of their uneven distribution, but conservation techniques and alternate fuels offer some promise for both industrialized and less industrialized regions.
> 4. Issues related to culture and society in the world's regions involving distribution of wealth, the political power of less industrialized versus industrialized regions, and societal changes affecting the role of men and women may create major changes in the cultural geography of the world.
> 5. Environmental issues associated with erosion, loss of the tropical rainforest, air and water pollution, and waste disposal will pose major challenges for the foreseeable future.

The issues facing the world at the beginning of the twenty-first century will be closely related to those facing today's world. They will just be more intense if we do not begin a worldwide effort to resolve problems of relative economic development, poverty, and resource allocation. Some observers suggest that the intense nationalism underlying less industrialized countries' demands for equitable distribution of resources and wealth represent the greatest challenge the world has faced. The fragile economies that are the basis of life for the world's poor cannot survive without a committed and concerted effort by the industrial, developed world to ensure worldwise economic improvement.

If the industrialized world is unwilling to make this commitment, the potential for world peace is remote. Extremes of poverty and wealth, hunger and affluence, cannot coexist forever, either nationally or internationally. With the majority of the world dominated by a single issue—simple survival—the pressure on the wealthy to curb their conspicuous consumption will continue to intensify. For the poorest people—the 1 in 5 in the mire of absolute poverty—the outlook is grimmer. Their destiny is deprivation—insecure livelihoods, malnutrition, hunger, disease, despair, and early death. The potential solutions to the issues of tomorrow lie with increasing acceptance of the reality of the interdependent nature of our world. The issues facing the less developed world are more life-threatening than those facing the industrial world, but the fate of the poor of the world is inextricably tied to that of the wealthy.

The issues themselves are increasing world population, the Malthusian dilemma posed by the need to feed the increasing population, the moral responsibility to ensure basic human rights in addition to food, the inevitable necessity for reallocation of world resources, the political and economic instability resulting from the world division between rich and poor, and environmental problems presented as we attempt to resolve the litany of issues facing tomorrow's world.

POPULATION: THE CENTRAL ISSUE

In spite of some changes in population growth rates, which provide optimism for the future, population will remain the central issue in tomorrow's world. The problems of development, provision of adequate food and shelter, political unrest, social change, and human development are closely related to the overriding issue of population. The population problem is an inseparable part of the larger, overall problem of development, but it is more than that. Some observers maintain that excessive population growth is the greatest single obstacle to economic and social advancement in the developing world. There are other problems that handicap development and improvement of the human lot, but none are as pervasive, intractable, or punitive. Population growth affects all other tasks that must be accomplished if an acceptable worldwide standard of living is to be achieved.

The magnitude of the population issue that will face the world at the turn of the century has been alluded to throughout this book. It is helpful to place the issue of tomorrow's population in perspective by examining where we are today. There were 5.3 billion people in 1990 and there will be an estimated 6.5 billion at the end of the century. Only one generation ago (1945) there were only 2.3 billion people. Since then we have added an additional 3 billion, and with a worldwide population growth rate of 1.7 percent yearly, in the 10 years between now and the end of the century we will add more than 1 billion more. In only half a century the world's population will have nearly tripled.

The problem is the impact of the additional population on the already strained resource base supporting humankind. The degradation or outright destruction of the physical environment that has resulted from worldwide population growth represents one of the most severe challenges facing the human population. Environmental deterioration ranges from destruction of tropical rain forest by increased numbers of slash-and-burn farmers, expansion of sedentary tropical agriculture, or demands for fuel for growing peasant populations, to the environmental degradation from industry and mining in the industrialized world. Depletion or exhaustion of mineral and fuel resources, declines in the volume of fish caught in the oceans, or the pillage of the world's cropland by soil erosion, expanding suburbia, or strip mining are threats to the continued existence of world civilization. Attempts to limit the rate of population growth address the issue of quality of life, but that quality is directly related to the impact of population on the quality of the environment.

The present rapid population growth will begin to decline as family sizes decrease, but significant change will probably not occur before the end of the century. The high rate of population growth that the world is experiencing now is *not* the result of high fertlity rates. In fact, fertility rates are probably lower than those throughout most of human history. Even in the less industrialized the average number of live births per woman is only 4.7, compared to an average of 2.3 in the industrialized world. For the world as a whole, the average number of births per adult woman is slightly more than 4, which is less than the average number of live children most women bore in the past.

The reason for the rapid population growth is the lower infant mortality rate in essentially all countries since the end of World War II. Global life expectancy is now 65 years, and even in the less developed world is approximately 60 years. Lower death rates with their resulting increase in life expectancy are the cause of the population increase. Ironically the achievement of eliminating unnecessary, tragic, and wasteful infant deaths is now viewed as contributing to the problem of overpopulation. Even so, lowering the death rate in the past half century should not be regarded as a disaster. It is probably the greatest achievement of this century. It allows most people to live a relatively full span of years without the fear of premature death. Women can bear children with reasonable expectation that they will live to adulthood. It is the first and most critical step toward a better life, but it creates the issue of dealing with the population that survives.

The issue for tomorrow centers around the momentum that population growth has attained, and the impact of increased population on the potential for improving the general human condition. Rapid population growth has a built-in momentum that requires several generations to get back to zero growth. In high-growth populations each succeeding age group is larger than the previous one. The number of women in the premarriageable ages is usually twice as large as that in their mothers' generation (Figure 18.1). Even when the **fertility** rate is halved, the **growth** rate remains the same because the number of mothers is doubled. For the majority of the less industrialized world, if the average family size were reduced to 2 surviving children per woman, the population would still increase an average of 66 percent before a stable population is achieved.[1]

The second factor handicapping attempts to curtail population growth in the hopes of improving the standard of life is *social inertia*. Limiting the number of children in a family has a profound and persistent effect on society in general. It affects kinship duties and relationships, it may challenge religious beliefs,

---

Fertility rate: The number of children born to women in the reproductive age group (15–44) as a percent of women in that group.

Growth rate: The net rate of population increase (births minus deaths) per year.

---

[1] Nathan Keyfitz, "On the momentum of population growth," *Demography* 8, No. 1 (February 1971). p. 80.

World priorities: world realities

Tension and violence are on the rise in a world dangerously overarmed and undernourished. Two superpowers, locked into cold-war antagonisms, were the driving force behind an arms buildup without parallel in history. It has created a nuclear menace on hair-trigger alert, threatening all life on earth. Local wars are increasingly internationalized and destructive. Half of the governments in developing countries are under military control.

An arms race no country can afford crowds out social needs. Government budgets give lower priority to deep-seated, debilitating problems than to military solutions and the latest technological advances in weapons. In the civilized world the disregard of human rights and human life is appalling.

The public outcry, focused principally against nuclear weapons, is stronger, and firmly backed by church councils and many other organizations. How soon it will affect government priorities is not yet clear but it has already become a healthy counterweight to official policy that has lost touch with reality.

In 25 countries young men are eligible to go to war at an earlier age than they are eligible to vote.

Every minute, 30 children die for want of food and and inexpensive vaccines, and every minute the world's military budget absorbs $1.8 million of the public treasure.

The U.S. now devotes over $300 billion a year to military defense against foreign enemies but 45 percent of Americans are afraid to go out alone at night within one mile of their homes.

Among 20 developing countries with the largest foreign debt, arms imports between 1976 and 1980 were equivalent to 20 percent of the increase in debt in that period.

The world's stockpile of nuclear weapons represents an explosive force over 5000 times greater than all the munitions used in World War II.

The cost of a single new nuclear submarine equals the annual education budget of 23 developing countries with 160 million school-age children.

The USSR, which spent $4.6 trillion between 1960 and 1988 for military power, now ranks 23rd among 142 countries in economic-social performance.

In a year when U.S. farmers were paid to take nearly 100 million acres of cropland out of production, 450 million people in the world are starving.

Source: Ruth Leger Sivard, *World Military and Social Expenditures*, 1988 (Washington: World Priorities, 1988) p. 5.

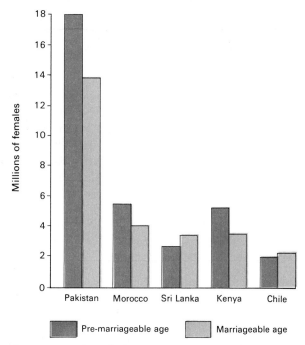

Figure 18.1 Growth in population of females of marriageable age.

and it creates changes in personal perception of wellbeing. Although it is easy to show that decreasing the number of births will benefit an entire country, it is more difficult to demonstrate to an individual farmer that a smaller family is in fact an economic gain. The farmer who has three children compared to his neighbor's six may not see any real benefit in total farm production, in assistance with the labor of the farm, in protection for his old age, in prestige, and in ability to accumulate capital through assigning differing tasks to family members. Even where social and economic changes associated with movement from rural to urban areas make the utility of children for labor marginal and increase the cost of providing the basics of life, engrained attitudes toward size of family can rarely be overcome in a single generation. The fear of being left without surviving children is so great in many less industrialized societies that even when there is no economic need for large families, the desire for many children persists (Table 18.1).

If the world is to attain an acceptable standard of living for everyone, one of the areas of change must be in population. It is not impossible to raise living standards with a rapid population growth, but it is much more difficult. Obviously fast-growing populations require a greater proportion of capital invest-

TABLE 18.1 Desired Number of Children Compared to Births per Woman (Total Fertility Rate)

Country	Desired Number of Children	Fertility Rate 1970	1987
Benin	7.6	6.9	6.5
Cameroon	8.0	5.7	7.0
Ghana	6.0	6.6	6.3
Ivory Coast	8.4	6.6	7.0
Kenya	7.2	7.8	7.7
Egypt	4.1	6.1	4.5
Turkey	3.0	5.3	3.7
Bangladesh	4.1	6.9	5.5
Pakistan	4.2	6.9	6.7
Indonesia	4.2	5.5	3.7
South Korea	3.1	4.3	2.1
Colombia	4.0	5.4	3.2
Peru	3.8	6.4	4.1
Venezuela	4.3	5.4	3.8
Mexico	4.4	6.6	3.6
Haiti	3.6	5.8	4.3
Jamaica	4.1	–	2.9
Austria	2.2	2.3	1.5
Italy	2.2	2.4	1.5
United Kingdom	2.4	2.4	1.8
West Germany	2.4	2.1	1.3

SOURCE: *International Family Planning Perspectives*, vol. 11, No. 2 (June 1985), p. 37; and *The World Bank Atlas, 1988*, pp. 6–9.

ment in facilities for additional educational opportunities, medical care, jobs, homes, and food and clothing for the incremental population growth. Countries with slower growth rates can use the same capital to increase worker productivity and hence standards of living. It is abundantly clear that industrialization that occurred in Europe in the nineteenth century took place without competing demands of population growth rates of 2 or 3 percent per year. The capital accumulation that Europe needed for industrialization was facilitated by population growth rates less than half those in less industrialized countries today. It may be possible to raise the standard of living of all residents of the world without curbing the population growth, but a rapidly increasing population will hinder the pace of improvement.

Assuming that population limitation will benefit the majority of the world, the issues for tomorrow center around methods. Substantial declines in growth rates have been recorded in China, South Korea, Taiwan, Singapore, Hong Kong, and a number of other countries where there was a concerted effort to provide information and techniques for limiting population growth. The issue is not whether population growth can be curtailed, but whether countries are willing to invest the capital necessary to make such a change. Many poor countries are suspicious of outside suggestions that they limit their population growth. They assume that the encouragement by industrialized nations for population control is a selfish attempt to protect Western industrial interests. Programs for population limitation work best when they come from the people of the country involved.

Nearly all countries have some type of population control program, but they suffer from limited funding. Less than 2 percent of foreign aid funds are channeled into population planning programs. In most less developed countries, less than one-half of 1 percent of national budgets go to population programs. Countries like China that recognize the interdependence of population growth and standard of living have made population planning a central part of their development effort. Until the poorest countries of the world recognize this fundamental relationship, population will be a central problem preventing rapid increase in quality of life for the world's poorest.

FOOD: THE MORAL DILEMMA

Closely related to the population issue in the coming decades will be food production and availability. For the world as a whole, both total food production and per capita production have increased in the last few years (Figure 18.2). Unfortunately this food has not been evenly distributed, and much of the growth has been the result of increased energy inputs in the industrialized world and the adoption of hybrid varieties in the developing world. The resulting increases in food supplies have not been uniformly distributed, as even the Green Revolution failed to be the panacea for world hunger that it promised. The issue facing the world of tomorrow is twofold: If world population increases at the rate estimated by the United Nations, food production must rise by 50 percent if there are no changes in consumption patterns. The second issue is the need to change the consumption patterns and resulting food distribution on a world basis. As has been noted earlier, the wealthy nations of the world consume an inordinate amount of the world's resources directly or indirectly. Even in consumption of food, the industrial nations have the world's lowest **direct** per capita consumption of cereals but consume

---

Direct consumption: Direct human consumption of cereal grains, and such, rather than meat or alcohol products.

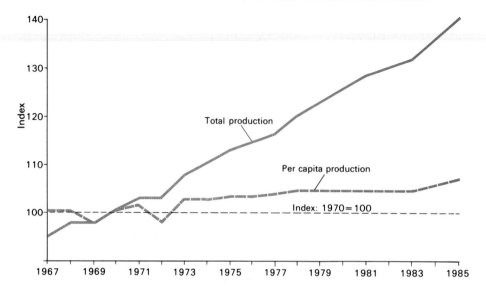

Figure 18.2 Per capita food production is growing at a slow rate because of rapid population growth.

nearly twice as much per capita indirectly as meat. The total annual consumption of grain for livestock feed in the rich countries is greater than the total human consumption of cereals in India and China combined.

There are other factors affecting the availability of food on a world scale. The majority of food exports are from the industrialized countries. Their high yields are associated with abundant use of fertilizer and mechanization of agriculture. The increasing cost of energy dictates that if the high yields of the United States, Japan, and Western Europe are to be maintained, the price paid to farmers must increase proportionally. Tragically, people in many importing countries do not have sufficient capital to pay the increased cost of the expensive food from these industrial nations. Faced with the prospect of higher cost and a limited market, farmers in the food-surplus areas will produce less, further handicapping attempts to increase the amount of food available to the people of the less industrialized world. Environmental problems of erosion, loss of soil fertility, and spread of insects and weeds further curtail the productivity of the earth's agricultural lands. Spread of urban areas as population increases takes additional acres from the agricultural resource base.

Solution of problems of food production will require a multifaceted approach. It would include changing the consumption habits of the rich minority population, land reform to place land in the hands of the peasant farmers, experimentation with new crops and new varieties, provision of farm credit programs, and agricultural education. Changing the food habits of the wealthy world could have a ripple effect in that as the market for high-priced beef is removed, the wealthy landowners of much of the less industrialized world may revert to their original practices of allowing individual peasant farmers to grow crops for their own consumption. With the present demand for meat and the wealth in industrial countries, the poor of the less industrialized world cannot compete for the products of the Green Revolution.

Another means of providing greater food production in the less industrialized world countries involves land reform. The coming decades will see an increasing demand on the part of the peasant majority for expropriation of landholdings of the wealthy minority of their countries. Such reforms may not result in absolute increases in food production, but they change the production ethic from production for export to production for local or regional consumption. Some observers maintain that land reform is essential if the world is to feed its population adequately by the year 2000.

These social changes require difficult changes in social organization and attitudes that are often implemented only with great human suffering. Other approaches to resolving the food production dilemma require only application of existing technology. The Green Revolution, which yielded the rice and wheat varieties that provided the cushion of food for the 1970s, has not been equally applied to other staple crops. Hybridization of such basic foods as manioc, yams, sweet potatoes, millets, or peanuts may yield equally spectacular improvements in productivity. The industrial world could supply adequate funding to ensure that the other staple foods of the world are researched to develop higher yielding varieties.

Alternative food sources

At a very rough estimate, 5 to 10 million species inhabit the Earth. The number is rough because a large proportion of the Earth's species—about 2.5 million—are yet undiscovered. Estimates are based on what is known about well-studied areas. About 1.5 million species have been scientifically described, named, and classified.

Only a tiny portion of the Earth's species have been used by people on a large scale. Just three cereal grain species—rice, wheat, and corn—serve most of the world's population as principal food sources. Over half the Earth's cropland is devoted to growing them. Humanity relies on less than 20 other plant species as major food sources. They include grains such as millet and sorghum; legumes such as peas, beans, soybeans, and peanuts; root crops such as potatoes, yams, sweet potatoes, and cassava; sugar cane and sugar beets; coconuts; and bananas.

Only a few dozen species of animals have been domesticated. Just nine domesticated animals supply almost all of the animal protein people consume. These are cattle, pigs, sheep, goats, water buffalo, chickens, ducks, geese, and turkeys. Poultry, beef, and pork in about equal amounts add up to some 90 percent of the meat people consume.

The only significant use of wild animal species for food is marine fish. Since 1971, the world fisheries catch has fluctuated between 70- and 75-million metric tons, supplying approximately 14% of the animal protein in the human diet worldwide. More indirectly, as a component in animal feeds, marine fish nourish man. As with plants, the possibilities of domesticating animals have hardly been explored. Aquaculture, the domestication of marine and freshwater fish, has only partially realized its enormous potential.

Despite human exploitation of relatively few plant and animal species, many other species of wild plants and animals have equal potential as food crops or serve other useful functions. Throughout the world, wild relatives of now-exploited crops and wild plant and animal species with unexplored value are being lost as natural habitats are transformed by development. Systematic investigation of the usefulness of plant species as medicines, for example, has barely begun. The thorough screening of even a single species for useful compounds is a difficult and laborious commercial enterprise requiring extensive commitment and investment.

Just as institutions such as the International Agricultural Research Centers devote research and training to the production of major crops and livestock of developing countries, other institutions should focus on domesticating new species and developing their economic potential.

Source: Caldwell, Hahn, "Environment & Development: Balancing Both Worlds." Horizons Agency for International Development, Fall 1984, p. 20–22.

Another approach to increasing food supply involves consideration of *alternative food* sources. There are many underexploited tropical food plants with high promise. Some specific crops that have not been utilized in the past but show high promise are such cereals as the *grain amaranths, echinochloa turnerana,* and *quinua*. The grain amaranths are fast-growing, cereal-like plants that produce a mass of seeds in heads similar to sorghum. They are among the most efficient photosynthesizers in the plant kingdom and are exceptionally high in protein. There are a number of amaranth varieties, some of which have a higher protein content than corn. Amaranths were cultivated and eaten in Latin America before the European conquest and subsequent expansion of corn throughout the continent. They are still important to Asian hill tribes and people in parts of Africa and Latin America. The plant is an annual, grows well under adverse conditions, is easily cultivated, and adapts well to rural farmers' small plots and mixed cropping techniques. Amaranth grain can be milled into a flour that is high in gluten and has excellent baking qualities. Research on the utility of amaranths is still in its infancy.

Quinua is a protein-rich grain that may prove to be a better food source than most true cereals. It is a staple for millions in the highlands of Bolivia, Chile, Ecuador, and Peru. It can be cultivated at altitudes of 8,000 to 13,000 feet (2,500 to 4,000 meters) and produces an abundance of white or pink seeds in large sorghum-like clusters. *Echinochloa turnerana* is a little-known wild Australian grass that has never been cultivated. It offers promise as a forage and grain crop for arid regions because only a single saturation of the soil by either heavy rains or irrigation is necessary for it to develop from germination to harvest. It produces a nutritious and palatable fodder, and most livestock will eat the grain.

In addition to these three cereal-like crops, there are several roots and tubers with great promise. A large number of root crops are underexploited because they have never received attention from geneticists. Among them is *arracacah*, a perennial that produces large, thick, edible, carrot-shaped, starchy roots. It is native to the Andean highlands and is cultivated in most Latin American Indian countries, usually in small gardens for local use. All parts of the plant can be

An Ecuadorian holds a dish of quinua. In South America, the grain is normally cooked whole rather than made into flour.

used for either livestock feed or human consumption, and in many areas it has replaced potatoes as a staple because it costs only half as much to grow.

In tropical areas, *cocoyams* are grown in a wide range of environments, but they have never been systematically examined to select the best varieties. Some varieties are capable of yielding 148 tons per acre (370 tons per hectare). The tender leaves can be used as a spinach-like vegetable, and although low in protein, they do provide minerals and vitamins. Cocoyams can be grown in a wide range of climates and are relatively disease-resistant and pest-free.

Certain vegetables hold great promise, the most widely known being the so-called winged bean. The winged bean is a tropical legume with a multitude of nitrogen-fixing nodules. It produces seeds, pods, and leaves that are all edible by humans or livestock. It is high in protein, produces edible oil from the seeds, and has a tuberous root that provides starch and protein. It rivals soybeans in oil and protein content with the added advantage of having protein-rich roots and edible foliage. It is a fast-growing perennial that grows in the wet tropics where protein deficiency is a serious problem that is difficult to overcome. The plant is a vine that grows to over 10 feet (3 meters) high when supported, producing pods with four "wings," each containing up to 20 seeds. The winged bean is cultivated mainly for its young, tender pods, which are picked 10 weeks after sowing. The plant continues to bear pods indefinitely, and young leaves and shoots may be eaten as a leafy vegetable. The immature roots of the plant are eaten like potatoes and contain more than 20 percent protein.

Another example of a food source that has been underutilized is the water buffalo. The water buffalo does well in tropical areas but has been virtually ignored in research. Water buffalo are of major importance in only a few countries, notably Egypt where it is the most important meat-producing domestic animal. It also produces 60 percent of India's milk needs. Butter derived from buffalo milk is the major source of cooking oil in India. Water buffalo are also useful for draft power and can be used in agriculture. Buffalo are highly adaptable, thriving in both dry and wet regions. The buffalo is related to cattle of the Northern Hemisphere, but is much more efficient in consumption of forage found in tropical and subtropical areas. It is at home in the hot, humid environments and is not susceptible to the diseases and heat there. Although crucial to Egypt's economy and food supply, buffalo are not used in any other country in Africa.

Cocoyams.

They have been used in the Amazon Basin and in Venezuela to some degree, but they are almost unknown in other parts of South and Central America. The limited distribution of the water buffalo illustrates the potential for increased food production by using alternate plant and animal life in the developing world. The immediate necessity is to begin a program of *serious* research and education to facilitate the adoption of alternate food sources.

Fuel needs of the residents of the less industrialized world and resulting deforestation may both be aided by adoption of new species of trees. The *Leucaena leucocephala* (commonly called *leucaena*) tree is an example of an alternative tree that holds great promise for resolving problems of fuel, reforestation, and food. There are several varieties of leucaena, but the most important is a single-trunked tree. In tropical areas it attains a height of 65 feet (20 meters) and a diameter of 16 inches (40 centimeters) in just five years. The leaves and young pods can be eaten in soups and salads. The wood is suitable for charcoal making, cooking, and pulp for paper. When properly cured, it can be used in construction. Leucaena is important because it helps to maintain soil fertility. Once leucaena is established, it regrows after being cut. It is being adopted in the Philippines and other areas of the tropics where it holds great promise.

PROVIDING OTHER BASIC HUMAN NEEDS

Provision of adequate food is essential for human life, but maintenance of the human spirit requires provision of other basic necessities. A continuing issue for the world will be the need to provide medical care, education, housing, employment opportunities, and freedom from fear for the evergrowing world population at the turn of the century. The tragedy of the modern interconnected world is that the industrialized nations have had only limited success in exporting their standard of living to the rest of the world. There is little reason to believe that the problems of providing basic needs to the world of 1990 will change much by the turn of the century. The fundamental problem is that we do not have a clear idea of how to resolve the issue of providing basic needs of individual rights and freedoms.

An examination of only one of the problems, providing jobs for the world population, indicates the staggering complexity and magnitude of the problem the world will face 20 years from now. In the industrialized region, 65 million jobs will have to be created between 1990 and the year 2000 to absorb those joining the work force and to eliminate existing unemployment. In the less industrialized world over 600 million new jobs would be needed to give each member of the work force an income adequate to meet minimum basic needs. The near impossibility of providing so many jobs independently dictates that meeting basic human needs is nearly impossible at the individual country level.

Providing medical care and housing are equally monumental tasks for less industrialized countries. The cost of providing Western-style medicine is prohibitive. Provision of housing modeled after the north and Western nations is equally difficult. Countries like China, who have succeeded in providing basic needs for one-quarter of the human population without even increasing the average per capita gross national product above $500 per year, may provide a more usable pattern.

It is estimated that at the beginning of 1990 there were at least 1.5 billion people whose basic human needs are unfulfilled. The irony for the world is that there is sufficient money to provide for the human needs of the entire population. The issue for tomorrow's world is ensuring that the wealth of the world is used to better the quality of life of its inhabitants rather than to destroy it. Military expenditures on a world basis equaled 1.8 million a minute in 1987.

The only place that water buffalo are used extensively in Africa is in Egypt, as in the Nile Delta where these buffalo are harnessed to a water wheel used to raise water for irrigation.

Analysis of government spending at the world level reveals that only one government in three spends as much on health care as it does on armaments. A single multiple-warhead nuclear missile costs as much as providing food for 50 million malnourished children, or building and stocking 65,000 local health care facilities, or building 340,000 simple primary schools in underdeveloped countries. The problem of providing for basic needs at a world scale is readily resolvable. The money required to provide adequate food, water, education, health care, and housing for everyone in the world is estimated at $17 billion a year. This is a huge sum of money, about as much as the world spends on arms every two weeks.[2]

The issue for tomorrow is how the wasteful use of world resources for destructive purposes can be changed to raise the standard of living of the entire human population. The total world military expenditure is approximately $500 billion a year, greater than the total income of the poorest half of humanity. Provision of adequate medical care can be accomplished through adopting models other than the traditional Western system. Smallpox has been eradicated from the world for all intents and purposes through massive worldwide concern and technology. Other diseases of less industrialized countries such as bilharzia, intestinal parasites, or yaws could also be eliminated if the medical profession, concentrated in the industrialized wealthy nations and focusing on the problems of the wealthy minority of the world, began to devote time to them. This may involve a radical departure in medical care delivery systems, like the Chinese barefoot doctors. The point is that with the technology and abilities available, it is essential to have the will to adopt the programs required to institute change. The simple issue of potable water for the world's population exemplifies the issue of tomorrow and of today. The waste of human effort involved in African women walking up to 6 miles (10 kilometers) to obtain water for their families, the human misery from unclean water sources, and the increasing demand for water as population increases illustrate the magnitude of this problem.

RESOURCE ALLOCATION AND ECONOMIC DEVELOPMENT

Another broad issue that will be the focus of increased attention at the turn of the century will be allocation of the world's finite resources. The present inequities in resource consumption by the wealthy nations will probably defeat all attempts at true equity in resource allocation, making this a perennial issue for the present generation. There may be more questions asked in the future on how to share world resources—questions concerning who has had what in the past, in what proportion to the total, to its population, and to its contribution. There may be a conscious trend toward redistributing resource opportunities and wealth at all levels. If the industrialized, developed world persists in its conspicuous consumption of resources, there may well be a tendency for increased world conflict, war, and social disorder. There is an irreversible trend toward equality in the distribution of resources, and it is in the best interest of the world's wealthy minority to facilitate an orderly transfer of resources to the less industrialized countries.

The difficulty of resource allocation lies in the conflicting views of the world as either a finite or an infinite storehouse of resources depending on technological changes. Those who support the view of limited resources argue that it is impossible for all people in the world to attain the level of life of the industrialized countries because resources are inadequate. These individuals accept the idea that there are absolute limits to the physical resources of the earth and that it is necessary to distribute those resources equitably. The conflicting view maintains that there are an infinite amount and variety of resources, based on the technological ability of the society of a place. These observers argue that the problem is not one of reallocation of the resources, but simply of producing more to allow the less wealthy of the world to increase their standard of living. Such conflicting views make it difficult for the world to grapple effectively with the issue of resource allocation. As long as a vocal group argues that the need to reallocate world resources is not the issue, it is impossible to restructure the method by which resources are made available.

ENERGY: A CHANGING RESOURCE

Some resources, particularly the fossil fuels, are available only in finite quantities. The increasing cost and scarcity of petroleum in the last two decades is indicative of the problems facing the world of the twenty-first century. The amount of energy available in presently used sources of petroleum is sufficient for only the next 15 years at present rates of consumption, but future discoveries will extend the supply into the twenty-first century. Clearly, however, the less industrialized countries must look to alternative energy sources if they are to increase their standard of living,

---

[2]*Development Forum*, January–February 1980 (New York: United Nations), p. 8.

and the wealthy must look to alternative sources to maintain their high standard of living.

In the less industrialized countries wood is a major energy source, and production of alternative tree crops such as the leucaena could allow them to provide adequate fuel for cooking. Leucaena does not lend itself well to the internal combustion engine, and even the relatively modest demands for petroleum for the less industrialized countries are going to be increasingly costly. The disproportionate share of wealth in the wealthy countries will allow them to control the majority of the world's energy. A major issue for the twenty-first century is the development of a program to ensure that the less industrialized countries have access to the energy necessary to industrialize if they so desire.

Another issue for tomorrow's world centers around developing alternative sources of energy. Not all countries of the world need turn to the industrialized, energy-profligate life-styles of Western Europe and North America. The move to develop alcohol from plant material as an alternative fuel, which is providing a significant portion of the fuel for automobiles in Brazil, may in fact benefit the less industrialized countries. The countries with a strong agricultural base such as those in the tropics may in the future be most able to acquire energy through their own efforts. Countries in the tropics and subtropics also may have an advantage in utilizing solar energy. India, for example, could provide for the energy needs of its 500 million peasant village dwellers through use of a simple solar cooker. Unfortunately, the Indian government has failed to recognize the utility of solar energy and is attempting to develop large electrical generating plants to provide electricity for cooking and heating. If the world problems of energy use are to be resolved, countries like India will have to utilize solar and other energy sources that they have in abundance rather than relying upon costly imported petroleum. Development of local energy resources would also free capital for investment to improve the quality of life of the Indian peasant.

Another alternative energy source used in China is methane gas, produced with a simple underground chamber where waste products are placed in an air-free (anaerobic) atmosphere and anaerobic bacteria convert the vegetative matter into methane gas (*biogas*). A simple, inexpensive biogas conversion project can be completed using local technology for under $100. The waste from a single cow and a family with less than an acre of ground produces enough gas to meet the energy needs for cooking and heating for the family. It is essential that solutions to tomorrow's issues be addressed through consideration of alternative energy sources as well as through better allocation of existing resources.

Another of the critical issues facing the world of tomorrow will be the necessity to conserve resources through changing life-styles in the industrial nations. Much of the consumption of resources in the industrial world is wasteful, and relatively simple conservation practices can result in major savings and resource consumption. Recycling of common products such as copper, steel, paper, aluminum, and tin not only free resources for use elsewhere in the world but they foster a changing attitude toward life-style. The world of tomorrow with two or three times the population of today will require everyone to utilize the resources of the earth in a responsible and frugal fashion.

The problem of providing energy for common daily needs such as cooking or heating water is often almost of crisis proportion in the less industrialized world. To peoples who rely on gathering free wood for fuel, the daily chore of obtaining enough for cooking and heating water for washing and bathing is time-consuming. An estimated one-half of the world's population relies on traditional fuels of wood, charcoal, dung, and agricultural waste for cooking and other energy needs. Growing populations in these countries are destroying forests for fuel or to clear land for ag-

Biogas unit being examined by U.N. observers in China.

riculture. Some estimates place the loss of tropical forest since 1950 at 37 percent for Latin America, 66 percent for Central America, 38 percent for Southeast Asia, and 52 percent for Central Africa. Often however, there are abundant energy supplies that are not utilized. In the Philippines the Filipino word for rice husk is "ipa," and ipa stoves were common before the introduction of kerosene, gas, and electricity. Ipa stoves burn rice husks that are otherwise wasted, but the desire to adopt Western ways has led to their abandonment. Increasing costs for energy on a world scale are fostering a return of the ipa stoves, which produce an almost smokeless heat. Ipa stoves consist of a simple metal hopper into which rice husks are dumped, a valve to control the amount of rice husks dropped into the fire chamber, and a pipe to carry out the smoke at the other end. One ipa stove built by a local craftsman provides all the cooking and heating needs for a hospital in a small Filipino community at a cost of only 10¢ a day.

In Kenya a new stove has been developed that improves the widely used charcoal brazier (jiko). It utilizes the same fuel, but consumes only half as much charcoal to produce equivalent amounts of heat. Throughout the less industrialized world the increasing cost of petroleum is causing local craftsmen and artisans to reexamine the traditional cooking implements and to increase their efficiency. If simple technical changes were made in the cooking stoves used in the less industrialized world, the demand for wood for fuel could be reduced by at least 50 percent. In simple terms, twice as many people in the rural less industrialized world could be provided with energy at present levels with no greater consumption of forest resources. Where alternative fuels such as straw, leaves, corn cobs, or other minerals that are presently wasted are utilized, the impact on the world's forests is reduced even more.

In the Himalayas the life of a typical woman illustrates the problem of energy for the world's majority. For a village woman life begins each day at 3:30 A.M. as she rises and splashes her face with water carried the night before. She then must begin threshing a supply of rice for the day, using a heavy wooden beam to power her millstones. It requires an hour to mill the day's flour and then another hour to walk to fetch water. Preparation for breakfast takes three or four logs for the fire, and she obtains the logs by walking to the remaining forest on the mountains and carrying back 70 to 90 pounds (30 to 40 kilograms) of wood. Wood is becoming increasingly difficult to obtain because the demand for firewood by an ever-growing population has pushed the edge of the forest up to the highest peaks, far from the villages, which 10 years ago were surrounded by woods. It takes two days per week to cut sufficient wood for a family's weekly fuel needs. Kerosene can be purchased, but the place in Nepal where road development allows kerosene to be delivered is a day's walk away. The price of kerosene is extremely high, resricting its use to dinnertime in the evening when the family has worked until dark in the fields. The government solution to the problem, based on Western technical advisers, is to build a $1.5 billion dam. The cost is seven and one-half times the total budget of the entire country. The dam will be 920 feet high and 227 feet

The lack of even simple technology such as a rope and pulley adds to the work of women of most less-developed regions. (Women of Mali drawing water.)

THE HUMAN DIMENSION

A Windmill and a Farmer

Bahri Yadav is a 45-year-old farmer of Deokali village in Ghazipur, Uttar Pradesh, India. He lives in an extended family with his adult brother and their mother. Besides two males and three female adults (wives and mother) in the family, the household has five school-age children.

With 1.6 hectares (2.95 acres) of land and four cattle (two bullocks, one cow, and a calf), the Yadav household is typical of small farmers in a backward district. There was no water source on their land which, in fact, was dry and of poor soil quality. With this meager resource base, the Yadav family has been merely eking out its livelihood by cultivating the 1.6 hectares and depending solely on the grace of nature. The main crops were paddy (rice) and sorghum in the rainy season with an average yield of 625 kilograms/hectare (560 lb./acre) and 500 kilograms/hectare (445 lb./acre), respectively. The yields were extremely poor. In the winter season the Yadavs would grow wheat. Wheat yields varied between 375 and 1000 kilos/hectare (890 lb./acre) depending on the amount of rainfall. Once in every three years the wheat crop failed completely. Without irrigation facilities the Yadavs could not cultivate any summer crops.

Farming being the only source of income, the dry-land cultivation of the Yadavs could barely meet their subsistence needs. In fact, their economic situation had worsened so much that since 1975 the family made serious attempts to have some of them migrate to an urban area. The Yadavs had been searching for jobs in the nearby towns without any success.

It was in this desperate situation that a bioenergy source—namely, a windmill—dramatically turned the fate of the Yadav family. It was a simple windmill of Dutch design, costing about 10,000 rupees. In 1981 a development group selected the Yadav family as a possible windmill testing site.

A 4-inch borewell was sunk in the Yadav farm in October 1981. The development officials helped the Yadavs in getting a 33 percent subsidy for the whole project from the Indian Small Farmers' Development Agency. An earthen water storage tank of 0.1 hectare (.247 acres) was constructed in February 1982.

Having installed the windmill, the Yadavs applied pyrite on the farm for soil improvement and changed the cropping pattern completely. First, on the 0.8 hectare (2 acres) paddy field, they started cultivating improved, high-yield varieties, which from the second year onward produced an average yield of 1700 kilos/hectare (1500 lb./acre) compared to 625 kilos/hectare with the previous varieties. This was more than sufficient for meeting the staple food needs of the family. In fact, a small part of the paddy output even became a marketable surplus, bringing some cash income for the family.

Since November to January are low-wind-velocity months, the Yadavs found the windmill energy inadequate for wheat cultivation. Instead, in 1983 they started experimenting with sugarcane cultivation on 0.25 hectares (.62 acres), partly in response to the ready market at a nearby sugar factory. However, the yield of sugarcane obtained was not very good, although the family received 400 rupees as cash income in the first year.

But what made the Yadavs augment the food energy nexus on the farm was their decision to cultivate year-round a variety of vegetables (chili, potatoes, peas, onions, and others). With 1 hectare (2.47 acres) of land under vegetable cultivation, the Yadav family started earning a net annual income of 3,049 rupees in addition to meeting their own requirements for vegetables.

Besides agriculture, the Yadavs had been keenly interested in rearing fish in the storage tank from the very beginning of the project. With the adoption of scientific fish-rearing methods, 0.1 hectare (.247 acres) of water area in the tank yielded a net annual income of 436 rupees.

Thus a simple bioenergy input in the form of a windmill on the Yadav's farm has substantially augmented the food productivity and family income. In monetary terms, the total net annual income of the Yadav family has been raised to 3,885 rupees ($243) as compared to less than the subsistence income of 1,000 rupees before the bioenergy intervention. It should be noted that the net annual income of 3,885 rupees excludes the family's subsistence needs of staple food, rice, vegetables, fish, and the cash income from the marketable surplus of rice (paddy).

With the increased food and energy created by the windmill, this small farmer is fairly confident that there is no necessity to migrate to an urban center. At least the family is no longer threatened with the specter of hunger and poverty.

This one example, of the improvement in the life of a single family, only indicates the potential of renewable energy sources in augmenting the food-energy nexus. Throughout the less industrial world where agriculture is the dominant economic activity, a host of similar opportunities exist for improving the quality of life by the addition of simple technology.

---

long (850 meters long and 210 meters high) and will take 10 years to build, produce 3,600,000 kilowatts of electricity, and irrigate 714,000 acres (300,000 hectares). The tremendous size of the project dwarfs the tiny country, and the majority of the power will be exported to India. An alternative is to develop the existing waterpower sources of the Nepalese people who have used primitive waterwheels for generations. Engineers from West Germany have worked in Nepal and in the 1970s developed a small turbine electric plant that can be installed at a local level. These generating plants are simple enough to be built in engineering shops in Nepal, require limited maintenance, use only materials obtainable in Nepal without foreign currency, are comparable in efficiency and cheaper than imported turbines, and have parts that can be moved by human porters with no part exceeding 155 pounds (70 kilograms). The turbine developed by the West German engineers is suitable for use in the existing watercourses that bring water to the terraced rice fields. The plants do work that is difficult and demanding for humans, including milling flour, pressing oil seeds, and husking rice. The installation is done by a team of three Nepalese: a fitter, a bricklayer, and an assistant. The promise for the future of the Nepalese experience is that small-scale projects utilizing local input are suitable for resolution of major energy problems.

CAPITAL TRANSFER: AID OR HINDRANCE

The problems facing the world today and tomorrow in the less industrialized areas of Africa, Latin America, and Asia require not only the application of existing expertise but capital to finance those changes and technology appropriate to the needs of the local populace. The so-called nonaligned nations, the 114 small countries that comprise the world's poor, have requested that industrialized countries transfer to the less developed economies the equivalent of 1 percent of their gross national product as foreign aid. Although a small percentage, few countries presently donate this much (Figure 18.3). This *capital transfer* represents but a small fraction of the money needed if relative inequities in world wealth are to be overcome.

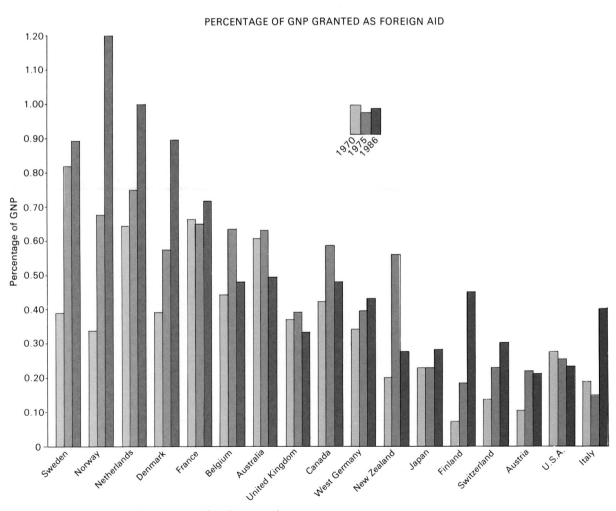

Figure 18.3 Percentage of GNP granted as foreign aid.

Even with the 1 percent, it is anticipated that the gap between the rich and poor countries will be wider 10 years from now than it is today.

There are some who argue that although the aid that is received is almost indispensable, it does not address the real problem of increasing standards of living. Aid grants allow a country to meet current responsibilities but do not assist in overcoming long-term problems. The problem is exacerbated by granting loans to less industrialized countries because repayment consumes an increasing portion of the poorer countries' gross national product. As of 1988, less industrialized countries owed $1,223 billion, and the interest and other service costs totaled $152 billion a year. (Figure 18.4).

Increased prices not only in the less industrialized countries but at a world level are causing worldwide inflation that threatens to jeopardize the minute advances the less industrialized countries have made in the previous decades and decrease the ability of the industrial countries to assist growth. In the less industrialized countries the growth rate was an average of 5.2 percent in the 1970s compared to less than 3 percent in the 1980s. The slowing of the growth rate coupled with rising inflation affects the entire world, including the centrally planned economies such as the Soviet Union.

Although the effects of inflation are widespread, disproportionate damage is concentrated in the less developed countries. In Peru, for example, between 1983 and 1989, the real buying power of workers was cut in half although wages increased. In Ghana, the capital city of Accra exemplifies the effects of inflation. The impoverished majority require one-quarter of their daily wages to pay the cost of traveling to and from work. It takes three days' wages to buy a kilogram (2.2 pounds) of meat and one day's wages for a can of condensed milk. (Figure 18.5).

Although past inflationary periods have affected only parts of the world, today's inflation is a worldwide phenomenon affecting all. The increasing interdependence of the world and the growing awareness of the finite nature of its resource base has pushed the inflationary spiral to crisis dimensions. The phenomenon of global inflation carries a grave message of utmost importance to humanity. Faced with exploding demand, rising cost for essential commodities, and stagnating productivity, many countries in the less industrialized world will suffer a serious decline in real standard of living. This decline will impact the poor of the world the most because of their absolute low level at the present. The issue for tomorrow is how to curb demand so that inflationary spirals do not continue. Inflation is a symptom of the imbalance be-

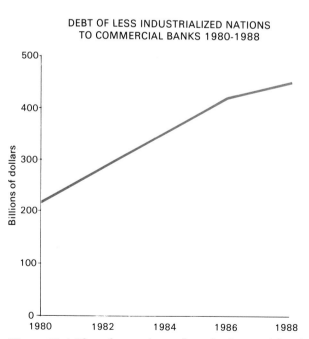

Figure 18.4 The substantial growth in the foreign debt of the developing nations handicaps development efforts.

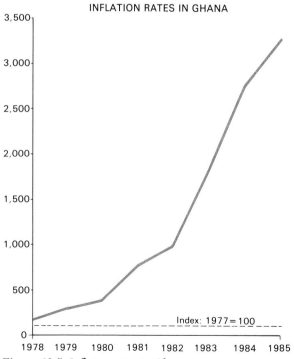

Figure 18.5 Inflation rates in Ghana.

tween soaring human demands and the physical resources to satisfy them.

TECHNOLOGY TRANSFERS AND TOMORROW'S POOR

Another issue is the degree to which Western technology is available or desirable for adoption in less industrialized countries. Exporting the technology for textile mills to India increases the productivity of a few thorough mechanization, but it deprives tens of thousands of their traditional work. The issue for tomorrow centers around whether it is in the best interest of less industrialized countries to model their economies and technologies after the industrialized, wealthy countries. The view that development and individual standard of life can be improved by simple adoption of Western technology and methods may well be false. The economic capabilities of many less industrialized countries are too small to launch a development policy on their own. They are too small in the same way that a factory of 50 workers is too small to compete with General Motors. They lack adequate resources, labor force, and skills to adopt the Western approach. The few large underdeveloped countries like India, China, or Indonesia suffer the reverse problem. They have more people than they need to adopt the Western model, and adopting the Western mechanized economy would destroy the livelihood of hundreds of millions of individuals. The problem becomes how to harmonize the individual countries' abilities and needs with the appropriate technology.

A final issue related to technology transfers is the potential for exploitation of economies in the less industrialized world by companies or corporations from the industrialized world. The General Motors Corporation has an annual value of production exceeded by only 22 nations of the world. Other major corporations rank only slightly behind General Motors, illustrating the problem faced by the small, less industrialized countries as they attempt to develop. Adoption of Western technology often simply means that Western industry establishes factories to capitalize on the low-cost labor of the host country. Tomorrow's issue will focus on harmonizing the need for jobs in the less industrialized world with the tendency of multinational companies to establish only industries reliant on cheap labor. As the less industrialized world's work force becomes more skilled, industries with higher wage rates may be attracted.

Technology designed to minimize labor requirements in recurrent tasks can provide free time to enable other developments. In this picture a simple grain grinder has been introduced into a village in Burkina Faso (Upper Volta), sharply cutting the time required for providing ground meal.

POLITICAL AND SOCIAL ISSUES IN TOMORROW'S WORLD

A number of political issues will be central to tomorrow's world. Crucial among these will be the respective roles of the traditional superpowers versus the traditionally powerless less industrialized countries. The influence and role of countries like the United States and the Soviet Union are declining as less industrialized countries recognize that they are not powerless and that they have a right to chart their own destiny. A major political issue in the future will be the extent and nature of the relationship between the industrialized and less industrialized world. The common relationship between the superpowers and the small nations in the past was essentially a client-supplier dyad. The rising expectations of the less industrialized world make it improbable that this asymmetrical relationship will be tolerated in the future. The oil-producing and exporting countries have shown that less industrialized countries can challenge the entire industrialized world (the United States, Europe, Japan) even though those countries could crush OPEC militarily. This change in political relationship will be one of the more difficult adjustments facing the world in the next decade.

Along with changes in relationship between less industrialized and industrialized countries will be political changes within individual countries. Demand by the majority poor for a voice in their government and a share in the prosperity of the country will lead to revolution and political instability in those less industrialized countries that are not progressing toward human equality or recognition of base human rights. An example of this occurred in China in 1989 when student demands for a free, open, democratic society resulted in the deaths of an estimated 2 to 3,000 students as the military opened fire on demonstrators in Beijing, and later executed students accused of plotting to overthrow the communist party. The need to provide stable governments that can allow individuals to lead peaceful lives will remain a critical issue at the turn of the century. The greatest threat is that some countries will attempt to capitalize on the political instability to increase their own sphere of influence and develop a new form of colonialism based on political control.

Closely related to the political issues of the future will be social issues. The social and political upheavals in portions of the Islamic world indicate the relationship of political and social changes. The demand by the world of Islam to develop using its own model rather than a Western one and to maintain its social values indicates the pressure of social change in these countries. One of the greatest issues will be the role of women in the world. Unless social changes involve the role of women, it is doubtful whether there can ever be a true improvement in the quality of life of the poor. Women in all societies, particularly in many less industrialized countries, have traditionally been relatively powerless. The issues growing out of the role of women in a changing, interdependent world are related to their historical role of second-class citizens. They have been disadvantaged because they are seen almost exclusively as childbearers and nurturers who are discriminated against when economic criteria are used to set priorities for national life.

Women's needs and concerns are often viewed as secondary to the real needs of the economy, and the Western model of development gives little consideration to their role and needs. The exclusion of women from the economic structure makes them invisible, especially in their nondomestic functions. Women do not receive a fair share of the benefits of a less industrialized society because of social attitudes limiting their role and ability to achieve personal development. There are now almost a half billion illiterate women in the world, 62 percent of all illiterates. This lack of education has a direct impact on women's ability to leave the home and its associated domestic work. Young women from rural areas in the less industrialized world migrate to the cities where they find only the most menial labor or are exploited through prostitution or other antisocial activities.

Women are often seen as problems to be corrected rather than as people who make intelligent choices in the context of their reality. For example, the problem of population is often viewed as a problem of women, but attempts to deal with the population issue rarely provide an alternative goal for women's personal achievement, security, or labor assistance now provided by children. The technological needs of women are largely ignored. Over 60 percent of the women in the world spend a disproportionate amount of their time preparing basic foods, especially hand grinding grains. Yet there has been no project to create an inexpensive, hand-powered grinder to meet the requirements of a small village. The social issues involved in the role of women in the world are only beginning to be addressed in the industrialized world and are essentially ignored in the less industrialized world; yet they affect 50 percent of the world's population.

Changes in the rural villages will be another major trend and issue of the twenty-first century. The increasing spread of economic development to the less industrialized countries is affecting the traditional villages in a wide variety of complex and subtle ways. Men will increasingly leave the village work to women while they migrate for a week or longer to industries in the cities (Figure 18.6). Where industry is close enough to allow daily commuting, village societies will reflect the change as extended families begin to disintegrate. The issues for tomorrow will

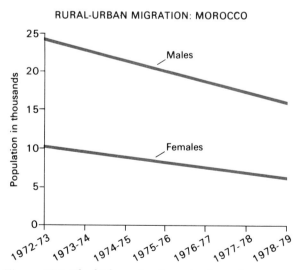

Figure 18.6 The higher migration rates for males disrupts traditional family and societal organization in developing countries.

THE ENVIRONMENTAL DIMENSION

The Greenhouse Effect and the Ozone Layer

Hypotheses concerning climatic change raise serious questions about the earth as the home of humans. The first hypothesis is related to the "Greenhouse Effect," and refers to the carbon dioxide essential to the existence of life on earth. The second hypothesis relates to the ozone layer that shields inhabitants of the earth from damaging ultraviolet radiation. The Greenhouse Effect occurs because of carbon dioxide. Heat from the sun is absorbed by water vapor and carbon dioxide in the atmosphere, making the earth habitable. The Greenhouse Effect is the cumulative warming of the earth's atmosphere due to increased carbon dioxide in the atmosphere caused by fossil fuel combustion.

Some climatologists theorize that in the past few million years there have been regular cycles in which ice ages roughly 100,000 years in length are separated by about 10,000 years of a warmer interglacial period. These climatologists conclude that increased carbon dioxide in the atmosphere (CO_2) is causing increased warming of the interglacial period the earth is presently experiencing.

Proponents of this model of the Greenhouse Effect rely on samples from glaciers in the Antarctic, which suggest that the carbon dioxide level was about 280 parts per million before 1850. In 1986, the carbon dioxide content measured in Hawaii (which has the longest continuous measurements) was 350 parts per million. The earliest record in Hawaii (1958) showed only 316 parts per million. Climatologists believe that the carbon dioxide content was about 250 parts per million during the glacial periods, and that present levels may reflect a level higher than at any time since the beginning of the most recent ice age—about 100,000 years ago.

Today's higher Co_2 content is related to the increased combustion of carbon-based fuels (coal, petroleum, natural gas, and wood) which adds carbon dioxide to the atmosphere. Some scientists hypothesize that about 58 percent of all carbon dioxide produced by burning (since monitoring began in 1958) has stayed in the air. In the past it was assumed that much of the carbon dioxide was transformed into oxygen by vegetation in the tropics. Continued loss of the tropical forests limits their role in the ecology of the earth, causing greater concentration of atmospheric CO_2.

The implications of climatic change for the geography of the world are significant. Global average temperature may increase between 1.5 and 4.5 degrees Celsius if the level of carbon dioxide increases to approximately 500 parts per million. Records suggest that temperatures increased .5 degrees Celsius in the last century.

The impact of global warming on human occupation of the earth varies from place to place. Climatologists suggest that high latitude regions will warm much more than the low latitude regions, and that the resultant climate changes will differ in terms of both temperature and precipitation. The industrialized countries of the higher northern latitudes will suffer most from the predicted climatic changes. The U.S. corn and grain belts will be particularly hard hit by drought, whereas increased precipitation will result in more stable crop production in the lower tropical and subtropical latitudes.

Another environmental change of growing importance relates to the ozone layer in the atmosphere. Ozone is present in minute amounts in the atmosphere, and concentrations vary with altitude. Ozone plays a major role in shielding the earth from ultraviolet radiation by absorbing wavelengths from the sun, which endanger the health of people, plants, and animals. The ozone layer of the atmosphere is affected by the use of chlorofluorocarbons (CFCs). CFCs are used in refrigerants and as propellants in aerosol sprays, such as hairspray. A molecule of CFC neutralizes up to 30,000 molecules of ozone, making continued addition of CFCs to the atmosphere a major concern. Declining ozone concentrations lead to increased skin cancers and climatic changes, because the radiation absorbed by the ozone is the primary source of heat for the stratosphere. The presence of a "hole" in the ozone layer over Antarctica was discovered by scientists in the late 1980s, intensifying demands for decreased reliance on fluorocarbons that began in the late 1960s. Although the industrial countries met to discuss ways of limiting CFCs in 1988, a solution to the ozone threat has not yet been achieved.

---

focus on maintaining the beneficial attributes for village life as the society changes. Preventing dissolution of the shared nurturing of the extended family, while allowing individual freedom and expression for its members, will require patience and time.

Adoption of high-yielding miracle grains causes breakdowns of the centuries-old tenant-landlord relationship with attendant potential for violence and suffering, and a major issue for tomorrow's world is the need to reform land tenure systems. The land must be distributed to ensure that cultivators have an opportunity to share in the increasing standard of living.

Just as there are changes and issues for the village, there are major issues affecting the community life of the industrial world. Social changes will result from increasing interdependence of the world and changes in life-style from conspicuous consumption to conservative use of resources. The Western, consumer-oriented society needs more products that are manufactured to last. We somehow feel that each year we should have more money and a higher standard of living, but actually the present high levels of the Western world will allow us to live a life of prosperity and ease. Social issues in the developed, industrialized world will also incude the conflict between rich and poor, as minority groups increasingly demand an equal share in the wealth of their countries. Should the effort of the individual be devoted to acquisition of more of the national product (increased per capita income for individuals) or to a reappraisal of what constitutes human happiness within the overall context of development? The issues for the developed world will involve the change in values that such a question implies.

ENVIRONMENTAL PROBLEMS

There are many environmental problems related to present and future development, including acid rains (increased acidity in precipitation caused by industrial development), landscape destruction from mining and resource extraction activities, loss of forests to fuel demands, or deterioration of the water and soil base of the world. Soil erosion is a major problem throughout the world, but adoption of mechanized cultivation techniques increases erosion dramatically. In the heartland of the Corn Belt of the United States, soil losses average more than 10 tons per acre per year, equivalent to a tax of two bushels of soil for each bushel of corn raised in Iowa. In the less industrialized realm, soil erosion may exceed these figures when Western technology and cultivation techniques are introduced.

The tropical rain forests are decreasing at a rate equal to the area of Massachusetts every month (Table 18.2). Loss of so much forest will have a number of direct and indirect effects on the world's occupants.

TABLE 18.2 World Deforestation

Region	Forest Area in Millions of Acres (hectares)	Annual Loss in Millions of Acres (hectares)
Central America	148 (60)	2.5 (1.0)
South America	1,300 (526)	22.0 (8.9)
West Africa	247 (100)	2.2 (0.9)
East Africa	217 (88)	2.0 (0.8)
West Asia	77 (31)	2.7 (1.1)
East Asia	815 (330)	17.5 (7.1)

SOURCE: *The Global 2000 Report* (Washington, D.C.: U.S. Interagency Task Force on Tropical Forests).

These environments are the repository of a host of plant and animal species not found elsewhere (Figure 18.7). They also play a role in the world's climatic system and act as key regulators of water supply at local and regional levels. The loss of tropical forests might influence weather farther away because of the area they cover. Of more critical concern is their role in maintaining the carbon dioxide in the atmosphere. Continued depletion of forests may well destroy the delicate atmospheric balance before the end of the twenty-first century. The issue facing the world as it considers such ecological disasters is that the loss of the tropical rain forest is not solely the result of hunger or population pressure. A major reason is the rush for land to raise cattle or cash crops such as coffee, which have a high market value in the wealthy industrial countries. Environmental problems will continue to plague the society of the world for the foreseeable future. The problem is to achieve worldwide conservation to ensure that all resources provided by the earth are used rationally and wisely to maintain essential life-support systems, to preserve genetic diversity, and to ensure the sustainable yield of the earth's biosphere. Overcoming environmental issues as the world continues to increase the standard of living of its population will require a major commitment on the part of all the earth's residents to live in harmony with the world around them.

WORLD CONFLICT OR WORLD PROSPERITY: THE ALTERNATIVE FUTURES

The range and intensity of issues that will face the people of tomorrow's world illustrate the necessity for true interdependence of the world community. Understanding of the importance of individual actions at a local and world scale is the first step toward changing the values and attitudes of the world's inhabitants. Because the inhabitants of the wealthy industrialized societies have a much greater impact in all of the issues discussed, changes in their attitudes and actions will also have more impact. If the residents of the industrialized world are willing to change from conspicuous consumption to studied conservation, the world will have taken the first giant step toward resolving the problems of inequitable distribution of world wealth and resources. If they accept the need to familiarize themselves with the poverty, deprivation, human suffering, and waste of human potential in the less industrialized world, it may be possible to begin the conomic changes necessary to integrate the world's poor into the world's society. Acceptance of the need

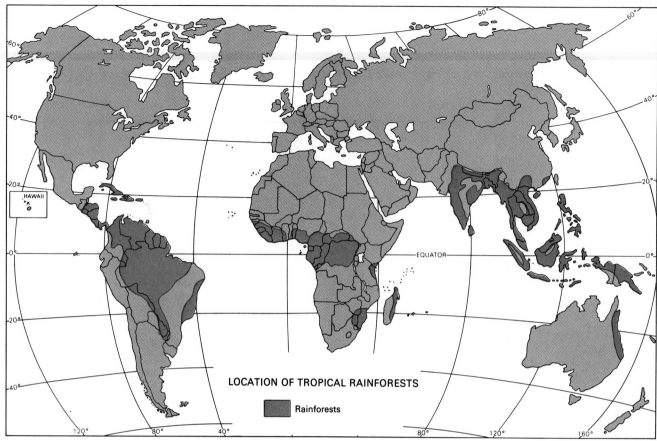

Figure 18.7 Location of tropical rain forests.

to provide people of the less industrialized world with an adequate market price for their products would be an important element in transfering excess wealth from the industrial countries to the less developed world. Only as changes in attitude take place in the industrial and less industrial countries will the world be able to overcome the problems that prevent all human kind from enjoying an adequate life-style, provided with the basic needs and free from unnecessary stress or fear.

FURTHER READINGS

ARAD, R. W. et al. *Sharing Global Resources*. New York: McGraw-Hill, 1979.

BROWN, L. R. *The Worldwide Loss of Cropland*. Washington, D.C.: Worldwatch Institute, 1979.

BROWN, LESTER R. *The Changing World Food Prospect: The Nineties and Beyond*. Washington, D.C.: Worldwatch Institute, 1988.

BROWN, SANDRA, et al. "Biomass of Tropical Tree Plantations and its Implications for the Global Carbon Budget." *Canadian Journal of Forest Research*, Vol. 16, No. 2, 1986.

CAMPBELL, K. O. *Food for the Future*. Lincoln: University of Nebraska Press, 1979.

CHAUHAN, SUMI KRISHNA. "Does Food Aid Help the Hungry?" *Earthscan* as reprinted in *World Press Review*, October 1982.

CHEN, ROBERT S., BOULDING, ELISE, and SCHNEIDER, STEPHEN H., eds. *Social Science Research and Climate Change: An Interdisciplinary Appraisal*. Boston: D. Reidel Publishing, 1983.

CHRISTIAN, SHIRLEY. "Pilots Fly over the Pole Into Heart of Ozone Mystery," *The New York Times*, September 22, 1987.

CLERC, JEAN-PIERRE. "The Militarization of Outer Space." *Le Monde* as reprinted in *World Press Review*, October 1983.

COLE, J. *The Development Gap: A Spatial Analysis of World Poverty and Inequality*. Chichester, England: Wiley, 1981.

COOMARASWAMY, RADHIKA. "A Third World View of Human Rights." *Unesco Courier*, August/September 1982.

ECKHOLM, E. *The Other Energy Crisis: Firewood.* Washington, D.C.: Worldwatch Institute, 1975.

FISHER, GLEN. "International Negotiation: Cross-cultural Perception." *The Humanist*, November/December 1983.

GILLAND, B. *The Next Seventy Years: Population, Food and Resources.* Tunbridge Wells, England: Abacus Press, 1979.

HALL, RAY. "Changing Fertility in the Developing World and its Impact on Global Population Growth." *Geography* 69, January 1984, pp. 19–27.

HARRISON, PAUL. "A Tale of Two Stoves," *New Scientist*, May 28, 1987.

LAPPE, F. M. et al. *Food First: Beyond the Myth of Scarcity.* Boston: Houghton-Mifflin, 1977.

LEWIS, FLORA. "The Habit of War." *The New York Times*, September 13, 1983.

MCCORMICK, JOHN. *Acid Earth.* (Washington, D.C.: International Institute for Environment and Development, 1985); *Acid Magazine*, Vol. 1, 1987.

MEMMI, ALBERT. "'Us' and 'Them'." *Unesco Courier*, November 1983.

MINTZER, IRVING M. *A Matter of Degrees: The Potential for Controlling the Greenhouse Effect.* Washington, D.C.: World Resources Institute, 1987.

MURDOCK, STEVEN, LEISTRITZ, F. LARRY, and HAMM, RITA R., eds. *Nuclear Waste: Socioeconomic Dimensions of Long-Term Storage.* Boulder, Colo.: Westview Press, 1983.

PACIONE, MICHAEL, ed. *Problems and Planning in Third World Cities.* New York: St. Martin's Press, 1981.

———., ed. *Urban Problems and Planning in the Developed World.* New York: St. Martin's Press, 1981.

RAMALINGASWAMI, V. "Children in Peril." *Unesco Courier*, April 1984.

RUSHER, WILLIAM A. "The Truth About Nuclear Weapons." *National Review*, November 26, 1982.

SALATI, ENEAS et al. "Amazon Rainfall, Potential Effects of Deforestation, and Plans for Future Research," in Ghillean T. Prance, ed., *Tropical Rain Forests and the World Atmosphere.* Boulder, Colo.: Westview Press, 1986.

SHEA, CYNTHIA POLLOCK. *Protecting Life on Earth: Steps to Save the Ozone Layer.* Washington, D.C.: Worldwatch Institute, 1988.

SIVARD, RUTH L. *World Military and Social Expenditures, 1987–88.* 12th ed. Washington, D.C.: World Priorities, 1987.

SMITH, D. M. *Where the Grass Is Greener: Geographical Perspectives on Inequalities.* London: Croom Helm, 1979.

SOBEL, L. A., ed. *Refugees: A World Report.* New York: Facts on File, 1979.

SUYIN, HAN. "The Woman Revolution." *Unesco Courier*, July 1982.

TAYLOR, WILLIAM J., Jr., and MAARANEN, STEVEN A., eds. *The Future of Conflict in the 1980s.* Lexington, Mass.: Lexington Books, 1982.

WELLBURN, ALAN. *Air Pollution and Acid Rain: The Biological Impact.* New York: Wiley, 1988.

XIANG, HUAN. "The West's Decline?" *Beijing Review* as reprinted in *World Press Review*, June 1982.

GLOSSARY

Absolute location: The geographic location defined by latitude and longitude and specific internal geographic characteristics of a place. Used interchangeably with site.

Accessibility: The ease with which interchange can occur between two places or people.

Acculturation: Changes occurring in a culture through intercultural borrowing. Changes may include technology, language, values, and so forth.

Acid rain: An acidified rainwater that severely damages plant and animal life. Caused by the oxides of sulfur and nitrogen that are released into the atmosphere when coal, oil, and natural gas are burned.

Afrikaans: The white population of South Africa who are descendants of the Boers (Dutch farmers).

Air mass: A mass of air similar in character covering a given area.

Alluvial: Refers to the mud, silt, and sand deposited by rivers and streams.

Alluvium: Sediment deposits found in stream channels and the floodplain of a stream.

Alps: An alpine chain of high rugged mountains in central Europe.

Alternative foods: Other species of wild plants and animals that have potential value as food crops.

Altitudinal zones: The climate changes with an increase in elevation, which is similar to the climate changes that occur with increasing latitude.

Agricultural revolution: The changes that have occurred in agriculture to increase yields and types of farming. The first was the change from gathering wild plants and hunting wild animals to plant and animal domestication. The second revolution was the improvement of equipment and types of seeds to increase yield and productivity.

Amazon Basin: The lowland, tropical rain-forest area of the Amazon river drainage region in South America.

Animists: People who worship naturally occurring phenomena such as mountains, trees, or animals.

Anthracite: A hard coal with a high carbon content that was formed under conditions of high pressure and temperature, which eliminated most impurities.

Apartheid: Literally, "apartness." The Afrikaans' term given to South Africa's policies of racial separation—and the highly segregated socio-geographical patterns they have produced.

Arabic: A principal language of the Semites usually associated with Arabs.

Arable: Land suited for farming.

Archipelago: A group or chain of islands in close proximity to one another.

Arid China: The drier western part of China.

Arithmetic growth: Refers to simple addition as a means of increase.

Assimilation: The process of integration of one culture into another so that the former loses its distinctiveness.

Asymmetrical relationship: One in which one participant benefits more than, or at the expense of, the others involved.

Atolls: Coral islands that are low, normally surrounding an open lagoon.

Bagasse: The fiber that remains after the juice has been removed from sugarcane.

Balkan Peninsula: A peninsula in southeastern Europe, including the countries of Greece, Albania, Yugoslavia, and Bulgaria.

Balkanization: The fragmentation of a region into smaller, often hostile, political units.

BAM: The Baikal-Amur Mainline railroad opened in 1984 is a 2000-mile track stretching across southern Siberia from Bratsk west of Lake Baikal to Sovetskaya-Gavan on the Pacific coast.

Barefoot doctors: Term used to refer to medical technicians in China who are roughly equivalent to practical nurses in the United States and provide basic medical services.

Barrios: The term for slums or shantytowns in Spanish-speaking Latin America.

Basin irrigation: A system of irrigation in which

floodwaters of a river are trapped in basins created by building small dikes around fields to retain the silt-rich floodwater.

Basque: An ethnic group in northern Spain and southern France with a unique language, dress, diet, religion, and economic standards.

Bilharzia: A disease that is carried by parasitic liver flukes. Common in tropical areas, it reduces a person's strength by as much as 80 percent.

Biomass: The amount of vegetative (organic) matter in an ecosystem in a designated region.

Birthrate: The annual number of births per 1000 individuals within a given population or country.

Bituminous: Soft high-grade coal. Some classes of bituminous coal can be converted to coking coal for use in iron and steel production.

Blue Nile: The Blue Nile's source is in Ethiopia. It joins the White Nile in Sudan to create the Nile.

Boer: Farmers of Dutch ancestry in South Africa.

Brahman: The highest group in the Hindu caste system.

Break-in-bulk: Location where large cargoes are broken down into smaller units.

Brittany Peninsula: The peninsula in western France occupied by the Bretons.

Buddhism: A religion derived from the teachings of Buddha, who attempted to reform the Hindu belief system.

Buffer zone: A country or number of countries separating two powerful countries that are adversaries.

Bustee: Any type of unsubstantial housing in India made from low-cost materials.

Calorie deficit: The total amount of food as measured in calories that would need to be added to provide minimum nutritional needs for the malnourished of the world.

Cape Coloured: An ethnic group in South Africa composed of descendants from a mixture of the African population and the early European visitors and immigrant Malays.

Capital transfer: The transfer of money from one nation to another.

Capitalism: An economic system based on the concept of private ownership of property and the freedom of the individual to engage in economic activity and to receive the benefits of that economic activity (profit).

Carpathians: A mountain chain in Eastern Europe centered in Romania.

Carrying capacity: The carrying capacity of an area is the number of animals, crops, or people it can support on a continual basis without degrading the environment. The carrying capacity varies with technology, land-use techniques, and geographic characteristics.

Caste system: The strict social segregation of people in Hindu society on the basis of ancestry and occupation.

Central place: Any place providing goods and services to a surrounding nonurban area.

Centrally planned economies: The political economic system in which the state controls the total economy.

Centrifugal forces: Forces that tend to weaken or destroy a country's unity.

Centripetal forces: Refers to forces that tend to unify a country or area.

Chernozem: A fertile black soil of the mollisol groups that have an abundance of humus and are among the most productive in the world.

China Proper: The name applied to humid eastern China.

City State: A state consisting of a city and its immediately surrounding territory, such as Singapore.

Clan: Clans consist of cultural groups based on extended family relationships.

Climatic types: The generalized regions based on variations in the complex interrelationships of the amount of heat, precipitation, and vegetation in a particular place.

Club of Rome: A group of about seventy individuals from 26 countries created the Club of Rome to assess or consider possible alternative futures of the world's environment.

Collectivization: The reorganization of a country's agriculture through expropriation of private holdings and creation of relatively large-scale units, which are farmed and administered cooperatively.

Colonialism: The political and economic control of a country by a foreign power.

Common Market: Name given to a group of 12 European countries (as of 1988) that belongs to a supranational association to promote their economic interests. Official name is European Economic Community (EEC), commonly shortened to European Community (EC).

Communes: Settlements based on collective ownership and use of property, goods, and the means of production.

Communism: The economic system where all factors of production are state owned in the name of the workers.

Compact state: A term used to describe a state that possesses a roughly circular, oval, or rectangular territory.

Complementarity: Production of goods or services by two or more places in a mutually beneficial fashion.

Complementary resources: Resources whose usefulness is enhanced by the presence of the other.

Complementary relations: Two or more regions exchanging goods and services that one lacks and the other has produced or developed.

Confucian ethic: Chinese and other societies that adopted the teachings of Confucious are thought to be more work and goal oriented by some observers.

Confucian meritocracy: Government characteristics developed in China based on the principles of Confucianism, which created a series of proper relationships between leader and follower, master and servant, father and son.

Coniferous: Cone-bearing trees—with needle leaves, commonly found in cool climates.

Conspicuous consumption: High levels of consumption of resources and goods associated with individuals or countries.

Continental drift: The movement of continents that led to the breakup of Pangaea, resulting in the present continental shields and ocean basin floors.

Continentality: The characteristic of continental climates, isolated from the moderating influences of the sea, resulting in seasonal extremes of temperature and marked differences in precipitation.

Contract system: The recent Chinese system of allowing individuals or members of a production brigade to keep the production or output required over the contract quota.

Conurbation: Extensive urban area formed by expansion of cities to form one continuous urbanized area.

Convectional: The rising of air due to heating of the land surface by the rays of the sun.

Cooperatives: In cooperatives a group of farmers pool their resources to purchase equipment and to market their products, but individuals farm their own land.

Copper belt: The area of Zambia and Zaire, which has a large concentration of copper.

Copra: The dried meat of the coconut used for making coconut oil.

Coral atolls: A coral island or islands with a reef surrounding a lagoon.

Corn belt: A region in the central United States where corn is the dominant crop.

Cottage industries: Cottage industries involve small-scale production using high labor inputs.

Crop rotation: The rotation of crops to maintain or increase the soil fertility and yield.

Cultural boundaries: Boundaries that are created as a result of cultural factors or variables.

Cultural changes: Changes that occur in the culture of an area.

Cultural convergence: The tendency of world cultures to become more alike as European culture and/or technology is dispersed, particularly to urban areas of less industrialized countries.

Cultural fragmentation: The presence in one region or country of a host of cultural traits as opposed to dominance of one trait over a large area.

Cultural pluralism: Variety of cultures within a country, but not necessarily occupying distinct territories.

Cultural realm: A group of countries sharing related cultural traits, as the Islamic realm.

Cultural revolution: A period of time in China when universities were closed and many youths were involved in trying to restore the communist revolution to a condition of altruism they perceived existed earlier.

Culture of poverty: A way of life created as a result of being poor.

Cyclonic: Frontal activity resulting from air mass movement into a region or area having an air mass that is different in character.

Cyrillic alphabet: The old Slavic alphabet used today in the Soviet Union, Bulgaria, and Serbia.

Death rate: The ratio between deaths and the number of people in a given population for a given time period.

Deciduous: Trees that have a dormant period in which they lose their leaves.

Delta: A flat, fertile lowland created by a river as it deposits its alluvium near the mouth when the water slows.

Democratic revolution: A development of free and open elections; freedom from taxation without representation; freedom from arbitrary acts of those in power; freedom of knowledge; freedom of public assembly and speech; and freedom to vote through use of a secret ballot.

Demographic transition: Model of population change that suggests countries move from a slow population growth stage with both high birth- and high death rates, to a stage of rapid population growth when death rates drop, to a stage of slow or negative population growth as birthrates also fall.

Dependency ratio: The proportion of a population that is either too young or too old to be economically self-sustaining or produce a surplus.

Desert pavement: Land surface composed of pebbles and rock fragments, often cemented together by calcium carbonate or other salts left at the surface through evaporation.

Desertification: The expansion of the desert to moister areas along its margins as plant cover and soils are destroyed through overgrazing and erosion or climatic change.

Devolution: In political geography, the disintegration of a nation-state, especially as the result of emerging or reviving regionalism.

Dew point: The temperature level or point at which condensation occurs in air containing water vapor as a result of being cooled.

Dharma: The Hindu concept of duty within society.

Direct consumption: Direct consumption of cereal grains and such, rather than meat or alcohol products.

Diurnal: Daily.

Domino theory: The political idea that if one nation of a region falls to a particular ideology, the other nations of the region will follow like falling dominos.

Double cropping: Planting two crops in succession in a field in the same year.

Dravidians: Members of an ancient Australoid race in southern India; also a family of languages spoken in that region.

Dry farming: A method of cultivation in which crops are alternated with a fallow period in which water accumulates in the soil.

Dual economy: An economy with both a small, intensive modern sector and a large, traditional sector. The modern system is tied to the broader world economy, whereas the traditional economy supplies local needs.

Dual societies: Societies in which both market and peasant exchange systems operate.

Dynasty: A ruler or ruling party that persists for an extended period of time.

Economic development: The nature of a country's economy in terms of level of industrialization and modernization.

Economic takeoff: The point at which the economy of a country begins to develop or improve rapidly.

Economies of scale: The savings that accrue from large-scale production whereby the unit cost of manufacturing decreases as the level of operation enlarges.

Ecumene: Refers to that portion of a region that is inhabited.

EEC: European Economic Community.

Effective National Territory: That part of a country that is controlled by the central government, contributes to the economic base of the country, and has allegiance to the country.

Ejidos: Cooperative farms in Mexico.

Elongated state: The shape of a state that is long and narrow, such as Vietnam.

Emergent landforms: Landforms exposed by a lowering of sea level or a rise of the crust, or both.

Encomendero: The Spaniard to whom an encomienda was granted.

Encomienda: A large grant of land including its Indian occupants. It was designed to "civilize" the Indians and protect them from European exploitation by placing them under the protection of a Spaniard who agreed to teach, feed, and clothe them in return for their labor.

Environmental determinism: The theory that the physical environment controls certain aspects of human action.

Environmental perception: The mental image of the physical environment held by an individual or a group.

Epicanthic: The epicanthic eyefold is an extra fold of skin in the inner corner of the eye, resulting in the distinctive eye shape of the Mongoloid people.

Erosion: The wearing down or away of the earth's suface by wind or water.

Estuary: The widening mouth of a river as it reaches the sea.

Ethnic religions: Local religions of groups that share beliefs and practices; membership comes through birth into the group.

Ethnocentricity: The idea that one's own ethnic group, race, or other group is in some way inherently superior to others.

Ethnocentrism: Belief in the superiority of one's own ethnic group.

Europort: A large port established on the Rhine downstream from Rotterdam to handle imports and exports of the Common Market nations of Europe.

Exotic stream: A river that originates in a humid region and flows through a dry region.

Exponential growth: Referring to the increase associated with increasing any number by some exponent or power. Malthus maintained that populations tended to double, thus the exponent is 2.

Expropriation: The process by which a country seizes the property of private citizens, businesses, or corporations. Ownership is then maintained by the government either directly or indirectly.

Extended families: Those family members beyond the nucleated family of husband, wife, and children.

Fall line: A line joining points where parallel-flowing rivers drop suddenly from a piedmont to a coastal plain in falls or rapids; the upstream limit of river navigation.

Fallow: The farming process of allowing land to lie in a natural state for a period of time, usually a year.

Far East: An area commonly including the Koreas, Japan, China, and the islands belonging to them.

Sometimes used to refer to all of Asia east of Afghanistan.

Federal government: A government in which groups or specific areas elect representatives to govern the country.

Fertile Crescent: Crescent-shaped zone of productive lands extending from near the southeast Mediterranean coast through Lebanon and Syria to the alluvial lowlands of Mesopotamia. One of the world's great source areas for agricultural innovations.

Fertility rate: The number of children born to women in the reproductive age group (15–44) as a percent of women in that group.

Filariasis: Infestation of parasites of the blood or tissues. Usually transmitted by mosquitoes.

Fjord: Narrow, steep-sided, elongated, and inundated coastal valley deepened by glacier ice that has since melted away, leaving the sea to penetrate.

Floodplain: The level, low valley floor bordering a river.

Four Modernizations: The policy in China involving modernization and mechanization of agriculture; modernization and expansion of industry; modernization and development of science, technology, and medicine; and modernization of the military.

Fragmentation: The continued division of land with each generation into smaller and smaller pieces.

Fragmented state: A state whose territory consists of several separated parts, not a contiguous whole.

Full employment: The official policy of the Soviet Union until 1988, stating that it is the government's responsibility to provide jobs for all.

Generative centers: See Generative city

Generative city: A city having complementary relations with its hinterland.

Geopolitics: The use of geographical relationships to justify a policy of national expansionism.

Ghettos: A distinct section of a city characterized by a particular ethnic composition.

Gradational forces: Forces that wear down the surface of the earth, such as wind, water, ice, gravity, and the sea.

Green Revolution: The development of higher-yield, fast-growing varieties of rice and wheat that has led to increased production per unit area and a temporary narrowing of the gap between population growth and food needs.

Gross National Product (GNP): Refers to the total value of all goods and services produced annually in a country.

Growth rate: The net rate of population increase calculated by subtracting deaths and emigrants from births and immigrants.

Hacienda: Another term referring to large estates in Latin America, commonly used in Mexico.

Hajj: The pilgrimage by Muslims to Mecca.

Harijans: The untouchables in India who are at the bottom of the caste system.

Head link: A city that serves as a link to the rest of the world for a country as imports and exports flow through it.

Headwaters: The source or beginning of a river.

Hejira: The flight of Muhammad, Prophet of Islam, from Mecca to Medina in A.D. 622; the date on which the lunar-based Muslim calendar begins.

Hinterland: The area a city serves and draws people from to purchase goods and services.

Holistic: The idea that the entirety of actions affects a given place.

Hollow frontier: A frontier that moves on without leaving permanent towns, farms, and industry, as in Brazil where the exploitive economy traditionally collapsed, creating hollow frontiers rather than permanent colonization.

Horizontal nomadism: The periodic movement of herds over long distances in search of pasture.

Humus: Dark-colored upper layer of a soil that consists of decomposed and decaying organic matter such as nutrient-rich leaves and branches, giving the soil a high fertility.

Hydraulic civilizations: Social organizations that developed in the Middle East and North Africa to ensure adequate water for the villages through a highly structured distribution system.

Iberian Peninsula: The 230,000-square-mile peninsula of southern Europe occupied by Spain and Portugal.

Import substitution: The protection of local industries by tariffs to prevent imports from competing with them.

Incipient industrial base: The railroads, technology, and skilled labor resulting from initial industrialization efforts by countries that speed subsequent rapid industrialization.

Industrialization: The development and increase in manufacturing and industrial activity.

Industrial Revolution: The Industrial Revolution involved the substitution of inanimate (machine) power for animate power sources, allowing production increases and a growth in demand for resources.

Infant mortality rate: The ratio between deaths in the first year of life and the number of people in a given population. Often expressed as the number of deaths per thousand per year.

Institutional framework: Refers to the customary or legal processes that set the parameters within which the economy of a country functions.

Insular: Referring to an island or locations having islandlike characteristics.

Insurgent state: Political and territorial control over part of a country by insurgent (guerrilla) movements that oppose the existing government.

Interconnections: The connections or linkage between countries of the world in communication, transportation, business, cultural values, and political arrangements.

Invisible exports: Sale of products or services for which no product leaves a country, such as tourism, banking, or other services.

Invisible trade: The flow of invisible exports and imports into and out of a country.

Irish curve: A rapid population growth followed by a precipitous drop in numbers and stabilization at a lower level.

Irredentism: Claims on lands culturally or historically related to a nation that are now occupied by a foreign government.

Irrigation: The distribution of water by building channels from source areas of water or storage areas such as lakes.

Islam: Religion founded by the prophet Mohammed (Muhammad) in Saudi Arabia around A.D. 624. Islam is the name of the religion and means submission to the will of one God (Allah). Muslim or Moslem refers to a member, one who submits himself or herself to the will of Allah.

Isolation: The condition of being geographically cut off or far removed from mainstreams of thought and action.

Japanese clans: In Japan these clans expanded in size and power to control Japan's government and economy in the feudal system.

Job security: A worker has a job for life with a particular company; common in Japan.

Junta: A group of military officers controlling the government, generally obtained by forceful seizure of the government and often exercised dictatorially.

Jute: A fiber of a plant that is used for burlap, twines, sacks, or carpet backing.

Karma: The concept of an accounting of good and evil in Hinduism.

Kibbutzim: A collective farm or settlement in Israel.

Kolkhoz: A collective farm in the USSR in which farming activities are performed by farm members who share in the profits after meeting quotas required by the central government.

Koran: Writings of the prophet Muhammad that Muslims accept as divine revelations from Allah.

Kwashiorkor: Severe malnutrition resulting in anemia, loss of skin pigmentation, hair loss or color change, and protruding stomachs.

Labor exploitation: When there is an abundant labor force or more workers than jobs, companies will pay a smaller wage, thus exploiting the worker for increased economic gain.

Labor specialization: The emergence of specific occupational categories that replaced general labor requirements of the early Neolithic period.

Landforms: The configuration of the land surface into distinctive forms such as hills, valleys, and plateaus.

Land tenure: Referring to the form of land ownership and control, including absentee landlordism, sharecropping, and tenancy.

Lateritic soils: Highly leached soils of the tropics that have high concentrations of iron and aluminum.

Latifundia: The latifundia are large estates in Latin America owned by individuals of European descent, who typically control most of the wealth in the individual countries.

Leaching: A process by which minerals and organic matter percolate downward or are entirely removed from the soil structure by intense and prolonged rainfall.

Lifeboat earth: The concept that there is only a finite quantity of resources; thus industrialized countries should concentrate only on their own well-being and not assist the less industrialized nations.

Lignite: Also called brown coal, a low-grade variety of coal somewhat higher in fuel content than peat but not nearly as good as the next higher grade, bituminous coal.

Lingua franca: Refers to use of a second language that can be spoken and understood by many peoples in regions with a diversity of languages.

Localized resources: Resources that are found only in specific locations.

Loess: Fine, very fertile soil that has been deposited by wind.

Malthusian equation: The early nineteenth-century hypothesis of Thomas Malthus who argued that population growth would always outrun the earth's capacity to produce sufficient food.

Mandarin: Dominant Chinese language, spoken mostly in northern China.

Maquis: Thick, scrub underbrush found along the shores of the Mediterranean Sea.

Maritime: A climate characterized by moderate temperature, medium to high rainfall, and high humidity, generally found along coasts.

Marxism: The economic doctrine that wealth and control of factors of production should be held by the workers.

Megalopolis: A group of large cities that has effectively become one large urban area through their growth.

Meiji Restoration: The calculated policy of modernization based on combining traditional Japanese values with Western technology.

Mercantilism: Protectionist policy of European states during the sixteenth to the eighteenth centuries, which promoted a state's economic position in the contest with other countries.

Meritocracy: The idea developed from Confucianism in China based on the concept that only the most qualified should have positions of authority.

Metropolis: A large urban center.

Migration: A change in residence intended to be permanent.

Migratory genocide: Term used to refer to destruction of a group through forced migration, under harsh conditions.

Minifundia: Tiny farms in Latin America that provide a subsistence existence upon which a significant portion of the Indian population depends.

Mir: A village community in Russia prior to 1905 characterized by joint ownership of the land.

Mixed farming: Farming that uses a combination of crop cultivation and animal husbandry.

Monoculture: The large fields of same crops as the great wheat fields or cornfields of the midwestern United States.

Monotheistic religion: A religion which believes that there is only one God.

Monsoon: It is described as a seasonal reversal of winds, but it also implies differences in precipitation.

Mortality rate: The number of deaths in a given period of time or place, normally given as number per 1000 population.

Moshav: A cooperative farm or settlement in Israel.

Nagana: A destructive disease of the African tropics spread by the tsetse fly.

Nation: Refers to a group of people with a distinct culture. The nation may or may not coincide with political boundaries.

Nationalism: The feeling of pride and/or ethnocentrism focused on an individual's home territory.

Nationalization: The process by which ownership of private property is assumed by the government.

Nation-state: A country whose population possesses a substantial degree of cultural homogeneity and unity and where the political unit coincides with the area settled by a nation.

Natural boundaries: Boundaries that are formed, which are natural either due to physical or cultural characteristics.

Neocolonialism: Control of former colonies by colonial powers, especially by economic means.

Neolithic revolution: The change from dependence on gathering wild plants and hunting wild animals to plant and animal domestication and sedentary farming. Also called the first agricultural revolution.

Nirvana: The concept of unity with the universal spirit in Buddhism.

Nomadism: Cyclic movement among a definite set of places. Nomadic peoples are mostly pastoralists.

Nonubiquitous resources: Resources that are not found everywhere.

North Atlantic Drift: The relatively warm currents of the Atlantic resulting from the Gulf Stream.

Nuclear families: Families that include only the immediate family of father, mother, and children.

Oil shale: A kind of rock containing oil in dispersed form.

OPEC: Cartel of oil-producing and exporting nations, the Organization of Petroleum Exporting Countries, formed in 1960.

Orient: The countries of the Asian continent excluding the Soviet Union.

Orographic precipitation: Precipitation caused by the forced rise of moist air over a mountain barrier.

Outback: The large dry interior of Australia.

Overdevelopment: The concept that some regions or countries have excessively industrialized and excessively urbanized, creating shortages in energy, natural resources, environmental pollution, congestion, and individual alienation.

Ozone depletion: The reduction of the ozone layer that shields and protects the earth's surface from the effects of ultraviolet radiation found in the sun's rays.

Pacific ring of fire: Zone of crustal instability along tectonic plate boundaries, marked by earthquakes and volcanic activity, that rings the Pacific Ocean basin.

Paddy: Field of wet rice, and term used to refer to unhusked rice.

Pangaea: The earth at one time was a single continental land mass known as Pangaea.

Parasitic city: A city that does not have a complementary relationship with its hinterland.

Peasants: Rural farmers in a stratified society. Peasants may or may not own land; the term refers to a way of life.

Peasant society: A society in which the roles are highly specific and usually established at birth or by living in a specific village.

Peasant systems: Individuals are born or participate in specific roles such as carpentry, farming, or production of a specific item.

Peoples communes: After the Chinese revolution, cooperatives were grouped together to form a massive farm with great labor potential.

Periodic markets: Markets that operate on some regular interval (as every Sunday, for example) rather than daily.

Permafrost: Permanently frozen water in the soil and bedrock, as much as 1000 feet in depth, producing the effect of completely frozen ground. Can thaw near surface during brief warm season.

Persian: A language group in the Middle East, spoken by Iranians.

Physical boundaries: Boundaries that result from physical characteristics such as lakes, rivers, or mountains.

Physical remoteness: Locations that are separated from other locations by great distances or rugged landforms.

Piedmont: Hilly, rolling land, lying at the foot of a mountain range and forming a transition between mountain and plain.

Plane of the ecliptic: An imaginary plane in which the earth's orbit lies.

Plantation: A large estate normally owned by a corporation concentrating on large-scale production of a cash crop for export from the tropics.

Plantation economy: An economy of a region or country that is based on upon plantation agriculture. Normally indicates a separation in occupation between local residents who provide labor and foreign administrators.

Plebiscite: A plebiscite is simply an election by the voters of a country or region to determine what course of action they will follow.

PLO: Palestine Liberation Organization founded in 1962 by displaced Arabs from Israel.

Plural society: In a plural society various ethnic groups live intermingled throughout a country, although often occupying distinct residential sections of cities. Each ethnic group recognizes its own distinctive cultural traits, icons, values, and life-styles.

Podzol: Acidic infertile soil common in coniferous forests of cool moist climates, as in large areas of Canada and the Soviet Union.

Point settlements: Point settlements are located in harsh environmental settings to capitalize on unique site features such as resources. They are generally extractive in nature with little or no hinterland.

Polder: Land reclaimed from the sea adjacent to shore by constructing dikes and pumping out the water.

Polyculture: A farming system in which numerous types of crops are grown together. Also known as intercropping.

Polyethnicism: Several ethnic groups within a country's borders, each group occupying a distinct territory.

Population density: The number of people in a given area, usually indicated per square mile or kilometer.

Population doubling time: The number of years it takes a given population, such as a country, to double.

Population explosion: The rapid growth of the world's human population during the past century, attended by ever-shorter doubling times and accelerating rates of increase.

Ports: A place where goods are brought into and out of a country.

Postindustrial society: The characteristics of a nation that has moved from an industrial society to one in which most of its employment is in the service sector.

Prairie wedge: The area in the midwestern region of the United States that constitutes the largest, most naturally fertile area of soil in the world.

Precipitation types: The various forms of precipitation, such as rain, snow, and hail.

Primary activities: Work or occupations involved with producing raw materials from farming, forestry, fishing, or mining.

Primate city: A city that dominates the urban network of a country, by definition at least twice as large as the next largest city.

Producer's cooperatives: China organized residents of a village of thirty to fifty households into a cooperative, in which they were able to combine their resources and their land in effort to increase productivity.

Prorupt state: A type of territorial shape that exhibits a narrow, elongated land extension leading away from the main body or territory.

Qanat: A gravity-fed subterranean irrigation tunnel in the Middle East and North Africa.

Quality of life index: A measure based on the standard of living indicated by key indices, such as literacy, infant mortality, and access to medical and other social services.

Rainforest destruction: The loss of tropical rain forest through burning, road and dam building, settlement, and air pollution.

Rain shadow: The relative dryness in areas downwind of mountain ranges.

Regionalism: A strong identity of feeling for a region.

Reincarnation: The belief that the souls of the dead return to life in new bodies or new forms.

Relative deprivation: Referring to relative levels of poverty rather than absolute poverty. A millionare may feel poor compared to a billionaire, but both are wealthy by absolute standards.

Relative location: Refers to location of a place or region with respect to other places or regions. Used interchangeably with situation.

Repatriation: The process of returning people to the country of their birth or where they hold citizenship.

Replicative cities: Cities that are similar or modeled after those in another culture.

Replicative societies: Societies that result from conscious or unconscious efforts to model themselves on another society.

Resource: The physical elements of the world that are perceived as useful by the population of a region or world.

Revolutionary changes: Changes that occur as a result of revolution.

Ritual purity: The purifying of the soul by ritual procedures or behavior, especially in Hinduism.

Riverine population concentrations: The growth of populations along the river basins of the world.

Romansch: A romance language spoken in Europe.

Rural-urban migration: The movement of people from rural areas to urban areas.

Sahel: Semiarid zone across most of Africa between the southern margins of the arid Sahara and the moister savanna and the forest zone to the south.

Salinization: The accumulation of salts in the upper levels of soil as salts carried by irrigation water are left at or near the surface as water evaporates.

Samurai: The warrior class of medieval Japan; also their rigid code of behavior.

Savanna: The wet-dry tropical regions of the world. Also, a grassland with scattered trees and bushes located on the edge of equatorial rain forests.

Sawah: Indonesian term for wet rice cultivation.

Schistosomiasis: A disease caused by parasitic worms affecting the intestines, urinary bladder, liver, and spleen.

Scientific revolution: The tremendous increase in scientific development and discoveries in health and technology of the nineteenth and twentieth centuries.

Secondary activities: Occupations or work that process raw material and transform them into products. Also referred to as manufacturing.

Sedentary: Permanently located in a particular area.

Seigneur: A feudal lord in French Canada who held land by feudal tenure.

Semitic: A cultural group in the Middle East and North Africa, The two major Semitic groups are Arabs and Jews.

Seventh approximation: A system for classifying the world's soils, also known as the Comprehensive Soil Classification System.

Shatter belt: Region located between stronger countries (or cultural-political forces) that is recurrently invaded and/or fragmented by aggressive neighbors.

Shifting agriculture: Crops cultivated in forest clearings that are abandoned in favor of newly cleared nearby forest land after a few years.

Shiite: An Islamic minority created after the death of Muhammad, who believe that the leader of Islam should be a direct descendant of Muhammad.

Sikh: An individual who follows the Sikh religion. They are characterized by adherence to the common surname of Singh; males have long hair, a full beard, wear a turban, and carry a dagger.

Sikhism: A religious group that combined elements of Islam and Hindu beliefs.

Site: The internal locational attributes of a place, including its local spatial organization and physical setting.

Situation: The external locational attributes of a place; its relative location with reference to other nonlocal places.

Slash-and-burn agriculture: (also known as shifting agriculture) Consists of clearing and burning the vegetation, planting crops for one or two years until new vegetation begins to overrun the fields and then moving on to a virgin area to repeat the process.

Small planet: The concept that the planet has a finite set of resources and all should work together to conserve and develop.

Social inertia: Social change is slow or is difficult to begin.

Social revolution: A change that takes place in a society so dramatically that it creates a new social order, as in Japan after the Meiji Restoration.

Socialism: A variety of political and economic theories and systems of social organization based on collective or governmental ownership and administration of the means of production and the distribution of goods.

Soil classifications: The various soils are divided into broad categories.

Solstices: The point in the ecliptic at which the sun is farthest from the equator both in the North in December and in the South in June.

South Island: One of the two major islands that comprise New Zealand.

Sovkhoz: A farm in the Soviet Union owned by the state; farm members are paid a set wage regardless of the profitability of the farm.

Spatial interaction: The relative amount of interaction between a place or region and other places or regions.

Special Economic Zones (SEZs): Specific cities where foreign firms are allowed to establish factories to capitalize on Chinese labor.

Spheres of influence: Regions or countries of the world in which the superpowers (such as the USSR or the United States) have influence.

State: The formal name for the political units we commonly call countries.

State-oriented capitalism: When the government regulates selected aspects of the economy, but lets other aspects function under laws of supply and demand.

Storm surge: Flooding caused by a storm that increases the height of the tide.

Street people: People who have no home.

Structural changes: Changes that are made in the formal structures of a country, such as land tenure, political organization, or economic relationships with other areas.

Submergent coast: Coastal areas that have sunk below the ocean.

Submergent landforms: Land areas that have sunk, usually along coastal areas or river mouths.

Subsistence systems: Agricultural systems in which most of the produce is consumed directly by the farm family rather than being traded or marketed.

Suffrage: The universal and free right to vote.

Sunnite: The major religion group of Islam.

Superimposed boundaries: Superimposed boundaries are those placed over the existing political and cultural patterns, fragmenting cultural groups and joining peoples of diverse backgrounds.

Symbiotic: Mutually beneficial relationships in which both parties in a relationship are better off than operating alone.

Taiga: The great northern coniferous forest; also referred to as the boreal forest.

Tariff: A tax placed on imports or exports into a country. Often used on imports to protect an economy from lower cost imported goods.

Tariff barriers: Methods such as tax, quotas, or health standards placed on exports or imports to retard the flow of goods into a country.

Tectonic forces: Forces created by pressure beneath the earth's crust, which results in folding or faulting of the earth's surface.

Terra incognita: Greek term, literally meaning unknown land.

Territorial shape: The form or shape of a country.

Tertiary activities: Occupations and work that engage in services, such as banking, tourism, transportation, retailing, finance, or education.

Tierra Caliente: The hot coastal lowlands and piedmont (usually below 2500 feet in elevation) in the Andes Mountains of Latin America.

Tierra Fria: The cool upland elevations (usually above 6000 feet) in the Andes Mountains of Latin America.

Tierra Templada: The temperate middle elevations (usually between 2500 and 6000 feet) in the Andes Mountains of Latin America.

Topography: The configuration of a land surface, including its elevations and the position of its natural and man-made features.

Transfer payments: The process of transferring wealth from one area to another. It takes many forms, from welfare within countries to economic aid or tourism between countries.

Transhumance: Seasonal movement of people and their livestock in search of pastures. This movement may be vertical—that is, into highlands during the summer and back to lowlands in winter—or horizontal, in pursuit of seasonal rainfall.

Treaty ports: Selected ports where Western powers established commercial enclaves from which they traded with China.

Triage: The concept of world population that suggests that the industrialized nations should help those that need only partial aid and let the extremely overpopulated nations fend for themselves.

Tribal religion: A religion associated with a specifc tribe or group and a specific place.

Tribalism: The practice of tribal religion.

Tropical cyclone: Another term for what Americans call a hurricane.

Tropical ecosystem: The collection of all living organisms in the tropical environment and their inner relationships.

Trust territories: A trust territory is a former colonial holding assigned by the United Nations to one of the industrialized nations for development assistance.

Tsetse fly: Insect pest in Africa that infects people with sleeping sickness and animals with nagana.

Tundra: A zone between the northern limit of trees and the polar region in North America, Europe, and Asia.

Ubiquitous elements: Those elements that are found in many areas.

Ubiquitous resources: Resources that are found in many areas.

Underemployment: The condition that exists when part or all of the labor force has only part-time or seasonal work (as in agriculture) or are engaged in labor that is only marginally productive.

Universalizing religion: One that has broken its ties to a specific space and maintains that its beliefs are appropriate for all people.

Untouchables: The lowest level of the caste system in India.

Urban morphology: The pattern and relationship of land use, economic activity, and population distribution in urban areas.

Urban shadow: Cities located near a larger city are frequently smaller than they normally would be if located away from the city, because the larger city retards growth and development of some economic activity.

Virgin Lands Project: The development of a new wheat- and grain-producing region in the Soviet Union in the steppe and humid continental climate region east of the Urals.

Westerlies: Prevailing surface winds in the midlatitudes that generally blow from the west to the east.

Wheat belts: Areas of the world where the major crop is wheat, such as in midwestern United States.

Xerophytic vegetation: Vegetation that is tolerant of arid conditions.

Zero population growth (ZPG): ZPG represents a stable population that basically maintains itself without increasing or decreasing.

Zionism: The desire for a Jewish homeland and its expansion.

Photo Credits

Chapter 1
Opener page 2 Franklin/ Sygma. *page 7* Sally and Richard Greenhill/ Stock, Boston. *page 18* Alain Keler/ Sygma. *page 20* Georg Gerster/ Rapho/Photo Researchers. *page 21* T. Matsumoto/Sygma. *page 23* Beryl Goldberg.

Chapter 2
Opener page 32 Bruno Barbey/ Magnum. *page 42* Peter Turmley/ Black Star.

Chapter 3
Opener page 64 Christa Armstrong/ Photo Researchers. *page 73 (top left)* Sebastiao Salgado Jr./ Magnum. *page 73 (top right)* Owen Franken/ Stock, Boston. *page 73 (bottom left)* Rick Smolan/ Stock, Boston. *page 73 (bottom right)* Barbara Pfeffer/ Peter Arnold. *page 83* Lily Solmessen/ Photo Researchers.

Chapter 4
Opener page 86 David Burnett/ Stock, Boston. *page 92* Skyviews Survey. *page 93* Stuart Cohen/ Comstock. *page 94* George Holton/ Photo Researchers. *page 102* Nogues/ Sygma. *page 104* Sebastiao Salgado Jr./ Magnum.

Chapter 5
Opener page 112 German Information Center. *page 121* Louis Goldman/ Ralpho/Photo Researchers. *page 130* Ulrike Welsch/ Photo Researchers. *page 132* Jan Lukas/ Photo Researchers. *page 143* Stuart Cohen.

Chapter 6
Opener page 150 Courtesy of German Information Center. *page 163* George Schaub/ Leo de Wys. *page 166* Cagatay/Gamma-Liaison. *page 173* Jacques Paylovsky/ Sygma.

Chapter 7
Opener page 176 Sam C. Pierson, Jr./ Photo Researchers. *page 185* Alex S. Maclean/ Landslides. *page 186* U.S. Army Corps of Engineers. *page 187* Georg Gerster/ Photo Researchers. *page 190* USDA. *page 192* USDA/ Soil Conservation Service. *page 196* Ira Kirschenbaum/ Stock, Boston. *page 197* Agriculture and Stabilization and Conservation Service. *page 200* J.P. Laffont/ Sygma. *page 201* Gans/ The Image Works.

Chapter 8
Opener page 214 Promotion Australia. *page 218 (left)* Australian Information Service. *page 218 (right)* Australian Information Service. *page 220* Jean Guichard/ Sygma. *page 224 (left)* Australian Tourist Commission. *page 224 (right)* Alex Webb/ Magnum. *page 226* New Zealand Consulate General.

Chapter 9
Opener page 234 Jim Harrison/ Stock, Boston. *page 250* A. Keller/ Sygma. *page 256* Tass from Sovfoto. *page 261 (top left)* Sovfoto. *page 261 (left center)* Tass from Sovfoto. *page 261 (right center)* Paolo Koch/ Rapho-Photo Researchers. *page 261 (bottom left)* Sovfoto. *page 261 (bottom right)* A. Kuzyarin/ Sovfoto. *page 263* Tass from Sovfoto.

Chapter 10
Opener page 268 Peter Menzel/ Stock, Boston. *page 278* Official U.S. Air Force Photo. *page 279* J.R. Heller/ Stock, Boston. *page 282* Georg Gerster/ Rapho/ Photo Researchers.

Chapter 11
Opener page 290 John Isaac/ U.N. Photo. *page 295* Keith Gunnar/ Photo Researchers. *page 296* Bernard Pierre Wolf/ Photo Researchers. *page 298* J.P. Laffont/ Sygma. *page 305* WHO Photo. *page 312* Bernard Pierre Wolf/ Photo Researchers. *page 313* Jules Bucher/ Photo Researchers.

Chapter 12
Opener page 324 Ira Kirschenbaum/ Stock, Boston. *page 333* Jill Hartley/ Photo Researchers. *page 336* Patrice Habans/ Sygma. *page 345* Baldev/ Sygma. *page 346* Baldev/ Sygma. *page 348* R. Darolle/ Sygma.

Chapter 13
Opener page 354 © Abbas/ Magnum Photos. *page 364* China Photo Service/ Eastfoto. *page 365* Owen Franken/ Stock, Boston. *page 367* WHO Photo. *page 371* Lehtikuva Oy/ Woodfin Camp and Associates. *page 376* Bruno Barbey/ Magnum. *page 381* Eastfoto. *page 384* J.P. Laffont/ Sygma. *page 389* Robin Laurance/ Photo Researchers. *page 391* Jonathan Wenk/ Black Star.

Chapter 14
Opener page 394 Ira Kirschenbaum/ Stock, Boston. *page 398* Courtesy of The United Nations. *page 400* Nick de Vore/ Bruce Coleman. *page 402* United Nations. *page 403 (top)* WHO Photo. *page 403 (center)* Courtesy of The United Nations. *page 403 (bottom)* Hawaii Visitors Bureau. *page 404* Gordon N. Converse/ Christian Science Monitor. *page 405* H. Null/ FAO Photo. *page 412* James R. Holland/ Stock, Boston. *page 413* Owen Franken/ Stock, Boston. *page 419* UNICEF Photo.

Chapter 15
Opener page 424 Owen Franken/ Stock, Boston. *page 429* Robert Azzi/ Woodfin Camp and Associates. *page 433* Bruno Barbey/ Magnum. *page 434* El Koussy/ Sygma. *page 437* Georg Gerster/ Rapho/Photo Researchers. *page 438* Victor Englebert/ Photo Researchers. *page 439* R. Steedman/ The Stock Market. *page 440* Georg Gerster/ Photo Researchers. *page 441* Courtesy of The United Nations. *page 453* M. Nasr/ Unrwa Photo. *page 454* Ingo Morath/ Magnum. *page 456* Chester Higgins, Jr./ Photo Researchers. *page 460* Bernard Pierre Wolf/ Photo Researchers. *page 462* Georg Gerster/ Rapho/Photo Researchers.

Chapter 16
Opener page 466 Owen Franken/ Stock, Boston. *page 473* Eugene Gordon/ Photo Researchers. *page 477* Ira Kirschenbaum/ Stock, Boston. *page 479* FAO Photo. *page 482* Georg Gerster/ Rapho/Photo Researchers. *page 485* Eugene Gordon/ Photo Researchers. *page 488* FAO Photo. *page 489 (left)* E. Boubat/ Photo Researchers. *page 489 (right)* Bernard Pierre Wolf/ Photo Researchers. *page 496* Lehtikuva Oy/ Woodfin Camp and Associates. *page 499* David Turnley/ Black Star. *page 500* Satour. *page 503* Chris Steele Perkins/ Magnum Photo.

Chapter 17
Opener page 508 Ellan Young/ Photo Researchers. *page page 518* Victor Englebert/ Photo Researchers. *page 520* Beryl Goldberg. *page 527 (top left)* Stuart Cohen/ Comstock. *page 527 (top right)* John Running/ Stock, Boston. *page 527 (center)* Ulrike Welsch. *page 527 (bottom left)* Stuart Cohen/ Comstock. *page 527 (bottom right)* Owen Franken/ Stock, Boston. *page 529* Carl Frank/ Photo Researchers. *page 534* Dick Davis/ Photo Researchers. *page 539* Sergio Penchansky/ Photo Researchers.

Chapter 18
Opener page 524 Lionel J.M. Delevighe/ Stock, Boston. *page 552 (top)* F. Mattiolo/ FAO. *page 552 (bottom)* IBPGR/ FAO. *page 553* George Holton/ Photo Researchers. *page 555* Eastfoto. *page 556* Georg Gerster/ Rapho/Photo Researchers. *page 560* Beryl Goldberg.

INDEX

Aberdeen, 155
Aborigines, 224
Absolute location, 24
Accessibility, 104, 105
Acculturation, 202
Acid rains, 140, 166, 167
Adelaide, 219
Afghanistan, 265, 451, 454
Africa, 30, 78, 79, 105, 468
   climate, 475
   colonial boundaries, 471
   development, 497
   disease, 478
   environment, 474
   population, 488
   resources, 490
   women, 485
Afrikaan, 501
Age of discovery, 70
Agrarian economies, 359
Agrarian society, 293
Agribusinesses, 207
Agricultural exports, 207
Agricultural land, 207
Agricultural Regions of North America, 207
Agricultural Revolution, 67, 68, 130
Agriculture:
   Asia, 318
   Denmark, 165
   Eastern Europe, 171
   India, 342
   Soviet Union, 247
   United Kingdom, 158
Agriculture surplus, 15, 221
Ahmadabad, 341
Aid, 84, 558
Ainu, 270, 284
Air, 210
Air pollution, 58, 166, 190, 210
Albania, 169, 170, 172
Albanians, 127, 129, 169
Alfisols, 52, 123
Algeria, 135
Alkaline soils, 371
Allah, 431
Alluvial, 49, 303
Alluvium, 371, 372
Alpine mountain, 121, 124
Alpine system, 119
Alps, 40, 124, 125, 168
Alsace, 146
Alsace–Lorraine, 124, 125, 133
Alternative food, 551
Altitudinal zones, 522
Alunite, 246
Amaranth, 551

Amazon Basin, 522, 539
Amazon River, 42, 522, 539
American:
   agriculture, 206
   dream, 198
   farms, 207
   Indians, 194, 195
   population, 202
Americans, 194
Amsterdam, 27
Amu Dar'ya, 240
Andes, 40, 520
Anglo-America, 178, 180, 182
   Chernozem, 187
   population, 194
Annamese people, 406
Anshan, 380
Anthracite coal, 189
Antwerp, 161
Aomen, 359
Apartheid, 500
Appalachia, 184
   fields, 189
   mountains, 40, 183, 203
Appennine Mountains, 121, 168
Applied geography, 29
Arabian peninsula, 446, 458
Arabic, 436
Arab people, 147
Archipelagos, 274, 397
Arctic, 35
Arctic Ocean, 126
Argentina, 536, 538
Arid, 218
   China, 370
Aridisols, 52, 372
Arithmetic growth, 76
Arracacah, 551
Asia, 30, 105, 292
   agricultural systems, 318
   climates, 305
   immigrants, 224
   language patterns, 305
   malnutrition, 296
   mineral resources, 307
   petroleum, 307
   population, 292, 293–299, 304, 305
   population growth, 299
   tin, 303
Asian minority, 201
Asians, 194, 197, 490
Assimilation, 201, 202
Assyrians, 430
Aswan Dam, 457, 461
Asymmetrical relationship, 332
Atacama, 42

Ataturk (Mustafa Kemal), 444
Athens, 168
Atlanta, 205
Atlantic coastal plain, 183
Atlantic Ocean, 123
Atolls, 417
Atomic Bomb, 271
Australia, 78, 216
Austria, 159, 160, 164
Austro-Asiatic, 399
Austroloid, 402
   peoples, 402
Austronesian, 401
Authoritarian government, 336
Automobile, 16, 210
Autumnal equinox, 43
Avalanches, 126
Aztec, 516, 519

Baader–Meinhof Gang, 148
Babylon, 430
Babylonians, 430
Bagasse, 532
Bahamas, 531
Bahrain, 457
Baikal-Amur Mainline (BAM), 254
Baku, 252
Balfour Declaration, 438
Balkanization, 129, 169
Balkanize, 129
Balkan peninsula, 118, 129
Baltic, 124
Baltimore, 203, 204
Bangalore, 341
Bangladesh, 324, 336, 338, 350, 351
Baotou, 372
Barcelona, 146, 168
Barefoot doctors, 367
Barley, 60
Barrios, 20
Basel, Switzerland, 124
Basin irrigation, 456
Basin and Range region, 185, 208
Basque, 146
   party, 146
Bass Straits, 222
Bauxite, 125, 522
Bazaar, 433
Beans, 60
Bedouin, 35, 442
Beef, 53
Beijing, 299, 361, 364
Belgium, 123, 127, 161
Beneficiation, 191
Bengali, 338
Bengali language, 338
Berlin, 133, 160

Berlin Conference, 492
Bessemer converter, 131
Bhutan, 327, 351
Biafra, 472
Bilharzia, 102, 455, 495, 554
Bilingual education, 201
Biogas, 555
Biomass, 45
Birmingham, England, 154, 155
Birth control, 301
Birthrate, 14, 19, 72, 73, 292, 301, 302, 349, 384, 413, 453
Bituminous coal, 187
Blacks, 180, 194, 199, 200, 211, 490
Black Sea, 124, 244
Boers, 499
Bolivia, 99, 104, 536
Bolshevik, 233, 234
Bombay, 299, 341
Botswana, 478
Boundaries, 471
   cultural, 471
   physical 471
   superimposed, 471
Bradford, 154
Brahman(s), 330
Brahmaputra-Ganges, 350
Brahmaputra River, 303, 326, 335
Brain drain, 136
Brasilia, 540
Brazil, 19, 99, 539, 540
Break-in-bulk, 3, 25, 139, 203, 204
Bretons, 126
Brisbane, 217, 218
British Commonwealth, 141
British East India Company, 332, 333
British Indian empire, 310
British Isles, 118, 123, 154
Brittany, 146
Brunei, 307, 401
Brussels, 162
Bubonic plague, 136
Buddha, 312
Buddhism, 311, 312, 313
Buddhist, 351
Buffer zone, 169
Bulgaria, 123, 169, 172
Bulgars, 129, 412
Burkina Faso, 21, 26
Burma, 400, 401
Burundi, 24, 108
Bushmen, 490
Bustees, 300, 347

Cacao, 519
Calcutta, 26, 28, 299, 300, 340

Calgary, 206
Caloric deficit, 3, 13
Campine, 124, 133
Canada, 29, 174–182, 191–195, 198, 199
  population, 202
Canal, 203
  boats, 203
Canberra, 218
Canterbury Plain, 225
Cantonese, 310
Cape Town, 26
Capital intensive, 207
Capitalism, 89, 90
Capitalist doctrine, 233
Capitalist Market System, 90
Capital-surplus, 449
  countries, 9, 11, 17, 18, 19, 20, 21, 401
  oil producers, 457, 463
Capital transfer, 566
Caribbean, 99, 510, 531, 532
Carpathians, 121, 126
Carrying capacity, 494
Cascade Range, 185
Caspian Sea, 240
Cassava, 60, 483, 519
Caste system, 105, 330, 333, 340
Catalonia, 146
Cataracts, 476
Catholic, 168, 201, 517, 529
Catholicism, 168, 518
Cattle production, 220
Caucasian, 194, 258
Caucasus Mountains, 244, 249, 257
Celtic, 127
  language, 127
Central America, 535
Central business district, 204
Central city, 199
Central lowland, 184
Central place, 3
Central planning, 250
  development pattern, 154
  economy, 19, 29, 30, 91, 169, 170, 171, 257, 264, 349, 391
  society, 390
Central Siberian Uplands, 237
Central Valley, 185, 208
Centrifugal forces, 127, 171, 173, 207, 218, 259, 336, 413, 491
  Canada, 199
Centripetal forces, 143, 202, 334
Cerrejon, 536
Chad, 24, 473
Chaing Kaishek, 361, 362, 363
Chang Jiang, 68, 303, 356, 361, 368, 370, 371, 378
Charlemagne, 129, 140
Chemical fertilizer, 377
Chernozem, 52, 242
Cherrapunji, 307
Chiatura, 246
Chicago, 186, 203, 204, 205
Children, 74, 107
  costs, 74
Chile, 536, 537
China, 30, 298–302, 356
  agriculture, 375
  climates, 369
  coal, 372
  deserts, 370
  family, 368, 385
  humid, 370
  industry, 378–381
  iron, 372
  petroleum, 372
  physical geography, 368
  population, 372, 373
  proper, 372
  resources, 372
  revolution, 361
  transportation, 382, 383
  urban, 386
    population, 383
    growth, 386
  women, 368, 384
China, People's Republic of, see People's Republic of China
China Proper, 372
Chinese, 314, 364, 365, 399
  boat people, 414
  communism, 89
  Communist Revolution, 356, 361
  Communists, 359, 361, 366, 367, 368, 372, 375
  culture, 270, 312
  family size, 383, 385
  female, 367
  Industrial Revolution, 364, 365
  Revolution, 361, 362, 367
  Society, 367, 374, 383–385
Chongqing, 366
Chongquin, 371
Christchurch, 226
Christian, 311, 350
Christianity, 336, 428, 430, 470, 493
Chrome, 448
Circulation, 131
Cities, 20, 25, 26, 202, 203, 232, 292
Civil service, 333
Clan, 95, 270
Classless society, 233
Climate, 41–49, 102, 182, 305, 369, 370, 397
  controls, 241
Club of Rome, 33, 35, 36, 37, 38, 60
Coal, 54, 55, 125, 189, 222, 244, 245, 338, 396
  fields, U.S., 189
Coarse grains, 207
Cocoyams, 552
Coke, 131
Collective farms, 169, 233, 246, 247
Collectivization, 233, 246, 247, 252, 364, 377, 498
Cologne, 159
Colombia, 19, 535
Colonial, 314, 324
  boundaries, 471
  China, 315
  empire, 132
  India, 315
  possessions, 270
  power, 315, 408
Colonialism, 134, 331, 332, 405, 412, 477, 493, 499, 528, 561
Colonies, 333, 470, 518
Colonization, 512
  Southeast Asia, 315
Colorado, 186
Coloured, 499
Columbia River, 186
Combines, 278, 279, 282
Commercial agriculture, 21, 487
Commercial crops, 521
Commitment to nation, 281
Common Market, 141–145, 166, 169, 173, 472
  parliament, 142
  Supreme Court, 142
Commonwealth of Nations, 141
Commune, 364, 365, 374, 384, 390
Communism, 89, 109, 127, 137, 232, 359, 361, 365
Communist, 169, 232, 234, 235
  economics, 106
  party, 232, 262, 263
  Revolution, 232, 233, 252, 253
Compact shape, 410
Complementarity, 3, 25
Complementary, 26
  relationships, 210
  resources, 104
Compound-ethnic, 311
Compulsory education, 278
Conflicts, 428, 430, 459, 471, 535, 563
Confucian:
  ethic, 358, 410, 416
  Mandarins, 358, 360
  meritocracy, 357
Confucianism, 311, 357, 389
Confucius, 357
Congo River, 474
Coniferous forest, 46, 123, 182, 242
Conservation ethic, 187, 192, 408
Conspicuous consumption, 28, 348, 361, 546, 555, 562
Consumer goods, 253
Continental drift, 474
  glaciers, 121
Continentality, 42, 236
Contraceptive, 301
Contract system, 376, 379
Conurbation, 280
Convectional precipitation, 44, 45
Cooperatives, 165, 363
Copper, 53, 480, 481
Copra, 343
Coral islands, 400
Core–periphery, 107
Corn, 52, 60, 206, 249, 519
  belt, 207, 208
  production, 207
Coromandel, 328, 339
Corporate farms, 207
Corsica, 146
Cortes, 516
Cottage industries, 339, 366
Cotton, 207, 444, 455
  belt, 207
Council of Indies, 518
Coup d'etat, 233
Craft industry, 251
Croats, 127, 129, 169
Crops, 49
  rotation, 130
  West Germany, 161
Cuba, 531, 532
Cultural:
  boundary, 471
  complexity, 129
  conflict, 145
  fragmentation, 152
  groups, 144
  pluralism, 202
  poverty, 195, 200
  realm, 178
  Revolution, 382
Customs unions, 146
Cyclonic precipitation, 44, 45
Czars, Czarist rule, 231, 232, 233, 234, 235, 251, 252, 264
Czechoslovakia, 127, 168, 169, 172
Czechs, 127

Dairying, 207
Dallas, 206
Danube, 121, 124
Danubian Plains, 121, 169
Dardanelles, 236
Dark Ages, 358
Death rate, 15, 324, 367, 513
Deccan Plateau, 328, 339, 343
Deciduous forests, 46, 123, 242
Deforestation, 553
Democracy, 194, 273
Democratic, 338
  government, 317
  political, 326
  Revolution, 130, 137, 148, 194
  society, 351
Demographic transition, 72, 73, 76, 92, 95, 136, 287, 297, 302, 385, 528
Demokratizatsiya, 248
Denmark, 164–166
Denver, 206
Dependency ratio, 297
Dependent economy, 445
Desert, 182, 218, 243, 303, 446
  climates, 48
  pavement, 427
Desertification, 478, 482, 494
Detroit, 205
Developed:
  countries, 10, 25, 29, 53, 74
  world, 106
Developing world, 293, 542, 545
Development, 28, 34, 481, 494
Devolution, 145
Dew point, 42
Dharma, 330
Diamonds, 480, 481, 503
Diffusion, 270
Dinaric Alps, 121
Direct consumption, 549
Disease(s), 76, 102, 396, 478, 516, 552, 554
Disposable income, 16
Distributaries, 27
Diurnal, 446, 476, 477
Division of Labor, 70
Dogger Banks, 126
Domesticated plants, 519
Domestication, 68, 430, 551
Domino theory, 407, 413
Donetsk, 255
  basin, 244
  –Krivoy Rog, 255
Dongbei, 303, 363, 364, 365, 366, 368, 378
  plain, 368
Dortmund, 159
Double cropping, 283, 307, 329, 378, 455
Double rate, 72
Doubling time, 72
Douro River, 124
Dravidians, 335

Drought, 350, 361, 366, 370, 378, 482, 494, 498
Dry farming, 441
Dual economies, 293, 410, 481, 504, 510, 526, 528, 533
Dual societies, 94, 348, 493
Duisburg, 159
Duluth, 186
Dusseldorf, 159
Dynasty, 357, 373

Earthquakes, 121, 126
Earth's population, 68
Eastern Europe, 29, 127, 152, 168
  agricultural systems, 169
Eastern Orthodox, 127
East Germany, 160, 169
Economic(s):
  development, 3, 4, 5, 27, 28, 108, 109, 230, 418, 442, 520
  growth, 3, 4
  North America, 206
  order, 80
  significance, 74
  system, 89
  takeoff, 409
  unification, 140
Economies of scale, 165
Ecuador, 536
Ecumene, 307
Edinburgh, 155
Edmonton, 206
Education, 263, 367, 383, 386, 514
EEC, 140–144, 161, 173
Effective national territory, 522
Egypt, 454, 455, 456, 457, 462, 492
  population, 455
Eire, 158
Ejidos, 533
Elbe River, 124, 160
Elongated shape, 410
Elongated state, 536
El Salvador, 529, 535
Emergent coastal plain, 183
Emerging countries, 137
Employment, 298
Enclaves, 305, 357, 509
Encomendero, 508
Encomienda, 515, 516
Energy, 554
  consumption, 192, 301
England, 123, 154
English Channel, 154
Entisols, 52
Environment, 79
  degradation, 139, 173, 192, 547
  destruction, 59
  deterioration, 547
  determinism, 33, 58
Environmental:
  exploitation, 58
  impact, 61
  issues, 192
  modification, 59
  perception, 33
  pollution, 139, 203, 210
  problems, 563
Epicanthic, 308
Equatorial areas, 44
Equinox, 42, 43
Erbo River, 124
Erie Canal, 203, 205
Erosion, 41, 80, 208, 401, 534, 547
Erzgebirge, 169

Essen, 159
Estuaries, 183
Ethiopia, 473, 494
Ethnic:
  minorities, 284
  prejudice, 202
  religion, 431
Ethnic composition, Anglo American, 194
Ethnic groups, 198, 259, 294
  America, 198–202
Ethnocentricity, 58, 101, 148, 356
Ethnocentrism, 321
Eugenics protection laws, 285
Euphrates River, 68, 454, 463
Eurodollar, 142
Europe, 29, 112
  population, 71
European:
  colonial powers, 293
  colonization, 70, 216
  core, 172
  cultural realm, 114
  environmental setting, 118
  expansion, 515
  explorers, 272
  lowland, 159
  plain, 121, 160
  river system, 124
  settlers, 121
European Coal and Steel Community, (ECSC), 141
Europort, 142, 162
Eurotunnel, 154
Euzkadi, 148
Exchange, 91
Exclaves, 307
Exotic stream, 239
Exponential increase, 68, 76
Exports, 114, 206, 220, 512, 532, 540
Expropriation, 233, 391, 406, 528, 532, 550
Extended family, 92, 93, 108, 293, 313, 462, 483, 486, 561

Factors of production, 91
Factors of wealth, 90
Factory acts, 132
Factory system, 131
Fall line, 184
Fallow, 283
Family planning, 348
Famine, 76, 233, 321, 324, 328, 338, 342, 361, 363, 376, 386, 496, 505
Far East, 292
Farm(s), 294
  consolidation, 283
  family, 209
  size, 207
Fascists, 140
Fatalism, 338
Favelas, 20
Federal government, 472
Fertility rate, 547
Fertilizers, 104, 209, 401
Feudal:
  China, 362
  lords, 272
  societies, 272
Feudalism, 272
Field patterns, 283
Fifth system, 441

Fiji, 470
Finland, 122, 164, 166
First World, 6, 106
Five-guarantees, 385
Five-year plan, 341
Fjords, 41, 123, 166
Flemish, 127, 146, 161
Flood, 371
  plain, 371
Food:
  crops, 60
  deficit countries, 14
  production, 50, 51, 549
  resources, 50
  sources, 551
  surplus, 187, 206, 207
Foreign aid, 84, 100
Foreign workers, 172, 458, 459
Forests, 44–46
Four Modernizations, 375, 382, 385
Four truths, 312
Fragmentation, 283, 343
Fragmented state, 410
France, 123, 125, 159, 160, 161
Free enterprise, 376
French, 195, 198, 199
French Canadians, 202
Friday Mosque, 433
Frontal precipitation, 44
Fuels, 53
Full employment, 257

Galicia, 146
Gambia River, 494
Gandhi:
  Indira, 349
  Mohandas, 349
Ganges:
  Delta, 334
  plains, 343
  region, 335
  River, 26, 68, 303, 326
General farming, 207
Generative centers, 217
Generative cities, 3, 25
Geography of terrorism, 82
Geopolitical, 278
Germanic, 164
  language, 127
  peoples, 126, 128
Germany, 123, 159, 160
Germ theory, 136
Ghats, 328, 329
Ghettos, 20, 199, 201
Glaciation, 41, 164, 183
Glasgow, 155
Glasnost, 248
Global Environmental Characteristics, 34
Global Geographic Characteristics, 4
Gluten, 60
Gobi, 42
Gold, 480, 481
Golden Horseshoe, 205
Gondwanaland, 474
Gorbachev, 246
Gosplan, 254
Gradational forces, 41
Grain exports, 207
Grand canal, 383
Grapes, 208
Grasslands, 47, 49, 477
Great Britain, 24

Great Hungarian Plain, 123, 169
Great Lakes, 185
Great Leap Forward, 366, 367, 383
Great Plains, 184, 185
Great Salt Lake, 185
Greece, 126, 167, 168
Greek Orthodox, 168
Greeks, 127
Green Revolution, 317, 318, 342, 343, 495, 550
Greenhouse, 562
Gross domestic product, 513
Gross national product (GNP), 3, 8, 39, 302, 387
Growth rate, 15, 324, 346, 381, 383, 547
Guadalquivir, 124
Guangzhou, 359, 364, 371, 373
Gulf Stream, 123
Gypsies, 169

Haarlem, 161
Haciendas, 517, 533
Hague, The, 161
Haiti, 98, 531, 534
Hajj, 434, 435
Hammurabi, 430
Han Chinese, 373, 374
Handicraft, 251
  industry, 333
Hapsburg Empire, 127
Hardwood, 46
Harijans, 311, 330
Hazards, 126
Head link, 133, 472
Health services, 79
Hebrew, 436
Hejira, 431
Hermit Kingdom, 390
Hierarcial order, 133
Highland Africa, 474
High latitude, 42
High-technology industries, 139, 211
High Veld, 499
Himalayas, 40, 292, 303, 326, 556
Hindi, 337, 348
Hindu, 311, 335
Hinduism, 311, 312, 330, 331
Hinterland, 25
Hiroshima, 273, 278
Hispanics, 180, 200
Histosols, 52
Hobard, 217
Ho Chi Minh, 299, 409
Hokkaido, 274, 281, 286
Holistic, 24
Hollow frontiers, 539
Homelands, 500
Hong Kong, 20, 301, 364, 382, 388, 389
Honshu, 274, 275, 281, 286
Hookworm, 479
Horizontal nomadism, 442
Household contract, 376
Houston, 206
Huabei Plain, 303, 361, 366, 368, 377
Huang He, 303, 357, 370, 371, 372
Hubei, 303, 305, 307
Hudson-Mohawk routeway, 205
Hudson River, 205
Human dimension, newly industrialized countries, 22

INDEX
583

Human equation, 65
Humid, 369
　continental climates, 48, 49, 60, 123, 169, 182, 274, 305, 369, 390
　　subtropical, 43, 123, 168, 182, 218, 243, 274, 329, 370, 390
Humid china, 370
Humus, 49
Hungarian language, 127
Hungary, 123, 169, 171
Hunger, 295, 296
Hungry millet, 60
Hwang Ho (Yellow River), 68
Hybrid:
　rice, 317
　varieties, 317
　wheat, 317
Hydraulic civilizations, 441
Hygiene, 72

Ice cap, 48
Iceland, 164, 165
Igarka, 240
Ile de la Cité, 159
Illegal aliens, 200
Illegal immigrants, 198
Illinois-Michigan Barge Canal, 186
Illiteracy, 210, 299, 493
　rate, 210, 513
Illyrian language, 127
Immigrants, 172, 180
Immigration, 198
　laws, 197
Imported minerals, 193
Imports, 117, 192
Import substitutions, 222
Incan, 519
Inca(s), 516, 519
Incentive, 250, 254, 259
Inceptisols, 52
Incipient industrial base, 253
Income transfers, 61, 180
India, 30, 105, 134, 302, 307, 323
　agriculture, 342
　cattle, 343, 344
　literacy rate, 347
　population, 324, 345–350
　　increase, 346
　　policy, 348, 349
　urbanization, 348
Indian(s), 194, 195, 518, 522, 526, 533, 535, 536, 538
　culture, 334, 399
　development, 338
　life-style, 331
　North American, 194, 195
　perimeter, 350
　population, 516, 517, 525, 534
　reservations, 195, 202
　Ruhr, 159
　subcontinent, 328, 334
　villages, 338
　women, 338, 340
Individualism, 194
Indochina, 406, 409
Indo-European language, 127, 335, 439
Indonesia, 77, 299, 305, 307, 315, 412
Indonesian archipelago, 406
Indus, 68, 303, 306, 328, 335, 350
Indus River, 303, 326, 350
Industrial, 203

base, 364
dominance, 270
power, 270
production, 365
regions, 209
Revolution, 71, 72, 90, 129–137, 178, 232, 233, 284, 359, 364, 365, 396
　Western Europe, 152
Technological Revolution, 130
world, 29, 50
Industrialization, 3, 4, 54, 55, 84, 148, 252, 334, 339, 453, 456, 503, 510, 524, 531
Industrialized:
　countries, 9, 10, 11, 23, 57, 73, 74, 88, 89, 96, 97, 381, 550, 559, 561
　realm, 6, 7
　world, 60, 61, 72, 96, 114, 216
Industries, 209
Infant mortality rate, 8, 21, 70, 88, 210, 293, 298, 547
In-Field, 130
Inflation, 531, 538, 559
Infrastructure, 285
Inland sea, 280
Institutional framework, 89, 525
Insular, 152, 284
Insurgent state, 413, 529
Intensive agriculture, 320
Interaction, 91
Interconnected world, 96–100, 206, 265, 486, 498, 505, 553
Interconnection, 16, 133, 209, 216
Intercropping, 319
Interdependence, 24, 133
Interfingering, 118
International Rice Research Institute, 317
Intestinal parasites, 27
Invisible exports, 163
Ipa, 556
Iran, 453, 454
Iranians, 198, 439
Iraq, 454
Ireland, 123, 158, 159
Irish, 127, 159, 196
　curve, 159
　Republican Army (IRA), 148
Iron industry, 252
Iron ore, 125, 154, 166, 191, 222, 246, 396, 401, 448, 480, 481, 522
Irrawaddy River, 303
Irredenta, 160, 314
Irrigated, 168, 187, 243, 249, 275, 318, 478, 533
　agriculture, 209
　land, 441
Irrigation, 49, 70, 238, 303, 305, 328, 329, 343, 350, 364, 371, 455, 461, 478, 495, 498, 534
　systems, 70, 366
Islam, 30, 169, 259, 311, 334, 428, 430, 431, 439
Islamic, 30, 169, 311, 335, 432, 433, 434, 440
　cities, 433, 439, 440
　culture, 310
　Republic, 350
　world, 561
Island arcs, 305
Isolation, 34, 104, 105, 180

Israel, 448–451
Italian peninsula, 118, 121
Italy, 118, 167
Ivory Coast, 494

Jainism, 336
Jamaica, 522
Jamestown, 194
Jamshedpur:
　iron, 341
　region, 340
Japan, 24, 29, 270
　agricultural land, 283
　climate, 274
　coal, 274
　feudalistic system, 272
　human resources, 278
　industrial core, 279
　industrial periphery, 281
　isolation, 271
　mechanization, 283
　migrants, 270, 281
　mineral resource, 275, 278
　peasants, 272
　population, 281, 284
　rural north, 281
　trading companies, 281, 282
Japanese, 271, 315
　archipelago, 274
　clans, 270
　culture, 270
　economy, 278
　industry, 280
　invasion, 364
　islands, 271
　job security, 281
　middle class, 286, 287
　miracle, 282, 284, 285
　society, 282, 284
Java, 305, 413
Jews, 436
Judaism, 428, 430
Jungle, 470, 476
Junta, 538
Jute, 340
Jutland peninsula, 118

Kaaba, 433
Kalahari, 473, 478
Kampuchea, 300, 316, 400, 412
Karachi, 350
Karaganda, 244
Karelia, 259
Karma, 330
Kashmir, 336
Kenya, 80, 556
Keynes, John Maynard, 90
Khaliston, 336
Khmer, 316
Khrushchev, Nikita, 250
Khymer Pass, 326
Kibbutzim, 451
Kinshasa, 26
Kiruna, 125
Kitakyushu, 280
Kobe, 280
Kolkhoz, 246, 247, 248
Köppen's climatic types, 44, 48
Koran, 350, 432, 437, 439, 461
Korea, 301, 389
Korean, 305
　Peninsula, 389
　War, 390
　workers, 458

Krivoy Rog, 246, 252, 255
Kshatriya, 330
Kuala Lumpur, 408
Kulaks, 232
Kurds, 439, 463
Kursk, 246
　Magnetic Anamoly, 246
Kuwait, 17, 457
Kuznetsk, 244, 245
Kwashiorkor, 484
Kyoto, 280
Kyushu, 274, 275, 277, 281
Kyushu industrial region, 281

Labor, 89
　exploitation, 91
　force, 194
　intensive, 318
　specialization, 68, 89
Laboring class, 233
Labrador-Quebec, 190
Lagos, 504
Lake:
　Baikal, 240, 254
　Edward, 474
　Erie, 210
　Maracaibo, 523, 535
　Michigan, 186
　Nassar, 457
　Nyasr, 474
　Tanganyika, 474
　Turkana, 474
　Yssel, 162
Land:
　forms, 40, 326
　reclamation, 366, 459
　reform, 232, 279, 283, 345, 363, 391, 456, 459, 535, 536, 537
　resource, 50, 51
　tenure, 279, 342, 441, 516, 517, 520, 536, 540, 562
Landless, 232, 363, 391
　tenants, 342, 406
Landlords, 362, 363, 378, 391
Landscape, 281
Laos, 400, 412
Laotians, 198
Lateritic, 476
Latifundia, 517, 528, 536, 537, 540
Latin, 127
Latin America, 30, 70, 105, 510–541
Law of the Sea, 417
Leaching, 49, 208
Least industrialized, 5, 6, 12, 13, 21, 22, 50, 54, 55, 79, 88
Le Duc, 188
Leeds, 133, 154
Lenin, 233, 256, 264, 384
　Heights, 237
Leningrad, 232, 235, 240, 242, 255, 260
Leninist, 375
Lentils, 60
Leprosy, 27
Less developed:
　areas, 281
　countries, 5, 6, 12, 13, 38, 50, 52, 74, 98, 339
　realm, 13
　world, 24, 78, 84, 95, 103, 106, 178, 206, 339
Less industrialized, 324
　areas, 148

INDEX

countries, 74, 75, 79, 81, 82, 83, 84, 88, 89, 95, 97–100, 105, 108, 114, 117, 412, 451, 530, 553, 558, 559
 lands, 445
Levant, 436, 446
Libya, 457, 492
Liege, 133
Lifeboat, Earth, 77
Life expectancy, 13, 73, 88, 154, 286, 302, 349
Life spans, 75
Lignite, 189
Lille, 159, 161
Limits of earth, 34
Limits of growth, 35
Lingua franca, 117, 310, 311, 491
Lisbon, 168
Literacy, 210, 273, 334
 rate, 9, 15, 21, 95, 154, 160, 165, 167, 264, 300, 324, 349, 412, 514
"Little Havana", 198
Liverpool, 133, 155
Localized resources, 39
Location, 24, 25
 absolute, 24
 holistic, 3
 relative, 3, 24
Locusts, 479
Loess, 41, 49, 122, 123
Loess soils, 367, 370
London, 118, 131, 133, 154
Long lots, 195
Long March, 362
Los Angeles, 200, 204
Louisiana, 188
Lowland Africa, 474
Low Veld, 478
Lutheran, 127, 164
Luxembourg, 161
Lyon, 161

Macedonians, 169
Madras, 299
Madrid, 118, 168
Maghreb, 426, 437, 442, 445, 446
Magnitogorsk, 246, 252, 256
Magyar Republic, 127
Magyars, 127, 129
Maize, 52
Malabar coasts, 328
Malaria, 102
Malay, 310
 Peninsula, 307
Malays, 314, 399, 403
Malaysia, 66, 314, 401, 413
Malnutrition, 58, 211, 276
Malthus, Thomas Robert, 75, 76
Malthusian dilemma, 76, 546
Malthusian equation, 76
Manchester, 133, 155
Manchuria, 273, 389
Mandarin, 310, 385
Mandarins, 358
Manganese, 246
Manhattan Island, 92, 205
Manioc, 60, 519
Manufacturing, 15, 19, 209, 210
Manutius, 130
Mao, 357, 362, 367
Mao Zedong, 356, 361, 362, 374
Maori, 225, 226
Maquis, 124

Marco, Polo, 359
Marine west coast, 46, 48, 123, 218, 226
 climate, 48
Maritime, 118
Market exchange economy, 91, 92
Marseille, 161
Marshal Tito, 170
Marx, Karl, 90, 91, 232, 233
Marxist economic theory, 232
Marxists, 108, 374, 535, 537
Materialism, 194
Mayas, 517
Meat, 206
Mecca, 432, 434, 435
Mechanization, 247, 248
 of agriculture, 318
Medical care, 70, 103
Mediterranean, 48, 123, 124, 127, 140, 153, 167
 agriculture ranges, 167
 climate, 48, 123, 167, 182, 219, 446, 478, 522, 537
 Sea, 123, 153
Megalopolis, 3, 27, 204, 205
Meiji Restoration, 272, 273
Mekong River, 303, 400
Melanesia, 416, 417, 418
Melanesoid peoples, 402
Melbourne, 217, 218, 223
Menam Chao Phraya, 303
Mercantilism, 90
Merchant, 272
Mercury poisoning, 285
Meritocracy, 312, 385
Mestizo, 105, 525, 527, 535, 536, 538
Metals, 139
Metallic minerals, 191
Mexican-Americans, 200
Mexico, 20, 22, 134, 178, 533, 534, 535
 City, 528, 533, 534
Micro-farms, 363
Micronesia, 416, 417, 418
Microstates, 473
Middle America, 510, 533
Middle Atlantic, 209
Middle class, 107, 137, 178, 231, 279
Middle East, 426, 427, 428
 and North Africa, 358
Middle latitudes, 42, 44
Middlesborough, 154
Midlands, 125, 139, 154, 172
Midwest, 207
Migrant(s), 222, 496, 538
 labor, 163
 rural, 80–82
Migration, 71, 222, 462, 493, 494, 496, 535
 Asia, 299, 300
 Europe, 136
 New World, 196
 rural to urban, 80–82, 514
Migratory genocide, 454
Milan, 168
Militarism, 278
Milk, 207
Millet, 60
Minamata, 285
Mineral resources, 38, 53, 54, 104, 338, 522, 523
 Anglo-America, 198

Mini-farms, 318
Minifundia, 517, 528
Minneapolis–St. Paul, 186
Mir, 232, 246
Mississippi, 184, 185
Missouri, 184
Mixed farming, 207
Mobility, 203
Mogul dynasty, 331
Mollisols, 52, 123, 242, 372
Mongoloid, 308, 309, 402
 peoples, 402
Mongols, 127, 358, 372
Monoculture, 319
Monopoly, 91
Monotheistic religion, 430
Monsoon, 306, 328, 369, 397
Montenegrins, 169
Montreal, 205
Moraines, 41
Moral restraints, 76
Morphology, 204
Moscow, 232, 236, 254, 260, 262
Moshav, 451
Mother Russia, 236, 246
Mount Everest, 295
Mount Fuji, 285
Mount Glittertinden, 121
Mount Narodnaya, 237
Muhammad, 431, 432, 433, 439
Mulatto, 535
Murmansk, 236, 240
Muslims, 169, 172, 331, 430, 435, 449, 461

Nagasaki, 273, 278
Nagoya, 280
Nancy, 159
Naples, 168
Narita, 280
Nation, 129
National:
 identity, 281
 minorities, 373
National Environment Protection Act, 210
Nationalism, 127, 414, 445, 473, 546
Nationalistic, 445
Nationalization, 158, 414, 528
Nationalized, 139, 536
Nation-state, 124
Natural gas, 126, 203
Natural resources, 34
NATO, see North Atlantic Treaty Organization
Nauru, 401, 422
Navajo, 58, 59
Near East, 426
Negrito peoples, 402
Negroid, 403, 468, 490
Nei Monggol, 370, 376
Neocolonialism, 96, 135, 265, 414
Neolithic Revolution, 67, 68, 430
Nepal, 327, 351, 556
Nephelite, 246
Netherlands, 25, 72, 108, 114, 124, 161
Net increase, 67
Newcastle-upon-Tyne, 154, 155
New Economic Policy (NEP), 233, 247, 252
New England, 209
New France, 195

Newly industrialized, 163
 countries, 10, 11, 12, 17, 18, 19, 20, 409, 412, 510
New Orleans, 186, 203
New York, 201, 203, 204, 205, 206
New Waterway, 27
New Zealand, 29, 215, 225
Nicaragua, 529, 531
Nickel, 307
Nigeria, 472, 504
Niger River, 474, 494
Nikopol, 246
Nile River, 68, 447, 455, 456, 474, 498, 554
Nirvana, 312
Nomadism, 482
Nomads, 442, 463, 482, 493, 496
Nonindustrial, 6
Nonproductive sector, 75
Nonreciprocal exchange, 293
Nonrenewable, 53
Nonubiquitous, 39
North Africa, 426, 427, 428. See also Middle East, and North Africa
North America, 74, 178
North Atlantic Drift, 123
North Atlantic Treaty Organization (NATO), 96, 162
North China, 381
Northern Ireland, 154, 159
North Island, 225
North Korea, 389–391
North Sea, 126, 141, 156, 166
Norway, 121, 123, 126, 164, 166
Nuclear family, 74, 108
Nuclear household, 91
Nutrients, 49
Nutrition, 296

Oases, 441, 446, 447
Oats, 60
Ob River, 237, 239, 256
Obsolescence, 139, 140, 156, 203
Obsolete equipment, 163
Oceania, 419
Ocean transport, 134
Odessa, 240
OEEC, 140
Ohio, 185
 River, 203
Oil, 401, 428
 exporting countries, 8, 9, 11, 13, 17, 19, 20, 399, 457
 producing countries, 88
 shales, 189
Oklahoma, 188
Old-age assistance programs, 74
Onchorcerciasis, 27
One-child family, 385
Ontario, 209
OPEC (Organization of Petroleum-Exporting Countries), 99, 447, 457, 560
Open hearth, 131
Opium, 332
 trade, 359
 Wars, 359
Ordos bend, 370
Ore Mountains, 124, 169
Organization of Coffee Exporting Nations, 99
Orient, 292–301, 304, 306, 308, 311, 317, 359

INDEX
585

Orient (*Continued*)
  cities, 294
  cultures, 301
  population, 295–302
    problem, 294
  religion, 311–313
Orinoco River, 524
Orkney, 155
Orographic precipitation, 44, 45, 329
Orthographic lifting, 123
Osaka, 280
Ottoman Empire, 442, 443, 444
Outback, 219
Overconsumers, 52
Overdevelopment, 17
Overpopulation, 28, 300, 547
Owner-operator, 283
Oxisos, 52
Ozone, 562

Pacific, 30
Pacific Islands, 400, 416
Pacific Ring of Fire, 41, 273, 303, 400
Paddy, 403
Pakistan, 80, 147, 305, 324, 334, 350, 431
Palermo, 168
Palestine, 449
Palestinian Liberation Organization (PLO), 450
Pamirs, 237
Pampas, 48, 538
Panama canal, 522
  zone, 528
Pangea, 40, 41, 474
Papuan, 399
Paraguay, 538
Paramos, 522
Parasitic, 3
  cities, 25, 26, 444, 472
  nature, 334
  relationship, 26, 324
Paris, 118, 133, 159, 160
Parliamentary government, 154
Peanuts, 518
Peas, 60
Peasants, 92, 106, 232, 233, 363, 366, 442, 526
  agriculture, 106
  economy, 92
  farmers, 377, 483
  life, Middle East and North Africa, 442
  societies, 92, 93
  women, 368
Peat bogs, 52
Pembina, 188
Peninsula, 116
Pennine, 154
People's communes, 366
People's Republic of China, 29, 71, 295, 300, 388
Per capita income, 114, 159, 216
Perception, 56, 58
Perestroika, 248
Periodic Markets, 94
Periods of Migration, 196
Peripheral, 226
Periphery, 226
Permafrost, 241
Persian, 439
  Gulf, 428, 448, 457, 459

Perth, 217
Peru, 21, 536
Pestilence, 76
Peter the Great, 251, 252, 255
Petroleum, 54, 55, 126, 155, 156, 163, 166, 169, 188, 245, 338, 401, 428, 448, 457, 480, 523
pH, 167
Philadelphia, 203, 206
Philippines, 305, 413
Phosphates, 448
Physical boundaries, 471
Physical quality of life index, 302, 539
Physical remoteness, 224
Piedmont, 184
Pilgrims, 195
Pindus, 121
Pinyin, 299, 365
Pizarro, 516
Plaid Cymru (Wales), 145
Planned society, 234, 264
Plantation, 318, 351, 407, 427, 487, 493, 525, 528, 536
Plateau, 368
Plebescite, 141, 409
Ploesti oil fields, 169
Plural societies, 201, 314
Pogroms, 136
Point settlements, 121, 182
Poland, 122, 168, 169, 172
Polar regions, 44
Polders, 121, 161
Polish, 127
Political fragmentation, 129, 490
Pollutants, 167
Pollution, 35, 53, 132, 139, 167
Polyculture, 319, 405, 484
Polyethnic, 315
Polyethnicism, 314, 338
Polynesia, 416, 417, 418
Polynesians, 403
Population, 50, 65, 488
  Africa, 488, 546
  Asia, 77, 294–298, 304, 305, 306
  control, 81
  density, 77, 303, 304
  distribution, 66, 70, 77
  explosion, 80
  growth, 36, 65–83, 294, 299, 383, 415, 416, 454, 455, 457, 489, 514, 546, 547, 548, 549
  history, 67
  movement, 70
  problem, 294
  pyramid, 76, 297
  riverine concentrations, 303
Pork, 52
Portugal, 124, 167
Postindustrial society, 211
Potato, 52, 60, 519
  Blight, 136, 159
Po Valley, 121, 123, 127, 168
Poverty, 5, 88–90, 101, 199, 537
  culture of, 199
PQLI, 302
Prairie, 47, 48
  wedge, 187
Precipitation, 43, 44, 49
Primate, 522
  cities, 411, 415, 528
Primary activities, 194
Private enterprise, 194
Private ownership, 195

Private plots, 245, 249, 384
Privatization, 139, 158, 226
Producer's cooperatives, 364
Production teams, 367
Productivity, 132
Prorupt state, 410
Protectionism, 139
Protestant, 158
Puerto Rico, 531
Punjab, 328
Punjabi, 336
  language, 336
Puritans, 196
Pygmies, 95, 490
Pyrenees, 121, 146

Qanat, 441
Qin dynasty, 357
Qings, 358
Qing Zang (Tibetan) Plateau, 303
Quality of life, 3, 8, 11, 12, 13, 38, 108, 302, 368, 547
Quebec, 198, 199, 202, 209
Quinling mountains, 368
Quinua, 551
Quotas, 99, 223, 250, 257
Quota system, 197

Racial strife, 147, 499
Racism, 147
Railroads, 131
Rails (guage), 217
Rain shadow, 44, 329, 397
Ramadan, 433
Randstaad, 27
Rajasthani, 338
Reciprocal exchange markets, 273
Reclamation, 209, 371, 461
Red Brigades, 148
Red River, 400, 409
Reforestation, 80, 124, 496
Refugees, 198, 388, 454, 504, 505
  Palestinian, 449
Regionalism, 144
Regionalization, 6, 144
Reincarnation, 330
Relative deprivation, 109, 200
Relative location, 3, 24, 105
Religion, 311–313, 431
Renaissance, 131
Renewable resources, 53
Repatriated, 222
Replica city, 226
Replicative societies, 216
Resource depletion, 192
Resources, 53, 104, 338, 554
Responsibility, system, 376
Revolutionary, 531
  changes, 107, 520, 531
Revolutions, 82, 106, 109, 445, 528
Rhine River, 27, 124, 140, 142, 160, 173, 196
Rhone River, 124
Rice, 52, 60, 320, 343, 378
  paddies, 364
  yields, 320
Rift Valley, 474
Rio de Janeiro, 514, 539, 540
Rising expectations, 23
Ritual purity, 330
Rocky Mountain field, 190
Rocky Mountains, 179, 180, 206
Role of women, 561
Roman Catholic Church, 127, 518

Romance, 127
Roman Empire, 70, 127, 144
Romania, 123, 127, 169
Romans, 126
Romansh language, 127
Rome, 168
Rostow, 106
Rotterdam, 25, 27, 124, 133, 142, 161, 162
Rubber, 519
Ruhr, 124, 133, 139, 159, 160
Rural:
  areas, 203
  ethic, 317
  landscape, 281
  population, 281
  poverty, 168
  -to-urban migration, 284, 415
Russia, 231, 248
Russian city, 260
Russian Plain, 237
Russians, 198, 258, 259
Russo-Japanese War, 273
Rwanda, 24, 108, 473
Ryukyu Islands, 273

Saar, 124
Sacramento River, 186
Sahara, 40, 42, 427, 446, 468, 473, 477, 478, 482, 488, 494, 498
Sahel, 102, 483, 494, 495, 496
Saigon, 300
St. Lawrence River, 185, 195, 207
St. Lawrence Seaway, 186
St. Louis, 203
Saint Petersburg, 255
Salinization, 328
Sambre-Meuse, 124, 133
Samoa, 420
Samurai, 272
San, !Kung, 95
Sanitation, 70, 72
Sao Paulo, 514, 539, 540
Saskatchewan, 188
Satellite countries, 29
Saudi Arabia, 17, 428, 431, 432, 436, 441, 442, 449, 457–459, 463
Sauer, Carl, 402
Sault Sainte Marie Canal, 185
Savanna, 217, 218, 303, 315, 444, 473
  climates, 48, 49, 319, 397, 401, 456
Savoy, 146
Sawah, 403, 404, 406
  agriculture, 403
Scandinavia, 123, 126
Scandinavian:
  countries, 91
  nations, 84, 148
  north, 148
  peninsula, 118, 121
Scarcity, 192, 212
Scenery, 126, 161
Schelde River, 124
Schistosomiasis, 102, 479
Schumann Plan, 140
Scientific Revolution, 130
Scientific-technological, 6
Scotland, 121, 154, 155
Scots, 127
Scottish migrants, 196
Scottish Nationalist Party, 145

Seattle, 206
Secondary, 88, 153
Second World, 6
Sedentary agriculture, 484
Sedentary cultivators, 480
Segregation, 195, 207
Seigneurs, 195
Seine River, 124
Semiarid, 219, 220
Semites, 434
Semitic, 434
Senegal, 494, 495
Separatism, 144, 145
Separatist party, 198
Serbs, 127
Serenghetti Plain, 40
Serfdom, 272
Serfs, 232, 251
Sertao, 539
Service industry, 17
Seventh approximation of soils, 52
Shaba, 481
Shanghai, 300, 359, 371, 381, 383
Shatter belt, 129, 169
Sheep, 219, 221
Sheffield, 154, 155
Sheikhs, 449, 457
Shenyang, 372, 381
Shenzhen, 382
Shetland, 155
Shifting agriculture, 95, 318, 403, 483
Shifting cultivation, 406, 483
Shiites, 431, 453, 463
Shikoku, 274, 281
Shintoism, 313
Shopping malls, 204
Siberia, 71, 243, 244, 249, 255, 264
Sichuan, 307, 365, 381
 basin, 371
Sicily, 168
Sickle-cell anemia, 478
Sierra Nevada, 185
Sikhism, 336
Sikhs, 336
Sikkim, 327
Silesia, 124, 169
Silicon Valley, 211
Silt, 41
Singapore, 20, 301, 302, 407, 410
Sinhalese, 351
Sino-Tibetan, 399
 languages, 310
Site, 323
 situation, 25, 26, 133, 285
Situation, 3, 25, 204
Slash and burn agriculture, 51, 95, 318, 404, 519, 547
Slave labor, 517
Slaves, 71, 468, 491, 492, 493, 525
Slavic, 126, 127, 169
 people, 258
Sleeping sickness, 478
Slovaks, 127, 129, 167
Slovenes, 127, 129, 169
Slums, 300, 347, 534, 539
Small planet, 77
Smallpox, 79
Smith, Adam, 90
Smog, 173
Smolensk ridge, 237
Social:
 conditions, 132
 inertia, 547

mobility, 178
organization, 105
problems, 286, 367
programs, 16
restraints, 81
revolutions, 136, 137, 272
status, 74
stratification, 105, 178, 311, 313
structure, 311
welfare, 235
Socialism, 89, 90, 139
Soil(s), 49, 50, 104, 243, 477
 erosion, 53, 404, 494, 547, 563
 fertility, 49
 vegetation in Europe, 123
Solar energy, 42
Solzhenitsyn, 236
Sorghum, 60
South Africa, 499–503
South America, 510
Southeast, 209
Southeast Asia, 30, 307, 309, 396
Southeast Asia and Pacific, 30, 396, 397
Southern Alps, 225
Southern Europe, 121, 127, 152, 167, 168
South Island, 225
South Korea, 20, 133, 389, 390, 391, 392
Soviet Central Asia, 238, 240, 249, 257, 259
Soviet Far East, 260
Soviet Union, 29, 121, 230
 agriculture, 249–251
 climate, 241–244
 demographic changes, 259
 desert climate, 243
 employment structure, 260
 ethnic groups, 258
 gross national product, 253
 humid continental climate, 242
 industry, 253, 255
 petroleum, 245
 population demographic characteristics, 259
 ports, 240
 resources, 244–246
 steppe climate, 242
 tundra, 241
 urbanization, 259
 women, 259, 263
Sovkhoz, 247, 375
Soybeans, 60
Spain, 124, 167
Spanish, 195, 200
Spatial interaction, 3, 24
Spheres of influence, 359
Special Economic Zones, 382
Spices, 331, 340, 347
Spinning frame, 131
Spinning jenny, 131
Spodosols, 52
Sprawl city, 204
Squatters, 20, 82
 settlement, 340
Sri Lanka, 89, 324, 351
Stalin, 233, 246, 247
Stalinism, 235, 264
Stalinist, 375
Standard of Living, 8, 11, 61, 79, 89, 91, 95, 97, 108, 132, 139, 148, 161, 168, 178, 201, 210,

216, 359, 510, 513, 547, 548, 553
State, 129
 farm, 375
 oriented capitalist societies, 91
Stavanger, 166
Steam:
 engine, 131
 locomotive, 131
Steamboats, 203
 era, 203
Steel-producing, 381
Stephenson, George, 131
Steppe(s), 47, 48, 218, 219, 370, 446
 climates, 182, 242, 306, 446, 477
 grassland, 372
Storm surge, 304
Street people, 347
Strikes, 232, 252, 282
Strip mining, 60, 190
Structural changes, 107, 520, 540
Subarctic, 241
 climate, 48, 182, 241
Submergent, 183
Subsistence:
 agriculture, 20, 418, 419, 536
 economies, 95, 96
 farmers, 22, 483
 societies, 95
Subtropical, 370
 climate, 49, 275, 305, 478, 522
 crops, 207
Suburban, 203
Sudan, 490, 498
Sudd, the, 498
Sudres, 330
Suez canal, 429, 449
Suffrage, 326
Sugar, 517, 528
Sugar cane, 222
Sumerians, 430
Sunflowers, 169
Sunnites, 432, 463
Sun Yat-sen, 361
Suqs, 461
Superimposed boundary, 405, 471, 472
Surplus(es), 208
 population, 20, 80
Surplus Capital Oil Exporting Nations, 399
Swahili, 490, 491
Sweden, 122, 164–166
Sweet potatoes, 60
Switzerland, 129, 159, 160, 162
Sydney, 217, 219, 223
Symbiotic relationship, 330, 405, 442
Syr Dar'ya, 240

Tahiti, 400, 420
Tahitian, 400
Tai, 373
Taiga, 242
Taipei, 385
Taiping rebellion, 359
Taiwan, 20, 299, 301, 385
 economy, 385
Takeoff, 101, 107
Tamils, 314, 351
Tangshan, 372
Taoism, 311
Tariffs, 140, 222, 226, 283, 463

Tarim Basin, 306, 368, 372
Tar sands, 189
Tashkent, 257
Tasmania, 222
Technical obsolescence, 211
Technological:
 development, 37, 68
 revolutions, 130, 203
 world, 14, 15
Technology, 109
 transfer, 560
Tectonic forces, 41
Tejos, 124
Temperatures, 48
Tenant:
 farmers, 318, 360
 laborer, 391
Terracing, 305
Terra incognita, 356
Terrorism, 148
Tertiary, 88, 154
 occupations, 194
Texas, 188
Textile, 139, 280
 industry, 131
Thailand, 400, 411, 412, 416
Thames River, 124, 140
Thar Desert, 336
Third World, 6, 106
Thunder Bay, 205
Tiananmin Square, 382
Tianjin, 381
Tibet, 40
Tien Shan, 237
Tierra caliente, 522, 535
Tierra fria, 522, 535
Tierra templada, 522, 535
Tigris, River, 68, 454, 463
Tin, 401, 408, 413, 522
Tobacco, 519, 522
Togliatti, 256
Tokugawa clan, 272
Tokyo, 273, 280
 metropolitan area, 280
Tonkin coalfield, 401
Topography, 40
Toronto, 205, 206
Totalitarian, 317
 governments, 140
Tourism, 531, 533
Tourists, 126
Traditional society, 106
Transfer payments, 420, 451, 458
Transhumance, 442, 463
Transportation, 203, 252, 396
 routes, 70
Transport costs, 419
Trans-Siberian railway, 252, 254
Treaty ports, 359
Triage, 77
Tribal, 311
 religions, 311, 430, 431, 468
Tribalism, 490
Tropical, 51
 climates, 48, 305, 419, 522, 536
 crops, 518, 522, 535, 540
 cyclone, 304
 ecosystem, 401
 forest, 45
 rainforest, 49, 319, 397, 476, 522, 539, 563
 savanna, 48, 329, 401, 477, 522
 soil, 401
Trotsky, 233

Trust territories, 418
Trypansomiasis, 478
Tsetse fly, 103, 478, 489
Tubers, 51
Tulsa, Oklahoma, 186
Tundra, 49, 182, 372
  climate, 123
Turin, 168
Turkey, 437, 451, 452, 453
Turkic, 258, 437
Turks, 127, 172
Tyranny of geography, 106
Tyumen, 245

Ubiquitous, 39
Uganda, 472
Ukraine, 245, 246, 250, 254, 255
Ultisols, 52
Underdevelopment, 4, 21, 106
  countries, 101
  economies, 333
Underemployed, 299
Undernourished, 50
Undeveloped, 6
Unemployment, 80
Union of Soviet Socialist Republic, 228
United Arab Emirates, 17, 457
United Kingdom, 126, 132, 154–159, 216, 324
United Nations, 97
United States, 9, 14, 29, 52, 98, 99, 178
Universalizing religion, 311, 430
Untouchables, 105, 330
Uralic, 127, 164, 258
Uralmash, 256
Ural Mountains, 246, 254
Urals, 237
Uranium, 53

Urban:
  agglomeration, 161
  America, 203
  areas, 194, 203, 232
  centers, 18
  functions, 25, 204
  growth, 82
  migration, 386
  morphology, 204
  places, 14, 207
  residents, 82
  services, 26
  sprawl, 188, 204
Urbanization, 70, 348, 462
Urdu, 335
Uruguay, 538
Ussuri, 240

Vaisyas, 330
Values, 194
Vancouver, 201
Varanasi, 312
Vegetation, 44–49, 328, 476
Veldt, 40
Venezuela, 535
Verkhoyansk, 243
Vernal equinox, 43
Vertisols, 52
Vietnam, 108, 135, 300, 315, 400, 412
Vietnamese, 198, 201, 403
Village life, 317, 338, 440
Villages, 440–442
Virgin Islands, 531
Virgin Lands Project, 250
Vladivostok, 235, 240
Volcanic soils, 305
Volcanism, 400
Volcano, 400
Volga Heights, 237

Volga River, 239, 256
Volga-Ural fields, 245
Volgograd, 256
Volta, 494
Voyages of discovery, 359
Voyages of exploration, 130, 331
Vulcanism, 41

Wahabis, 432
Wales, 154, 155
Wallachian Plain, 123, 169
Walloon, 127, 146, 161
War, 77
Washington, 204
Water, 347
  pollution, 173, 210
  resources, 186
  shortages, 186
Watt, James, 131
Wealth, 88, 90, 101
Wealthy elite, 526
Welfare intervention, 87
Welsh, 127
West Coast, 209
Westerlies, 42
Western Europe, 71, 74, 123, 143, 154, 245
West Germany, 159
West Siberian Lowland, 237
West Siberian Plain, 245
Wheat, 52, 60, 207
  belt, 208
  exports, 207
White only, 222
Wind-chill, 243
Windmill, 557
Windward, 44
Women, 259, 263, 340, 368, 376, 385, 432, 485, 486, 487, 488, 496, 503, 561

  agriculture, 263
Wood, 80
Wool, 219, 221
Wool and mutton, 226
Work ethic, 194
Workers, 91
  guest, 172
Working class, 232
World:
  population, 303, 546
    growth, 66–81
  trade, 280
  urbanization, 82
World Health Organization, 100, 102
World War II, 253, 315, 408, 421

Xerophytic, 48
  vegetation, 123, 478
Xi Jiang, 371, 377
Xizang (Tibet), 370, 377

Yams, 56
Yangtze, 68, 359
Yenisey, 239, 240
Yorkshire field, 125
Yugoslavia, 121, 123, 129, 170
Yugoslavs, 173
Yungui, 368, 370

Zaire, 481, 490
Zambia, 478
Zambos, 525
Zero population growth, 385
Zhengzhou, 365
Zhiguli, 257
Zhiguli Hills, 237
Zimbabwe, 136, 478, 501
Zionists, 444, 449
Zuider Zee, 162